FLUORINE CHEMISTRY AT THE MILLENNIUM

FASCINATED BY FLUORINE

Related Titles of Interest

BOOKS

FILLER, KOBAYASHI & YAGUPOLSKII: Organofluorine Compounds in Medicinal Chemistry and Biomedical Applications
NAKAJIMA, ZEMVA & TRESSAUD: Advanced Inorganic Fluorides: Synthesis, Characterization and Applications

Tetrahedron Organic Chemistry Series

CLARIDGE: High-Resolution NMR Techniques in Organic Chemistry
HASSNER & STUMER: Organic Syntheses based on Name Reactions and Unnamed Reactions
McKILLOP: Advanced Problems in Organic Reaction Mechanisms

Major Reference Works

BARTON, NAKANISHI, METH-COHN: Comprehensive Natural Products Chemistry
BARTON & OLLIS: Comprehensive Organic Chemistry
KATRITZKY & REES: Comprehensive Heterocyclic Chemistry I CD-Rom
KATRITZKY, REES & SCRIVEN: Comprehensive Heterocyclic Chemistry II
KATRITZKY, METH-COHN & REES: Comprehensive Organic Functional Group Transformations
SAINSBURY: Rodd's Chemistry of Carbon Compounds, 2nd Edition
TROST, FLEMING: Comprehensive Organic Synthesis

Journals

JOURNAL OF FLUORINE CHEMISTRY

Also of interest
BIOORGANIC & MEDICINAL CHEMISTRY
BIOORGANIC & MEDICINAL CHEMISTRY LETTERS
HETEROCYCLES (distributed by Elsevier)
IL FARMACO
TETRAHEDRON
TETRAHEDRON: ASYMMETRY
TETRAHEDRON LETTERS

Full details of all Elsevier Science publications, and a free specimen copy of any Elsevier Science journal, are available on request from your nearest Elsevier Science office or from the Elsevier Science web site www.elsevier.nl

FLUORINE CHEMISTRY AT THE MILLENNIUM

FASCINATED BY FLUORINE

Edited by

R. E. Banks

Emeritus Professor of Fluorine Chemistry
The University of Manchester Institute of
Science and Technology (UMIST), UK

2000

ELSEVIER

AMSTERDAM – NEW YORK – OXFORD – SHANNON – SINGAPORE – TOKYO

ELSEVIER SCIENCE Ltd
The Boulevard, Langford Lane
Kidlington, Oxford OX5 1GB, UK

First edition 2000

Library of Congress Cataloging-in-Publication Data

Fluorine chemistry at the millennium : fascinated by fluorine / edited by R.E. Banks.–
1st ed.

p.cm.
Includes bibliographical references and indexes.
ISBN 0-08-043405-3 (hc)
1. Fluorine. 2. Fluorine compounds. I. Banks, R.E. (Ronald Eric), 1932-

QD181.F1.F54 2000
546'.731–dc21

00-026273

British Library Cataloguing in Publication Data
A catalogue record from the British Library has been applied for.

ISBN: 0-08-043405-3

⊗ The paper used in this publication meets the requirements of ANSI/NISO Z39.48-1992 (Permanence of Paper).

Printed in The Netherlands.

CONTENTS

FOREWORD

This volume brings together contributions by leading researchers covering a wide scope so characteristic of fluorine chemistry. It was in 1986 that we celebrated the centennial of the isolation of fluorine by Moissan, and much was said at the time about the impact of fluorine chemistry during its first century. Nearing the new millennium we can only guess what remarkable new results and progress the next century will bring. Fluorine continues to intrigue chemists who overcome the challenge of handling this remarkable halogen and work to develop varied methods for synthesizing fluorine compounds. The unique properties of these materials, ranging from inert perfluorocarbons and fluoropolymers to a multitude of other fascinating fluorinated compounds, firmly established their place amidst the technological achievements of the 20th Century. Fluorine substitution in biologically active compounds has become particularly significant in recent years.

Fluorine is the most electronegative of the elements and forms very strong bonds with other elements, contributing to the unique properties of its compounds. The extreme stability of some organofluorine compounds can, however, also become detrimental – as in the case of the highly persistent nature of chlorofluorocarbons in the atmosphere. In many cases (high-performance polymers, inert fluids, etc.), however, this represents a great advantage. Fluorine chemistry was, of course, essential to the separation of U-isotopes through volatile UF_6, and the related development of highly-resistant fluorinated materials also contributed to the emergence of the atomic age. It was fluorine chemistry which turned 'inert' gases into 'noble' ones by making possible the preparation of many of their remarkable compounds, and the increasingly important area of superacid chemistry, which I particularly enjoy, also is based on the use of fluorinated systems.

When giving a lecture on some of my work in fluorine chemistry years ago I referred to a French proverb: 'One always returns to his first love'. I still feel the same way. Even though my interests have carried me into other areas, I still believe that fluorine chemistry is and will continue to be one of the most fascinating and dynamic areas of all chemistry.

Eric Banks as Editor is to be congratulated on having assembled such an excellent group of authors. Their contributions make this volume a testimonial of fluorine chemistry in the second half of the 20th Century and an inspiration for future generations of researchers in the field.

GEORGE A. OLAH

EDITOR'S PREFACE

About this book

Fascinated by Fluorine is a *Festschrift* in honour of all who have contributed directly to the massive developments in fluorine chemistry and technology witnessed by the past 50 years. It is neither a textbook nor a structured exposition of fluorine chemistry as it stands at the beginning of the new millennium,[1] but a monograph of historical character comprising personalized accounts of progress and events in areas of particular interest to me, written almost without exception by fluorine chemists I've 'interacted' with during my research lifetime. As explained below, this relates to my drive to celebrate in literary fashion the centenary of Moissan's isolation of fluorine in 1986; in turn, that effort stemmed from the particular pleasure I'd derived over many years from reading anecdotal accounts of the history of chemistry.

Readership

This monograph should appeal not only to all dedicated fluorine chemists worldwide (lapsed and active alike) but also to many who perhaps spent only their PhD training years discovering that fluorine – *the superhalogen, the ultimate combiner, the enabler* – is full of surprises.[2] There is also much to interest and instruct chemists from other disciplines, too, since a good proportion of the chapters contain a considerable amount of 'hard' referenced information relating to modern organic, organoelemental and inorganic chemistry. Historians of chemistry and technology will no doubt be tempted to dip into this book, and surely whoever addresses the task of commemorating Moissan's achievement at the 150-years stage will bless us all in some measure for its existence.

---

[1] Readers seeking recent (1990s) in-depth accounts of fluorine chemistry in monograph form should find the Kirk-Othmer Encyclopedia Reprint *Fluorine Chemistry: A Comprehensive Treatment* (ed. M. Howe-Grant), John Wiley, New York, 1995 (ISBN 0-471-12031-6), a useful broad-based starting point. Fluoroorganic Chemistry is particularly well endowed with single-tome learned expositions in the forms of *Organofluorine Chemistry: Principles and Commercial Applications* (ed. R. E. Banks, B. E. Smart and J. C. Tatlow), Plenum, New York, 1994 (ISBN 0-306-44610-3) and *Chemistry of Organic Fluorine Compounds II* (ed. M. Hudlicky and A. E. Pavlath), ACS Monograph 187, 1995 (ISBN 0-8412-2515-X), which are quite complementary; also the encyclopaedic *Houben-Weyl Vol. E10: Organo-fluorine Compounds* (Georg Thieme Verlag, Stuttgart) has just (1999) become available.

[2] 'Once a fluorine chemist, always a fluorine chemist [at heart]', my one-time boss at ISC (Avonmouth), A. K. ('Joe') Barbour, asserts – and there is much more than a grain of truth in that, judging from the tenor of conversations I've enjoyed with many of my former research students since they left UMIST.

Origins

The idea of producing a book that would help me in my old age to re-remember friends, colleagues and personalities associated with events and progress in the world of fluorine chemistry since I made my first C—F bonds in 1953, and to do so in time to celebrate the new millennium, came up during a holiday conversation with my wife, Linda, in France in September 1996. Teasing me about my 'pseudo' retirement, she enquired whether I was at last going to put pen to paper and write a novel about getting to the top (→ → → knighthood) via fluorine chemistry, after the style of *The Struggles of Albert Woods* by William Cooper [3] – a challenge that I had been threatening to tackle for quite a while.

As usual, I was able to identify a literary task of higher priority – this time to pick up where I'd left off in the early 1980s with work on a book describing the lives and contributions of British scientists and technologists to fluorine chemistry. Conceived as a component of a multi-volume series entitled *In Praise of Moissan*, its partners being contributions from countries then viewed as having a major stake in fluorine chemistry (America, France, Germany, Italy, Japan and Russia), that endeavour necessarily came to fruition in completely modified form: *Fluorine – The First Hundred years (1886-1986)*, ed. R. E. Banks, D. W. A. Sharp and J. C. Tatlow (Elsevier Sequoia, Lausanne and New York, 1986).[4] This left me holding piles of documentation relating to the stillborn *In Praise of Moissan*, including taped (voice) recordings of interviews Dr Kathleen R. Farrar (of UMIST's then Department of the History of Science and Technology) and I conducted with, among others, the notable British pioneers of fluorine chemistry H. J. Emeléus, H. R. Leech, B. C. Saunders and M. Stacey – all deceased now.[5] As I listened to those tapes again, I recalled the names of a fair number of other eminent fluorine chemists from around the globe I'd connected with but who also had died in recent years;[6] and that led me to contemplate the imminent or actual retirement of senior figures in the field of fluorine chemistry globally. Viewing it, therefore, as a now-or-perhaps-never situation, I started to mail invitations to write for *Fascinated by Fluorine* to enthusiasts who, between them, would create a book with a British bias yet certainly not lacking a substantial international flavour.

As regards finding a publisher, I was confident initially through my work as Editor of *Fluorine Technology Bulletin*[7] that *Fascinated by Fluorine* could be published in-house.

---

[3]The battered, much-read paperback copy I own is the 1966 Penguin Book (2409) version. William Cooper is the pseudonym of H. S. Hoff, who, at one point in his career, was Personnel Consultant to the United Kingdom Atomic Energy Authority.

[4]Reproduced in *J. Fluorine Chem.*, 33 (1986) 1-399.

[5]The Emeléus tape is particularly good and I hope one day to get around to submitting a written version for publication in *J. Fluorine Chem.* I provided a short excerpt to be played during the oral tribute to Harry at the 15th ISFC (International Symposium on Fluorine Chemistry, Vancouver, 1997) but that was not done. Excerpts played at a meeting of Fluorine Technology Bureau (UMIST) shortly after Harry died in 1993, aged 90, were much appreciated. He had such a distinctive voice – and an excellent sense of humour.

[6]*E.g.* Charles B. Colburn, 1923-1988; Lev German, 1931-1994; Nobuo Ishikawa, 1926-1991; Ivan L. Knunyants, 1906-1990; John M. Tedder, 1926-1994. Short accounts of the lives and work of these deceased fluorine chemists (and of 17 others) – all based on talks given at a 'retrospective symposium' organized by Milos Hudlicky at the 15th ISFC, can be found in *J. Fluorine Chem.*, 90 (1998) 151-211.

[7]*F T Bulletin* is the magazine issued by Fluorine Technology Ltd., a small Manchester-based organization I co-founded in 1988 with Roger Benn (then Director of UMIST's Chemserve unit) and Dr Mike Stevenson of Fluorochem. Until recently, it was produced by a small 'multimedia' organization run by Dr Basil T. Abdo, an ex-PhD student and postdoc of mine.

Unforeseen events, coupled with a fortuitous conversation between myself and Dr Guido Zosimo-Landolfo (of Elsevier Science SA, Lausanne) in Vancouver on the evening of August 5, 1997, while strolling across UBC's superb campus (location of the 15th ISFC[5]) following the dinner at Cecil Green Park for members of the Editorial Board of *J. Fluorine Chem.*,[8] led about a year later to finalization of the Agreement under which Elsevier Science contracted to publish my 'work' under its present (extended) title.[9]

About the Chapters

It was never my intention to sectionalize this book, hence the numerical ordering of chapters on an alphabetical-sequence-of-authors'-names basis – except in my own case that is, since a book-ends rôle appealed on several counts, not least eliminating the need to compose a closing section! Some may argue that in terms of content and style the chapters can be divided into two broad categories: career lifestories of selected fluorine chemists written in freestyle mode and more formal accounts of specific topics. But that's too simplistic in my view because a continuum of sorts runs between Karl Christe's contribution, say, and Malcolm Atherton's.

Authors were asked to write in a light yet authoritative style and to dwell on important discoveries and events, providing as much information about themselves, their careers and interactions with other chemists as circumstances (topic under discussion, 'censorship requirements', etc.) would allow and personal inclinations tolerate. Anecdotes were flagged as perfectly acceptable (provided they were not libellous!) and those dealing with research-based careers were encouraged to restrict detailed presentation of results to major discoveries. Naturally, I've exercised pragmatic flexibility where these guidelines are concerned.

For a variety of reasons not every contribution I planned for can be found here. The saddest of these is the premature cancer-related death of Clay Sharts (San Diego State University, USA) in January 1999; Clay, of Sheppard and Sharts' fame,[10] was the official photographer of the Fluorine Division of the ACS, and his tragic departure eliminated the opportunity to display here a selection of photographs from the collection he amassed while pursuing his favourite hobby at fluorine symposia over many years. Health problems intervened to prevent two former 'bosses' of mine, Joe Barbour (at ISC) and Bob Haszeldine (at UMIST), from finalising major contributions; however, Joe has provided an Appendix for my chapter and Bob hopes to complete his memoirs for publication in *J. Fluorine Chem.*

[8] At that time, Guido was a Publishing Editor at Elsevier Science and responsible for the *Journal of Fluorine Chemistry* (JFC); very shortly afterwards, he became an Associate Editor at Elsevier and Dr Adrian Shell (based in Oxford, UK) took over as Publishing Editor responsible for JFC. Guido, a great admirer of *Fluorine: The First Hundred Years (1886-1986)*, wrote to me after our conversation in Vancouver saying: 'Regarding your book project, I also find it seducing and agree with your conditions'.

[9] The sub-title 'Fascinated by Fluorine' was conjured up by my wife, Linda, during our holiday conversation in 1996: she had always liked the title *The Fascination of Fluorine* used by Professor R. N. Haszeldine for a public lecture we attended soon after my arrival in Manchester in 1958, and through such a simple modification neatly pinpointed a feature common to at least the great majority of this book's contributors.

[10] W. A. Sheppard and C. M. Sharts, *Organic Fluorine Chemistry*, W. A. Benjamin, New York, 1969. For details of the origins of this book, see Clay's appealing reminiscences about his friendship with Bill Sheppard (1928-1978) in *J. Fluorine Chem.*, 90 (1998) 197-199.

Chapters planned to include information about the French and German fluorochemicals industries and one centred on American accomplishments in the fluoropolymers area are currently in the 'lost' category, but moves are afoot to produce a special issue of *J. Fluorine Chem.* devoted to industrial fluorochemistry on a worldwide basis. A special issue of the *Journal* devoted to Japanese contributions to fluorine chemistry should also appear – a happy outcome, hopefully, to a problem I caused concerning the nature of the original contribution I received (via the good offices of Takashi Abe) from Japan. A few other contributions had to be transferred to the distinctive 'Centennial Issue' (vol. 100) of *J. Fluorine Chem.*, which also carried six chapters from this book – as a taster for potential customers.

An item which did not elude me – as it did when I was working on the 'Moissan book' in the early-to-mid 80s – is the 'Yvonne Désirant story', brought to us now by Dirk Tavernier, who is not a fluoro-organic chemist (hence we've never actually met, only corresponded) but pursues the history of chemistry as a hobby. Professor emeritus Heinz G. Viehe (Université catholique de Louvain), whose personal fascination with fluorine, he reminded me, '... started with the finding in 1964 of spontaneous oligomerization of *t*-butylfluoroacetylene to aromatics via isolable valence bond isomers...', kindly and efficiently located Dirk for me, via Professor Pierre De Clercq (Universiteit Gent). Only when Dirk sent me the final version of his chapter did he reveal why he was glad of the chance to investigate the Swarts-Désirant-hexafluorobenzene connection: 'I had a special motivation in that as a 17 year old I browsed through the 1958 issue of the *Bulletin des Sociétés Chimiques Belges* and noticed Désirant's full hexafluorobenzene paper. I remember my amazement at reading that the work had been done in 1934-36, and wondered what had happened. Now I know.'

Acknowledgements

Producing this book has been a complex exercise involving more than 200 pieces of outgoing correspondence; and a great many people, in one way or another, have made a contribution – including, to my delight, Professor George Olah. My thanks go to all of them, but pre-eminently to the contributing authors: their professionalism, enthusiasm, good humour and friendship will long remain a highlight for me. I'm truly grateful also to my research associate at UMIST, Dr Mohamed K. Besheesh, who (with support from Dr Max J. Parrott) did me a tremendous service – well beyond the call of duty – by modifying electronic versions of manuscripts I'd adjusted; the knowledge he gained by doing so obviously gave him great pleasure, which augurs well concerning the appeal of *Fascinated by Fluorine* to the younger end of the fluorine fraternity. Last – but by no means least – we all should thank Drs Guido Zosimo-Landolfo and Adrian Shell of Elsevier Science for their enthusiastic support and help; they too, I've noticed, are enchanted by the world of fluorine chemistry.

My wife, Linda, has been enormously supportive – as ever – during this project. And she hasn't mentioned that novel again ... yet!

ERIC BANKS

ABOUT THE EDITOR

Emeritus Professor **Ronald Eric Banks** is well known internationally for his wide-ranging contributions to fluorine chemistry, most of them stemming from his researches carried out at the University of Manchester Institute of Science and Technology (UMIST). An Englishman, born in Stoke-on-Trent on Guy Fawkes' Day 1932, he graduated in chemistry from Durham University [BSc (first-class honours), 1953; PhD, 1956], his doctoral work on halogenation of benzene with chlorine trifluoride being performed in W. K. R. Musgrave's laboratory. In lieu of compulsory military service, he then worked on a Ministry of Supply contract for two years (1956-58) in A. K. Barbour's group at Imperial Smelting Corporation, Avonmouth (Bristol), doing R & D connected with high-performance fluoropolymers of interest to the Royal Aircraft Establishment, Farnborough; studies there on perfluorocyclohexadienes led him to discover by chance the basis of the first commercial route to hexafluorobenzene.

Moving to the Manchester College of Science and Technology (now UMIST) in 1958, he began a fruitful collaboration with R. N. Haszeldine which lasted until 1974 and did much to ensure the award of a DSc degree by the Victoria University of Manchester in 1971 for 'original contributions to knowledge of fluorocarbon chemistry'.

Since 1974, Eric has continued to serve fluorine chemistry as a researcher and educator, particularly through activities designed to benefit industry, such as his work for Fluorine Technology Bureau, a Manchester-based 'fluorine club'. He retired early (31 December 1993) from his Chair in Fluorine Chemistry to allow time to taper off his research career in an organized manner, and to be free to come and go to UMIST as he pleased. Currently an Honorary Visiting Professor at UMIST, he received the American Chemical Society's Award for Creative Work in Fluorine Chemistry in 1993, the year in which he was cited as the Inventor of Air Products' *Selectfluor*^{TM} electrophilic fluorinating agent F-TEDA-BF_4 when the company received an *R & D 100* prize in the US for developing this technologically significant product. In addition to his work on *Fascinated by Fluorine*, he has edited or co-edited *Organofluorine Chemicals and Their Industrial Applications* (Horwood, 1979) and its companion volume *Preparation, Properties and Industrial Applications of Organofluorine Compounds* (1982), *Fluorine in Medicine in the 21st Century* (UMIST Chemserve, 1994), *Fluorine in Agriculture* (Fluorine Technology; 1995), *Fluorine: The First Hundred Years (1886-1986)* (Elsevier Sequoia, 1986), and *Organofluorine Chemistry: Principles and Commercial Applications* (Plenum, 1994); also, he was senior co-reporter for all three volumes of The Chemical Society's biennial specialist periodical reports *Fluorocarbon and Related Chemistry* (*1*, 1969-70; *2*, 1971-72; *3*, 1973-74), and wrote the classical two-edition text *Fluorocarbons and their Derivatives* (Oldbourne, 1964; Macdonald, 1970). Not unexpectedly, he is a member of the Editorial Board of the *Journal of Fluorine Chemistry*. In all, his publications (research papers, patents, reviews and books) exceed 300, and he has been involved with the training of well over 100 research students.

Eric lives in Hazel Grove, Cheshire (about 10 miles from UMIST) with his wife Linda (née Raine) – now a retired schoolteacher – whom he met when they were students at adjacent colleges in Durham; they have 3 grown-up children (Christopher, Nicholas and Philippa – in order of birth) and 2 grandchildren (via Christopher). Eric's hobbies are following chemistry and sport (naturally, he is a Manchester United fan), although he has been known to shoot 'clays', paint in oils, and play the odd round of golf; in his youth he was a good cricketer and he played competitive field hockey regularly for 25 years. There is no real sign that the considerable gardening skills possessed by Eric's paternal grandfather (a professional gardener) and father have been passed down to him, but Linda demonstrably has enough talent in that direction for two. None of their children earns a living through chemistry: Christopher ('No way, the number of hours you put in for such low pay!'), the oldest, is a company director; Nicholas – who flirted with chemistry in his late teens – is an ex-policeman now trading as a precision engineer ('One should put a high price on job satisfaction'); and Philippa, a financial consultant, is well known in the UK's brass band fraternity ('Where's Dad? What's he doing? . . . Not working on *another* book!!').

Chapter 1

FOLLOWING FLUORINE IN NUCLEAR FUEL MANUFACTURE AT BNFL

MALCOLM J. ATHERTON

BNFL, Springfields, Salwick, Preston, Lancashire, PR4 0XJ, UK

Introduction

The genesis of British fluorine cells is described by Powell in the chapter dealing with ICI's contribution to fluorine chemistry.[1] Following ICI's withdrawal from elemental fluorine manufacture in 1976, the development of the British design of fluorine cell depended solely on British Nuclear Fuels plc (BNFL) – in effect a reversal of roles, as many of the development studies on uranium manufacturing processes used in the early days of the British nuclear industry stemmed from work carried out at ICI facilities.

The requirements of uranium processing dictated that BNFL[2] needed access to fluorine technology to establish the base for an indigenous nuclear fuel cycle. Indeed, much of the early research and development work in the UK related to piecing together the nuclear fuel cycle was conducted by ICI, via the Tube Alloys Directorate.[3] It was natural in the early days of the Cold War that the British Government should look to existing technology within British industry to fulfil its requirements – hence the links to ICI's fluorine programme. ICI Widnes had been the home of the Tube Alloys Project during World War II, and it was there that the initial research and development associated with the manufacture of pure uranium metal was carried out. However, in 1946 it was decided to move operations to a site acquired by the Ministry of Supply – a disused former ICI munitions and poison gas establishment at Springfields, located some five miles north-west of Preston, Lancashire. Sir Christopher Hinton, the former head of the atomic energy production organisation, remarked in 1957[4] that that decision had been reached because: '... the contribution of ICI to the development of atomic energy (in the UK) could never be sufficiently realised.'

The decision to use Springfields for uranium metal manufacture was made early in 1946 by Hinton and Springfields became the first production plant to be established, in March 1946, as part of Britain's atomic energy programme. The movement of equipment,

---

[1] Chapter 22.

[2] BNFL was formed in 1970 from the Production Group of the United Kingdom Atomic Energy Authority (UKAEA – fondly known as 'ukulele'), which in turn had been formed from the Department of Atomic Energy of the Ministry of Supply in 1954.

[3] 'Tube Alloys' was the code name for a division of the Government's Department of Scientific and Industrial Research (DSIR) set up in the autumn of 1941 to handle all aspects of Britain's nuclear effort. Wallace Akers, the Research Director of ICI, and his assistant Michael Perrin ran this organisation [1].

[4] During the prestigious Melchett Lecture, given at the Institute of Fuel, London on 19 February 1957.

resources and personnel from Widnes and elsewhere to Springfields was to be accomplished as soon as possible. At essentially the same time it was decided to use batch processing at the proposed Springfields chemical plants, the principal reason being that there was insufficient time to develop continuous processes to meet the demands of the Windscale piles programme. The Springfields facility's prime purpose was to be the manufacture of fuel elements for early British experimental reactors – particularly the 'Windscale piles', plutonium producing reactors – and feed materials for uranium enrichment facilities.[5]

A chronological account of operations at Springfields does not make a lot of sense out of context, so here I've chosen to weave the role and fate of fluorine in the nuclear cycle into the wider story of the history of the Springfields chemical manufacturing facilities. The element fluorine enters the story at two distinct points: firstly in the form of hydrogen fluoride (HF) which is used to convert uranium dioxide (UO_2) to uranium tetrafluoride (UF_4), a key intermediate for the manufacture of both uranium metal and uranium hexafluoride (UF_6); and secondly, as fluorine itself or chlorine trifluoride (ClF_3)[6], needed to convert uranium tetrafluoride to uranium hexafluoride. Fluorine leaves at two distinct points, too: during the production of metallic uranium from uranium tetrafluoride, and via the conversion of 'enriched' uranium hexafluoride[7] to uranium dioxide.

Fluorine IN: the manufacture of uranium tetrafluoride

The first uranium plant [2]

A plant to convert uraninite ore (pitchblende) to UF_4 at Springfields was fully operational soon after Hinton's decisions, *i.e.* by mid-1948. The imported ore comprised essentially of uranium oxides (UO_2, U_3O_8) containing impurities such as thorium, zirconium, lead, vanadium and radium, was converted to UF_4 by a basic nine-stage process designed to process low quality ore (Fig. 1.1). Steps **8** and **9** were known collectively as the Dryway Process. ICI had developed a wet method for accomplishing the conversion of ammonium diuranate (ADU) to UF_4 but its projected operating costs and requirement for large quantities of platinum catalyst meant that the Dryway Process, although operating only on a laboratory scale in 1946, was preferred.

Only sketchy information on the hydrofluorination stage (step **9**, Fig. 1.1) survives; it was a batch process in which hydrogen fluoride, entrained in nitrogen, was passed over UO_2 contained in trays in a heated, cylindrical reactor. This procedure and others tried at the time was both inefficient and expensive and plagued by significant downtime owing to equipment failure, caused mostly by corrosion. The lessons learned from this experience were applied to the second UF_4 plant built in the late 1950s.

---

[5]The first British experimental reactor, GLEEP (Graphite Low Energy Experimental Pile), started operations at Harwell, Oxfordshire in August 1947 using uranium metal elements made by ICI from American-produced uranium oxide, sourced from Canadian ore.

[6]The use of chlorine trifluoride for this purpose was discontinued by BNFL in 1965.

[7]As is very well known, the volatility of uranium hexafluoride (m.p. 64.5 °C at 765 torr; *cf.* UF_4, m.p. 1036 °C) lies at the heart of the solution devised by American and British scientists and engineers working on the Manhattan Project during World War II to separate uranium-235 (fissionable) from uranium-238. Uranium occurs in nature as three isotopes: U-238 (99.28% abundance), U-235 ($<$0.72%), and U-234 ($\sim$0.0055%).

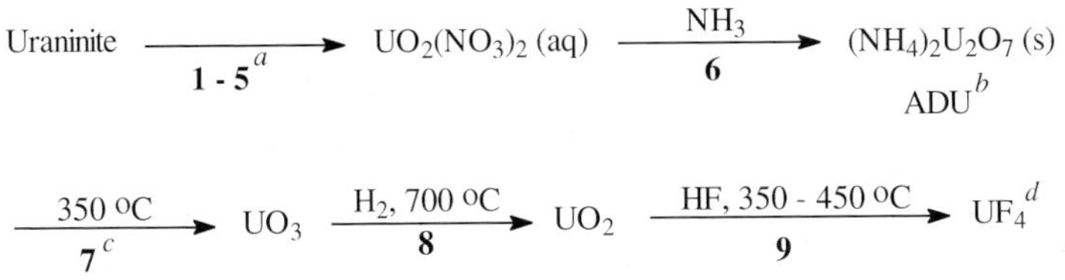

Fig. 1.1. Initial route used at Springfields to convert uranium ore to uranium tetrafluoride.

a STEPS: **1**, prepare a slurry of crushed ore in water; **2**, add HNO_3-H_2SO_4 to produce a solution of uranyl salts, then add $BaCl_2$ to co-precipitate radium with $BaSO_4$, filter; **3**, add H_2O_2 (the original batches were purchased from war surplus propellant stocks) to precipitate the uranium as uranium peroxide ($UO_4 \cdot xH_2O$, $x = 3, 4$); **4**, dissolve the uranium peroxide in HNO_3 and evaporate the solution to provide crude uranyl nitrate, $UO_2(NO_3)_2 \cdot 6H_2O$; **5**, solvent extract with diethyl ether and back extract the organic phase with weakly acidified water to produce pure uranyl nitrate solution.

b Ammonium diuranate (ADU) is a polymerized, hydrated polymer, for which exact formulation is difficult. A more accurate expression than the above is $UO_3 \cdot xNH_3 \cdot yH_2O$, where $x = 0.5$, $y = 1.5$ approximately.

c Using an Inconel (Ni-Cr-Fe) vessel.

d Reduced to uranium metal with calcium initially and, later, with magnesium.

Process research and development continued throughout the operating lifetime of the first uranium plant. Advances, which would be incorporated into new processes ultimately operated at 10 000 tonnes U per annum, included: adaptation of dissolution and purification processes to purer, higher-grade chemical concentrates, which now became commercially available from the mines; replacement of the inherently hazardous ether solvent in the $UO_2(NO_3)_2$ extraction step (**5**, Fig. 1.1) by a less volatile and less flammable mixture of 20% tributyl phosphate in odourless kerosene (TBP/OK) – a change actually instituted in the first uranium plant around 1955; use of counter-current mixer-settler units; replacement of the Dryway units (steps **8**, **9**, Fig. 1.1) by fluidized-bed reactors (the first development rigs to test this improvement began operation in 1950-51); and replacement of ADU precipitation (step **6**, Fig. 1.1) and subsequent calcination (step **7**, Fig. 1.1) by a thermal denitration operation providing uranium trioxide directly from uranium nitrate hexahydrate.

The second uranium plant [3]

In the late summer of 1953 civil construction work had commenced on the world's first civil nuclear power station at Calder Hall in Cumbria. To meet the increased demand for fuel, the design and construction of a new facility at Springfields became a high priority issue. A 1955 Government White Paper recommended the creation of the First UK Nuclear Power Programme, of 5 GW capacity, all the reactors of which would be based on the Calder Hall 'Magnox'[8] natural, metal-fuelled design. To meet this unprecedented demand for metal fuel, the construction of a complex of new facilities began at Springfields

[8] The low U-235 content (<0.72%) of natural uranium means that a high flux of thermal neutrons is required to obtain continuous nuclear fission. The fuel cans must ideally absorb as few thermal neutrons as possible. They are made from an alloy (99% magnesium, 1% aluminium) which has a very low neutron absorption cross-section and does not react with carbon dioxide, the reactor cooling medium (hence the name Magnox – 'magnesium, no oxidation'). The term 'Magnox' has also become the generic name applied to all first-generation British gas-cooled, graphite-moderated nuclear reactors.

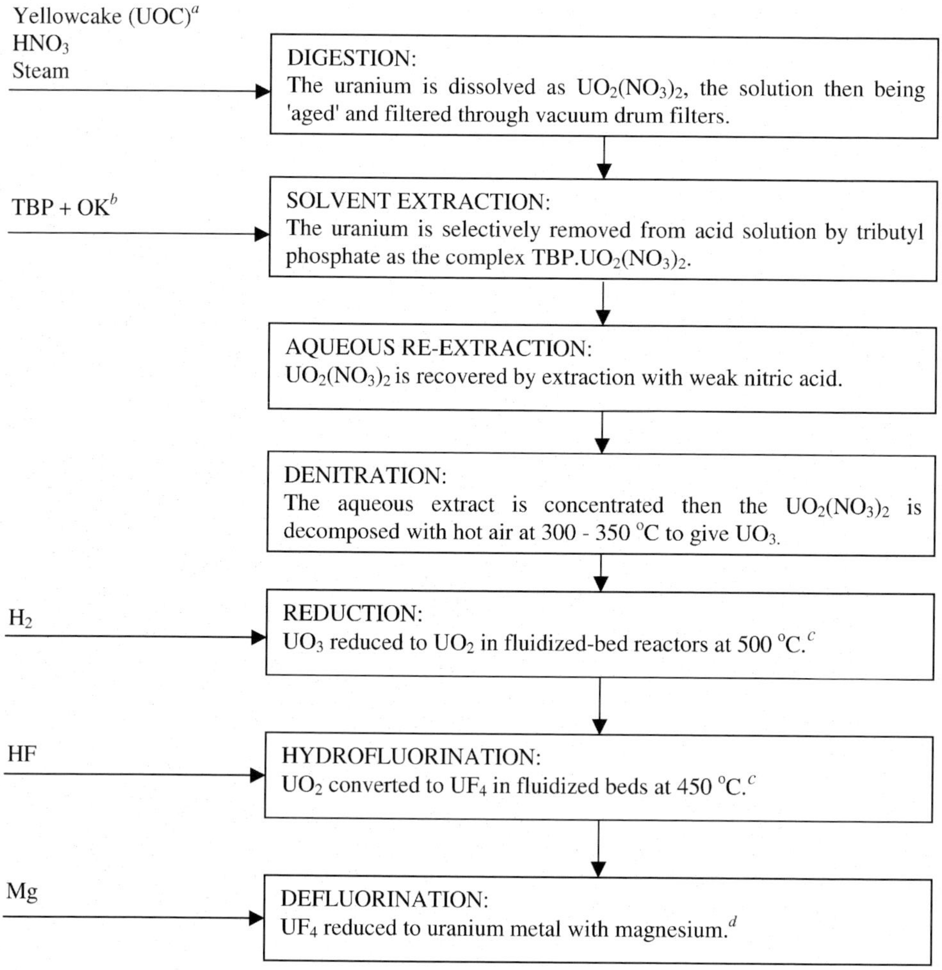

Fig. 1.2. Flowchart for Springfields' second uranium plant.

a Uranium ore concentrate (UOC) is also known as yellowcake because of its colour. Primary treatment of the ores is carried out at the mines. The ore is crushed, dissolved in nitric acid and then further treated to produce a powder containing between 60 and 90% uranium in a variety of chemical forms. The commercial ores are pitchblende, uranophane, uraninite and carnotite. Yellowcake may contain U_3O_8, $(NH_4)_2U_2O_7$, or the sodium, calcium or magnesium salts of uranium.

b Odourless kerosene, b.p. 175-325 °C.

c Rotary kiln reactors were introduced in 1978 (see the text).

d See later for details (Fluorine OUT section).

in May 1955; known collectively as 'The Second Uranium Plant'; they became operational in 1958. The set of processes employed in the new plant (Fig. 1.2) operated largely unchanged throughout the 1960s and 1970s, the only significant alterations being associated with increases in capacity to meet market demands.

Whereas all processes up to the formation of UO_2 were continuous, its hydrofluorination to UF_4 was operated initially on a batch basis, since the reaction kinetics leading to complete conversion of UO_2 (essential for later processes) were such that the necessary temperature profiles and residence times within the reactor could not be achieved on a continuous basis. By 1973 ten Inconel fluidized-bed hydrofluorinators were in operation with a combined capacity of 5000 tonnes U per annum.

In one cycle for a hydrofluorination reactor, 8 tonnes of uranium dioxide was preheated to 400 °C whilst being fluidized with nitrogen; when the bed temperature had stabilized, superheated AHF (anhydrous hydrogen fluoride) was introduced into the nitrogen stream. Bed temperature dictated the concentration of AHF in the fluidizing gas flow, and had to be kept below 450 °C to prevent sintering of the product. The AHF flow was maintained until the bed samples showed a composition containing $<1\%$ by weight of UO_2. Reactor off-gases were distilled after condensation; the head product was AHF, which was recycled, and the bottoms comprised an HF-water azeotrope[9], which was sold on as a by-product.

The kiln plant

The expansion of world markets in the early 1970s prompted by the growth of civil nuclear power programmes offered an opportunity for BNFL to develop a business based on the toll conversion of UOC to uranium hexafluoride. The existing dissolution, purification and denitration technologies of the second uranium plant clearly could be upgraded and expanded to accommodate the new business; by contrast, increasing the scale of the fluidized-bed reactors used for reduction of UO_3 and hydrofluorination of UO_2 was economically unacceptable. Following a technological review, rotary kiln technology was chosen as the basis for plant expansion.

BNFL had had experience of operating rotary kilns since 1971 (see below), and development through the 1970s demonstrated that a continuous low-pressure counter-current, gas-solid reaction system was both practical and economic. The 'rotary kiln UF_4 plant' installed converted pure uranium trioxide to uranium tetrafluoride in three stages:

(i) Hydration of UO_3 with acidified (nitric acid) water, a 'wetting' agent and air (this 'activates' the oxide and increases its surface area) in a conventional continuous mixer which delivers hydrated UO_3 as a dry free-flowing powder to a reduction kiln.

(ii) Reduction of hydrated UO_3 to UO_2 at ~480 °C by passing it through a horizontal rotating (3 rpm), stainless steel kiln (10 m long × 1.2 m dia.) against a counter-current flow of H_2.

(iii) Hydrofluorination of the UO_2 by screw-feeding it to a horizontal, rotating Inconel kiln (16.7 m × 1.2 m) with a series of heated zones along its length, the counter-current gaseous HF feed being generated and maintained from a self-regulating flash boiler.

Uranium tetrafluoride (a green solid, m.p. 1036 °C) produced in this manner either forms feedstock for metal production or is routed to the Hex (uranium hexafluoride) Plant. The kiln plant was commissioned in 1978 and for some years afterwards it operated in tandem with the fluidized bed processes.

[9] An HF-water mixture containing ~37% wt/wt HF, b.p. ~111 °C.

Fluorine IN: the manufacture of uranium hexafluoride

As early as 1952 proposals were made to produce 235U-enriched uranium dioxide at Springfields, the compound being needed to make fuel for the prototype Windscale Advanced Gas-cooled Reactor (WAGR). The 'First Uranium Hexafluoride Plant' was handed over for commissioning in November 1951 and produced its first uranium hexafluoride in February 1952. All the fuel for WAGR, and indeed all the uranium hexafluoride manufactured at Springfields prior to 1968, except in development rigs, was made in this plant. The process used chlorine trifluoride, manufactured at ICI's Rocksavage Works at Runcorn in Cheshire (see below), as the fluorinating agent.

Uranium hexafluoride from chlorine trifluoride

The 'First Uranium Hexafluoride Plant' was based on batch process operation, and by 1955 seven units were operating at a total throughput of 700 tonnes U per annum. By 1959, contract requirements exceeded 800 tonnes U per annum; this necessitated plant expansion to encompass nine units, each with an enhanced processing capacity. The maximum processing capacity reached by the plant was 1200 tonnes U per annum.

The direct reaction between UF_4 and ClF_3 to produce UF_6 and Cl_2 gas was carried out in a sealed nickel reactor, known as a retort – a designation that seems to have carried through from earlier research and development. Each retort consisted of a vessel [jacketed to allow for R-11 ($CFCl_3$) cooling] containing a direct-drive, multi-bladed stirrer. It was topped by an angled reflux condenser containing perpendicular cooling tubes through which R-11 could be passed at temperatures between −35 and +80 °C. Because of the configuration of the components, the vessels were known as 'swan-necked reactors'.

Operations were carried out batch-wise using an excess of ClF_3, a typical cycle taking approximately 18 hours to complete, with the actual reaction taking 5-6 hours. With the original reactor configuration, 310 kg batches of UF_4 (requiring 72 kg of ClF_3 for complete fluorination) were used; following upgrading, the batch size rose to 414 kg of UF_4 (needing 96 kg of ClF_3 for conversion to UF_6). Once the green UF_4 powder had been loaded, each reactor was sealed, purged with N_2 and leak checked, then, with the reflux condenser set to its lowest temperature (−35 °C), the stirrer was started. Chlorine trifluoride (b.p. −11.8 °C) was added as a liquid feed at a nominal rate of 1 kg per minute, during which time the retort was held between 85 and 105 °C. Pressure was continually monitored during the reaction and the condenser off-gas (Cl_2 and the excess ClF_3) was vented to a scrubbing system at a pressure of 7 kg cm^{-2}. Experience showed that after approximately one-third of the ClF_3 had been added the load on the stirrer increased dramatically due to condensed UF_6 re-entering the retort and causing the remaining UF_4 powder to agglomerate. As the reaction continued the load decreased again as the solids inventory fell. When fluorination was complete, each retort was cooled to 50 °C (at which temperature UF_6 is a solid) before standard procedures were used to recover the UF_6 and allow it to distil into evacuated transit cylinders held at 5 °C in water baths. Over a period of time, performance deteriorated due to the accumulation of non-reactive residues and unreacted UF_4. Typically, such residues could be completely removed by treatment with a half-batch (~40 kg) of ClF_3 at 90 °C. In operation radiological problems were encountered from the accumulation of active residues (plutonium and fission products) in retorts. Thus every 160 batches (~40 te U), or approximately every four months, the retort bases were replaced.

The overall process efficiency was good: 99.3% for uranium and 77.8% for chlorine trifluoride, and the product had a typical residual chlorine content of just 0.03%. This plant operated until 1965, when production was halted to allow the construction of 'Line 1' – the first commercial-scale direct fluorination plant.

Uranium hexafluoride from fluorine

'Line 1' commenced operation in 1968. Originally rated at 2000 te U pa, the operating capacity was in excess of 3000 te U pa by 1973, and processing continued until 1982. 'Line 2', rated at 6500 te U pa, commenced operation in 1973 and ran until 1992, when it was replaced by a new plant of similar capacity.

The reaction between UF_4 and nitrogen-diluted fluorine to form UF_6 is carried out in fluidized bed reactors containing inert beds of powdered calcium fluoride. The UF_4, fed to a reactor at a controlled rate by screw feeder, combines with gaseous fluorine in a highly exothermic manner and the calcium fluoride acts as a diluent and prevents UF_4 sintering within the reactor. Bed temperature is maintained around 450 °C by forced-air cooling. The uranium hexafluoride produced passes, together with the excess of fluorine and diluent nitrogen, through sintered nickel filters to remove entrained solids and then to a condensing system, cooled to below −35 °C by R-11. The efficiency for the conversion of UF_4 to UF_6 is greater than 99% [4]. Material produced as described above is stored either at Springfields or a customer's facility, prior to shipment to an enrichment facility where the isotopic enrichment process for partially separating U-235 from U-238 is carried out.

A Digression: fluorine cells and ClF_3 production

Fluorine generation [5]

The first fluorine generator operated at the Springfields site was a medium-temperature ICI 1 kA cell (electrolyte: $KF \cdot 2HF$ maintained at ~80 °C), which fed a 3-inch diameter, continuous fluidized-bed 'hex' (uranium hexafluoride) development unit that functioned between October 1954 and March 1955. Only limited operations were carried out with the cell at this time because the downstream manufacturing unit was shut down later that year: development priority had been switched to developing a UF_4 fluidized-bed process.

Laboratory-scale work using F_2 had been carried out from just prior to the installation of the 1 kA cell, using cylinders containing 4 lb weight (at 400 psig) of 'pure' F_2 gas obtained from ISC Chemicals (Avonmouth). The bulk of this laboratory work was in support of a direct fluorination route to UF_6, although some other inorganic fluorides were synthesised as part of the research programme. Direct fluorination of UF_4 to UF_6 was incorporated into the design of the Prototype Hex Plant in 1959 as part of the Civil Advanced Gas-cooled Reactor (CAGR) programme; cell options being considered for the provision of F_2 to feed the plant were: ICI 4.4 kA; USAEC (United States Atomic Energy Commission) 5 kA; and Allied Chemical Corporation 5 kA units.

The UKAEA entered into a dialogue aimed at purchasing cells of ICI or USAEC design, and a conceptual design based on the Allied's cell was produced. Eventually all three avenues of approach were shelved in favour of ICI's suggestion that supplies of refrigerated liquid fluorine could be made available from their Rocksavage works at Runcorn in Cheshire. This offer was made possible by a technology exchange between ICI and Allied, which lead to the installation of a fluorine liquefaction plant at Rocksavage. Thus, a purpose-built demountable triple-skinned tank cooled with liquid nitrogen (b.p. −196 °C; F_2 boils at −188 °C), and based on Allied Chemical Corporation's designs, was manufactured for the trade in liquefied fluorine which ensued. The tank held a maximum inventory of 500 kg of F_2 and a total in excess of 5 tonnes of F_2 was transported across Lancashire (a distance of about 45 miles) in the days before the M6 motorway existed, to provide feedstock for the initial operations of the Prototype Hex Plant between 1961 and 1963.

By 1963, the Prototype Hex Plant had been satisfactorily demonstrated at the design throughput of 150 te U pa, and it was obvious that the potential existed to double that throughput using the same design. Following a technology review, the decision was taken to rebuild the hex plant and incorporate on-site F_2 generators supplied by ICI directly from their ClF_3 plant. Thus, a termination agreement for the current trade in chlorine trifluoride between ICI and the UKAEA was drafted which included the statements: 'ICI hereby grant to the Authority a royalty-free non-exclusive non-transferrable right to the use of ICI know-how relating to the construction and operation of fluorine and chlorine trifluoride plants together with the requisite royalty-free non-exclusive non-transferrable licence under the following British Patents ...' (5 patents were listed); and 'ICI shall also make available to the Authority free of charge information to enable the Authority to set up and operate a pilot chlorine trifluoride plant.' Initially, three 4.4 kA fluorine cells were transferred from Rocksavage to Springfields, complete with two DC motor generator sets and HF absorption equipment; these cells were operational by 1964, although they were initially down-rated to 3 kA 'for practical reasons'. Three more cells were received at Springfields in early 1966. In 1969, Union Carbide offered for sale redundant fluorine cells from the Oak Ridge K-25 Plant. These were of nominal 5 kA capacity, and one was purchased and installed, for test purposes, in the Prototype Hex Plant, but after about two months its use was discontinued due to non-availability of spares.

The ICI 4.4 kA cell design (nominal F_2 production rate 3 kg per hour) formed the basis for the first cellroom installation for the commercial-scale manufacture of UF_6; known as 'Line One', this comprised 26 units and was built in 1968. A second batch of 24 cells, 'Line Two', was built in 1972/73 and was based around the same cell design, uprated to 5 kA and provided with a significantly modified electrical supply system. Following experience with a 7.5 kA prototype (operated briefly in 1971), seven 11 kA cells were added as 'Line 2A' in 1978, followed by seven 15 kA units ('Line 2B') in 1979. The larger capacity units, which incorporated features such as recirculating electrolyte and external electrolyte cooling, did not prove to be as successful in operation as hoped for, and they were replaced by smaller capacity (5 kA) units during major refurbishment of the F_2 production plant in the early 1990s.

The essential features of the ICI/BNFL design cells are: provision for the ease of removal and replacement of anodes and anode assemblies, enabling cells to be maintained at high current efficiency and minimum power consumption; the use of non-polarising porous carbon anodes, attached to nickel and mild steel hangers by means of a nickel spraying technique; a cathodic body, permitting a cheaper mild steel construction, by comparison with Monel-bodied USAEC cells; and the operating characteristics of individual anodes can be monitored. Details of the design and operational features of the cells can be found in reference [6], and the ICI pioneer Rudge [7] has discussed the scientific background related to the development and performance of these cells. An informative comparison between British fluorine cells and other designs has been written by Ring and Royston [8].

Chlorine trifluoride manufacture

Uranium isotope enrichment is carried out in the UK at Capenhurst near Chester. Prior to the commercial application of centrifugation, which replaced gaseous diffusion, as the enrichment process, there was a requirement for ClF_3 as a diffusion membrane cleaning and passivation agent. The process that BNFL operated to manufacture ClF_3 was developed by ICI.

Small-scale production of ClF_3 had commenced at ICI's Weston Point plant in 1941, and production expanded later in the decade since the compound was considered to be an easy way to ship fluorine. ICI supplied all the requirements of Britain's fledgling nuclear industry from 1946, and the process used was scaled up before 1947 (between 1949 and 1951, a 1.5 kg ClF_3 h^{-1} unit was operated intermittently at ICI's Rocksavage plant). By 1959, ClF_3 required for the 'First Uranium Hexafluoride Plant' at Springfields was being manufactured by ICI in a plant fed by 14 $\times$ 1.4 kA fluorine cells. Following the closure in 1965 of the first UF_6 plant and the switch to F_2 as the means for converting UF_4 to UF_6 (see earlier), the remaining trade, involving use at the enrichment plant at Capenhurst, was not a viable proposition for ICI. In preparation for this eventuality, the 1964 technology transfer agreement referred to earlier allowed the UKAEA to prepare to manufacture ClF_3 for its own applications. Construction of a plant capable of producing at least 10 te pa was approved in 1967. Competition between various options, including proposals for a 100 te pa unit, resulted in considerable delay between approval and operation, but the commercialisation of centrifugation[10] reduced the requirement back to a nominal 10 te pa plant; that facility commenced operation in 1973. The plant was simple in principle: a mixture of Cl_2 (fed from cylinders purchased from ICI) and F_2 (generated in-situ from two 1 kA cells) in three-to-one stoichiometric ratio was passed through an all-nickel double reactor system, the two reactors being maintained at temperatures of 320 and 270 °C respectively. The reaction proceeded in two stages: $Cl_2 + F_2 \rightarrow 2ClF$; $ClF + F_2 \rightarrow ClF_3$, and the lower temperature of the second reactor helping to minimise thermal dissociation of the required product, which condensed in nickel product receivers cooled to -40 °C by R-11. Operationally, although rated at 10 te pa, the plant ran smoothly for short periods at production rates in excess of 15 te pa. The shutdown of BNFL's U-235 diffusion enrichment plants at Capenhurst ended the internal trade in ClF_3, so by 1980 the plant was shut down and dismantled, although the 1 kA F_2 cells continued to operate for several years as test beds for cell development programmes.

[10]In centrifugation, gaseous uranium hexafluoride is fed continuously at low pressure into the rotor of a high-speed centrifuge. As the gas rotates the heavier molecules concentrate at the rotor wall whilst the lighter molecules concentrate near the axis of the rotor. This produces two product streams, which can be continuously withdrawn from the centrifuge. The effect is small and commercially many centrifuges have to be connected together in series ('cascaded') to achieve significant changes in the isotopic content.

Fluorine OUT: the manufacture of uranium metal and uranium dioxide

Uranium metal billet production

Britain's first commercial power reactors, the Magnox stations, required uranium metal, which was made initially by reducing uranium hexafluoride with calcium. From the start of operations in the 'first uranium plant' it was important to produce bulk uranium metal that was not prone to failure due to brittle fracture. This meant operating a process that achieved complete separation of the metal product from the slag (calcium fluoride) produced. In practice enough heat had to be generated by the reaction to melt both the metal and the slag so that their differing densities affected a complete separation. Calcium chips were added to drums of uranium tetrafluoride powder held in enclosed cabinets. The reactants were thoroughly mixed under a blanket of inert gas (argon) and then transferred to a vertical, calcium fluoride-lined crucible. The crucible was sealed into a 'firing chamber' where reaction was initiated by inserting an ignited pellet composed of a mixture of potassium nitrate and lactose.

When the reaction was complete a uranium metal billet was mechanically separated from the calcium fluoride slag and sent to the metallurgical and canning plants for further processing. The slag was sent to a recovery plant for crushing and leaching to remove residual uranium. The last calcium reduction at Springfields was carried out in 1957, the year when magnesium began to be used. The falling price of high-purity magnesium and the comparative ease with which it could be handled had prompted the replacement of calcium by magnesium in the metal-forming reduction step. The reaction between magnesium metal and UF_4 is significantly less exothermic than that of calcium and necessitates external heating of the reactants. With magnesium, a more-intimately-mixed reaction mixture is needed, as is a higher purity grade of UF_4.

To feed the magnesium reduction process, UF_4 and magnesium turnings are intimately mixed and compressed into 'compacts' weighing about 3 kg. Compacts are stacked and loaded into a graphite liner, which in turn is loaded into a stainless steel reactor vessel. The reactor vessel is heated inside an electric furnace, evacuated and purged with argon. When the temperature reaches about 650 °C the charge fires, and the temperature is then maintained for sufficient time for the metal/slag mixture to separate under gravity. Molten uranium metal falls into a catchpot in the base of the reactor vessel from whence a 350 kg billet can be separated from the magnesium fluoride slag prior to cleaning. Residual uranium is recovered from the fluoride slag by acidic leaching.

Deconversion of uranium hexafluoride to uranium dioxide

In order to manufacture ceramic-grade, enriched (up to 5% U-235) uranium dioxide a process was required to 'deconvert' enriched UF_6. Springfields' original process, operated between 1953 and 1965 to provide fuel for Windscale's Advanced Gas Reactor (WAGR), was batch reactor based, complex and inefficient, and determined to be unsuitable for scale-up to commercial operation. By 1965, however, a much-improved continuous-operation deconversion plant with a throughput of 300 te U pa had been commissioned. The processes employed (Fig. 1.3) produced high-density, ceramic grade UO_2 suitable for fabrication into fuel pellets.

The Integrated Dry Route (IDR) process

Once process development work at Springfields had demonstrated the versatility and adaptability of rotary kilns for multi-stage, continuous chemical processing, the so-called wet route (Fig. 1.3) from UF_6 to UO_2, was replaced in 1971 by the IDR process, which is well-documented [2]. Briefly, UF_6 is vaporised and metered into a rotary kiln where it reacts with a concurrent flow of steam before meeting a countercurrent flow of hydrogen; the reaction with steam forms particulate uranyl fluoride (UO_2F_2), which is transported through the kiln by a scroll feeder; further reaction with steam and hydrogen produces high quality UO_2 powder. The kiln's off-gas stream contains HF, which is scrubbed prior to dis-

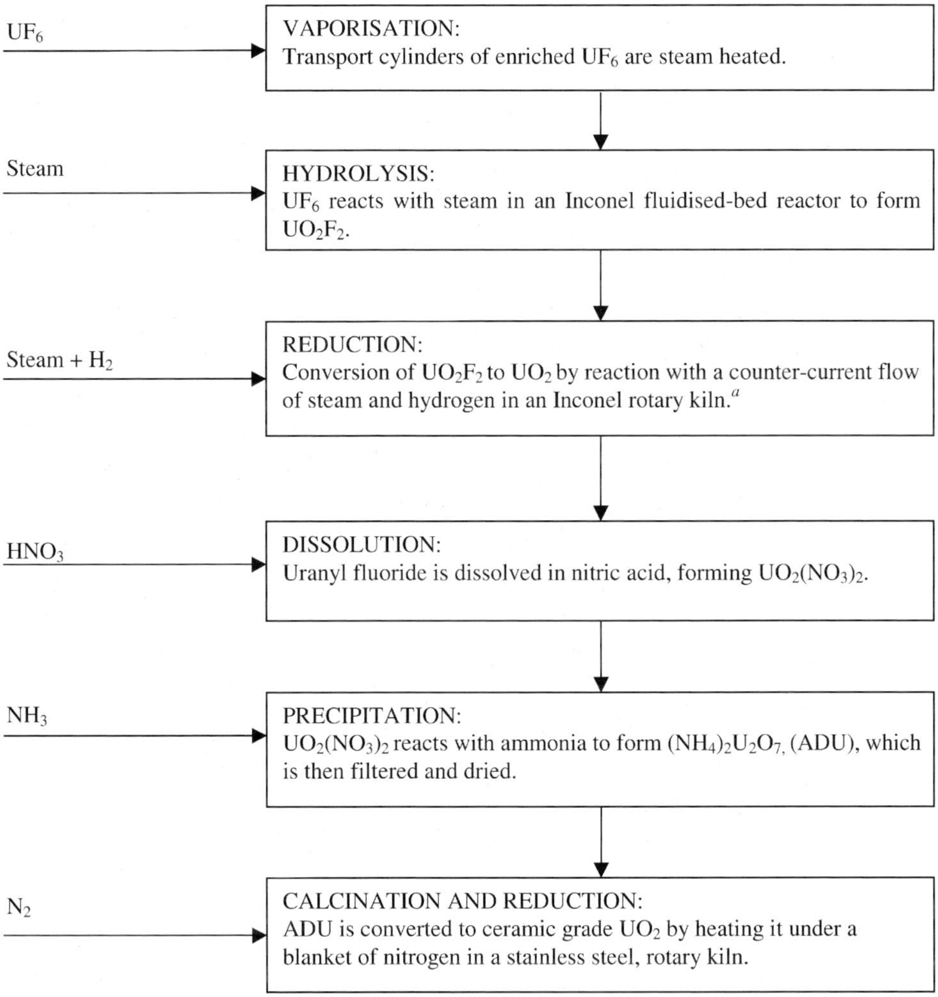

Fig. 1.3. Flowchart for Springfields' continuous-operation deconversion plant.

a The physical properties of the product powder made it unsuitable for manufacturing ceramic-grade fuel pellets and necessitated further processing.

tillation. In 1994, six rotary kilns with a total throughput of 650 te U pa were in operation. The IDR process now forms the heart of Springfields' new Oxide Fuels Complex, recently built to replace the earlier facilities.

Fluorine Recovery Plant (FRP)

Currently BNFL is researching and developing a route by which elemental fluorine can be recovered from 'tails depleted hex', *i.e.* UF_6 product from the isotope enrichment process in which the fissile isotope (U-235) content has been reduced significantly below the natural abundance of 0.72%. A commercial-scale demonstrator is currently being built at Springfields and BNFL's immediate objective is to close the fluorine cycle within the nuclear industry by reusing the FRP fluorine product to make new natural UF_6.

Diversification: F2 Chemicals Limited

Industry in general has made many enquiries of BNFL, and formerly UKAEA, concerning the applicability of its fluorine technology to the manufacture and processing of non-nuclear materials. It became quite evident that there were commercial opportunities beyond the nuclear area which could be exploited by the application of aspects of BNFL's direct fluorination expertise. This led, eventually, to the exploration of diversification opportunities based on the exploitation of elemental fluorine manufacturing and handling capabilities. A subsidiary company, F2 Chemicals Limited (formerly BNFL Fluorochemicals Limited), operating from the Springfields site was set up in 1992. F2 Chemicals currently operates a 150 te pa electrolytic fluorine plant and other smaller cells for development purposes. The fluorine produced is used to manufacture perfluoroalkane fluids via cobalt fluoride fluorination, a technology acquired from Rhône-Poulenc. New direct fluorination technologies developed by F2 Chemicals to manufacture pharmaceutical and agrochemical intermediates are currently in the development stage at pilot plant scale.

Acknowledgements

I wish to thank British Nuclear Fuels plc and F2 Chemicals Limited for granting me permission to write this chapter. My thanks go also to my mentors, who set me on the path to true enlightenment – a career with fluorine, and to all my colleagues, friends and acquaintances, both past and present, who have helped and guided me over the years.

References

1 R. W. Clark, *The Greatest Power on Earth: The Story of Nuclear Fission*, Sidgwick & Jackson, London, 1980.
2 A more thorough review of the entire manufacturing process is contained in a series of articles by D. H. Locke and J. R. Smith, *The Nuclear Engineer, 34* (3, 4) (1993) and references contained therein.
3 H. Rogan, 'Fuel Manufacturing Technology and Production Facilities at BNFL Springfields', BNFL, Warrington, 1977.
4 H. Rogan, 'Facilities and Technology needed for Nuclear Fuel Manufacture', International Atomic Energy Agency, Vienna, 1973.
5 H. R. Cartmell, private communication.

6 J. F. Ellis and G. F. May, in R. E. Banks, D. W. A. Sharp and J.C. Tatlow (eds.), *Fluorine: The First Hundred Years (1886-1986)*, Elsevier-Sequoia, Lausanne, 1986, pp. 133-148.
7 A. J. Rudge, 'Production of Elemental Fluorine by Electrolysis', in A. T. Kuhn (ed.), *Industrial Electrochemical Processes*, Elsevier, Amsterdam, 1971, pp. 1-70.
8 R. J. Ring and D. Royston, 'A Review of Fluorine Cells and Fluorine Production Facilities', Report AAEC/E281, September 1973.

BIOGRAPHIC NOTE

Malcolm J. Atherton was born in Liverpool, England in 1952 and received his chemical training at the University of Newcastle-upon-Tyne where he was introduced to fluorine chemistry by Drs Bob Dobbie and Bruce Tattershall. In 1974 he moved to the University of Leicester where, under the supervision of Dr John Holloway he worked on chalcogenide fluorides of molybdenum (VI) and tungsten (VI) for his PhD. He joined BNFL Fuel Division in 1977 where his career path became inextricably linked with fluorine and uranium hexafluoride manufacture. A founder member of the team which became BNFL Fluorochemicals Limited, and subsequently F2 Chemicals Limited where he was R&D Manager until 1997, he rejoined BNFL in May 2000 to work on the Fluorine Recovery Plant.

Malcolm J. Atherton

Chapter 2

LOOKING IN ON FLUORINE CHEMISTRY IN RUSSIA AND UKRAINE

DEREK L. AVERRE

Centre for Russian and East European Studies, School of Social Sciences, The University of Birmingham, Edgbaston, Birmingham B15 2TT, UK

Introduction

In the Soviet period a lot of information about R & D and production of fluorine compounds was classified because of their high-technology military-related applications. Scholarly research by Soviet chemists was published in Russian and Western journals, but in general travel and direct collaboration with overseas scientists were restricted. Nevertheless there is a history of contacts between Soviet (now, of course, Russian and Ukrainian) and Western fluorine chemists, dating from the late 1950s, which established the former as very much part of the global fluorine community.[1] The political and economic changes of the past decade have opened up the field to investigation, so we now know more about the scale of the fundamental research effort in fluorine chemistry and a lot more about applied research and the industrial production of fluorine compounds in Russia and Ukraine, as institutes and manufacturers try to forge business links with their Western counterparts. Also many leading fluorine chemists from these countries have spent periods working in universities or companies abroad, and some have emigrated.

One British expert has told me that the USSR was probably the world's leading nation in basic fluorine chemistry in terms of the number of chemists involved and quality of research. The eminence achieved by the leading research institutes in this sphere can be partly explained by the high priority given to their work by the Soviet government and the Academy of Sciences, whose generous funding allowed them to concentrate on pure research and to a large extent shielded them from demands to apply their findings in a commercial environment. As a result many talented chemists were attracted into the field. In Soviet times the Institute of Organic Chemistry (IOC) in Novosibirsk and the A. N. Nesmeyanov Institute of Organoelement Compounds (Russian acronym INEOS) in Moscow were (and may still be) two of the biggest centres in the world in terms of staff directly employed on organofluorine research, and the Russian Research Centre (RRC) 'Kurchatov Institute' is a world leader in the development of inorganic fluorine compounds. Generous funding was not the only attraction to Soviet chemists, however. The pride and prestige attached to scientific work in a country where the very idea of science was revered, and the intense intellectual interest in the rapidly developing field of fluorine chemistry in the

[1] The chapter in this volume by Gerald Brooke about the Novosibirsk school gives a more personal colouring to this relationship.

post-WW2 period, meant that the fluorine community in the USSR produced many distinguished chemists.

Developments in science, technology and industry since the break-up of the USSR have had a far-reaching and largely deleterious effect on R & D and production, however, which fluorine has unfortunately not escaped. Without the substantial state support previously afforded them, institutes are facing financial difficulties, in some cases extreme. The sharp drop in funding from the defence budget – an important source of finance for many of the leading fluorine institutes – has hit both military and civilian-related research projects. Chemicals, equipment and even literature are often difficult to get hold of. The Russian fluorine chemicals industry, itself suffering from economic and organisational problems and a lack of investment, is unable or unwilling to finance R & D. Due to priority accorded to defence-related chemistry, there was relatively limited commercial development in the Soviet economy of certain areas in which fluorine chemicals are widely used in the West, and it is difficult for Russian companies to break into well-established Western markets for a variety of reasons.

My extensive discussions with representatives of institutes and industrial enterprises suggest that there is unlikely to be any improvement on the current situation in the foreseeable future. The lack of any thoroughgoing reform of science means that finance is still centrally allocated through the Russian Academy of Sciences or, in the cases of the RRC 'Kurchatov Institute' and the Russian Scientific Centre (RSC) 'Applied Chemistry', directly from the government; together with the availability of grants (particularly from the Soros Foundation in 1994-96) and overseas contracts for research, therefore, this has meant that many fluorine research programmes and staffing have so far been maintained. The longer-term fundamental problems already described – the depressed domestic market, difficulties in breaking into international trade and lack of investment – make it likely that significant losses will occur over the next few years, however.

INEOS is in a relatively healthy state at present, mainly due to its high reputation (bringing grants for priority research from the Russian Academy of Sciences (RAN) and international programmes) and to the breadth of its research, which serves a number of sectors and so has a wider potential market in industry; its location in Moscow is also a positive factor. Even at INEOS the outlook is uncertain, however, as it is still heavily dependent on central funding; the majority of its funding in 1996–97 came from the RAN, with only around 30% from contract work. A similar situation exists at the Kurnakov Institute of General and Inorganic Chemistry, 70% of whose funding is currently provided by the RAN. The RRC 'Kurchatov Institute' and the RSC 'Applied Chemistry' are still receiving enough central funds to keep most research going, although they have lost some, particularly younger, research staff. The institutes in Novosibirsk, Ekaterinburg and Irkutsk are in a less favourable situation; they are having less success in attracting central funding and grants (Ekaterinburg has to find 70% from commercial work) and their organofluorine research base might well shrink over the next few years. Fluorine chemistry in Ukraine, where research is concentrated in the Kiev-based Institute of Organic Chemistry and the Institute of Bioorganic Chemistry and Petrochemistry (both now part of the Ukrainian Academy of Sciences), is in similar difficulties.

Various ways of minimising research cutbacks are being tried; forms of cooperation with overseas companies include offering product, R & D services, or licences for technology to manufacture fluorine compounds. This is done either in the name of the institute itself or by commercial companies set up within institutes and using their facilities. There have been successes; for example, the independent firm 'P & M' was set up by INEOS chemists in 1990 to research and develop novel methods of synthesis and to produce organofluorine chemicals for marketing abroad (production facilities have been established both at INEOS and at the premises of other institutes and industrial manufacturers). However, there have been problems in putting these agreements on a sound commercial footing. There has not yet been any large-scale direct investment by Western companies in Russian fluorine materials production, although some are known to be seeking investment opportunities.

Basic organofluorine research

In Moscow

One does not have to go very deeply into the history of Soviet fluorine chemistry to come across the name Ivan Lyudvigovich Knunyants. A Professor at INEOS and Major-General at the S. K. Timoshenko Military Academy of Chemical Defence, Knunyants literally inspired generations of Russian fluorine chemists[2]. Born in 1906, he was a pupil of A. E. Chichibabin and his achievements are by no means limited to organofluorine chemistry, a field he began to develop in 1941. Much of his early work concentrated on probing and rationalising the unusual chemical behaviour of fluoro-olefins (particularly perfluorinated examples such as $CF_3CF{=}CF_2$ and $(CF_3)_2C{=}CF_2$) and related compounds (this was reviewed in 1974 [2]), and his sizeable team also carried out much of the fundamental chemistry underpinning the development of fluorocarbons, on which a large industrial effort in the Soviet period was subsequently based. With more than 1300 publications (including about 200 patents) to his credit, the majority dealing with organofluorine compounds [1, 3], it is perfectly understandable why Knunyants[2] is viewed as one of the giants of modern fluorine chemistry.

Knunyants could, as one of his pupils told me, 'inspire either love, hatred or fear' – usually the first of these emotions – but never indifference. The source of the inspiration he provided to so many lay in the fact that, as Knunyants himself said, 'he was above all preoccupied with the psychology of fluorine as much as its anatomy or physiology' – in other words the behaviour of a then still largely unexplored element. He was very much the model of the best type of Soviet scientist, combining imagination with a classically rigorous and systematic approach to research – even those of his pupils who revere him still recall coming under his the lash for such misdemeanours as not adequately recording their experiments – and broader intellectual interests, in Knunyants' case chess and picture collecting and restoring. But perhaps his defining feature was that, despite his reputation and influence as an Academician and sometime President of the Mendeleev Chemical Society (as well as chief consultant to the military on chemical matters), Knunyants was a liberal figure, free of the conceit which characterised some senior people in the Soviet scientific hierarchy. The members of his school were relatively free from ideological pressures and could choose their research under his patronage. His death in 1990, which coincided with the final months of the Soviet era, might be seen as the closing of a chapter of the history of Russian science.

[2]For a succinct account of the life and work of I. L. Knunyants, see ref. [1].

One of Knunyants' early postgraduate students, Aleksandr Vasil'evich Fokin, emerged as a leading fluorine chemist in his own right. Born in 1912, Fokin spent most of the earlier part of his career at the Chemical Defence Academy before becoming the deputy scientific secretary of the Presidium of the USSR Academy of Sciences and (from 1974) a member at INEOS, of which he was Director from 1980-88. In the post-war period, he began (together with Knunyants) to carry out seminal synthetic and mechanistic studies on organofluorine compounds, particularly fluoro-olefins. Like Knunyants his work in organic chemistry extended beyond fluorine, into such areas as industrial routes to ethylene oxide and propylene oxide and the synthesis and production of new phosphorus- and sulphur-containing materials for use in the recovery and purification of non-ferrous metals; also like Knunyants, he was widely respected by his colleagues. Professor Fokin died in mid-1998, just as I was finishing this article. Though evidently infirm, he generously agreed to speak to me during a visit to INEOS in 1997 and asked to be remembered to his British colleagues, some of whom still recall his impressive singing voice heard during one of his visits to the West.

Today there are between 30 and 40 fluorine scientists working at INEOS, mainly in the Laboratory for Organofluorine Compounds (around 20 chemists) and the Laboratory for Physiologically-Active Organofluorine Compounds (8 chemists). The former is now headed by Professor Georgii Belen'kii; he arrived from the Mendeleev Institute via a spell in industry in the mid-1960s and took over on the death of Lev German in 1994 [4], another name well-known to Western fluorine chemists. Research in Belen'kii's group is mainly directed towards fluoroaliphatic compounds – developing new methods of synthesis and investigating their reactivity – and fluoromonomers. Fluorine compounds for 'blood substitution' and a number of other medical applications have also been developed in his laboratory. The Laboratory for Physiologically-Active Organofluorine Compounds is headed by Professor Nikolai Chkanikov, who took over from Dr A. Kolomiets in 1995. The main research interest there lies in strongly electrophilic organofluorine compounds suitable as precursors of new biologically-active substances; this follows on from INEOS' work in the field of bioactive organophosphorus compounds for application in medicine and plant protection. Chkanikov himself is a specialist in the use of biologically-active fluorine compounds for organic synthesis; fluorine-containing amino acids and peptides represent a new field of research for his team, which is collaborating with a group at Leipzig University. Two other fluorine-related research groups are active at INEOS: the Team for Fluorocarbons, which studies synthetic methods and structure-property relationships in organofluorine chemistry, and the Team for Technical Chemistry, which investigates methodology for the production of organofluorine compounds.

Prior to my 1997 visit mentioned above, I was at INEOS in 1996 as a member of a UK DTI/Chemical Industries Association delegation on a science and technology mission. During these two visits I was fortunate to interview most of the chemists mentioned above, and hence share their recollections of Knunyants and early developments in the fluorine field. I also spoke to, or heard about, several others well-known to fluorine chemists internationally. Lev Gervits (who came from Moscow State University in 1971) and Yurii Zeifman (who arrived as long ago as 1961 to work under Natal'ya Gambaryan, who is now retired) are still working in Professor Belen'kii's lab.; other veterans of what Belen'kii described as the 'golden years' of the 1960s and 1970s, such as Yurii Cheburkov and Sergei Sterlin (now editor of the patents section of the *Journal of Fluorine Chemistry*), are also still active. Evgenii Rokhlin, Viktor Cherstkov and Vera Popkova (who worked with Rafael Bekker until his death in 1983 [5]) are among the better-known of the younger generation. The Institute is striving to carry on the fluorine tradition through a scheme designed to fund younger chemists.

Some research is also carried out by the Scientific Industrial Association 'P & M', whose director, Dr Sergei Igumnov (another of Knunyants' pupils), heads the Team for Technical Chemistry at INEOS. This organisation concentrates on developing commercial outlets for different groups of products, including perfluoroalkyl halides.

In Novosibirsk

Exploits in fluorine chemistry at the Novosibirsk Institute of Organic Chemistry (NIOC), founded by N. N. Vorozhtsov in the late 1950s and famous for its massive contributions to knowledge of polyfluoroarene chemistry [6], are covered in this book by Gerald Brooke, so I shall be brief here.

NIOC is part of the Siberian Division of the RAN, and its principal research activity still lies in the field of polyfluoroaromatic chemistry. Its research developed separately from the INEOS school, although Professor Vorozhtsov and his contemporaries maintained good scientific links with Professor Knunyants and his group.

I paid a very short visit to the NIOC in 1996 while on the DTI/Chemical Industries Association mission in the company of Dr Tony Bastock (who studied fluorine chemistry at Birmingham University) and Bill Denison (a Durham University graduate and latterly MD of BNFL Fluorochemicals). During the visit two things became clear. The first was that the high scientific standards associated with the Institute were being maintained. The second was that despite its distinguished research record, NIOC is facing greater problems than INEOS. The remoteness of Novosibirsk (Moscow appears to attract a greater share of grants and commercial contracts) and its relatively narrow specialisation are largely responsible for this, but the generation factor may also play a part; most of the principal fluorine chemists have been there since the early days and are now in their sixties. Nevertheless the fluorine team survives and was very generous with its time and hospitality. When we left Novosibirsk, Professor Platonov (now on the editorial board of *the Journal of Fluorine Chemistry*) insisted on getting up at 5 am after a pleasant dinner the previous evening in the House of Scientists to accompany us to the airport; he seems to have the energy of a much younger man.

In Ekaterinburg

Moscow and Novosibirsk are unquestionably the main centres of basic organofluorine chemistry in Russia. But another institute extensively involved in the field might qualify as the 'Urals school': the Institute of Organic Synthesis (IOS) in Ekaterinburg, formerly Sverdlovsk, part of the Urals Division of the RAN. In 1920, Isak Postovskii, who worked in Munich with Emil Fischer, was awarded the chair of organic chemistry at the Urals Polytechnical Institute, which became the cradle of organic chemistry in the Urals. In 1948-49 Postovskii (at what was by then known as the Institute of Chemistry) was asked to develop fluorocarbon technology for the nuclear industry: subsequently he worked with Boris Lundin (now retired), Sergei Sokolov (latterly of VNIISK – see below) and Vladimir Kazakov on the development of organofluorine chemistry.

Since 1976 (the IOS split off from the Institute of Chemistry in 1993) Professor Kazimir Pashkevich has headed the fluorine effort at Ekaterinburg which concentrates on fluorinated complexing agents, metal chelates, and functionally-substituted carbonyl compounds for use in the synthesis of fluoroalkyl-containing chemicals, the IOS' strongest field. Research has been done on per- and poly-fluoroalkyl halides, sulfides and related compounds (in recent times there has been no funding for this) as well as on ring-fluorinated aromatic compounds for pharmaceutical applications, but that work also has suffered since it was done in collaboration with the RSC 'Applied Chemistry' branch in Perm', which is facing difficulties. Pashkevich's students have included V. I. Saloutin, L. E. Deev, and, more recently, V. P. Filyakova.

In Irkutsk

Fluorine chemistry has been practised at the IOC in Irkutsk, one of the largest industrial centres in Siberia, but now appears to be under threat; its leading figure, Professor Mikhail Voronkov, is a full member of the Russian Academy of Sciences and has a distinguished research record, but is now in his late 70s and the Institute's funding has been drastically cut. The main research interests seem to lie in fluorinated organosilanes, polyfluorinated vinyl ethers, epoxides and polymers.

In Chernogolovka

The Institute for Physiologically Active Substances (IPAS) in Chernogolovka (Moscow Region), which was formerly a closed institute, was founded in 1979; early fluorine work there was headed by Professor Ivan Martynov, who worked at INEOS. Under Dr Valerii Brel' the group is now occupied with research on physiologically-active heterocyclic compounds. Part of this includes studies in fluorine and phosphorus chemistry for medicinal applications, and research on fluorine-containing sugars and amino acids (a new field in Russia). Although pure fluorine chemistry is of less importance these days, the laboratory for organofluorine compounds still survives.

There are two other institutes in Chernogolovka with an interest in fluorine compounds. The Institute of Chemical Physics is active in several areas of fluorine-related research, most prominently via studies by Drs L. T. Eremenko and G. N. Nestelenko on fluorinated nitroalkanes and fluorine-containing heterocycles. At the Institute of Energy Problems of Chemical Physics, Dr Aleksandr Kharitonov and his colleague Dr Yurii Moskvin carry out both fundamental and applied research on surface modification of polymers via direct fluorination (with F_2 or F_2-inert gas blends), oxyfluorination (with F_2-O_2 blends) and post-oxyfluorination graft copolymerization [7].

Chernogolovka is a very pleasant and quiet little town (about an hour's drive from Moscow) where I have spent several pleasant afternoons in the company of Sasha Kharitonov and Valerii Brel', sampling the local fresh produce and (inevitably!) sundry beverages; I'm looking forward to taking up an outstanding invitation to pick mushrooms and eat *shashliki* in the forest surrounding the town.

In Kiev

Moving a few hundred miles westwards into Ukraine, the Institute of Organic Chemistry (IOC) and the Institute of Bioorganic Chemistry and Petrochemistry (IBCPC) in Kiev are the only former Soviet centres outside Russia extensively involved in fluorine research.

Their main chemists are well-known internationally. Former director Professor L. M. (Lev) Yagupol'skii, who though retired is still active, pioneered work at the IOC on the synthesis of fluorinated dyestuffs and the use of fluorosulfuranes as fluorinating agents. As a young man Lev Yagupol'skii was encouraged to work in fluorine chemistry by the then director of the IOC, Professor A. I. Kipriyanov, who, however, warned him that it would be a lonely occupation (at the time research at INEOS in Moscow was still in its early stages). The fluorine programme is now directed by his son, Professor Y. L. (Yurii) Yagupol'skii – giving a literal meaning to the phrase 'the fluorine family!' – who is now a member of the editorial board of the *Journal of Fluorine Chemistry*. The roll call of Lev Yagupol'skii's other pupils is extremely long – 7 doctors of science and 72 candidates of science! – and I hope he will forgive me for not listing them all; among the better-known names are S. V. Pazenok, A. A. Kolomeitsev, I. I. Gerus, V. A. Soloshonok, Yu. M. Pustovit, I. I. Maletina, N. V. Ignat'ev, R. Yu. Garlyauskaite and A. N. Rechitskii. Another branch of the Kiev fluorine school was represented by another former IOC director, Professor Leonid Markovskii (who sadly died while this account was being written), a pupil of Academician A. V. Kirsanov; this branch is being continued by Markovskii's pupils Yu. Shermolovich and V. Pashinnik. Work on organofluorine-based intermediates and process technology relating to plant protection agents and pharmaceuticals have traditionally been, and remain, an important part of the IOC's research programme.

The IBCPC was formed from the IOC several years ago; indeed its Director, Professor Valerii Kukhar' (a vice-president of the Ukrainian Academy of Sciences), was another pupil of Academician Kirsanov. The main interests there centre on medical applications of fluorinated amino acids and fluoropolymers. Professor Kukhar's team includes A. E. Sorochinskii, V. A. Soloshonok and A. M. Kornilov. Dr Gerus (a pupil of Lev Yagupol'skii) has been studying β-alkoxyvinyl polyfluoroalkyl ketones at IBCPC, and Professor O. Kolodyazhnyi is working on phosphorus-fluorine compounds.

Applied organofluorine chemistry

Russia has a substantial applied research effort in the field of organofluorine chemistry. Fluorine monomers for the production of fluoropolymers are mainly the province of the RSC 'Applied Chemistry' (formerly the State Institute of Applied Chemistry, and often referred to by its former acronym GIPKh) in St. Petersburg, which is also the head institute for chlorofluorocarbons (CFCs) and develops numerous other fluorine compounds. R & D in the sphere of fluorinated plastics and rubbers is carried out at two other St. Petersburg-based institutes, the Scientific Research Institute for Polymerised Plastics (NIIPP) within the Plastpolimer association and the Academician S. V. Lebedev Scientific Research Institute for Synthetic Rubber (SP NIISK, formerly VNIISK)], and at the N. D. Zelinskii Institute of Organic Chemistry and the N. N. Semenov Institute of Chemical Physics, both in Moscow. The RSC 'Applied Chemistry' is a major development centre for organofluorine chemicals which has its own large experimental plant in St. Petersburg, and does the main process engineering work for fluorine technologies at the large industrial plants. The main interests lie with 'freons' for use as propellants, refrigerants, foaming agents and in power engineering, fluorine-containing gases for fibre optics and microelectronics applications, perfluorinated dielectrics useful as heat-transfer agents, and fluorinated surfactants.

The senior fluorine chemists are Professor Boris Maksimov and Dr Valerii Barabanov, a 'freons' specialist.

One of the programmes at SP NIISK concerns research into and the development of fluorine-containing rubbers (elastomers), and the provision of technical know-how for the organisation of industrial production. Professors Sergei Sokolov (who, as was mentiond earlier, spent part of his early career in the Urals school) and Viktor Gubanov, a specialist in fluoroacrylates, head fluorine-linked research at this institute. Together with N. V. Veretennikov, Gubanov has developed fluoroelastomers for a range of applications. Large-scale industrial manufacture of fluoroelastomers has been established at the Kirovo-Chepetsk Chemical Combine based on SP-NIISK work. The drop in funding for science and technology throughout Russia is, of course, hindering progress at SP NIISK; however, up to 1996 there had not been any cutbacks in staffing, although the Institute was having to survive mainly on contract earnings.

In the post-war period, NIIPP, which became part of the Plastpolimer association, pioneered the Soviet research effort on fluorovinyl polymers on which the 'ftoroplast' series of fluoroplastics is based (*Ftoroplast 4F* is PTFE). Industrial production is located at 'Galogen' Perm' and the Kirovo-Chepetsk plant (see below). The head of the fluoropolymer department is V. Konovalenko.

Chemists at the Zelinskii Institute of Organic Chemistry are developing fluoropolymer materials for various applications, mainly perfluoropolyether oligomers and related polymers. From the 1960s, the institute was involved in research for military purposes (primarily aerospace) under Professor Vasilii Ponomarenko; the Laboratory for Polymer Chemistry is now headed by Professor Stanislav Krukovskii, and a team there under Professor Svyatoslav Shevelev is working on fluoronitro-compounds (in collaboration with the Institute of Chemical Physics in Chernogolovka – see above), although its primary focus is not fluorine chemistry. Dr Igor Dalinger is working on fluorine-containing heterocyclic compounds, especially *N*-(difluoroamino)azoles – a new class of *N*-substituted azoles.

Researchers at the Semenov Institute of Chemical Physics have also developed fluoropolymer materials and process technology; the work is currently lead by Dr Andrei Lyapunov. Studies on the kinetics of radical polymerization and the synthesis of fluorine-containing compounds originated in the mid-1970s, and now centre on radical-initiated graft polymerization, including grafting of fluorinated surface-active substances onto hydrocarbon polymers. Other fluorine-related research at the Semenov Institute includes using elemental fluorine to study *N*-fluorination of asymmetric nitrogen compounds (*e.g.* aziridinecarboxylates) under Professor Remir Kostyanovskii (a pupil of Knunyants from some time ago), and fluorine-containing bio-active compounds, although again the primary focus of these areas is not fluorine chemistry (the fluorine input is provided by INEOS).

Other centres have been involved in developing fluorine-related technologies. For example, research into the synthesis of fluoroacrylates as the basis for polymeric materials was initiated in the early 1970s at the Kargin institute in Dzerzhinsk in collaboration with the Orgsteklo industrial plant in the same town, but was recently discontinued due to lack of funding. Kazan' State University is investigating fluorine-containing phosphorus compounds in collaboration with the Arbuzov Institute of Organic and Physical Chemistry (located in the same town) and with Ul'yanovsk State University.

Inorganic fluorine chemistry

Fundamental research on inorganic fluorine compounds is mainly the province of the Institute of Applied Chemical Physics at the Russian Research Centre 'Kurchatov Institute', headed by Professor Boris Chaivanov. The main work of the Kurchatnik, as it is commonly known, has centred on noble gas chemistry, the use of atomic fluorine in chemical synthesis, fluorides of transition elements, and modification of polymer surfaces using fluorine compounds. Major contributions to the field have been made by the Kurchatnik, especially where the production of high-quality xenon difluoride and the generation of atomic fluorine from molecular fluorine are concerned.

Research in the USSR on inorganic fluorides for the atomic industry originated in the early 1950s under Vladimir Prusakov and Nikolai Tarasenko. Noble gas chemistry was pioneered in the early 1960s – there were few links with teams in the West – by Academician Valerii Legasov, who was subsequently part of the team which reported to the IAEA on the Chernobyl accident. Professor Chaivanov studied under Legasov from 1964, developing knowledge of XeF_2; in the 1970s he joined Professor Vladimir Sokolov to work on atomic fluorine. Sokolov's early fluorine-related work (also starting in the 1960s) included research into krypton difluoride and collaboration with the plutonium chemist Vitalii Serik and with Dr Sergei Spirin, who has worked since the mid-1970s on the chemistry of fluorides in anhydrous solutions and atomic fluorine. Recent new fields of investigation concern fluorofullerenes, fluoride-containing gas mixtures for chemical lasers, polymer surface modification using fluorine-containing compounds – under Dr Georgii Barsamyan – and applied work on fluorides in microelectronics and optics.

The Kurchatnik offers commercial quantities of most of the compounds it synthesises (sometimes far in excess of quantities usually sold on the world market!). Despite some difficulties in adjusting to the lower level of demand for its expertise, its fluorine research programmes appear to be surviving. Direct funding from the state budget and a commercial income of around $3m per annum has secured the Kurchatnik as a whole, and sales of XeF_2 and other compounds, coupled with the transfer of inorganic fluorides technology to other countries, contributes to this. The Institute's work was formerly so secret that its location in the suburbs of Moscow was not even represented on street maps; but in these more open days Professor Chaivanov and his team extend a warm welcome to visitors, as I found during a recent visit.

The other main centres involved with research into inorganic fluorine chemistry are the N. S. Kurnakov Institute of General and Inorganic Chemistry in Moscow, where this field has been extensively developed since the 1950s, and the Institute of Inorganic Chemistry (IIC) in Novosibirsk. The inspiration for the early inorganic fluorine work at the Kurnakov Institute, again linked with applications in the atomic industry, was provided by the outstanding analytical chemist Academician Professor I. V. Tananaev; starting in the 1930s, he developed (partly in collaboration with the Kurchatnik) what was then the new field of transition metal fluorides. Initially, progress was more or less independent of Western work in the field, although later there were good links with Professor Harry Emeléus in the UK. Major current fluorine research programmes at the Kurnakov (under Academician Professor Yurii Buslaev in association with Professors Evgenii Il'in, Yurii Kokunov and S. P. Petrosyants) focus on high-purity glass based on heavy metal fluorides, electrolytes, ion-selective electrodes, and high-temperature lubricants. At the IIC Novosibirsk, Professor Igor Igumenov and Dr Aleksandr Mishchenko are investigating fluorocomplexes of noble metals and fluorine compounds for use in chemical lasers. They have also worked on polymeric graphite fluorides, in connection with cathode materials for lithium batteries produced at Angarsk.

Industrial manufacture of fluorine compounds

Most of Russia's manufacturing capacity for organic fluorine compounds, mainly built up during the 1950s and 1960s, is located at the Kirovo-Chepetsk Chemical Combine and two enterprises in the major chemical-producing town Perm' – namely the Perm' branch of the RSC 'Applied Chemistry', which is a small-scale industrial manufacturer, and the 'Galogen' Joint-Stock Company. Facilities at these sites manufacture basic starting materials, including hydrofluoric and fluosilicic acid, fluoroaromatic compounds, perfluorinated alcohols and acids, 'freons', fluoromonomers and fluoride salts. A number of other major enterprises engage in large volume manufacture of fluorine compounds, but this is mainly confined to commonly-used CFCs [especially 11 ($CFCl_3$) and 12 (CF_2Cl_2)]. Small-scale industrial production exists for specialised fluorine-based materials such as perfluoropolyethers, fluoroacrylates, perfluoro-olefins [mainly C_2 and C_3, with some C_4 and oligomers (C_6 and C_9) of hexafluoropropene].

Production capacity for CFCs in the USSR at the end of the 1980s was large-scale – perhaps as high as 150 000 tonnes. However, the fall in domestic demand, the lack of investment in recent years, and the need to comply with the Montreal protocol in phasing out ozone-depleting substances (ODS) mean that these figures are likely to have decreased substantially (annual output was down to around a third of former levels by 1995). CFC production facilities in the USSR successor states can be found at a number of sites in the Russian Federation (the only site I identified outside Russia is the Tajikkhimprom enterprise in Yavan, Tajikistan, which was producing CFCs 11 and 12 but now appears to have stopped). The most important manufacturers are Galogen Perm' and the Kirovo-Chepetsk Chemical Combine, but a number of other plants still produce 'freons'; the RSC 'Applied Chemistry' facility in Perm' manufactures some of the less common types and is also the site of a semi-industrial unit for the non-ODS CFC-12 replacement HFC-134a (CF_3CH_2F). There is also large capacity at the Kaustik and Khimprom plants in Volgograd for CFCs 11 and 12, and at the 'Altaikhimprom' company in Slavgorod, and smaller capacity at the Redkino Experimental Plant. CFCs were used in the USSR mainly as aerosol propellants (65% of demand), refrigerants (22%), solvents and cleaning fluids (8%), and foam-blowing agents (3%).

As indicated above, Russia is establishing its own production of well-established non-ozone-depleting fluorocarbons like HFC-134a using technology developed by the RSC 'Applied Chemistry' and based on existing facilities. Small industrial-scale units have already been set up at or are planned for a number of sites. The RSC 'Applied Chemistry' is involved with work on some of the newer fluorine-containing substances, such as fluorinated ethers and 2*H*-heptafluoropropane designed to replace traditional 'freon' refrigerants and Halons (bromofluorocarbons) respectively. The cost of fully adapting production to non-ODS has been estimated to be in excess of $100 million; although financial aid has been provided by the Global Economic Fund to cover part of this sum, the rest is being financed by Russian manufacturers, including companies producing goods in which freons are used. The RSC 'Applied Chemistry' has therefore concentrated its efforts on the development of compounds which can directly substitute ODS without major hardware modifications ('drop-in' substitutes).

The Kirovo-Chepetsk Chemical Combine, whose capacity is in the order of tens of thousands of tonnes, and to a lesser extent the 'Galogen' plant in Perm are the principal manufacturers of organofluorine monomers which support fluoropolymer production in Russia. These two plants are also, as might be expected, the main industrial producers of domestic fluorine-based plastics and elastomers. The Okhtinskii Experimental Plant attached to 'Plastpolimer' in St. Petersburg also manufactures fluorine plastics, although capacity is estimated to be no more than several hundred kg. A similar capacity exists at an experimental unit in Borislav, Ukraine; this is the only known manufacturing site for fluoropolymers in the USSR successor states outside Russia. Most of the known commercial fluoroplastics are produced in Russia: the homopolymers poly(tetrafluoroethylene), poly(chlorotrifluoroethylene), poly(vinyl fluoride), and modified poly(vinylidene fluoride); copolymers of tetrafluoroethylene with hexafluoropropene, vinylidene fluoride, ethylene, trifluoroethylene and perfluoro(propyl vinyl ether); and the copolymer of chlorotrifluoroethylene with ethylene.

Apart from small-scale experimental production at SP NIISK, it appears that virtually all fluoroelastomer production in the USSR successor states is at the Kirovo-Chepetsk plant. The main types produced are copolymers of vinylidene fluoride with chlorotrifluoroethylene (under the tradename *SKF-32)* and with hexafluoropropene (*SKF-26*). Both are produced by emulsion copolymerization techniques using ammonium persulphate as initiator, alkali metal salts of perfluorocarboxylic acid as emulsifiers, and appropriate alcohols, hydrocarbons or bromo/iodo-perfluoroalkanes as molecular weight regulators. Capacity for speciality fluoroelastomers also exists at Kirovo-Chepetsk. Due to the sharp drop in domestic demand actual current production of fluorinated plastics and elastomers is but a fraction of total plant capacity; however, an export market for some production has been found (the Kirovo-Chepetsk plant is reported to be selling substantial quantities of its fluoropolymers overseas via a joint-venture agreement with DuPont).

Fluorine-based oils and greases are produced in substantial amounts at the Perm' branch of the RSC 'Applied Chemistry', with some production at Kirovo-Chepetsk (along with some other speciality compounds). Perfluoropolyether fluids are manufactured at the 'Orgsteklo' plant in Dzerzhinsk under the trade name *Porafol*. Other specialised compounds are also produced in the USSR successor states – again, almost exclusively in Russia – despite the fact that the domestic market is generally less well-developed than in Western industrialised countries. Fluorine compounds for medical applications are manufactured at the Kirovo-Chepetsk Chemical Combine and at 'Galogen' Perm'. Small-scale capacity for fluorine-based (mainly PTFE) chemical fibres exists at the Serpukhov Chemical Fibres Plant near Moscow and at Khimvolokno in Klin (which one source in the industry has told me may now be the only surviving producer). Organofluorine compounds units have been established at plants in Angarsk and Tomsk, based on freed capacity for fluorine which was formerly used to produce inorganic fluorides for the nuclear and aerospace programmes.

Capacity for inorganic fluorides exists at the Kurchatnik (see above), whose products are manufactured at its own facilities in Moscow, and at the Siberian Chemical Combine in Tomsk and the Angarsk Electrolytic Chemicals Combine, both of which were producing substantial quantities. Inorganic fluorine compounds are also produced at 'Galogen' Perm.

Concluding remarks: a future for fluorine?

Although my chapter contains much less chemistry than others in this book, I hope I have conveyed some idea of the breadth, depth and vitality of fluorine research and development in Russia and Ukraine over the past few decades. The distances involved, coupled with financial problems, mean that fluorine chemists and technologists from these countries are less frequent visitors to other leading fluorine chemistry nations than they might be; but they continue to be extensively represented in the literature, and are in my experience always eager to welcome visitors to their institutes and exchange scientific findings.

It would be an understatement to say that science is facing a tough time in Russia and Ukraine; initiatives to assist science in these countries does exist at UK government level, and individual companies and universities do provide help to their Russian and Ukrainian counterparts, but more could be done. Perhaps a special programme, however modest initially, could be set up to develop and foster links specifically in the fluorine field? It would be in everybody's interest to capitalise on the existing expertise in these countries and to assist the emergence of a new generation of Russian and Ukrainian fluorine chemists at the beginning of the new millennium.

Acknowledgements

I was helped enormously in the preparation of this chapter by a number of Russian and Ukrainian fluorine chemists, most of whom are mentioned in the text, and would like to acknowledge their contributions to the account. I am also in debt to the following: Dr Petr Valetskii, Deputy Director of INEOS; Professors Vladislav Vlasov and Vladimir Starichenko, Deputy Directors of the Institute of Organic Chemistry Novosibirsk; Academician Professor Nikolai Zefirov, Chair of the Department of Organic Chemistry, Moscow State University and Director of the Institute for Physiologically Active Substances, Chernogolovka; Professors Miron Lozinskii and Anatolii Sinitsa, respectively Director and Deputy Director, Institute of Organic Chemistry, Kiev; Academician Professor V. A. Tartakovskii, Director, N. D. Zelinskii Institute of Organic Chemistry; Academician Professor Fedor Kuznetsov, Director, Institute of Inorganic Chemistry, Novosibirsk; Dr Boris Mislavskii, General Electric International (Moscow); Mr Igor Uklonskii, General Director, 'Galogen' Perm'; Mr Yurii Burushkin, Kirovo-Chepetsk Chemical Combine; Dr Richard Powell, ICI Chemicals & Polymers (UK); Mr Ken Johns, Chemical & Polymer (UK); Mr Peter Whitehead, Apollo Scientific (UK); Mr Derek Ronksley, Mindeco International (UK).

References

1 L. M. Yagupolskii, *J. Fluorine Chem.*, *90* (1998) 181.
2 I. L. Knunyants, 'Development of the Chemistry of Organofluorine Compounds in the Academy of Sciences in the USSR', *Izv. Akad. Nauk SSSR, Ser. Khim.*, (1974) 1098 (Engl. transl., p. 1046).

3 See Appendix 11.4 in R. E. Banks and J. C. Tatlow, 'A Guide to Modern Organofluorine Chemistry', *J. Fluorine Chem.*, *33* (1986) 227.
4 See G. G. Belen'kii, *J. Fluorine Chem.*, *90* (1998) 173 for a thoughtful, concise account of Lev German's life at INEOS.
5 V. Ya. Popkova, *J. Fluorine Chem.*, *90* (1998) 153.
6 V. D. Shteingarts, *J. Fluorine Chem.*, *90* (1998) 203.
7 A. P. Kharitonov, 'Commercial Applications of the Direct Fluorination of Polymer Surfaces', *Fluorine Technology Bulletin*, No. 25 (1997/98) 38.

BIOGRAPHIC NOTE

Derek Lambton Averre, the only non-chemist represented in this volume, was nevertheless probably fated to contribute to it: born only a few miles from Durham, one of the centres of UK fluorine chemistry, he graduated in Russian Studies from Manchester University, a stone's throw away from Eric Banks' territory at UMIST; he also completed his doctoral thesis at Manchester. After several years working in industry, trading with what was then the Comecon bloc, he took up a Research Fellowship at Birmingham University, home of yet another prominent school of fluorine chemists, in the Centre for Russian and East European Studies. His main research interests are science, technology and industry in the USSR successor states and Russian security and disarmament issues.

Derek Lambton Averre

Chapter 3

FORTY YEARS OF FLUORINE CHEMISTRY: KING'S COLLEGE, NEWCASTLE (1954-57); THE UNIVERSITY OF BRITISH COLUMBIA (1958-66); PRINCETON UNIVERSITY (1966-69); AND THE UNIVERSITY OF CALIFORNIA AT BERKELEY (1969-98)

NEIL BARTLETT

College of Chemistry, University of California, Berkeley, CA 94720, USA

Early education and King's College, Newcastle, 1954-57

The most fortunate event in my life was to pass the eleven-plus examination, a test taken at that time in England by children aged 11 or 12 to select suitable candidates for grammar schools. That meant that I had the benefit of an educational system (at Heaton Grammar School in Newcastle upon Tyne) that I believe has had no superior. Immediately, I was aware that there were many of my own age who were at least as clever as I, and some much cleverer. From the beginning we were ranked on the basis of our scholastic performance, and the instruction (provided by school masters who wore academic gowns, except in laboratory classes, when they wore lab coats) was at a high academic level. From the first year (when I was just short of twelve) we had separate classes in chemistry and physics. That included an hour of work in the laboratory each week. I vividly recall an early lab experience (it may have been the first) in which we added aqueous ammonia to a solution of copper sulfate in water. We were asked to record all of our observations and explain (!) them. Incidentally we set the deep blue cuprammonium solution aside on a clock glass until the following week. At that next lab session we found beautiful well-formed crystals. I was hooked! Fortunately my widowed mother was highly tolerant and I soon had a home laboratory. From my pocket money, and sundry earnings, I bought a wide variety of lab equipment from the School Furnishing Company on Northumberland Road, Newcastle. This included a retort, a still head, flasks, beakers, a Liebig's condenser, and so on. I used to make alcohol by fermentation, followed by fractional distillation. I had to buy my sulfuric acid, but I prepared my own fuming nitric acid using it, and my retort. Because of these interests, I developed a strong background in chemistry, and with that strength as its basis, I eventually won a State Scholarship that underwrote my undergraduate education in chemistry.

In my later schooldays I had decided that I would try to become a biochemist. My advisors said that I should first become a natural products chemist, and since that branch of chemistry was moderately strong at the local university (King's College in the University of Durham) that is where I applied for admission. Fortunately I had excellent teachers for chemistry at my school, so when I went up to University I already had a good general knowledge of chemistry, at least as keyed to the Periodic Table. Before going up to University, I already owned and had read N. V. Sidgwick's *Electronic Theory of Valence* as well as his two-volume opus *The Chemical Elements and Their Compounds* the latter being a school book prize. In fact the teaching of natural product chemistry proved not to be inspiring and perhaps not surprisingly, the Inorganic aspects were always easy to assimilate, and were interesting. When I graduated (1954) I was persuaded by members of the Inorganic Chemistry research group, run by the Reader in Inorganic Chemistry, P. L. Robinson, that I should consider doing research in inorganic chemistry. They pointed out that there were many exciting opportunities for research in that sub-department, including Fluorine Chemistry. Without quite knowing what I was doing I jumped aboard!

I did not immediately begin to do fluorine chemistry research, but it was going on all around me and I quickly learned what each person in the sub-department was doing. In fact the first problem I was given to tackle could not have been better in preparing me for my future research.

Research beginnings

P. L. Robinson had some research money from the UK Atomic Energy Authority, for research on a homogeneous nuclear reactor system. I do not know who specified what, but I was asked by P. L. Robinson to look into the possibility of using liquid sulfur trioxide (the trimeric form of SO_3) as solvent for uranyl sulfate and potassium sulfate mixtures. I very quickly discovered that liquid S_3O_9 polymerized with astonishing ease, to give the polymeric 'asbestos form', which was of course useless for my purposes. If I were to be effective in pursuing my research I had to learn how to manipulate S_3O_9 under thoroughly anhydrous and grease-free conditions. By taking care to rigorously dry all apparatus by flaming out under vacuum it became possible to work easily enough with the S_3O_9 but it proved to be a disappointingly poor solvent, so I switched to 100% sulfuric acid in its place. After about one year of work on the three component system $H_2SO_4/UO_2SO_4/K_2SO_4$ the phase diagram was fairly well mapped out and P. L. Robinson asked if I would like to take over the problem of making and purifying sulfur tetrafluoride. I accepted with alacrity. It is probably of interest to add a few comments on Fluorine Chemistry at Newcastle at this point.

After World War II, A. J. Rudge at ICI Runcorn, had approached his old research director with a proposal that Robinson undertake research in fluorine chemistry for which ICI would supply and maintain the fluorine generators. Robinson accepted, and R. D. Peacock became the first student to engage in that research. Peacock was one of my instructors in my undergraduate years and on occasions he would tell me of the research that he was carrying out. Indeed, it was hard to avoid seeing something of the research since much of it was done in the same large room in which we received our laboratory training.

The research laboratory in which I worked was separated from the large teaching laboratory by a central aisle in which several fume hoods were also located. There were also fume hoods ranged along the end walls of the room, and it was on narrow tables placed in front of such 'hoods', that my fluorinations were carried out. A large capacity fluorine cell stood on the floor in a corner immediately in front of a fume hood. The DC power supply control panel was mounted on the wall above the cell and one adjusted the fluorine supply (diluted with nitrogen gas flowing through the anode-head chamber) by adjusting a large spring-loaded copper lever which made electrical contact with an arc of copper studs, the amperage indicating the rate of fluorine generation. We monitored the HF content of the electrolyte by keeping a log in which the amp-hours of use were recorded. This did not always work as it should, and on occasions the electrolyte level in the cell would drop to the level where the fluorine would get into the hydrogen of the cathode compartment – then there were explosive 'coughs' from the cell and we knew it was time to add HF! This we did by joining an HF tank with a polyethylene tube to the cell, the tank standing on a scale, so that a known weight of HF could be added. To heat the HF tank we had to borrow P. L. Robinson's electric fire from his office!

As I have indicated fluorine was always carried in nitrogen. We also made most of the fluorinations in Pyrex glass apparatus, although occasionally fused-silica, sintered-alumina or nickel tubes were joined by graded seals or neoprene-gasketed compression fittings, to the Pyrex. To protect such apparatus from attack by HF, the F_2 in N_2 was passed through a tube packed with finely powdered NaF. In addition the apparatus was always flamed out under vacuum, and a vacuum pump mounted on a trolley was brought up to the work place for that purpose. Mobile glass-blowing carts were also brought to the fluorination site only when needed. In this way the place of the fluorination was kept relatively uncluttered. Because of the need to completely eliminate HF and to prevent ingress of moisture, any preparative apparatus had a fore-trap and an end trap that were cooled in liquid O_2 (liquid N_2 condenses F_2 at partial pressures above 300 torr).

My broad research directive from P. L. Robinson was to work on improving the preparation of SF_4 and to look out for lower fluorides of sulfur. As it turned out I was kept entirely occupied with the characterization of SF_4, SeF_4 and TeF_4.

SF_4 was made by fluorinating sulfur at dry-ice temperature after the method [1] devised in the previous year by Brown and Robinson. This always generated a mixture of SF_4, SF_6, SOF_2 and S_2F_{10} and Brown had built a Podbielniak fractional distillation column for the separation of this mixture. A hazardous aspect of this distillation was the necessity of taking off the SF_6 fraction (the first) at two atmospheres pressure (to liquefy the SF_6). This was done by building the pressure in the still against a two-atmosphere column of mercury. Unfortunately, it was common for the sulfur fluorides in the still to super-heat after the fractionation had been under way for some hours, then the liquid mixture would boil explosively, producing a local mercury geyser, followed by a cloud of sulfur fluorides, both of which, of course, enveloped the operator of the still. I soon resolved to find a chemical purification for SF_4! This quickly came to hand.

Since SF_6 was known to be a thermodynamically favorable and kinetically inert molecule I was confident that SF_4 would be a good reducing agent. M. A. Hepworth, a senior member of the Robinson group at that time, suggested the synthesis of PdF_2 from PdF_3, the latter being well characterized magnetically [2] as well as structurally [3]. The former had been claimed by Ruff and Ascher [4] to have a rutile structure. If they were correct, PdF_2 should be paramagnetic. No paramagnetic Pd[II] compound was then known, even PdO having a low-spin d^8 configuration. Indeed, when SF_4 was passed over PdF_3 in a hot Pyrex tube at ~300 °C, we observed that reduction to PdF_2 did occur, the product having the rutile structure as claimed by Ruff and Ascher, but this reduction also produced some Pd metal. Immediately SeF_4 was used in place of SF_4 as the reducing agent, and high purity PdF_2 obtained [5], from which structural [6] and magnetic [7] information were derived to fully characterize the material. This excursion into transition-metal fluorine chemistry was to have a profound impact on my subsequent career, but that first experiment with SF_4 and PdF_3 had another important consequence.

It was noticed, in that experiment, that when the Pyrex apparatus was hot, the SF_4 appeared to attack it. Simultaneously a colorless crystalline solid appeared on the cooler walls downstream. This colorless solid sublimed on heating and transferred to the traps under vacuum. I conjectured that the SF_4 had fluorinated the glass by fluorine-for-oxygen exchange (this was long before the group at DuPont de Nemours Inc., demonstrated [8] the utility of SF_4 in the fluorination of oxygen-containing organic compounds):

$$2SF_4(g) + SiO_2(s) \rightarrow 2SOF_2(g) + SiF_4(g)$$
$$3SF_4(g) + B_2O_3(s) \rightarrow 3SOF_2(g) + 2BF_3(g)$$

the BF_3, but not the SiF_4, then combining with the SF_4:

$$SF_4(g) + BF_3(g) \rightarrow SF_4 \cdot BF_3(s)\downarrow$$

To check this I prepared both BF_3 and SiF_4 and mixed each with SF_4. Tensimetry established that the solid was a 1:1 adduct, $SF_4 \cdot BF_3$, and provided the heat and entropy of formation from the constituent molecules (vapour density measurements having established that the adduct was fully dissociated in the vapour phase) [9, 10]. The immediate benefit, however, was that I now had a chemical way of separating SF_4 from the other sulfur fluorides. I simply combined my crude sulfur fluoride mixture with an abundance of BF_3 then removed all volatiles at −78 °C; at which temperature the $SF_4 \cdot BF_3$ was involatile. The SF_4 was recovered by displacement [11] with SeF_4 or [9] by passing the adduct through a tube packed with alkali fluoride:

$$SF_4 \cdot BF_3(s) + SeF_4(l) \rightarrow SF_4(g) + SeF_4 \cdot BF_3(s)$$
$$SF_4 \cdot BF_3(s) \rightarrow SF_4(g) + BF_3(g); \quad BF_3(g) + NaF(s) \rightarrow NaBF_4(s)$$

These discoveries had however revealed something of even greater importance, namely the 'basic' character of the chalcogenide tetrafluorides. In addition, the marked difference in the acceptor properties of BF_3 and SiF_4 aroused a desire to quantify this difference, which was the basis of much work, later, on the assessment of gaseous F^- affinities.

I quickly established [9, 10] that SF_4, SeF_4, and TeF_4 would each form a 1:1 adduct with the well known fluoroacids AsF_5 and SbF_5. We did not have access in those days to infrared spectroscopy since we still had to learn about AgCl as a suitable window material, so we had no evidence except one X-ray powder photograph on which to conjecture the nature of the adducts. That photograph, of $SF_4 \cdot SbF_5$, showed it to be simple cubic. This inclined me to the view that these adducts were probably the salts of SF_3^+, but although, with the help of K. H. Jack, much effort was spent in trying to solve the structure of the SF_4SbF_5 from the powder data, we failed. While I was teaching chemistry at the Duke's School in Alnwick (see below), Seel and his co-worker Detmer published [12] infra-red evidence, which proved that these and a variety of other related compounds, were indeed fluoronium salts. A number of years on, the structure of the $SF_4 \cdot BF_3$ adduct was solved in my Berkeley laboratories [13] and indeed was shown to be $SF_3^+BF_4^-$. But the next quest in my doctoral research was for PtF_2. That search was to have unforeseen important consequences.

Although PtF_2 and PtF_4 had been described by Moissan [14], there was no firm support for PtF_2's existence in 1956. Our confirmation of paramagnetic rutile-structure

PdF_2 however gave me some hope that SF_4 or SeF_4 could be effective mild reductants for the conversion of PtF_4 to PtF_2. The tetrafluoride had been confirmed a few years before by the work [15] of Alan Sharpe at Cambridge. But as Sharpe was careful to point out, the PtF_4 made in his way, via a BrF_3 complex, was always contaminated with bromine. In order to get rid of the bromine from PtF_4 prepared in the Sharpe way, I undertook to fluorinate the PtF_4 in a stream of fluorine, with the intention of converting the bromine to BrF_5, which was known to form weaker complexes than BrF_3. The crucial experiment was carried out on the evening of November 5, 1956 (Guy Fawkes night). As I heated the PtF_4 (which was contained in a nickel boat, within a Pyrex glass tube) in the fluorine stream I was puzzled to see a deep red vapour above the PtF_4. My first thought was that the fluorine supply had been cut off and that I was observing Br_2 but this proved not to be the case, and as I raised the temperature the red vapours increased dramatically, and a deep-red solid collected on the cooler glass downstream. I was immediately sure that this red solid must represent a new high oxidation state of platinum, but I had also noted that the red vapours appeared at the same time as the fluorine was attacking the Pyrex glass, and therefore conjectured that the material was probably an oxyfluoride of platinum.

In the spare time from my thesis work on the chalcogenide tetrafluorides, I attempted to settle the nature of this new platinum oxyfluoride material; but it proved to be extremely difficult to analyze and characterize, so eventually the problem had to be set aside. I took it up in a full time way later at the University of British Columbia, but even in Newcastle I was able to show that oxygen was necessary to the synthesis of the sublimable red solid. When PtF_4 was fluorinated in a nickel tube a sticky low-volatility product was obtained, which my fluorine and platinum analyses indicated was probably PtF_5. These however were days of high confusion in the fluorine chemistry of the platinum metals. It was not as clear as now, that this low melting solid had to be a pentafluoride. I will simply point out that the fluoride then known [16] as OsF_8 was to be shown [17] in 1957 by Weinstock and his co-workers at the Argonne National Laboratory, to be OsF_6. The then OsF_6 was to be shown by Hargreaves and Peacock [18] to be OsF_5. The material that had been identified by two groups [19, 20] as iridium tetrafluoride, IrF_4, was eventually shown [21] by my co-worker P. R. Rao at UBC to be IrF_5. Rao also prepared (independently of John Holloway) [22] RhF_5 and genuine [23] IrF_4.

I left the Robinson research group in July 1957 in order to take up a post not far from Newcastle as Senior Chemistry Master at the Duke's School in Alnwick (Northumberland), this job excusing me from military service. A little over one year later I was offered and accepted an appointment at the University of British Columbia in Vancouver, where I was to spend the next eight years.

Independent research beginnings

By the time I wrote my PhD thesis in early 1958, the popularization of Crystal (and Ligand) Field Theory, particularly by Griffith and Orgel [24], and the expansion of the Sidgwick and Powell valence electron pair repulsion rules by Gillespie and Nyholm [25], had provided a good basis for the understanding of the relationship between valence shell electron configuration and molecule and crystal geometry. What was far less clear was what the valence limits for a given element were. My laboratory research experiences had already indicated that Pt[V] and possibly Pt[VI] were attainable, with fluorine as the oxidizer and ligand. Indeed, Weinstock and his co-workers had just reported [26] PtF_6, as well as showing [17] that OsF_8 was in fact OsF_6. I pondered on the possible existence of AuF_6 and PdF_6, but also recognized, from my experience with PdF_2, that reducing fluorides such as SF_4, and SeF_4 could give access to lower fluorides of the transition elements. This had the attractive prospect of fluorine chemistry providing a wide range of oxidation states, at least for the later transition series elements. Not only were the platinum fluorides of interest to me, but even the seemingly well described PdF_3 had a puzzling aspect. Such problems were uppermost in my mind as I set off for Vancouver.

Perhaps the most important aspect of my new life in Vancouver was that I felt entirely free to do what was most important to me. C. A. McDowell, the Head of the Chemistry Department, saw to it that I had a small but adequate starting research budget. He also assisted in a number of other ways to launch my independent research career. Another aspect which at the time was a burden, but which was to provide a valuable learning experience, was the immediate need to assist in graduate course teaching. This I did in concert with Drs H. C. Clark and W. R. Cullen. I chose to instruct the graduates in the use of X-ray powder diffraction for structural analysis, of which I then had but a smattering of knowledge. The thorough understanding that this gave me proved to be invaluable in the research I was to carry out.

It was my great good fortune to be preceded by one year in Vancouver by another chemist with interests in fluorine chemistry, Howard C. Clark who had been trained in H. J. Emeléus' lab in Cambridge. Clark had collected the necessary hardware for a fluorine supply by the time I arrived in early September. But this involved using fluorine under pressure ($\sim$450 psi) with which neither of us had previous experience! It was with great trepidation that we eventually opened the valve of the giant cylinder, which was even more impressive behind its wall of bricks – all, incidentally, in a large walk-in fume hood located in the north west corner of the top floor of the old Chemistry Building. But, apart from an initial leak of fluorine gas, the result of our over-cautious tightening of the fittings, all went well, and by November 1958 fluorine chemistry was launched at UBC!

Pd(II) Pd(IV)F_6, and PdF_4

My immediate personal research involved the preparation of the red oxyfluoride of platinum, but when an eager recent graduate (J. W. Quail) indicated that he would like to do some fluorine chemistry under my direction, I set him to work on the easier task of defining the conditions for Pd[IV] versus Pd[III], in fluoro solvents such as BrF_3 and SeF_4. Sharpe had shown [15] that from the dissolution of $PdBr_2$ in BrF_3 one obtained a complex, $BrF_3 \cdot PdF_3$, vacuum pyrolysis of which gave PdF_3. But he also showed [27] that when alkali fluorides were present in the BrF_3 solution one obtained PdF_6^{2-} salts. Clearly BrF_3 was not a sufficiently good base itself to facilitate the formation of Pd[IV] but I thought it likely that SeF_4 would do so, and it did. Addition of SeF_4 to $BrF_3 \cdot PdF_3$ led to immediate oxidation [28] of the Pd to $(SeF_3)_2PdF_6$. There was, however, an even more important problem associated with palladium fluoride chemistry and that was the nature of PdF_3 itself. A structure by K. H. Jack and his co-workers [3], based on a high quality X-ray powder diffraction pattern (XRDP), had established the trifluoride to be rhombohedral, the site symmetry for each of the two Pd atoms in the cell being D_{3d} and essentially octahedral. This contradicted the indications from the magnetic data [2] of Sharpe and Nyholm that the electron configuration was low spin d^7. A low spin d^7 configuration should have produced a gross Jahn-Teller distortion [24]. The answer came two years further on, when, with my student P. R. Rao, we carried the Quail findings to their logical conclusion, and with acidified BrF_3 thereby limited the palladium oxidation to Pd[II]. In this way we prepared [29] several Pd[II]MF_6 salts (M = Sn, Ge, Pt) that proved to be isostructural with PdF_3, *e.g.*

$$PdBr_2(s) + SnBr_4(s) + 2BrF_3(l) \rightarrow PdSnF_6(s) + 4Br_2(g)\uparrow$$

Indeed the magnetic data, for these salts and PdF_3, proved that the last must be the mixed oxidation state material Pd[II]Pd[IV]F_6. (Some years later, a powder neutron diffraction study established the existence [30] of two Pd-F distances at 2.17 and 1.90 Å.) This work on Pd[II]Pd[IV]F_6 prompted Rao and myself to search for a tetrafluoride of palladium, which we soon synthesized [29], by the fluorination of $PdGeF_6$. The liberation of gaseous GeF_4 helped to bring about the oxidation of all Pd[II] to Pd[IV]:

$$PdGeF_6(s) + F_2(g) \rightarrow PdF_4(s) + GeF_4(g)\uparrow$$

Incidentally, we found also a new route to PdF_2, namely pyrolysis of $PdGeF_6$:

$$PdGeF_6(s) \rightarrow PdF_2(s) + GeF_4(g)\uparrow$$

The MF_4 structures (M = Pd, Pt, Ir, Re, Rh and Os)

The XRDP of PdF_4 [29] proved to be very similar to that of PtF_4 [31]. Eventually, IrF_4 [23], RhF_4 [23], OsF_4 [32] and ReF_4 [32] were all found to have similar patterns. The structure was worked out [33] in collaboration with Alain Tressaud while I was on sabbatical leave at the University of Bordeaux in 1974; it is essentially that of rutile with half of the metal atoms omitted in ordered array.

GeF_2

The gaseous nature of GeF_4 prompted me to seek the lower fluoride, GeF_2, as a reducing agent, since a report [34] by Dennis and Laubengayer indicated that a white solid formed by passing GeF_4 over hot Ge could be the difluoride. Little else was known, and since I also had an interest in the steric activity of the non-bonding valence electron pair in such fluorides, I set a senior undergraduate student (K. C. Yu) the task of verifying the Dennis and Laubengayer synthesis, which he quickly did [35]. My indexing of the XRDP indicated the unit cell to be a distorted form of that reported [36] for its isoelectronic relative SeO_2. This conjecture eventually proved to be correct, when my graduate student co-worker M. Akhtar was able to grow a single crystal of GeF_2, making possible a full structure determination [37]. Akhtar also examined the reducing properties of the difluoride, which proved to be disappointingly mild [38].

PtF_5, '$PtOF_4$', and $O_2^+PtF_6^-$

In the summer of 1959 J. W. Quail left to do doctoral work with R. J. Gillespie, but D. H. Lohmann, fresh from a masters degree at Queen's University, Ontario, joined me for doctoral work. Since I was making little headway with the platinum fluorides investigation, owing to the pressures of my other duties, I introduced him to that problem, and he accepted it with enthusiasm. Indeed, because Lohmann was inexperienced in handling fluorine much of the early preparative work and some of the analysis was done jointly. Very quickly we obtained reproducible analytical evidence to support my King's College findings that the low melting red solid, obtained by heating PtF_4 in F_2 in the absence of oxygen, was PtF_5. Analysis of the more volatile oxyfluoride posed many more difficulties. In those days it was my habit to do the analyses of fluorides by first hydrolyzing them in concentrated aqueous alkali. The PtF_5 interacted well, most of the platinum passing into solution as PtF_6^{2-} and the remainder precipitating as the hydrated dioxide. Both the Pt and F analyses were routine. Interaction of the oxyfluoride with the aqueous alkali was very different.

When the red solid made contact with the solution, there was an immediate flash of light, ozone smelling gases were released, and a nearly black precipitate that contained some elemental platinum was thrown down. However, some of the platinum also appeared in solution as PtF_6^{2-}. We had much difficulty in gaining reproducible analyses. At this stage I was not attempting to do oxygen analysis, the formulation resting on Pt and F alone. In addition I was sure that this oxyfluoride had to be $PtOF_4$ since other MOF_4 species ($M = Mo, W, Re$) were known to be more volatile than related pentafluorides. So when we obtained F and Pt values in rough agreement with the $PtOF_4$ formulation, I sent off a communication [39] announcing PtF_5 and $PtOF_4$. By the time that this note appeared in print we were already very uneasy about the formulation $PtOF_4$.

Try as we might, we could never confirm our earlier analyses for $PtOF_4$; indeed when we did the Pt analysis in one step, by very slowly hydrolyzing the solid with slightly moist air, followed by reduction in hydrogen, we obtained consistent results, but these were far too low for our material to be $PtOF_4$. At this point we saw that we would have to devise an analysis for oxygen, as well as a more reliable one for fluorine. Indeed, we were afraid that we might be loosing some fluorine as F_2O, during the dissolution in aqueous alkali. Since low-valent bromine oxides and oxyfluorides appeared to lose their oxygen as O_2, I decided that dissolution of the oxyfluoride in BrF_3 might provide the oxygen content. With this approach we found that one mole of O_2 was liberated for each mole of Pt in the oxyfluoride! To ensure that no fluorine could be lost, as F_2O or other volatile fluorides, we arranged to react the oxyfluoride with sodium in a Parr bomb. These F analyses, with our O and Pt values summed to 100%, but gave the astonishing composition PtO_2F_6! Since the Pt atom has ten valence electrons, it was at that time feasible to conclude that this was an oxyfluoride of ten valent platinum, but I had already made one mistake in the assessment of this material, and was determined not to make another. In the next few months we amassed structural magnetic and reaction chemistry data that, taken together, unlocked the secret of this oxyfluoride.

Unfortunately, Lohmann was not able to participate in the final resolution of the problem because he had an assistant lectureship at University College, Dublin, that he had to take up in early October 1961. This left me without experienced help. I had however been fortunate to have two new doctoral students, N. K. Jha, and P. R. Rao, join me as Lohmann left. Jha and Rao, at this stage, however, were inexperienced, and as yet without the necessary glass working and other skills essential to full immersion in this sort of work.

A very helpful piece of information was, however, obtained by Jha. Lohmann and I had obtained XRDP's of PtO_2F_6. These XRDP's showed it to be cubic. It was also clear that the Pt atom arrangement (Pt being the dominant X-ray scatterer in such a material) had to be a simple cubic one, with a closest Pt-Pt distance of 5.016 Å. This suggested the possibility of salt formulations such as $O_2^+PtF_6^-$ or even $PtF_6^+O_2^-$, as well as *trans* $Pt(O)_2F_6$, all in the Ia3 space group required by the observed X-ray reflections. To provide Jha with training tasks, I had him prepare OsF_6 and mix it with nitric oxide to make what I was confident would be $NO^+OsF_6^-$. The preparation went as expected and the XRDP, as I had hoped, proved to be almost identical (except for a small shift in line positions) to that of the PtO_2F_6. Indeed, the similarity of the relative intensities of those X-ray powder lines to which Pt or Os atoms made no contribution, were so close, that I was now sure that the light atoms as well as the heavy had to occupy essentially the same positions in the two materials. Calculation of the X-ray powder pattern assuming a salt formulation was very persuasive but could not differentiate between $O_2^+PtF_6^-$ and $PtF_6^+O_2^-$; for that I resorted to thermodynamic and chemical arguments, which made a convincing case for the former. Up to this point I had been careful to avoid doing any work with PtF_6, because this would have been an infringement on the territory of Weinstock and his co-workers. Now, however, I had come to the point where I had no alternative but to make PtF_6 since, if my deductions were correct, it ought to oxidize O_2 to O_2^+! It seemed strange that in the four years that had elapsed, that the Argonne National Laboratory chemists would not have observed that remarkable capability in PtF_6. Nevertheless they had not. When I mixed PtF_6 with O_2, I obtained my familiar red solid:

$$O_2(g) + PtF_6(g) \rightarrow O_2^+PtF_6^-(s)$$

This was not only the first O_2^+ salt, but also the first PtF_6^- salt. Implications of the energetics, for the oxidizing power of PtF_6 were extraordinary.

When, finally, all was in place and I had written the communication [40] that would announce the discovery, I presented the work as a research seminar to the department at UBC. My seminar was very well attended and there were a few friendly questions, which I had no difficulty in responding to, but after the talk one of my senior colleagues made the observation that 'maybe something had been overlooked'. This was an allusion to my conclusion [41] that the electron affinity of PtF_6 had to exceed 160 kcal mol^{-1} for it to have the capability to oxidize O_2 to O_2^+. This skepticism was certainly a challenge to which I was compelled to respond, but in addition it was clear that through this oxidizer more remarkable chemistry should be possible. The immediate consequence of this was the launching of noble gas chemistry, but it also inspired much of the remaining chemistry of my career, including that of recent years.

Noble-gas chemistry begins

$XePtF_6$ and $Xe(PtF_6)_2$

In early February 1962, as I was preparing a lecture, I noticed in a textbook the familiar plot of the first ionization potential as a function of atomic number. This immediately suggested a new test for the oxidizing power of PtF_6. The plot reminded me that the effective nuclear charge of an atom decreases markedly with increasing atomic number, down any group in the Periodic Table. I rapidly checked the values for the heavier noble gases, and saw that the ionization potential of xenon was essentially the same as that of O_2, radon

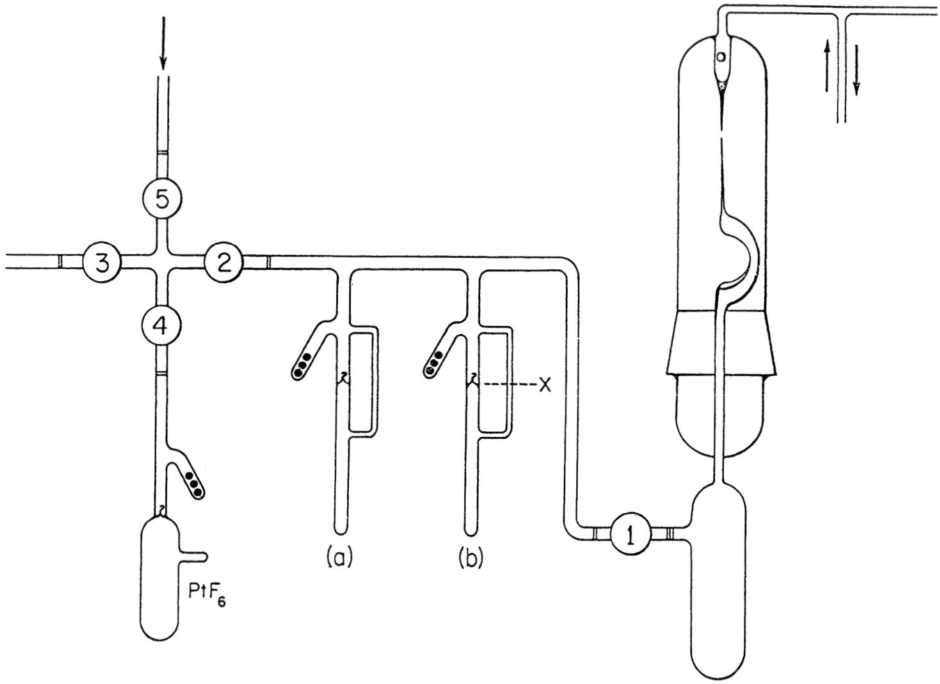

Fig. 3.1. The Xe + PtF_6 experiment. A small sample of PtF_6 was transferred to the quartz sickle gauge, and was allowed to vaporize in the gauge, closed by the metal valve. Following pressure measurement it was transferred to (b) via the break-seal by-pass which was then sealed at X. Xenon was admitted to the gauge to the same pressure as the PtF_6 sample. The sample of Xe from the gauge was condensed in (a) at −196 °C and valves 2 and 1 closed to ensure a small volume. Both the Xe and PtF_6 were vapourized, then the break-seal separating them was broken with nickel balls, moved within the system by means of an external magnet. The interaction of the gases, to produce a yellow solid, was immediate and the gauge showed that the residual pressure in the system was low.

being even lower. Because the anion in such a salt as $O_2^+PtF_6^-$, or the speculated $Xe^+PtF_6^-$, is relatively large, the small increase in cation size associated with the latter formulation would, I estimated, only lower the lattice energy by about 10 kJ mol^{-1}. Since work with radon was out of the question for me, I immediately ordered xenon and prepared some PtF_6. The first experiment was carried out in dry glass and quartz apparatus, as illustrated in Fig. 3.1.

Because my co-workers at that time (March 23, 1962) were still not sufficiently experienced to help me with the glassblowing and the preparation and purification of PtF_6 necessary for the experiment, I was not ready to carry it out until about 7 pm on that Friday. When I broke the seal between the red PtF_6 gas and the colorless xenon gas, there was an immediate interaction, causing an orange-yellow solid to precipitate (see Fig. 3.2). At once I tried to find someone with whom to share the exciting finding, but it appeared that everyone had left for dinner!

A period of feverish activity followed. First, I satisfied myself that the orange-yellow solid was not a product derived from gross impurities in the xenon. To do this, I warmed

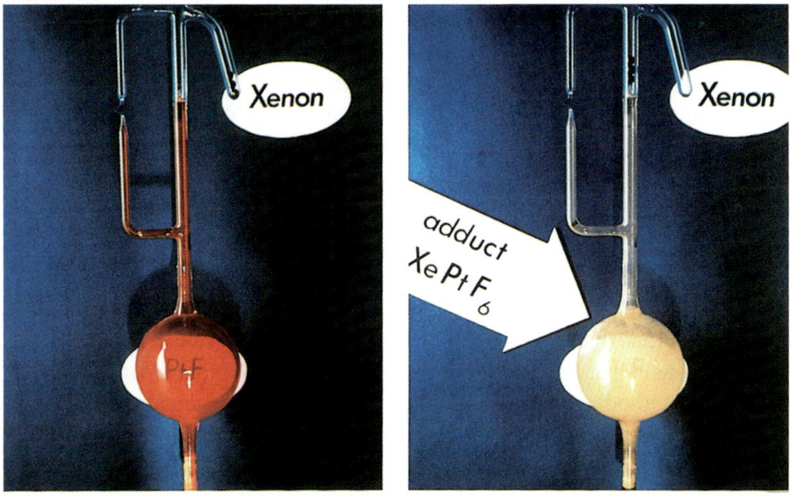

Fig. 3.2. The oxidation of Xe by PtF_6.

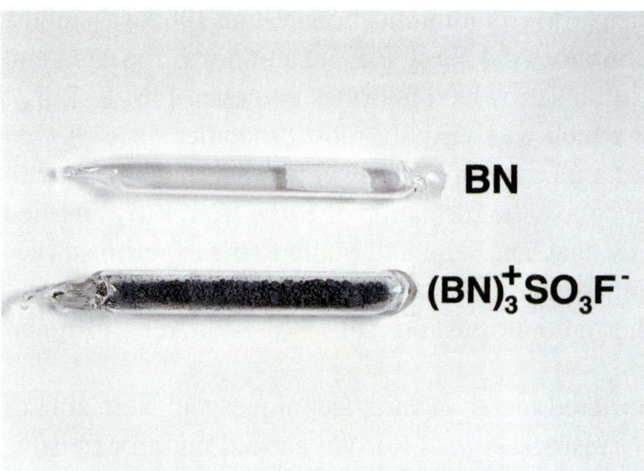

Fig. 3.3. The blue metallic intercalation compound of hexagonal boron nitride, $(BN)_3SO_3F$, compared with colourless BN (an insulator). (See p. 44.)

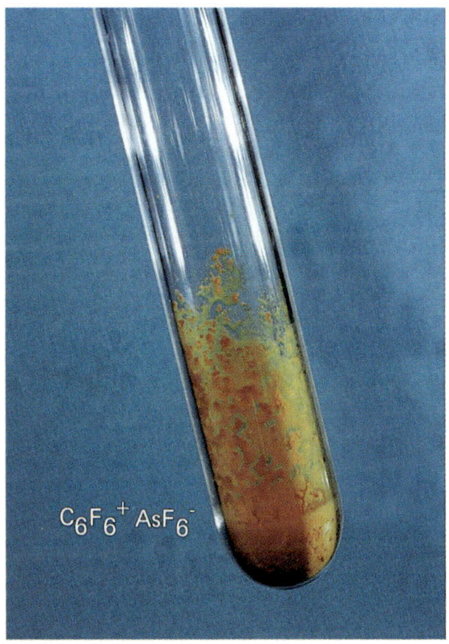

Fig. 3.5. $C_6F_6^+AsF_6^-$ (see p. 46).

the solid in a vacuum (some subliming) then hydrolysed it by adding liquid water (done by vacuum distillation). The gases from the hydrolysis were collected and were analyzed by my colleague, Professor David Frost, by mass spectrometry. He found xenon and oxygen, the latter being attributed to the oxidation of the water by the Pt(V) and the oxidized xenon. Although the tensimetry of the PtF_6 + Xe reaction indicated that more PtF_6 was consumed than Xe, this was what could be expected if the PtF_6 (which was prepared in a quartz flow reactor) contained some SiF_4, and had also reacted, in part, with the metal valves, glass and quartz of the reactor. Indeed it was already clear, at this moment of triumph, that I was in a very vulnerable position as compared to the group at the Argonne National Laboratory. It was evident to me that, if I was to work effectively with oxidizing hexafluorides, I would need to convert to the Argonne mode of working, which was in metal vacuum lines, Kel-F® traps and the like. The loss, on Tuesday March 27, by my carelessness, of my delicate quartz sickle gauge (I had left it under a dynamic vacuum overnight and during the night the main line had broken) also indicated that I should change my methods. The technicians of the mechanical shop at UCB were magnificent in their support and the Argonne National Laboratory (via H. H. Hyman) also sent me Kel-F traps and other valuable equipment. Within 7 months my students (in new laboratory space in the new North Wing at UBC) were able to follow the methods of Weinstock and his co-workers in making PtF_6 [26], RuF_6 [42] and RhF_6 [43]. But by then my original publication on $XePtF_6$ [44] had produced a cascade of events.

Since our $O_2^+PtF_6^-$ note had already appeared before I carried out my xenon oxidation with PtF_6, my chief concern had been that the chemists of the Weinstock group would also have the idea to oxidize xenon, and would immediately apply all of the hexafluorides that they had at hand. Indeed, as I sent off the first announcement of the xenon oxidation by PtF_6 on April 2, 1962[1] I had already formulated an effort to oxidize krypton using rhodium hexafluoride. The Argonne chemists had shown [43] that this was the least stable of the hexafluorides and on this basis I conjectured [45] that it might be the most powerfully oxidizing of all.

The impact of the publication of the $XePtF_6$ communication in June 1962 was almost immediate. At the Argonne National Laboratory the Xe + PtF_6 reaction was repeated and quickly extended to the other late transition series hexafluorides and even [46] to PuF_6! It was the interaction of RuF_6 with Xe which was visually most dramatic, since the red RuF_6 gas, in its interaction with Xe, gave a mixture of a nearly colorless solid and a green material, the latter looking like RuF_5. Of course the formation of RuF_5 from RuF_6 implied fluorination of xenon and this quickly led Claassen, Selig and Malm to the experiment [47] in which they produced XeF_4 by heating xenon and fluorine in a nickel can at 400 °C! Publications from the Argonne, and other laboratories, on XeF_2 [48, 49], XeF_6 [50] and $XeOF_4$ [51] quickly followed.

In the meantime, Jha and I had mastered metal vacuum techniques and were able to produce clean PtF_6 by the method of Weinstock *et al.* [26]. We carried out an extensive series of Xe + PtF_6 reactions. These studies [52] showed that the solid product of the interaction varied in composition between $XePtF_6$ and $Xe(PtF_6)_2$, depending on the relative proportions of the reacting gases. Even at this early stage it was noted that the sticky red product of composition $Xe(PtF_6)_2$ was characterized by a distinctive XRDP which was not present in materials close to the $XePtF_6$ composition. This pattern was subsequently shown by Sladky [53], at Princeton, to be that of $XeF^+PtF_6^-$ (a paramagnetic solid), which he prepared from the 1:1 interaction of XeF_2 with PtF_5 in BrF_5 solution. Jha showed [54] also that the reaction product of composition $Xe(PtF_6)_2$, when warmed gently ($\leqslant 60$ °C) became

[1] A 'letter' to *Nature*, announcing $XePtF_6$ was sent from Vancouver on April 2, but its receipt was acknowledged by a postcard sent sea-mail which arrived in Vancouver in June! I withdrew the 'letter' in the interim.

friable, did not change in composition, and gave a new XRDP. Sladky demonstrated that this XRDP was characteristic of $XeF^+Pt_2F_{11}^-$ (made from 1:2 XeF_2 plus PtF_5). The initial $Xe(PtF_6)_2$ product could therefore be formulated as $XeFPtF_6$ + PtF_5. (The latter like most metal pentafluorides crystallizes with difficulty, especially when impure.) Evidently the 1:1 product was being fluorinated by additional PtF_6:

$$XePtF_6 + PtF_6 \rightarrow XeFPtF_6 + PtF_5$$

Jha found [52] that $Xe(PtF_6)_x$ $(1 < x < 2)$ always gave PtF_6^- salts with alkali fluoride in IF_5 solution therefore it was concluded that the platinum was quinquevalent. (In the light of subsequent experiments [54] it appears that PtF_6^{2-} salts should have been produced also, but the weakly basic IF_5 solutions cannot have been condusive to their formation.) He also discovered (November, 1962) that when $Xe(PtF_6)_x$ was heated in a quartz tube above 165 °C it decomposed to yield XeF_4, which crystallized on the cooler reaches of the tube, the solid red residue being diamagnetic and analyzing as $XePt_2F_{10}$. This formation of XeF_4 was unfortunately subsequent to the Argonne work. It was also puzzling. Why should $XePtF_6$ (or $XeF^+Pt_2F_{11}^-$) yield XeF_4 together with $XePt_2F_{10}$ and not XeF_2? These questions were largely clarified by our later work.

$XeF_5^+PtF_6$

When my first post doctoral associate Don Stewart arrived from T. A. O'Donnell's lab in early 1965, I had him attempt a clean preparation of $XePtF_6$ from O_2PtF_6 with high pressure Xe (using sodium elsewhere in the pressure vessel as 'getter' for the O_2). The projected reaction failed. When a combination of Xe with F_2 was applied to PtF_5 to attempt to form the crystalline product of the Xe + PtF_6 reaction, he found that he could easily (F_2 at 10 psi at 180-220 °C) make Xe(VI) complexes, one of which was isolated in single crystal form and proved [55] the formulation $XeF_5^+PtF_6^-$:

$$Xe + 3F_2 + PtF_5 \rightarrow XeF_5^+PtF_6^-$$

The poor F^- donor properties of XeF_4

There was never a hint of XeF_4 complexes of Pt(V) in this work of Stewart, but the existence of $Xe(PtF_6)_2$ proved that XeF_2 derivatives of Pt(V) should exist. Sladky's work at Princeton clarified these issues. He showed [53] the existence of $Xe_2F_3^+PtF_6^-$, $XeF^+PtF_6^-$ and $XeF^+Pt_2F_{11}^-$ and confirmed that XeF_4 would not form complexes with PtF_5, or the other noble metal pentafluorides, or AsF_5. This provided a basis for the chemical purification [56] of XeF_4 from its admixture with XeF_2 and XeF_6 and proved that XeF_4 was a poorer F^- donor than either XeF_2 or XeF_6. The detailed findings also suggested that XeF_6 was a superior F^- donor to XeF_2 and pointed to the crowded ligancy of XeF_6 as a contributing cause. Berkowitz and his co-workers were later to give [57] quantitative support for this sequence of the F^- donor properties of the xenon fluorides. This work of Sladky also provided a partial explanation for Jha's observation that pyrolysis of $Xe(PtF_6)_x$ gave XeF_4. The intricate redox reactions that underlay that reaction were revealed some years later by the work of Žemva and Graham at UC Berkeley [58].

Oxidation of Pt(IV) by Xe(II), and Xe(II) by Pt(V)

Žemva and Graham [58] showed that liquid XeF_2 at ~140 °C would oxidize Pt(IV) to Pt(V):

$$5XeF_2(c) + 2PtF_4(c) \xrightarrow{140\,°C,\ XeF_2} 2Xe_2F_3^+PtF_6^-(c) + Xe(g)\uparrow$$

The $Xe_2F_3^+$ salt lost XeF_2 in a vacuum:

$$Xe_2F_3^+PtF_6^-(c) \xrightarrow{70\,°C\ \textit{in vacuo}} XeF^+PtF_6^-(c) + XeF_2(g)\uparrow$$

and pyrolysis of the $XeF^+PtF_6^-$ obtained, close to the temperature at which Jha had observed XeF_4 production from $Xe(PtF_6)_x$, gave XeF_4 and diamagnetic $XePt_2F_{10}$:

$$2XeF^+PtF_6^-(c) \xrightarrow{150\text{-}160\,°C} XePt_2F_{10}(c) + XeF_4(g)\uparrow$$

thus establishing that Pt(V) was able to oxidize Xe(II) to Xe(IV). The remarkably stable Pt(IV) material, $XePt_2F_{10}$ (probably $XeF^+Pt_2F_9^-$), did not decompose until 430 °C:

$$XePt_2F_{10}(c) \xrightarrow{430\,°C} 2PtF_4(c) + XeF_2(g)\uparrow$$

The summation of these four reactions amounts to the disproportion of XeF_2:

$$2XeF_2 \rightarrow Xe + XeF_4$$

for which ΔG°_{298} is slightly unfavorable, even for solid XeF_2 and XeF_4. Undoubtedly the poor fluorobasicity of XeF_4, relative to XeF_2, helps to drive the third reaction.

To account for Jha's observations on the pyrolysis of $Xe(PtF_6)_x$ it is first necessary to note that his material, which gave XeF_4, was an approximate 1:1 mixture of $XePtF_6$ and $Xe(PtF_6)_2$. To account for the absence of XeF_2 (only XeF_4 being observed) it is necessary to accept that the acidic component of $Xe(PtF_6)_2$ (*i.e.* the PtF_5 or $Pt_2F_{11}^-$ species) would abstract XeF_2 from the $XePtF_6$ to generate $XeF^+PtF_6^-$, in that process the $XePtF_6$ being converted to $XePt_2F_{10}$. Above 150 °C the $XeF^+PtF_6^-$ would also yield $XePt_2F_{10}$, as shown above.

The nature of $XePtF_6$

Material of composition close to $XePtF_6$ was prepared by Graham [54] by mixing PtF_6 (diluted with SF_6) with a large molar excess of Xe. The mustard-yellow solid, amorphous to X-rays, and insoluble in aHF, was never quite diamagnetic, probably because of contamination by $XeF^+PtF_6^-$ and PtF_5 (see p. 40–41) but always a much weaker paramagnet than $XeF^+PtF_6^-$. In addition, it was found [54] that PtF_4 dissolved in aHF, when it contained a large molar excess of XeF_2 (over the PtF_4). That solution was found, from ^{19}F n.m.r. spectra, to contain PtF_6^{2-}. Removal of aHF and excess XeF_2, *in vacuo* at room temperatures, gave an amorphous, mustard-yellow, diamagnetic solid, which, like the product of the Xe + PtF_6 reaction, was insoluble in aHF. This had the composition $XePtF_6$. It did dissolve in aHF made basic with alkali fluoride, and from such a solution a PtF_6^{2-} salt was obtained. It can therefore be safely concluded that the $XePtF_6$ obtained from the XeF_2 + PtF_4 reaction in aHF is $XeF^+PtF_5^-$ – a Pt(IV) salt.

The behavior of the XeF_2 interaction with PtF_4 in aHF, stands in sharp contrast to that of molten XeF_2 with PtF_4, where oxidation to Pt(V) occurred [58]. This may derive from the XeF^+ and $Xe_2F_3^+$ cations being solvated (and hence less oxidizing) in aHF, but not in the molten difluoride.

The insolubility of $XePtF_6$ in aHF and the recurrent failure to secure crystalline material, are compatible with the anion of $XeF^+PtF_5^-$ being a polymeric entity (the Pt being six-coordinated by F, with two of the ligands shared with another Pt). The $XePtF_6$ made in the Xe + PtF_6 reaction must be the same. When this material interacts with more PtF_6, that electrophile must attack the anion, capturing the electron to liberate $(PtF_5)_n$ oligomer, with simultaneous formation of $XeF^+PtF_6^-$ – a Pt(V) compound!

Redox reactions in the XeF_2/PdF_3 system

A related study by Žemva and Graham [58] of the XeF_2/PdF_3 system, uncovered parallel redox reactions to those of the XeF_2/PtF_4 system, and established the compounds $XePdF_6$ and $XePd_2F_{10}$, the latter from its XRDP clearly very close structurally to $XePt_2F_{10}$. $XePdF_6$ was found to be diamagnetic, and on the basis of vibrational evidence was conjectured to be $XeF^+PdF_5^-$. This is probably a structural relative of $XeF^+PtF_5^-$.

Just as in the Pt system the interaction of XeF_2 in aHF was different from that of the melt. In aHF, XeF_2 in part reduced PdF_4 to Pd_2F_6, although some dissolution of PdF_4 (to give $XePdF_6$ on removal of volatiles) did occur.

Attempted oxidation of krypton

By the end of 1962, Jha had succeeded in preparing a small quantity of RhF_6 by the Argonne method [43]. It did not oxidize krypton [52]. The interaction with Xe was, however, strikingly parallel to that of PtF_6. The latter being so much easier to prepare, we concentrated on it, to understand such reactions.

Noble-gas cation geometries

The work that arose from attempts to understand the Xe + PtF_6 reaction helped greatly in defining the F^- donor properties of the xenon fluorides. Single-crystal structures illustrating the $Xe_2F_3^+$ [59], XeF^+ [60], XeF_5^+ [61], $Xe_2F_{11}^+$ [62] and even XeF_3^+ [63] were obtained, the observed shapes conforming well to a modified valence-shell repulsion model. A crystallographic and vibrational spectroscopic study [64] of XeF_2, in a wide variety of complexes, defined the ionization pathway ($XeF_2 \rightarrow XeF^+ + F^-$), from one-electron bonding in the Xe-F bonds of the molecule, to the electron-pair bonding of the cation.

Ligands other than fluorine for xenon

Interaction of BCl_3 with XeF_4 at −78 °C was carried out by Jha [65] in late 1962, in an attempt to make $XeCl_4$ or $XeCl_2$. Xe, Cl_2 and BF_3 were quantitatively formed. I was, however, sure that a xenon oxide could be derived from XeF_4 by interaction with water, even though I also recognized that the great strength of the bonding in O_2 would mean that such an oxide would be thermodynamically unstable. P. R. Rao took on the task with enthusiasm, and soon obtained a small amount of colorless solid from such a reaction, the xenon analysis of which (by the Dumas method for N_2!) indicated a formula weight greater than that required for XeO_2 ($XeO_2 \cdot H_2O$ or $XeO_2 \cdot 2H_2O$ fit better). Throwing caution to the winds, I had him prepare a larger sample. My late night intervention in Rao's experiment of January 27, 1963, led to disaster. As Rao removed the last of the HF and water from his reaction mixture we noted beautiful colorless crystalline needles of product. Since it was in my mind that this could be a hydrated oxide, I had an intense interest in seeing if these crystals would fall to a powder under vacuum. Because I could not see well enough through my plastic visor, I incautiously threw it back. Rao did likewise. At that point the oxide sample detonated and we received eye and other injuries which put us in hospital for a month. We warned the scientific community through a note to *Science* [66], but the pace of events was such that on the day of my hospital release I received a preprint of a crystal structure (from Allan Zalkin, at Berkeley) which indicated that the hydrolysis product of XeF_4 was the trioxide, XeO_3 [67].

By the time of my move to Princeton in the summer of 1966, it was clear that effective bonding to radon, xenon and krypton would only be achieved with small highly electronegative atoms (such as an oxygen ligand in $-OSO_2F$ or $-OClO_3$). Fortunately, my skilful postdoctoral associates F. O. Sladky and M. Wechsberg took up the suggestion and quickly prepared [68] FXeL and XeL_2 (L = OSO_2F or $OClO_3$) from interaction of the strong anhydrous acids (HL) with XeF_2:

$$XeF_2 + HL \rightarrow FXeL + HF$$
$$XeF_2 + 2HL \rightarrow XeL_2 + 2HF$$

and single crystals of $FXeOSO_2F$ obtained by Wechsberg provided a structure [69]. These studies were extended by Dr R. Mews [70] and by Dr K. Seppelt [71] when they were with me in Berkeley. $Xe(ClO_4)_2$ proved to be an even more potent detonator than XeO_3, so its study has been left for braver souls than I. The decomposition of $Xe(SO_3F)_2$ on gentle heating, however, gives high purity $S_2O_6F_2$:

$$Xe(SO_3F)_2 \rightarrow Xe + S_2O_6F_2$$

and we have always used this reaction as a safe route to this very useful oxidizing and fluorosulfonating reagent, it having been shown by Dudley and Cady to be a source of $FSO_3\cdot$ radicals [72]. My co-workers at Berkeley, using $S_2O_6F_2$, were able to intercalate BN, at ~20 °C, to produce [73] a blue metallic material of composition $(BN)_3SO_3F$ (see Fig. 3.3, p. 39).

Other fluorine chemistry of high oxidation states

The hexafluorides of the transition elements

After the discovery of the $O_2^+PtF_6^-$ formulation, I quickly attributed [45] the remarkable electron affinity of PtF_6 (~8 eV) to the high effective nuclear charge of platinum, the antibonding π symmetry t_{2g}^4 *d* electrons being almost non-bonding. This simple model (see Fig. 3.4) predicted that the electron affinity should rise steadily along each of the second

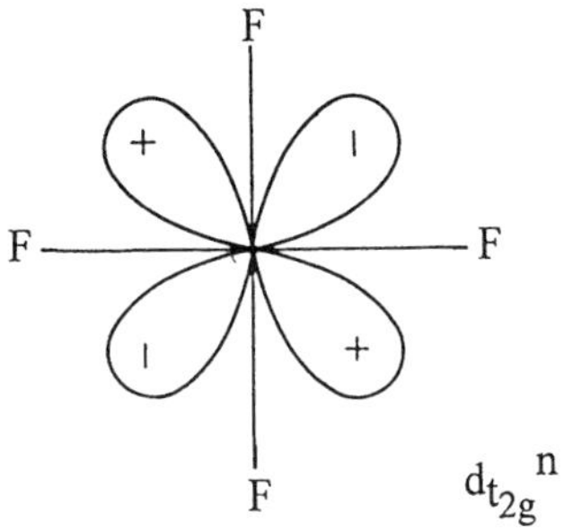

	WF_6	ReF_6	OsF_6	IrF_6	PtF_6	AuF_6
n =	0	1	2	3	4	5
E(eV)	3.5	4.5	5.6	6.7	7.8	8.9

Fig. 3.4. Electron configurations (t_{2g}^{n}) for the 3rd transition series hexafluorides and their estimated electron affinities (eV).

and third transition series for each set of MF_6 molecules. We concentrated our attention on the more extensive third transition series set.

S. P. Beaton (who joined my group at UBC in 1963), together with Jha, made a series of experiments [74] with Cl_2, SF_4 and the strongly basic gaseous fluoride ONF which differentiated these oxidizers, and clearly showed [75] that the electron affinity increases steadily in the MF_6 sequence M = W < Re < Os < Ir < Pt. Moreover, it was judged that the electron affinity must increase by ~1 eV for each unit increase in nuclear charge. This meant that AuF_6 should be ~1 eV superior to PtF_6. Much effort was expended by Rao to prepare this molecule. Although without success in its prime objective, this work did give us our first hints of Au(V) (which was later prepared, in the first AuF_6^- salts, by K. M. Leary in Berkeley [76]) and fortuitously gave us single crystals of AuF_3 for a structure determination [77]. An important and unforeseen aspect of the MF_6/ONF studies, however, was the discovery of ONF_3.

Nitrogen oxide trifluoride, ONF_3

C. J. Willis and his co-workers published [78] a preparation of CF_3O^- salts from which I conjectured that the molecule ONF_3 should exist. Fortunately, J. Passmore had joined me in the Fall of 1964, and he undertook the task of searching for ONF_3. This began with a direct approach of attempting the fluorination of ONF in nickel bombs. Coincidentally, Beaton was investigating the interaction of ONF with IrF_6. I soon realized that the intense absorption bands that occurred in the IR spectra of the gaseous products of Beaton's reaction were one and the same with a product of Passmore's pyrolysis of $(NO)_2NiF_6$, the latter having been a product of the F_2 + ONF interaction, with the nickel containers. Passmore quickly found methods of purifying the ONF_3 (which is kinetically stable enough to withstand a wash with water) and went on to demonstrate that it is a moderately good fluoride-ion donor, giving a ONF_2^+ salt even with BF_3 [79]. It was only after I had delivered a lecture at the 1965 IUPAC Congress in Moscow, USSR, in which the ONF/IrF_6 reactions were mentioned that I was informed, by a senior American inorganic chemist, that the ONF_3 molecule was 'known' but was in the classified literature (such irony!). We were left bitterly disappointed and indeed aggrieved, since rumors circulated that I had been privy to the existence of ONF_3. This was not so.

$OsOF_5$ and IOF_5

Since OsO_4 [80] and OsO_3F_2 [81] were easy to make and yet the highest fluoride then (1964) appeared to be OsF_6 [17], I encouraged Jha, as an exercise in preparing and characterizing a new molecular species, to take up the tasks of making $OsOF_5$ or OsO_2F_4. At that point, I had come to the conclusion that the coordination of high valent osmium was unlikely to exceed six. Some support for that supposition had come from Jha's own findings that although the interaction of ONF with OsF_6 gave $ON^+OsF_7^-$, it was very easily reduced to $ON^+OsF_6^-$; I therefore had him fluorinate OsO_2, since I knew from unpublished work of M. A. Hepworth that this furnished a volatile oxyfluoride (or oxyfluorides). Jha was able to prove that the major product of that fluorination was $OsOF_5$, then the first Os(VII) compound [82]. Although he and, later, Dr Nygen-Nghi at Princeton were to seek OsO_2F_4, they did not succeed [83]. It was not until the work of Christe and Bougon [84], using KrF_2 as oxidant, that the latter molecule became known. The successful preparation of $OsOF_5$ stimulated a search for IOF_5.

The existence of IOF_5 was established independently in three different laboratories almost simultaneously [85-87]. (The synthesis of $XeOF_4$ from XeF_6 may have had a role in this coincidence.) My co-worker in the UBC discovery [86] was an undergraduate student, L. E. Levchuk, who made it by treating IF_7 with various oxides, including water, and showed that all previous structural studies of IF_7 were suspect, as a consequence of the close physical similarity of IOF_5 to IF_7. The evidence of published spectroscopic data [88] on IF_7 showed heavy contamination by IOF_5. It is of interest that the widely accepted D_{5h} symmetry of IF_7 at this time, was based on such spurious vibrational data!

$C_6F_6^+$ and related perfluoroaromatic-cation salts

I had noted in 1963 that the first ionization potential of C_6F_6 (9.97 eV [89]) was sufficiently low for oxidation by PtF_6, and perhaps by earlier members of the series, but attempts by Jha to prepare $C_6F_6^+$ using PtF_6 and by Richardson using IrF_6, although they gave promising bright orange-yellow solids at low temperature, always resulted in decomposition to higher fluorocarbons on being warmed up to ~20 °C. Eventually, it occurred to me that the O_2^+ salts of the kinetically stable AsF_6^- and SbF_6^- anions should provide a better route and in T. D. Richardson's hands this proved to be so [90], giving golden yellow solids (see Fig. 3.5, p. 39):

$$C_6F_6 + O_2AsF_6 \rightarrow C_6F_6AsF_6 + O_2\uparrow$$

Initially the solvent was liquid WF_6, but subsequently both SO_2ClF and aHF proved to be more useful [91]. $C_6F_6^+AsF_6^-$, which has a primitive rhombohedral unit cell, indicative of a rhombohedral variant of the CsCl structure, proved to be a simple paramagnet. It is a valuable reagent [90, 91] in the quantitative preparation of higher aromatic cation salts (Fig. 3.5.), *e.g.*

$$C_6F_6^+AsF_6^- + C_{10}F_8 \rightarrow C_{10}F_8^+AsF_6^- + C_6F_6$$

since it can be used in slight excess, that excess then being eliminated by simply raising the temperature to ~20 °C, at which temperature the $C_6F_6^+AsF_6^-$ decomposes cleanly to C_6F_6

and a cyclohexadiene, 1,4-C_6F_8. The same products are formed when $C_6F_6^+$ is attacked by F^- (from an alkali fluoride).

The chemistry of these salts is still largely unexplored (especially the consequences of attack by anions other than F^-) but it was the realization of the ease of oxidation of the higher aromatics that first drew my attention to graphite intercalation chemistry [92]; this in turn focused my attention again on thermodynamic aspects of fluorine chemistry [93, 94].

Fluorine chemistry exploiting anhydrous liquid hydrogen fluoride

AgF_3, NiF_4 and NiF_3

A synthesis, in 1983, by Bougon *et al.* [95], of what was reported to be 'AgF_3', puzzled me greatly. The 'trifluoride' was described as a simple, yet weak, paramagnet. It was evident from the reported XRDP data, however, that the binary fluoride was neither related structurally to RhF_3 [3] nor to AuF_3 [77]. Indeed, I had expected AgF_3 to have the same structure as AuF_3, and like it to be diamagnetic (*i.e.*, low-spin d^8). Although since the middle seventies I had spent much of my effort on research in graphite and related chemistry, I had maintained a connection with higher oxidation-state chemistry through a joint programme with the group under Boris Žemva's leadership, at the Jožef Stefan Institute in Ljubljana, Slovenia. In 1986, our joint-project renewal proposal was due to be submitted, and for that purpose I arranged to spend a few days in Ljubljana following the International Fluorine Meeting in Paris. This meeting celebrated the Centennial of Moissan's isolation of fluorine, and there K. O. Christe described his chemical route to elemental fluorine [96]. This involved thermal decomposition of MnF_4, which was derived from K_2MnF_6 and SbF_5 (the K_2MnF_6 being preparable without the need of elemental F_2). This gave me the idea that one could probably prepare thermodynamically unstable (yet kinetically stable) binary fluorides, such as AgF_3, *by precipitating them from aHF, at dry ice temperature*, using gaseous fluoroacids such as BF_3, PF_5 and AsF_5. I already knew from joint work with the Jožef Stefan group [97] that $XeF_5^+AgF_4^-$ was soluble and unsolvolyzed in aHF. We also knew that K_2NiF_6 was highly soluble, since we had long used that reagent to dry aHF (by oxidation of the water). Žemva shared my optimism. We embodied the idea in our new proposal and we made plans to begin the programme. For the summer of 1987, he arranged a one month stay in Ljubljana for myself and my wife. Because it is always more exciting to do something entirely new, we made the synthesis of NiF_4 our first target [32, 98].

By passing AsF_5 through a solution of $(Xe_2F_{11})_2NiF_6$ in aHF below −60 °C, a permanent precipitate was obtained after four molar aliquots of AsF_5 had been passed for each molar aliquot of NiF_6^{2-} salt:

$$(Xe_2F_{11})_2NiF_6 + 4AsF_5 \xrightarrow{\text{aHF at} \leqslant -60\,^\circ\text{C}} 4XeF_5AsF_6 + NiF_4\downarrow$$

The NiF_4 proved to be thermally unstable above −60 °C, hence it has yet to be characterized structurally or magnetically. Its controlled decomposition, however, yielded NiF_3:

$$NiF_4(c) \xrightarrow{\text{aHF } 0\,^\circ\text{C}} NiF_3(c) + 1/2\ F_2(g)\uparrow$$

three forms of which have been structurally and magnetically defined [98]. They are valuable fluorinating agents and much of my time during my last four years at the bench has been spent in the application of the nickel trifluorides, NiF_4, and NiF_6^{2-} salts to the fluorination of organic molecules at 20 °C or lower temperatures. It was only after much of the early work had been completed that we rediscovered the papers of Court and Dove [99]. They had anticipated the 1986 idea, at least for MnF_4 and NiF_3 generation, although they had not succeeded in isolating NiF_3 pure enough to structurally characterize it. Nor had they found evidence for NiF_4.

The new synthetic approach also worked for AgF_3 [32, 100]. It was quickly found that this was best made by precipitation using BF_3 at 0 °C:

$$AgF_4^-(\text{solv}) + BF_3(\text{g}) \xrightarrow{\text{aHF, 0 °C}} BF_4^-(\text{solv}) + AgF_3(\text{c})\downarrow$$

Unlike the 'trifluoride' of Bougon *et al.*, however, this bright red solid proved to be diamagnetic, and it was also isostructural with AuF_3. Overnight in aHF at ~20 °C, however, the solid evolved F_2:

$$3AgF_3(\text{c}) \xrightarrow{\text{aHF, } \sim 20 \text{ °C}} Ag_3F_8(\text{c}) + 1/2\, F_2(\text{g})\uparrow$$

The XRDP of this product proved to be the same as that reported by Bougon *et al.* [95] for their 'trifluoride.' Moreover, like their solid, this was a simple paramagnet. The formulation $Ag^{2+}(AgF_4^-)_2$ was shown to be appropriate, with its synthesis from Ag^{2+} and AgF_4^- salts:

$$Ag^{2+}(\text{solv}) + 2AgF_4^-(\text{solv}) \longrightarrow Ag_3F_8(\text{c})\downarrow$$

and this has recently been structurally established with the indexing of the XRDP data, using unit cell parameters close to those of $Ag^{2+}(AuF_4^-)_2$, the single-crystal structure of which has been reported [101] by B. Müller and his co-workers.

Room temperature fluorination of organic molecules by NiF_3, NiF_4 and NiF_6^{2-}

In the Fall of 1994 and 1995, working with R. D. Chambers group at Durham University, UK, I undertook investigation of NiF_6^{2-}, NiF_3 and NiF_4 (the latter made *in situ*) in aHF, at or below room temperatures, as fluorinators [102] for a variety of organic compounds chosen by Chambers. In this, I was much assisted in the separation and characterization of products by A. J. Roche and R. C. H. Spink. This work is exemplified below.

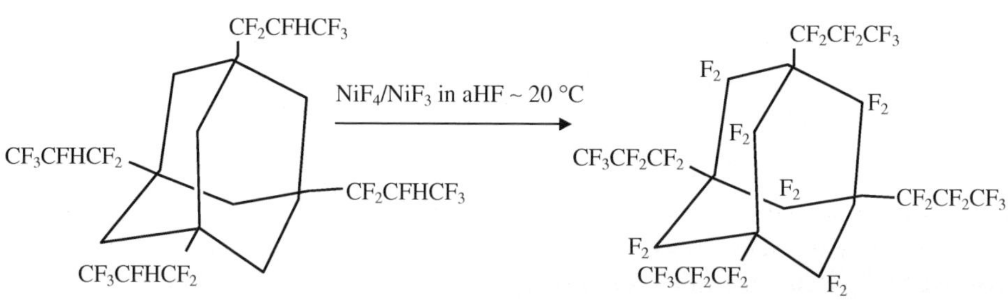

It is probable that these oxidizers first take an electron from the substrate molecule, which is then attacked by solvated F^-. Of the hydrocarbons only CH_4 (I = 12.6 eV) [89] has a higher ionization potential than Xe.

Subsequently, it was found that the NiF_6^{2-} species itself (in aHF) is also an effective fluorinator, although less potent [102] than NiF_3 or NiF_4. In one category, however, *i.e.* the fluorination of cationic species, the NiF_6^{2-} anion is especially valuable, since elemental fluorine (an electrophile) is not effective in such fluorinations. My Berkeley co-workers, L. C. Chacón and J. M. Whalen, used NiF_6^{2-} solutions in aHF, to fluorinate $(CH_3)_4N^+$, to obtain a variety of fluorinated cationic derivatives, the most fluorinated of which proved [103] to be $(CHF_2)_3CH_3N^+$. It is evident that the cationic nature of the ammonium species must be effectively maintained throughout a sequence of oxidative fluorine transfers from NiF_6^{2-} anions.

The advent of NiF_4, NiF_3 and AgF_3 and especially their valuable fluorinating properties, led me to ponder on more efficient routes to the synthesis of NiF_6^{2-} and AgF_4^- salts. These had first been made nearly fifty years ago in Münster by Wilhelm Klemm [104] and Rudolph Hoppe [105] and their co-workers, using fluorine under pressure in metal reactors above 300 °C.

Room temperature synthesis of high oxidation-state anion salts in basic aHF

Early experience with palladium chemistry in BrF_3 had taught me that a high oxidation state is much more easily attained in an anion. This I had rationalized as a consequence of the lower electronegativity of that oxidation state in the electron-rich environment of an anion. This led to the room temperature conversion of gold and the platinum metals (except Rh) to make AuF_4^- and OsF_6^-, RuF_6^-, IrF_6^-, PtF_6^{2-} and PdF_6^{2-} salts [106]. This discovery (except for the gold oxidation) was simultaneously also made by Holloway and his co-workers [107]. We found it impossible to similarly oxidize AgF_2, even with very strongly basic aHF and F_2. It occurred to me, however, that the photo-dissociation of F_2, to atoms, could provide the essential condition for AgF_4^-, and this proved to be so [108] (see Fig. 3.6):

$$AgF_2(c) + F^-(solv) + F^{\cdot}(solv) \xrightarrow{\text{aHF}, \sim 20\ ^\circ\text{C}} AgF_4^-(solv)$$

PtF_6^- and AuF_6^- salts can be made similarly [109]. A photochemical synthesis of NiF_6^{2-} was also achieved [110]:

$$NiF_2(c) + 2F^-(solv) + 2F^{\cdot}(solv) \xrightarrow{\text{aHF}, \sim 20\ ^\circ\text{C}} NiF_6^{2-}(solv)$$

With this approach, using LiF as the alkali fluoride, the first syntheses of $LiAgF_4$, $LiAuF_6$, $LiPtF_6$ and Li_2NiF_6 have also been made and their structures characterized [108-110]. It is, therefore, now possible to easily prepare NiF_6^{2-} and AgF_4^- salts, and to thereby recover their reduction products (NiF_2 and AgF_2) as the salts.

Room temperature synthesis of Pd^{2+} and Au^{2+} salts from the elements in acidified aHF

In aHF acidified by strong F^- acceptors, such as AsF_5, SbF_5, or BiF_5, metals usually dissolve to give cationic species. In cations, the electronegativity of a given oxidation state

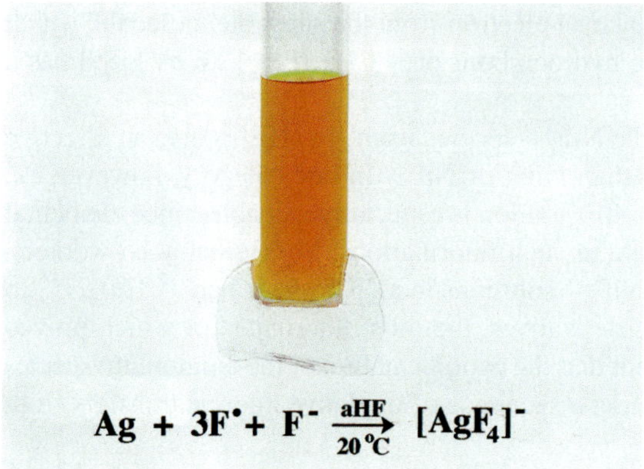

Fig. 3.6. Room temperature preparation of an AgF_4^- salt.

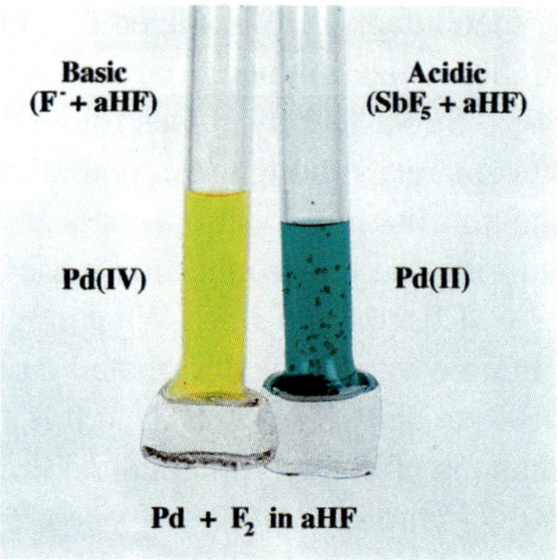

Fig. 3.7. Oxidation of Pd by F_2 in basic and acidic aHF.

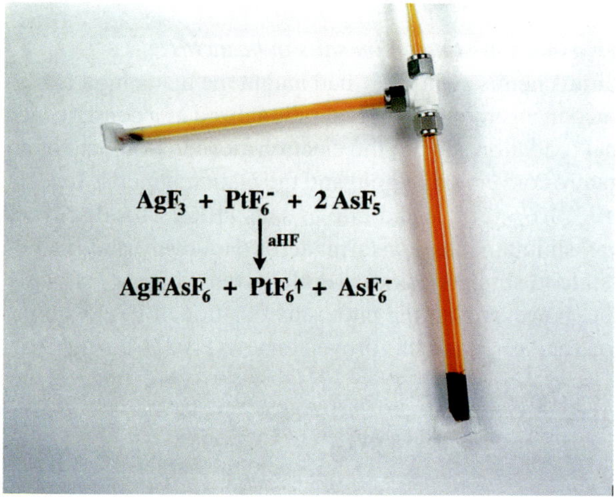

Fig. 3.8. The room temperature oxidation of PtF_6^- to PtF_6 (the red vapour) by Ag(III) in acidic aHF.

is at its highest, whereas in an anion it is at its lowest. It is, consequently, often more difficult to attain a given higher oxidation state in acidified aHF than in neutral or basic aHF. Palladium provides an excellent example, illustrated in Fig. 3.7. As I have already mentioned, the metal, with F_2, easily dissolves in aHF, made basic with alkali fluoride, to give yellow solutions of a PdF_6^{2-} salt. Pd(IV) is not accessible in acidified aHF, where the oxidation only proceeds to formation of the aquamarine Pd^{2+}(solv) [106]:

$$\mathrm{Pd(c)} + \mathrm{F_2(g)} + 2\mathrm{H^+(solv)} \xrightarrow{\text{aHF, } \sim 20\ ^\circ\mathrm{C}} \mathrm{Pd^{2+}(solv)} + 2\mathrm{HF(solv)}$$

The insights that this chemistry provided led to the preparation [111] of the first fluoro derivative of Au(II), $Au(SbF_6)_2$:

$$Au(c) + F_2(g) + 2SbF_6^-(solv) + 2H^+(solv) \xrightarrow{aHF,\ \sim 20\ ^\circ C} Au(SbF_6)_2(solv) + 2HF(solv)$$

We were, however, unsuccessful in our attempts to derive AuF_2 from $Au(SbF_6)_2$. A large relativistic effect at gold [112-114] so stabilizes the metal, and simultaneously provides for relatively easy oxidation of 5*d* electrons, that AuF_2 disproportionates:

$$4AuF_2(c) \longrightarrow Au_3F_8(c) + Au(c)$$

The Au_3F_8 is structurally similar [115] to Ag_3F_8, and therefore is $Au(II)\{Au(III)F_4\}_2$.

Ag^{2+} oxidation of O_2 to O_2^+ in acidified aHF

To illustrate the higher electronegativity of a cationic oxidation state, it is instructive to consider Ag^{2+}(solv) in aHF. The beautiful blue solutions of the latter were found to be quickly generated (with F_2 evolution) when AgF_3 was dissolved in acidified aHF [100]. In searching for evidence of Ag(III) solution species, it was found that even the Ag^{2+}(solv) had the remarkable ability to oxidize xenon gas at ~20 °C [116]. Puzzlingly, we were not able to oxidize O_2(g) (which has the same ionization potential as Xe [89]). It was my student co-worker, W. J. Casteel, Jr., who showed that this is because the (hard, low polarizability) O_2 molecule has low solubility in aHF (in contrast to the highly polarizable Xe atom). When the solutions of Ag^{2+}(solv) were cooled to −78 °C, O_2 was quickly taken up and oxidized, colorless $O_2^+AsF_6^-$ and $AgAsF_6$ precipitating [117]:

$$Ag^{2+}(solv) + O_2(g) + 2AsF_6^-(solv) \xrightarrow{aHF,\ -78\ ^\circ C} O_2AsF_6(c)\downarrow + AgAsF_6(c)\downarrow$$

Raising the temperature (which increases the positive value of TΔS for the reverse reaction) regenerated O_2 and Ag^{2+}.

Room temperature oxidation of MF_6^- (M = Pt, Ru, Rh) by Ag(III) and Ni(IV) in acidified aHF

From the early sixties, I had been seeking a route to AuF_6. The remarkable oxidizing capability of Ag^{2+}(solv) in aHF clearly presaged even more potent capabilities for Ag(III) and in that species, in strongly acid aHF, I therefore had a real hope for the oxidation of AuF_6^- to AuF_6. In that, I was to be disappointed, but such solutions were shown [116], by my student co-worker G. M. Lucier, to be able to oxidize the monoanions of the most powerfully oxidizing transition element fluorides:

$$Ag^{3+}(solv) + MF_6^-(solv)\ (M = Pt, Ru, Rh) \xrightarrow{aHF,\ \sim 20\ ^\circ C} Ag^{2+}(solv) + MF_6(g)\uparrow$$

The reaction is illustrated in Fig. 3.8 for the PtF_6 case. Here, the yield is better than 70% of the theoretical one. This indicates that the $Ag^{3+}_{(solv)}$ is sufficiently long lived, to encounter at least 70% of the PtF_6^- species present in solution.

We later discovered [117] that NiF_4, or NiF_3 with acidified aHF would also liberate these fluorides (but not AuF_6 from AuF_6^-!). With acid Ag(III) or Ni(IV) in aHF, *we therefore have the strongest oxidizers known to us so far*. If cationic Cu(IV) can be generated in aHF, it may be able to generate AuF_6. Failing that, access to AuF_6 via electrolysis of a low-melting salt might yet be possible.

Acknowledgements

Because this account has concentrated on only the most novel (and sensational?) aspects of my career in Fluorine Chemistry, many of my excellent co-workers have received no mention; this is especially so for the work involving graphite systems and thermodynamic interests, which are described elsewhere [92-94]. It is a matter of deep personal regret that one of the great innovators in transition-element fluorine chemistry, Bernard Weinstock, did not live to see the fulfillment of the revolution that he had initiated. He, I am sure, would have been delighted with the room temperature generation of PtF_6, RuF_6, and RhF_6.

I have been especially fortunate in finding generous support for my research. Initially, this came from the National Research Council of Canada and The Research Corporation. In Princeton, the National Science Foundation supported me; and since my move to Berkeley I have always enjoyed total funding of my 'exotic' chemistry from the US Department of Energy (previously the Atomic Energy Commission).

References

1 F. Brown and P. L. Robinson, *J. Chem. Soc.*, (1955) 3147.
2 R. S. Nyholm and A. G. Sharpe, *J. Chem. Soc.*, (1952) 3579.
3 M. A. Hepworth, K. H. Jack, R. D. Peacock and G. J. Westland, *Acta Cryst.*, *10* (1957) 63.
4 O. Ruff and E. Ascher, *Z. anorg. Chem.*, *183* (1929) 211.
5 N. Bartlett and M. A. Hepworth, *Chemistry and Industry*, (1956) 1425.
6 N. Bartlett and R. Maitland, *Acta Cryst.*, *11* (1958) 747.
7 P. R. Rao, R. C. Sherwood and N. Bartlett, *J. Chem. Phys.*, *49* (1968) 3738.
8 W. C. Smith, C. W. Tullock, R. D. Smith and V. A. Engelhardt, *J. Am. Chem. Soc.*, *82* (1960) 551.
9 N. Bartlett and P. L. Robinson, *Chemistry and Industry*, (1956) 1351.
10 N. Bartlett and P. L. Robinson, *J. Chem. Soc.*, (1961) 3417.
11 N. Bartlett and P. L. Robinson, *Proc. Chem. Soc.*, (1957) 230.
12 F. Seel and O. Detmer, *Z. anorg. Chem.*, *301* (1959) 113.
13 D. D. Gibler, C. J. Adams, M. Fischer, A. Zalkin and N. Bartlett, *Inorg. Chem.*, *11* (1972) 2325.
14 H. Moissan, *Le Fluor et ses Composes*, Steinbeil, Paris, 1900, p. 216.
15 A. G. Sharpe, *J. Chem. Soc.*, (1950) 3444.
16 O. Ruff and F. W. Tschirch, *Ber.*, *46* (1913) 929.
17 B. Weinstock and J. G. Malm, *J. Amer. Chem. Soc.*, *80* (1958) 4466.
18 G. B. Hargreaves and R. D. Peacock, *J. Chem. Soc.*, (1960) 2618.
19 O. Ruff and J. Fischer, *Z. anorg. Chem.*, *179* (1929) 161.
20 G. J. Westland and P. L. Robinson, *J. Chem. Soc.*, (1956) 4481.
21 N. Bartlett and P. R. Rao, *Chem. Commun.*, (1965) 252.
22 J. H. Holloway, P. R. Rao and N. Bartlett, *Chem. Commun.*, (1965) 306.
23 P. R. Rao, A. Tressaud and N. Bartlett, *J. Inorg. Nucl. Chem. Supplement*, (1976).
24 J. S. Griffith and L. E. Orgel, *Quart. Revs. Chem. Soc.*, *XI* (1957) 381.
25 R. J. Gillespie and R. S. Nyholm, *Quart. Revs. Chem. Soc.*, *XI* (1957) 339.

26 B. Weinstock, H. H. Claassen and J. G. Malm, *J. Am. Chem. Soc.*, *79* (1957) 5832.
27 A. G. Sharpe, *J. Chem. Soc.*, (1953) 197.
28 N. Bartlett and J. W. Quail, *J. Chem. Soc.*, (1961) 3728.
29 N. Bartlett and P. R. Rao, *Proc. Chem. Soc.*, (1964) 393.
30 A. Tressaud, M. Wintenberger, N. Bartlett and P. Hagenmuller, *C.R. Acad. Sci. Paris*, *282*, serie C (1976) 1069.
31 N. Bartlett and D. H. Lohmann, *J. Chem. Soc.*, (1964) 619.
32 B. Zemva, K. Lutar, A. Jesih and W. J. Casteel, Jr., *J. Chem. Soc., Chem. Commun.*, (1989) 346.
33 N. Bartlett and A. Tressaud, *C.R. Acad. Sci. Paris*, *278* (1974) 1501.
34 L. M. Dennis and A. W. Laubengayer, *Z. physik. Chem.* (Leipzig), *130* (1927) 530.
35 N. Bartlett and K. C. Yu, *Can. J. Chem.*, *39* (1961) 80.
36 J. D. McCullough, *J. Am. Chem. Soc.*, *59* (1937) 790.
37 J. Trotter, M. Akhtar and N. Bartlett, *J. Chem. Soc.*, *A*, (1966) 30.
38 M. Akhtar, M.Sc. Thesis, U.B.C., 1965.
39 N. Bartlett and D. H. Lohmann, *Proc. Chem. Soc.*, (1960) 14.
40 N. Bartlett and D. H. Lohmann, *Proc. Chem. Soc.*, (1962) 115.
41 N. Bartlett and D. H. Lohmann, *J. Chem. Soc.*, (1962) 5253.
42 H. H. Claassen, H. Selig, J. G. Malm, C. L. Chernick and B. Weinstock, *J. Am. Chem. Soc.*, *83* (1961) 2390.
43 C. L. Chernick, H. H. Claassen and B. Weinstock, *J. Am. Chem. Soc.*, *83* (1961) 3165.
44 N. Bartlett, *Proc. Chem. Soc.*, (1962) 218.
45 N. Bartlett, *Chemistry in Canada*, *15* (1963) 33.
46 C. L. Chernick *et al.*, *Science*, *138* (1962) 136.
47 H. H. Claassen, H. Selig and J. G. Malm, *J. Am. Chem. Soc.*, *84* (1962) 3593.
48 J. L. Weeks, C. L. Chernick and M. S. Matheson, *J. Am. Chem. Soc.*, *84* (1962) 4612.
49 R. Hoppe, W. Dähne, H. Mattauch and K. M. Rödder, *Angew. Chem.*, *74* (1962) 903.
50 J. Slivnik, B. S. Brcic, B. Volavsek, J. Marsel, V. Vrscaj, A. Smalc, B. Frlec and A. Zemljic, *Croat. Chem. Acta*, *34* (1962) 253; J. G. Malm, I. Sheft and C. L. Chernick, *J. Am. Chem. Soc.*, *85* (1963) 110; F. B. Dudley, G. Gard and G. H. Cady, *Inorg. Chem.*, *2* (1963) 228; E. E. Weaver, B. Weinstock and C. P. Knop, *J. Am. Chem. Soc.*, *85* (1963) 111.
51 Ref. [47] and D. F. Smith, *Science*, *140* (1963) 899; J. Shamir, H. Selig, D. Samuel and J. Reuben, *J. Am. Chem. Soc.*, *87* (1965) 2359.
52 N. Bartlett and N. K. Jha, 'The Xenon-Platinum Hexafluoride Reaction and Related Reactions', in H. H. Hyman (ed.), *Noble Gas Compounds*, University of Chicago Press, 1963, p. 23.
53 F. O. Sladky, P. A. Bulliner and N. Bartlett, *J. Chem. Soc.*, (1969) 2179.
54 L. Graham, O. Graudejus, N. K. Jha and N. Bartlett, *Coordination Chemistry Reviews*, *197* (2000) 321.
55 N. Bartlett, F. Einstein, D. F. Stewart and J. Trotter, *Chem. Commun.*, (1966) 550; *J. Chem. Soc., A*, (1967) 1190.
56 N. Bartlett and F. O. Sladky, *J. Am. Chem. Soc.*, *90* (1968) 5316.
57 J. Berkowitz, W. A. Chupka, P. M. Guyon, J. H. Holloway and R. Spohr, *J. Phys. Chem.*, *75* (1971) 1461.
58 N. Bartlett, B. Žemva and L. Graham, *J. Fluorine Chem.*, *7* (1976) 301.
59 F. O. Sladky, P. A. Bulliner, N. Bartlett, B. G. DeBoer and A. Zalkin, *Chem. Commun.*, (1968) 1048.
60 N. Bartlett, M. Gennis, D. D. Gibler, B. K. Morrell and A. Zalkin, *Inorg. Chem.*, *12* (1973) 1717; A. Zalkin, D. L. Ward, R. N. Biagioni, D. H. Templeton and N. Bartlett, *Inorg. Chem.*, *17* (1978) 1318.
61 K. Leary, D. H. Templeton, A. Zalkin and N. Bartlett, *Inorg. Chem.*, *12* (1973) 1726.
62 K. Leary, A. Zalkin and N. Bartlett, *Inorg. Chem.*, *13* (1974) 775.
63 D. E. McKee, A. Zalkin and N. Bartlett, *Inorg. Chem.*, *12* (1973) 1713.
64 B. Žemva, A. Jesih, D. H. Templeton, A. Zalkin, A. K. Cheetham and N. Bartlett, *J. Am. Chem. Soc.*, *109* (1987) 7420.
65 Reported in N. Bartlett, *Endeavour*, *23* (1964) 3, and ref. [46].
66 N. Bartlett and P. R. Rao, *Science*, *139* (1963) 506.
67 D. H. Templeton, A. Zalkin, J. D. Forrester and S. M. Williamson, *J. Am. Chem. Soc.*, *58* (1963) 817.
68 N. Bartlett, M. Wechsberg, F. O. Sladky, P. A. Bulliner, G. R. Jones and R. D. Burbank, *Chem. Commun.*, (1969) 703.
69 N. Bartlett, M. Wechsberg, G. R. Jones and R. D. Burbank, *Inorg. Chem.*, *11* (1972) 1124.
70 M. Wechsberg, P. A. Bulliner, F. O. Sladky, R. Mews and N. Bartlett, *Inorg. Chem.*, *11* (1972) 3063.

71 L. K. Templeton, D. H. Templeton, K. Seppelt and N. Bartlett, *Inorg. Chem.*, *15* (1976) 2718.

72 F. B. Dudley and G. H. Cady, *J. Am. Chem. Soc.*, *79* (1957) 513.

73 N. Bartlett, R. N. Biagioni, B. W. McQuillan, A. S. Robertson and A. C. Thompson, *J. Chem. Soc. Chem. Commun.*, (1978) 200; C. Shen, S. G. Mayorga, R. Biagioni, C. Piskoti, M. Ishigami, A. Zettl and N. Bartlett, *J. Solid State Chem.*, *144* (1999) (in press).

74 N. Bartlett, S. P. Beaton and N. K. Jha, *Chem. Commun.*, (1966) 168.

75 N. Bartlett, *Angew Chem. Int. Ed. Engl.*, *7* (1968) 433; N. Bartlett, *Angew. Chem.*, (1968) 453.

76 K. Leary and N. Bartlett, *J.C.S. Chem. Commun.*, (1972) 902; N. Bartlett and K. Leary, *Revue de chimie minerale*, *13* (1976) 82.

77 F. W. B. Einstein, J. R. Rao, J. Trotter and N. Bartlett, *J. Chem. Soc., A*, (1967) 478.

78 D. C. Brandley, M. E. Redwood and C. J. Willis, *Proc. Chem. Soc.*, (1964) 416.

79 N. Bartlett, J. Passmore and E. J. Wells, *Chem. Communs.*, (1966) 213.

80 G. Brauer (ed.), *Handbuch der Preparativen Anorganischen Chemie*, F. Enke Verlag, Stuttgart, 1981, p. 1745.

81 M. A. Hepworth and P. L. Robinson, *J. Inorg. Nuclear Chem.*, *4* (1957) 24.

82 N. Bartlett, N. K. Jha and J. Trotter, *Proc. Chem. Soc.*, (1962) 277; N. Bartlett and N. K. Jha, *J. Chem. Soc.*, (1968) 536.

83 Nguyen-Nghi and N. Bartlett, *C.R. Acad. Sci. Paris*, *269*, serie C (1969) 756.

84 K. O. Christe and R. Bougon, *J. Chem. Soc., Chem. Commun.*, (1992) 1056.

85 R. J. Gillespie and J. W. Quail, *Proc. Chem. Soc.*, (1963) 278.

86 N. Bartlett and L. E. Levchuk, *Proc. Chem. Soc.*, (1963) 342.

87 L. G. Alexakos, C. D. Cornwall and S. B. Peirce, *Proc. Chem. Soc.*, (1963) 341.

88 R. C. Lord, M. A. Lynch, W. C. Schumb and E. J. Slowinski, *J. Am. Chem. Soc.*, *72* (1950) 522.

89 R. D. Levin and S. G. Lias (eds.), *Ionization Potential and Appearance Potential Measurements, 1971-81*, NSRDS-NBS 71, U.S. Dept. of Commerce.

90 T. J. Richardson and N. Bartlett, *J. Chem. Soc. Chem. Communs.*, (1974) 427.

91 T. J. Richardson, F. L. Tanzella and N. Bartlett, *J. Am. Chem. Soc.*, *108* (1986) 4937.

92 N. Bartlett and B. W. McQuillan, 'Graphite Chemistry', in M. Stanley Whittingham and Allan J. Jacobson (eds.), *Intercalation Chemistry*, Academic Press, New York, 1982, p. 19-53.

93 N. Bartlett, F. Okino, T. E. Mallouk, R. Hagiwara, M. Lerner, G. Rosenthal and K. Kourtakis, in M. K. Johnson *et al.* (eds.), *ACS Advances in Chemistry Series, No. 226* (1990) 20.

94 C. Shen, R. Hagiwara, T. E. Mallouk and N. Bartlett, *ACS Symposium Series, Inorganic Fluorine Chemistry*, *555* (1994) 26.

95 R. Bougon and M. Lance, *C.R. Acad. Sci.*, *297*, serie C (1983) 117; R. Bougon, T. Bailtuy, M. Lance and H. Abazli, *Inorg. Chem.*, *23* (1984) 3667.

96 K. O. Christe, *Inorg. Chem.*, *25* (1986) 3721.

97 K. Lutar, A. Jesih, I. Leban, B. Žemva and N. Bartlett, *Inorg. Chem.*, *20* (1989) 3467.

98 B. Žemva, K. Lutar, L. Chacón, M. Fele-Buermann, J. Allman, C. Shen and N. Bartlett, *J. Am. Chem. Soc.*, *117* (1995) 10025.

99 T. L. Court and M. F. A. Dove, *J. Chem. Soc., Chem. Commun.*, (1971) 726; and *J. Chem. Soc. Dalton Trans.*, (1973) 1995.

100 B. Žemva, K. Lutar, A. Jesih, W. J. Casteel, Jr., P. Wilkinson, D. E. Cox, R. B. Von Dreele, H. Borrmann and N. Bartlett, *J. Am. Chem. Soc.*, *113* (1991) 4192.

101 D. Koeler, R. Fischer and B. G. Müller, *Abstr. P(2)116, 15th Int. Symp. Fluorine Chem.*, Vancouver, B.C., Canada, Aug. 2-7, 1997.

102 N. Bartlett, R. D. Chambers, A. J. Roche, R. C. H. Spink, L. Chacón and J. M. Whalen, *Chem. Commun.*, (1996) 1049.

103 J. M. Whalen, L. Chacón and N. Bartlett, *Electrochem. Soc. Proceedings, 1997, 97-15* (1997) 1.

104 W. Klemm and E. Huss, *Z. anorg. Chem.*, *258* (1949) 221.

105 R. Hoppe, *Z. anorg. Chem.*, (1957) 292; R. Hoppe and R. Homann, *Z. anorg. Chem.*, *379* (1970) 193.

106 G. Lucier, S. H. Elder, L Chacón and N. Bartlett, *Eur. J. Solid State Inorg. Chem.*, *33* (1996) 809.

107 J. H. Holloway, E. G. Hope and C. D. Puxley, *Eur. J. Solid State Inorg. Chem.*, *33* (1996) 821.

108 G. M. Lucier, J. M. Whalen and N. Bartlett, *J. Fluorine Chem.*, *89* (1998) 101.

109 O. Graudejus, S. H. Elder, G. M. Lucier, C. Shen and N. Bartlett, *Inorg. Chem.*, *38* (1999) 2503.

110 J. M. Whalen, G. M. Lucier, L. Chacón and N. Bartlett, *J. Fluorine Chem.*, *88* (1998) 107.

111 S. H. Elder, G. M. Lucier, F. J. Hollander and N. Bartlett, *J. Am. Chem. Soc.*, *119* (1997) 1020.

112 K. S. Pitzer, *Accnts. Chem. Res.*, *12* (1979) 271.
113 P. Pyykkö and J.-P. Desclaux, *Accnts. Chem. Res.*, *12* (1979) 276.
114 N. Bartlett, *Gold Bulletin*, *31* (1998) 22-25.
115 O. Graudejus, A. P. Wilkinson and N. Bartlett, *Inorg. Chem.*, *39* (1999) 1545.
116 B. Žemva, R. Hagiwara, W. J. Casteel, Jr., K. Lutar, A. Jesih and N. Bartlett, *J. Am. Chem. Soc.*, *112* (1990) 4846.
117 G. Lucier, C. Shen, W. J. Casteel, Jr., L. Chacón and N. Bartlett, *J. Fluorine Chem.*, *72* (1995) 157; and P. Botkovitz, G. M. Lucier, R. P. Rao and N. Bartlett, *Acta Chim. Slov.*, *46* (1999) 141.

Chapter 4

CONTRIBUTION TO THE PERPETUATION OF MOISSAN'S MEMORY: THE MOISSAN PRIZE

ROLAND BOUGON

Institution du Prix Moissan, 28300 Champhol, France

Background

The isolation of fluorine

'At lunchtime on June 26, 1886, Henri Moissan's wife and small son, Louis, were walking in the Rue Michelet bordering the buildings where Moissan, a young associate professor at the Ecole Supérieure de Pharmacie, was working in a makeshift laboratory provided by Professors Friedel and Debray, who had very quickly noted the outstanding research talent displayed by Moissan in his experiments. Moissan was confined to a makeshift laboratory because he had been evicted from the laboratory where he had previously worked by Debray's assistant, who criticized him for monopolizing the platinum equipment and frosting the glassware. Moissan had to content himself with using, outside classroom hours, the amphitheatre in the Sorbonne annexe on the Rue Michelet.

Fully aware of the importance Moissan attached to the results of the experiment he was conducting that day, his wife, a constant help and support throughout his scientific career, was obviously not just a casual passer-by. She anxiously went up to the small window that had just opened in the amphitheatre and heard him call out "it's working".

On the other side of the window, witnessed by Rigault, the assistant of Troost and Friedel, who had just entered the room, Moissan observed the release through a tube connected to the anode section of his electrolytic cell of a gas upon contact with which a handful of silicon crystals burst into flame. Moissan had postulated that, given the stability of silicon tetrafluoride, a substance with which he had often worked, fluorine would attack silicon and that the heat produced would cause incandescence. The long-awaited demonstration of the isolation of fluorine had been made... This crowning point of Moissan's scientific work earned him the Nobel Prize for Chemistry twenty years later, on December 10, 1906' [1].

The origin of the Moissan Prize

The Moissan Prize originates from the International Symposium held in Paris in August 1986 to commemorate the 100th anniversary of the isolation of fluorine by Moissan. Apart from the usual scientific and industrial aspects of this symposium, in which all fields involving fluorine were represented, several special events took place, including:

- Moissan Fluorine Centennial Medallions (17) were awarded (Table 4.1).
- An exhibition devoted to Henri Moissan took place at the Faculté de Pharmacie on the very spot where Moissan isolated fluorine.

- Documents and laboratory equipment which belonged to Moissan were displayed.
- Plenary lectures were given by the late Professor Claude Fréjacques, member of the French Academy of Sciences, and Professor Jean Flahaut from the Faculté de Pharmacie de Paris on the isolation of fluorine and on the life and works of Henri Moissan, respectively.
- A press conference was held in the presence of the late Dr R. J. Plunkett, inventor of Teflon®, and the organizers of the Symposium.
- A brochure entitled *Henri Moissan, his Life and Work* written by Professor Claude Viel from the Faculté de Pharmacie de Chatenay-Malabry was distributed to the participants.

The Moissan Symposium attracted a large international audience, comprising 621 persons (459 participants and 162 accompanying persons) representing 19 countries (an apposite number!).

The Organizing Committee comprised P. Hagenmuller and P. Plurien (Chairmen), R. Bougon (General Secretary), J. Andrioly, A. Cambon, B. Cochet-Muchy, M. Ramanadin, R. Romano, N. Roux, A. Tressaud, C. Wakselman (Members), and P. Barberi (Consultant). Generous sponsorship was obtained from French government institutions and companies, and from companies in the USA, Japan, Italy, Germany, and Great Britain; and this fact, coupled with the unexpected large attendance and significant profits from capital investments, meant that despite charging reasonable registration fees and providing a memorable free cruise for all on the River Seine, the organizers were left in a very happy financial position once the Symposium was concluded. A consensus quickly emerged from discussions between the Chairmen and the General Secretary about the use of the residual capital, and it was decided to give continuation to the Moissan Symposium by instituting a Moissan Prize. As far as the monetary viewpoint is concerned, it was assumed that investment of the capital would afford perpetuation of the Prize.

Selecting a Moissan Prize winner

The Prize was created to commemorate Henri Moissan's isolation of elemental fluorine in 1886, and to stimulate research in the field of fluorine chemistry. An international Selection Committee chosen by the Institution du Prix Moissan, which administers the Prize, nominates the candidates and selects the winner. The composition of the Selection Committee is reconsidered before each Prize, the final choice of members being guided by the prestige of each fluorine chemist discussed and the need to establish a balance in respect of nationality and subject discipline.

In the first round of the election, each member of the Selection Committee proposes a maximum of five candidates. The nominees for the Prize are those candidates who receive a significant number of ballots and the list (six names in 1997) is sent to all the members of the Selection Committee, who anonymously select a maximum of two names from this list. The winner is the candidate who receives the largest number of suffrages, but the Prize may be shared if two candidates get an equal number of votes. Nominations can be based on the candidate's entire career or a recent outstanding contribution to the field of fluorine chemistry.

The Prize, which is presented every three years at the International Symposium on Fluorine Chemistry (ISFC), carries a diploma and a monetary award (100 000 French francs through 1997).

The Moissan Laureates

The recipients prior to the year 2000 were GEORGE CADY (Professor Emeritus, University of Washington, Seattle, USA), who shared the prize with NEIL BARTLETT (Professor at the University of California, Berkeley, USA) (1988); HARRY J. EMELÉUS (Professor Emeritus, University of Cambridge, UK) (1991); ROBERT N. HASZELDINE (formerly Professor of Chemistry and Principal at the University of Manchester Institute of Science and Technology, UK) (1994); and PAUL HAGENMULLER (Professor Emeritus, University of Bordeaux I, France) (1997). The Awards Ceremonies took place at the 12th (Santa Cruz, California), 13th (Bochum, Germany), 14th (Yokohama, Japan), and 15th (Vancouver, Canada) ISFCs respectively.

Henri Moissan

TABLE 4.1
Recipients of the Moissan Fluorine Centennial Medallion

NEIL BARTLETT	*University of California, Berkeley, USA*
YURI A. BUSLAEV	*Institute of General and Inorganic Chemistry, Academy of Sciences, Moscow, Russia*
GEORGE H. CADY	*University of Washington, Seattle, USA*
HARRY J. EMELÉUS	*University of Cambridge, UK*
RONALD J. GILLESPIE	*MacMaster University, Hamilton, Ontario, Canada*
OSKAR GLEMSER	*Institut für Anorganische Chemie der Universität Göttingen, Germany*
PAUL HAGENMULLER	*University of Bordeaux I, France*
RUDOLF HOPPE	*Institut für Anorganische und Analytische Chemie der Justus-Liebig Universität, Giessen, Germany*
IVAN L. KNUNYANTS	*Institute of Organo-Element Compounds, Academy of Sciences, Moscow, Russia*
YOSHIRO KOBAYASHI	*Tokyo College of Pharmacy, Tokyo, Japan*
JOHN G. MALM	*Argonne National Laboratory, Argonne, Illinois, USA*
WILLIAM T. MILLER	*Cornell University, Ithaca, New York, USA*
ROY J. PLUNKETT	*DuPont Company, USA*
PIERRE PLURIEN	*Commissariat à l'Energie Atomique, Centre d'Etudes Nucléaires de Saclay, France*
COLIN TATLOW	*University of Birmingham, UK*
NOBUATSU WATANABE	*Kyoto University, Faculty of Engineering, Kyoto, Japan*
WEI-YUAN HUANG	*Institute of Organic Chemistry, Academia Sinica, Shanghai, China*

Neil Bartlett[1]

Neil Bartlett was born in Newcastle upon Tyne, England, in 1932 and was educated at Heaton Grammar School and King's College, University of Durham. He was a faculty member of the University of British Columbia in Vancouver BC, Canada, from 1958 to 1966. In 1966 he was appointed Professor of Chemistry at Princeton University and was simultaneously a member of the scientific staff at Bell Telephone Laboratories, Murray Hill, New Jersey, USA. In 1969, he took up appointments as Professor of Chemistry at the University of California, Berkeley, USA (Emeritus from 1998) and as Principal Investigator at the Lawrence Berkeley Laboratory.

Professor Neil Bartlett

Neil Bartlett's major interests have been in the nature of the chemical bond, and particularly the highest oxidation-state limits to bonding, but all from an experimental approach. Emphasis has been placed on the chemistry of elements which are most resistant to oxidation, such as the noble metals and the noble gases. Because fluorine is the most electronegative of the elements and excites the greatest range of oxidation states, most of his chemistry has been centred on fluorine. In many of the high-oxidation systems of interest to him, molecular oxygen and elemental xenon are reducing agents at room temperatures.

Some of his work has been oriented towards possible practical applications. Electrical conductors derived from graphite relatives of boron and/or nitrogen were prepared because of their possible use in electrical energy storage or as anodes in electrochemistry. Metallic high oxidation-state metal fluorides have also been sought for related reasons.

Professor Bartlett is best known for his synthesis of the first true compound of a noble gas, in which he used platinum hexafluoride to oxidize xenon. He has remained a leader in the chemistry of the noble gases, both in enlarging the range of compounds and in exploiting them as chemical reagents. By exploiting the solvent and basic properties of xenon hexafluoride he and his co-workers were able to prepare the first quinquevalent gold compound, $Xe_2F_{11}^+AuF_6^-$. The first O_2^+ salt was made in his laboratory; and the first Os(VII) compound. New binary fluorides such as PdF_4 and RhF_5, were also prepared and structurally characterized; and in (1989) with B. Zemva and his co-workers, a new way was found to synthesise thermodynamically unstable high-oxidation-state fluorides such as NiF_4 and AgF_3. In 1974, a general synthesis of salts containing perfluoroaromatic cations, such as $C_6F_6^+$ and $C_{10}F_8^+$, was established. His group has also prepared new synthetic metals from graphite or graphite-like boron nitride and more recently has undertaken the synthesis of novel B/C, C/N and B/C/N relatives of graphite.

Professor Bartlett's work has brought him wide recognition. He is member or corresponding member of many scientific Academies, and has received worldwide many rewards and honorary degrees.

[1] See Chapter 3 for a detailed account of Neil Bartlett's researches.

George Cady (1906-1993)

George Cady was first introduced to fluorine chemistry as a graduate student with Professor Joel H. Hildebrand at the University of California at Berkeley, where he studied the vapour pressure and critical temperature of liquid fluorine. As an instructor at MIT, he published two key papers. The first concerned a definitive study of freezing points and vapour pressures in the KF-HF system, thus providing for the first time the optimum conditions for the electrolytic generation of fluorine and the basis on which all fluorine cells operate. The second reported the reaction of fluorine with nitric acid to form a gas that was later shown to be O_2NOF, the first of many hypofluorites to result from the Cady school. After a brief period in industry where, through his study of hypochlorous acid and its anhydride, dichlorine oxide, an industrial process for the production of calcium hypochlorite was developed, Dr Cady joined the faculty at the University of Washington, Seattle, in 1938. With the exception of a period spent on the Manhattan Project at Columbia University during WWII where, as group leader, he played a key rôle in the production and study of the first fluorocarbon lubricating oils, he spent the remainder of his professional life at Washington.

Professor George H. Cady

The very productive years that followed were dedicated almost entirely to compounds which contained fluorine. New hypofluorites were prepared [ClO_3OF, FSO_2OF, SF_5OF, SeF_5OF, $SeF_4(OF)_2$, $FC(O)OF$, $CF_3C(O)OF$, $CF_3CF_2C(O)OF$, CF_3OF and the first bis(hypofluorite), $CF_2(OF)_2$], and their reactions studied; CF_3OF, the first member of the new class of perfluoroalkyl hypofluorites, was shown to have an exceptionally broad chemistry and has had quite an impact on modern electrophilic fluorinations. Numerous compounds that contain the SF_5 group, including the first known derivative of sulfur hexafluoride, CF_3SF_5, were discovered and characterized.

Another key chapter of Cady fluorine chemistry is that of fluorinated peroxides, such as FSO_2OOSO_2F, F_5SOOSF_5, $F_5SeOOSeF_5$, F_5SOOSO_2F, $CF_3SO_2OOSO_2CF_3$, $FC(O)OOC(O)F$, CF_3OOCF_3, CF_3OOSO_2F, $FC(O)OOCF_3$, and F_5SOOCF_3. The first of these proved to be especially versatile in its reactions due to its reversible thermal dissociation to $FSO_2O^{\cdot}$ radicals from which a wide range of novel fluorosulfates were prepared, including $FOSO_2F$, $ClOSO_2F$, $BrOSO_2F$, $Br(OSO_2F)_3$, $IOSO_2F$, $I(OSO_2F)_3$, and $KI(OSO_2F)_4$, demonstrating the pseudohalogen nature of the radical. Acting as a radical source, FSO_2OOSO_2F underwent hydrogen abstraction reactions and effected saturation of C=C bonds in fluoroalkenes. Oxidative addition with SF_4 [$\rightarrow SF_4(OSO_2F)_2$], radical-radical coupling involving N_2F_4 [$\rightarrow NF_2OSO_2F$], and simple oxidation of PF_3 [$\rightarrow OPF_3$] exemplify the three modes of behaviour exhibited by this peroxide. The displacement of CO from metal carbonyls to form dioxodi(fluorosulfates), such as $MoO_2(OSO_2F)_2$ from $Mo(CO)_6$, as well as reactions with metals and metal oxides show further the breath of the chemistry of this reagent. Cady explored synthesis via electrochemical fluorination and determined the solvent properties and solution chemistry of anhydrous hydrogen fluoride and trifluoroacetic acid. Also, he shared with workers in three other laboratories the independent discovery of XeF_6 and explored a number of its properties.

George Cady's achievements were not exclusively in the synthetic field. His research included thermochemistry, kinetics, electrochemistry, nonaqueous systems, and detailed nuclear magnetic resonance studies of inorganic and organic compounds. He was gifted with great chemical insight, coupled with exceptional skill as an experimentalist; and not least, he was a dedicated and completely unselfish teacher and human being who successfully passed on his enthusiasm and fascination with fluorine to those who were privileged to work with him.

Harry Julius Emeléus (1903-1993)

Harry Emeléus graduated in chemistry from Imperial College, London, then stayed on to work for the PhD degree, his thesis dealing with the luminescent oxidation of phosphorus being examined in 1925. In the following years he continued to widen his interest in physical chemistry at Princeton University, USA, where he worked with H. S. Taylor on the photosensitisation of ethene polymerization. Prior to his time in Princeton, his first exposure to inorganic chemistry came during a stay in the laboratory of Alfred Stock at Karlsrühe, Germany; there he acquired the techniques for handling volatile, extremely air-sensitive compounds, and together with E. Poland he synthesized $B_{10}H_{14}$, a key compound in a unique and unprecedented series of polyhedral cluster compounds. The collaboration with Stock profoundly influenced Harry's research career, and after returning from America to Imperial College in 1931 he continued to pursue his interest in the inorganic chemistry of main group elements through preparative studies on new volatile hydrides such as MH_3 (M = P, As, Sb) and H_3Si-derivatives. During this period Emeléus and his colleague J. S. Anderson from Oxford published (in 1938) *Modern Aspects of Inorganic Chemistry*, a collection of essays which played a major part in the revival of inorganic chemistry worldwide.

Professor Harry J. Emeléus

During the Second World War (1939-45), Emeléus worked at Imperial College on defence-related topics which included the preparation of a range of highly reactive interhalogen compounds, such as ClF, ClF_3, BrF_3 and BrF_5. Much of that work was not published at the time. His stay at Imperial College (1931-1945) was interrupted in early 1944 when he was sent to the USA to work at Oak Ridge, Tennessee, on topics related to the Manhattan Project. On his return to Britain in 1945 he moved to Cambridge, first as University Reader and then Professor of Inorganic Chemistry, and subsequently established there a famous school with wide-ranging interests in inorganic chemistry. During his Cambridge period (1945-1970), Emeléus attracted PhD students, postdoctoral researchers, and guest scientists from all over the world. Many of the members leaving his school entered academic life and continued very successfully to develop main group element chemistry. His main interest at Cambridge lay in aspects of inorganic fluorine chemistry, including perfluoroalkyl derivatives of main group elements. His synthetic approach to perfluoro-organoelement chemistry had its basis in the large-scale synthesis of CF_3I, which became a key compound in this area. His broadly-based studies on fluorinated organoelement compounds were complemented by investigations into the behaviour and utility of non-aqueous solvents, and the chemistry of interhalogens and polyhalides, as well as work on metal fluorides and pseudohalides.

The perfluoroalkyl-element chemistry pioneered by Emeléus and his school at Cambridge was later extended to such an extent by his former students and colleagues all over the world that since 1973 the Gmelin Institute has published 18 volumes of the *Handbook of Inorganic and Organometallic Chemistry* dealing with perfluorohalogen-organic compounds of main group elements; and all of them are dedicated to Harry Emeléus, a man who was universally liked and an inspirational fluorine chemist.

Robert Neville Haszeldine

Professor Robert N. Haszeldine

After graduating (PhD) from the Birmingham school of organofluorine chemistry in 1947, Robert ('Bob') Haszeldine moved to the University of Cambridge where he rose to be an Assistant Director of Research (1956) before becoming Professor of Chemistry (Organic and Inorganic) in the Faculty of Technology at the University of Manchester (UMIST, as it is known today). At Cambridge he not only pursued his personal interest in organofluorine chemistry but also enjoyed a fruitful collaboration with H. J. Emeléus which lead to significant advances in knowledge of organoelemental compounds of the perfluorocarbon class. At Manchester he greatly extended his research repertoire – particularly in organic areas of perfluorocarbon chemistry – with the help of a sizeable world-class team of collaborators, postdoctoral fellows, PhD students, and technicians. His contributions to chemistry through his fluorine researches have been recognized in the UK by the award of the Meldola Medal (1953), the Corday-Morgan Medal and Prize (1960), a Tilden Lectureship (1968), and election to the Fellowship of The Royal Society (1968). He has published around 500 papers, officiated as a consultant to many chemical companies throughout the world, and served as a member of British Government committees. He was Principal of UMIST from 1976 to 1982.

Paul Hagenmuller

Paul Hagenmuller is one of the founders of modern solid-state chemistry. After receiving his PhD degree at Paris University (1950) he has been, successively, Assistant Professor at Paris Sorbonne, CNRS Fellow, Assistant Professor in Hanoi and Saigon (1954-56), and full Professor in Rennes (1956-60) and then Bordeaux, where he has built up the CNRS Solid State Chemistry Laboratory, a French institution largely open to international cooperation and foreign scientists. Paul and his scientific school have produced hundreds of PhD graduates and published more than a thousand papers in international journals dealing with various areas of materials research. Many of his co-workers are now involved in university or industrial research in France (*e.g.* Portier, Tressaud, Grannec, Réau, Doumerc, de Pape, Ferey), Germany, USA, Japan, China, Brazil and Morocco. A member or corresponding member of many scientific Academies, Paul has received worldwide many rewards and honorary degrees. As a young man during WWII, he was active in the French resistance and is one of the few survivors of the European concentration camps.

Professor Paul Hagenmuller

Particular interest attaches to inorganic fluorides due to the high electronegativity of fluorine, which as a rule leads to M-F bonds more ionic than those of homologous chlorides, oxides or sulfides. Paul Hagenmuller has shown with his Bordeaux research school, often in a predictive way, that this feature results in quite singular properties. A particularly significant example is the 'white bronzes' K_xFeF_3 and K_xCrF_3 with the same structure as the tungsten bronzes K_xWO_3 but insulating in behaviour. An exceptionally high F^- anion conductivity may appear in fluorite- or tysonite-type fluorides as a consequence of F^- excess or vacancy leading to cluster formation. Interesting optical properties have been achieved such as intense Nd^{3+} laser emission with fluoride glasses (a field later developed by Jacques Lucas in Rennes), strong fluorescence in the blue or the green thanks to monochromatic excitation by Eu^{2+} doping in $BaY_2F_8:Er^{3+}$ or Tb^{3+}, and optical UV windows. Ferrimagnetic fluorides have been prepared, and these respect Néel's molecular field law much better than the corresponding oxides because the weakness of magnetic interactions allows one to neglect couplings involving the second cationic neighbours. Hagenmuller's group has also studied the influence of partial oxygen-fluorine substitution on the physical properties of many oxides.

Acknowledgements

The author would like to express his thanks to Eric Banks for help with the text concerning Bob Haszeldine, to Neil Bartlett and Paul Hagenmuller for providing him with information on their scientific careers, and particularly to Jean'ne Shreeve and Alois Haas for producing outlines of the careers of George Cady and Harry Emeléus, respectively.

Note added at the final proof stage

Karl Christe (see Chapter 8) is the winner of the first Moissan Prize of the new millennium. Roland Bougon announced this result and presented the Prize to Karl on 17 July 2000 during the interval at an evening concert given in Durham Cathedral as part of the 16th ISFC (16-21 July; University of Durham, UK).

Eric Banks

Reference

1 Extract from the text of 'The Isolation of Fluorine' by B. Cochet Muchy and C. Fréjacques which was presented by C. Fréjacques at the Faculté de Pharmacie de Paris, on August 28, 1986.

BIOGRAPHIC NOTE

After graduation (PhD) at the Faculté des Sciences d'Orsay in 1962, Roland Bougon entered Dr Pierre Plurien's department at the Centre d'Etudes Nucléaires de Saclay (Commissariat à l'Energie Atomique) where he spent his working life (*i.e.* till retirement in 1994) pursuing studies in preparative inorganic fluorine chemistry and associated topics in physical and applied chemistry. He was the Scientific Secretary of the International Symposium on Fluorine Chemistry held in Avignon, France, in 1979 and General Secretary of the 'Moissan Symposium' (Centenaire de la Découverte du Fluor) in Paris in 1986; he has been in charge of the 'Institution du Prix Moissan' since its inception in 1988.

Chapter 5

ORGANOFLUORINE CHEMISTRY IN NOVOSIBIRSK (SIBERIA)

GERALD M. BROOKE

Department of Chemistry, University of Durham, Science Laboratories, South Road, Durham DH1 3LE, UK

Introduction

Final-year chemistry courses inevitably focus the minds of undergraduates on the future. In particular, one must decide whether to continue as a 'professional' chemist (perhaps undertaking research leading to a Masters or Doctoral qualification before seeking employment) or simply to collect the Bachelor's Degree and then pursue a completely different career. In late 1958 when I graduated (BSc) at the University of Birmingham, scientists were considered by the UK Government to be of value to the country, hence conscription could be deferred indefinitely, and there seemed to be no problems in securing employment in the chemical industry. The way was open, therefore, for me to fulfil my long-held ambition to become a professional chemist immediately on graduation.

In the University of Birmingham at that time, Professor Maurice Stacey was the Head of the Department of Chemistry, and he made all final-year students aware of the different areas of research work that were being pursued in the Department. An undergraduate lecture course on *organofluorine chemistry* given by Dr Robert Stevens first alerted me to one emerging area. The subject seemed to be connected mainly with aliphatic chemistry, an impression which was further emphasised by the contents of the Methuen Monograph *Fluorine and Its Compounds* by Haszeldine and Sharpe, the only book available at the time to impecunious students. Only six pages were devoted to ring-fluorinated aromatic compounds; but C_6F_6 and $C_6F_5CF_3$ were documented [1]. Having heard that polyfluoroaromatic chemistry was beginning to be developed in the Department, it became my desire to be involved in studying the chemistry of such novel compounds; after all, I reasoned, aromatic chemistry was already a pretty big area, and the possibility of being involved in creating a new field based not on aromatic C—H bonds but on C—F bonds was truly exciting. It came as a great relief, therefore, to be accepted to work in the Fluorine Team directed by Dr (later Professor) J. C. Tatlow.

Dr James Burdon was my immediate supervisor (I was his first PhD student) when I began work in October 1958. Earlier that summer I had written to the then Dr Colin Tatlow to find out more precisely the topic on which I would be engaged. It became clear that work on nucleophilic substitution reactions of hexafluorobenzene and its derivatives (C_6F_5X) would be the main area for investigation, the reactions of the derivatives being of major interest, since the selective preference for replacement of fluorine sited *ortho*-, *meta*- or *para*- to the different X groups needed to be established (and then rationalised) – all pretty basic chemistry. Relatively large amounts of C_6F_6 and C_6F_5H (five to ten of grams at a time) were becoming available (somewhat irregularly) within Tatlow's group via the recently discovered dehydrofluorination reactions of polyfluorocyclohexanes [2, 3], so it was clear that a vast area for investigation was opening up.

Bob Stevens and Maurice Stacey had made us aware as undergraduates that in the years during and after World War II (1939-1945) other countries besides the UK were interested in organofluorine chemistry. Indeed, work actually started in Birmingham during the war under the direction of Professor Stacey and Dr Fred Smith on the preparation of fluorocarbon fluids, oils and greases required for use in plant handling highly reactive UF_6 for uranium isotope separation. After the war, further research was conducted, and among those taking PhDs were W. K. R. Musgrave, J. C. Tatlow, R. N. Haszeldine, the late P. G. Harvey, the late J. M. Tedder and A. K. Barbour [4]. The fact that chemists in the then USSR were deeply involved in fluorine chemistry was particularly fascinating to me, and I viewed the country as shrouded in mystery and quite inaccessible owing to the prevailing political situation. However, it clearly was not totally so, because Professor Stacey paid a visit to Moscow in 1957 as a member of a party of British scientists from the Royal Society, headed by Lord Adrian; how we enjoyed the 'slide show' which he gave to the Department on his return!

New recruits in the Birmingham F-Team in 1958 soon realised that a milestone in fluorine chemistry was coming the following year: the *1st International Symposium on Fluorine Chemistry*, actually to be hosted by our very own Department. The prospect of having one's own research results presented at an international meeting was a tremendous encouragement to get things done in the laboratory, and by summer 1959 quite a number of postgraduate students were gathering data in the general area of polyfluoroaromatic chemistry for Professor Tatlow's plenary lecture.

As a diversion from first-year benchwork, yet at the same time potentially of very great use in the future, I began a course of study in Russian for Scientists put on by the Department of Modern Languages. There were only about ten participants at the weekly class, but it did provide an opportunity to learn something of the language. Suffice it to say that it didn't prove too difficult to learn the cyrillic alphabet, but thereafter, coming to terms with the grammar was a different matter. Nevertheless, with the possibility of Russians coming to Birmingham in the summer of 1959, there was an incentive to struggle along and learn a few Russian phrases.

Two senior chemists from the USSR were expected at the Birmingham Symposium: Professor I. L. Knunyants from Moscow and Professor N. N. Vorozhtsov from Novosibirsk (Siberia). There was always some doubt in those early days whether chemists who had signified that they were coming from the USSR to a meeting to give a paper would *actually* turn up, and it did not take long for future conference organisers to schedule such speakers at the end of morning or afternoon sessions so that the failure of individuals to appear did not unduly affect the smooth-running of the programmes.

Professors Knunyants and Vorozhtsov *did* arrive on time, and with *some* knowledge of the language. Colin Tatlow assigned me as a young enthusiast to guide them around the campus. I have one very clear recollection of taking them to the Student Union Building for a cup of tea. The visitors' grasp of English was only marginally greater than mine of Russian, but we got on fine in our own way and I was presented with a packet of Russian cigarettes by Professor Knunyants, which I said I would pass on to my father (I've always been a non-smoker). My acquaintance with Professor Vorozhtsov was later renewed at the *3rd International Symposium* in Munich (W. Germany) in 1966. He greeted me with much warmth and gave me a set of photographs taken in Academgorodok, the suburb of Novosibirsk which is the centre of the Siberian Division of the Russian Academy of Sciences. Little did I know then that one day I would take up temporary residence there.

One of the great privileges of being a member of the academic scientific community is this possibility of meeting up with scientists from other countries who share the same interests. Inevitably, there is a measure of professional rivalry, which seems important at the time, but which in later years is of no great historical consequence; no person, group or country has the monopoly of great ideas! Nevertheless, those of us in the UK with a particular interest in polyfluoro-aromatic and -heteroaromatic chemistry became keen followers of publications emanating from the Institute of Organic Chemistry in Novosibirsk. The first personal contacts we had with some of the scientists involved came in July 1971 at the *6th International Symposium* in Durham, home territory for me by then. The delegates, 11 in number, were mainly from Moscow and Novosibirsk; friendships forged at that time with the Siberian members (the late George Yakobson[1], Vladislav Vlasov[2], Slava Platonov[3] and Tamara Petrova[4]), were nurtured at subsequent International Symposia and happily remain to the present day (see later).

The political climate in the USSR began to change in the late 1980s with 'glasnost' and 'perestroika', and in August 1991 President Yeltsin came to power. Contact between UK and Russian chemists had been increasing during that period, principally because fuller use was made of a joint agreement between the Royal Society of Great Britain and the Russian Academy of Sciences which had been set up in 1956 to promote exchange visits and encourage collaborative work. Less senior people have taken advantage of the opportunities to travel in more recent years, and visits to the UK by fluorine chemists from the USSR took place in 1988, 1990 and 1993; a small party from Britain went to the 1st UK-Russian Conference on Fluorine Chemistry in July 1991, in Novosibirsk.

---

[1] Memorial Issue to G. G. Yakobson, *J. Fluorine Chem.*, *28* (1985).

[2] *Vladislav M. Vlasov* is a true Siberian, born in Novosibirsk in 1936. He graduated from the Engineering Chemical Technology Faculty, MCTI in 1958 and worked in the Urals before being awarded a postgraduate studentship at NIOC in 1962; he obtained his PhD degree in 1965. In 1983 he was awarded the degree of Doctor (for work which concerned physico-chemical aspects of C-, N- and O-centred anions) in the manner singularly characteristic of the USSR (and former USSR) – by a public 'defence' of his thesis. Securing the prestigious degree of Doctor (roughly equivalent to a British DSc) entitles the holder to be addressed as 'Professor'. Currently he is a Vice-Director of NIOC. When I met him in 1996, he owned a *very* big dog and I was advised not to attempt to pat the beast.

[3] *Vyacheslav E. Platonov* was born in 1937 in Leningrad (now St. Petersburg) and moved east with his family during a difficult period at the time of the Second World War. He graduated from the Ural Polytechnic Institute, Sverdlovsk (now Ekaterinburg) in 1959 and then worked at the Aniline Dyes Factory in Kemerovo. He secured a postgraduate studentship in the Laboratory of Halogen Compounds at NIOC in 1961, and was awarded his PhD degree in 1965 and a Doctorate (see footnote 2) in 1979. Professor Platonov has made many contributions to the chemistry of polyfluoroaromatic compounds, not least the independent Russian discovery of the 'dry' conversion of C_6Cl_6 to C_6F_6 (Scheme 5.3), and also some unique thermolytic reactions.

[4] *Tamara D. Petrova* was born in 1935 in Korosten, Ukraine. She graduated from the Intermediates and Dyes Section, MCTI in 1957. She then continued work at the same Institute but moved to NIOC in 1961 and completed her PhD degree in 1962. Professor Petrova successfully defended her thesis for the Doctor's degree in 1995 at the Institute of Organic Chemistry of Ufimsky, Ufa, which is affiliated to the Russian Academy of Sciences. Tamara has carried out extensive research work in polyfluoroaromatic chemistry, in particular intramolecular nucleophilic displacement of fluorine and the synthesis and chemistry of polyfluoroaryl carbimidoyl dichlorides.

Through the support of my colleagues in the Chemistry Department at Durham, the University granted me sabbatical leave from October 1995 to September 1996, and it was my great pleasure to spend almost three months of that (March to May 1996) as a guest of the Russian Academy of Sciences at the Institute of Organic Chemistry in Novosibirsk, situated in a pleasant suburb of the city called Academgorodok. There I was able to study all the papers devoted to polyfluoro-aromatic and -heteroaromatic chemistry which had been published by the Institute's researchers and so gather information for incorporation in a comprehensive review of global development of the subject, published in late 1997 [5]. My stay in Novosibirsk also provided information for the general account of the fluorine group there which follows. The Siberian Division of the Academy of Sciences has many institutes in Academgorodok, developed from a forest area. It also has its own hotel, the Golden Valley Hotel, an eight-storey building which was my temporary home during my stay.

The first section which follows deals with the establishment and staffing of the Institute of Organic Chemistry, and I am indebted to Professor Tamara Petrova[4] for her help with it.

The Novosibirsk Institute of Organic Chemistry (NIOC)

The early days

The region of the world known as Siberia conjures up in the minds of those not acquainted with it as a place which is decidedly unpleasant because of the harshness of the climate in winter and its remoteness – all influenced by writings of authors dealing with an earlier political age. It is well known that 'science' was highly valued in the days of the USSR, and to become a scientist was a noble achievement. With such a vast territory available for development, two Academicians of the Academy of Sciences, M. A. Lavrentiev and S. A. Christianovich, put forward a proposal to the Council of Ministers of the USSR to establish a major scientific centre in Siberia. This was approved on May 18th, 1957. Thus, the Siberian Division of the Academy of Sciences of the USSR was conceived, and on June 27th, 1958 Academician N. N. Vorozhtsov Jr. was appointed as the first Director. While the building of the Institute was in progress, the first appointees started their work early in 1961 at the Aniline-Dyes Factory in Kemerovo, a city situated about 300 km north east of Novosibirsk; within a short time, the group moved to their permanent home, in 1962.

Prior to moving to Novosibirsk, Professor Vorozhtsov was Head of the Intermediates and Dyes Section at the Moscow Chemical Technological Institute (MCTI). His particular interest was nucleophilic substitution reactions of aromatic halogen derivatives, particularly chlorine compounds. Under his supervision, G. G. Yakobson, a graduate of the Institute, carried out part-time research work there during his spare time (his main employment was as a chemistry teacher at a school in Moscow). In 1954, Yakobson was appointed as a full staff member of MCTI and was encouraged by his supervisor to explore new routes to *fluoro*aromatic compounds as alternatives to the classical Balz-Schiemann reaction involving diazonium tetrafluoroborates. The fruit of this work was the discovery that an activated chlorine group can be replaced by fluorine under the action of potassium or caesium fluoride in the absence of solvents (Scheme 5.1) [6]. Yakobson was awarded the degree of PhD in 1958 for his thesis entitled *Interaction of halo- and dinitro-compounds with metal*

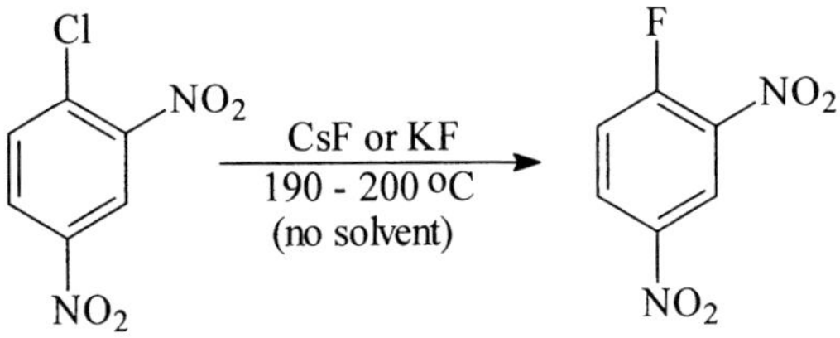

Scheme 5.1.

$$C_6Cl_5F \xrightarrow[200\ ^{o}C]{30\%\ NH_3\ aq.} C_6Cl_5NH_2$$

Scheme 5.2.

$$C_6Cl_6 \xrightarrow[450\ -\ 500\ ^{o}C]{KF,\ autoclave} C_6F_6\ (21\%) + C_6F_5Cl\ (20\%) + C_6F_4Cl_2\ (14\%) + C_6F_3Cl_3\ (12\%)$$

Scheme 5.3.

fluorides. Vorozhtsov's interest in nucleophilic aromatic substitution reactions in general expanded to include pentachloro derivatives (Scheme 5.2) [7], and a new laboratory was set up in the Intermediates and Dyes Section.

In 1961, Professor Vorozhtsov moved his research group east, from Moscow to the temporary accommodation in Kemerovo. Among this group was a young postgraduate student, V. E. Platonov[3], who was directed to study the reaction of hexachlorobenzene with potassium fluoride at high temperatures in the absence of solvents. The successful production of hexafluorobenzene at this time (Scheme 5.3) [8] marked the beginning of research in polyfluoroaromatic chemistry and the group moved into its new building in Novosibirsk the following year.

As Director of the new Institute of Organic Chemistry in Novosibirsk, Professor Vorozhtsov appointed George Yakobson as Head of the Laboratory of Halogen Compounds; he was accompanied by a number of other postgraduates from MCTI in Moscow, namely Tamara Petrova[4], Lyuba Kobrina[5], Vitalii Shteingarts[6] and Victor Odinokov.

---

[5] *Lyuba S. Kobrina* comes from Alma-Ata, Kazakhstan where she was born in 1937. She graduated from the Intermediates and Dyes Section of MCTI in 1959 and moved to NIOC in 1961 where she studied nucleophilic reactions of C_6Cl_5X derivatives, gaining her PhD in 1964 and Doctor's degree in 1984. She then worked with polyfluoroaromatic systems, making major contributions to the understanding of their reactions with free radicals. She is married to Vitalii Shteingarts[6] and they enjoy outings in their car (a rare possession) and watersport.

I enjoyed two visits with LSK and VS to 'datcha-land' near Novosibirsk in April/May 1996 to experience something of rural life in Siberia. The gardens associated with datchas are used largely to grow vegetables and fruit, which are stored partly for winter consumption. Tea drinking is important in Russia and to drink tea made traditionally in a samovar was quite an event on one of these visits.

[6] *Vitalii D. Shteingarts* is a Muscovite, born in 1937, and a graduate of the Intermediates and Dyes Section, MCTI (1959). After a short period in industry in Moscow, he joined NIOC and was awarded his PhD degree in 1965 for work on the reactions of polyfluorinated aromatic compounds with nitric acid. In 1978 he defended his thesis for the Doctor's degree, which dealt with polyfluorinated arenonium ions and some early work with radical cation species.

From 1963 to 1972, polyfluoroaromatic chemistry at NIOC was carried out in two laboratories, one under George Yakobson, the other headed by Professor Vorozhtsov and V. A. Barkhash[7].

The recruitment of young active chemists to work in Siberia after life in Moscow did not seem to present any difficulties to the people involved. The climate in West Siberia with its centre in Novosibirsk is one of the warmest places in the whole of Siberia, and in any case, the people there are accustomed to coping with adverse conditions. Tamara Petrova wanted to finish her PhD work, started at MCTI under Vorozhtsov and Yakobson. Vitalii Shteingarts worked in industry in Moscow after graduating from MCTI, but grasped the opportunity to engage in fundamental research work once again when invited to join the group destined for NIOC. For each recruit, the move did represent a higher job 'profile' and a modest increase in salary, but the main reason for the move east was purely scientific – the prospect of working under the inspired leadership of Vorozhtsov and Yakobson in the largely unexplored area of polyfluoro-aromatic and -heteroaromatic chemistry. It really was not too hard to change one's life completely, spurred on by the attraction of new laboratories, with excellent facilities (equipment and materials) and the opportunity to exchange ideas with scientists engaged in other areas of organic chemistry in the Institute. [N. B. While organofluorine chemistry was a major activity in the early days of the Institute's foundation, other areas of research included the chemistry and technology of heterocyclic compounds, natural products (primarily 'forest' chemistry) and nucleic acids, and to this day a broad range of topics are studied.] Nevertheless, as a safeguard, all recruits from Moscow were assured that if they did not settle to life in the new territory they would be allowed the option of returning to the capital; the authorities did appreciate that it could be quite stressful for members of families to be separated from loved-ones 'back home'. It is a testimony to the success of the whole venture that nobody made use of this opportunity to return to Moscow.

Some fascinating facets of fluorine chemistry from NIOC

Inevitably, the ground covered here stems from my own personal interests and is related to chemistry pursued post-1958 in Professor Tatlow's laboratories in Birmingham.

Vitalii, the husband of Lyuba Kobrina[5] enjoys water-based activities and now uses his small motor-boat on the Ob Sea (close to Novosibirsk) – which is less taxing than his former exploits in a canoe.

On returning to Academgorodok from a visit to datchaland with Lyuba and Vitalii on a warm afternoon in May 1996, the attention of the author was drawn to a male sunbather who had just risen from the grass at the edge of the forest and was making a thorough examination of his body. Vitalii explained that the fellow was looking for signs that his skin might have been penetrated by the dreaded *kleshch* (a 'tick'). These parasites are sources of encephalitis, and if one is found the host must go straight to hospital for a blood test, which, if positive, necessitates an injection of gamma globulin. In May 1996, one of Lyuba's co-workers, Dr V. N. Kovtonyuk, picked up a tick in the garden of a relative and went to a hospital for treatment – one of over 200 patients in that day for the same reason. I was strongly advised by several friends independently to keep to the main paths in the forests as the warmer days came along. It seemed wiser to avoid the forests completely!

[7] *Vladimir A. Barkhash* was born in Moscow in 1933. He graduated from the Intermediates and Dye Section of Moscow Chemical Technological Institute (MCTI) in 1955 and obtained his PhD degree from Moscow State University in 1960 and was awarded his Doctor's degree in 1977. He began work in fluorine chemistry in NIOC in 1963, but in 1972 became interested in non-classical carbocations, a topic which was very popular in the West at that time. Nevertheless, polyfluorobenzocyclenes were at the centre of this work.

Tatlow's group needed pentafluoronitrobenzene for a study of its reactions with nucleophiles, and although work in the late 1950s had demonstrated that C_6F_5H could be halogenated and sulphonated (Scheme 5.4) [9], perhaps surprisingly, no really satisfactory nitration procedure had been found; it was only during the mid-60s that an efficient procedure was discovered (use of fuming nitric acid in sulfolane saturated with boron trifluoride) [10]. In the meantime, Dr R. D. Richardson had oxidised $C_6F_5NH_2$ with peroxytrifluoroacetic acid to make the nitro-compound [11] (Scheme 5.5).

Significantly, while research on the reactions of polyfluoroaromatic compounds with nucleophiles (S_NAr reactions) was proceeding both inside and outside the USSR, the reactions of *electrophiles* with *per*fluoroaromatic compounds was being undertaken essentially exclusively in Novosibirsk: attack at a C−F bond (*ipso* attack) occurred to give *addition products*. The nitrofluorination reaction of hexafluorobenzene is shown in Scheme 5.6 [12]. What is truly amazing, however, is the behaviour under similar conditions of compounds with ring hydrogen also present: both nitration *and* nitrofluorination occur with C_6F_5H [12, 13], whereas with 2*H*-heptafluoronaphthalene only nitrofluorination takes place [14] (Scheme 5.7), the aromatic C−H bond being untouched! Surely such a course of reaction would have been totally unpredictable? But what fun to establish these equations! Even more fascinating chemistry has been described involving carbon electrophiles and polyfluoroaromatic compounds which formally require the loss of F^+ to account for the products; two of these are shown in Schemes 5.8 [15] and 5.9 [16]. The mechanisms of these reactions are still unexplained.

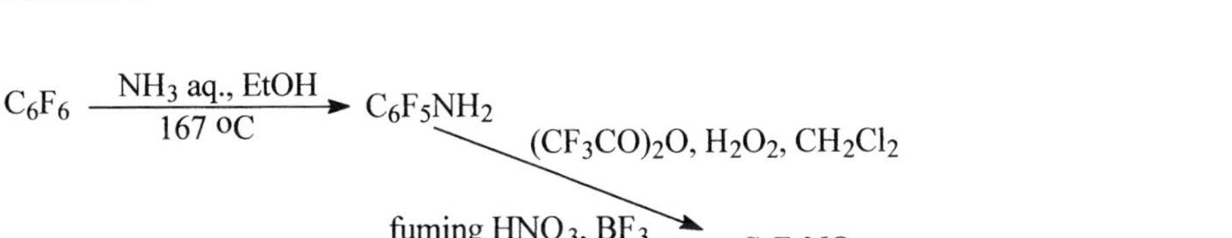

Scheme 5.4.

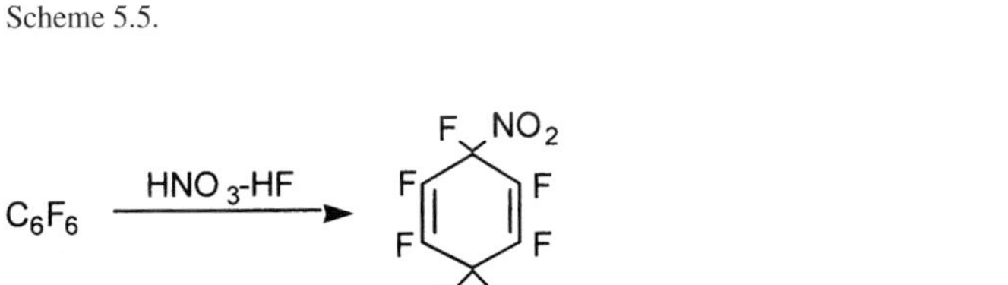

Scheme 5.5.

Scheme 5.6.

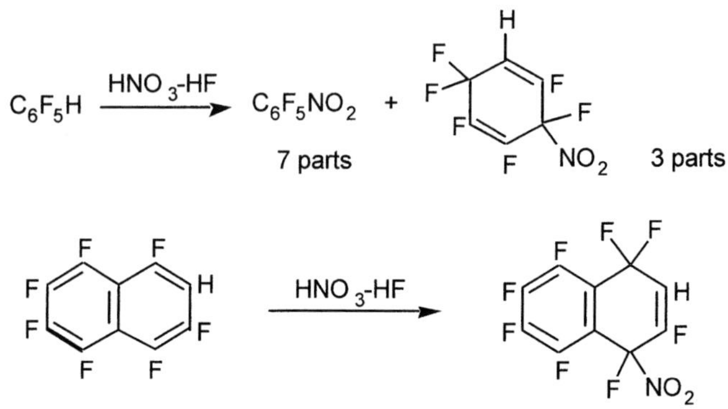

Scheme 5.7.

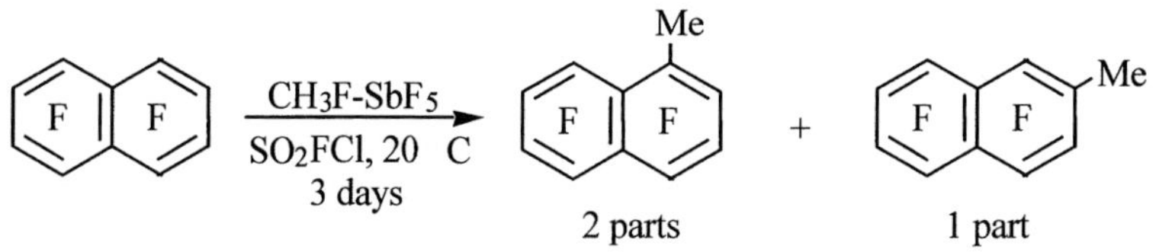

Scheme 5.8.

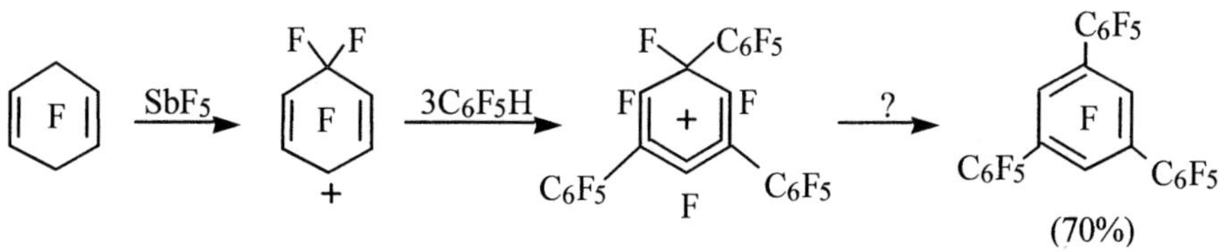

Scheme 5.9.

Thiols are powerful sulfur nucleophiles, and as such have been used extensively as their alkali-metal salts in reactions with polyfluoroaromatic compounds [17]; lead(II) benzenethiolate has also been used to synthesise $C_6H_5SAr_F$ compounds [18]. A spectacular transformation from nucleophilic to electrophilic sulphur takes place in the reaction reported by Furin[8] *et al.* of lead thiocyanate with pentafluorobenzene in SbF_5, the powerful oxidising properties of which enable the NCS^- ion to be converted into the potential NCS^+ species (Scheme 5.10) [19].

A recent exploitation of SbF_5 has been demonstrated in the remarkable isomerisation of one tetralin derivative to another (Scheme 5.11) [20].

The first S_NAr orientation reaction I carried out as a PhD student involved treating C_6F_5H with $LiAlH_4$ to give mainly replacement of the fluorine *para* to the hydrogen [21].

[8] *George G. Furin* was born in 1939 in Novorossiisk, Russia, graduating from Dnepropetrovsk State University in 1962, when he began work at NIOC. Nucleophilic substitution of C_6F_5X compounds was the basis of his PhD studies, obtained in 1969. In 1983 he was awarded the degree of Doctor. His main achievements have been connected with the use of SbF_5 and superacids in fluorine chemistry.

$$C_6F_5H + Pb(SCN)_2 \xrightarrow{SbF_5} C_6F_5SCN$$

Scheme 5.10.

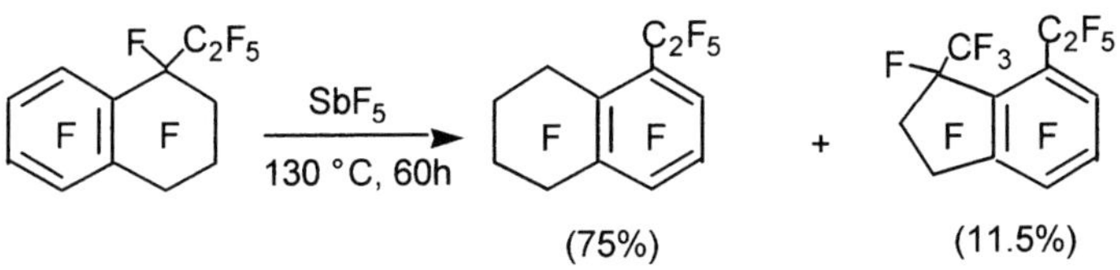

Scheme 5.11.

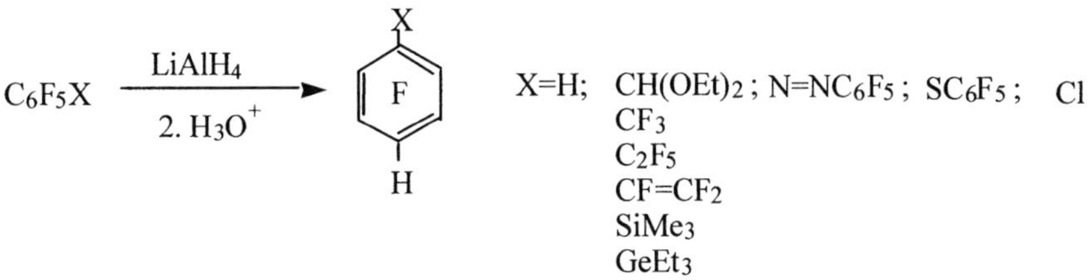

Scheme 5.12.

As it turned out later, the major type of substitution product from attack on a vast number of C_6F_5X compounds by $LiAlH_4$ is the *para* isomer (Scheme 5.12) [5]. So how exciting it must have been for Gerasimova[9] *et al.* [22] to find that for X = CO_2H and CH(R)OH, treatment with $LiAlH_4$ led exclusively to replacement of *ortho*-F by H! It turned out that this selectivity was maintained with other metal-containing reagents, and was rationalised in terms of metal chelation with the heteroatom in the functional group X. Scheme 5.13 shows a reaction involving a nitrogen-centred nucleophile [23].

High-temperature reactions carried out at NIOC between polyfluoroaromatic compounds and tetrafluoroethylene have established novel routes to a variety of materials, the reactions involving both difluorocarbene (from the C_2F_4) and fluorinated radicals. For example, controlled insertion of :CF_2 into the C−F bond of C_6F_6 gives perfluorotoluene, which in turn can be converted into xylene derivatives (Scheme 5.14) [24]. With pentafluoronitrobenzene or pentafluorobenzenesulfonyl chloride as substrates, perfluorotetralin is formed, the mechanism proposed being shown in Scheme 5.15 [24].

Perfluoroindane is the main product when C_6F_5OH, C_6F_5OMe, or C_6F_5SH is heated with C_2F_4, the best yield being achieved with the thiol (Scheme 5.16) [25, 26].

The chemistry of non-fluorinated sulphur-nitrogen compounds was a major research topic in the Chemistry Department at Durham over the period 1967-1997 [27], so the synchronized incorporation of these two elements in polyfluoroheterocyclic compounds by

[9] *Tatiana N. Gerasimova* comes from St. Petersburg, where she was born in 1934. After graduating from MCTI in 1957, she worked with Professor Vorozhtsov in the same Institute and moved to NIOC in 1962. Professor Gerasimova secured her PhD degree in 1965 and Doctor's degree in 1981, based on studies of the reactions of polyfluoraromatic carbonyl compounds with organometallic compounds containing magnesium and lithium.

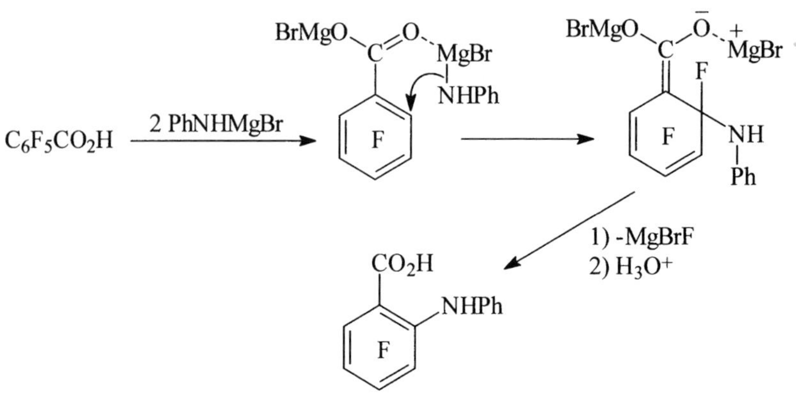

Scheme 5.13.

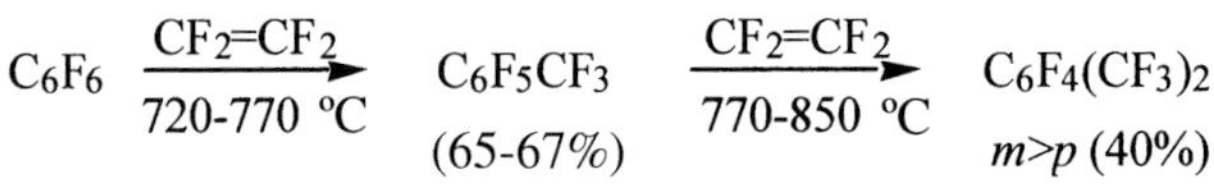

Scheme 5.14.

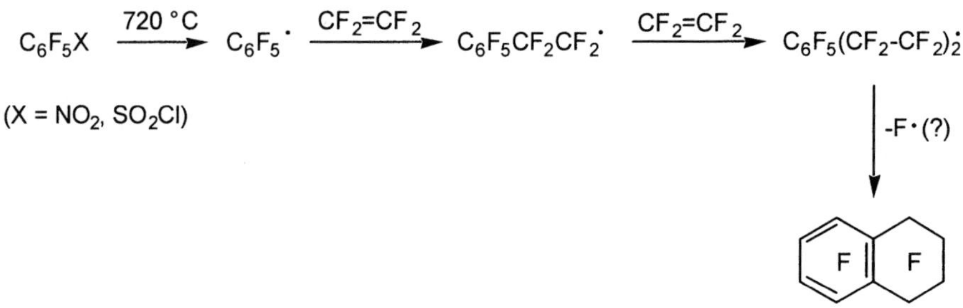

Scheme 5.15.

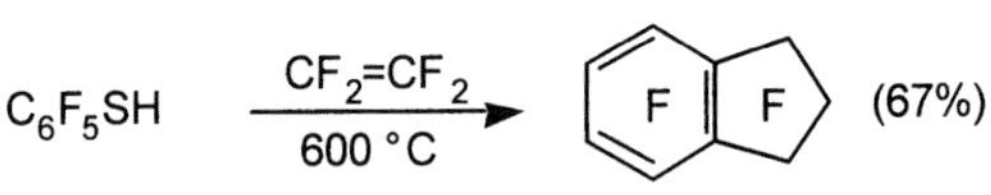

Scheme 5.16.

Zibarev[10] *et al.* in the NIOC group was of particular interest to us, not least from the point of view of the orientation in the case of the 2-naphthalene system (Scheme 5.17) [28].

---

[10] *Andrei V. Zibarev* was born in Tomsk (Siberia) in 1952, where he graduated from the State University in 1975. He has been working at NIOC since 1976, and obtained his PhD degree in 1981, and successfully defended his thesis for the Doctor's degree ('Azetines involving aromatic substituents of opened or closed topology') in 1996. His main work in fluoroorganic chemistry has been the fusion of S- and N-containing rings to polyfluoro-benzene and -naphthalene ring systems. Andrei likes to take his dog for a walk. (Academgorodok is a *very* doggy place: with forests separating Institutes from residential areas, it is a dog's delight to roam and sniff.)

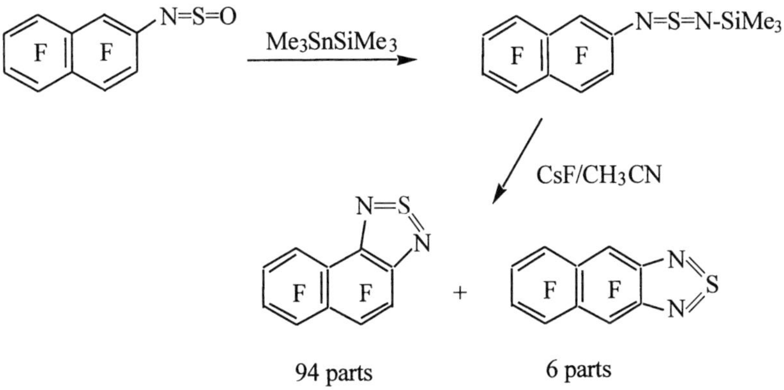

Scheme 5.17.

$$C_6Cl_5NHCOCCl_3 \xrightarrow[160\,^\circ C]{PCl_5} C_6Cl_5N{=}C(Cl)CCl_3$$

Scheme 5.18.

$$C_6F_5NHCOCCl_3 \xrightarrow[120\,^\circ C]{PCl_5} C_6F_5N{=}PCl_3$$

Scheme 5.19.

The subtle effects of polyfluoroaromatic groups on the course of a reaction expected to proceed in a certain way, and which in fact followed a completely *different* pathway must continue to be a source of fascination to all organic chemists. Many examples are known, but one will suffice here: we found that just as anilides react with PCl_5 to give arylimidoyl chlorides, so the trichloroacetyl derivative of pentachloroaniline can be converted into the corresponding acetoimidoyl derivative (Scheme 5.18) [29]; however, when the NIOC group reacted the related C_6F_5 derivative in the same way, the trichloroacetyl group was eliminated (Scheme 5.19) [30].

The present position at NIOC

During my 3 months stay at NIOC (March to May in 1996) gathering material for my fluoroaromatic review [5], I was able to make a fair assessment of fluorine chemistry in progress. Considering the existing conditions, research work was continuing at a healthy pace, and I understand that it is still continuing (1998).

In the UK, access to a huge variety of chemicals is taken for granted, with next-day deliveries after ordering being commonplace. In the days of the former USSR, materials produced *in* the country were relatively cheap: the only problem was the time taken for delivery. By contrast, apart from chemicals produced in socialist countries, importation from abroad was virtually impossible, and so compounds had to be synthesised from whatever was available locally. The present situation is the reverse: 'Western' chemicals are completely accessible – provided there is money available for their purchase. In reality, however, there is no money! Moreover, even chemicals formerly produced in the USSR have become hopelessly expensive or have completely disappeared. Thus, even such common entities like diethyl ether, chloroform, mineral acids and the like are now rarities. Inevitably this situation has a negative effect on morale, and furthermore, maintaining high-tech equipment in good working order is not without its headaches: difficulties associated with purchasing spare parts can leave equipment out of action for months, and sometimes experiments are interrupted because the local electricity supply company decides to make a saving by stopping the generation of power.

During the days when the 'Iron Curtain' stretched around the USSR, usually only Senior Scientists (Academicians, Institute Directors etc.) enjoyed visits abroad at the State's expense; other scientific workers were denied these privileges – presumably because the State feared that a delegate might 'defect' and seek political asylum, thereby casting doubt that everybody in the country was happy with the *status quo* at home[11]. Times have changed dramatically: during the earlier International Fluorine Symposia delegates from the USSR were viewed with a sense of awe – as though they came from a different planet! – but nowadays Russian scientists are enjoying freedoms completely unheard of until not so very long ago. Owing to the worsening economic state of the country since 1991 and its effect on scientific activity, many scientists have sought work abroad, and the Chemistry Department in Durham, for example, is benefitting enormously from the skills and insights of a number of brilliant crystallographers from Russia who have joined Professor Judith Howard's group. The NIOC group remains dedicated to keeping organofluorine chemistry as one of the major research groups at the Institute. Nevertheless, short-term placements outside the country have been enjoyed by a number of the NIOC group, including some of the younger scientists [Vadim Bardin[12] (in Duisberg, Germany with Professor Frohn) and Victor Karpov[13] (in Bochum, Germany with Professor Haas)].

---

[11]When the 11 delegates arrived for the Durham Symposium in 1971, we often wondered what they really thought, and were they, in the eyes of the USSR, 'reliable'? The four people from NIOC (George Yakobson, Tamara Petrova, Slava Platonov and Vladislav Vlasov) came at considerable personal expense, but there was some help for their scientific tour from the Academy of Sciences and Intourist. All four returned to NIOC, none the worse for an escapade in London when they went walking one evening in Hyde Park. As it grew dark they made their way to the main gates only to find them locked! Shouts for help drew no attention, so they walked around the perimeter of the park until they found a section of railing which seemed surmountable and managed to climb over. They still laugh about the incident, surmising on what would have happened if a policeman had caught them escaping from the park in such an ungainly fashion!

[12]*Vadim V. Bardin* is a Siberian, born in Kemerovo in 1952; he graduated from Novosibirsk State University in 1970. In 1982 he was awarded his PhD degree at NIOC, and since joining the staff there, he has studied the fluorinating properties of VF_5 with polyfluorinated materials and the use of $P(NR_2)_3$ reagents to remove halogens other than fluorine.

[13]*Victor M. Karpov* was born in 1948 in Mogocha, in the Chitinsky region of Siberia, near Lake Baikal. He graduated from Novosibirsk State University in 1970, joined the staff at NIOC and produced his PhD thesis in 1976. Victor successfully defended his thesis for the Doctor's degree in 1996, which brought together his work on polyfluorinated benzocycloalkenes and indenes.

References

1 R. N. Haszeldine and A. G. Sharpe, *Fluorine and its compounds*, Methuen & Co. Ltd., London, 1951.

2 J. A. Godsell, M. Stacey and J. C. Tatlow, *Nature, 178* (1956) 199; *Tetrahedron, 2* (1968) 195.

3 R. Stephens and J. C. Tatlow, *Chem. Ind.*, (1957) 821.

4 J. C. Tatlow, *J. Fluorine Chem., 73* (1995) vii.

5 G. M. Brooke, *J. Fluorine Chem., 86* (1997) 1.

6 N. N. Vorozhtsov and G. G. Yakobson, *Zh. Obshch. Khim., 27* (1957) 1672; N. N. Vorozhtsov and G. G. Yakobson, *Khim. Nauka i Prom., 3* (1958) 403.

7 N. N. Vorozhtsov, G. G. Yakobson and T. D. Rubina, *Dokl. Akad. Nauk SSSR, 134* (1960) 821.

8 N. N. Vorozhtsov, V. E. Platonov and G. G. Yakobson, *Izv. Akad. Nauk SSSR, Ser. Khim.*, (1963) 1524.

9 E. Nield, R. Stephens and J. C. Tatlow, *J. Chem. Soc.*, (1959) 166.

10 P. L. Coe, A. E. Jukes and J. C. Tatlow, *J. Chem. Soc. C*, (1966) 2323.

11 G. M. Brooke, J. Burdon and J. C. Tatlow, *J. Chem. Soc.*, (1961) 802.

12 A. A. Shtark and V. D. Shteingarts, *Zh. Org. Khim., 12* (1976) 1499.

13 V. D. Shteingarts, O. I. Osina, G. G. Yakobson and N. N. Vorozhtsov, *Zh. Vses. Khim. Oba, 11* (1966) 115; A. A. Shtark and V. D. Shteingarts, *Zh. Org. Khim., 22* (1986) 831.

14 V. D. Shteingarts, O. I. Osina, N. G. Kostina and G. G. Yakobson, *Zh. Org. Khim., 6* (1970) 833.

15 P. N. Dobronravov and V. D. Shteingarts, *Zh. Org. Khim., 17* (1981) 2245.

16 Yu. V. Pozdnyakovich and V. D. Shteingarts, *Zh. Org. Khim., 13* (1977) 1911.

17 K. R. Langille and M. E. Peach, *J. Fluorine Chem., 1* (1971/72) 407.

18 M. E. Peach and K. C. Smith, *J. Fluorine Chem., 27* (1985) 105.

19 G. G. Furin, S. A. Krupoder, M. A. Fedotov and G. G. Yakobson, *Izv. Sib. Otd. Akad. Nauk SSSR, Ser. Khim. Nauk, 5* (1982) 120.

20 V. M. Karpov, T. V. Mezhenkova and V. E. Platonov, *J. Fluorine Chem., 77* (1996) 101.

21 G. M. Brooke, J. Burdon and J. C. Tatlow, *J. Chem. Soc.*, (1962) 3253.

22 T. N. Gerasimova, N. V. Semikolenova and E. P. Fokin, *Zh. Org. Khim., 14* (1978) 100.

23 T. N. Gerasimova, N. N. Semikolenova, N. A. Orlova, T. V. Fomenko and E. P. Fokin, *Izv. Sib. Otd. Akad. Nauk SSSR, Ser. Khim. Nauk, 14* (1975) 54.

24 V. E. Platonov and G. G. Yakobson, in M. Vol'pin (ed.), *Soviet Scientific Reviews, Section B, Chemistry Reviews*, Vol. 5, 1984, p. 297.

25 V. E. Platonov, G. G. Furin, N. G. Malyuta and G. G. Yakobson, *Zh. Org. Khim., 8* (1972) 430.

26 N. G. Malyuta, V. E. Platonov, G. G. Furin and G. G. Yakobson, *Tetrahedron, 31* (1975) 1201.

27 J. M. Rawson, A. J. Banister and I. Lavender, *Adv. Heterocycl. Chem., 62* (1995) 137; A. J. Banister, N. Bricklebank, I. Lavender, J. M. Rawson, C. I. Gregory, B. K. Tanner, W. Clegg, M. R. J. Elsegood and F. Palacio, *Angew. Chem. Int. Ed. Eng., 35* (1996) 2533.

28 I. Yu. Bagryanskaya, Yu. V. Gatilov, A. O. Miller, M. M. Shakerov and A. B. Zibarev, *Heteroatom Chem., 5* (1994) 561.

29 G. M. Brooke, R. D. Chambers, W. K. R. Musgrave, R. A. Storey and J. Yeadon, *J. Chem. Soc., Perkin Trans., 1* (1976) 162.

30 T. D. Petrova, V. E. Platonov, T. I. Savchenko and L. M. Meshalkina, *Izv. Akad. Nauk SSSR, Ser. Khim., 11* (1978) 2635.

BIOGRAPHIC NOTE

Gerald M. Brooke

Gerald M. Brooke, a Yorkshireman born in Dewsbury in 1937, was both an undergraduate and postgraduate student at Birmingham University. He returned to the North of England in 1962 to join the staff at Durham University where he is now Reader in Chemistry. Polyfluoro-aromatic and -heteroaromatic chemistry has always been his 'first love' in research, and even in the days when non-classical carbonium ions were in vogue during the 1960s, he carried out some solvolysis studies on tetrafluorobenzonorbornenyl derivatives with Herbert C. Brown at Purdue University in the USA (1968/69; the work was never published). Since 1991 he has broadened his interests and engaged in the synthesis of a variety of pure long-chain oligomeric materials: alkanes (*e.g.* $C_{390}H_{782}$) and various nylons urgently needed by UK polymer physicists to serve as models for the corresponding commercial high-molecular-weight polymers. Despite his teaching and administrative duties in Durham, he has always spent much time at the bench doing personal research.

Chapter 6

THE IOWA CONNECTION

DONALD J. BURTON
Department of Chemistry, University of Iowa, Iowa City, IA 52242, USA

Introduction

The research initiated in my group at the University of Iowa was influenced by my associations with Professors W. T. Miller and H. C. Brown, which aroused my interest in fluoro-olefin chemistry and organoborane chemistry respectively. As the reader will note, we have developed synthetic methodology in several different areas, some which are interrelated while others stemmed from current trends in organofluorine chemistry. In order to provide some coherence, I have organized the chemistry into general topics rather than adopt a strict chronological presentation. I have explained why we pursued particular synthetic topics and described how each topic developed from its conception until the present or until we lost interest in it. Because of space limitations, it's not been possible to discuss every research topic pursued during the past 36 years, and I apologize to any co-workers whose work is not presented here.

Ylide chemistry

My initial research topic as a graduate student in Professor Miller's laboratory was the preparation of 2-phenyl-substituted perfluoroallyl halides and a mechanistic study of S_N2' reactions of such halides with fluoride ion. At that time (1956), however, the lack of suitable routes to this class of perfluoroallyl halides caused us to postpone the work. Instead, I pursued the aluminum halide-catalyzed rearrangement and disproportionation of fluorohaloalkanes and fluorohaloallyl halides as a PhD topic [1]. However, my initial thesis topic continued to be of interest to me, and a stubborn determination to conquer this topic was never abandoned by my subconscious, even though it took us another 15 years to succeed!! The key to this research problem was found through our studies on fluorinated phosphorus ylides.

Fluorohalomethylene ylides from fluorohalomethanes and tertiary phosphines

My initial interest in ylide chemistry was piqued by Frank Herkes, one of the first graduate students to join my research group at Iowa in 1962. He pointed out to me that Fuqua and co-workers [2] had prepared 1,1-difluoro-olefins from aldehydes by heating an aldehyde, triphenylphosphine, and sodium chlorodifluoroacetate in a glyme solvent. This approach, however, failed with ketones. The ylide $[Ph_3P^+{-}CF_2^-]$ was proposed as a transient species in this transformation, formed via capture of $[:CF_2]$ (from thermolysis of the acetate) by the nucleophilic tertiary phosphine. Since neither one of us knew anything about ylides at that time, Frank and I naively assumed that this ylide was insufficiently nucleophilic to react with standard ketones. However, we anticipated that it would attack the more electrophilic carbonyl group in ketones of the type $ArC(O)R_F$, and thus established a new route to β-phenyl-substituted perfluoro-olefins [3, 4]:

This venture initiated our work in ylide chemistry, which continues to the present time. We were fortunate that the tertiary phosphine, Ph_3P, did *not* react either with the perfluoroalkyl ketones or with the olefinic product. Mechanistic experiments suggested that the ylide was not formed by trapping of [$:CF_2$] but was formed via decomposition of a phosphobetaine salt [4]:

$$Ph_3P + CF_2ClCO_2Na \longrightarrow NaCl + [Ph_3\overset{+}{P}\text{-}CF_2\text{-}C(=O)\text{-}O^-] \longrightarrow [Ph_3\overset{+}{P}\text{-}\bar{C}F_2]$$

With ketones containing a carbanion stabilizing group, it became apparent that significant amounts of fluoride ion were formed in the generation of the ylide, since *both* the olefin and its HF-addition product were formed, *e.g.*

$$p\text{-}ClC_6H_4C(O)CF_3 \xrightarrow[\text{diglyme, 100 °C}]{Ph_3P,\ CF_2ClCO_2Na} \underset{(32\%)}{p\text{-}ClC_6H_4C(CF_3){=}CF_2} + \underset{(37\%)}{p\text{-}ClC_6H_4CH(CF_3)_2}$$

Also, with ketones containing perfluoroalkyl groups of more than one carbon, a mixture of the terminal olefin and internal olefin (isomerized product) was formed [5, 6], *e.g.*

$$C_6H_5C(O)C_2F_5 + [Ph_3\overset{+}{P}\text{-}\bar{C}F_2] \longrightarrow cis/trans\text{-}C_6H_5C(C_2F_5){=}CF_2 \xrightarrow{F^-} C_6H_5C(CF_3){=}CFCF_3$$

HF-Addition and fluoride ion-catalyzed isomerization could be avoided by using lithium chlorodifluoroacetate in DMF to generate the ylide [6]. Attempts to pre-generate the ylide were unsuccessful [4].

Fuqua and co-workers had attempted to prepare 2-phenylpentafluoropropene via the more nucleophilic ylide [n-$Bu_3P^+{-}CF_2^-$] [7]. However, when trifluoroacetophenone was heated with n-Bu_3P and sodium chlorodifluoroacetate in N-methyl-2-pyrrolidone (NMP), they obtained 1,1,1-trifluoro-2-phenyl-2-hexene in low yield. To account for this, these workers proposed that the difluoromethylene ylide formed initially rearranged before being trapped by ketone. However, Ken Klabunde in our laboratory demonstrated that the products isolated by Fuqua *et al.* [7] were actually formed via direct reaction of the tertiary trialkylphosphine with the perfluoroalkyl aryl ketone [8]. This work provided the first reported examples of reactions between a perfluoroalkyl ketone and a trialkylphosphine:

$$RC(O)CF_3 + (n\text{-}C_4H_9)_3P \xrightarrow[\text{reflux}]{\text{hexane}} cis/trans\text{-}CF_3C(R){=}CH(CH_2)_2CH_3 \quad (43\text{-}64\%)$$

($R = C_6H_5$, p-$CH_3C_6H_4$)

This initial work was extended by Henry Krutzsch, who established that the mixed halogen ylide $[Ph_3P^+-CFCl^-]$ could be generated by two different methods [9, 10], as outlined below

$$Ph_3P + CFCl_2CO_2Na + {>}C{=}O \xrightarrow[85\ ^\circ C]{\text{triglyme}} {>}C{=}CFCl + Ph_3PO + NaCl + CO_2$$

$$Ph_3P + CHFCl_2 + KO^tBu \xrightarrow[0\ ^\circ C]{\text{heptane}} [Ph_3P{=}CFCl] \xrightarrow{{>}C{=}O} {>}C{=}CFCl + Ph_3PO$$

The similarity in the *cis/trans* ratio of fluorochloro olefinic products from the two methods suggested that the mixed ylide was a common intermediate. In contrast to the difluoromethylene ylide, the mixed fluorochloro ylide exhibited some stability and could be pre-generated (albeit in moderate to low yields). Coincident with our work, Ando and coworkers reported the same mixed ylide, formed via the reaction of methyl dichlorofluoroacetate with triphenylphosphine and sodium methoxide [11].

The unavailability of *meta-* and *para-*substituted trifluoroacetophenones, with substitutents such as NH_2, CN, Br, and I, seriously impeded the preparation of β-phenyl-substituted perfluoro-olefins. However, Klabunde developed a simple, useful route to the amino-substituted ketones, which could then be readily transformed into other derivatives by the Sandmeyer reaction [12]:

$R = CF_3, C_2F_5, n\text{-}C_3F_7$

These ketones were subsequently utilized by Klabunde for the preparation of phenyl-substituted hexafluoropropanes, $CF_3CHArCF_3$. The acidity of these propane derivatives was studied by base-catalyzed hydrogen, deuterium and tritium isotope exchange methods to probe fluorine hyperconjugation, intermediate carbanion stability and carbanion geometry [13]. None of this work would have been possible without the ylide route to the olefin precursors required for the 2-hydropropane compounds. Similarly, the availability of fluorochloro-olefin products derived from the mixed halogen ylide provided us with the opportunity to study the stereochemistry of nucleophilic displacement of chloride ion in β-substituted-1-chloroperfluoro-olefins [14]; in all cases, retention of configuration was observed.

Even though the use of lithium chlorodifluoroacetate in DMF circumvented some of the problems with substituted aryl perfluoroalkyl ketones [6], it did not solve them all. For example, with *m*-bromophenyl trifluoromethyl ketone, only the H-F addition product was isolated:

$$m\text{-}BrC_6H_4C(O)CF_3 + Ph_3P + CF_2ClCO_2Li \xrightarrow[100\ ^\circ C]{\text{DMF}} m\text{-}BrC_6H_4CH(CF_3)_2 \qquad (65\%)$$

Similarly, with *m*-bromophenyl pentafluoroethyl ketone, the internal olefin m-$BrC_6H_4C(CF_3){=}CFCF_3$ was the major product. With other good carbanion stabilizing groups, such as Br, I, CN, NO_2 in the perfluoroalkyl ketone, similar results were observed with CF_2ClCO_2Li. Clearly, an alternative fluoride-ion free route to these reactive β-substituted-1,1-difluoro-olefins was required.

Rabinowitz had suggested that 1,1-difluoro-olefins could be formed via the interaction of dibromodifluoromethane with Ph_3P and RCHO. However, no details were provided or subsequently published. Doug Naae subsequently developed a procedure whereby a perfluoroalkyl aryl ketone was treated with CF_2Br_2 and Ph_3P at 70 °C in glymes or DMF to give excellent yields of 1,1-difluoro-olefins [15]. No H-F addition products or isomerized olefins were observed, *e.g.*

$$m\text{-}BrC_6H_4C(O)CF_3 + 2Ph_3P + CF_2Br_2 \xrightarrow[\text{diglyme}]{70\ ^\circ C} m\text{-}BrC_6H_5C(CF_3){=}CF_2 \qquad (85\%)$$

$$m\text{-}BrC_6H_4C(O)C_2F_5 + 2Ph_3P + CF_2Br_2 \xrightarrow[\text{diglyme}]{70\ ^\circ C} m\text{-}BrC_6H_4C(C_2F_5){=}CF_2 \qquad (87\%)$$

In addition, the yields of olefinic products are generally higher than via the chlorodifluoroacetate method. However, like the acetate route, the CF_2Br_2 route failed with ketones that did not contain an activating perfluoroalkyl group, *e.g.* acetophenone.

Naae resolved this problem by employing the more nucleophilic ylide $[(Me_2N)_3P^+\text{-}CF_2^-]$, generated from $(Me_2N)_3P$ and either CF_2Br_2 or CF_2Cl_2:

$$C_6H_5C(O)CH_3 + 2(Me_2N)_3P + CF_2Br_2 \xrightarrow{\text{RT}} C_6H_5C(CH_3){=}CF_2 \qquad (68\%)$$

Ketones, such as cyclohexanone, other non-activated ethyl phenyl ketones and diethyl ketone also gave good yields of the corresponding 1,1-difluoro-olefins [16, 17]. Naae was able to isolate and identify the bromodifluoromethylphosphonium salts formed initially from the reaction of the tertiary phosphine and CF_2Br_2. Mechanistic experiments demonstrated that these salts were formed via a difluorocarbene route (Scheme 6.1). A second halophilic attack on the phosphonium salt gave the corresponding ylide:

$$[R_3P^+CF_2Br]Br^- + R_3P \rightleftharpoons [R_3\overset{+}{P}-\overset{-}{C}F_2] + R_3PBr_2$$

Extension of this methodology to the bromofluoromethylene ylide by Bill Vander Haar [18] provided a new route to bromofluoromethylene olefins directly from aldehyde or ketone precursors Unfortunately, the bromofluoro ylide olefination route is not stereospecific and gives *cis/trans* mixtures of olefins. The phosphonium salt $[Ph_3PCFBr_2]^+Br^-$ can be isolated when $CFBr_3$ and Ph_3P are used in equimolar amounts.

$$CFBr_3 + 2Ph_3P + C_6H_5C(O)CF_3 \xrightarrow[70\ ^\circ C]{\text{THF}} C_6H_5C(CF_3){=}CFBr \quad E/Z = 54/46 \qquad (82\%)$$

Similar work by Mike Van Hamme [19] provided access to the phosphonium salt, $[(Me_2N)_3PCFCl_2]^+Cl^-$ [via reaction of $(Me_2N)_3P$ with $CFCl_3$] and hence the ylide $(Me_2N)_3\overset{+}{P}-\overset{-}{C}FCl$. He also demonstrated that Ph_3P, $CFCl_3$, $>C{=}O$ and zinc dust in DMF at 60 °C could provide chlorofluoromethylene olefins in one step [20].

$$R_3P + CF_2Br_2 \longrightarrow [R_3\overset{+}{P}Br][CF_2Br]^-$$

$$[CF_2Br]^- \longrightarrow [:CF_2] + Br^-$$

$$R_3P + [:CF_2] \longrightarrow [R_3\overset{+}{P}\text{-}\bar{C}F_2]$$

$$[R_3\overset{+}{P}\text{-}\bar{C}F_2] + [R_3\overset{+}{P}Br] \longrightarrow [R_3\overset{+}{P}CF_2Br] + R_3P$$

Overall reaction: $R_3P + CF_2Br_2 \rightarrow [R_3P^+CF_2Br]Br^-$

$(R = Ph, Me_2N)$

Scheme 6.1.

The phosphonium salts prepared in our ylide work provided convenient sources of other transient intermediates, such as dihalocarbenes and trihalomethide ion. Naae demonstrated that bromodifluorotriphenylphosphonium bromide was readily cleaved by methoxide or fluoride ion to give difluorocarbene [21]. He developed this route as a one-pot procedure to produce difluorocyclopropanes via *in situ* formation of the phosphonium salt precursor, *e.g.* [21]

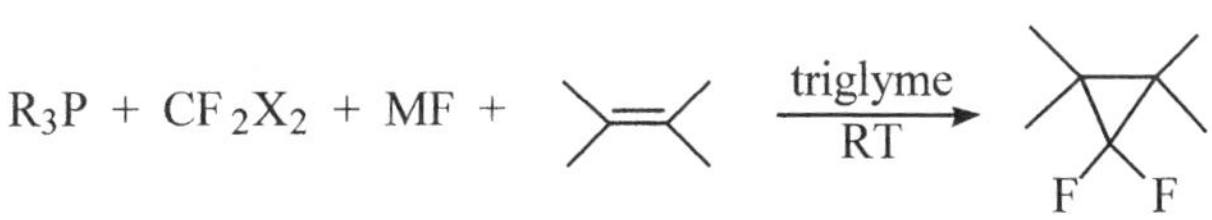

$R = Ph, Me_2N$; $X = Br, Cl$; $M = K, Cs)$

This facile formation and capture of difluorocarbene has been extensively utilized by many workers over the past 25 years and has become the most widely employed mild, non-basic route to difluorocarbene.

The carbene is formed via initial cleavage of the phosphonium salt to give a trihalomethide ion which collapses via α-elimination to the carbene:

$$[R_3P^+CF_2Br]Br^- \xrightarrow{F^-} [R_3P^+F][CF_2Br^-]$$

$$[CF_2Br^-] \rightleftharpoons [:CF_2] + Br^-$$

Jack Kesling utilized this methodology to capture $[CF_2Br]^-$ with electrophiles such as alcohols, iodine, fluoro-olefins and acyl fluorides [22]. In related work, Greg Wheaton demonstrated the generation and capture of $[CF_2Cl]^-$ with electrophiles, such as trifluoromethyl ketones, pentafluoro pyridine and fluoro-olefins [23, 24]. Previous to these reports, there was no concrete evidence in the literature for the existence of halodifluoromethide ions. This work stimulated others to re-investigate earlier work and to confirm the finite lifetime of these intermediates. Subsequent work by Kesling and Seiji Shin-Ya generated $[CF_2Br]^-$, $[CF_2Cl]^-$ and $[CFCl_2]^-$ from analogous phosphonium salts [25].

In collaborative work with the Department of Nuclear Medicine (National Institutes of Health), Rick Flynn and Steve Hansen studied the hydrolysis of $[Ph_3PCF_2Br]^+Br^-$ and $[Ph_3PCFBr_2Br]^+Br^-$ in the presence of a radioactive isotope of bromine. With $[Ph_3PCF_2Br]^+Br^-$ unequivocal evidence was obtained that demonstrated that the mechanism of hydrolysis proceeds through a difluorocarbene intermediate and that decomposition of bromodifluoromethide is reversible [26]: $[CF_2Br^-] \rightleftharpoons [:CF_2] + Br^-$; with $[Ph_3PCFBr_2]^+Br^-$, hydrolysis of the salt proceeds via $[CFBr_2^-]$ and not [:CFBr] [27].

When Kesling [28] attempted to prepare $[n\text{-}Bu_3PCF_2Br]^+Br^-$from CF_2Br_2 and n-Bu_3P, he discovered that the initial reaction between these reactants to form $[n\text{-}Bu_3PCF_2Br]^+Br^-$ was slower than the subsequent reaction of the phosphonium salt with a second equivalent of the trialkyl phosphine. The main product (the exclusive product if two equivalents of n-Bu_3P were utilized) was a bis(phosphonium) salt [28]:

$$2n\text{-}Bu_3P + CF_2Br_2 \longrightarrow [n\text{-}Bu_3P^+CF_2P^+Bu_3\text{-}n] \quad 2Br^-$$

Bis(phosphonium) salt formation was not observed with two equivalents of the triarylphosphine Ph_3P. Kesling rationalized bis(phosphonium) salt formation via nucleophilic attack of the intermediate ylide on dibromotri-n-butylphosphorane (Scheme 6.2). We assumed that the reaction stopped at this stage because any further halophilic reactions would involve abstraction of F^+ from the bis(phosphonium) salt. However, if abstraction of X^- ($X = Cl$, Br, I) were available, we predicted that the reaction would continue to form a phosphoranium salt, $[R_3P^+\text{-}^-CF\text{-}P^+R_3]Br^-$. However at that time (~1974) our NMR capabilities were poor and we could not observe the phosphoranium salt spectroscopically. Consequently, we postponed further work in this direction for a few years until our NMR capabilities improved and allowed Daryl Cox (in the early 1980s) to verify our earlier expectation of phosphoranium salt formation [29]:

$$3R_3P + CFX_3 \longrightarrow [R_3\overset{+}{P}-\overset{-}{C}F-\overset{+}{P}R_3]X^- + R_3PX_2$$

(R = n-Bu, n-octyl; (90-95%)
X = Cl, Br)

The phosphoranium salts exhibited excellent reactivity with perfluoroacyl fluorides. Ylides generally undergo acylation with acid halides; however, Cox demonstrated that perfluoroacyl fluorides undergo a stereoselective Wittig olefination reaction with the phosphoranium salts to give a (Z)-perfluorovinyl phosphonium salt; subsequent basic hydrolysis gave *only* the (E)-1-hydroperfluoro-olefin [30] (Scheme 6.3). This work illustrates one of the rare examples of a Wittig olefination reaction with an acyl halide and provided the first example of the preparation of a fluoro-olefin directly from an acyl halide.

Subsequent work by Cox and Narayanasamy Gurusamy [31] demonstrated that aldehydes also react with the fluorine-containing phosphoranium salts, derived from n-Bu_3P, to produce a fluorine-containing vinylphosphonium salt, basic hydrolysis of which gives the corresponding 1-fluoroalkenes with retention of configuration. The stereochemistry of the vinyl phosphonium salt is aldehyde dependent: with aliphatic aldehydes, the major product is the (E)-vinylphosphonium salt, which gives the (Z)-1-fluoro-alkene exclusively on hydrolysis, while aryl aldehydes give the (Z)-phosphonium salt and hence the (E)-1-fluoroalkene [31] (Scheme 6.4). The stereochemistry, in the case of aromatic aldehydes, is

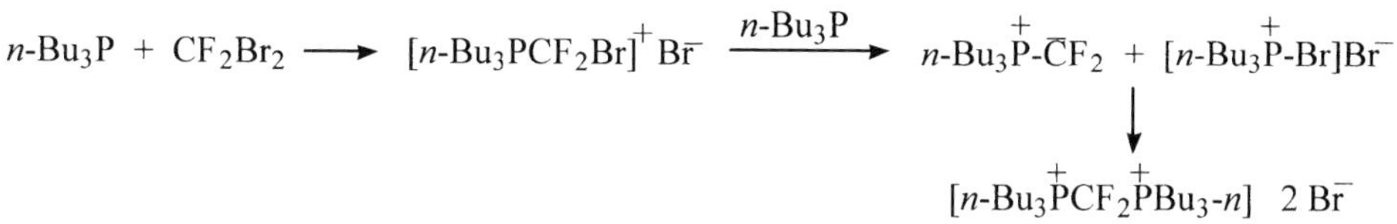

Scheme 6.2.

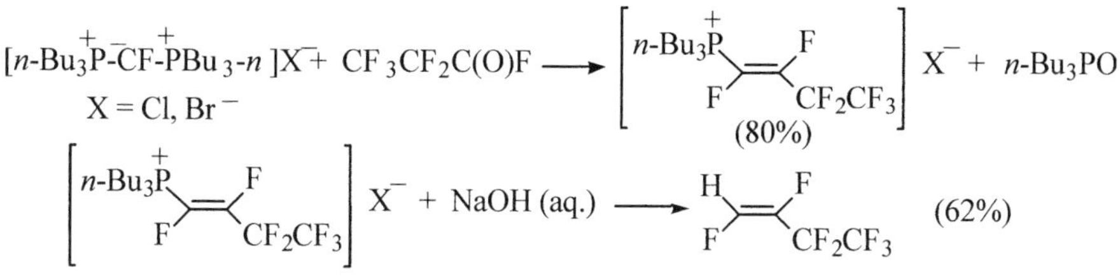

Scheme 6.3.

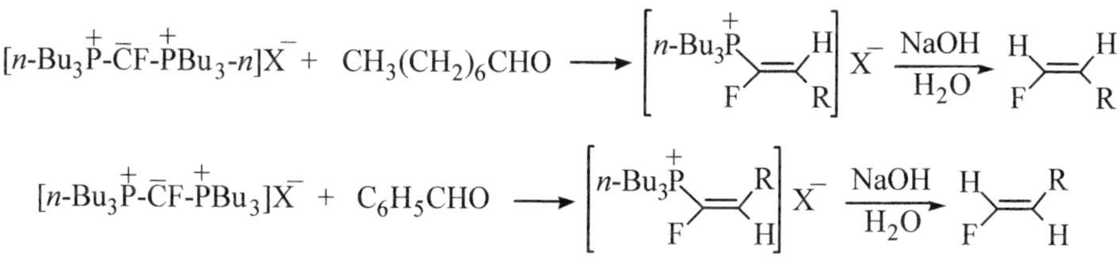

Scheme 6.4.

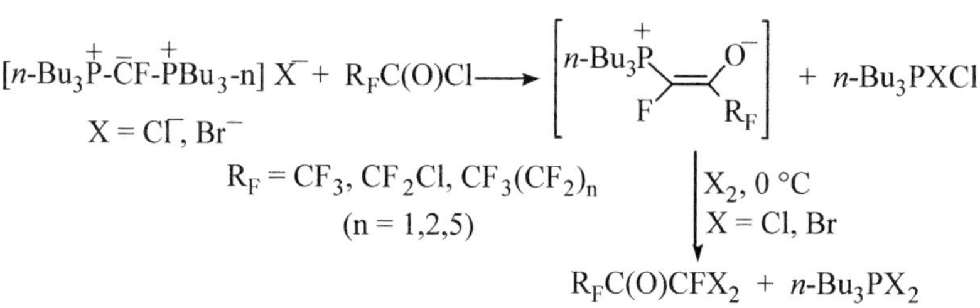

Scheme 6.5.

judged to be controlled by formation of a through-space charge-transfer complex between the aromatic ring and the positively charged phosphonium center during formation of the oxaphosphetane intermediate [31].

In contrast to the Wittig olefination reaction noted above, In Howa Jeong and Cox [32] observed normal acylation of the fluorine-containing phosphoranium salts with perfluoroacyl chlorides. Jeong subsequently developed this methodology into a regiospecific preparation of α,α-dihalofluoromethyl perfluoroalkyl ketones (Scheme 6.5) [33]. Treatment of $R_FC(O)CFCl_2$ with SbF_5 provided a useful route to α-chloroperfluoro-2-alkanones and perfluoro-2-alkanones [34].

Thus, what began as a simple preparative route to 1,1-difluoro-olefins eventually led us to bis(phosphonium) salts and eventually to the novel fluorine-containing phosphoranium salts. We developed an in-depth understanding of the mechanisms of these reactions, which aided us in the development of subsequent phosphonate chemistry. This work provided some unique and novel chemistry and several useful precursors for our studies in organometallic chemistry (see later).

Ylides via metal dehalogenation of phosphonium salts

The fluoromethylene ylide $[Ph_3P^+\text{-}CHF^-]$ was initially generated by Pete Greenlimb under typical Witting conditions from $[Ph_3P\text{-}CH_2F]^+I^-$; however, it was not very stable [35], and a better method was developed which involved Zn(Cu)-dehalogenation of an α-halophosphonium salt [36]. Although only modest success was achieved in this work by Greenlimb, *e.g.*

$$[Ph_3P\text{-}CHFI]^+I^- + Zn(Cu) + C_6F_5CHO \xrightarrow[0\,°C]{DMF} E/Z\text{-}C_6F_5CH{=}CHF \qquad (65\%)$$

the metal dehalogenation approach was extended successfully to other ylides. For example, Van Hamme employed this route to form stable ylide complexes with Group IIB metals [37]:

$$[(Me_2N)_3\overset{+}{P}{-}CFCl_2]Cl^- + M \xrightarrow[60\,°C]{THF} [(Me_2N)_3\overset{+}{P}{-}CFClMCl]Cl^- \quad (M = Zn, Cd, Hg)$$

These complexes exhibited surprising stability in ethereal solvents and served as effective chlorofluoromethylene transfer agents in Wittig reactions via dissociation into the corresponding ylide and metal chloride, *e.g.*

$$[(Me_2N)_3P^+\text{-}CFClZnCl]Cl^- \rightleftharpoons (Me_2N)_3\overset{+}{P}{-}\overset{-}{C}FCl + ZnCl_2$$

When the metal was zinc, a solution of the complex could be generated on a large scale and utilized in Wittig olefination reactions over a period of weeks.

When the difluoromethylene ylide, $[R_3\overset{+}{P}{-}\overset{-}{C}F_2]$, were generated from $[R_3P^+CF_2Br]Br^-$ and a tertiary phosphine (R_3P) in the presence of substrates such as C_6F_5CHO, $C_6F_5C(O)CF_3$, $C_6H_5C(O)CF_2Cl$ and $C_6F_5C(O)CF_2Cl$, low yields of the 1,1-difluoro-olefins were obtained since the ketones reacted with the tertiary phosphine used. Kesling and Naae, however, found that when the ylide was generated via dehalogenation of the phosphonium salt in the presence of these substrates, good yields of 1,1-difluoro-olefins were obtained [38], *e.g.*

$$[Ph_3P^+CF_2Br]Br^- + C_6H_5C(O)CF_2Cl \xrightarrow[RT,\ Cd]{Triglyme} F_2C{=}C(C_6H_5)CF_2Cl \qquad (60\%)$$

Similar dehalogenation of $[Ph_3PCFBr_2]^+Br^-$ provided a useful route to $[Ph_3P^+\text{-}^-CFBr]$ [39].

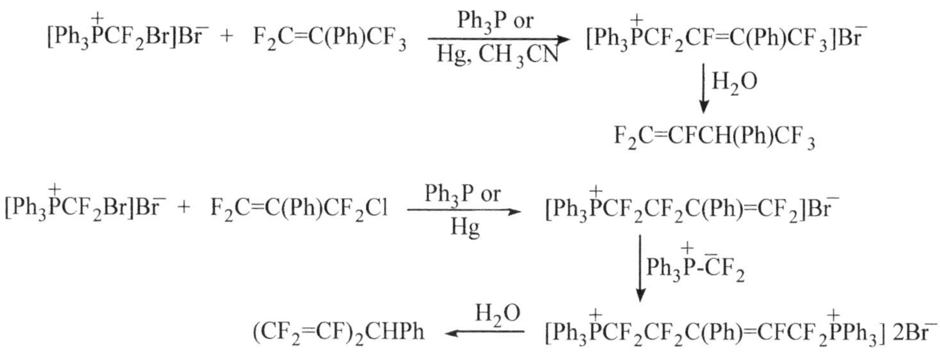

Scheme 6.6.

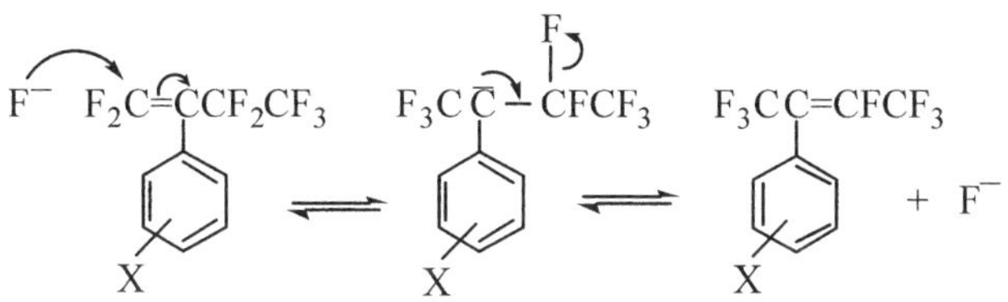

Scheme 6.7.

Chain-extension reactions

Yoshio Inouye and Jim Headley demonstrated that nucleophilic difluoromethylene ylides reacted with 1,1-difluoro-olefins to give chain-extended phosphonium salts, which on hydrolysis gave chain-extended fluorinated alkenes or alkadienes [40]. With an allyl halide derivative, the initial S_N2' product reacted with a second equivalent of the ylide to give chain-extension by two CF_2 units (Scheme 6.6).

Fluoride ion-catalyzed isomerization of 2-aryl-F-butenes

Our success with the difluoromethylene ylide chemistry permitted us to prepare a series of β-phenyl-substituted perfluoro-1-butenes. This *finally* enabled us to accomplish my initial PhD thesis topic, namely to investigate the mechanism of fluoride ion-involved catalyzed S_N2' reactions in such systems. Headley [41] carried out a detailed kinetic study of the isomerization. The resultant Hammett plot was non-linear with a concave downward break near $\sigma = 0$, and a two-step mechanism involving formation of a carbanionic intermediate best explained the data (Scheme 6.7). A change in the rate-limiting step caused the break in the Hammett plot.

Ylide-carbene chemistry

In an attempt to prepare 1,1-difluoro-olefins without formation of Ph_3PO (the main by-product in Wittig chemistry), Wheaton developed a novel reaction between non-stabilized ylides and chlorodifluoromethane [42, 43]. The idea was to utilize the basic ylides in a two-fold way: (i) as a base to generate the electrophilic difluorocarbene *in situ*; and (ii), to trap the electrophilic carbene with the nucleophilic ylide

$$Ph_3\overset{+-}{PC}R^1R^2 + CHF_2Cl \longrightarrow [Ph_3\overset{+}{P}CHR^1R^2]Cl^- + [:CF_2]$$

$$Ph_3\overset{+-}{PC}R^1R^2 + [:CF_2] \longrightarrow F_2C{=}CR^1R^2 + Ph_3P:$$

Overall Reaction:

$$2\ Ph_3\overset{+-}{PC}R^1R^2 + CHF_2Cl \longrightarrow F_2C{=}CR^1R^2 + Ph_3P: + [Ph_3\overset{+}{P}CHR^1R^2]Cl^-$$

Scheme 6.8.

(Scheme 6.8). The triphenylphosphine can be readily recovered and utilized to prepare more $[Ph_3PCHR^1R^2]^+Cl^-$ ylide precursor). The $[Ph_3PCHR^1R^2]^+Cl^-$ formed in the reaction can be recovered and recycled to prepare more ylide. Consequently, olefin formation involves one equivalent of phosphonium ylide and one equivalent of CHF_2Cl. No Ph_3PO is formed. This reaction can also be accomplished with polymer-supported phosphonium ylide, which makes recycling of the tertiary phosphine and phosphonium salt more efficient [44]. This approach to 1,1-difluoro-olefins is an example of Trost's atom economy concept.

In the above discussion of the highlights of our ylide chemistry, many details and examples were of necessity omitted. For a more thorough presentation of the work, the reader should consult our *Chemical Reviews* article on this topic [45].

Cyclic fluorinated ylides

In addition to the fluoromethylene ylides, we also pioneered the preparation and identification of *stable* fluorinated cyclic ylides. Our interest in this subject was stimulated by a report by Stockel and co-workers, who proposed that the 1:1 adduct between triphenylphosphine and perfluorocyclobutene was a rapidly equilibrating set of 1,3-dipolar species [46] (Scheme 6.9). Our knowledge of fluorinated carbanions (in 1970) led us to suspect that these 1,3-dipolar species would be unstable and easily lose fluoride ion. Consequently, we prepared the 1:1 adduct and, finding its ^{19}F NMR spectrum ambiguous, carried out an X-ray analysis and unequivocally demonstrated the compound to be the *stable* cyclic ylide **1** [47], formed as shown in Scheme 6.10. (Perfluorocyclopentene and related four and five-membered perfluorinated cycloalkenes behave similarly [45].)

When the cyclic ylide **1** is treated with BF_3, the corresponding vinyl phosphonium tetrafluoroborate is formed and can be isolated. However, treatment of the tetrafluoroborate salt with fluoride ion reforms the ylide [45, 48, 49] (Scheme 6.10). Cyclic ylides like **1** are stabilized by the phosphonium centre and the beta fluorines. It's not necessary to invoke any stabilization by 'd' orbitals on phosphorus. In fact, trialkylamines form analogous stable ammonium ylides, as demonstrated by work of Dick Howells and Paul Vander Valk in our laboratory [50, 51]. Later work by Vander Valk demonstrated the applicability of this method to analogous arsonium ylides [52] (Scheme 6.11). Note that this work provided not only the first stable fluorinated ylides but also the first examples of stable fluorinated carbanions. Subsequent work by others has demonstrated the generality of this seminal work.

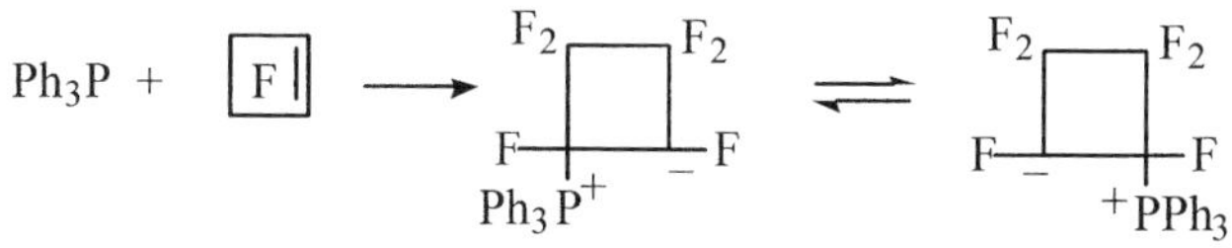

Scheme 6.9.

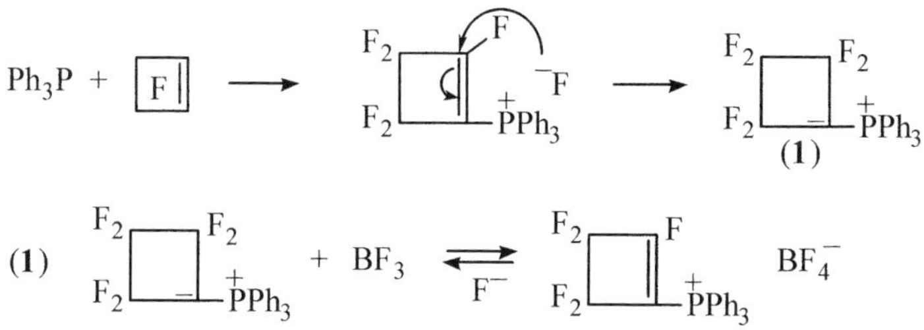

Scheme 6.10.

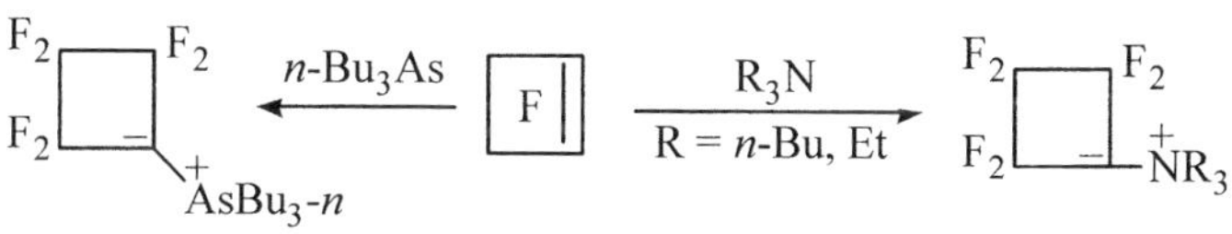

Scheme 6.11.

Extension of this work to acyclic olefins by Seiji Shin-Ya and Howells [53] demonstrated that stable ylide formation was possible when *only* β-fluorines are present in the ylide. Thus, perfluoro-2-butene reacts with n-Bu_3P to give a stable ylide (in solution):

$$CF_3CF{=}CFCF_3 + n\text{-}Bu_3P \xrightarrow{Et_2O} CF_3CF_2\text{-}\bar{C}(\text{-}CF_3)\text{-}\overset{+}{P}Bu_3\text{-}n$$

However, when both α- and $\beta-$fluorines are present in the ylide, the ylide is unstable and collapses to a vinylphosphorane [53]. Hydrolysis of the vinylphosphorane derived from hexafluoropropene gives only (E)-1-hydropentafluoropropene [54, 55]:

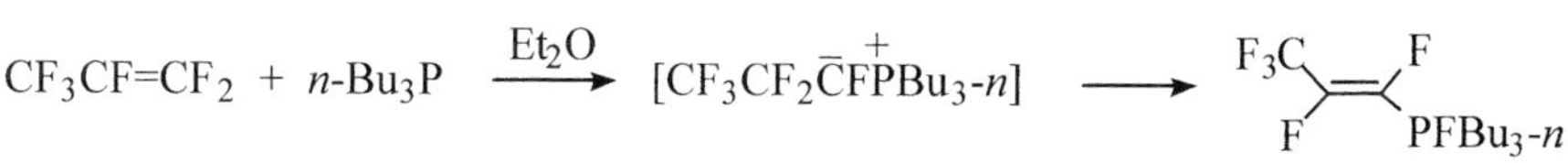

Shin-Ya and Howells demonstrated the generality and utility of this methodology as a stereoselective reductive route to polyfluorinated olefins. We and others have employed this method as a one-pot route to many 1-hydroperfluoro-olefins. The vinylphosphonium tetrafluoroborates are also readily cleaved by KF/I_2 or Na_2CO_3/I_2 to give perfluorovinyl iodides with retention of geometry [56]. This methodology has also been used by others to prepare the analogous SF_5 derivatives.

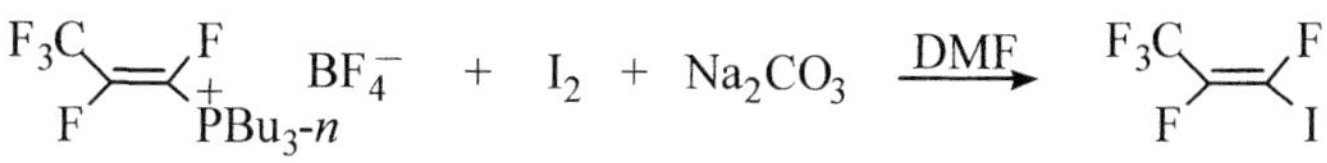

Ylide functionalization

In addition to the typical Wittig reaction to directly give olefinic products, we became interested in more complex *in situ* functionalization of fluorine-containing ylides. The main architect of this work in our laboratory was Alagappan Thenappan. Initially he developed a route to α-fluoroalkanoates via alkylation of (fluorocarboalkoxymethylene)tri-n-butylphosphorane [57]. The conversion can be carried out as a one-pot reaction:

$$[n\text{-Bu}_3\overset{+}{P}\text{CHFCO}_2\text{R}']\text{Br}^- \xrightarrow[\substack{\text{THF} \\ -78\ °\text{C} \\ 2)\ \text{RX}}]{1)\ \text{BuLi}} [n\text{-Bu}_3\overset{+}{P}\underset{\substack{| \\ \text{R}}}{\text{C}}\text{FCO}_2\text{Et}]\text{X}^- \xrightarrow[\text{RT}]{\text{NaHCO}_3\ \text{(aq.)}} \underset{(40\text{-}60\%)}{\text{RCHFCO}_2\text{Et}}$$

Similarly, Thenappan developed a convenient route to α-fluoro-β-ketoesters via acylation of (fluorocarboethoxymethylene)tri-n-butylphosphorane [58]:

$$[n\text{-Bu}_3\overset{+}{P}\text{CHFCO}_2\text{Et}]\text{Br}^- \xrightarrow[\substack{\text{THF} \\ -78\ °\text{C} \\ 2)\ \text{RC(O)X}}]{1)\ \text{BuLi}} [n\text{-Bu}_3\overset{+}{P}\text{CF(CO}_2\text{Et)C(O)R]X}^- \xrightarrow{\text{NaHCO}_3\ \text{(aq.)}} \underset{(40\text{-}77\%)}{\text{RC(O)CFHCO}_2\text{Et}}$$

The R group in the acyl halide can be alkyl, aryl, perfluoroalkyl or cycloalkyl, and the phosphonium bromide starting material is easily made from n-Bu_3P and $BrCHFCO_2Et$. This mild method can be carried out as a one-pot reaction via sequential addition of the reagents.

This methodology was subsequently modified by Thenappan and Hou-Jen Tsai to provide a one-pot synthesis of unsymmetrical and symmetrical tetrasubstituted α-fluoro-α,β-unsaturated esters from α-fluorophosphonates, as illustrated below [59]:

$$\text{(EtO)}_2\text{P(O)CHFCO}_2\text{Et} \xrightarrow[\substack{2)\ \text{RC(O)Cl} \\ 3)\ \text{R}'\text{M}/-78\ °\text{C to RT}}]{1)\ \text{BuLi}/-78\ °\text{C}} \text{RR}'\text{C=CFCO}_2\text{Et} \qquad (52\text{-}63\%)$$

$$\text{(EtO)}_2\text{P(O)CHFCO}_2\text{Et} \xrightarrow[\substack{2)\ \text{ClC(O)Cl} \\ 3)\ \text{R}'\text{M}/-78\ °\text{C to RT}}]{1)\ \text{BuLi}/-78\ °\text{C}} \text{R}'_2\text{C=CFCO}_2\text{Et} \qquad (50\text{-}55\%)$$

When alkyloxalyl chlorides were used as the acylating agent, the above procedure provided an expedient synthesis of α-fluoro-α,β–unsaturated diesters [60, 61]:

$$\text{(EtO)}_2\text{P(O)CHFCO}_2\text{Et} \xrightarrow[2)\ \text{ClC(O)CO}_2\text{R}]{1)\ \text{BuLi/THF}/-78\ °\text{C}} \text{(EtO)}_2\text{P(O)CF(COCO}_2\text{R)CO}_2\text{Et} \qquad (\sim 90\%)$$

$$\text{(EtO)}_2\text{P(O)CF(COCO}_2\text{R)CO}_2\text{Et} \xrightarrow[-78\ °\text{C to RT}]{\text{R}'\text{MgX}} E/Z\text{-R}'\text{(CO}_2\text{R)C=CFCO}_2\text{Et} \qquad (50\text{-}68\%)$$

An additional modification introduced by Thenappan was the low-temperature *in situ* reduction of an ester to an aldehyde with di-isobutylaluminum hydride, followed by *in situ* reaction of the aldehyde with a fluorine-containing phosphonate anion to give α-fluoro-α,β-unsaturated esters [62, 63]. The ester's R group can be alkyl, aryl, perfluoroalkyl or fluoroalkyl. With ethyl formate this procedure provides a facile preparation of ethyl 2-fluoroacrylate [64].

$$\mathrm{RC(O)OR'} \xrightarrow[\substack{1)\ \mathrm{DIBAL} \\ 2)\ [(\mathrm{EtO})_2\mathrm{P(O)^-CFCO_2Et}]\mathrm{Li}^+}]{\mathrm{THF}/-78\ ^\circ\mathrm{C}} E/Z\text{-}\mathrm{RCH{=}CFCO_2Et} \quad (45\text{-}66\%)$$

A series of fluorine-containing aldehydes was prepared by Don Wiebe via Wittig chemistry [65], *e.g.*

$$\mathrm{CF_3C(O)Ph + Ph_3P^+C^-HOCH_3} \xrightarrow[5\ ^\circ\mathrm{C\ to\ RT}]{\mathrm{Et_2O}} E/Z\text{-}\mathrm{CF_3C(Ph){=}CHOCH_3} \quad (89\%)$$

$$E/Z\text{-}\mathrm{CF_3C(Ph){=}CHOCH_3} \xrightarrow[65\ ^\circ\mathrm{C}]{\mathrm{H_2SO_4\ (aq.)}} \mathrm{CF_3C(Ph)CHCHO} \quad (83\%)$$

Halogenation of these aldehydes could be regioselectively controlled via the enol or free-radical halogenation at the formyl hydrogen:

$$\mathrm{CF_3C(Ph)CHCHO + Cl_2} \xrightarrow{h\nu} \mathrm{CF_3C(Ph)CHC(O)Cl} \quad (70\%)$$

$$\mathrm{CF_3C(Ph)CHCHO + Cl_2} \xrightarrow{\mathrm{CH_3CO_2H}} \mathrm{CF_3C(Ph)CClCHO} \quad (74\%)$$

$$\mathrm{(CF_3)_2CHCHO + Br_2} \xrightarrow{h\nu} \mathrm{(CF_3)_2CHC(O)Br} \quad (57\%)$$

$$\mathrm{(CF_3)_2CHCHO + BrCl} \xrightarrow{\mathrm{CH_3CO_2H}} \mathrm{(CF_3)_2CBrCHO} \quad (48\%)$$

This selective halogenation work provides a facile route to either α-halopolyfluorinated aldehydes or polyfluorinated acyl halides, and demonstrates the effect of CF_3 groups on the formation of free-radical sites at an adjacent carbon atom.

Inouye developed a simple *in situ* preparation of $\mathrm{Ph_3\overset{+}{P}{-}\overset{-}{C}(CF_3)_2}$ from tetrakis(trifluoromethyl)-1,3-diethetane, which is readily prepared from hexafluoropropene. *In situ* capture of this ylide with aldehydes gave good yields of the bis(trifluoromethyl)olefins [66]. Ketones do not capture this ylide.

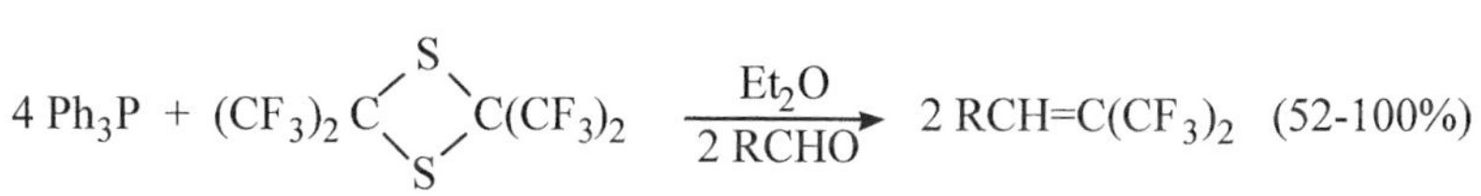

Organometallic chemistry

Our initial ventures in organometallic chemistry were quite straightforward and provided useful but not especially novel results. In the early 1970s, Jerry Hahnfeld explored some fluorinated vinyl-lithium reagents and demonstrated that these retained their configuration – contrary to an earlier report in the literature [67]. As part of our interest at that time, he also explored the low-temperature preparation and utility of trihalomethyl-lithium reagents such as $CFCl_2Li$ and $CFBr_2Li$ as fluorochloro- and fluorobromo-carbene precursors [68]. Hahnfeld also

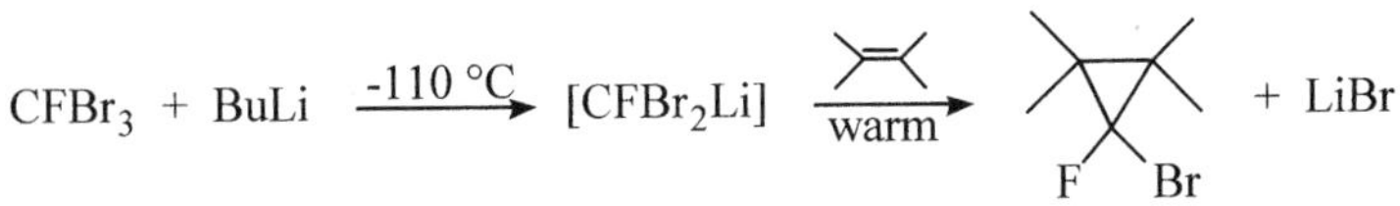

developed a photochemical route to fluorocarbene from di-iodofluoromethane [69].

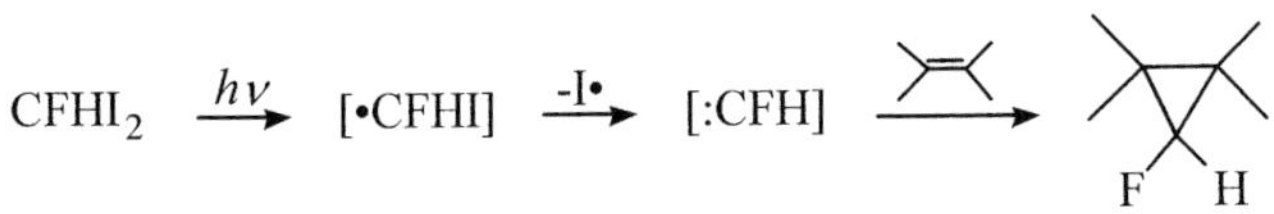

Perfluoroalkyl, perfluoroallyl and perfluoroaryl organometallics

Our interest in carbene chemistry [21] and the frustration of dealing with exotherms during scale-up of recipes involving low-temperature lithium reagents provided the impetus for our more novel work on organometallic reagents. From our previous ylide work in which we utilized phosphonium salts as difluorocarbene precursors, Jack Kesling developed a new route to trifluoromethyl derivatives of mercury and tin [28, 45]. He reasoned that the fluoride ion utilized in carbene generation (see p. 85) could reversibly trap $[:CF_2]$ to form trifluoromethide, hence if decomposition of the phosphonium salt was triggered with F^- in the presence of an electrophile rather than a nucleophilic olefin, capture of $[CF_3^-]$ would produce a trifluoromethylated product:

$$[Ph_3P^+CF_2Br]Br^- + KF \longrightarrow Ph_3PFBr + [CF_2Br]^- \longrightarrow [:CF_2] + Br^-$$
$$[:CF_2] + F^- \rightleftharpoons [CF_3^-]$$

The two electrophilic reagents successfully utilized by Kesling were phenylmercuric chloride and chlorotrimethyltin:

$$[Ph_3P^+CF_2Br]Br^- + KF + PhHgCl \longrightarrow PhHgCF_3$$
$$[Ph_3P^+CF_2Br]Br^- + KF + Me_3SnCl \longrightarrow Me_3SnCF_3$$

This important advance by Kesling provided the impetus for later work on the preparation of trifluoromethyl zinc, cadmium and copper reagents.

As part of our phosphonate chemistry, Ryutaro Takei and Shin-Ya prepared a *stable* cadmium reagent from diethyl (bromodifluoromethyl)phosphonate:

$$Cd + (EtO_2)P(O)CF_2Br \longrightarrow [(EtO)_2P(O)CF_2CdBr]$$

We chose to explore cadmium for two reasons: (i) we were not aware of the use of cadmium in any coupling (dimerization) reactions; and (ii) we needed a way to demonstrate the

$$M + CF_2X_2 \longrightarrow M^+[CF_2X_2]^{\dot{-}} \longrightarrow X^- + M^{2+} + [CF_2X^-]$$

$$[CF_2X^-] \rightleftharpoons [:CF_2] + X^-$$

$$[:CF_2] + Me_2NCH{=}O \longrightarrow CO + Me_2NCHF_2 \rightleftharpoons [Me_2\overset{+}{N}{=}CHF]F^-$$

$$F^- + [:CF_2] \rightleftharpoons [CF_3^-] \xrightarrow{MX_2} CF_3MX + (CF_3)_2M$$

Scheme 6.12.

formation (in solution) of the organometallic reagent, being well aware (from literature precedent) that cadmium NMR could assist our structural assignments [70].

In later work, Takashi Ishihara and Masamichi Maruta [71] prepared the analogous zinc reagent: $(EtO)_2P(O)CF_2Br + Zn \longrightarrow [(EtO)_2P(O)CF_2ZnBr]$. In the course of investigating the chemistry of this reagent, we often encountered an unknown singlet in the ^{19}F NMR spectrum at *ca.* 44 ppm upfield from $CFCl_3$. Denise Wiemers proposed that this signal could be a trifluoromethylzinc reagent. Hence we devised an experiment to generate difluorocarbene from dihalodifluoromethanes with concomitant formation of a metal halide, arguing that if the *in situ* generated $[:CF_2]$ could be captured by fluoride ion, trifluoromethide would become available to attack the *in situ* generated metal halide. Thus was born our simple preparation of trifluoromethyl cadmium and zinc reagents from dihalodifluoromethanes [72], *i.e.*

$$2M + 2CF_2X_2 \xrightarrow{DMF} CF_3MX + MX_2 + CO + [Me_2N^+{=}CFH]X^-$$
$$(M = Cd, Zn; \quad X = Br, Cl) \quad (80\text{-}95\%)$$

DMF is not a neutral partner in this reaction: it serves as both the solvent and the reactant needed to produce the fluoride ion required for trifluoromethide formation from difluorocarbene produced via an electron transfer route (Scheme 6.12). Detailed mechanistic experiments provided evidence for the electron transfer step, the formation of $[:CF_2]$, CO and Me_2NCHF_2 [72, 73].

Thus, the earlier work of Kesling, Takei, Shin-Ya, Ishihara and Maruta culminated in the development of this novel route to trifluoromethylated organometallic reagents [72, 74].

In subsequent work, Wiemers utilized her reagents, particularly the cadmium reagent, in

$$[CF_3MX + (CF_3)_2M] \xrightarrow{CuY} [CF_3Cu] \qquad (90\text{-}100\%)$$
$$Y = I, Br, Cl, CN$$

exchange reactions with Cu(I) salts to achieve the first pre-generation and spectroscopic detection of trifluoromethylcopper [75]. Using dihalodifluoromethanes, $[CF_3Cu]$ could be pre-generated via a one-pot reaction:

$$2M + 2CF_2X_2 \xrightarrow{DMF} [CF_3MX + (CF_3)_2M] \xrightarrow{CuY} [CF_3Cu]$$

Several interesting observations were made during this work. Firstly, following the exchange reaction with Cu(I)Y, we detected three singlets in the ^{19}F NMR spectrum of the product – suggesting the presence in the DMF solution of three different copper reagents! Obviously, we were surprised and puzzled at this observation. When oxygen was vigorously excluded, one of these signals disappeared, suggesting that it belonged to an oxidation product of $[CF_3Cu]$. Later work by Monica Willert-Porada demonstrated that the oxidation product was in fact a perfluoroalkyl copper(III) species, $[(CF_3)_4Cu^-]$, [76]. Subsequent X-ray work by Nauman confirmed this. Secondly, when the initially formed $[CF_3Cu]$/DMF solution was stored overnight, the reagent was converted to $[CF_3CF_2Cu]$ – an observation made earlier by Kobayashi in the preparation of $[CF_3Cu]$ from CF_3I [77]. However, when HMPA was added to stabilize the $[CF_3Cu]$ reagent, coupling reactions with aryl iodides could be effected at 70 °C without formation of trifluoroethylcopper, *e.g.*

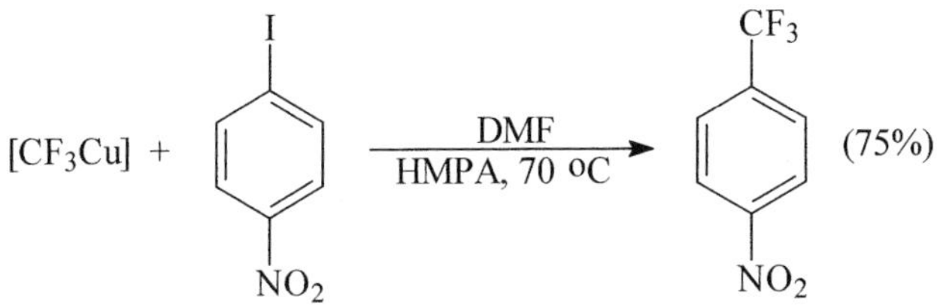

Based on related work by Willert-Porada and Haridassan Nair, the three ^{19}F signals mentioned above were assigned to the species $CF_3Cu \cdot L$ (L = metal halide), $CdI^+[(CF_3)_2Cu]^-$ and $CdI^+[(CF_3)_4Cu]^-$ [78, 79].

The spontaneous conversion of $[CF_3Cu]$ to $[CF_3CF_2Cu]$ in DMF at room temperature accelerated at higher temperatures, and surprisingly the oligomerization did not stop at the perfluoroethylcopper stage. Thus at 85-95 °C (in DMF), chains of 14 to 16 carbons were formed, all possible values of n (to 14) being found – *both* odd *and* even values. Hence chain growth

$$[CF_3Cu] \xrightarrow[85\text{-}95\ ^\circ C]{DMF} [CF_3(CF_2)_nCF_2Cu], \quad n = 0 \text{ to } 14$$

must occur by insertion of CF_2 units, not by dimerization of $[:CF_2]$ to $F_2C{=}CF_2$ (TFE) followed by oligomerization of TFE with $[CF_3Cu]$. This chain-extension process is not unique to trifluoromethyl copper; for example, $(EtO)_2P(O)CF_2Cu$ undergoes a similar process when heated [79]:

$$[(EtO)_2P(O)CF_2ZnBr] \xrightarrow[DMF]{CuX} [(EtO)_2P(O)CF_2Cu] \xrightarrow{warm} (EtO)_2P(O)(CF_2)_nCu, \quad n = 2, 3$$

This initial work by Wiemers on the oligomerization of $[CF_3Cu]$ was subsequently taken up by Jerry Easdon, who also prepared a similar oligomeric copper reagent directly from dibromodifluoromethane or bromochlorodifluoromethane and Cu^0 [80]: $CF_2XY + Cu^0$ (X, Y = Br, Cl) (in DMF at 85-95 °C) → $[CF_3(CF_2)_nCF_2Cu]$, where $n = 0$ to 14. The distribution of oligomeric chains was determined by coupling of the oligomeric copper reagent with iodobenzene and subsequent analysis of the (perfluoroalkyl)benzenes formed by GLPC and mass spectrometry. This unprecedented oligomerization process complimented the oligomerization of TFE as an entry to longer chain perfluoroalkyl derivatives, since the copper reagent is easily functionalized. TFE oligomerization reactions can give only even numbered products, whereas the copper reagent gives both odd and even products. The University of Iowa patented this novel process [81], which delayed its announcement, although it was revealed in several conference presentations. Easdon also developed a protocol for the preparation of trifluoromethylated aromatics via direct reaction of dihalodifluoromethanes, copper metal, and aryl iodides [80]. Initial experiments indicated that oligomerization of the initially formed [CF_3Cu] competed with coupling of trifluoromethylcopper with iodobenzene, as perfluoroalkyl chains from 0 to 9 carbons were detected in the product. After a search for an additive which would effectively suppress the rate of oligomerization relative to the rate of coupling of trifluoromethylcopper with the iodoaromatic, Easdon found that fluoride ion worked best. Thus, when coupling was carried out in the presence of KF or CsF, 95% of the coupled product was the trifluoromethyl derivative [80, 81]:

$$CF_2BrCl + Cu^0 + C_6H_5I + KF \xrightarrow[85\text{-}95\ ^\circ C]{DMF} \underset{95}{C_6H_5CF_3} + \underset{:\ 5}{C_6H_5CF_2CF_3}$$

Since the initial discovery of the formation of [CF_3M], where M = Cd, Zn or Cu, from difluorodihalomethanes and the elucidation of the oligomerization of the copper derivative mechanism by Wiemers, several other workers have prepared [CF_3Cu] by utilizing a different source of [$:CF_2$] [82-84]. The seminal work of Wiemers led the way.

The ease with which [CF_3M] oligomerized suggested that it might be employed as a [CF_2] transfer agent. Work by Zhen-Yu Yang and Wiemers [85] demonstrated that in a competition between pentafluorophenylcopper and trifluoromethylcopper, selective transfer (insertion) of [CF_2] occurred with pentafluorophenylcopper at low temperature:

$$C_6F_5Cu + 2CF_3Cu \xrightarrow{-30\ ^\circ C\ \text{to RT}} C_6F_5CF_2CF_2Cu \qquad (70\text{-}80\%)$$

The perfluorobenzylcopper presumably forms first but is more reactive than C_6F_5Cu where [CF_2] insertion is concerned. This perfluoro(phenethyl)copper could also be directly prepared from the trifluoromethyl cadmium reagent [85]. Interestingly, multiple insertions of [CF_2] units into $C_6F_5CF_2CF_2Cu$ were not observed, even when a large excess of [CF_3Cu] was utilized at room temperature. Perfluoro(phenethyl)copper is readily functionalized with allyl halides and halogens, couples smoothly with vinyl halides and iodoaromatics, and readily undergoes SO_2 insertion at the C−Cu bond [85].

Our success with trifluoromethylcadmium halides as synthetic reagents prompted us to explore other compounds of this class. Pam Heinze studied the direct preparation of perfluoroalkyl cadmiums from perfluoroalkyl iodides [86], in DMF at room temperature: $R_FI + Cd \rightarrow R_FCdI + (R_F)_2Cd$; yields were good to excellent when $R_F = CF_3$, C_2F_5, or n-C_3F_7 and moderate for higher homologues. The reaction also worked well with bromopentafluorobenzene. The reaction was later exploited by Kathy MacNeil, who generated pentafluorophenylcopper via copper(I) halide exchange with the [C_6F_5CdBr] produced [87].

Greg Hartgraves prepared difluoromethylcadmium via similar methodology [88] and used ^{113}Cd NMR to characterize the mono and bis reagents formed:

$$CH_2FBr + 2Cd \xrightarrow[50\ °C]{DMF} CH_2FCdBr + (CH_2F)_2Cd + CdBr_2 \qquad (65\text{-}75\%)$$

These difluoromethylcadmium reagents react with Cu(I) bromide to give difluoromethylcopper [89]; both types of reagent and perfluoroalkylcopper species were used by Hartgraves to synthesize a range of fluoroalkylated allenes [90, 91], *e.g.*

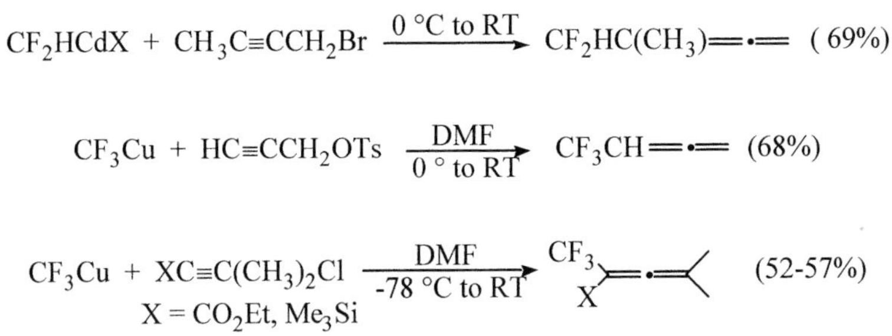

Heinze, with Yasuo Tarumi, used perfluoroallyl cadmium and copper reagents from $CF_2{=}CFCF_2I$ [92], while Yang and MacNeil [93] and Ba Van Nguyen [94] extended the work to the aromatic series (Scheme 6.13).

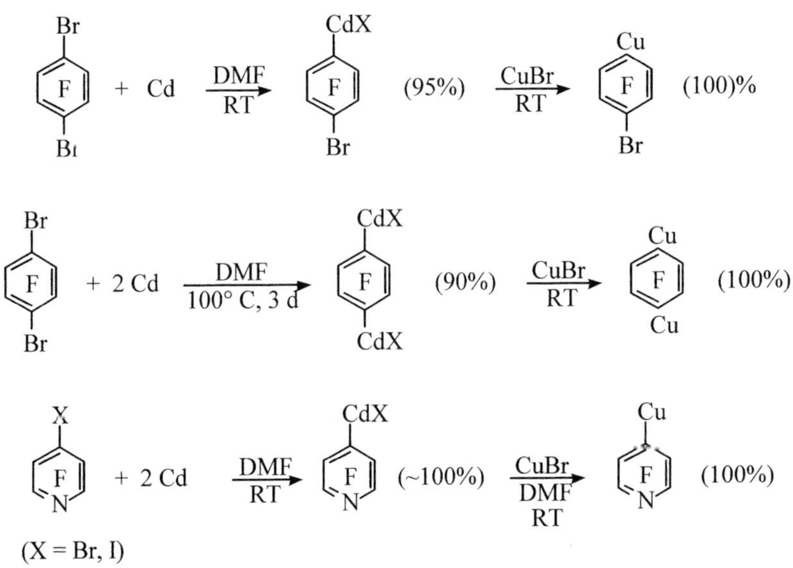

Scheme 6.13.

Perfluorovinyl zinc, cadmium and copper reagents

Steve Hansen and Terry Spawn developed a general, direct stereospecific preparation of *stable* fluorinated vinylzinc reagents from the corresponding fluorinated vinyl iodides or bromides and zinc metal [95]. Solvents such as DMF, THF, DMAC, CH_3CN and glymes were employed.

$$CF_2{=}CFX + Zn + \xrightarrow[\geqslant RT]{\text{solvent}} CF_2{=}CFZnX + (F_2C{=}CF)_2Zn + ZnX_2 \qquad (72\text{-}97\%)$$

(X = Br, I)

$$(Z)\text{-}CF_3(CF_2)_4CF{=}CFBr \xrightarrow[\text{DMF, 60 °C}]{Zn} (Z)\text{-}CF_3(CF_2)_4CF{=}CFZnBr \qquad (77\%)$$

$$F_2C{=}CBr_2 + Zn \xrightarrow[RT]{DMF} F_2C{=}CBrZnBr \qquad (97\%)$$

The corresponding vinylcadmium reagents were prepared by Hansen using similar methodology [96]. These vinylcadmium reagents exhibit excellent thermal stability, and we have isolated $(CF_2C{=}CF)_2Cd$ and $(Z)\text{-}(CF_3CF{=}CF)_2Cd$ as DMF or triglyme solvates via distillation [97].

$$F_2C{=}CFI + Cd \xrightarrow[RT]{DMF} F_2C{=}CFCdI + (F_2C{=}CF)_2Cd + CdX_2$$

The bis-cadmium species are low-melting, moisture sensitive white solids. The fluorovinyl zinc and cadmium reagents are very useful synthons. Spawn, for example, developed a high yield, general route to trifluorovinyl ketones via the copper(I) mediated acylation of trifluorovinylzinc reagents [98]; these ketones are excellent Michael acceptors.

$$[F_2C{=}CFZnBr] + RC(O)Cl \xrightarrow[CuBr]{\text{glyme}} F_2C{=}CFC(O)R$$

Although numerous literature reports describe attempts to pregenerate fluorinated vinyl copper reagents, success was not achieved until Hansen demonstrated that a copper(I) metathesis reaction of the corresponding vinyl cadmium and/or zinc reagents proceeds stereospecifically to give *stable* polyfluorinated vinylcopper reagents in excellent yields [99]. His work provided the first *unequivocal* route to fluorinated vinylcopper reagents.

$$F_2C{=}CFM \xrightarrow[DMF]{CuBr} [F_2C{=}CFCu] \qquad (M = Zn, 72\%; M = Cd, 99\%)$$

$$(Z)\text{-}CF_3CF{=}CFCdX \xrightarrow[DMF]{CuBr} (Z)\text{-}CF_3CF{=}CFCu \qquad (92\%)$$

$$\underset{E/Z = 90/10}{CF_3CF{=}C(Ph)CF{=}CFZnX} \xrightarrow[RT]{CuBr} \underset{E/Z = 90/10}{CF_3CF{=}C(Ph)CF{=}CFCu} \qquad (63\%)$$

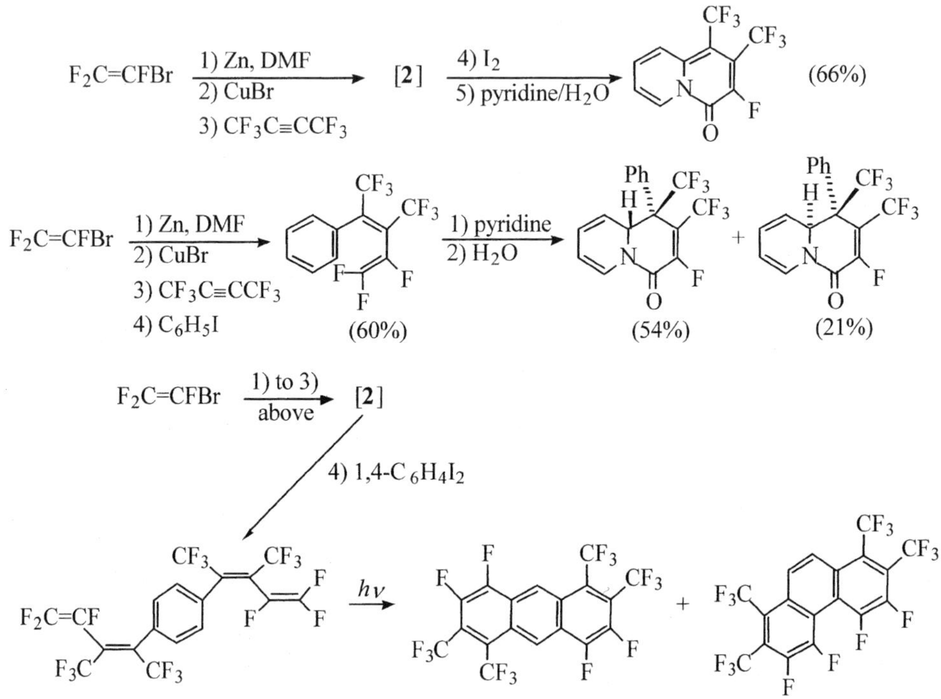

Scheme 6.14.

Subsequent work by Hansen demonstrated that these vinyl copper reagents add stereospecifically to perfluoro-2-butyne [97], *e.g.*

$$[F_2C{=}CFCu] + CF_3C{\equiv}CCF_3 \xrightarrow{DMF}$$

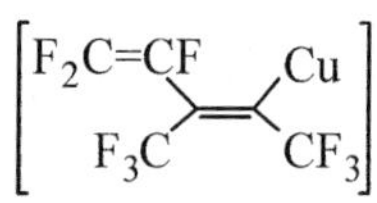

(**2**)

The synthetic utility of the prototypical *syn* adduct **2** is exemplified in Scheme 6.14 (one-pot techniques are used). These few examples demonstrate the wealth of chemistry now available via these copper reagents. As expected, these fluorinated vinylcopper reagents are readily acylated, allylated, alkylated, benzylated and stereospecifically coupled with vinyl halides, as demonstrated by Hansen [99-101]:

$$(Z)\text{-}CF_3CF{=}CFCu + (E)\text{-}CF_3C(Ph){=}CFI \longrightarrow (E,Z)\text{-}CF_3CF{=}CFCF{=}C(Ph)CF_3 \quad (54\%)$$

In contrast to the stability of the perfluorinated vinyl zinc and copper reagents, those containing an α-halogen other than fluorine exhibit diminished stability, especially the copper reagents. Thus, when $CF_3C(Ph){=}CBrZnX$ (stable) was treated with Cu(I)Br, Peter Morken found

$F_2C{=}CFCu$	$F_2C{=}CBrCu$	$CF_3C(Ph){=}CFCu$	$CF_3C(Ph){=}CBrCu$
stable	unstable	stable	unstable

that the product immediately dimerized to a mixture of (*E*)- and (*Z*)-butatrienes [102-104] which are readily separated by silica gel chromatography:

$$CF_3C(Ph){=}CBr_2 \xrightarrow[\text{2) CuBr}]{\text{1) Zn/DMF}} \underset{E/Z}{CF_3C(Ph){=}C{=}C{=}C(Ph)CF_3} \qquad (72\%)$$

Morken also demonstrated that zinc dehalogenation of the isomeric trienes provides a convenient entry to divinyl acetylenes derived from $C_2F_5C(Ph){=}CBr_2$ and its higher R_F homologues [105, 106], *e.g*

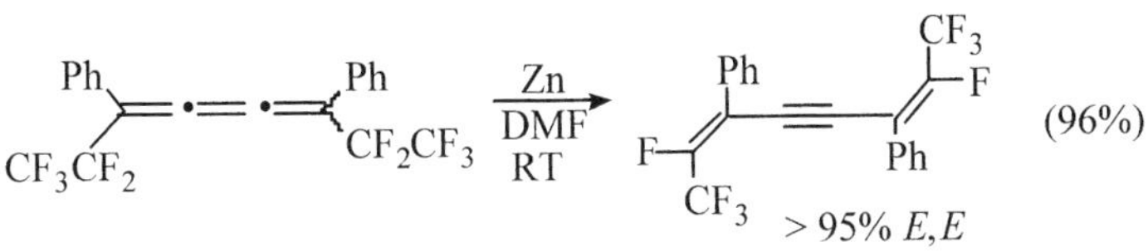

Heinze recognized that the perfluorovinylzinc reagents could be utilized in palladium-catalyzed coupling sequences with aryl iodides and vinyl halides, and she developed this methodology to provide a useful, mild route to trifluorovinyl styrenes [107, 108]:

$$\underset{(X = Br, I)}{F_2C{=}CFX + Zn} \xrightarrow[\text{RT}]{\text{DMF}} [F_2C{=}CFZnX] \xrightarrow[\text{Pd(PPh}_3)_4,\ 60\text{-}80\ ^\circ\text{C}]{\text{ArI}} F_2C{=}CFAr \qquad (61\text{-}81\%)$$

A wide variety of functional groups are tolerated in this approach, and the mild conditions prevent cyclodimerization of the styrene product. This route to (trifluorovinyl)benzenes has received wide acceptance and has been utilized industrially on a large scale. With longer-chain perfluorovinylzinc reagents, the coupling reaction proceeds stereospecifically, *e.g.* [108-110]

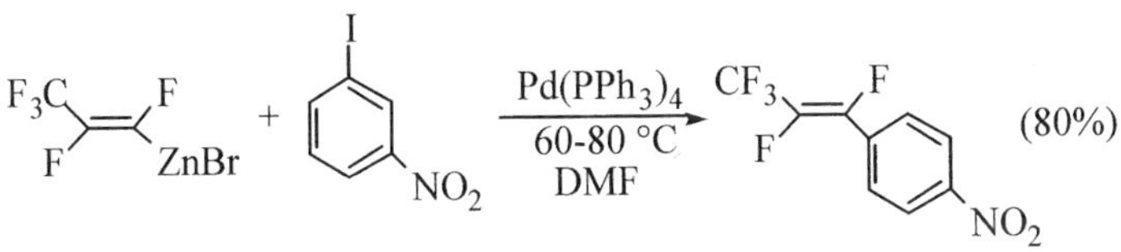

This approach was extended by Nguyen for the facile preparation of β,β-difluorostyrenes and (2,2,2-trifluoroethyl)benzenes from $F_2C{=}CHI$ [111]:

$$F_2C{=}CHI \xrightarrow[\text{DMF}]{\text{Zn}} [F_2C{=}CHZnI] \xrightarrow[\text{Pd(PPh}_3)_4,\ \text{DMF/60}\ ^\circ\text{C}]{\text{ArI}} F_2C{=}CHAr \xrightarrow[\text{DMSO}]{\text{KF/H}_2\text{O}} CF_3CH_2Ar$$

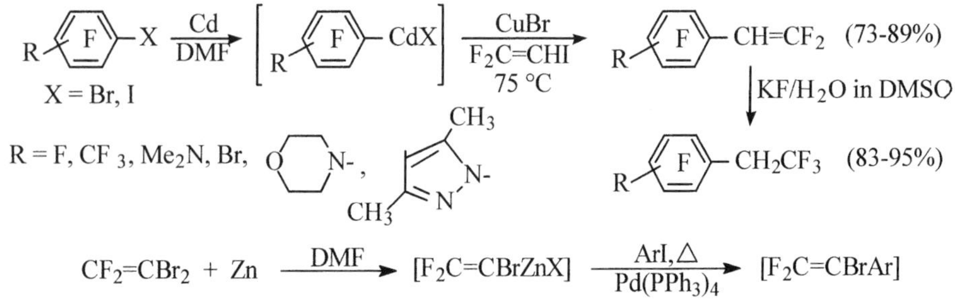

Scheme 6.15.

The corresponding bromoethylene is not suitable for conversion to the zinc reagent: using $CF_2{=}CHBr$, Nguyen discovered one of the first examples of an acid-base reaction during vinylzinc formation, *i.e.*

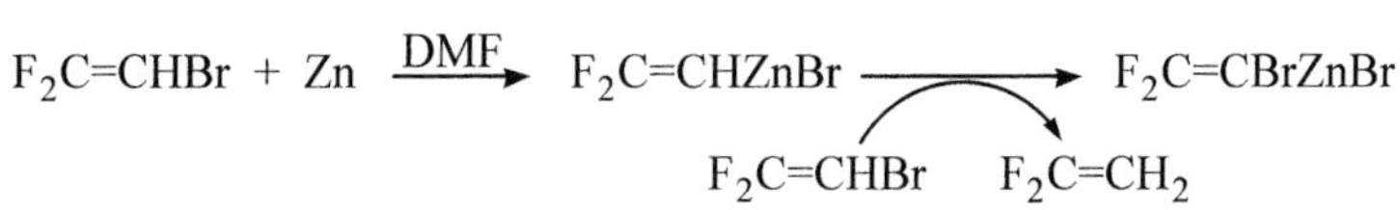

Useful extensions of this work are outlined in Scheme 6.15 [111, 112].

Charles Davis developed a stereospecific route to (*Z*)-α,β-difluorostyrenes via the vinylzinc/palladium coupling methodology. The requisite zinc reagent was prepared from (*E*)-$HFC{=}CFSiEt_3$, which was prepared by a novel photochemical isomerization developed by

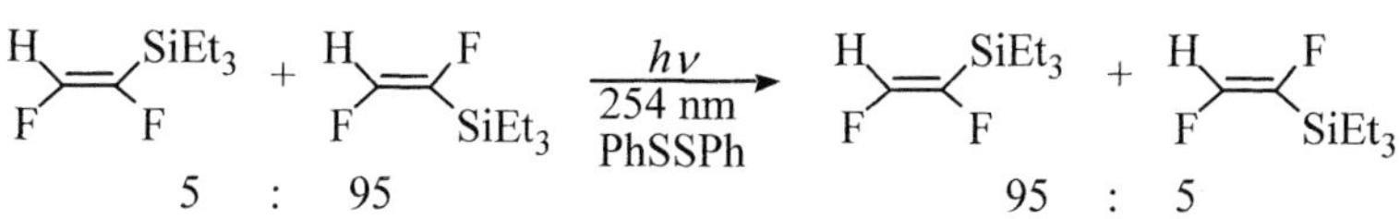

Simonetta Fontana and Davis in our laboratory [113]. This work provided the first general route to *cis*-1,2-difluoro-olefin precursors. On a large scale, the *E*/*Z*-vinyl silanes can be separated by fractional distillation; iododesilylation then provides the requisite (*E*)-vinyl iodide for zinc reagent formation [114]:

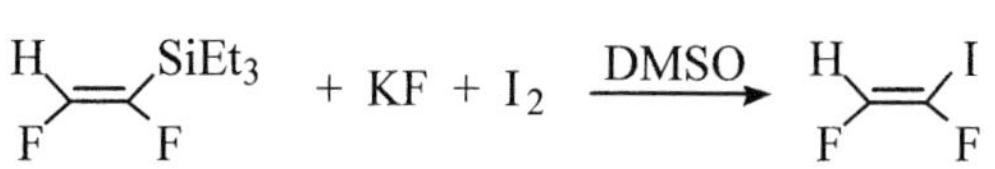

Subsequent formation of the zinc reagent followed by palladium catalyzed coupling with aryl iodides stereoselectively gave the (*Z*)-α,β-difluorostyrenes [115]. The styrenes can be readily

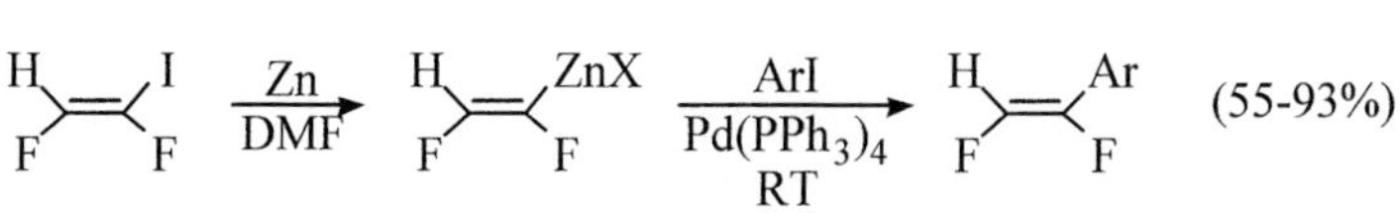

converted to (*E*)-α,β-difluoro-β-iodostyrenes or the analogous vinylstannanes, useful precursors for elaboration into other *cis* analogues.

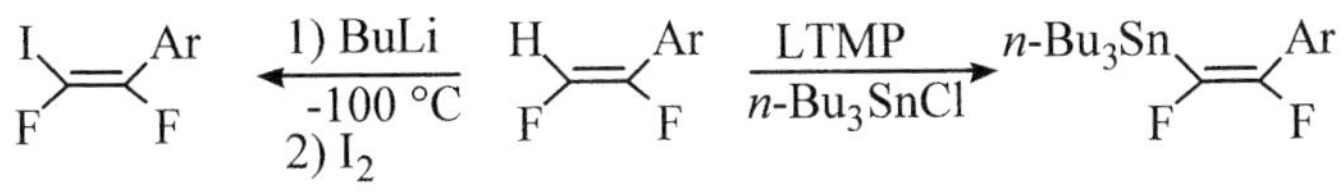

One of the most useful synthetic transformations in our laboratory was developed by Ling Xue, who demonstrated that 1,2-difluorovinylsilanes can be stereospecifically converted into the corresponding 1,2-difluorovinyl stannanes [116, 117]:

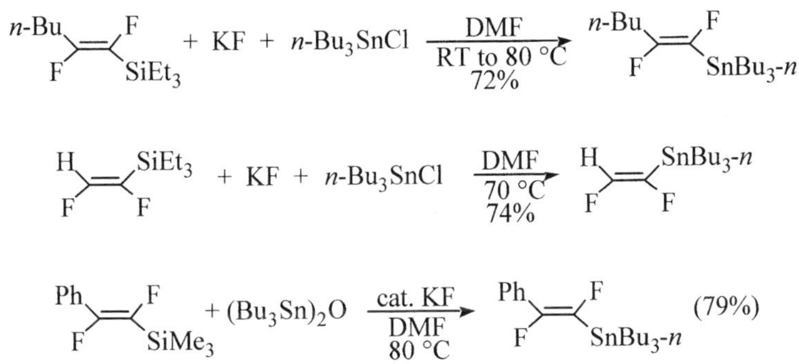

This work readily permits the preparation of fluorinated vinylstannanes from the corresponding fluorinated vinylsilanes, which are readily prepared from trifluorovinylsilanes via addition-elimination reactions:

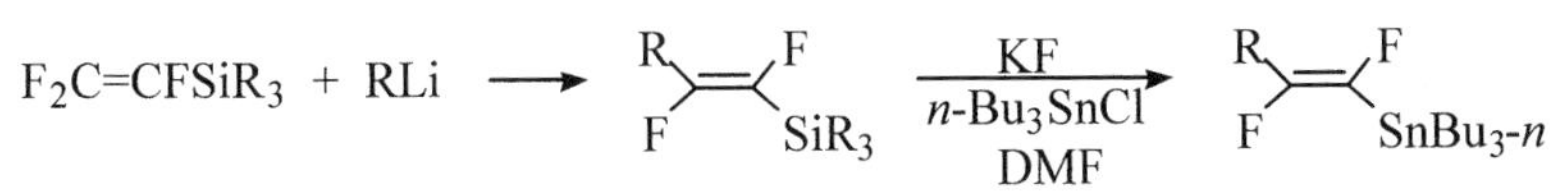

The corresponding vinylstannanes cannot be easily prepared via similar methodology, since the stannane group easily exchanges with the alkyl lithium reagent: $F_2C{=}CFSnR_3 + R'Li \rightarrow [F_2C{=}CFLi] + R_3SnR'$. This approach permits the facile preparation of almost any (*E*)- or (*Z*)-fluorinated vinylstannane, which readily participates in palladium-catalyzed coupling reactions under Liebeskind conditions, as demonstrated by Long Lu [118], *e.g.*

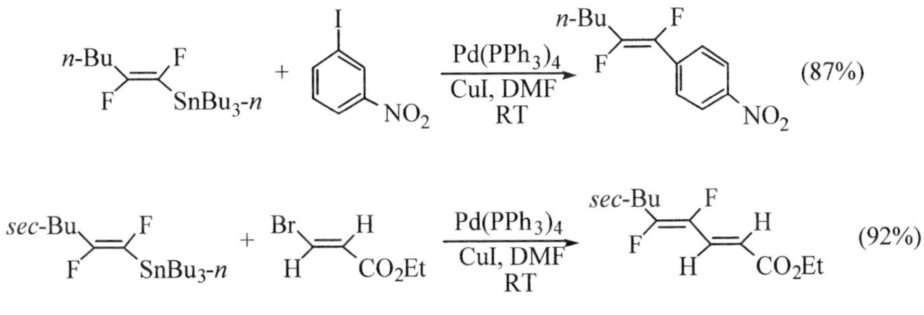

We believe that this silane/stannane conversion discovered in our laboratory and its extensions will become a general stereospecific entry to fluorinated dienes, trienes and polyenes.

The vinylcopper chemistry and the palladium-catalyzed coupling chemistry of fluorinated vinylzinc reagents in our laboratory led to a fruitful collaboration with Professor Bill Dolbier's group at the University of Florida. In earlier work of Guan-Su Shaw in our laboratory, we had noted that the ring closure and ring opening of fluorinated dienes and fluorinated cyclobutenes did not appear to follow the usual selection rules promulgated for the analogous hydrocarbon analogues. When this work was presented at a Winter ACS Fluorine Conference in 1983, a serious and protracted discussion on the interpretation of our results ensued [119]. Current physical-organic orbital symmetry control arguments (in 1983) regarding the ring opening of *cis*- and *trans*-1,2,3,4-tetramethylcyclobutenes attributed the direction of ring opening to probable repulsive steric interactions that would be present in the transition state due to the two methyl groups rotating simultaneously inwards. This argument had been uniformly applied in the interpretation of virtually all electrocyclic ring-opening reactions of hydrocarbon cyclobutenes. In order to resolve the argument produced in our presentation, Bill graciously offered to collaborate with us on this problem. We carried out the synthesis of model dienes and Bill and his co-workers carried out the gas-phase reactions. This collaboration resulted in several seminal papers on this topic [120 - 122] and unequivocally demonstrated that trifluoromethyl groups (CF_3) in the 3,4-position of cyclobutenes preferentially moved inwards on ring-opening:

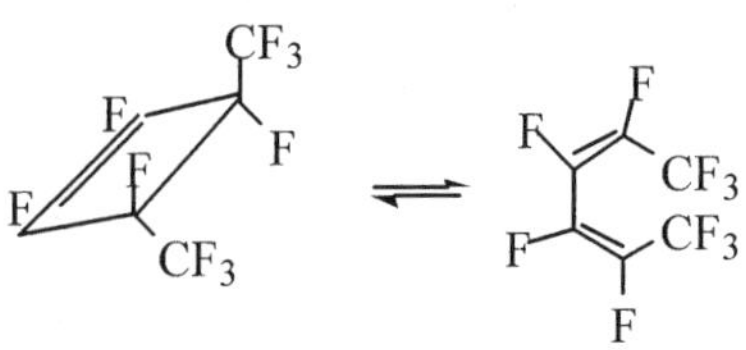

The cyclobutene opens to the *Z,Z*-diene with an activation energy ~18 kcal *less* than when it ring opens to the *E,E*-diene. Similar results were found with the (*E*)- and (*Z*)-1,3-pentadiene system. It was found that the (*Z*)-diene underwent cyclization much more readily than the (*E*)-diene.

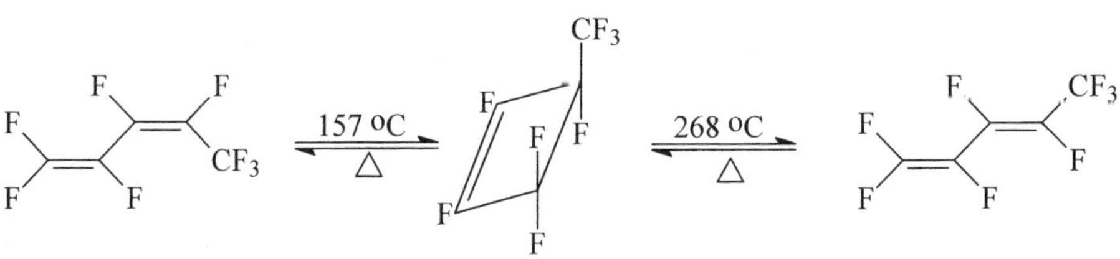

It was gratifying to see that our copper and zinc chemistry not only contributed to useful and novel synthetic methodology but also to the advancement of mechanistic organic chemistry. The role of Bill Dolbier was crucial in this work and contributed to its ultimate success. The collaborative effort was indeed greater than the sum of its parts!

I would be remiss if I did not comment on the work of MacNeil and its contribution to our more recent ventures. She combined the earlier work of Heinze and Hansen to prepare [C_6F_5Cu] from [C_6F_5CdX] and studied its *syn* addition to fluorinated alkynes [87]. The resultant adduct was readily functionalized with electrophiles, such as H^+, I_2, allylic halides, alkyl halides, aryl halides, vinyl halides, and acyl halides. With an unsymmetrical alkyne, regiospecific addition was observed:

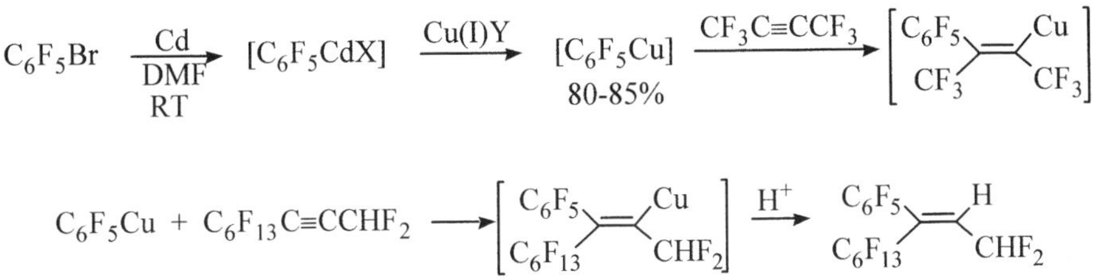

Fluoride ion-catalyzed isomerization of the acylated product, and PhSSPh/*hν* catalyzed the isomerization of the protonated copper reagent. In a few cases, the initial vinylcopper adduct could undergo a second *syn* addition to a second equivalent of perfluoro-2-butyne [87, 123], *e.g.*

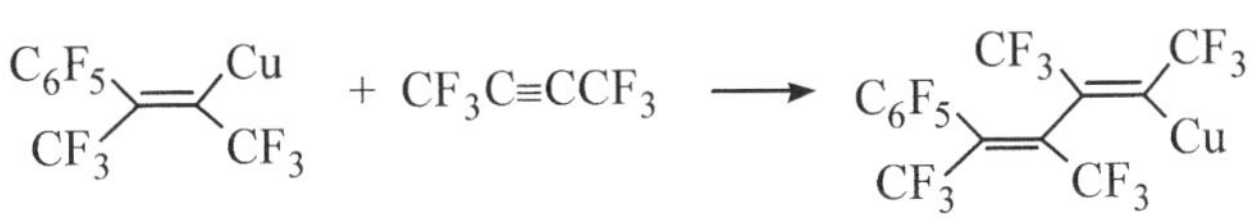

When the product from coupling with perfluorovinyl iodides was treated with fluoride ion, sequential isomerization occurred; initially at the C_6F_5 terminus and then at the perfluorovinyl terminus to give the *cis* propenyl group:

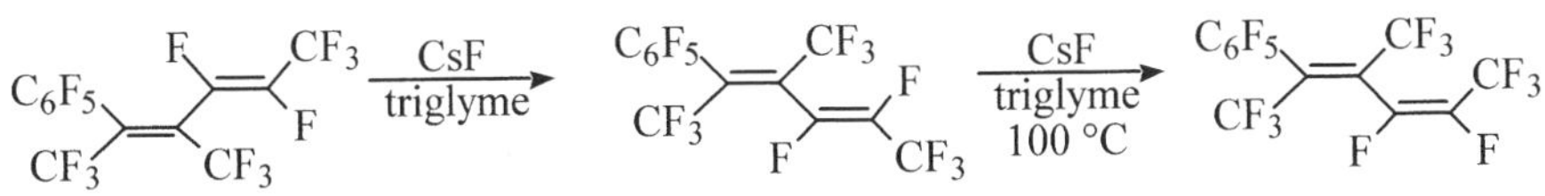

This isomerization reaction challenged our previous assumption of the thermodynamic stability of *cis*- and *trans*-fluoro-olefinic groups and prompted current investigations in our research group on the thermodynamic stability of 1,2-difluoro-olefinic groups [124].

Enynes and alkynes

Conjugated enynes provide an efficient method for the assembly of multifunctional molecules and one would anticipate that fluorine-containing enynes would be useful building blocks to partially fluorinated multifunctional molecules or partially fluorinated natural products. Our interest in the generality of palladium-catalyzed coupling reactions of fluorinated vinyl and aryl iodides directed us to investigate the utility of these precursors in enyne and/or alkyne preparations. Yang developed an efficient, mild, stereospecific route to fluorinated enynes via coupling of fluorinated vinyl iodides with terminal alkynes [125-127]:

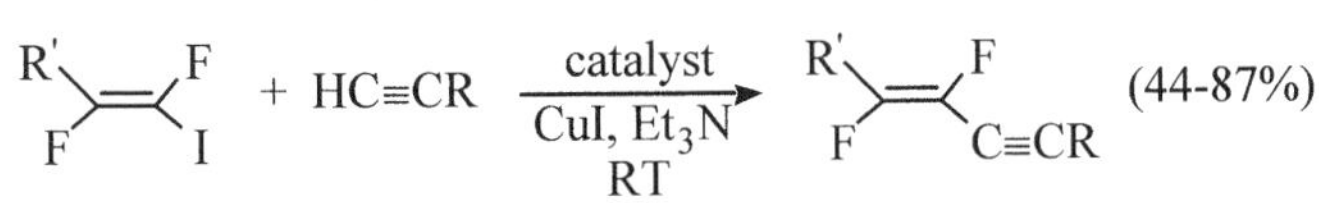

Catalysts, such as $Pd(PPh_3)_4$, $Pd(PPh_3)_2Cl_2$ and $Pd(PPh_3)_2(OAc)_2$ were found to be equally effective; and the R′ group in the vinyl iodide can be F, perfluoroalkyl, aryl, or $(RO)_2P(O)$ while the R group in the alkyne can be aryl, alkyl, Me_3Si, or RC≡C. Generally, the reactions were carried out in an excess of Et_3N as solvent, but DMF, HMPA, CH_3CN, dioxane, benzene and hexane can also be utilized, *e.g.*

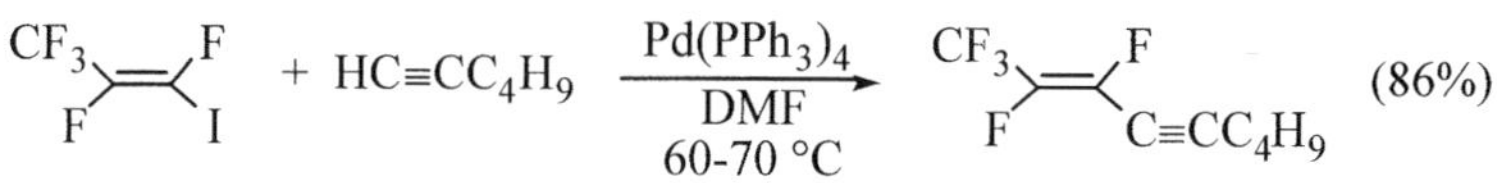

Similar palladium-catalyzed coupling of 4-substituted tetrafluorophenylhalides with 1-alkynes provides a facile route to functionalized fluorinated phenyl alkynes [128, 129]. Functional groups, such as alkyl, alkenyl, phenyl, vinyl ether, alkoxy, phenoxy, hydroxy, amine, trimethylsilyl and cyano are tolerated. With dialkynes, bis(fluorophenyl)dialkynes are formed in good yields. With 1,4-dibromotetrafluorobenzene, the bis-alkynlated tetrafluorobenzenes are obtained in good yields. Nguyen and Yang developed this methodology into a practical approach to fluorinated phenyl functionalized alkynes.

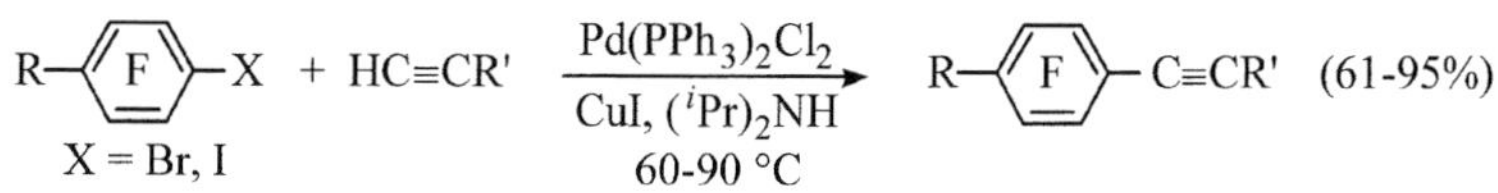

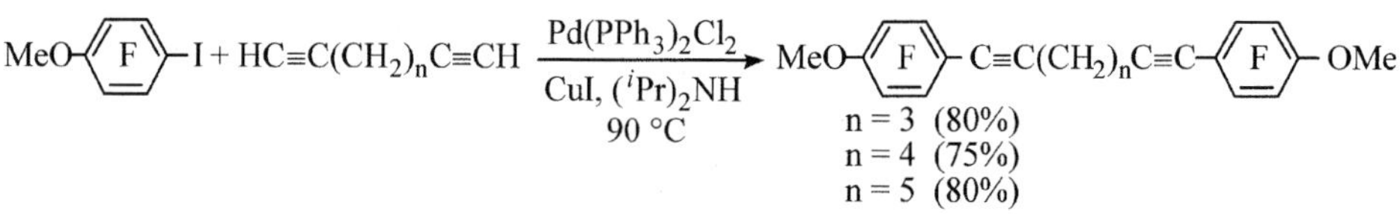

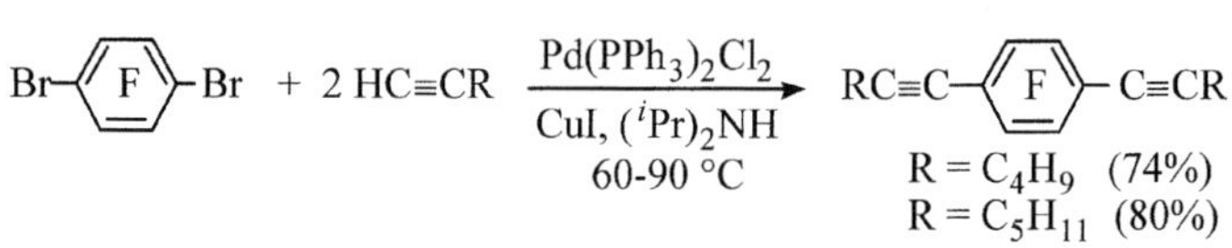

Our initial venture into organometallic chemistry was a diversion to circumvent the frustration of dealing with the low stability of fluorinated vinyl-lithium reagents. This diversion eventually took on a life of its own and guided us into unknown territory (at least to us), such as zinc, copper, and cadmium chemistry and eventually into palladium catalysis chemistry. It has been an enjoyable and delightful excursion, no doubt fuelled by the creative work of numerous students and postdocs. Our work in this area continues even today with new studies on palladium-catalyzed reactions involving fluorinated stannanes, polyenes and phosphonates.

Phosphonate chemistry

The chemical literature in the mid-1970s was almost devoid of examples of fluoroalkyl-phosphonates, especially halo-*F*-methyl derivatives. Reviews on phosphorus chemistry in that era noted that fluoromethanes such as CF_3Cl, CF_3I, and $CFCl_3$ were inert to trialkylphosphites. Of course, the premise for the Michaelis-Arbuzov reactions at that time was based (mechanistically) on a succession of S_N2 events. Organofluorine chemists were cognizant of the lack of activity of CF_3Cl, CF_3I, and $CFCl_3$ in S_N2 processes, hence the failure of the Michaelis-Arbuzov reaction with these methanes was not totally unexpected.

Our interest in phosphonate chemistry was stimulated by our earlier work with difluoromethylene ylides. The key to formation of the halofluorophosphonium salts was halophilic attack by the tertiary phosphine on the halogen atom of a halomethane with concomitant formation of a fluorinated methide ion (or carbene). Our ylide work with methanes, such as CF_2Br_2, $CFBr_3$, $CFCl_3$, and CF_2BrI convinced us that the earlier reported failures were dictated by improper choice of the methane precursor. Since we were convinced that fluorinated phosphonates would be useful building blocks, we decided to pursue the preparation of this important, yet unexplored, class of phosphorus derivatives.

Richard Flynn initially chose to investigate halofluoromethanes that we knew were suceptible to halophilic attack by phosphorus (tertiary phosphines). We anticipated that the halofluoromethide ion would be produced and lead to the corresponding fluorinated phosphonate. In retrospect, this study was more successful than we could have anticipated. Flynn discovered that merely refluxing trialkyl phosphites with dibromodifluoromethane or tribromofluoromethane in ether or triglyme gave good to excellent yields of the respective fluorinated phosphonates [130]. The reactions are exothermic, but can be easily and safely carried out in solvents like ether

$$(EtO)_3P + CF_2Br_2 \xrightarrow[\text{reflux}]{\text{ether}} (EtO)_2P(O)CF_2Br \qquad (95\%)$$

$$(EtO)_3P + CFBr_3 \xrightarrow[\text{reflux}]{Et_2O} (EtO)_2P(O)CFBr_2 \qquad (78\%)$$

or triglyme, and are readily scaled-up. Subsequent work by Flynn demonstrated that analogues, such as $Ph_2P(O)CF_2Br$ and $EtO(Ph)P(O)CF_2Br$ could be prepared by similar methodology from Ph_2POEt and $PhP(OEt)_2$ respectively [45, 131]. The mechanism of the formation of these phosphonates was investigated by Flynn [131] and is dependent on the halofluoromethane precursor. With CF_2Br_2, the mechanism is similar to that of difluoromethylene ylide formation from this precursor, and is outlined in Scheme 6.16. The main difference between phosphonate formation and ylide formation is that the dealkylation step is irreversible and shifts all equlibria to product. Other CF_2XY analogues behave similarly.

In the $CFBr_3$ case, the initial step is similar (halophilic attack). However, subsequent steps, as expected [27], do not invoke carbene intermediates and the mechanism is more reminiscent of the Michaelis-Arbuzov reaction (Scheme 6.17). Consequently, the key to success is the presence of a polarizable halogen in the methane and the formation of a stabilized fluoromethide ion or carbene. With precursors such as CF_3I and C_6F_5I, only one of these requirements is met, consequently they are unreactive towards trialkylphosphites. However, Flynn developed a photochemical procedure for the preparation of $(EtO)_2P(O)CF_3$ and $(EtO)_2P(O)C_6F_5$ in moderate yields [132].

Later collaborative work with Bruce Smart's group at DuPont, corroborated our proposal that ylide formation is reversible and produces [:CF_2]. Thus, when the reaction of

$$(EtO)_3P + CF_2Br_2 \xrightarrow[\text{attack}]{\text{halophilic}} [(EtO)_3\overset{+}{P}Br] + [CF_2Br]^-$$

$$[CF_2Br]^- \rightleftharpoons [:CF_2] + Br^-$$

$$(EtO)_3P + [:CF_2] \rightleftharpoons (EtO)_3\overset{+}{P}\text{-}\overset{-}{C}F_2$$

$$(EtO)_3\overset{+}{P}\text{-}\overset{-}{C}F_2 + (EtO)_3\overset{+}{P}Br \text{ (or } CF_2Br_2) \rightleftharpoons [(EtO)_3\overset{+}{P}CF_2Br] + (EtO)_3P$$

$$Br^- + [(EtO)_3\overset{+}{P}CF_2Br] \xrightarrow[\text{dealkylation}]{S_N2} (EtO)_2P(O)CF_2Br + EtBr$$

Overall Reaction: $(EtO)_3P + CF_2Br_2 \rightarrow (EtO)_2P(O)CF_2Br + EtBr$

Scheme 6.16.

$$(EtO)_3P + CFBr_3 \xrightarrow[\text{attack}]{\text{halophilic}} [(EtO)_3\overset{+}{P}Br] + [CFBr_2]^-$$

$$[CFBr_2]^- + (EtO)_3\overset{+}{P}Br \xrightarrow{S_N2} [(EtO)_3\overset{+}{P}CFBr_2] + Br^-$$

$$Br^- + [(EtO)_3\overset{+}{P}CFBr_2] \xrightarrow[\text{dealkylation}]{S_N2} (EtO)_2P(O)CFBr_2 + EtBr$$

Overall reaction: $(EtO)_3P + CFBr_3 \longrightarrow (EtO)_2P(O)CFBr_2 + EtBr$

Scheme 6.17.

$[Ph_3PCF_2Br]^+Br^-$ with triphenylphosphine is carried out in a large excess (21 equivalents) of tetramethylethylene, the corresponding difluorocyclopropane is formed [133]:

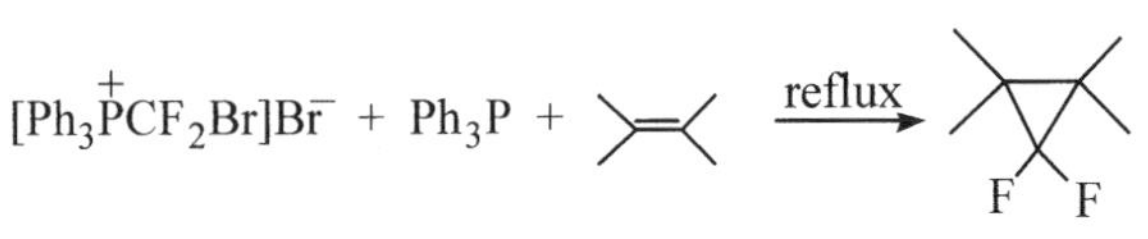

(35%)

Similar conclusions had been earlier reached by Naae [45, 134], who proposed a dissociation mechanism in the exchange of the $[CF_2Br]$ group when bromodifluoromethylphosphium salts were treated with a more nucleophilic tertiary phosphine (Scheme 6.18). Flynn observed a similar exchange with trialkylphosphites [131, 133]:

$$[Ph_3PCF_2Br]^+Br^- + (EtO)_3P \longrightarrow (EtO)_2P(O)CF_2Br + Ph_3P + EtBr$$

The dealkylation step rapidly shifts all equilibria to the phosphonate product [133]. We became intrigued by the mechanism of phosphonate formation from CF_2Br_2: why didn't the

$$[Ph_3\overset{+}{P}CF_2Br]Br^- + (Me_2N)_3P \rightleftharpoons Ph_3\overset{+}{P}\text{-}\bar{C}F_2 + [(Me_2N)_3\overset{+}{P}Br]Br^-$$

$$Ph_3\overset{+}{P}\text{-}\bar{C}F_2 \rightleftharpoons Ph_3P + [:CF_2]$$

$$(Me_2N)_3P + [:CF_2] \rightleftharpoons [(Me_2N)_3\overset{+}{P}\bar{C}F_2]$$

$$[(Me_2N)_3\overset{+}{P}Br]Br^- + (Me_2N)_3\overset{+}{P}\text{-}\bar{C}F_2 \longrightarrow [(Me_2N)_3\overset{+}{P}CF_2Br]Br^- + (Me_2N)_3P$$

$$\text{Overall reaction:} [Ph_3\overset{+}{P}CF_2Br]Br^- + (Me_2N)_3P \longrightarrow [(Me_2N)_3\overset{+}{P}CF_2Br]Br^- + Ph_3P$$

Scheme 6.18.

phosphonate anion $[(EtO)_2P(O)CF_2]^-$ not attack the $[(EtO)_3PBr]^+$ cation and form a bisphosphonate? Presumably, we deduced, halogen abstraction from the cation or CF_2Br_2 occurs faster than bisphosphonate formation. However, Flynn devised an *in situ* reaction, whereby phosphonate anion formation simultaneously produced the corresponding $(EtO)_2P(O)Br$ so that *in situ* phosphorylation would give the bisphosphonate. The simultaneous generation of the two intermediates required was tested by the following reaction:

$$(EtO)_2P(O)CF_2Br + (EtO)_2P(O)Na \xrightarrow{(EtO)_2P(O)H} (EtO)_2P(O)CF_2H \qquad (75\%)$$

Abstraction of positive halogen gave the phosphonate anion and phosphoryl halide. Rapid quenching of the anion by diethylphosphite gave the observed difluoromethylphosphonate. When the reaction is repeated in the *absence* of the dialkyl phosphite, *in situ* phosphorylation becomes the dominant reaction and bisphosphonate formation is observed [135]. Our initial yields were not spectacular – 47% in hexane. However, later work using toluene gave significantly better results (60% for the diethyl analogue) [136, 137]. Flynn also demonstrated that the bisphosphonate could be obtained via direct reaction of CF_2Br_2 with an excess of sodium dialkylphosphite [135]. Nevertheless, the design of the synthetic approach was demonstrated very nicely by Flynn and provided the first synthesis of this class of bisphosphonates.

The preliminary work by Flynn to produce the phosphonate anion from the bromodifluoromethylphosphonate via reaction with a dialkyl phosphite anion also demonstrated again the ease of dissociation of the difluoromethylphosphonate anions. Thus, when the groups on phosphorus were different, rapid exchange of $[CF_2]$ occurred to give a mixture of bisphosphonates via scrambling of $[:CF_2]$ among all possible phosphite anions in solution [138]:

$$\begin{aligned}(BuO)_2P(O)Na + (EtO)_2P(O)CF_2Br \longrightarrow\ & (EtO)_2P(O)CF_2P(O)(OEt)_2 \\ & + (BuO)_2P(O)CF_2P(O)(OEt)_2 \\ & + (BuO)_2P(O)CF_2P(O)(OBu)_2\end{aligned}$$

Hydrolysis of the bisphosphonates is readily accomplished via the silylesters and the bisphosphonic acid **3** is formed in good yield [139]:

$$(EtO)_2P(O)CF_2P(O)(OEt)_2 + 4\ Me_3SiBr \xrightarrow[2)\ 50\ ^\circ C]{1)\ RT} (Me_3SiO)_2P(O)CF_2P(O)(OSiMe_3)_2\ (\sim 100\%)$$

$$\downarrow H_2O$$

$$(HO)_2P(O)CF_2P(O)(OH)_2\quad (\sim 100\%)$$

(**3**)

We investigated this acid as a chelating agent for calcium in collaboration with Professor Donald Pietrzyk, one of my Analytical Chemistry colleagues [140]. Dorothy Rowe in our Dental School also investigated **3** as a chelating agent for calcium in bone [141]. Blackburn and co-workers proposed that **3** can mimic pyrophosphate in biological substrates and touched off an explosion of activity with its analogues.

In our laboratory, we tried to develop methodology to bisphosphonates (and acids) of the type $(RO)_2P(O)(CF_2)_nP(O)(OR)_2$, where $n > 1$. Initial success, utilizing a solution free-radical approach achieved some success [142], but the chemistry was difficult and yields were variable. Our interest in these compounds at this time were their potential utility as fuel cell electrolytes or electrolyte additives. One of these compounds, $(HO)_2P(O)CF_2CF_2P(O)(OH)_2$, did show good activity in a H_2/O_2 fuel cell in collaborative work with Professor Ernest Yeager's group at Case Western Reserve University [143].

Later work by Nair utilized a photochemical approach to the bisphosphonates [144, 145] which also proved applicable to perfluoroalkyl phosphonate analogues [145], *e.g.*

$$(EtO)_2POP(OEt)_2 + R_FI \xrightarrow{h\nu} (EtO)_2P(O)R_F \qquad (35\text{-}80\%)$$

$R_F = CF_3$, C_2F_5, $CF(CF_3)_2$, C_4F_9, C_6F_{13}, C_6F_5, $CF_2{=}CFCF_2$, $CF_2ClCFClCF_2CF_2$, $FSO_2(CF_2)_2OCF_2CF_2$, $FSO_2(CF_2)_4$, CF_3CCl_2

This photochemical approach is generally milder and more efficient (higher yields) than previous approaches to these compounds. Nair also showed (surprisingly) that when substrates, such as $BrCF_2CF_2I$ and/or $ClCF_2CF_2I$, are photolysed in the presence of trialkylphosphites under photochemical conditions, surprisingly the iodotetrafluoroethyl phosphonate was obtained [146]; under thermal free-radical conditions, only the bromo or chloro analogues are formed:

$$(RO)_3P + BrCF_2CF_2I \xrightarrow{h\nu\ (254\ nm)} (RO)_2P(O)CF_2CF_2I \qquad (42\text{-}48\%)$$

$(R = Et,\ i\text{-}C_3H_7)$

$$\underset{(X = Br,\ Cl)}{(EtO)_2POP(OEt)_2 + XCF_2CF_2I} \xrightarrow[\substack{CF_2ClCFCl_2 \\ 125\text{-}130\ ^\circ C \\ 2)\ Me_3CO_2H}]{1)\ Me_3COOCMe_3} \underset{(X = Br,\ Cl)}{(EtO)_2P(O)CF_2CF_2X,\ 52\text{-}62\%}$$

In the photochemical approach, the reaction proceeds via $F_2C{=}CF_2$; thus, by proper choice of methodology one can easily access either of the $XCF_2CF_2P(O)(OEt)_2$ building blocks.

The photochemical methodology can be utilized to prepare perfluorovinyl ether monomers containing phosphonate terminal groups. Thus, monomers such as $CF_2{=}CFO(CF_2)_nP(O)(OEt)_2$ ($n = 2, 3$) and $F_2C{=}CFOCF_2CF(CF_3)O(CF_2)_2P(O)(OEt)_2$ have been prepared and co-polymerized with TFE or terpolymerized with TFE and $F_2C{=}CFOC_3F_7$ [147, 148] to provide membranes for evaluation in CH_3OH/O_2 fuel cells. Clearly Nair's work was the key to this project.

$(RO)_2P(O)CF_2ZnX$ and $(RO)_2P(O)CF_2CdX$

Zinc and cadmium readily insert into bromodifluoromethyl phosphonate to give the respective zinc and cadmium complexes [70, 71]; these reagents are stable, readily prepared on a large scale, and easily functionalized by many electrophiles (see Scheme 6.19 for examples). Exchange with Cu(I) salts gives the corresponding copper reagent, which participates in useful coupling reactions. The zinc reagent was initially functionalized with acyl halides [71]; subsequent extension of this work by Lee Sprague to chloroformates and related derivatives provided a useful route to difluorophosphonacetic acid and its derivatives [149, 150]. Sprague developed a convenient route to 1,1-difluoro-3-alkenephosphonates [151], exemplified in Scheme 6.19, which also shows one of the coupling reactions carried out with aryl iodides by Weiming Qiu [152].

More recently our attention has turned to α-fluorophosphonates. Xin Zhang and W. Qiu prepared several analogues of this class of phosphonates via two approaches: (i) free-radical addition of $(EtO)_2P(O)CFBr_2$ to alkenes, followed by reduction of the addition adducts with Bu_3SnH; (ii) Cu^0 or $Pd(PPh_3)_4$ catalyzed addition of $(RO)_2P(O)CFHI$ to 1-alkenes followed by reduction of the iodine with $Zn/NiCl_2 \cdot 6H_2O$ [153]:

$$({}^i PrO)_2P(O)CFHI + H_2C{=}CHR \xrightarrow[2)\ Zn/NiCl_2 \cdot 6H_2O]{1)\ Pd(PPh_3)_4} ({}^i PrO)_2P(O)CFHCH_2CH_2R \quad (60\text{-}80\%)$$

Our studies in phosphonate chemistry have produced several compounds or reagents that have been utilized by many other workers with great success. It has been particularly gratifying to me to see the success of the synthetic work by talented students and postdocs in my laboratory utilized in many biological applications.

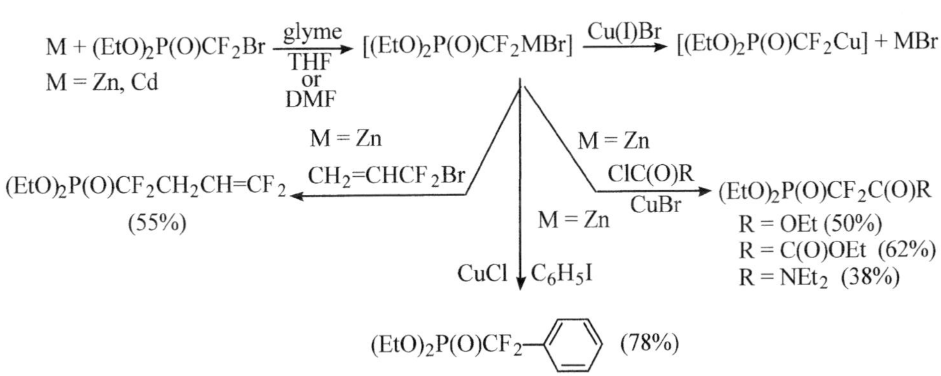

Scheme 6.19.

Electron-transfer chemistry

Our work with metal-initiated reactions began in the late 1980s with the *gem*-difluoroallylation of aldehydes and ketones by Yang [154]:

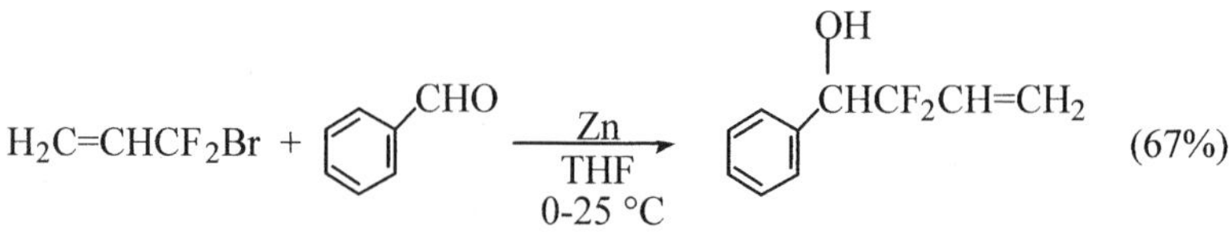

The reaction was easy to carry out experimentally, avoided low temperature reactions with unstable allylic intermediates (lithium reagents), and gave the addition product regiospecifically with the CF_2 terminus bonded to the carbonyl carbon, similar to earlier observations by Seyferth and Hiyama. Other metals, such as Cd or Sn, also catalyzed the reaction. Interestingly, in the presence of aluminum and catalytic amounts of tin dichloride, the reaction was successful in protic solvents; for example, with benzaldehyde in ethanol instead of THF, the benzyl alcohol shown above was obtained in 56% yield.

Yang also utilized copper powder to initiate the addition of iododifluoroacetates to alkenes. The reaction is suppressed by *p*-dinitrobenzene and di-*tert*-butyl nitroxide and gives cyclized products with diallyl ether, which is consistent with an SET mechanism. Reductive deiodination of the adduct was achieved with $Zn/NiCl_2 \cdot 6H_2O$ [155, 156]:

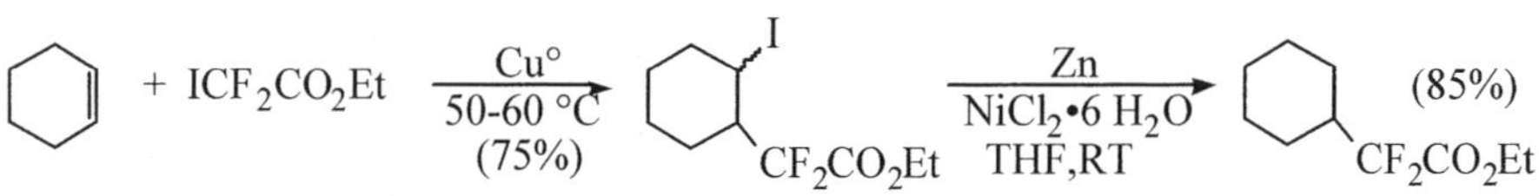

$$ICF_2CO_2Et + (EtO)_2P(O)CH_2CH{=}CH_2 \xrightarrow[\substack{50\text{-}60\ ^\circ C \\ (77\%)}]{Cu^\circ} (EtO)_2P(O)CH_2CHICH_2CF_2CO_2Et$$

$$\xrightarrow{\text{reduction (as above)}} (EtO)_2P(O)(CH_2)_3CF_2CO_2Et \quad (77\%)$$

Yang [157, 158] also demonstrated that zinc in the presence of nickel chloride hexahydrate in moist THF could give the α,α-difluoroesters in a one-pot reaction. Zinc reduces the nickel chloride to Ni^0, which catalyzes both the addition and reduction reactions. A wide variety of functional groups,

$$Me_3SiCH{=}CH_2 + ICF_2CO_2R \xrightarrow[Zn/THF]{NiCl_2 \cdot 6H_2O} Me_3SiCH_2CH_2CF_2CO_2R \quad (65\%)$$

$$CH_3C(O)CH_2CH_2CH{=}CH_2 + ICF_2CO_2R \xrightarrow[Zn/THF]{NiCl_2 \cdot 6H_2O} CH_3C(O)(CH_2)_4CF_2CO_2R \quad (76\%)$$

such as OH, OAc, ketone, ester, and silyl, are tolerated in this reaction, and the overall reaction is a facile entry to functionalized α,α-difluoroesters. Mechanistic experiments are in agreement with an SET pathway. Extension of this approach to iodofluoroacetates ($ICHFCO_2R$) gives α-fluoroesters in moderate to good yields [159].

Use of diethyl iododifluoromethylphosphonate by Yang provided an excellent route to α,α-difluorofunctionalized phosphontes [160, 161] from alkenes containing functional groups, such as Me_3Si, OH, epoxy, OAc, ketone and ester; dienes gave the corresponding bisphosphonates. Both copper metal and $Pd(PPh_3)_4$ initiated the reactions. Mechanistic evidence was again in agreement with an SET mechanism.

$$HO(CH_2)_8CH{=}CH_2 + ICF_2P(O)(OEt)_2 \xrightarrow[\text{RT}\ (78\%)]{Pd(PPh_3)_4} HO(CH_2)_8CHICH_2CF_2P(O)(OEt)_2$$

$$\xrightarrow{\text{Zn}\ |\ NiCl_2{\bullet}6\,H_2O} HO(CH_2)_{10}CF_2P(O)(OEt)_2\ (75\%)$$

Yang also developed a route to α,α-difluoromethylene functionalized sulfones via sulfination of bromodifluoroacetate or acetamide with sodium dithionite, followed by cuprous bromide catalyzed allylation [162, 163]:

$$CF_2BrC(O)Y + Na_2S_2O_4 \xrightarrow[CH_3CN/H_2O,\ 50\ ^\circ C]{NaHCO_3} NaO_2SCF_2COY\ (60\text{-}68\%)$$

$Y = OCH(CH_3)_2$
$Y = NEt_2$

$$NaO_2SCF_2COY \xrightarrow{CH_2{=}CHCH_2Br,\ CuBr,\ DMF} H_2C{=}CHCH_2SO_2CF_2COY$$

$Y = OCH(CH_3)_2$ (80%)
$Y = NEt_2$ (87%)

The aryl analogues were prepared by reaction of the bromodifluoroester or amide with appropriate mercaptides to give sulfides, which were oxidized to give the sulfones, *e.g.*

$$PhSCF_2CO_2Et \xrightarrow[CH_3CO_2H,\ 65\ ^\circ C,\ 20\ h]{30\%\ H_2O_2} PhSO_2CF_2CO_2H \quad (79\%)$$

Nguyen and Yang [164, 165] utilized SET chemistry to achieve perfluoroallylation of olefins, *e.g.*

$$Me_3SiCH{=}CH_2 + F_2C{=}CFCF_2I \xrightarrow[50\ ^\circ C]{Cu^\circ} Me_3SiCHICH_2CF_2CF{=}CF_2 \quad (78\%)$$

$$\xrightarrow{Zn/NiCl_2{\bullet}6\,H_2O} Me_3SiCH_2CH_2CF_2CF{=}CF_2 \quad (64\%)$$

As expected, a wide variety of functional groups are tolerated in this approach. The initial adducts readily form a zinc reagent, which can be elaborated further via reaction with electrophiles, *e.g.*

$$F_2C{=}CFCF_2CH_2CHICH_2C_6H_5 \xrightarrow[\text{2) CuBr, } CH_2{=}CHCH_2Cl]{\text{1) Zn/DMF}} F_2C{=}CFCF_2CH_2CH(CH_2CH{=}CH_2)CH_2C_6H_5 \quad (69\%)$$

$$F_2C{=}CFCF_2CH_2CHICH_2(CH_2)_3CH_3 \xrightarrow[\text{2) CuBr, } PhC(O)Cl]{\text{1) Zn/DMF}} F_2C{=}CFCF_2CH_2CH(C(O)Ph)CH_2(CH_2)_3CH_3 \quad (80\%)$$

This work demonstrates that the SET approach can not only provide ready access to a variety of iodine-free functionalized α- or α,α-difluoro derivatives, but also that the initial adducts can be further elaborated via a zinc derivative.

Zai-Ming Qiu utilized an SET approach to develop a general route to α,α-difluoroketones from iododifluoromethyl ketones [166, 167]; as usual, this method tolerates many functional groups, *e.g.*

$$RC(O)CF_2I + H_2C{=}CHR' \xrightarrow[\text{RT}]{Pd(PPh_3)_4} RC(O)CF_2CH_2CHIR' \quad (50\text{-}93\%)$$

R = alkyl, Ph
R' = alkyl

$$RC(O)CF_2CH_2CHIR' \xrightarrow{Zn/NiCl_2\cdot 6\,H_2O} RC(O)CF_2CH_2CH_2R' \quad (70\text{-}80\%)$$

In related work, he studied the photochemical addition of iododifluoromethyl ketones or perfluoroalkyl iodides to electron-deficient olefins [168-170], and utilized the products to develop a clever route to α-substituted β-fluoropyrroles [171, 172], *e.g.*

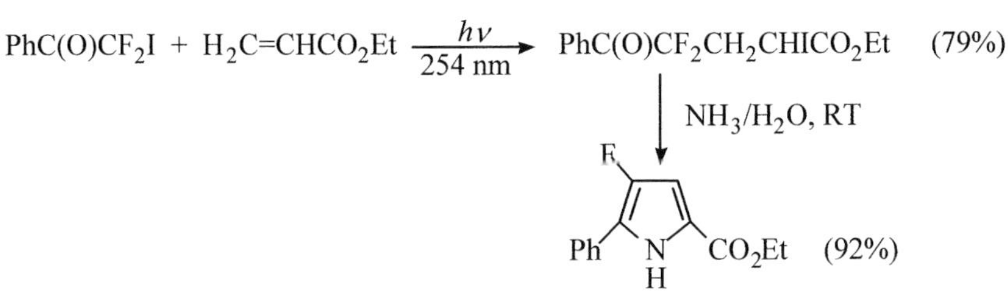

When the corresponding silyl derivatives are utilized in the reaction with aqueous ammonia, the reaction sequences stop at the 1-pyrroline stage.

Metal hydride chemistry

From my experience in Professor H. C. Brown's laboratory, I carried an interest in metal hydride and borane chemistry with me to the University of Iowa. At that time (1962), lithium aluminum hydride (LAH) was the reagent generally employed for reduction of polyfluoro-olefins. However, it was difficult to prevent over-reduction with this reagent, so product separation problems often arose. Richard Johnson then introduced sodium borohydride as a selective reagent to accomplish the olefin reduction process [173]. Although

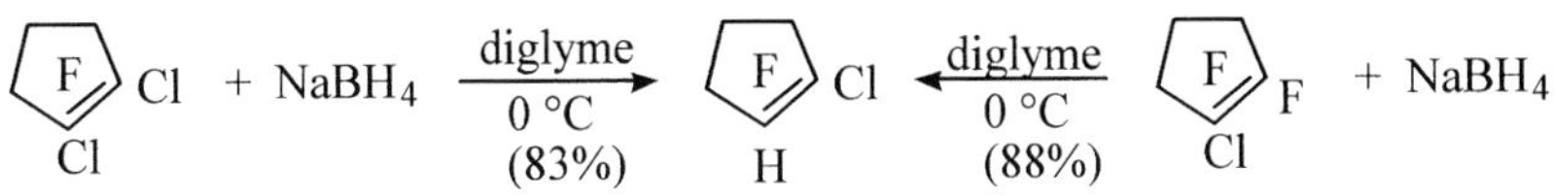

fluoro-olefins were inert to diborane under normal hydroboration conditions, diborane/MF solutions readily reduced polyfluoro-olefins, via *in situ* formation of the $[BH_3F]^-$ ion [174]. An article in *Fluorine Chemistry Reviews* [175] summarizes our initial work with $NaBH_4$ and the earlier work with LAH.

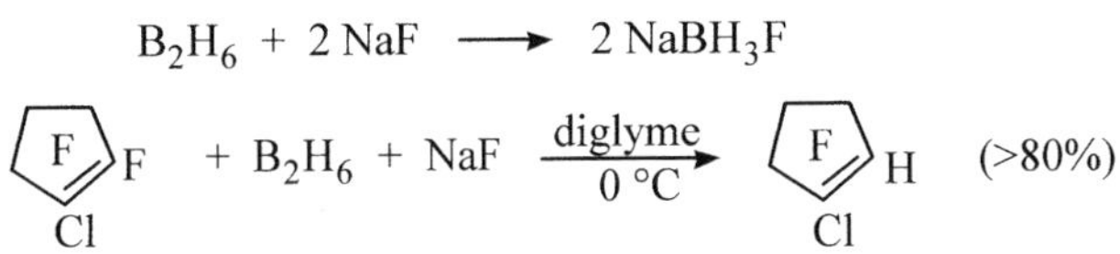

In later work with LAH, we found that when vinylic iodine was present, attack by the reducing agent occurred at iodine – not at carbon. We proposed that a *stable* aluminum complex was formed and in one case, (Scheme 6.20) attempted to isolate the complex. However, attempts to remove the last vestiges of solvent caused an explosion, presumably via elimination of the aluminum complex to form a cycloalkyne [176]. Since the reduced olefin could be easily removed under vacuum before hydrolysis, addition of D_2O to the aluminum complex gave a simple route to the deuterated olefin. Subsequent work by Frank Mettile with sodium aluminum hydride in diglyme provided an improved procedure for the preparation of the deuterated olefin [177].

Lynn Anderson developed a useful method for the hydrogenation of fluorinated ethylenes with sodium borohydride in protic media [178]. Since sodium borohydride is soluble (and stable) in water or alcohols, the ethylenes could be treated with $NaBH_4$ in diglyme with added water, ethanol or *t*-butyl alcohol. The major product does *not* arise via displacement of vinylic halogen but through the addition of the elements of H_2 to the

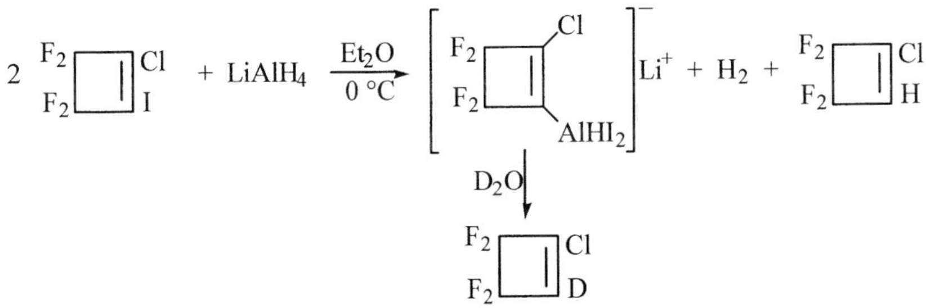

Scheme 6.20.

fluorinated olefin. The reaction can be utilized to add HD to

$$F_2C{=}CXY + NaBH_4 \xrightarrow[\substack{H_2O \\ 0\,^\circ C}]{\text{diglyme}} CHF_2CHXY \quad (X = F, Cl, Br; \quad Y = Cl, Br)$$

fluorinated ethylenes regiospecifically, since one of the hydrogens comes from the borohydride reagent and the other from the protic solvent.

$$F_2C{=}CFCl + NaBH_4 + D_2O \xrightarrow[0\,^\circ C]{\text{diglyme}} CHF_2CFClD \qquad (69\%)$$

$$F_2C{=}CCl_2 + NaBH_4 + D_2O \xrightarrow[0\,^\circ C]{\text{diglyme}} CHF_2CCl_2D \qquad (88\%)$$

Epilogue

In this review of my group's research activities in Iowa, I have attempted to present the genesis of ideas conceived and developed during the period 1962-1997. Any success we have achieved is due to the hard work, dedication, talent and creativity of the students and postdoctoral associates with whom it has been my pleasure to have shared many chemical adventures. They have made my laboratories a pleasant and exciting place, and have constantly initiated, challenged, extended and developed research ideas. Over the years, the students and postdocs in our laboratory continuously change, and each new group has brought its own ideas and enthusiasm to bear on the research effort. I would be remiss, however, not to acknowledge the one co-worker who has been the constant in all our endeavours and to whom I dedicate this article – my wife, Margaret. Over the past forty years, Margaret has assumed numerous duties and responsibilities in order that I could devote my time and efforts to my mistress, chemistry. She has used her talents and energy unstintingly to assist the co-workers in my research group in numerous ways, and without her help, understanding and sacrifices there is no possible way that I could have attempted and/or accomplished the many endeavours in synthetic methodology described above.

Acknowledgements

I am indebted to the late Professor W. T. Miller for stimulating my interest in fluorine chemistry and developing my passion for chemical research. No research programme today exists without financial support and I am indebted to the National Science Foundation, the Air Force Office of Scientific Research, 3M, and the National Institutes of Health for support of many of our research endeavours. It is also a pleasure to acknowledge Halocarbon, DuPont, 3M and Allied Chemical for providing us with many research chemicals. And last, but certainly not least, I acknowledge my colleagues, Don Pietrzyk, Bruce Friedrich, Dwight Tardy, Norm Baenziger, Dale Swenson, Dan Quinn and Bill Bennett for their collaborative work with us; Bill Dolbier, Gary Gard, the late Nobuo Ishikawa, Bruce Smart and Heinz Koch for their gracious and important collaborative efforts; and Paul Resnick and Dick Chambers for much chemical advice and pleasant rounds of golf.

References

1 D. J. Burton, PhD Thesis, Cornell University, 1961.
2 S. A. Fuqua, W. G. Duncan and R. M. Silverstein, *Tetrahedron Lett.*, *23* (1964) 1461.
3 D. J. Burton and F. E. Herkes, *Tetrahedron Lett.*, *23* (1965) 1883.
4 F. E. Herkes and D. J. Burton, *J. Org. Chem.*, *32* (1967) 1311.
5 D. J. Burton and F. E. Herkes, *Tetrahedron Lett.*, *50* (1965) 4509.
6 D. J. Burton and F. E. Herkes, *J. Org. Chem.*, *33* (1968) 1854.
7 S. A. Fuqua, W. G. Duncan and R. M. Silverstein, *J. Org. Chem.*, *30* (1965) 2543.
8 D. J. Burton, F. E. Herkes and K. J. Klabunde, *J. Am. Chem. Soc.*, *88* (1966) 5042.
9 D. J. Burton and H. C. Krutzsch, *Tetrahedron Lett.*, *1* (1968) 71.
10 D. J. Burton and H. C. Krutzsch, *J. Org. Chem.*, *35* (1970) 2125.
11 H. Yamanaka, T. Ando and W. Funaska, *Bull. Chem. Soc. Jap.*, *41* (1968) 757.
12 K. J. Klabunde and D. J. Burton, *J. Org. Chem.*, *35* (1970) 1711.
13 K. J. Klabunde and D. J. Burton, *J. Am. Chem. Soc.*, *94* (1972) 820.
14 D. J. Burton and H. C. Krutzsch, *J. Org. Chem.*, *36* (1971) 2351.
15 D. G. Naae and D. J. Burton, *J. Fluorine Chem.*, *1* (1971/72) 123.
16 D. G. Naae and D. J. Burton, *Synth. Commun.*, *3* (1973) 197.
17 D. G. Naae, H. S. Kesling and D. J. Burton, *Tetrahedron Lett.*, *44* (1975) 3789.
18 R. W. Vander Haar, D. J. Burton and D. G. Naae, *J. Fluorine Chem.*, *1* (1971/72) 381.
19 M. J. Van Hamme and D. J. Burton, *J. Fluorine Chem.*, *13* (1979) 407.
20 M. J. Van Hamme and D. J. Burton, *J. Fluorine Chem.*, *10* (1977) 131.
21 D. J. Burton and D. G. Naae, *J. Am. Chem. Soc.*, *95* (1973) 8467.
22 H. S. Kesling and D. J. Burton, *Tetrahedron Lett.*, *39* (1975) 3355.
23 D. J. Burton and G. A. Wheaton, *J. Am. Chem. Soc.*, *96* (1974) 6787.
24 G. A. Wheaton and D. J. Burton, *J. Org. Chem.*, *43* (1978) 2643.
25 D. J. Burton, S. Shin-Ya and H. S. Kesling, *J. Fluorine Chem.*, *20* (1982) 89.
26 R. M. Flynn, R. G. Manning, R. M. Kessler, D. J. Burton and S. W. Hansen, *J. Fluorine Chem.*, *18* (1981) 525.
27 D. J. Burton, R. M. Flynn, R. G. Manning and R. M. Kessler, *J. Fluorine Chem.*, *21* (1982) 371.
28 H. S. Kesling, PhD Thesis, University of Iowa, 1975.
29 D. G. Cox and D. J. Burton, *J. Org. Chem.*, *53* (1988) 366.
30 D. J. Burton and D. G. Cox, *J. Am. Chem. Soc.*, *105* (1983) 650.
31 D. G. Cox, N. Gurusamy and D. J. Burton, *J. Am. Chem. Soc.*, *107* (1985) 2811.
32 I. H. Jeong, D. J. Burton and D. G. Cox, *Tetrahedron Lett.*, *27* (1986) 3709.
33 D. J. Burton and I. H. Jeong, *J. Fluorine Chem.*, *62* (1993) 259.
34 D. J. Burton and I. H. Jeong, *J. Fluorine Chem.*, *65* (1993) 153.
35 D. J. Burton and P. E. Greenlimb, *J. Org. Chem.*, *40* (1975) 2796.
36 D. J. Burton and P. E. Greenlimb, *J. Fluorine Chem.*, *3* (1973/74) 447.
37 M. J. Van Hamme and D. J. Burton, *J. Organomet. Chem.*, *169* (1979) 123.
38 D. J. Burton, H. S. Kesling and D. G. Naae, *J. Fluorine Chem.*, *18* (1981) 293.
39 R. W. Vander Haar, PhD Thesis, University of Iowa, 1973.
40 D. J. Burton, Y. Inouye and J. A. Headley, *J. Am. Chem. Soc.*, *102* (1980) 3980.
41 D. J. Burton and J. A. Headley, *J. Fluorine Chem.*, *18* (1981) 323.
42 G. A. Wheaton and D. J. Burton, *Tetrahedron Lett.*, *12* (1976) 895.
43 G. A. Wheaton and D. J. Burton, *J. Org. Chem.*, *48* (1983) 917.
44 G. A. Wheaton, unpublished results, University of Iowa.
45 D. J. Burton, Z-Y. Yang and W. Qiu, *Chem. Rev.*, *96* (1996) 1641.
46 R. F. Stockel, F. Megson and M. T. Beachem, *J. Org. Chem.*, *33* (1968) 4395.
47 M. A. Howells, R. D. Howells, N. C. Baenziger and D. J. Burton, *J. Am. Chem. Soc.*, *95* (1973) 5366.
48 R. D. Howells, PhD Thesis, University of Iowa, 1973.
49 P. D. Vander Valk, PhD Thesis, University of Iowa, 1974.
50 D. J. Burton, R. D. Howells and P. D. Vander Valk, *J. Am. Chem. Soc.*, *99* (1977) 4830.
51 N. C. Baenziger, B. A. Foster, M. A. Howells, R. D. Howells, P. D. Vander Valk and D. J. Burton, *Acta Cryst.*, *B33* (1977) 2327.

52 D. J. Burton and P. D. Vander Valk, *J. Fluorine Chem.*, *18* (1981) 413.
53 D. J. Burton, S. Shin-Ya and R. D. Howells, *J. Am. Chem. Soc.*, *101* (1979) 3689.
54 D. J. Burton, S. Shin-Ya and R. D. Howells, *J. Fluorine Chem.*, *15* (1980) 543.
55 D. J. Burton, T. D. Spawn, P. L. Heinze, A. R. Bailey and S. Shin-Ya, *J. Fluorine Chem.*, *44* (1989) 167.
56 P. L. Heinze, T. D. Spawn, D. J. Burton and S. Shin-Ya, *J. Fluorine Chem.*, *38* (1988) 131.
57 A. Thenappan and D. J. Burton, *Tetrahedron Lett.*, *30* (1989) 3641; A. Thenappan and D. J. Burton, *J. Org. Chem.*, *55* (1990) 2311.
58 A. Thenappan and D. J. Burton, *Tetrahedron Lett.*, *30* (1989) 6113; A. Thenappan and D. J. Burton, *J. Org. Chem.*, *56* (1991) 273.
59 H-J. Tsai, A. Thenappan and D. J. Burton, *Phosphorus, Sulfur, and Silicon*, *105* (1995) 205.
60 H-J. Tsai, A. Thenappan and D. J. Burton, *Tetrahedron Lett.*, *33* (1992) 6579.
61 H-J. Tsai, A. Thenappan and D. J. Burton, *J. Org. Chem.*, *59* (1994) 7085.
62 A. Thenappan and D. J. Burton, *Tetrahedron Lett.*, *30* (1989) 5571.
63 A. Thenappan and D. J. Burton, *J. Org. Chem.*, *55* (1990) 4639.
64 A. Thenappan and D. J. Burton, *J. Fluorine Chem.*, *48* (1990) 153.
65 D. A. Wiebe, PhD Thesis, University of Iowa, 1973.
66 D. J. Burton and Y. Inouye, *Tetrahedron Lett.*, *36* (1979) 3397.
67 J. L. Hahnfeld and D. J. Burton, *Tetrahedron Lett.*, *10* (1975) 773.
68 D. J. Burton and J. L. Hahnfeld, *J. Org. Chem.*, *42* (1977) 828.
69 J. L. Hahnfeld and D. J. Burton, *Tetrahedron Lett.*, *22/23* (1975) 1819.
70 D. J. Burton, R. Takei and S. Shin-Ya, *J. Fluorine Chem.*, *18* (1981) 197.
71 D. J. Burton, T. Ishihara and M. Maruta, *Chem. Lett.*, (1982) 755.
72 D. J. Burton and D. M. Wiemers, *J. Am. Chem. Soc.*, *107* (1985) 5014.
73 D. M. Wiemers, PhD Thesis, University of Iowa, 1987.
74 D. J. Burton, *L'actualité Chimique*, (1987) 142.
75 D. M. Wiemers and D. J. Burton, *J. Am. Chem. Soc.*, *108* (1986) 832.
76 M. A. Willert-Porada, D. J. Burton and N. C. Baenziger, *J. Chem. Soc., Chem. Commun.*, (1989) 1633.
77 Y. Kobayashi and I. Kumadaki, *J. Chem. Soc., Perkin I*, (1980) 661.
78 M. A. Willert-Porada, unpublished results, University of Iowa.
79 H. K. Nair, unpublished results, University of Iowa, *cf.* ref. [145].
80 J. C. Easdon, PhD Thesis, University of Iowa, 1987.
81 D. J. Burton, D. M. Wiemers and J. C. Easdon, *U.S. Pat. 4,582,921* (1986); D. J. Burton, D. M. Wiemers and J. C. Easdon, *U.S. Pat. 4,650,887* (1987); D. J. Burton, D. M. Wiemers and J. C. Easdon, *U.S. Pat. 4,749,802* (1988); D. J. Burton, D. M. Wiemers and J. C. Easdon, *U.S. Pat. 4,895,991* (1990).
82 D. J. Burton and Z-Y. Yang, *Tetrahedron*, *48* (1992) 189.
83 Q-Y. Chen and S-W. Wu, *J. Chem. Soc., Chem. Commun.*, (1989) 705.
84 Q-Y. Chen and S-W. Wu, *J. Chem. Soc., Perkin I*, (1989) 2385.
85 Z-Y. Yang, D. M. Wiemers and D. J. Burton, *J. Am. Chem. Soc.*, *114* (1992) 4402.
86 P. L. Heinze and D. J. Burton, *J. Fluorine Chem.*, *29* (1985) 359.
87 K. J. MacNeil and D. J. Burton, *J. Org. Chem.*, *60* (1995) 4085.
88 G. A. Hartgraves and D. J. Burton, *J. Fluorine Chem.*, *39* (1988) 425.
89 G. A. Hartgraves, PhD Thesis, University of Iowa, 1988; G. A. Hartgraves and D. J. Burton, Third Chemical Congress of North America, Toronto, Canada, June 1988, Abstract FLUO #30.
90 D. J. Burton and G. A. Hartgraves, *J. Fluorine Chem.*, *49* (1990) 155.
91 D. J. Burton, G. A. Hartgraves and J. Hsu, *Tetrahedron Lett.*, *31* (1990) 3699.
92 D. J. Burton, Y. Tarumi and P. L. Heinze, *J. Fluorine Chem.*, *50* (1990) 257.
93 D. J. Burton, Z-Y. Yang and K. J. MacNeil, *J. Fluorine Chem.*, *52* (1991) 251.
94 B. V. Nguyen and D. J. Burton, *J. Fluorine Chem.*, *67* (1994) 205.
95 S. W. Hansen, T. D. Spawn and D. J. Burton, *J. Fluorine Chem.*, *35* (1987) 415.
96 D. J. Burton and S. W. Hansen, *J. Fluorine Chem.*, *31* (1986) 461.
97 S. W. Hansen, PhD Thesis, University of Iowa, 1984.
98 T. D. Spawn and D. J. Burton, *Bull. Soc. Chim. Fr.*, *6* (1986) 1.
99 D. J. Burton and S. W. Hansen, *J. Am. Chem. Soc.*, *108* (1986) 4229.
100 M. Yamamoto, D. J. Burton and D. C. Swenson, *J. Fluorine Chem.*, *72* (1995) 49.

101 M. Yamamoto, D. C. Swenson and D. J. Burton, 11th ACS Winter Fluorine Conference, St. Petersburg Beach, FL, January 1993, Abstract #P55; D. C. Swenson, M. Yamamoto and D. J. Burton, *Acta Cryst. (C)*, in the press.
102 P. A. Morken, N. C. Baenziger, D. J. Burton, P. C. Bachand, C. R. Davis, S. D. Pedersen and S. W. Hansen, *J. Chem. Soc., Chem. Commun.*, *8* (1991) 566.
103 D. C. Swenson, P. A. Morken and D. J. Burton, *Acta Cryst.*, *C53* (1997) 946.
104 P. A. Morken, P. C. Bachand, D. C. Swenson and D. J. Burton, *J. Am. Chem. Soc.*, *115* (1993) 5430.
105 P. A. Morken, D. J. Burton and D. C. Swenson, *J. Org. Chem.*, *59* (1994) 2119.
106 D. C. Swenson, P. A. Morken and D. J. Burton, *Acta Cryst.*, *C52* (1996) 2349.
107 P. L. Heinze and D. J. Burton, *J. Fluorine Chem.*, *31* (1986) 115.
108 P. L. Heinze and D. J. Burton, *J. Org. Chem.*, *53* (1988) 2714.
109 P. A. Morken, J. Lu, A. Nakamura and D. J. Burton, *Tetrahedron Lett.*, *32* (1991) 4271.
110 P. A. Morken and D. J. Burton, *J. Org. Chem.*, *58* (1993) 1167.
111 B. V. Nguyen and D. J. Burton, *J. Org. Chem.*, *62* (1997) 7758.
112 B. V. Nguyen and D. J. Burton, *J. Org. Chem.*, *63* (1998) 1714.
113 S. A. Fontana, C. R. Davis, Y-B. He and D. J. Burton, *Tetrahedron*, *52* (1996) 37.
114 C. R. Davis and D. J. Burton, *Tetrahedron Lett.*, *37* (1996) 7237.
115 C. R. Davis and D. J. Burton, *J. Org. Chem.*, *62* (1997) 9217.
116 L. Xue, L. Lu, S. D. Pedersen, Q. Liu, R. M. Narske and D. J. Burton, *Tetrahedron Lett.*, *37* (1996) 1921.
117 L. Xue, L. Lu, S. D. Pedersen, Q. Liu, R. M. Narske and D. J. Burton, *J. Org. Chem.*, *62* (1997) 1064.
118 L. Lu and D. J. Burton, *Tetrahedron Lett.*, *38* (1997) 7673.
119 D. J. Burton, G. S. Shaw and S. W. Hansen, 6th ACS Winter Fluorine Conference, Daytona Beach, FL, February 1983, Abstract #36.
120 W. R. Dolbier, Jr., H. Koroniak, D. J. Burton, A. R. Bailey, G. S. Shaw and S. W. Hansen, *J. Am. Chem. Soc.*, *106* (1984) 1871.
121 W. R. Dolbier, Jr., H. Koroniak, D. J. Burton, P. L. Heinze, A. R. Bailey, G. S. Shaw and S. W. Hansen, *J. Am. Chem. Soc.*, *109* (1987) 219.
122 W. R. Dolbier, Jr., H. Koroniak, D. J. Burton and P. L. Heinze, *Tetrahedron Lett.*, *27* (1986) 4387.
123 K. J. MacNeil and D. J. Burton, *J. Org. Chem.*, *58* (1993) 4411.
124 L. Lu and C. A. Wesolowski, unpublished results, University of Iowa.
125 Z-Y. Yang and D. J. Burton, *Tetrahedron Lett.*, *31* (1990) 1369.
126 Z-Y. Yang, P. A. Morken and D. J. Burton, *J. Fluorine Chem.*, *52* (1991) 443.
127 Z-Y. Yang and D. J. Burton, *J. Fluorine Chem.*, *53* (1991) 307.
128 B. V. Nguyen, Z-Y. Yang and D. J. Burton, *J. Fluorine Chem.*, *50* (1990) 265.
129 B. V. Nguyen, Z-Y. Yang and D. J. Burton, *J. Org. Chem.*, *58* (1993) 7368.
130 D. J. Burton and R. M. Flynn, *J. Fluorine Chem.*, *10* (1977) 329.
131 R. M. Flynn, PhD Thesis, University of Iowa, 1979.
132 D. J. Burton and R. M. Flynn, *Synthesis*, *8* (1979) 615.
133 D. J. Burton, D. G. Naae, R. M. Flynn, B. E. Smart and D. R. Brittelli, *J. Org. Chem.*, *48* (1983) 3616.
134 D. G. Naae, PhD Thesis, University of Iowa, 1972.
135 D. J. Burton and R. M. Flynn, *J. Fluorine Chem.*, *15* (1980) 263.
136 R. M. Flynn, unpublished results, University of Iowa.
137 D. J. Burton and R. M. Flynn, *U.S. Pat. 4,330,486* (1982); D. J. Burton and R. M. Flynn, *U.S. Pat. 4,393,011* (1983); D. J. Burton and R. M. Flynn, *U.S. Pat. 4,478, 761* (1984).
138 D. J. Burton, T. Ishihara and R. M. Flynn, *J. Fluorine Chem.*, *20* (1982) 121.
139 D. J. Burton, D. J. Pietrzyk, T. Ishihara, T. Fonong and R. M. Flynn, *J. Fluorine Chem.*, *20* (1982) 617.
140 T. Fonong, D. J. Burton and D. J. Pietrzyk, *Anal. Chem.*, *55* (1983) 1089.
141 D. J. Rowe, D. J. Burton and D. J. Pietrzyk, *J. Dent. Res.*, *60* (1981) 596.
142 H. K. Nair, R. D. Guneratne, A. S. Modak and D. J. Burton, *J. Org. Chem.*, *59* (1994) 2393.
143 K. Kanamura, A. Tanaka, D. Gervasio, V. Kennedy, R. Adzic, E. B. Yeager, D. J. Burton and R. D. Guneratne, *J. Electrochem. Soc.*, *143* (1996) 2765.
144 H. K. Nair and D. J. Burton, *Tetrahedron Lett.*, *36* (1995) 347.
145 H. K. Nair and D. J. Burton, *J. Am. Chem. Soc.*, *119* (1997) 9137.
146 H. K. Nair and D. J. Burton, *J. Am. Chem. Soc.*, *116* (1994) 6041.
147 S. D. Pedersen, W. Qiu, Z-M. Qiu, S. V. Kotov and D. J. Burton, *J. Org. Chem.*, *61* (1996) 8024.

148 S. V. Kotov, S. D. Pedersen, W. Qiu, Z-M. Qiu and D. J. Burton, *J. Fluorine Chem.*, *82* (1997) 13.
149 D. J. Burton, L. G. Sprague, D. J. Pietrzyk and S. H. Edelmuth, *J. Org. Chem.*, *49* (1984) 3437.
150 D. J. Burton and L. G. Sprague, *J. Org. Chem.*, *53* (1988) 1523.
151 D. J. Burton and L. G. Sprague, *J. Org. Chem.*, *54* (1989) 613.
152 C. R. Davis and D. J. Burton, *Fluorinated Organozinc Reagents*, in the series 'Practical Approach in Chemistry', Volume: *Organozinc Reagents: A Practical Approach*, Oxford University Press, edited by P. Knochel and P. Jones, 1999, p. 57.
153 X. Zhang, W. Qiu and D. J. Burton, *J. Fluorine Chem.*, *89* (1998) 39.
154 Z-Y. Yang and D. J. Burton, *J. Org. Chem.*, *56* (1991) 1037.
155 Z-Y Yang and D. J. Burton, *J. Fluorine Chem.*, *45* (1989) 435.
156 Z-Y. Yang and D. J. Burton, *J. Org. Chem.*, *56* (1991) 5125.
157 Z-Y. Yang and D. J. Burton, *J. Chem. Soc., Chem. Commun.*, *3* (1992) 233.
158 Z-Y. Yang and D. J. Burton, *J. Org. Chem.*, *57* (1992) 5144.
159 Y. Wang, Z-Y. Yang and D. J. Burton, *Tetrahedron Lett.*, *33* (1992) 2137.
160 Z-Y. Yang and D. J. Burton, *Tetrahedron Lett.*, *32* (1991) 1019.
161 Z-Y. Yang and D. J. Burton, *J. Org. Chem.*, *57* (1992) 4676.
162 Z-Y. Yang and D. J. Burton, *J. Chem. Soc., Perkin Trans. 1*, *8* (1991) 2058.
163 Z-Y. Yang and D. J. Burton, *Heteroatom Chem.*, *3* (1992) 261.
164 Z-Y. Yang, B. V. Nguyen and D. J. Burton, *Synlett*, *2* (1992) 141.
165 B. V. Nguyen, Z-Y. Yang and D. J. Burton, *J. Org. Chem.*, *63* (1998) 2887.
166 Z-M. Qiu and D. J. Burton, *Tetrahedron Lett.*, *34* (1993) 3239.
167 Z-M. Qiu and D. J. Burton, *J. Org. Chem.*, *60* (1995) 5570.
168 Z-M. Qiu and D. J. Burton, *Tetrahedron Lett.*, *35* (1994) 1813.
169 Z-M. Qiu and D. J. Burton, *J. Org. Chem.*, *60* (1995) 3465.
170 Z-M. Qiu and D. J. Burton, *J. Org. Chem.*, *60* (1995) 6798.
171 Z-M. Qiu and D. J. Burton, *Tetrahedron Lett.*, *35* (1994) 4319.
172 Z-M. Qiu and D. J. Burton, *Tetrahedron Lett.*, *36* (1995) 5119.
173 D. J. Burton and R. L. Johnson, *J. Am. Chem. Soc.*, *86* (1964) 5361.
174 R. L. Johnson and D. J. Burton, *Tetrahedron Lett.*, *46* (1965) 4079.
175 F. J. Mettille and D. J. Burton, *Fluorine Chem. Reviews*, *1* (1967) 315.
176 D. J. Burton and F. J. Mettille, *Inorg. Nucl. Chem. Lett.*, *4* (1968) 9.
177 D. J. Burton and F. J. Mettille, *J. Fluorine Chem.*, *20* (1982) 157.
178 A. L. Anderson, R. T. Bogan and D. J. Burton, *J. Fluorine Chem.*, *1* (1971/72) 121; A. L. Anderson, PhD Thesis, University of Iowa, 1971.

BIOGRAPHIC NOTE

Donald J. Burton

Donald J. Burton was born in Baltimore, Maryland (USA) and received his BS degree in 1956 from Loyola College of Baltimore. His PhD thesis work was carried out at Cornell University (Ithaca, New York) under the direction of Professor W. T. Miller, and after graduation (1961) he spent nineteen months doing postdoctoral research on organoboranes in the laboratory of Nobel Laureate H. C. Brown at Purdue University (Lafayette, Indiana). In August 1962, he commenced his academic career as an Assistant Professor in the Department of Chemistry, University of Iowa, where he has risen through the ranks to his present position of Carver/Shriner Professor of Chemistry.

During his long research career as an organofluorine chemist, Don has been a Fellow of the Japan Society For The Promotion of Science and a Visiting Professor and Lecturer in Japan, Korea, Russia, China and Taiwan. In the US, his contributions to fluorine chemistry have been recognized by the presentation of the ACS Fluorine Division's Award for Creative Work in Fluorine Chemistry, the Governor's Science Medal for Scientific Achievement, and the American Chemical Society Midwest Award. Don's research interests include synthetic methodology, organometallic chemistry, organofluorine/phosphorus chemistry and mechanistic aspects of organofluorine chemistry.

Chapter 7

ORGANOFLUORINE CHEMISTRY IN THE UNIVERSITY OF DURHAM, UK

RICHARD D. CHAMBERS

Department of Chemistry, University of Durham, South Road, Durham DH1 3LE, UK

Introduction

Richard D. Chambers

Organofluorine chemistry is unique and those not in the field do not always appreciate the attractions and excitements of the subject. For example, achieving the synthesis of a new compound such as pentafluoropyridine or a novel perfluorinated alkene is like starting an essay with a clean sheet of paper for the workers involved. The chemistry of such fluorinated systems then emerges through a combination of imagination, diligence, observation and recognition of the unusual, followed by exploitation of the latter. The excitement of the unusual is always present, and the opportunities for synthesis of unexplored systems is truly unique in organic chemistry.

Equally unique is the international kinship that exists within the global community of fluorine chemists, and this is one of the many pleasures derived from the subject and much appreciated by the Durham group. The international collaborations and friendships we have enjoyed are much too extensive to do justice to here and apologies are offered to our many friends not referred to. It is particularly important that fluorine chemistry is presented *outside* the fluorine community, and we try our best to do this. Indeed, it gives me great satisfaction to observe that since the 1950s the introduction of fluorine into an organic compound has gradually emerged from being viewed by many as a specialist interest, to the present situation where fluorine substitution is recognised as an important *general* tool. This change has stemmed partly from recognition of the fact that fluorine chemistry has made significant contributions to a whole range of products in the chemical and pharmaceutical industries; but equally important to this change, is the fact that the chemistry of organofluorine compounds has gradually become understood to a level where it is now integrated mechanistically with the rest of modern organic chemistry. Nevertheless, the process of reaching this stage has been only a gradual one, and it is important that the reader appreciates the changing situations that researchers in the field have worked under.

The studies and projects that have been initiated in Durham may be regarded as parts of a number of themes which have run through our research over many years. Inevitably, we have been concerned to develop methodology that involves the formation of C—F bonds, while our carbanion and free-radical chemistry has been directed towards the formation of C—C bonds from systems already fluorinated, and to developing 'understanding' that links fluorine chemistry to wider concepts in organic chemistry. In all aspects of our work we have enjoyed close relations with industry to foster applications, and happily we have often been influenced by industry.

We are particularly fortunate in Durham in having a tradition of excellent co-workers and support staff who are a pleasure to collaborate with. They have carried out all of the experimental work, and it is good to have the opportunity of paying tribute to their skill and dedication. What follows is an account of developments in Durham from my personal perspective; my colleagues Jim Feast, Gerald Brooke and David O'Hagan have been kind enough to provide brief accounts of their research activities, and these can be found in appendices at the end of this chapter.

The early days

Fluorine Chemistry began in Durham with the arrival of the then Dr Ken Musgrave (WKRM) in 1947, who started his research career at the University of Birmingham (UK), where he took his PhD with Dr Fred Smith (a sugar chemist who later went to the USA). Ken believes that he was the first person to study organofluorine chemistry in a UK University [1], and his work was, of course, associated with the Manhattan Project, which took him to Canada. There he met Professor F. Paneth, who was then Professor of Chemistry in Durham and persuaded him to take up a Lectureship here, replacing W. A. Waters who had gone to Oxford following his now classic work with D. H. Hey on free-radical additions to alkenes. (It is not generally appreciated that this pioneering work was carried out while Waters was in Durham.) For WKRM it was a return to the North East of England, the region of his birth and schooldays – at Stanley Grammar School (years later I also attended this school and was taught by the same chemistry teacher, Mr J. Scott).

WKRM began in Durham by exploring the reactions of chlorine trifluoride [2] and the chemistry of functional fluorine-containing compounds [3], work which must have been extremely difficult considering that only the most rudimentary equipment was available to him at that time. He had several graduate students prior to 1960, including Eric Banks (PhD, 1956) and me (PhD, 1959); in fact, during the academic year 1955-56 I shared a laboratory with Eric while pursuing my final-year undergraduate research project. Projects initiated by WKRM in this period included the fluorination of benzene and hexachlorobenzene with chlorine trifluoride or cobalt trifluoride, oxidation of aromatics with peroxytrifluoroacetic acid [4] (Scheme 7.1), and fluoropolymer chemistry.

After completing my PhD degree, I spent a year (1959-60) as a postdoctoral fellow in UBC, Vancouver, working with Howard C. Clark (who eventually became President of Dalhousie University), and sharing a laboratory with Chris J. Willis (now in London Ontario) and Ron Cavell (now in Edmonton). Also in the laboratory was the embryonic group of Neil Bartlett (justly world-renowned for his ground-breaking work on xenon fluorides) beginning the now classical chemistry with PtF_6. Neil's experiments were the first examples of the use of elemental fluorine that the author witnessed – burning platinum in

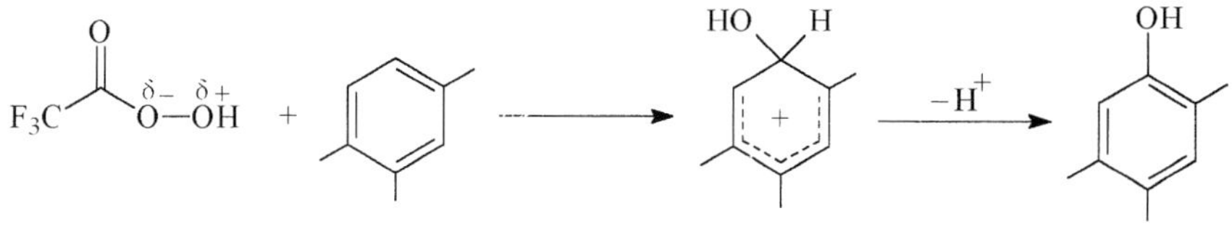

Scheme 7.1. Electrophilic aromatic hydroxylation.

$$Me_3Sn{-}SnMe_3 \xrightarrow{h\nu} 2Me_3Sn\cdot$$
$$Me_3Sn\cdot + CF_3I \longrightarrow Me_3SnI + CF_3\cdot$$
$$CF_3\cdot + Me_3Sn{-}SnMe_3 \longrightarrow Me_3SnCF_3 + Me_3Sn\cdot \longrightarrow \text{etc.}$$

Scheme 7.2. A free-radical reaction of hexamethylditin.

fluorine had a strong psychological impact on a young chemist from a country still emerging from post-war economic stringencies! That period in Vancouver, working on organotin compounds (*e.g.* the very early free-radical reaction shown in Scheme 7.2 [5]), fostered in me an interest in fluorinated organometallic compounds which I took back to Durham in 1960, where the tradition of research in organometallic chemistry had been initiated by Professor Geoffrey Coates in the mid-1950s.

1960-1970

This was a remarkable period in the development of the Chemistry Department at Durham during which Gerald Brooke, Jim Feast, David Clarke and I were all appointed to Lecturerships and made many contributions to organofluorine chemistry. The Department was accommodated in new buildings, with excellent facilities for the time, and therefore progress was rapid.

Fluorinated alkenes were identified as important 'building-blocks' in organofluorine chemistry and therefore an ongoing interest in the chemistry of these systems began. The development of vacuum-line techniques acquired from inorganic chemists (Coates in Durham and Clark in Vancouver) proved essential to us in this area, and we often wonder why organic chemists in general are strangely reluctant to adopt these very satisfying techniques. Addition of 'IF' (stoichiometric mixtures of IF_5 and I_2) to fluorinated alkenes was developed as a route to fluorocarbon iodides [6] (Scheme 7.3). This methodology proved to be very important commercially and plants have been built outside the UK to utilise this process. However, it is an interesting insight into the times that we were unable to obtain support to patent the process.

My work on pentafluorophenylated organometallic compounds began with studies on boron, aluminium, mercury, and tin compounds [7, 8]. It is interesting that the tetrakis(pentafluorophenyl)borate anion has become quite important as a non co-ordinating anion in catalysts for polymer synthesis [9] and that the first two graduate students who worked on this project have become very well known, albeit in different fields: Professor Tris Chivers (University of Calgary, Canada) is well known for his contributions to sulfur-nitrogen chemistry and the Rt. Hon. Dr Jack Cunningham was attracted from chemistry to

$$\text{'IF'} + CF_2{=}CFX \longrightarrow CF_3CFXI \ (e.g.\ X{=}F,\ CF_3)$$

Scheme 7.3. A new synthesis of fluorocarbon iodides.

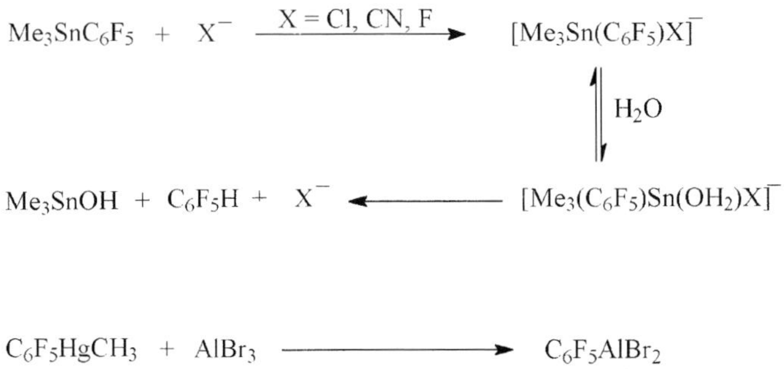

Scheme 7.4. Pentafluorophenylated organometallics.

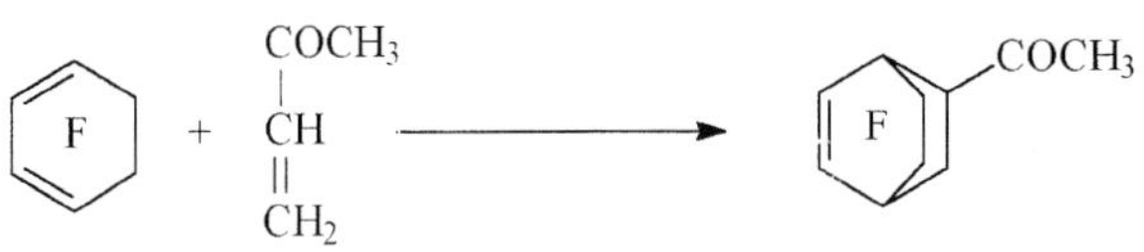

Scheme 7.5. A Diels-Alder reaction of a perfluorinated diene.

$$(CF_3)_2CFI + n\,CH_2{=}CF_2 \longrightarrow (CF_3)_2CF(CH_2CF_2)_nI \xrightarrow[Hg]{h\nu} [(CF_3)_2CFCH_2CF_2]_2$$

Scheme 7.6. Model compounds related to Viton® A.

Parliament and has developed a very high-profile career in UK national politics (at the time of writing he is Minister of Agriculture and wrestling with the problems of BSE in cattle). Examples from their researches are shown in Scheme 7.4 [7, 10]; the tin compounds are susceptible to both nucleophilic cleavage (providing an early example of fluoride-induced cleavage) and electrophilic attack, leading to the first stable fluorocarbon aluminium compounds. One of the first examples of a Diels-Alder reaction of a perfluorinated diene was encountered by D. A. Pyke in the early '60s [11] (Scheme 7.5).

Other members of the Durham group of that period who are now well-known in the fluorine field were John Hutchinson and Fred Drakesmith (known for his later work on electrochemical fluorination). Early work by Hutchinson on the synthesis of model compounds for DuPont's *Viton*® A elastomer was extended many years afterwards (see later) [12] (Scheme 7.6).

Fluoroaromatic compounds were obvious targets during the '60s, and further influential pioneering work by John Hutchinson [13, 14] (Scheme 7.7), then a Research Fellow, led to a programme that continues to this day, exploring synthesis and chemistry in what is now a considerable field – that of fluorinated heterocyclic systems. At the time that this work was developing, the Banks-Haszeldine group at UMIST was pursuing a similar path, so keen competition developed. (It is worth noting that such competition is an ex-

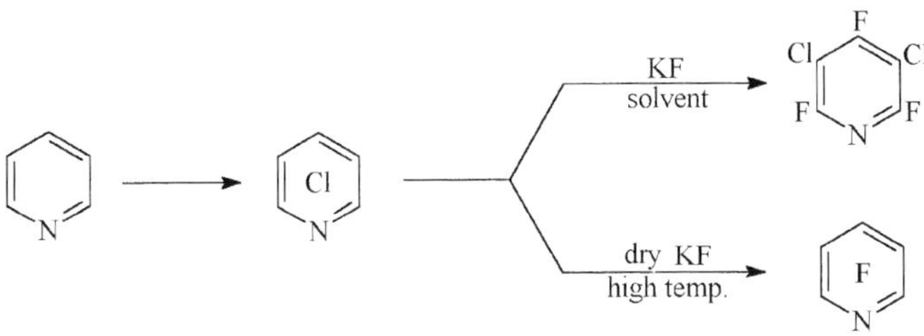

Scheme 7.7. Synthesis of fluorinated pyridines.

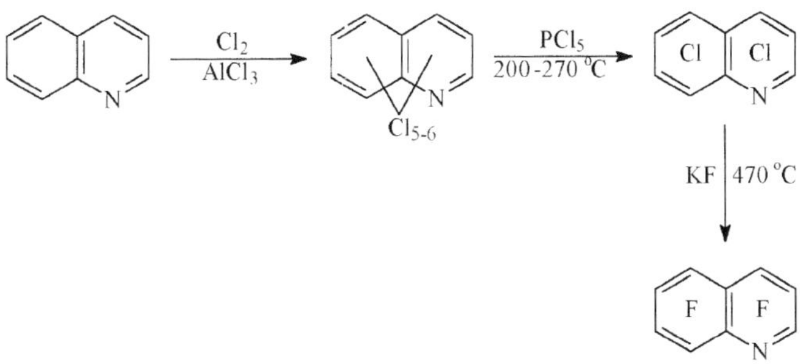

Scheme 7.8. Perchloro- and perfluoro-quinoline.

tremely effective way of stimulating graduate students to work unreasonable hours!) The general approach to the synthesis is relatively simple, but a considerable effort went into the preparation of the perchloro-compounds, most of which were themselves unknown. Nevertheless, various procedures were developed over time, and once perchlorination had been achieved, fluorination could progress.

An important development allowed us to synthesize perchloroquinoline and perchloroisoquinoline and hence their perfluoro counterparts [15, 16] (Scheme 7.8). A graduate student, Mike Hole, worked on this development, and Dr Brian Iddon, who was a temporary Lecturer at the time, collaborated with us. Brian went on to a career in heterocyclic chemistry at the University of Salford but, remarkably, like Jack Cuningham, he too is now in national politics, being a Member of Parliament for Bolton. (Note that Brian and Jack overlapped in Durham, lest any political influence is attributed to me!)

Eventually, John Hutchinson was recruited by ICI, and Dr Hugh MacBride joined our group; this took us into the synthesis of various fluorinated diazines (**1**-**5**) [17]. We also developed our first close interaction with the fluorine community in the then Soviet Union through a six-months visit by Dr Yuri Cheburkov [18] (now with 3M in Minnesota) to Durham. This was the beginning of lasting friendships and memorable visits to fluorine laboratories in Moscow and Novosibirsk.

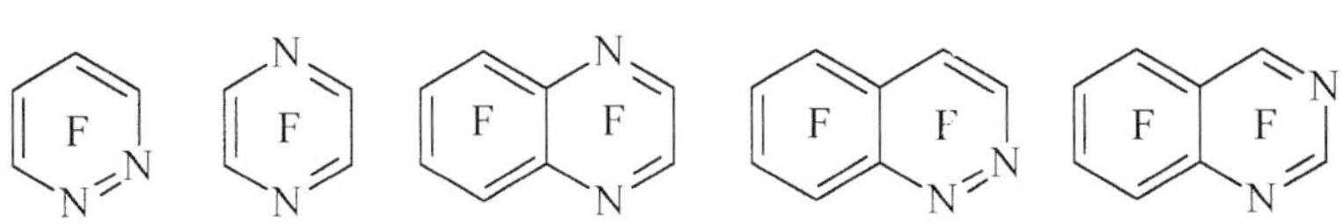

(1) (2) (3) (4) (5)

$$F^- + CF_2{=}CFCF_3 \rightleftharpoons CF_3\bar{C}FCF_3$$

$$(\text{cf. } H^+ + CH_2{=}CHCH_3 \rightleftharpoons CH_3\overset{+}{C}HCH_3)$$

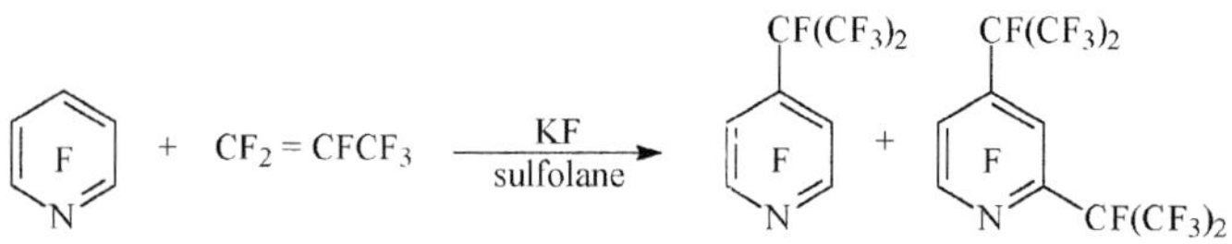

Scheme 7.9. Negative Friedel-Crafts reactions.

Use of fluoride-initiated processes for the formation of carbon-fluorine bonds is a continued interest of our group, but it was the analogy between fluoride ion in reactions with unsaturated fluorocarbons, and proton in reactions with unsaturated hydrocarbons (Scheme 7.9) that took us into really new chemistry. The late Professor W. T. (Bill) Miller (Cornell University, USA)[1] was the pioneer of fluorocarbanion chemistry, and our work was influenced by his early publications. These led us to develop the concept of 'negative Friedel-Crafts' reactions – a very simple way of introducing sterically-demanding groups into an aromatic system [19, 20] (*e.g.* Scheme 7.9).

Professor George Olah was situated at Case-Western Reserve in Cleveland (USA) at the beginning of our carbanion work, and his spectacular carbocation work was developing rapidly. He was kind enough to accept me in his department as a Visiting Fulbright Fellow and Lecturer for the period 1966-67, and this inspired some 'positive thinking' (see later for comments on fluorinated carbocations) [21]. Moreover, he persuaded me to 'write-up' my graduate course on organofluorine chemistry given in Cleveland as a book. It was, however, several years before this task, begun in Cleveland, was completed. It's easy to forget that in those days organofluorine chemistry was definitely *not* integrated with organic chemistry because the mechanistic basis of the subject was really just emerging from contributions by a range of workers, both 'aficionados' of the subject and otherwise. It was a difficult but satisfying task trying to put all this together.

WKRM established the first strong links with the University of Florida at Gainesville through his close friendship with Professor Paul Tarrant, and a series of postdoctoral workers from the Durham group went to the US to collaborate with Paul. I paid my first visit to Gainesville in 1967, and there began another enduring friendship and link which continues to this day; and I feel greatly honoured to have been invited to be the first (1999) Tarrant Visiting Professor at Gainesville. The Durham connection goes even further because Professor Alan Katritzky, who holds the distinguished Kenan Chair in Gainesville, claims that his transfer there from the University of East Anglia in the UK stemmed from a suggestion by Ken Musgrave that he should be a Visiting Professor in Gainesville. Alan subsequently decided that he liked it in Florida!

---

[1]Bill Miller was a good friend of mine over many years, and I was saddened by his death in 1998, quite soon after I had submitted this article for publication. It is a tribute to Bill that he was able to cope with such lively graduate students as Don Burton, Heinz Koch and Paul Resnick simultaneously! Coincidentally, I have enjoyed a long friendship also with each of these well-known fluorine chemists: Don and Paul continue to beat me at golf, and Heinz is still enjoying drinking my wine when he visits Durham.

1970-1980

This was a period of maturing for the Durham group. 'The book' [22] finally emerged, and fluoride-ion chemistry developed rapidly and in exciting ways. The analogies with proton-induced processes extended to rearrangements [23], and reactions involving hexafluorobut-2-yne showed the ability of this system to oligomerize or polymerize [24]. The structure of poly(hexafluorobut-2-yne) was established [25] in collaboration with David Clark, whose emerging ESCA techniques were fascinating.

The poly(perfluoroalkylation) studies gave crowded systems in a simple way, and it was becoming apparent that the stability of the C—F bond would allow us to look at skeletal rearrangements of heterocyclic sysems with less complications from side-reactions than occurred with corresponding hydrocarbon systems. Azaprizmane rearrangements [26] were mapped out for the first time, and the remarkable pyridazine-pyrazine rearrangement provided a unique case in photochemistry where substituent labelling and isolation of valence-isomer intermediates allowed progress of the rearrangement to be fully mapped [27] (Scheme 7.10).

Durham has a tradition of strength in reaction kinetics, and a collaboration with Lyn Williams proved to be extremely fruitful. The factors affecting orientation and reactivity in polyfluoroaromatic compounds undergoing nucleophilic attack were emerging. Dr Jim Burdon at the University of Birmingham (UK) had already made an important contribution to understanding by drawing attention to the I_π-repulsive effect of a *para*-fluorine. We subsequently established that in a process of nucleophilic aromatic substitution, the activating order of fluorine as a substituent is *meta* > *ortho* > *para* with respect to the C—F bond displaced, with *para-F* being not very different from hydrogen. Eventually, the gross features affecting orientation and reactivity were discussed in a series of papers [26, 28]. We also questioned whether 'negative hyperconjugation' has a major effect on reactivity [29].

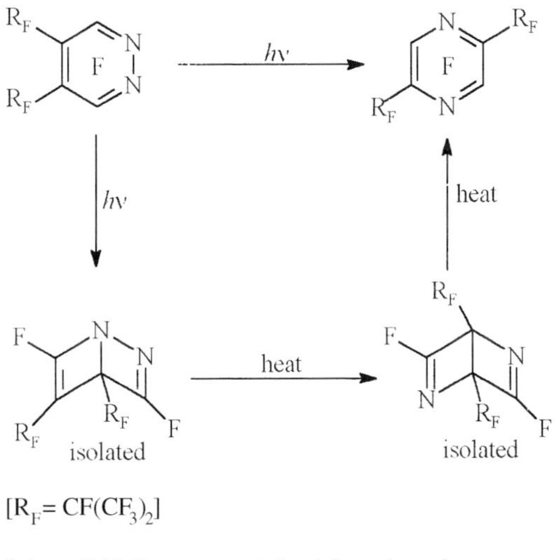

[R_F= $CF(CF_3)_2$]

Scheme 7.10. Rearrangments involving valence isomers.

The positive influence of George Olah eventually came through when we were able to observe some highly-fluorinated allyl cations (**6**, **7**) for the first time [30] and even more remarkable systems were to come.

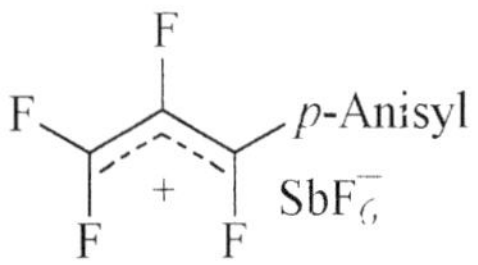

(**6**)

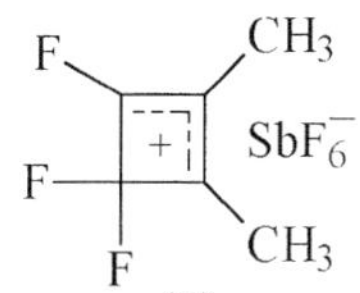

(**7**)

At the end of the 1960-70 decade an important organizational change occurred involving WKRM. In various capacities Ken had been playing a major role in the development of the University, and his skills were recognised when he was appointed Acting Vice-Chancellor (Chief Executive of the University) in 1979, a post that he occupied for two years. In essence this took him irreversibly away from the Chemistry Department.

1980-1990

The early part of this decade saw some far-reaching changes: WKRM retired and David Clark left for a senior position in ICI. Later in the decade, Jim Feast took up a leading role in a new centre for polymer synthesis, and David O'Hagan was appointed to a Lectureship, bringing biochemical skills to the Department and becoming 'fluorinated' in the process [31, 32]. I took over the Chairmanship for a period, and a rapid expansion in both personnel and facilities then began, through a series of Chairmen, which has continued to this day.

Our fluoride-ion chemistry finally yielded one of its major objectives – the generation of observable carbanions (*e.g.* **8**-**11**), derived in the first instance from perfluorobicyclobutylidine [33, 34]. Interesting rearrangements that caused some debate were also revealed [35]. The chemistry of oligomers of tetrafluoroethylene was yielding fascinating results, and the effects of angle-strain on reactivity were revealed by reactions of bicyclobutylidine [36, 37]. Early attempts to defluorinate oligomers of tetrafluoroethylene over iron gave new dienes [38]. An intriguing thermal rearrangement of a fluorinated pyridazine to a pyrimidine was finally unravelled by nitrogen-labelling experiments [39] (Scheme 7.11).

$\overset{+}{Cs}\,\overset{-}{C}(CF_3)_3$

(**8**)

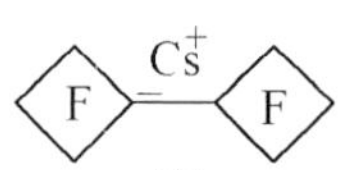

(**9**)

$\overset{+}{Cs}\,(CF_3)_2\overset{-}{C}(CF_2)_2CF_3$

(**10**)

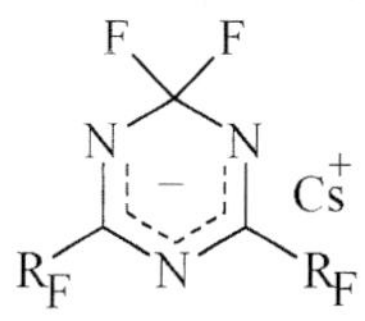

[R_F = F or $CF(CF_3)_2$]

(**11**)

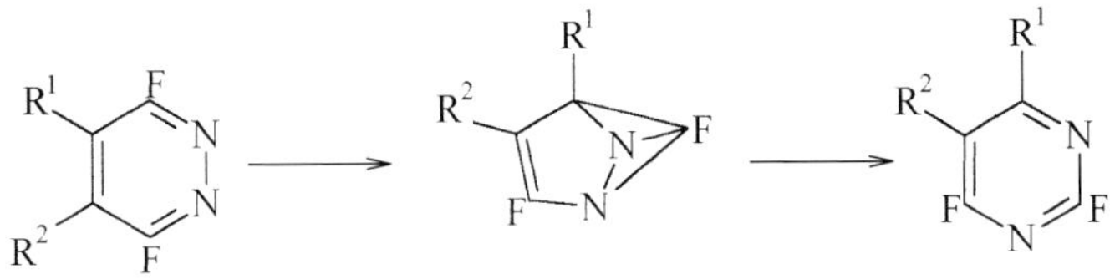

$R^1 = R^2 = CF(CF_3)_2$;
$R^1 = R^2 = C_2F_5$;
$R^1 = CF(CF_3)_2$, $R^2 = F$

$R^1 = R^2 = CF(CF_3)_2$;
$R^1 = R^2 = C_2F_5$;
$R^1 = CF(CF_3)_2$, $R^2 = F$

Scheme 7.11. Thermal rearrangements of pyridazines to pyrimidines.

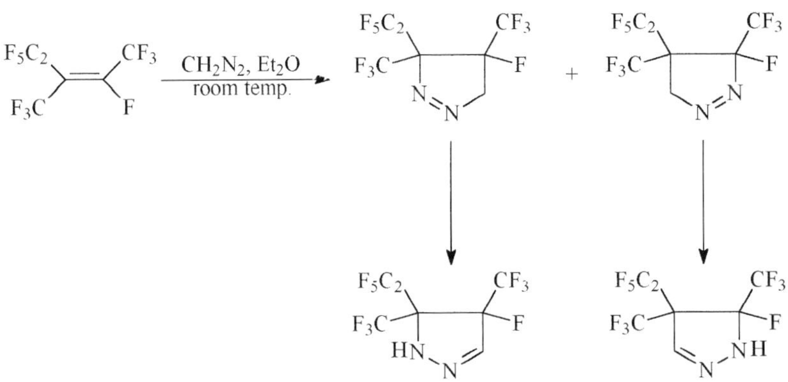

Scheme 7.12. Addition of diazomethane to *F*-alkenes.

Factors affecting the reactivity of fluorinated alkenes towards 1,3-dipoles were not understood, and we made efforts to elucidate the effects of perfluoroalkyl (activating, relative to hydrogen) and fluorine (comparable to hydrogen) directly attached to the double bond [40] (Scheme 7.12). It was concluded that steric effects of perfluoroalkyl groups limit reactivity towards 1,3-dipoles.

Our interest in the C—H bond as a functional group developed during the '80s, and we are fortunate enough to have a facility to house a high-intensity ^{60}Co γ-ray source which has been a splendid tool for promoting our work in free-radical chemistry. Homolytic additions of ethers to fluorinated alkenes, especially hexafluoropropene, were particularly significant in the later synthesis of perfluorinated ethers [41]; additions of cyclic ethers clearly illustrated stereoelectronic effects, where oxolane was more reactive than oxane in competition experiments [42, 43]. Studies in photochemistry continued to produce exciting results: azetes were generated from 1,2,4-triazines; and even the direct observation of azetes

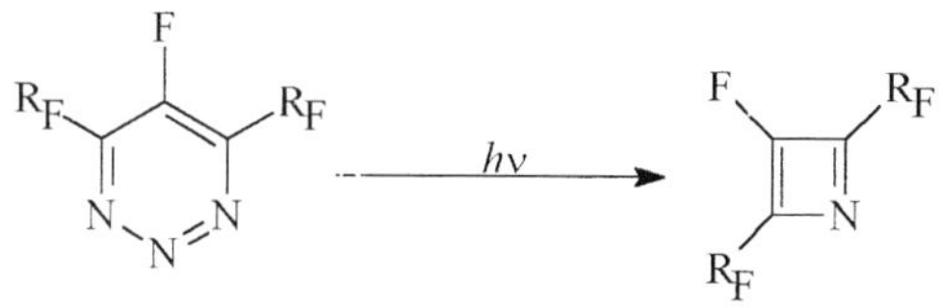

Scheme 7.13. Generation and observation of azetes.

was possible at low temperatures, using matrix isolation techniques [44] (Scheme 7.13).

Our studies in electrophilic fluorination began during the 1980s, initially using caesium fluoroxysulphate and fluorine to cleave organometallic derivatives in order to achieve site-specific fluorination [45].

1990-1999

The UK research grading system (grades 1-5) had an impact on all chemistry departments in this era and Durham emerged with a grade 5 (high). Jim Feast and I were elected as Fellows of the Royal Society in successive years (1996 and 1997, respectively) and this was recognition of the efforts of all the people who worked in our groups over the years, as well as the splendid contributions of support staff in the Department. At the beginning of this period, Bill Dennison of BNFL (the UK's nuclear energy company) called on me at Durham and, it is alleged, I 'burned his ear' a little about the fact that the company generated fluorine on a large scale but had not invested in uses other than making UF_6. This coincided with enlightened discussions taking place within BNFL on the exploitation of core expertise for non-nuclear uses. An outcome of these discussions was the setting up of a company now called F2 Chemicals Ltd, with Bill Dennison as Chief Executive and with the Durham fluorine group collaborating very closely, especially in the early stages. One of the employees was Dr John Hutchinson who retired from ICI to complete the circle of his professional career by returning to Durham. The other pleasant outcome was the return to Durham of Dr Graham Sandford (an ex-graduate student of mine who then did postdoctoral work with George Olah). Graham returned on a fellowship funded by BNFL, during the tenure of which he was awarded a Royal Society University Fellowship, despite strong competition.

The 'elemental fluorine project' has been highly successful, and our own perceptions of the viability of the element for selective fluorination have changed dramatically since the project began in 1991. Indeed, it is very satisfying to see fluorinations first carried out in Durham on the bench scale, now performed at multi-kilo levels by F2 Chemicals. That part of the development is, of course, entirely the work of the excellent team at the company, and it demonstrates what is possible for the future. I have enjoyed close collaboration with a number of companies, especially ICI, over many years, and it was a pleasure to be invited to join the Board of F2 Chemicals as a Non-executive Director – the process of education never ceases! Some examples from our work on the use of elemental fluorine are shown in Scheme 7.14 [46-52].

During our development of fluoride-ion chemistry, we have stressed the importance of processes that generate new carbon skeletons from readily-available fluorinated 'building-blocks' via simple procedures. The production of oligomers via

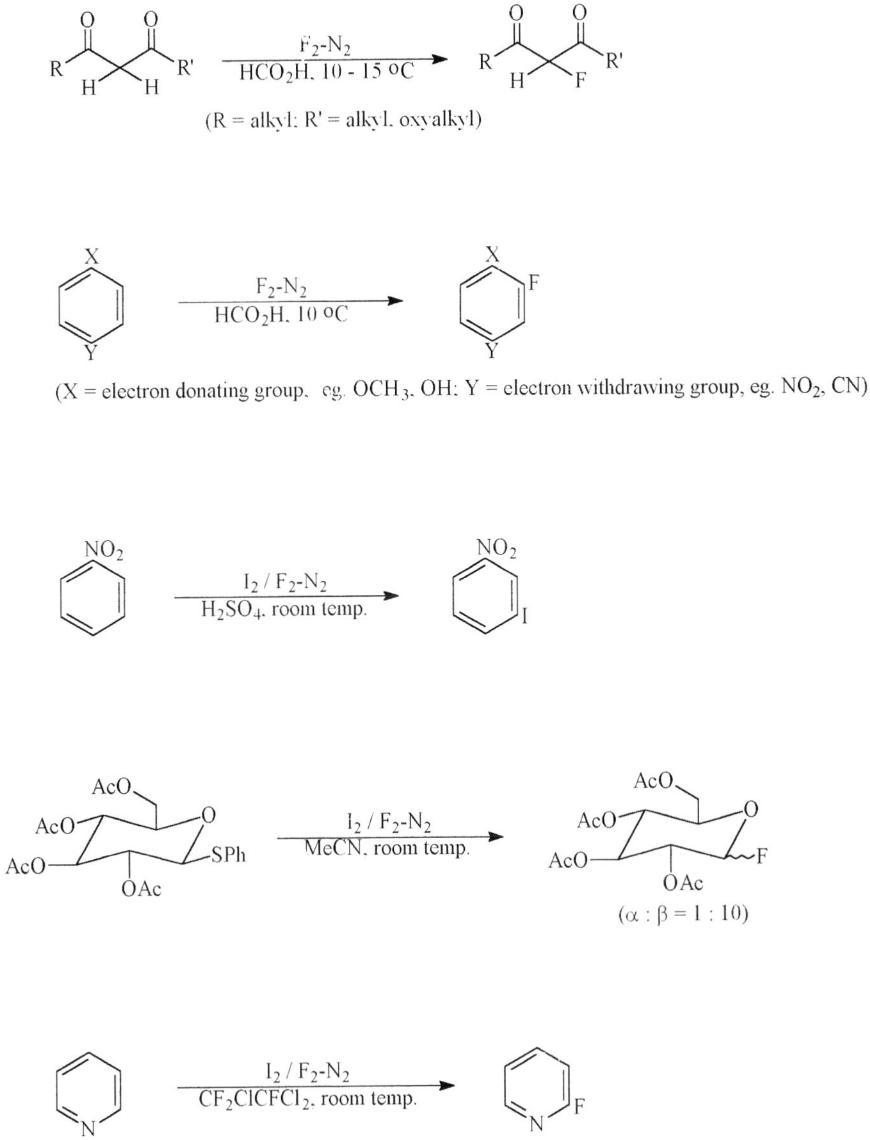

Scheme 7.14. Some uses of elemental fluorine.

fluoride-induced reactions of fluorinated alkenes, followed by defluorination to produce new dienes, nicely exemplifies this approach. Alkali-metals were used successfully initially, but defluorinations using tetrakis(dimethylamino)ethene (TDAE) proved to be very efficient and safe to scale up [53] (Scheme 7.15). Indeed, this reagent proved to be remarkably effective in promoting fluoride-induced solvent-free oligomerizations and polyfluoroalkylations [54], making many of these procedures synthetically very useful. Consequently, new electron-deficient dienes became available, and

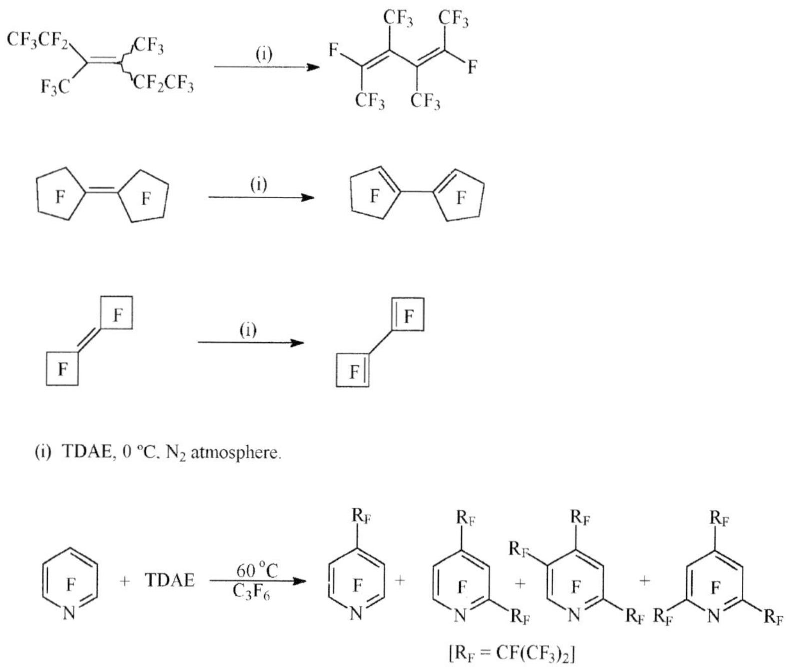

Scheme 7.15. Reactions induced by TDAE.

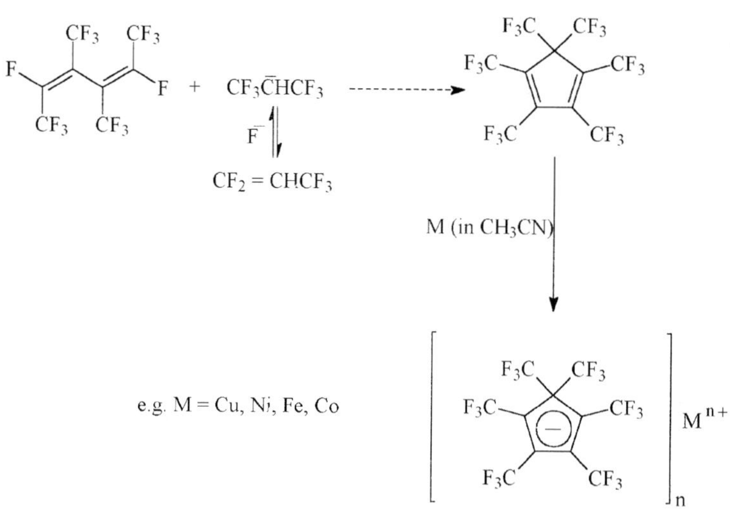

Scheme 7.16. Hexakis(trifluoromethyl)cyclopentadiene chemistry.

syntheses of pentakis(trifluoromethyl)cyclopentadienide salts and the novel hexakis-(trifluoromethyl)cyclopentadiene were established. The latter gave salts with a variety of metals directly, although π-bonded derivatives have, so far, eluded us [55] (Scheme 7.16).

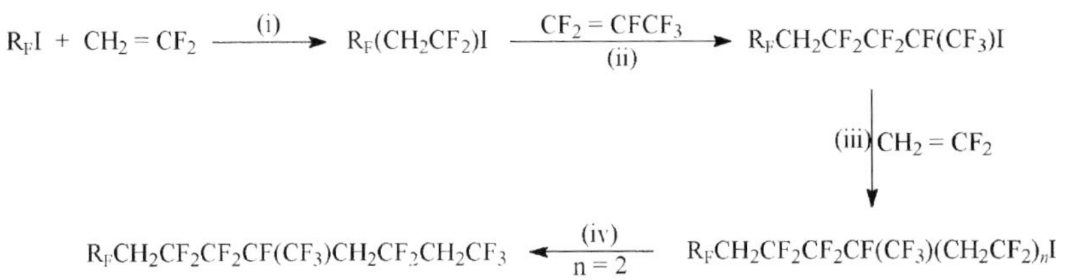

Scheme 7.17. Synthesis of polymer models. $R_F = CF(CF_3)_2$; i and ii, 185 °C, autoclave; iii, 200 °C autoclave; iv, SbF_5, 0 °C.

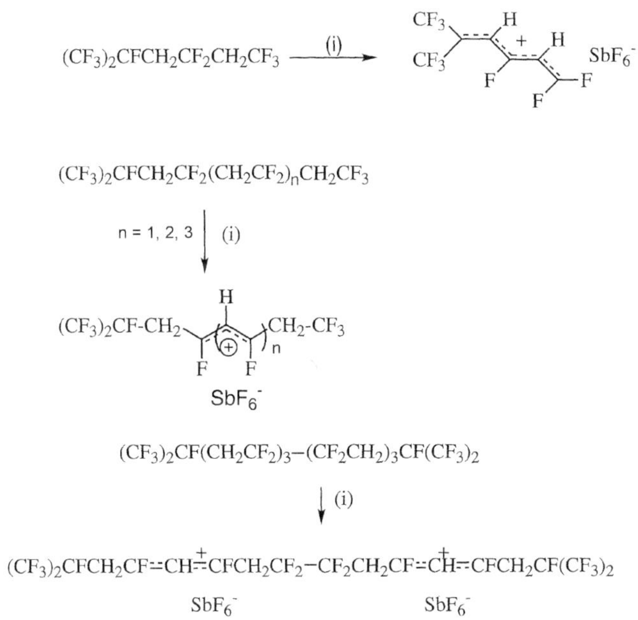

i, excess SbF_5, room temperature.

Scheme 7.18. Formation of conjugated carbocations.

A long-term collaboration we enjoyed with Ausimont was based on the synthesis of polymer models for the purpose of pursuing cross-linking studies [56] (Scheme 7.17). The work also led to some remarkable fluorinated carbocations [57] (Scheme 7.18).

Concluding remarks

It will give me a great pleasure to see fluorine chemistry continuing in Durham in the splendid hands of David O'Hagan and Graham Sandford (now appointed Lecturer in Durham). Between them they have a very diverse range of skills and interests. The year 2000 is my formal retirement year, so this article provides a timely opportunity to express my gratitude to so many people in the fluorine field for their help and friendship, but especially to my co-workers past and present.

References

1 W. K. R. Musgrave and F. Smith, *J. Chem. Soc.*, (1949) 3021.
2 J. F. Ellis and W. K. R. Musgrave, *J. Chem. Soc.*, (1950) 3608.
3 F. Brown and W. K. R. Musgrave, *J. Chem. Soc.*, (1952) 5049.
4 R. D. Chambers, P. Goggin and W. K. R. Musgrave, *J. Chem. Soc.*, (1959) 1804.
5 R. D. Chambers, H. C. Clark and C. J. Willis, *Chem. and Ind.* (London), (1960) 76.
6 R. D. Chambers, W. K. R. Musgrave and J. Savory, *J. Chem. Soc.*, (1961) 3779.
7 R. D. Chambers and T. Chivers, *J. Chem. Soc.*, (1964) 4782.
8 R. D. Chambers and T. Chivers, *Organometallic Chem. Rev., 1* (1966) 279.
9 W. E. Piers and T. Chivers, *Chem. Soc. Rev., 26* (1997) 345.
10 R. D. Chambers and J. A. Cunningham, *J. Chem. Soc. (C)*, (1967) 2185.
11 R. D. Chambers, W. K. R. Musgrave and D. A. Pyke, *Chem. and Ind.* (London), (1965) 564.
12 R. D. Chambers, J. Hutchinson, R. H. Mobbs and W. K. R. Musgrave, *Tetrahedron, 20* (1964) 497.
13 R. D. Chambers, J. Hutchinson and W. K. R. Musgrave, *Proc. Chem. Soc.*, (1964) 83.
14 R. D. Chambers, J. Hutchinson and W. K. R. Musgrave, *J. Chem. Soc.*, (1964) 3573.
15 R. D. Chambers, M. Hole, B. Iddon, W. K. R. Musgrave and R. A. Storey, *J. Chem. Soc. (C)*, (1966) 2328.
16 R. D. Chambers, M. Hole, W. K. R. Musgrave, R. A. Storey and B. Iddon, *J. Chem. Soc. (C)*, (1966) 2331.
17 R. D. Chambers, J. A. H. MacBride and W. K. R. Musgrave, *J. Chem. Soc. (C)*, (1968) 2116.
18 C. G. Allison, R. D. Chambers, Y. A. Cheburkov, J. A. H. MacBride and W. K. R. Musgrave, *Chem. Comm.*, (1969) 1200.
19 R. D. Chambers, R. A. Storey and W. K. R. Musgrave, *Chem. Comm.*, (1966) 384.
20 R. D. Chambers, J. A. Jackson, W. K. R. Musgrave and R. A. Storey, *J. Chem. Soc. (C)*, (1968) 2221.
21 G. A. Olah, R. D. Chambers and M. B. Comisarow, *J. Amer. Chem. Soc., 89* (1967) 1268.
22 R. D. Chambers, *Fluorine in Organic Chemistry*, Wiley-Interscience, New York, 1973.
23 R. D. Chambers, R. P. Corbally and W. K. R. Musgrave, *J. Chem. Soc., Perkin Trans. 1*, (1972) 1281.
24 R. D. Chambers, S. Partington and D. B. Speight, *J. Chem. Soc., Perkin Trans. 1*, (1974) 2673.
25 R. D. Chambers, D. T. Clark, D. Kilcast and S. Partington, *J. Polymer Science, Polymer Chem. Ed., 12* (1974) 1647.
26 R. D. Chambers and R. Middleton, *J. Chem. Soc., Perkin Trans. 1*, (1977) 1500.
27 R. D. Chambers, J. R. Maslakiewicz and K. C. Srivastava, *J. Chem. Soc., Perkin Trans. 1*, (1975) 1130.
28 R. D. Chambers and R. Middleton, *Chem. Comm.*, (1977) 154.
29 R. D. Chambers, J. S. Waterhouse and D. L. H. Williams, *Tetrahedron Letters*, (1974) 743.
30 R. D. Chambers, A. Parkin and R. S. Matthews, *J. Chem. Soc., Perkin Trans. 1*, (1976) 2107.
31 R. D. Chambers, R. Jaouhari and D. O'Hagan, *J. Fluorine Chem., 44* (1989) 275.
32 R. D. Chambers, R. Jaouhari and D. O'Hagan, *J. Fluorine Chem., 45* (1989) 5101.
33 R. D. Chambers, R. S. Matthews and G. Taylor, *J. Chem. Soc., Perkin Trans. 1*, (1980) 435.
34 R. D. Chambers, A. E. Bayliff, M. R. Bryce and G. Taylor, *J. Fluorine Chem., 35* (1987) 65.
35 R. D. Chambers, J. R. Kirk, G. Taylor and R. L. Powell, *J. Chem. Soc., Perkin Trans. 1*, (1982) 673.
36 R. D. Chambers, J. R. Kirk, G. Taylor and R. L. Powell, *J. Fluorine Chem., 22* (1983) 393.
37 A. E. Bayliff, M. R. Bryce, R. D. Chambers, J. R. Kirk and G. Taylor, *J. Chem. Soc., Perkin Trans. 1*, (1985) 1191.
38 R. D. Chambers, A. A. Lindley, H. C. Fielding, J. S. Molliet and G. Whittaker, *J. Chem. Soc., Perkin Trans. 1*, (1981) 1064.
39 R. D. Chambers, W. K. R. Musgrave and C. R. Sargent, *J. Chem. Soc., Perkin Trans. 1*, (1981) 1071.
40 M. R. Bryce, R. D. Chambers and G. Taylor, *J. Chem. Soc., Perkin Trans. 1*, (1984) 509.
41 R. D. Chambers and B. Grievson, *J. Fluorine Chem., 29* (1985) 323.
42 R. D. Chambers, B. Grievson and N. Kelly, *J. Chem. Soc., Perkin Trans. 1*, (1985) 2209.
43 R. D. Chambers and B. Grievson, *J. Chem. Soc., Perkin Trans. 1*, (1985) 2215.
44 R. D. Chambers, T. Shepherd, M. Tamura and P. Hoare, *J. Chem. Soc., Perkin Trans. 1*, (1990) 983.
45 R. D. Chambers, M. R. Bryce, S. T. Mullins and A. Parkin, *Bull. Soc. Chim. France*, (1986) 930.
46 R. D. Chambers, J. Hutchinson, C. J. Skinner and J. Thomson, WO 95/16649.
47 R. D. Chambers, M. P. Greenhall and J. Hutchinson, *Tetrahedron, 52* (1996) 1.
48 R. D. Chambers, J. Hutchinson, A. S. Batsanov, C. W. Lehmann and D. Y. Naumov, *J. Chem. Soc., Perkin Trans. 1*, (1996) 2271.

49 R. D. Chambers, C. J. Skinner, J. Hutchinson and J. Thomson, *J. Chem. Soc., Perkin Trans. 1*, (1996) 605.

50 R. D. Chambers, C. J. Skinner, M. J. Atherton and J. S. Moilliet, *J. Chem. Soc., Perkin Trans. 1*, (1996) 1659.

51 R. D. Chambers, G. Sandford, M. E. Sparrowhawk and M. J. Atherton, *J. Chem. Soc., Perkin Trans. 1*, (1996) 1941.

52 R. D. Chambers and G. Sandford, WO 96/19456.

53 M. W. Briscoe, R. D. Chambers, S. J. Mullins, T. Nakamura and J. F. S. Vaughan, *J. Chem. Soc., Perkin Trans. 1*, (1994) 3119.

54 R. D. Chambers, W. K. Gray and S. R. Korn, *Tetrahedron, 51* (1995) 13167.

55 R. D. Chambers, W. K. Gray, J. F. S. Vaughan, S. R. Korn, M. Médebielle, A. Batsanov, C. W. Lehmann and J. A. K. Howard, *J. Chem. Soc., Perkin Trans. 1*, (1997) 135.

56 G. C. Apsey, R. D. Chambers, M. J. Salisbury and G. Moggi, *J. Fluorine Chem., 40* (1988) 261.

57 R. D. Chambers, M. J. Salisbury, G. C. Apsey and G. Moggi, *J. Chem. Soc., Chem. Commun.*, (1988) 680.

Appendix 7.1

Fascinated by fluoroaromatic chemistry

GERALD M. BROOKE
Department of Chemistry, University of Durham, UK

The early 1960s marked the beginning of a big expansion in British Universities. I arrived in Durham in September 1962 to join the teaching staff and reinforce departmental research interests in fluorine chemistry under the leadership of Ken Musgrave. He immediately invited me to assist in some current projects being undertaken along with Dick Chambers; he also encouraged me to pursue independent work in heterocyclic chemistry based on highly-fluorinated benzenoid substrates, compounds which I had had considerable experience with during my time in the Fluorine Group in Birmingham[1]. Most of my personal fascination with fluorine chemistry has arisen directly or indirectly from this work, four examples of which are given below.

(i) The *potential mechanism* of the one-pot reaction between C_6F_6 and the sodio derivative of ethyl acetoacetate to produce a 4,5,6,7-tetrafluorobenzo[*b*]furan derivative [1, 2] provided the inspiration for one of my heterocyclic syntheses. Scientists in the UK envisaged an initial attack by oxygen followed by ring closure with a carbanion [1], but Russian workers later established unequivocally that the reverse sequence of bond-forming reactions took place: the final cyclization proceeded via attack by the heteroatom [3]. About that time syntheses of partially-fluorinated benzo[*c*]thiophene derivatives were being investigated in Durham and it was decided to explore nucleophilic addition of sulphur to an acetylene to produce a *carbanionic* intermediate for possible cyclization; it was gratifying to realise a positive reaction (Scheme 1.1) [4]. This reaction is one of the very few proceeding via this mode of cyclization: it is far more usual to exploit a heteroatom as the nucleophilic species in a substrate containing a carbon or a deactivating heteroatom already bonded to the aromatic ring [5].

(ii) The relatively ready availability of C_6F_5-derivatives in the 1970s provided yet another incentive to seek novel routes to fluorinated heterocyclic compounds. One of these involving the Claisen rearrangement reaction was eventually realised by carrying out the reaction in the presence of fluoride ion as a base (Scheme 1.2) [6]. Conditions designed to promote loss of HF pyrolytically turned out to be even more interesting in that following the initial 3,3-sigmatropic shift, one (**A**) of two possible intramolecular Diels-Alder reaction products (**A**, **B**) was formed and underwent further isomerization to the final carbobicyclic product (Scheme 1.3) [7]. Under milder conditions, the *other* internal Diels-Alder adduct, **B**, was the first ever adduct of this kind actually to be isolated from any Claisen rearrangement reaction involving prop-2-enyl ethers [8]; the electron-poor diene and the electron-rich dieneophile provided the ideal situation for this reaction.

[1] See Chapter 5 for details of that experience and my biographical data.

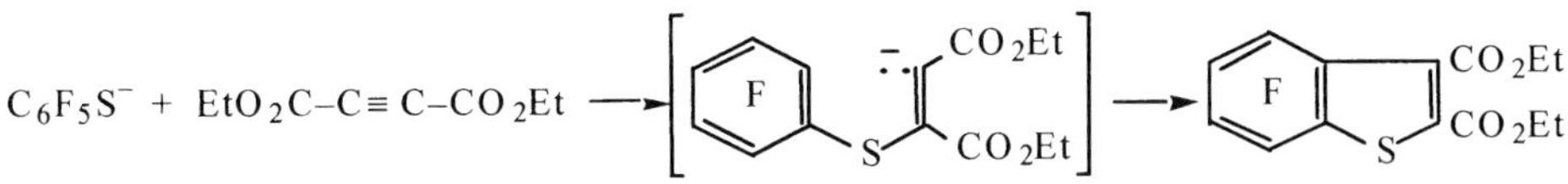

Scheme 1.1.

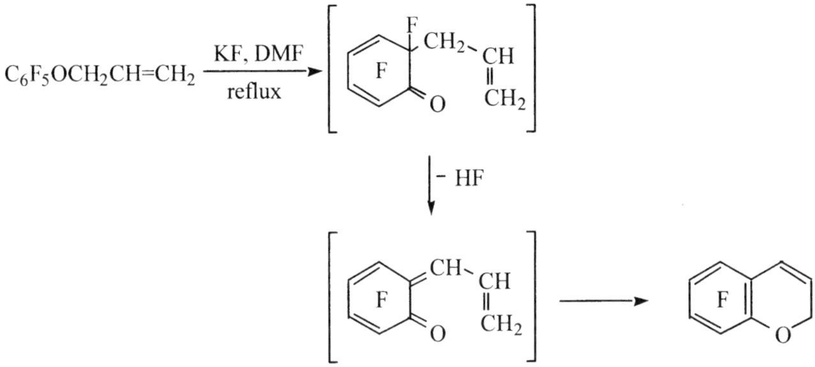

Scheme 1.2.

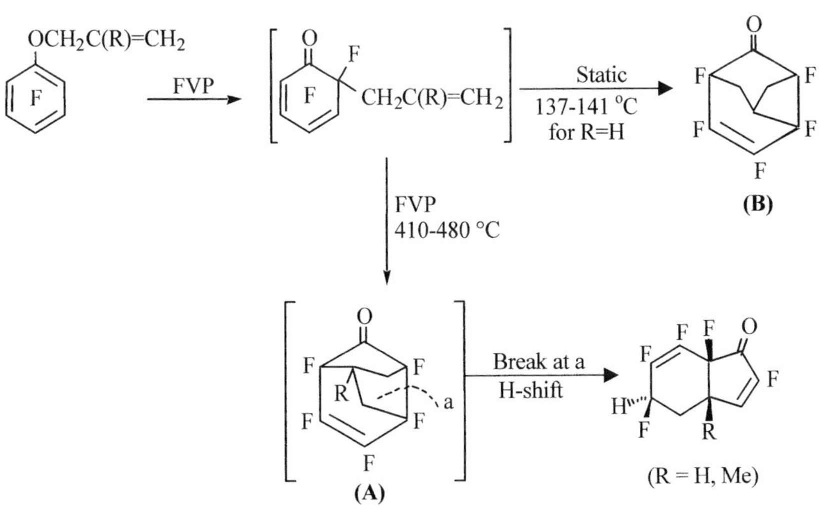

Scheme 1.3.

(iii) Further Claisen reactions, carried out under high-vacuum conditions in the vapour phase with prop-2-ynyl ethers [9] and thioethers [10], again resulted in isomerisation of the starting materials, the products being 2-fluoromethyl derivatives of ring-fused furans and thiophenes, respectively. It is remarkable that a formal 1,4-migration of fluorine is required to rationalize the formation of these products (Scheme 1.4).

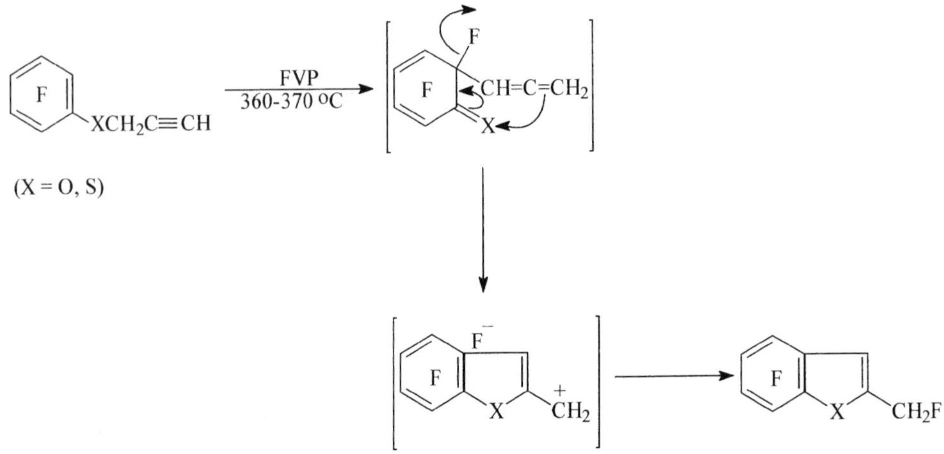

Scheme 1.4.

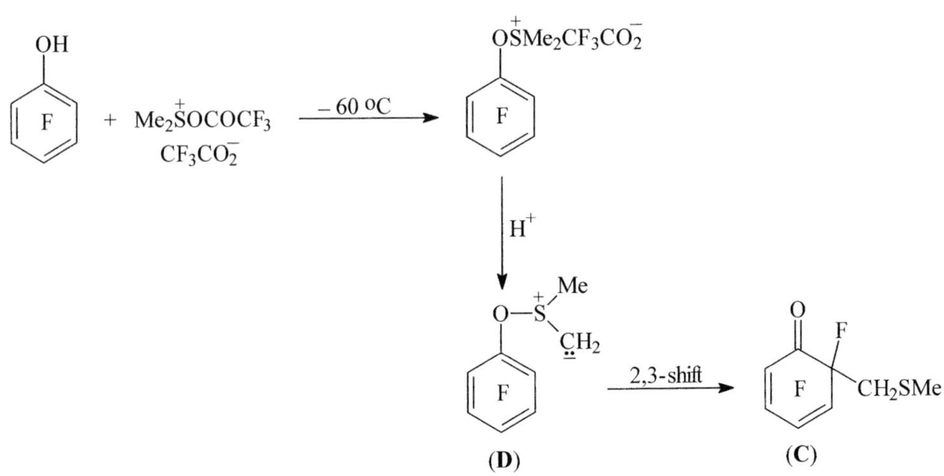

Scheme 1.5.

(iv) Following the work on Claisen rearrangements (3,3-sigmatropic reactions), 2,3-sigmatropic reactions (Sommelet-Hauser rearrangements) were studied and found to result in the formation of de-aromatised products, *e.g.* **C** (Scheme 1.5) [11]; intriguingly, *no analogous product* formed with the corresponding C_6Cl_5-derivative.

In an attempt to create another potential 2,3-sigmatropic shift reaction, I treated the sulfoxide $C_6F_5CH_2S(O)CH_3$ with BuLi with a view to forming the ion **E** – which is not too dissimilar to species **D** in Scheme 1.5. To my surprise, the product isolated after an aqueous work-up was 4,5,6,7-tetrafluorobenzo[*c*]thiophene (**F**) [12]. Currently there is much interest in conjugated polymers as potential conducting materials, and poly(benzo[*c*]thiophene) (**G**) made by Wudl *et al.* was particularly exciting because it is had a very small band gap of 1 eV [13]; however, it is totally insoluble in all known solvents. By contrast, poly(4,5,6,7-

tetrafluorobenzo[*c*]thiophene) (**H**) [14] is freely soluble in chlorinated solvents, rendering it manipulable! Unfortunately, this very unusual and desirable property is not matched by any other favourable properties of interest to the physics community.

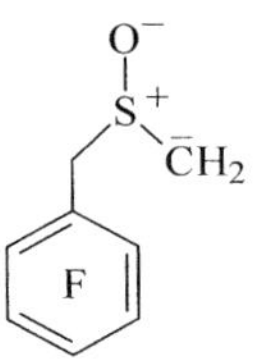

(**E**)

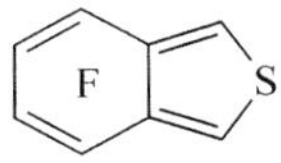

(**F**)

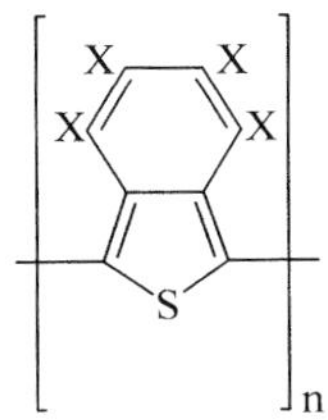

(**G**) X = H
(**H**) X = F

References

1 E. H. P. Young, 3rd International Symposium on Fluorine Chemistry, Munich, September 1965.
2 G. G. Yakobson, T. D. Petrova, L. I. Kann, T. I. Savchenko, A. K. Petrov and N. N. Vorozhtsov, *Dokl. Akad. Nauk S.S.S.R., 158* (1964) 926.
3 T. D. Petrova, L. I. Kann, V. A. Barkhash and G. G. Yakobson, *Khim. Geterotsikl. Soedin., 5* (1969) 778.
4 G. M. Brooke and Md . Abul Quasem, *J. Chem. Soc. C*, (1967) 865.
5 G. M. Brooke, *J. Fluorine Chem., 86* (1997) 1.
6 G. M. Brooke, *J. Fluorine Chem., 22* (1983) 483.
7 G. M. Brooke, *J. Chem. Soc., Perkin Trans. 1*, (1974) 233; G. M. Brooke and D. H. Hall, *J. Fluorine Chem., 10* (1977) 495.
8 G. M. Brooke and D. H. Hall, *J. Fluorine Chem., 20* (1982) 163.
9 G. M. Brooke and D. I. Wallis, *J. Chem. Soc., Perkin Trans. 1*, (1981) 1417.
10 G. Brooke and J. R. Cooperwaite, *J. Chem. Soc., Perkin Trans. 1*, (1985) 2643.
11 G. M. Brooke and J. A. J. K. Ferguson, *J. Chem. Soc., Perkin Trans. 1*, (1987) 2091.
12 G. M. Brooke and S. D. Mawson, *J. Chem. Soc., Perkin Trans. 1*, (1990) 1919.
13 F. Wudl, M. Kobayashi and A. J. Heeger, *J. Org. Chem., 49* (1985) 417.
14 G. M. Brooke, C. J. Drury, D. Bloor and M. J. Swann, *J. Mater. Chem., 5* (1995) 1317.

Appendix 7.2

Adventures with fluorinated dienes

W. J. FEAST[1]

Chemistry Department and IRC in Polymer Science and Technology, University of Durham, Durham DH1 3LE, UK

W. J. Feast

After undergraduate training in the exciting environment of the Chemistry Department at Sheffield University (1957-60), my first research experience was obtained in the Fluorine Group led by Colin Tatlow at Birmingham University. I worked for my PhD degree under the direction of Bob Stephens on a project which involved the synthesis, purification and characterisation of a series of fluorinated cycloalkanes, cycloalkenes and cycloalkadienes. My first solo research experiment was the reduction of 1,2-dichlorohexafluorocyclopentene with lithium aluminium hydride, a fortunate choice for a tyro scientist since it gave an unanticipated product which I was able to isolate, identify and explain before reporting to my supervisor [1]. This early experience gave me an appreciation of the excitement of chemistry and a taste for working in unpredictable areas and with novel materials.

My interest in polymers was aroused during my postdoctoral work on a Research Contract at Birmingham. The work was tedious and involved repetitive monomer synthesis and purification with almost no feedback from the sponsors, so I undertook some freelance polymer synthesis on a 'teach yourself' basis. I found the exercise fascinating and, although the results were never published, the experience paved the way for future research interests.

My appointment to a lectureship in Durham (1965) was made on the understanding that I would contribute to the existing activity in organic fluorine chemistry there. Initially, work was concentrated on thermal and photochemical cycloadditions and isomerizations of fluorinated alkenes and dienes (1965 was, of course, the year when Woodward and Hoffmann's first papers on the conservation of orbital symmetry appeared, so there was a high level of international interest in this general area). Several research papers resulted from our studies [2-5]. We established some routes to functionalized fluoroaromatics and heteroaromatics, and made several novel polycyclic polyenes. An example of the work from

[1]Professor Feast has been the Director of the Leeds-Bradford-Durham Interdisciplinary Research Centre for Polymer Science and Technology since 1994, having been Assistant Director since 1989, the year in which he was appointed to the Courtaulds Chair of Polymer Chemistry at Durham. He was elected to the Fellowship of the Royal Society in 1996. (Ed.)

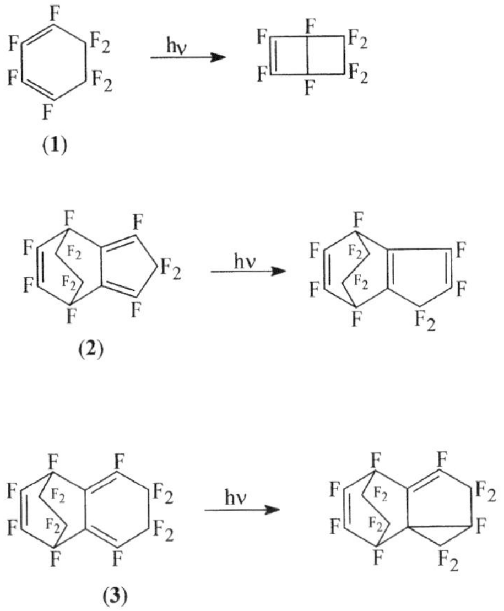

Scheme 2.1.

this period is summarised in Scheme 2.1. The three perfluorinated 1,3-dienes 1-3 were all useful in the synthesis of fluoroaromatics, but the feature of their chemistries which proved particularly interesting to me was their vapour-phase photoisomerizations, since simple product analysis clearly demonstrated that these superficially similar dienes behaved differently.

Diene **1** has the option of a 4π-closure or a 6π-opening and follows the 4π-route in 100% yield, presumably because the preference for locating F on a saturated carbon, combined with the C—C bond strengthening effect of fluorination at the CF_2-CF_2 bond, dominates over other considerations. Diene **2** gives a product which could have arisen via an initial 4π-closure followed by a $[_{\sigma}2_s + {}_{\sigma}2_a]$ process or via a 1,5-sigmatropic F migration. In the event, the isomerization of diene **3**, which can only be rationalized as the result of an initial 6π-opening followed by a photo-Diels-Alder reaction or $[4_s + 2_a]$ cycloaddition, shows that a 4π-closure in **2** is (commensurate with reasonable expectation) very unlikely, implying that the isomerization observed involves a fluorine migration. At that time this concept was unacceptable to referees, but subsequently several other well-documented examples of fluorine migrations have been established. Just when this part of our work was running into heavy weather with referees, I was developing a strong interest in the challenges presented by synthetic organic polymer chemistry and had spent a year's leave of absence (1968-69) in the laboratories of the late Georges Smets in Universiteit te Leuven, working in collaboration with Frans DeSchryver on the development of step-growth photopolymerization. The Chairman of the Chemistry Department at Durham, Ken Musgrave, approved my decision to reduce my activities in fluorine chemistry so that I could develop polymer synthesis as the main theme of my research; I am grateful to him for encouragement during the fairly difficult years immediately following this decision.

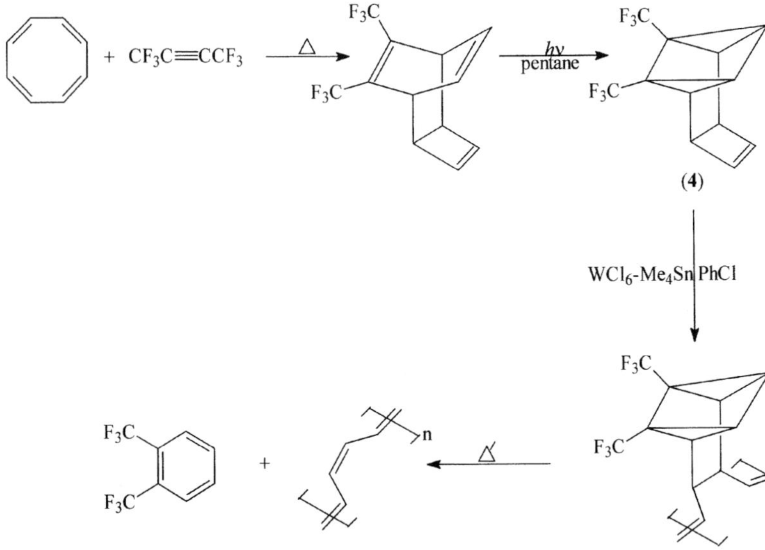

Scheme 2.2. Synthesis of poly(acetylene) via ring-opening metathesis polymerisation of 3,6-bis(trifluoromethyl)-pentacyclo[$6.2.0.0^{2,4}.0^{3,6}.0^{5,7}$]dec-9-ene (**4**).

Ring-opening metathesis polymerization of fluorinated monomers

We were initially attracted to the possibilities for polymer synthesis via ring-opening metathesis polymerization (ROMP) as a result of thinking about how to solve a perceived materials need. One of the perennial problems in various parts of technology is the desire of design engineers for materials embodying apparently mutually incompatible properties. In this case the need of aerospace engineers was (and still is!) for thermally-stable elastomers; such materials should have low glass transition temperatures and high thermal stabilities, and should be resistant to degradation by lubricants, hydraulic fluids and fuels (including liquid oxygen). We argued that since a ROMP polymer from cyclopentene, *cis*-poly(pentenamer), has a T_g of -114 °C, which is the lowest value known for a hydrocarbon polymer, it is possible that a fluorinated analogue, $[-CH{=}CH(CF_2)_3]_n-$, might well have a fairly low T_g and also be moderately stable both thermally and towards the fluids indicated above. While we believe that this idea remains valid, we have not yet succeeded in making this polymer and testing it; however, our attempts to do so led us into very fruitful areas.

ROMP is initiated at an active centre, a metallocarbene or cyclobutane, on a transition metal, and it has similarities with Ziegler-Natta polymerization in as much as the initiators are generated via the interaction between a transition metal compound (often a halide) and an alkylating agent (often an alkyl aluminium halide). At the time we started to work in this area the received opinion in the field was that functionalized monomers could not be polymerized by such initiator systems. The argument ran *'the active site on the transition metal will be poisoned by any lone pair donor in the system'*.[2] In the case of olefin metathe-

[2]Primarily as a result of the pioneering work of Professor R. H. Grubbs (Cal. Tech.), we now know that this view was nonsense, but at that time the very concept was heretical. Now, of course, it is possible to conduct living polymerization of functionalized monomers in water using transition metal carbenes as initiators.

sis we have demonstrated that this analysis is invalid; Brian Wilson (PhD 1978) was the first to describe ROMP of halogenated monomers, and his work led to a very active theme over several years and the synthesis of many novel fluoropolymers [6-11]. Indeed, many of our themes in polymer synthesis and properties still derive benefit from the grounding in fluorine chemistry provided by Bob Stephens and Colin Tatlow at Birmingham and my colleagues in the Durham Fluorine Group during my early research years; for example, our studies on the 'Durham Route' to polyacetylene [12-15, (Scheme 2.2)] and stereoregular polar polymers [16] are two fruitful areas which depended on fluorine chemistry.

References

1 W. J. Feast, D. R. A. Perry and R. Stephens, *Tetrahedron*, *22* (1966) 433-439.

2 L. P. Anderson, W. J. Feast and W. K. R. Musgrave, *J. Chem. Soc., Chem. Commun.*, *22*, 1433; *J. Chem. Soc. (C)*, (1969) 211-217 and 2559-2564.

3 W. J. Feast, W. K. R. Musgrave and R. G. Weston, *J. Chem. Soc. (D)*, (1970) 1337.

4 W. J. Feast and W. E. Preston, *Tetrahedron* (1972) 2805-2812 and *J. Chem. Soc., Chem. Commun.*, (1974) 985-986.

5 W. J. Feast, R. R. Hughes and W. K. R. Musgrave, *J. Fluorine Chem.*, *9* (1977) 271-278 and *10* (1977) 585-604.

6 W. J. Feast and B. Wilson, *Polymer, 20(10)* (1979) 1182-1183.

7 A. B. Alimuniar, P. M. Blackmore, J. H. Edwards, W. J. Feast and B. Wilson, *Polymer, 27(8)* (1986) 1281-1288.

8 W. J. Feast and L. A. H. Shahada, *Polymer, 27(8)* (1986) 1289-1295.

9 P. M. Blackmore and W. J. Feast, *Polymer, 27(8)* (1986) 1296-1303.

10 P. M. Blackmore and W. J. Feast, *J. Mol. Cat., 36(1-2)* (1986) 145-152.

11 G. Bazan, E. Khosravi, R. R. Schrock, W. J. Feast, V. C. Gibson, M. B. O'Regan, J. K. Thomas and W. M. Davis, *J. Amer. Chem. Soc.*, *112* (1990) 8378-8387.

12 J. H. Edwards and W. J. Feast, *Polymer, 21* (1980) 595-596.

13 J. H. Edwards, W. J. Feast and D. C. Bott, *Polymer, 25(3)* (1984) 395-398.

14 W. J. Feast and J. N. Winter, *J. Chem. Soc., Chem. Commun.*, (1985) 202.

15 P. I. Clemenson, H. Cramail, P. W. Dyer, S. Feast, W. J. Feast, V. C. Gibson, E. Khosravi, D. Parker and J. N. Winter, Chapter 13 in W. R. Salaneck, I Lindstrom and B. Rånby (eds.), *Conjugated Polymers and Related Materials: The Interconnection of Chemical and Electronic Structure*, Oxford University Press, 1993, pp. 171-184.

16 G. R. Davies, H. V. St. A. Hubbard, I. M. Ward, W. J. Feast, V. C. Gibson, E. Khosravi and E. L. Marshall, *Polymer, 36* (1995) 235-243.

Appendix 7.3

Fluorinated bio-organic compounds

DAVID O'HAGAN[1]
Department of Chemistry, University of Durham, Durham DH1 3LE, UK

David O'Hagan

Fluorine chemistry has flourished at Durham for very many years, and my entry into this field has been greatly facilitated by the culture and reputation of the Chemistry Department here as a centre of excellence for this subject. The main areas of expertise which have been developed over the years – not only by Chambers but also by Brooke (fluoroaromatics) and Feast (fluoropolymers) – concern the synthesis of perfluoro-organic compounds, mechanisms of fluorination, and the behaviour and properties of fluorinated materials. When I arrived in 1986, I began to explore the influence of fluorine in *bio-organic* chemistry, an area which had not really developed at Durham in a sustained way. It seemed appropriate to branch out in that direction since it broadened the scope of organofluorine chemistry in the Department, and also selectively-fluorinated compounds were enjoying an increasing profile in the pharmaceutical and bio-product industries – a trend which has continued undiminished to this day.

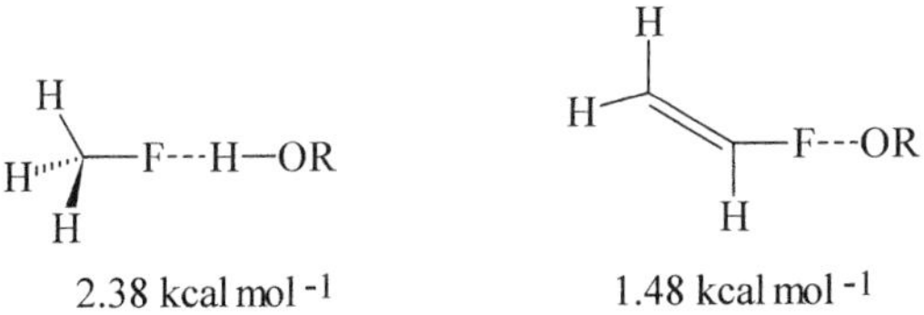

Our general goal was to try to assess the magnitude of perturbation which results from the introduction of fluorine substitution into a biologically relevant compound, in a given situation [1]. It was becoming increasingly popular to substitute F for OH, F for H, or CF_2 for O to generate fluorinated analogues of enzyme substrates, so we began to study systematically the binding to enzymes and stereoelectronic influence of a fluorine atom in enzymatic reactions after such modifications. Our approach was largely experimental:

[1] E-mail: david.o'hagan@durham.ac.uk

synthesising relevant fluorinated analogues such as the phosphonate series shown below (Scheme 3.1) and assessing their kinetic parameters (K_m, V_{max}) with appropriate enzymes [2]. However, many of our conclusions were also backed up by theoretical calculations and by analysing structures of fluorinated compounds in the Cambridge Crystallographic Data Base, in collaboration with Howard at Durham and Rzepa at Imperial College, London. It was calculated for example that the maximum strength of a CF· · ·HO bond is 2.38 kcal mol^{-1}, less than half that of a typical hydrogen bond to oxygen (5.0 kcal mol^{-1}). The strength of the F· · ·H bond is weaker (1.48 kcal mol^{-1}) again if the fluorine is attached to an sp^2 hybridised carbon [3]. In other studies on enzyme systems we have been able to reveal that the '*cis* fluorine-effect' controls the stereospecificity of fluoroacetyl-CoA condensation by citrate synthase to generate a single stereoisomer of fluorocitrate [4], and we have rationalised the ability of hydrolases to resolve α-fluoroacids and amides on the basis of stereoelectronic effects rather than the more conventional steric model [5].

A study on the nature of selectively-fluorinated stearic acids (Fig. 3.1) has been carried out over several years [6, 7] to assess what influence the replacement of F for H or CF_2 for CH_2 has on the physical properties on hydrocarbons and fats. Using LB methods, DSC, scanning tunnelling microscopy (STM to observe the fluorine atoms individually), and X-ray powder diffraction, it emerges that the CF_2 substitution has a significant effect and introduces considerable conformational disorder into such chains relative to the parent hydrocarbon. The origin of this phenomenon can be traced to $C{-}CF_2{-}C$ bond angle (115°) widening relative to $C{-}CH_2{-}C$ (109°).

Another major area which has developed during my time at Durham, and continues to gain momentum, is biological fluorination. Almost all organofluorine chemistry is

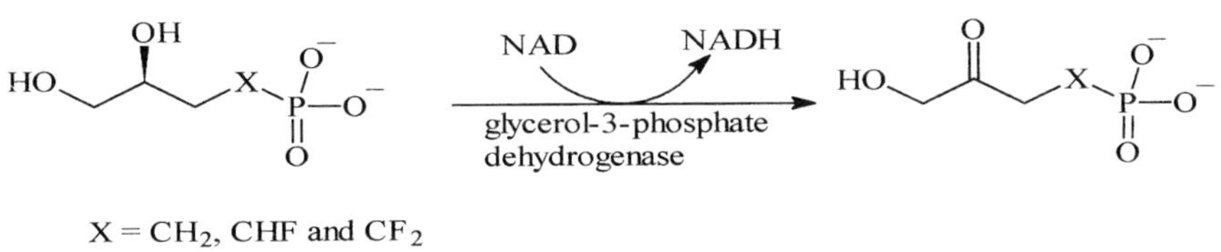

Scheme 3.1.

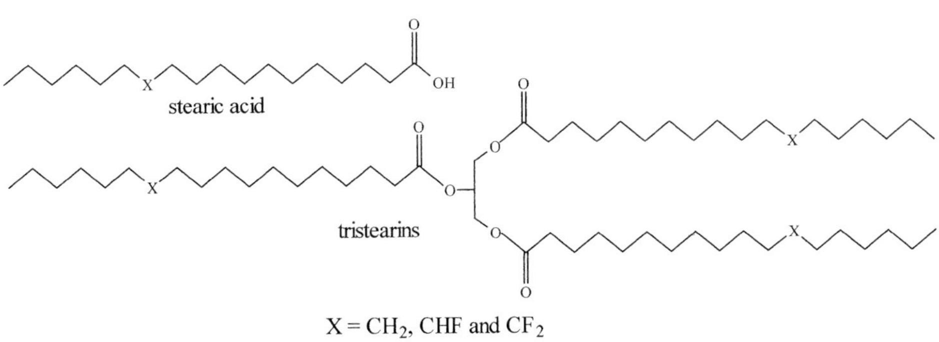

Fig. 3.1.

anthropogenic in origin and Nature has hardly developed a biochemistry around fluorine; however, there are a few rare organisms which do elaborate organofluorine compounds. Our research, being carried out in collaboration with a group at the Queen's University of Belfast, is directed towards discovering an enzyme capable of catalysing formation of the carbon-fluorine bond. Some early studies on the biosynthesis of organofluorine compounds from plants were carried out by Sir Rudolph Peters at Cambridge during the 1960s and 70s, but these studies never revealed the nature of fluorination. Our first efforts in the early 1990s focused on the fluoroacetate-producing plant *Dichapetalum cymosum* from South Africa [9]; however, as in Peters' studies, plants and tissue culture proved very difficult to work with, so over the last five years we have turned to the bacterium *Streptomyces cattleya*, a micro-organism which produces fluoroacetate (**1**) and 4-fluorothreonine (**2**) [10, 11]. We have mapped the pathway to the fluorometabolites through isotopic labelling studies and I would like to think that Durham (with Belfast) will be the place where the first fluorinating enzyme will be characterised in the near future. Our work is now being extended to fluorocarbohydrate synthesis, with aim of generating novel fluorinated antibiotics by biotransformation.

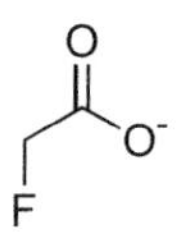

(**1**)

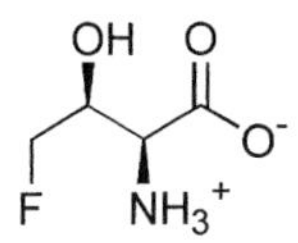

(**2**)

References

1 D. O'Hagan and H. S. Rzepa, *Chem. Commun.*, (1997) 645.
2 J. Nieschalk, A. S. Batsanov, D. O'Hagan and J. A. K. Howard, *Tetrahedron, 52* (1996) 165.
3 J. A. K. Howard, V. J. Hoy, D. O'Hagan and G. T. Smoth, *Tetrahedron, 52* (1996) 12613.
4 D. O'Hagan and H. S. Rzepa, *Chem. Commun.*, (1994) 2029.
5 D. O'Hagan and H. S. Rzepa, *J. Chem. Soc., Perkin Trans. 2*, (1994) 3; J. W. Banks, A. S. Batsanov, J. A. K. Howard, D. O'Hagan, H. S. Rzepa and S. Martin-Santamaria, *J. Chem. Soc., Perkin Trans. 2*, (1999) 2409; J. W. Banks and D. O'Hagan, *J. Fluorine Chem., 102* (2000) 235.
6 L. Dasaradhi, D. O'Hagan, M. C. Petty and C. Pearson, *J. Chem. Soc., Perkin Trans. 2*, (1995) 221.
7 D. O'Hagan, I. Kumadaki, M. Petty, H. Takaya and C. Pearson, *J. Fluorine Chem.*, in press.
8 R. A. Peters and M. Shorthouse, *Nature, 231* (1971) 123.
9 J. J. M. Meyer and D. O'Hagan, *Phytochemistry, 31* (1992) 2669.
10 J. Nieschalk, J. T. G. Hamilton, C. D. Murphy, D. B. Harper and D. O'Hagan, *Chem. Commun.*, (1997) 799.
11 J. T. G. Hamilton, C. D. Murphy, M. R. Amin, D. O'Hagan and D. B. Harper, *J. Chem. Soc., Perkin Trans. 1*, (1998) 759.

Chapter 8

NEVER SAY NO TO A CHALLENGE

KARL O. CHRISTE

Air Force Research Laboratory, Propulsion Sciences and Advanced Concepts Division, Edwards AFB, California 93524-7001, USA and Loker Hydrocarbon Research Institute, University of Southern California, Los Angeles, California 90089, USA

Preface

Writing about oneself is difficult because in their own minds most men greatly overestimate their significance. I've broken down this biographical tale into five parts: my childhood and high school years; the college years; the work at Stauffer Chemical; the years at Rocketdyne; and my present work at the Air Force Research Laboratory and the Loker Institute of the University of Southern California.

The early years

I was born in 1936 in Ulm, Germany, as the third child having two sisters. Ulm, a beautiful historic town on the Danube in Southern Germany, is famous for its gothic cathedral with the highest church tower in the world, and as the birthplace of Albert Einstein. My father, who was a high school chemistry teacher, blessed me with a logical mind and self discipline, and my mother, who came from a book, art and music shop, gave me a fair share of creativity and the ability to improvise. My father's family had come many generations earlier from the French Jura in Switzerland and probably was of Huguenot descent.

Prompted by the outbreak of World War II and the vicinity of our house to the Magirus truck plant, an excellent bombing target, my parents decided to move in 1939 to Bad Mergentheim, a small resort on the 'Romantic Road' in Franconia and my mother's hometown. Despite being 41 years old, my father was drafted in 1942 because of his open opposition to the Third Reich and was killed in action in 1944 on the Western Front. Living in a small resort town, crowded with many army hospitals, we were spared most of the physical devastation of WW II, except for a narrow escape in 1945, when our residence was destroyed by two heavy artillery shells half an hour after we had decided to seek protection in an air raid shelter.

I attended elementary and high school in Bad Mergentheim. The austere living conditions during and in the first years after the War strongly influenced our upbringing. We learned to make do with what we had, which was not very much, to conserve all resources and to rely on mutual help and support. Growing up in a family with only a mother and two sisters, I became rather independent at an early age. I still vividly remember those long summer vacation trips, when I rode my bicycle (mainly alone because I could not find a similarly crazy person to join me) through Germany, Austria, Switzerland, Belgium, the Netherlands, France, Britain and Italy, averaging about 200 km a day while carrying 40 pounds of supplies. The philosophy we grew up with and were strongly influenced by was pure existentialism, with extensive reading of Sartre, Camus, Kafka, and Freud, which seemed very fitting for the war and postwar periods.

For the first eight years in high school I minimized studying, but still maintained good enough grades to keep me within the top of my class. The sciences, particularly, were easy subjects; and I had the good luck to be instructed by an excellent chemistry teacher, Dr Otto Bayer. His teaching of the subject was entirely based on logic and understanding, rather than memorizing facts, which later prompted me to choose chemistry as my career. His lectures were filled with exciting experiments, and his teaching was proactive. He also conducted a voluntary afternoon chemistry lab course where we were allowed to choose the experiments ourselves. I still recollect the preparation of nitroglycerine, a reasonably safe experiment when carried out on a small scale. However, when Dr Bayer discovered that I had without his knowledge scaled up the prep by a factor of 10, he retreated wisely to an adjoining room and calmly and confidently kept giving me instructions through a small opening in the wall. After completing the prep, I carried the sample of several grams carefully to our small river and threw it off the bridge. To my great joy, it made a big bang when it exploded. My second career choice would have been architecture, because I liked its creative aspects and enjoyed geometry, art and sculpture.

Most of my time in high school was spent on sports. I competed in gymnastics, fencing, bicycle racing, and swimming, and despite living in a small town without good facilities or professional coaching, I managed to finish seventh in the under-16 age category National Championships in bicycle racing and to become the under-21 State Champion in fencing. Not being a naturally gifted athlete, I learned quickly that the secret of success was not so much unusual talent but hard work and, above all, determination.

When I finshed high school in 1955 university admission was still restricted, and was based on high school grades. Knowing this, I studied hard for about nine months and pulled up my average by more than one full grade, which some of my teachers never quite forgave: the fact that for eight years I had taken school very lightly and had underperformed scholastically did not sit well with them. The straw which broke the camel's back was the the final maths exam, a tough four-hour written test. I mistakenly thought that the exam would start at nine in the morning and left home at 8:45. On my way to school, a young student came running up to me, yelling that the exam had started already at eight. So I hurried to school, but after looking at the test and realizing how easy it was, I decided to first eat a couple of sandwiches which I had brought along. To keep a long story short, I finshed the test 40 minutes early and correctly solved not only the mandatory five problems but also a sixth voluntary one. I thought this was very cool, but my maths teacher apparently thought otherwise. He accused me of ridiculing the entire educational system and promptly retaliated. In spite of a straight A average for the whole year, including the written test, I had to take an oral exam to determine whether I should be given an A or a B. Needless to say, the outcome was predetermined and, to add insult to injury, I was lectured afterwards by him that this should serve me as a lesson for the rest of my life. So much for the German school system and tolerance!

College years

I followed my first love – chemistry, and acting on the good advice of Professor Ebert (a family friend and chemistry professor at the Technical University of Karlsruhe), enrolled in 1955 at the Technical University of Stuttgart, which at that time had an excellent reputation in chemistry. I was in for a rude awakening. The German university system in those days was very loosely structured and it was entirely up to the individual student to fight his way through it. Furthermore, there were long waiting lists for all laboratory courses, without which one could not take the final exams. Fortunately, the waiting list was ordered on a performance basis, *i.e.*, from the results of a comprehensive chemistry test. This provided the chance to beat out other students who had studied chemistry already for several years and were still waiting. Luckily, I was able to get one of those coveted lab spaces in my first semester by placing in the top 3%. If I had not already been familiar with Darwin's 'Survival of the Fittest', I certainly would have learned it quickly during those years.

Plate 8.1. Working as an unskilled labourer during semester breaks.

During semester vacations, I worked as an unskilled labourer on numerous construction and road building jobs to cover some of my college expenses. A faithful companion on many of these jobs was Manfred Engelhardt, an old high school friend and also a chemistry student at the University of Stuttgart, who went on to become an executive at Merck in Darmstadt. We quickly discovered the blessings of having a job that you can truly enjoy, and that digging ditches for a living was not our vocation!

My financial burden during semesters was also greatly eased by a caring couple, Dr Bernhard Schoeninger, a retired food chemist, and his wife. They were renting out one room of their flat to a chemistry student (so that Dr Schoeninger had somebody to talk shop with) and I had seen their ad on a University bulletin board, but by the time I got there, the room had already been rented to another student. Again, luck struck. When talking to Mrs Emilie Schoeninger, she noticed my last name, which is not very common in Germany, and asked me if, by any chance, I was from Schwaebisch Hall, a small town in Hohenlohe. I was not, but my grandfather and father had lived in Schwaebisch Hall for many years. It turned out that her father and my grandfather had been colleagues at the same post office and that she had known my grandfather quite well. She immediately adopted me like her long lost son and offered me an unheated attic storage room for free. In return, I took care of the coal-fired central heating for the whole house in winter and assisted them whenever they needed help. Being very tired from my usual evening fencing workouts, getting up dependably every day in winter at five in the morning to stoke up the furnace required as many as three alarm clocks, and for many years afterwards I still had occasional nightmares about the tenants banging on their radiators in order to wake me up.

During one semester break, I also learned another important lesson in my life: never accept a job through nepotism. If you are not being hired for your qualifications and your employer does not need your services, don't take the job. I had gotten a summer job with the telephone cable plant of Standard Electric in Stuttgart through an uncle of mine. The head of the Standard Electric Personnel Department had opposed my hiring, but was told that he had to take me for the summer. It turned out that they really did not have any meaningful assignment for me and tried to keep me out of their hair, so they told me to study potential work rationalization methods for their plant. I did not realize that this meant 'please do nothing', and I eagerly went to work. I studied for one week all their manufacturing steps and proudly came back with an extensive list of how they could combine, simplify or eliminate numerous jobs and steps. This was the wrong thing for an inexperienced and unwelcome summer student to do, and I was instantaneously reassigned to another task for which there was no solution. However, the Personnel Director was waiting for his chance to ambush me and he got it. During lunch breaks, I had taken pieces of coloured telephone wire from the trash barrels and, with the help of coloured tape, had modelled them into a series of beautiful animals. Since there was an art competion at a local gallery, I decided to take my 'pets' home and enter them. As I was leaving the plant, however, an overzealous (what else could you expect from a German in a uniform) guard at the exit gate noticed my models and insisted on reporting me for removing, without an official permit, items consisting of telephone wire. Next morning, the Personnel Director had me fired for theft without any chance for a hearing, taking advantage of the fact that his boss was on a four-week vacation. My 'animals', by the way, won a prize in the exhibition, with credit given to Standard Electric. I made sure that the Personnel Director got the Award Certificate from his boss, together with my letter expressing gratitude for his support of a struggling student.

I completed my 'Vordiplom' in the minimum allowable time of two years and then faced a second hurdle – a one-year waiting list for the organic laboratory courses. This time the waiting list was done on a first-come first-served basis, using the completion date of the Vordiplom. When I found out that in Austria there were no waiting lists and that work carried out at Austrian universities was accredited in Germany, it was off to the University of Vienna for me. After my arrival in Vienna, I discovered that one of my former colleagues from Stuttgart, Volker Dorn, had come there for the same reason, and we became close friends.

The year in Vienna (1957/1958) gave me a more balanced outlook on life. It exposed me to a mentality and culture very different from that of my native Suebia, which can best be summarized as 'schaffe, schaffe, Haeusle baue, sterbe,' which roughly translates as 'work, work, build your own little house, and then die.' I found out that there were other things in life besides hard work, and I learned to enjoy them. I spent much time in the theatre and opera and, in the summer, sailing on the Old Danube. As far as chemistry was concerned, I was again lucky. One of my first assignments was to prepare some crazy phenothiazines via a ring-closure reaction in molten sulfur, following some intentionally vague French patent literature. Staff members and some other students had tried repeatedly to run this reaction, but without success. On my first attempt I obtained, using my intuition, the desired compound in high yield as beautiful blue crystals with the correct melting point. As a result, several co-students were assigned to me as helpers, enabling me to spend more time on my newly found hobbies.

Having completed the required organic chemistry courses, I returned to Stuttgart in 1958. Since polymer chemistry seemed to be the field of the future, I tried to get an appointment with the organic chemistry professor to arrange for a Diplomarbeit (the equivalent of a Master's thesis). He declined to see me and sent me to one of his assistants who, with similar Teutonic arrogance, told me that Austrians do not know chemistry and that he might possibly consider me as a graduate student, provided that I first take all the oral exams for the Diplom in Stuttgart and get excellent grades in organic chemistry. The only problem was that in order to take the exams, I first needed to enroll in their own lecture series – which came around on a two-year cycle. Since I would have lost the time that I had so craftily gained by going to Vienna, I decided to look for a more sympathetic professor, and went to see Professor Josef Goubeau, the director of the Inorganic Institute – a decision which I have never regretted, even for one minute of my life. Goubeau was an outstanding chemist, a gifted teacher, and one of the finest gentlemen I've ever met. He had trained and retained a number of outstanding young scientists at his Institute, including H. Becher, K. Dehnicke, A. Allenstein and W. Bues, making the Technical University of Stuttgart a powerhouse in synthetic inorganic chemistry. Professor F. Seel, a well-established fluorine chemist, and his group had also joined the Institute, but they had great difficulty in attracting coworkers and were outclassed by Professor Goubeau, a Bavarian of Huguenot descent who received me with open arms, gave me a Diplomarbeit and – recognizing my financial needs – a teaching assistant position. He had very broad interests and gave me a wide range of topics to choose from, the two most fascinating being the synthesis of the NF_4^+ cation and an improved synthesis of diborane by catalytic gas-phase hydrogenation of trimethylborate. Since there were no existing facilities for the fluorine project, which was later unsuccessfully pursued by Eberhard Jacob, I opted for the diborane topic. A thermodynamic analysis of the hydrogenation of trimethylborate revealed that the the breakage of the C—O bond was favoured over that of the B—O bond, and this was also quickly verified by experiment. Therefore, I replaced trimethylborate by the equally cheap BCl_3 and developed in this manner a nice synthesis for boranes. Unfortunately, we were too slow to publish our results and were pre-empted by several US Patents, in which the same concept was described.

Up to this stage of my college career I had still intensely pursued my athletic ambitions. I had concentrated on fencing, and through endless hours of training and competition had reached a level where I could hold my own against almost anybody. My main goal was to make the German Team for the 1960 Olympic Games in Rome. Germany had a powerful young team that would go on to win two medals in Rome. In the six months preceding the Games, I had reached the finals in three World Cup tournaments and had taken second in the Europe Cup. I had made the German squad, survived the elimination tournament against East Germany, and everything looked rosy. But then disaster struck. I completely separated my right hamstring in a fencing accident and, because of an incompetent medical decision made in a Specialty Sports Clinic, I was not immediately operated on but sent home. Living by myself in the attic without a telephone, it took three days until my fencing training partner and close friend, Michel Bodemer, found me there unable to move, with my leg black from heel to hip and having a 90 degree bend at the knee. At this point it was too late for any surgical repairs. The main concern was to avoid an embolism and to get the leg straightened out, using a 200 pound bag of lead. I was told that I would never fence again and should be glad if I could learn to walk with a cane.

While in hospital, I found out through an aunt that my mother had been admitted to a clinic at the University of Wuerzburg and undergone exploratory brain surgery. She was diagnosed as having terminal brain cancer and given three more days to live. Since my hospital would not release me, I asked a friend to park my little Fiat outside the hospital then escaped at night and drove to Wuerzburg. It was quite a challenge to drive this car with an unsynchronized manual transmission requiring double declutching, using only one foot. I was very grateful to have reached my mother, with whom I had a very close relationship all my life, in time to be with her in her last hours.

The next few months were rough by any standards since, in addition to all the other problems, I also had to take my Orals for the Diplom. Within nine months I had sufficiently rehabilitated my leg by swimming every morning in an unheated outdoor pool, even through the nippy German winter. After one year I fenced again and in my first tournament, an international epee (the only weapon left for a crippled fencer) event in Switzerland, and managed to take a totally unexpected third place without making a single attack. Out of necessity, I had learned to rely exclusively on intelligence and flawless technique, since I no longer possessed superior speed and aggression. Had I fully recognized these principles before my accident, I might have been unbeatable!

After completion of my Diplom, Professor Goubeau wanted me to continue working on the synthesis of boranes for my PhD, but I preferred to change topics because of what happened to a fellow student named Palandt, who had survived four years on the Russian Front then eight years in a Siberian POW labour camp. While working in the same laboratory as me with trimethylborane, which, like diborane, is hypergolic, he had a tragic accident – almost dying from severe burns and becoming permanently crippled. One of the main causes of the accident was that owing to monetary constraints, we had to cool our cold traps with liquid air, which upon prolonged storage became almost pure liquid oxygen. Without the availability of liquid nitrogen, I was not interested in continuing my work on hypergolic materials, so Professor Goubeau offered me another challenging topic. This involved Dr W. Wilborn, a former student of his who had joined Farbwerke Hoechst and was looking for help with problems associated with a new class of inorganic high polymers derived from silicophosphates. Although no one in our Institute had any expertise in this area, I accepted the job because it paid 50 Marks more per month and freed me from my teaching obligations. With most of my other obsessions, such as fencing, greatly reduced, I concentrated hard on my work and completed the assigned task in about four months.

Realizing that even with the most understanding and benevolent thesis advisor in the world I could not get away with this for a PhD thesis, I made an appointment with Professor Goubeau and asked him what he expected me to accomplish in my work. He promptly turned the tables on me and wanted to know what I considered reasonable. So I told him about all my results, without actually mentioning that the topic was already complete. He was duly impressed and promised that if I could accomplish all this work, he would let me go. To save everybody's face, I fed him my results over several months, took a long camping trip with my friend Peter Kuhnle through the Balkan countries, Greece, and Turkey, and delivered the final data by the end of the year. I reminded Professor Goubeau of his earlier promise, and as a true gentleman, he honoured his word. I moved quickly and took my PhD oral exam two days before Christmas, thus completing my chemical education in the record time of 12 semesters, in spite of my many extracurricular activities and interests.

Professor Goubeau tried to persuade me to habilitate under Professor Hans Siebert in Clausthal-Zellerfeld, but I had different ideas; I preferred an industrial career – but not in Germany. Having worked as an unskilled labourer in the chemical industry during semester vacations and having done my PhD research for Hoechst, I had seen enough, actually more than enough, and I knew that rigid German hierarchies were incompatible with my Huguenot spirit of freedom and independence. I still remember my first visit to Hoechst. I had arranged to see a Professor W. Teske, so I located his office and told his secretary that I had an appointment with 'Herrn Teske'. She informed me that she did not know such a person. Since I was confident that I was at the right place, I waited for approximately ten minutes then repeated my request, which she dignified with the same contrived answer. So, I sat for another ten minutes in front of her desk, after which time she told me that they did have a Professor Doctor Doctor Teske, if, by any chance, I would like to speak to *him*. After I had finally earned my audience with the 'Herr Professor Dr Dr', the first thing he explained to me was that he hated people with beards. Needless to say, I was sporting a beard.

Overall, my years in Stuttgart and Vienna were – to steal a line from Frank Sinatra – very good years. Professor Goubeau had instilled in me an everlasting love and enthusiasm for chemistry, and since then I have considered chemistry as a paid hobby rather than work. Through my fencing activities, I had travelled all over Europe and developed many long-lasting friendships. Above all, I met my dear wife Brigitte at the University of Stuttgart, where she worked at the 'Studentenwerk'. We have been happily married since 1962 and together have raised three wonderful children. My upbringing and training, based on a 'sink or swim' philosophy, had prepared me well for my next move: the pursuit of the American dream.

Coming to the United States, and the Stauffer experience

Within four weeks of completing my PhD degree, I had married Brigitte and also managed to obtain an unrestricted immigration visa (green card) to the United States. Around 1962, there was a shortage of scientists in US Government Laboratories, so they were actively recruiting qualified people through 'Action Paperclip'. Transportation expenses were paid, but in return one was obliged to work for several years for an Agency chosen from a very limited list. I did not like this aspect of buying a pig in a poke. Unfortunately, all I had at that time were negative assets, and it required skilful negotiations with the American Consulate to obtain an unrestricted visa. The next hurdle was transportation. Through friends, my sister got me a free ride on a coal freighter from Hamburg to Norfolk in Virginia. My only obligation was to play 'Skat' (a German card game) every day with the captain and his officers. It certainly beat washing the dishes in the kitchen and, since the loser had to buy, it also provided me with plenty of free beer. Again, my rigorous training from my college and fencing days had paid off. The boat ride was rough, as we ran through the centre of two winter storms, but eventually I disembarked safely in Norfolk carrying all my belongings in two suitcases and took my first step on American soil. Contrary to all these touching stories about new immigrants falling on their knees and kissing the soil, my experience was quite different: my first step was on wet loose coal dust and I sank in almost to my knees. What a mess that was! I had expected a different kind of welcome.

Since I was severely undercapitalized, to put it mildly, I had purchased in Germany a $99/99-day unlimited Greyhound Bus ticket and avoided accommodation expenses by taking night buses whenever possible. My first scheduled stop was Wilmington in Delaware to take a look at DuPont. I arrived on a Sunday afternoon in beautiful downtown Wilmington. What an eye opener that was, compared to European cities, even shortly after the war. The only nice looking building was the Hotel DuPont. Gambling that it actually belonged to DuPont and that the company would cover my bill, I checked in, knowing very well that I would be unable to pay if DuPont declined. Through Reinhard Schmutzler, whom I knew from Stuttgart, I got an interview, my first job offer from a US company, and also a moderate travel cost reimbursement for the Norfolk-Wilmington leg. For the next two months I crisscrossed the country several times by bus, stopping in major cities and calling at chemical companies to request the favour of an interview. I quickly learned the ropes and managed to get offers from most companies, which somehow seemed to be impressed by my 'take no prisoners' attitude and willingness to tackle any kind of job.

Since it was winter, I concluded that California had by far the best climate and offered the best quality of life. Consequently, I made up my mind to become a California chemist. The only problem was that there was not much of a chemical industry on the West Coast. Having learned my lesson about nepotism well, I declined the help of Brigadier General William Kunzig, a former fencing team-mate from Stuttgart who was by then stationed at the San Francisco Presidio. I got two offers from California companies: one from Standard Oil as an X-ray crystallographer and one from the Western Research Centre of Stauffer Chemical in Richmond as an organic fluorine chemist. I accepted the second one and became fascinated with fluorine chemistry.

My group leader at Stauffer was Attila Pavlath, a Hungarian and one of George Olah's first graduate student's back in Budapest. In 1956, he had fled Hungary after the revolution and had come to this country. He and his wife Kata kindly and patiently helped me to get settled and started. By May 1962, I had raised enough money for an airline ticket for Brigitte and she joined me in California. We adjusted quickly, moved out of a seedy neighbourhood to a better area of Berkeley, had our first son, Ralf, and two years later our second son, Mark.

Stauffer was a major producer of chlorinated hydrocarbons and HF and, as far as we could judge, in an excellent position to enter the fluorocarbon market. My major contributions to the fluorocarbon area were the development of novel processes for making freons [1] using SiF_4 (a waste gas from Stauffer's fertilizer plants) and for introducing fluorine into aromatic rings [2]. After our Stauffer patents on these processes expired, they were commercialized by Bayer in Germany and ICI in England.

Stauffer was also interested in the production of inorganic fluorine oxidizers for rocket propulsion; among other ventures, they had a pilot plant for making N_2F_4 from NF_3 and carbon in a fluidized-bed reactor. Attila had submitted a proposal to the Office of Naval Research (ONR) to investigate the possible synthesis of $ClF_2^+NO_3^-$ and $ClF_2^+ClO_4^-$. I was very pleased when this proposal was funded, as it provided for me an opportunity to utilize my inorganic chemistry background and also allowed us to hire a second inorganic chemist, Dr Jacques Guertin from McGill University. We rather quickly demonstrated that the desired ClF_2^+ salts were thermally unstable and, in the process, discovered several new halogen fluoride ions, such as ClF_2^- [3]. At this point, it had become clear that ONR would drop our contract, unless we had some spectacular results. Without the blessing of Attila, who feared that we were violating the work statement of the ONR contract, Jacques and I started to work on synthesizing the NF_4^+ cation. At that time, this species was the 'Holy Grail' of nitrogen fluoride chemistry. Earlier calculations by a group of chemists at Shell Development had predicted that this cation could not exist as a stable species. Furthermore, for several years a team under Wes Tolberg at Stanford Research Institute (SRI) had unsuccessfully attempted to prepare $NF_4^+HF_2^-$ from NF_3, F_2, and HF at elevated temperatures. None of this deterred us. The exceptional stability of isoelectronic CF_4 was a sufficient incentive to pursue NF_4^+, and we were lucky. We built a low-temperature glow-discharge reactor and in our first experiments, using NF_3, F_2 and AsF_5 as starting materials, prepared $NF_4^+AsF_6^-$ [4]. The successful synthesis of a stable NF_4^+ salt was a major scientific breakthrough, but it also taught me some bitter lessons about human nature and greed.

Several months before our NF_4^+ synthesis, I had urged Wes Tolberg at a conference in Palo Alto to abandon his experimental approach and instead use NF_3, F_2, and SbF_5 at elevated temperature and pressure. Shortly after our successful synthesis and identification of $NF_4^+AsF_6^-$, he phoned me to say that, following my advice, he had obtained a stable white solid. I told him that, in the meantime, we had made $NF_4^+AsF_6^-$ and characterized it by vibrational spectroscopy. We agreed to take our sample to SRI and to record and compare the ^{19}F NMR spectra of both compounds. It turned out that they exhibited a common NMR signal and, therefore, were both NF_4^+ salts. Our Stauffer contract under ONR sponsorship was unclassified and we were ready to publish, whereas the SRI contract under Air Force sponsorship was classified. We assured Wes that we would hold our manuscript until he could get his work unclassified to give him a chance to publish his results back-to-back with ours in *Inorg. Nucl. Chem. Letters*. Instead of being grateful, Wes contacted ONR to get our work classified while getting his own work declassified, and then submitted his paper for publication. Fortunately, I found out about this through one of the reviewers of the Tolberg paper and was able to persuade Joe Katz, the editor of *Inorg. Nucl. Chem. Letters*, to slow down Tolberg's paper until we could get ours unclassified again; we barely made the same journal issue as Tolberg.

Greed about the NF_4^+ success regrettably also destroyed my close friendship with Attila. He wanted to exploit the perceived NF_4^+ goldmine all by himself with the help of Jacques Guertin, and forebade me to continue working on this programme. The loss was his, however, because in the two years from 1965 to 1967 he produced nothing I viewed as useful on this topic.

The work at Stauffer came to a fiery end in 1967, when Ferenc Pallos, a Hungarian organic chemist from our group, had a mishap while trying to copolymerize vinyl chloride and carbon disulfide in a glass autoclave at elevated temperature and pressure. The autoclave ruptured and the overheated carbon disulfide caused a secondary fuel-air explosion, resulting in a fire that destroyed the laboratory. Fortunately, Ferenc escaped unharmed.

At about the same time, we were pushing for the commercialization of our newly developed fluorocarbon processes, which required the approval of Hans Stauffer, the President of Stauffer Chemicals. When he found out that we wanted to compete against DuPont in the fluorocarbon market, he was flabbergasted. Stauffer Chemical had a large plant for making chlorinated hydrocarbons from chlorine and natural gas in Louisville, Kentucky and the feasibility of this operation depended entirely upon the sale of the HCl by-product to a neighboring DuPont neoprene plant. Stauffer clearly could not risk losing this operation by entering a competition with DuPont in fluorocarbons. This raised the obvious question of why had Stauffer carried out research in this area for the past ten years with a sizeable group if they had no inclination to enter this field, and their answer was to get rid of this group of pyromaniacs. Ironically, if we had been less successful, we could probably have done fluorocarbon research at Stauffer forever by not forcing top management to make a decision!

Although I was given the opportunity by Stauffer to stay and carry out process development on methane oxychlorination, I decided to look for more exciting work. I received offers from numerous companies, including Kaiser Aluminum, Kerr McGee, Cowles, the Naval Research Lab, and IIT Research Institute, but my dream company was Rocketdyne, the undisputed leader in rocket propellant research, employing several hundred first-class scientists working on all aspects of the subject. I had applied to Rocketdyne twice, only to be told that there were no openings; but before accepting employment elsewhere, I decided to try for the third time within three months, and this time I succeeded.

Again, as so often in my career, I was lucky. Dr Fred Bauer, a member of the Rocketdyne fluorine group, had just accepted the position at Cowles which I had originally been offered but, because I was not impressed by the Company, I had asked for double my Stauffer salary. This outrageous demand had prompted them to look for a more affordable employee and they found Dr Bauer. This freed up a position at Rocketdyne which I accepted for a mere 20% salary increase. It was a good choice, because a year later Cowles was bought by Stauffer. Again my German stubbornness and unwillingness to take no for an answer had paid off, and I enjoyed 27 years of highly productive and challenging work at Rocketdyne.

The Rocketdyne years

We moved in 1967 from Berkeley to Los Angeles and bought our own house and furniture. (After our first visit back to Germany in 1965, Brigitte and I had come to realize that Germany was a nice place, but only for visits.) Also in 1967, I became a US citizen – a status required of those working on classified government contracts.

In 1967 Rocketdyne was at its peak. It was and today still is the world's leading manufacturer of liquid-fuel rocket engines. It was instrumental in the Apollo Programme that culminated in 1969 with the first manned moon landing. In my personal opinion, this feat is the greatest technological achievement of mankind. My direct boss at Rocketdyne was Dr Don Pilipovich, a highly creative and free spirit who was a master in motivating people and delegating. I had the pleasure to work with excellent colleagues, particularly Drs Carl Schack, Dick Wilson, Ross Wagner, Earl Curtis, Lou Grant, Walt Maya, Art Axworthy, Howard Rogers, Frank Gunderloy, John Hon, Chuck Lindahl, and Milt Fraenkel, and I greatly improved my laboratory skills. The working climate at Rocketdyne was extremely competitive, not only towards other companies, but also among and within the individual groups. With the successful landing on the moon in 1969, the national goal of beating the USSR in the space race had been accomplished, so the US Government started to cut back drastically its funding for space programmes, and by 1970 the Rocketdyne workforce had shrunk from more than 25 000 in 1965 to a mere 3000. Again, I was very fortunate. I had been hired last and had the least seniority in our department, but the lay-offs were done exclusively on a performance basis. We were left with a very small but extremely competent group of fluorine chemists, with Carl Schack, Dick Wilson and myself being the core of this unit. The lay-offs were carried out mercilessly: if a person did not have a valid charge number, he was laid off with two weeks notification. We were lucky to have relatively stable, though modest, support from the Office of Naval Research which kept us going. In return, we were highly productive. For example, our final ONR report [5] covering a two-man effort from 1970 to 1978 contained 86 technical publications in refereed journals and 18 patents, in addition to unpublished work. During this period, we discovered many novel compounds, such as ClF_3O, ClF_3O_2, ClF_6^+, $ClF_2O_2^+$, ClF_2O^+, ClF_4O^-, $ClF_2O_2^-$, halogen perchlorates, the first stable OH_3^+ salts, H_2OOH^+, SH_3^+, $NH_2F_2^+$, $H_2PF_4^-$, NF_3^+, and a large body of NF_4^+ salts.

The quality and quantity of work generated during these years had established our Rocketdyne group as a major innovator and driving force in fluorine and energetic materials chemistry. However, we were in the same position as the last horse buggy makers after the discovery of the automobile. It does not help you – even if you make the best buggy in the world – if nobody buys horse buggies anymore. Fortunately for us, a new game – chemical lasers for the Strategic Defense Initiative (SDI or 'Star Wars') – had arrived in town. The two most promising systems were the HF/DF laser and the singlet delta oxygen/iodine (COIL) laser. However, the HF/DF laser used elemental fluorine or nitrogen fluorides and hydrogen and deuterium as supply gases, and the use of high pressure or cryogenic gases under battlefield conditions was considered unacceptable; therefore our Rocketdyne group proposed and patented the use of solid propellant F_2/NF_3 and H_2/D_2 gas generators, and was securely funded for about a decade under this programme. We used the pyrolysis of our NF_4^+ salts for the F_2/NF_3 generators, generating the required heat by burning a small percentage of these salts with some aluminum powder which was added to the formulations. The ultimate goal was to increase the NF_3/F_2 yields by making the counterions as light as possible, using multiple-charge anions, and employing high-oxidation-state anions which on decomposition would also yield F_2. We prepared a large number of novel NF_4^+ salts, including $(NF_4)_2NiF_6$, which, on a volume basis, contains 12% more usable fluorine in the form of F_2 and NF_3 than liquid fluorine itself at its boiling point [6]. We also developed processes for making NF_4^+ salts on a large scale and produced 40 kg of NF_4BF_4, formulated it into solid propellants and successfully test fired them on a 10 kg scale. The somewhat improved funding situation during this period allowed us to hire one new member, for our group, Dr William Wilson, who became a key player.

We learned to become wary of certain government in-house researchers, who, by giving us a token small contract, gained access to our concepts and results and, behind our backs, obtained patents on our work in their own names. We also learned that sometimes there are penalties for being fast and highly efficient: the parallel effort of solid propellant H_2/D_2 gas generators, an easier task, was progressing much more slowly and, as a result, was funded for several years longer. It shows that survival as a contractor has many pitfalls and is a lifelong learning process.

Other drawbacks of working in a government-controlled and highly-regulated industry included much unnecessary red tape and security regulations. It was a continuous game of beating the system in order to get a job done. This frequently brought us into conflict with our security people, who welcomed any opportunity to nail us. They controlled all our incoming and outgoing mail, stole the stamps from our letters, and distributed incoming personal letters throughout the Company. However, every dog has its day, and we got our revenge. In 1969, I had qualified for the World Championships in fencing as a member of the US National Team (my citizenship, granted in 1967, allowed this); the Championships were held in Cuba, which at that time topped the list of countries forbidden to US citizens, and we needed special permits from the US State Department and a visa from the Swiss Embassy in Mexico City. From Havana, I sent an open postcard to my boss, Don Pilipovich, at Rocketdyne, thanking him profusely for all the secret reports on 'TIHS' he had sent me, and signed it 'Fidel'. Our beloved security officer promptly intercepted this card and immediately ran with it to our Company President to show him the proof that this darn Pilipovich was spying for Cuba. Fortunately, the IQ of our Company President was much higher than that of the security person, and he not only quickly realized the hoax but also joined the game by asking the security officer whether he had tried to break the code word. When the officer responded that he had tried but not succeeded, he was told to read it backwards. You can imagine the embarrassment and humiliation of our security friend, and from then on he left us in peace, fearing further retaliation. I even managed to bring my close friend Dr Roland Bougon of the French Atomic Energy Commission into our laboratories every working day during 1984 using a daily visitor's badge. Roland was spending a one-year sabbatical with me, and the original plan to set him up in Professor Bill Kaska's lab at UC Santa Barbara proved unsatisfactory.

One of the achievements during my Rocketdyne years which received perhaps the most publicity was the first chemical synthesis of elemental fluorine. It was done in 1986, the year in which a special symposium was held in Paris to celebrate the centenary of Henri Moissan's isolation of elemental fluorine. As general secretary of the organizing committee, Roland Bougon had spent a full year helping to make sure that this 100-year jubilee would be a most impressive and memorable event, and he invited me to give a main lecture on the chemical generation of elemental fluorine. Attempts to synthesize elemental fluorine by chemical means predated Moissan's electrochemical synthesis in 1886 by almost 100 years, and every textbook maintained that it was impossible to prepare fluorine by chemical means because it is the most electronegative element. When I started to prepare slides for my lecture a week before the Moissan Symposium, I realized that my presentation would be rather anticlimactic if all I could say was that you cannot make fluorine by chemical means. Thinking some more about the problem, it occurred to me that I should be able to make fluorine in the same manner I had used for my solid propellant NF_3/F_2 gas generators. The only thing I needed was a high-oxidation-state transition metal fluoride anion which could be prepared without the use of elemental fluorine and which upon treatment with a strong Lewis acid would form a thermodynamically unstable fluorocation capable of decomposing spontaneously to a lower fluoride and fluorine. I quickly realized that all the required compounds were well known, and actually had been known in Moissan's time. Excited, I rushed to the lab, prepared K_2MnF_6 in aqueous HF solution from potassium permanganate using H_2O_2 as the reducing agent, as already described by Weinland and Lauenstein in 1899 [7], and heated it with SbF_5, which can be readily obtained from $SbCl_5$ and HF. And voilà! – in three days I had demonstrated a fair-yield chemical synthesis of elemental fluorine via thermal decomposition of MnF_4. I incorporated it in my presentation for Paris: the timing and occasion were perfect, and the work [8] received much attention. Our contributions to fluorine chemistry were also recognized by the bestowment of the 1986 ACS Award for Creative Work in Fluorine Chemistry in New York City, in connection with another celebration of the 100-year jubilee of Moissan's discovery. Again, the timing could not have been better.

However, it was not all fun and games; there were some major bumps in the road. One in particular was a harbinger of the eventual demise of our group. Ross Wagner, a member of my group, was working under the direct guidance of my director at the Santa Susana Field Laboratory on a new monopropellant, azido(dinitro)ethane. He had prepared half a pound of this material, but its unexpectedly high impact sensitivity indicated that it might contain a more sensitive impurity. He brought all the material to our Canoga Park Laboratories and distilled it to isolate the suspected impurity. While transferring one of the fractions with a syringe, friction set off the material and it exploded, initiating the detonation of the bulk of the sample. Miraculously, Ross survived this blast, which had the explosive power of one pound of TNT, but he was badly injured. He lost both eardrums and fingers on both hands, was severely burned, and needed many months of reconstructive surgery and skin grafts. Fortunately, Ross was an early bird and the explosion occurred half an hour before the start of the regular shift; otherwise the consequences could have been much worse, as several people normally worked in that area.

A glass vacuum line in the other corner of the laboratory contained about 30 grams of Ross' propellant that had survived the explosion, and this material had to be removed and desensitized before the accident investigators could be allowed to enter the laboratory. Since the responsible managers and my director were unwilling to touch this material, the task of removing it fell to me. Rather predictably, one of the fractions in the vacuum line exploded in my hands, breaking one of my fingers, in spite of the steel-reinforced safety gloves I was wearing. Our internal company doctor took X-ray photographs and placed a splint on the broken finger. To conceal it from the outside investigators and the public, my mishap was never documented, nor was I sent to a regular doctor. To add insult to injury, an internal investigating team consisting of three vice-presidents unsuccessfully attempted to make me the scapegoat for Ross' accident. Only after my threat of a lawsuit did they put the blame where it belonged – on the responsible managers. Ross exhibited exemplary fortitude throughout his ordeal: he never complained and after his rehabilitation rejected a company offer of early retirement on full pay. Instead he returned to Rocketdyne to resume work on energetic materials until he reached his regular retirement age of 65 and now, at age 70, he still works half-time with me at USC.

Because we had managed to solve the NF_3/F_2 gas generator problems so efficiently, we had to find new support and this time it came from the Air Force, enabling us to start an interesting programme aimed at identifying and developing High Energy Density Matter (HEDM) for propulsion purposes. This allowed us to continue our research on new energetic materials, and also resulted in significant fundamental contributions to chemistry. By relying heavily on the synergism between theory and synthesis, and collaborating more and more with excellent theoreticians such as Dr David Dixon, we made significant contributions to high-coordination-number chemistry (particularly for cases exceeding CN 6), and our discovery of a method for making truly anhydrous tetramethylammonium fluoride (often referred to in the literature as 'naked fluoride') resulted in a renaissance of the subject. We prepared and characterized exciting ions such as ClF_6^- [9] and, in collaboration with Professor Gary Schrobilgen's group, worked on the elusive PF_4^- anion [10], and the unique XeF_5^- anion [11] – the first example of a pentagonal planar AX_5 species. In collaboration with Dave Dixon, we developed the first quantitative scale for the strength of oxidizers [12].

Although internal funding became increasingly tighter as Rocketdyne was readied by its new president, Paul Smith, for sale to Boeing, the overall situation was reasonably stable until 1994. Then disaster struck twice. First, there was the big Northridge Earthquake, which destroyed our laboratories. It took us about half a year to rebuild the labs, but shortly after completing the reconstruction, two colleagues with whom I shared offices, Drs Larry Pugh and Otto Heiney, were killed in a tragic explosion while burning large amounts of outdated explosives at the Santa Susana Field Laboratory. Although my research group again had nothing to do with this operation, Rocketdyne's new president decided that all chemistry was evil, and within two weeks he took away all our chemicals and refused to sign my Air Force renewal contract. The message was clear: there was no future left for chemistry at the 'Rocket Factory', and the time had come to move on. Although I was concerned that at the age of 58 I might have difficulties finding a new home, it turned out that my fears were unfounded: within one week I had five offers to choose from.

Plate 8.2. The four musketeers from Rocketdyne in 1990. From left to right: Dick Wilson, Carl Schack, KOC, and Bill Wilson.

Overall, my 27 years at Rocketdyne had been highly productive and gratifying. I had had the pleasure of working with a tightly-knit group of friends, notably Carl Schack, Dick Wilson, Bill Wilson, and Ross Wagner, who respected, appreciated, and protected each other and were absolutely dependable; together, we had endured many hardships and lived by the motto of the three musketeers: 'all for one and one for all'. We survived the behaviour of our director who, for the past 15 years, had made life very difficult for us. As the responsible line manager of our group, I had to take the brunt of his abuse and was careful not to pass it on to my people. With the support of my family and friends, I overcame a bout with colon cancer and a partial colon removal in 1982. It made me realize that we are living only on borrowed time and that we must never compromise our principles and goals in life, even if it involves personal hardship and occasional controversy.

The 27 years at Rocketdyne also provided a stable basis for raising our children and sending them through college. (Our daughter, Tina, was born in 1969 and has been an absolute joy.) I still maintained some of my hobbies, including fencing, tennis and scuba diving. Although I had not been training anymore seriously, I still managed to win seven West Coast Individual Championships and two National Team Championships in epee fencing and I fenced on the US National Team in the World Championships. After finishing my active fencing career at the age of 42, I started teaching the sport and coached one of my students to a National Championship; also, I was the coach of the 1993 Women's National Epee Team at the World Championships.

The Air Force Research Laboratory and Loker Hydrocarbon Research Institute

Of the five job offers mentioned above, the two that were the most intriguing offered the best opportunity to continue my research in fluorine chemistry and energetic materials and did not force me to uproot my family. Of these, I decided to accept a position as a Senior Staff Advisor through an on-site contractor (Hughes STX) at the Air Force Research Laboratory (AFRL) at Edwards Air Force Base in the Mojave Desert, about two hours drive from Los Angeles. AFRL had funded my HEDM programme at Rocketdyne and had already set aside the money for my programme. I was also able to bring along Bill Wilson, which greatly facilitated the transition. Rocketdyne gracefully allowed us to take along most of our chemicals and laboratory equipment to AFRL, which immensely helped our start at Edwards, where funding had also become very tight.

The second offer came from an old friend, Professor George Olah of the Loker Institute of the University of Southern California, with whom I had previously coauthored a number of papers. George wanted me to join USC on a full-time basis, but when I told him about my preference for AFRL where I could carry out risky work more safely, he persuaded me to join Loker on a part-time basis as an Adjunct Professor and generously provided me with a laboratory and seed money to help me get started. AFRL agreed to let me spend 20% of my time at USC, and provided some additional seed money. With the help of Ross Wagner, who had retired from Rocketdyne and accepted a half-time job with me at USC, we built a well-equipped laboratory using old Rocketdyne equipment. I obtained an NSF grant that allowed me to hire through my German connections some outstanding postdoctoral collaborators, such as Drs Robert Gnann, Xiongzhi Zhang, Berthold Hoge, Joachim Hegge, Thorsten Schroer and Stefan Schneider, and we have developed a viable

Plate 8.3. If you are too old to win your own battles, teach.

research programme at USC and enjoy the collaboration with Professors Olah and Prakash. At Edwards, I hired an excellent postdoc under an NRC grant, Dr Greg Drake.

In the four years since the exodus from Rocketdyne, we have successfully reestablished ourselves in fluorine and energetic materials chemistry. Collaborating extensively with numerous excellent theoreticians, including Drs Jeff Sheehy and Jerry Boatz from AFRL, Dr David Dixon from Pacific NorthWest National Laboratory, Dr Anatoli Korkin and Professor Rodney Bartlett from the University of Florida, Dr Golam Rasul from USC, and Dr Harvey Michels from United Technologies, we have been highly successful in predicting, synthesizing, and characterizing numerous novel and interesting compounds. Typical examples include the POF_4^- anion [13], the trimethylperoxonium cation [14], the triazidocarbenium dinitramide and perchlorate salts [15], the NO_4^+ cation [16], $N(N_3)_3$ and the $N(N_3)_2^-$ and $N(N_3)_4^+$ ions [17], the IF_5^{2-} [18] and IF_5O^{2-} [19] anions, and the first heptacoordinated pnicogen anions, SbF_7^{2-} and BiF_7^{2-} [20]. We have also found new methods for chemically oxygenating NF_3 to NF_3O [21] and achieving stepwise fluorine-oxygen exchange in halogen fluorides using the dinitramide anion as a reagent [22]. The most recent discovery, which has received coverage even in the *New York Times* and *London Times* [23], is the single-step synthesis of the N_5^+ cation in essentially quantitative yield, work done in collaboration with Bill Wilson, Jeff Sheehy and Jerry Boatz [24]:

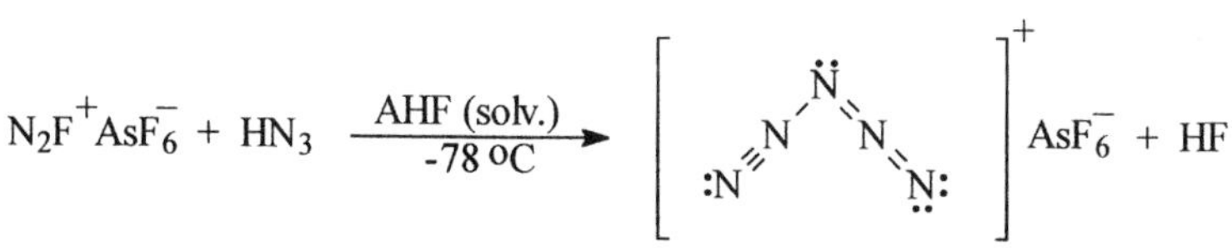

The N_5^+ cation is only the third homoleptic polynitrogen species that has been isolated in bulk and may provide the basis for new high-energy-density materials.

Concluding remarks

If I were given the choice to change any of the major decisions in my life, would I do so? The answer to this question is an emphatic NO. I have never said no to a challenge, and I have fought for my principles. It has hurt at times, but I have never regretted it. From my athletic career I have learned that you cannot and will not win all the time, but you certainly can fight for it. If you gave it your best shot, you will be at peace with yourself, no matter what the outcome is. My cancer ordeal also has taught me not to waste time on trivia and the importance of setting yourself goals and living every day as if it were your last.

Finally, I would like to comment briefly on happiness – a major goal in life. The choice to be either happy or miserable is entirely ours. Having a glass of delicious wine which is filled to the middle, we can either be happy that it is half-full or feel miserable that it is half-empty. How we feel about it does not change the level in the glass, but it certainly determines our happiness. Furthermore, we should always keep in mind that there is no permanent state of happiness, as we sense only changes but not steady states. Because upswings cannot last forever, we must learn to accept occasional depression as a prerequisite for happiness.

References

1 K. O. Christe and A. E. Pavlath, *J. Org. Chem., 29* (1964) 3007.
2 K. O. Christe and A. E. Pavlath, *J. Org. Chem., 30* (1965) 3170; *30* (1965) 4104; *31* (1966) 559.
3 K. O. Christe and J. P. Guertin, *Inorg. Chem., 4* (1965) 905; *4* (1965) 1785.
4 K. O. Christe, J. P. Guertin and A. E. Pavlath, *Inorg. Nucl. Chem. Letters, 2* (1966) 83; *Inorg. Chem., 5* (1966) 1921; *6* (1967) 533.
5 K. O. Christe, *Final Report, Inorganic Halogen Oxidizer Research, RI/RD 79-165.* 16 Feb. 1979, Rocketdyne Division of Rockwell International, Canoga Park, CA 91304.
6 K. O. Christe, *Inorg. Chem., 16* (1977) 2238.
7 R. F. Weinland and O. Lauenstein, *Z. Anorg. Allg. Chem., 20* (1899) 40.
8 K. O. Christe, *Inorg. Chem., 25* (1986) 3721.
9 K. O. Christe, W. W. Wilson, R. V. Chirakal, J. C. P. Sanders and G. J. Schrobilgen, *Inorg. Chem., 29* (1990) 3506.
10 K. O. Christe, D. A. Dixon, H. P. Mercier, J. C. P. Sanders, G. J. Schrobilgen and W. W. Wilson, *J. Am. Chem. Soc., 116* (1994) 2850.
11 K. O. Christe, E. C. Curtis, D. A. Dixon, H. P. Mercier, J. C. P. Sanders and G. J. Schrobilgen, *J. Am. Chem. Soc., 113* (1991) 3351.
12 K. O. Christe and D. A. Dixon, *J. Am. Chem. Soc., 114* (1992) 2978.
13 K. O. Christe, D. A. Dixon, G. J. Schrobilgen and W. W. Wilson, *J. Am. Chem. Soc., 119* (1997) 3918.
14 G. A. Olah, G. Rasul, A. Burrichter, M. Hachoumy, G. K. S. Prakash, R. I. Wagner and K. O. Christe, *J. Am. Chem. Soc., 119* (1997) 9572.
15 M. A. Petrie, J. A. Sheehy, J. A. Boatz, G. Rasul, G. K. S. Prakash, G. A. Olah and K. O. Christe, *J. Am. Chem. Soc., 119* (1997) 8802.
16 A. A. Korkin, M. Nooijen, R. J. Bartlett and K. O. Christe, *J. Phys. Chem. A, 102* (1998) 1837.
17 H. H. Michels, J. A. Montgomery, K. O. Christe and D. A. Dixon, *J. Phys. Chem., 99* (1995) 187.
18 K. O. Christe, W. W. Wilson, G. W. Drake, D. A. Dixon, J. A. Boatz and R. Z. Gnann, *J. Am. Chem. Soc., 120* (1998) 4711.

19 K. O. Christe, W. W. Wilson and D. A. Dixon, *J. Am. Chem. Soc., 121* (1999) 3382.
20 G. W. Drake, D. A. Dixon, J. A. Sheehy, J. A. Boatz and K. O. Christe, *J. Am. Chem. Soc., 120* (1998) 8392.
21 K. O. Christe, *J. Am. Chem. Soc., 117* (1995) 6136.
22 K. O. Christe and W. W. Wilson, *J. Fluorine Chem., 89* (1998) 97.
23 New York Times, Feb. 2, 1999; London Times, Feb. 10, 1999.
24 K. O. Christe, W. W. Wilson, J. A. Sheehy and J. A. Boatz, *Angew. Chem., Int. Edit. Engl., 38* (1999) 2004.

Chapter 9

THE ANIONIC SIDE OF FLUORINE CHEMISTRY

JAMES H. CLARK

Department of Chemistry, University of York, York YO1 5DD, UK

Starting out in London

I started my research career like many other English chemists at the tender age of 21 when, without any particular planning or thought, I found myself doing postgraduate research at Kings College, University of London. John Emsley[1] can take the credit for my introduction to fluorine chemistry even though he was not a fluorine chemist himself! John had one of those wonderful old academic offices equipped with a fume cupboard (not, I imagine, something we would expect to see in architects' plans for modern chemistry buildings!) and like all good scientists, he dabbled in his 'spare time'. John had made the remarkable observation that the heavier alkali metal fluorides and a few others readily dissolved in acetic acid [1]. The solubilities were not only very high in some cases but dissolution was accompanied by a lot of heat (causing the acids to boil in the more dramatic cases!). John had been awarded a grant from the UK Science Research Council to progress this work, and I was tempted by a postgraduate position that offered a salary considerably greater than a PhD grant.

My first research year at Kings was certainly one of the two or three most influential in my career so far – it was also one of the most enjoyable, thanks to many factors – including tremendous freedom in my research, lots of interesting results, and a great bunch of postgraduates who sensibly recognised that you could do good research and still enjoy other things in life, such as bridge and soccer! When I was given the chance to switch to a PhD grant with retrospective registration (from 1972), I jumped at the chance – the loss of income now seemed a small penalty.

As my PhD progressed, so did my research experiences. Having dabbled in various analytical techniques [2] (and found an interest in NMR spectroscopy that would stay with me for many years), I 'discovered' organic chemistry and, just as significantly, I encountered organofluorine chemistry. In a rather ambitious attempt to cover as many different areas as possible in my PhD work, I started to look at the reaction chemistry of fluoride solutions. My knowledge of organic chemistry was quite limited and I knew nothing of C—F chemistry (I was blissfully ignorant of all the pioneering organofluorine work that had been going on in the UK). We did realise, however, that dissolving ionic fluorides was one of the limiting factors in making C—F bonds by nucleophilic substitution – a situation that has changed little in the intervening years. Unfortunately, we were not as knowledgeable about

[1] John Emsley was a Lecturer then Reader in Chemistry at Kings College London before becoming Science Writer in Residence, initially at Imperial College London, and now in the Chemistry Department at the University of Cambridge.

the properties of the products we sought to make, and one of our first targets, monofluoroacetamide, is a compound best known for its considerable toxicity! Fortunately, our first attempts to make this fluoroamide were unsuccessful – KF in acetic acid proved not to be a source of fluoride but a source of the acetoxy group! Ironically, we managed to make this dangerous relative of monofluoroacetic acid by simply heating an intimate mixture of KF and monochloroacetamide and it was only then that a somewhat amused senior colleague told us of the compounds' claim to fame! We soon realised that the effects of the very strong $F^- \cdots H{-}O$ hydrogen bonding were not only manifest in the physical and spectroscopic properties of the solutions we were using, but also in the chemical properties [3]. I finished my PhD degree with a burning desire to learn more about the effects of hydrogen bonding on chemical properties and about the reactivity of fluorides.

Pursuing fluorides in Canada

My postdoctoral research with Jack Miller at Brock University in Canada (1975-1977) again gave me a lot of freedom to indulge my interests and allowed me to explore new areas, particularly in organic synthesis. By realising that fluoride activated carboxylic acids, we were able to exploit this phenomenon in a wider range of applications. Metal fluorides such as KF are not soluble in the more weakly acidic protic compounds such as phenols and anilines, but quaternary ammonium fluorides are. This was my first real experience with quaternary ammonium fluorides – a relationship that has lasted to the present day! These highly-soluble ionic compounds are excellent fluoride sources but have two major drawbacks – they are extremely hygroscopic and they have poor thermal stability. The combination of these makes it effectively impossible to completely dry them; indeed, there appears to be only one realistic claim to a dry onium fluoride and that is for the smallest member of the family, tetramethylammonium fluoride [4]. By carrying out parallel studies on the hydrogen bonding properties of fluorides and the synthetic utility of fluoride systems, we were able to use our understanding of one area to the benefit of the other. Thus we realised that it should be possible to remove the water from quaternary ammonium fluorides under mild conditions if the fluoride was offered an alternative protic molecule to which it could hydrogen bond. In this way we were able to form a variety of hydrogen bonded complexes such as $ArOH \cdots F^-$, $ArNRH \cdots F^-$, and $ArSH \cdots F^-$. The presence of the hydrogen bond was easily observed through shifts in the OH, NH or SH stretching bands in the IR spectra. We later showed for the first time how the then new technique of FAB mass spectrometry could be used to study and prove hydrogen bonds such as these. These complexes proved to be powerful sources of the conjugate anions of the organic molecules so that the fluoride systems could be used to functionalise phenols, anilines and thiols among other compounds [5] (Scheme 9.1).

The mechanistic aspect of the chemistry proved to be particularly interesting – it was clear that reactions did not simply occur by proton transfer to form the nuclephilic organic anion which then attacked the electrophile. Thus the methodology provided an in-situ source of the anion which we represented as the hydrogen bonded complex $ArZH \cdots F^-$. Independently, Emsley and others were following up our earlier spectroscopic and calorimetric studies on fluoride-carboxylic acid systems with the first ab initio calculations on very strong hydrogen bonds of this type. The results of these theoretical studies clearly showed that such anionic hydrogen bonded complexes were very stable and that the bulk

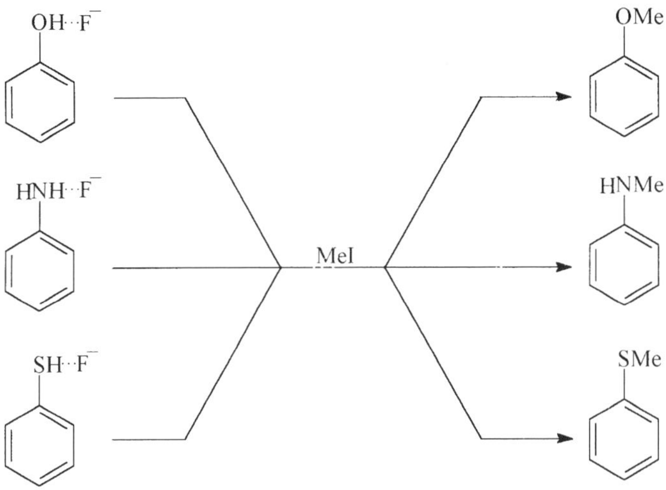

Scheme 9.1. Use of hydrogen bonded complexes as nucleophilic reagents.

of the negative charge could reside on the electronegative atom of the protic molecule (Z) [6]. This provided excellent supportive evidence for our theory. Remarkably, the calculations showed that the amount of charge on Z could exceed that in the free anion ArZ^-, *i.e.* the hydrogen bonded complex could be a more reactive source of ArZ^- than ArZ^- itself! This example of *hydrogen bond inhibition of resonance delocalisation of charge* has since been demonstrated in gas-phase experimental studies and in a other strongly hydrogen bonded systems [7]. Interestingly, later calculations of hydrogen bonded complexes between fluoride and nucleic acid bases such as uracil showed that water could easily disrupt these hydrogen bonds, causing proton transfer to occur (either back to organic-H + F^- or to organic$^-$ + HF). We had directly observed the consequences of this back at Kings in the early 70s when we had found that added water destabilised the KF-AcOH system, leading to the formation of quantitative amounts of KHF_2.

Fluorides back in the UK

After returning to England in 1977 I spent a short but very happy period working independently in the Chemistry Department at Exeter University. There I again benefited from a generous and supportive mentor, Eddie Abel, who gave me a lab and a small budget and let me indulge myself in speculative research. It was here that my insatiable love affair with supported reagents started ironically with the materials that have proven to be as complex as any I have since worked with – *supported* fluorides. Having experienced first-hand the difficulties of working with the highly hygroscopic tetrabutylammonium fluoride (TBAF), I decided to apply the relatively new methodology of supported reagents to the problem. My logic was that since it had proven possible to remove the water from such fluorides by offering them an alternative protic molecule with which to hydrogen bond to, the heavily hydroxylated surface of a chromatographic silica gel should achieve the same effect. This proved to be the case, at least to some extent, and I reported TBAF-silica as a dry and reasonably active soluble fluoride source for various base-catalysed reactions [8]. I was pleased to discover subsequently that TBAF-silica had become a laboratory reagent, notably for desilylation reactions.

After moving north-east to my first academic staff position at the University of York in 1979, I continued to work on supported fluorides as solid bases; and in particular we started to study supported KF – a much cheaper fluoride than any of the more reactive quaternary ammonium fluorides. KF-alumina was proving to be a popular supported reagent with several research groups around the world and we demonstrated its very high activity in Michael reactions in particular, where it is considerably more active than other popular fluoride sources, such as KF-18-crown-6 [9]. Analytical studies on KF-alumina proved its suprising complexity. There appeared to be much more to it than $F^{-}\cdots HO$ hydrogen bonds and it is now realised that the surface of KF-alumina contains a number of different species including AlF_6^{3-}, surface-O^-, and HO^- [10]. Part of this work was carried out during a number of enjoyable and productive NATO-sponsored work visits back to Brock University in Canada. On one of these trips we also decided to investigate the cause of the suprisingly low activity of fluoride-crown ether systems – their activity was not as high as would be expected for a soluble and 'naked' fluoride source. Serendipity provided the answer: we were surprised to see very wide lines in the ^{19}F NMR spectra of crown ether-aprotic solvent solutions of metal fluorides, and we rationalised that these must be due to ion aggregates in solution, thus disproving the then popular – if unsubstantiated – concept of 'naked' fluorides [11].

It took us several years of studies on the physical chemistry of ionic fluorides, their solutions and complexes before we started a serious attempt to exploit our knowledge in the area of nucleophilic fluorinations. It was clear that the familiar problems of low solubility for metal fluorides, and high hygroscopicity and low thermal stability for quaternary ammonium fluorides, were as much a problem to this area as any other. Owing to hydrogen bonding, as well as direct reaction with the highly fluorophilic silicon and alumina sites, fluorides supported on hydroxylated surfaces such as those of silica gels and aluminas had little if any residual fluoride activity. When metal fluoride did dissolve it was a result of strong hydrogen bonding leading to solvent activation but fluoride deactivation. Even 'tricks' such as using crown ethers had proved to be of little real value. Our experience in supported reagents had taught us the fundamental benefits of dispersing a reagent over a high surface area support, and it seemed that in order to exploit this we must find a high surface area material that is inert to F^-. Of the many materials that we screened then and since then, the one that has proven to be most useful is precipitated calcium fluoride. Its inactivity is legendary (should ordinary calcium fluoride ever prove to be an active fluoride source then it would have a dramatic effect on the economics of nucleophilic fluorination!) and it has no significant surface hydrogen bonding activity. Remarkably, we discovered at the point of publication that our research on the preparation and application of KF-CaF_2 (Scheme 9.2) had been going on in parallel to very similar work in Japan, and two consecutive publications appeared in *Chemical Communications* reporting KF-CaF_2 as an active reagent for nucleophilic aliphatic fluorinations [12, 13].

Subsequently, we turned our attention to novel soluble fluoride reagents. While much had been reported on quaternary ammonium fluorides, little was known of the analogous phosphonium fluorides. While desolvation of the former could only lead to ion aggregates (with the major mismatch in ion sizes being at the heart of their instability), the latter

$$RBr \xrightarrow[\text{MeCN or sulfolane}]{\text{KF-CaF}_2} RF \ (71\text{-}92\%)$$

R = $PhCH_2$, $C_{12}H_{25}$, PhCO, MeCO

Scheme 9.2. Use of KF supported on CaF_2 as a reactive source of fluoride ion.

could go through to fluorophosphoranes. Compounds of the formula R_4PF could exist in ionic and molecular forms; if this was tunable then it might be possible to prepare the compound in the molecular and presumably non-hygroscopic form, and then use that dry soluble compound as an in-situ source of F^-, *e.g.* by adjusting the polarity of the medium. The methodology proved to be partly successful, especially with the tetra-arylphosphonium fluorides, which have excellent thermal stability and good solubility in polar aprotic solvents. One unexpected twist to the story is that having struggled to avoid the formation of the hydrogen difluoride, Ar_4PHF_2, it proved to be highly active in nucleophilic fluorinations [14] – a possibility I had discounted having been brought up on the concept of strong hydrogen bonds suppressing the activity of F^-. A thorough spectroscopic study revealed the reasons behind this: the P^+ centre distorted the normally symmetrical HF_2^- ion, making it an in situ source of F^- [15]. The HF released was picked up by another molecule of the hydrogen difluoride. Unfortunately this meant that only 25% of the fluorine introduced to the system ended up in the organic product – economically unsatisfactory and, in these environmentally-conscious days, very wasteful!

Our work on nucleophilic fluorination also involved the spectroscopic study of Meisenheimer complexes. The high activity of TBAF resulted in the formation of a variety of anionic complexes on reaction with nitroaromatic substrates [16]. In doing this work we also discovered the potential of using the more reactive, onium fluorides for nucleophilic aromatic fluorodenitration – a reaction that has been known since the 1950s but for some 30 years had been regarded as little more than a side-reaction to halex. Through studying the reagents, the intermediates and the solvent (including the recognition that the key role of polar aprotic solvents in such reactions is the stabilisation of those intermediates [17]), we were in a good position to maximise selectivity in the reactions of many useful aromatic substrates that could undergo both fluorodechlorination and fluorodenitration. Attempts to efficiently fluorodenitrate nitroaromatics are often frustrated by low product yields due to the formation of phenols and ethers as side-products. These are believed to result from hydrolysis of the fluoroproduct and back-attack by displaced nitrite ions. Tetramethylammonium fluoride can be dried and gives relatively high product yields in at least some fluorodenitrations (see Scheme 9.3) – a result of its low water content and its apparent ability to capture and render nitrite inactive [18]. Product yields measured by gas chromatography can be very deceptive in this area of chemistry, however, and we have shown that large mass losses can occur in these reactions, presumably due to the formation of ethers as well as the occurrence of reactions involving the polar aprotic solvent (a consequence of high fluoride nucleophilicity going hand-in-hand with high basicity) [19]. Efficient fluorodenitration remains an elusive goal in many reactions, although we have been able to learn the significance of water, the solvent, the substrate and the cation, so that we can at least maximise the product yield and predict likely major problems from side-reactions.

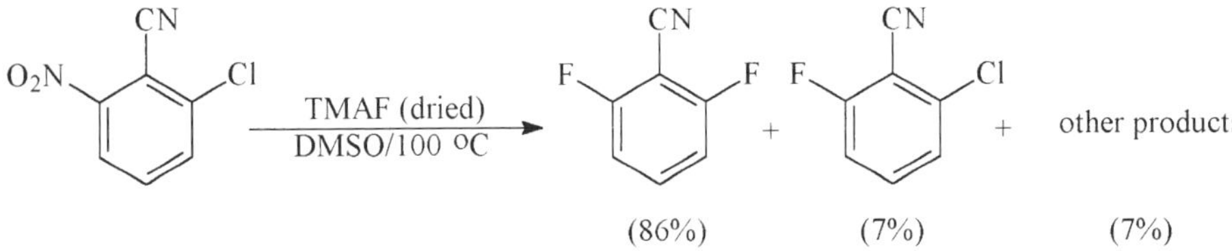

Scheme 9.3. Fluorodenitration of 2-chloro-6-nitrobenzonitrile using tetramethylammonium fluoride.

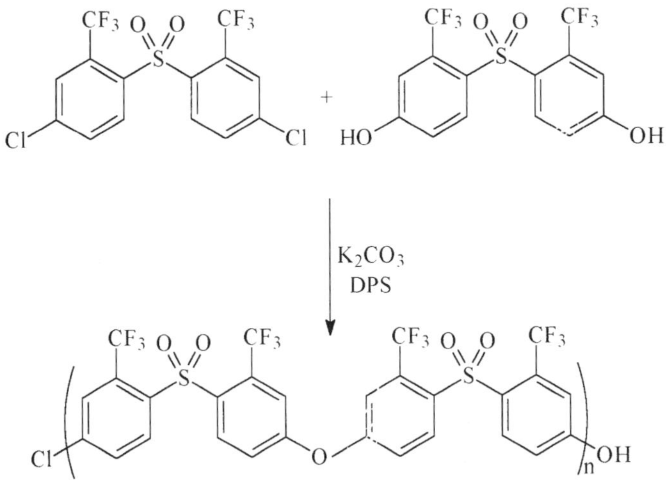

Scheme 9.4. Preparation of a trifluoromethylated aromatic polysulfone.

While fluoride ion and its chemistry has remained at the centre of the fluorine chemistry research in my laboratory, we have also looked further afield, notably to nucleophilic trifluoromethylation and trifluorothiomethylation. Our early work in these areas involved the study of the reagents $CuSCF_3$ and the elusive $CuCF_3$. Our continued interest in supported reagents encouraged us to study the effects of support materials on these reagents; $CuSCF_3$ is easily supported with some benefit to its activity [20], and while we were unable to succesfully trap $CuCF_3$ on a support, we were able to demonstrate an advantage in running $CuCF_3$ reactions in the presence of some support materials [21]. Our interest in CF_3 and CF_3S groups extended to the preparation of partially fluorinated benzophenone and diphenylsulphone monomers that we were able to incorporate into novel polymers [22] (*e.g.* Scheme 9.4).

In our on-going search for new and effective ways of introducing valuable substituents such as F and CF_3 into organic molecules, we have turned our attention recently to organosilanes and discovered some new and quite remarkable reactions: in the presence of KF, Ruppert's reagent (Me_3SiCF_3) will displace NO_2 and CN groups from activated aromatic substrates such as 2-chloro-6-nitrobenzonitirile [23] (Scheme 9.5). The substitution of nitro groups by trifluoromethyl is unusual, but the direct substitution of CN groups is without precedent. Dechlorination does not occur, and we believe that this reflects the

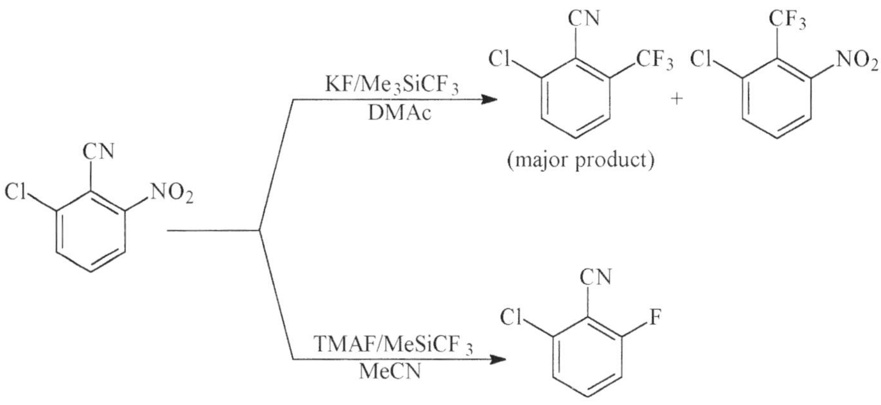

Scheme 9.5. Reactions of $\overline{F}/Me_3SiCF_3$ with 2-chloro-6-nitrobenzonitrile.

relative affinities of the leaving groups for silicon, with nitro and cyano forming stronger bonds than chloro. At the time of writing, product yields are low but there are clearly some exciting possibilities in this new area of chemistry. Suprisingly, different chemistry occurs when TMAF is used instead of KF. Here, fluorodenitration dominates and the CF_3 group is lost as CF_3H (Scheme 9.5). It seems that the solvent (MeCN) is deprotonated under these highly basic conditions. Once again, basicity has proven to be a highly influencial factor in nucleophilic fluorination systems. After 25 years working with this remarkable anion, F^- continues to surprise and to delight; most of all, though, it continues to challenge.

References

1 J. Emsley, *J. Chem. Soc. (A)*, (1971) 2511.
2 J. H. Clark and J. Emsley, *J. Chem Soc., Dalton Trans.*, (1973) 2154; (1974) 1125.
3 J. H. Clark, PhD Thesis, Kings College, London, 1975.
4 K. O. Christie, W. W. Wilson, R. D. Wilson, R. Bau and J. Feng, *J. Am. Chem. Soc., 112* (1990) 7619.
5 J. H Clark and J. M. Miller, *J. Am. Chem. Soc., 99* (1977) 498.
6 J. Emsley, O. P. A. Hoyte and R. E. Overill, *J. Chem. Soc., Perkin Trans. 2*, (1977) 2079.
7 J. H. Clark and C. W. Jones, *J. Chem. Soc., Chem. Commun.*, (1990) 1786.
8 J. H. Clark, *J. Chem. Soc., Chem. Commun.*, (1978) 789.
9 J. H. Clark, D. G. Cork and M. S. Robertson, *Chem. Lett.*, (1983) 1145.
10 T. Ando, S. J. Brown, J. H. Clark, D. G. Cork, T. Hanafusa, J. Ichihara, J. M. Miller and M. S. Robertson, *J. Chem. Soc., Perkin Trans. 2*, (1986) 1133.
11 J. M. Miller and J. H. Clark, *J. Chem. Soc., Chem. Commun.,*, (1982) 1318.
12 J. H. Clark, A. J. Hyde and D. K. Smith, *J. Chem. Soc., Chem. Commun.*, (1986) 791.
13 J. Ichihara, T. Matsuo, T. Hanafusa and T. Ando, *J. Chem. Soc., Chem. Commun.*, (1986) 793.
14 S. J. Brown and J. H. Clark, *J. Chem. Soc., Chem. Commun.*, (1985) 672.
15 S. J. Brown, J. H. Clark and D. J. Macquarrie, *J. Chem. Soc., Dalton Trans.*, (1988) 277.
16 J. H. Clark and D. K. Smith, *Tetrahedron Lett., 26* (1985) 2233.
17 J. H. Clark and D. J. Macquarrie, *J. Fluorine Chem., 35* (1987) 591.
18 N. Boechat and J. H. Clark, *J. Chem. Soc., Chem. Commun.*, (1993) 921.
19 D. J. Adams, J. H. Clark, H. McFarland and D. J. Nightingale, *J. Fluorine Chem.*, in the press.
20 J. H. Clark, C. W. Jones, A. P. Kybett, M. A. McClinton, J. M. Miller, D. Bishop and R. J. Blade, *J. Fluorine Chem., 48* (1990) 249.

21 J. H. Clark, M. A. McClinton, C. W. Jones, P. Landon, D. Bishop and R. J. Blade, *Tetrahedron Lett., 30* (1989) 2133.
22 J. H. Clark and J. E. Denness, *Polymer*, (1994) 5124.
23 D. J. Adams, J. H. Clark, L. B. Hansen, V. C. Saunders and S. J. Tavener, *J. Chem. Soc., Perkin Trans. 1*, (1998) 3081.

BIOGRAPHIC NOTE

James Clark

James Clark was appointed to the new Chair in Industrial and Applied Chemistry at the University of York in 1994, and currently holds a Royal Academy of Engineering-EPSRC Clean Technology Fellowship. He was awarded the RSC John Jeyes Medal and Lectureship in 1993 and the SCI Environment Medal in 1996. His research interests are quite diverse and include the use of heterogeneous catalysts in organic chemistry, the preparation of chemically modified mesoporous solids and the preparation of new materials, as well as new methods of fluorination (he is the author/coauthor of over 150 research articles, several reviews and three books in these areas). In 1998 he moved into new purpose-built laboratories in the *Clean Technology Centre* at York where he leads a research group of more than 20 graduates and postdoctoral researchers. James is also the Director of the new *RSC Green Chemistry Network*, the Scientific Editor of the new RSC journal *Green Chemistry*, and the series editor of the *RSC Clean Technology* monographs.

Chapter 10

LAPORTE AND ITS FLUORIDE BUSINESSES

ALAN E. COMYNS
Alan E. Comyns and Associates, 21 Churchward Close, Chester CH2 2BG, UK

Introduction

Laporte[1] is one of the only three UK companies presently (1999) making inorganic fluorides, the other two being ICI and Rhône-Poulenc. Laporte's involvement with fluorides began in 1959 when it acquired James Wilkinson & Sons Ltd, a long-established producer of fluorides in Rotherham, Yorkshire, and Glebe Mines Ltd, a Derbyshire fluorspar mining operation. The Sheffield Chemical Company, also acquired by Laporte at that time, had produced hydrofluoric acid (aqueous HF) in its Don Vitriol Works in the 1920s, but by 1959 it was making only sulfuric acid. Laporte in 1959 was a medium-sized chemicals producer whose main products were hydrogen peroxide, titanium dioxide, and fullers earth; the three new acquisitions created a fourth business area for the company – inorganic fluorine chemicals.

The inorganic fluorides business

James Wilkinson & Sons had been making acids in Sheffield for the steel industry for over a hundred years[2], and since 1925 had been making hydrofluoric acid and a range of fluoride salts. The hydrofluoric acid was used in the metallurgical and glass industries, and for 'general' cleaning. Railway companies, for example, used it for cleaning the windows of carriages, and members of the general public bought buckets of it at one shilling (12 old pence) per pint for cleaning the glazing of their greenhouses. The fluoride salts had a very wide variety of uses, many of them metallurgical. After World War II the chemical plant in Sheffield, situated in an unsuitable residential district near an elementary school (whose windows had to be replaced annually because of etching!), was moved to a more suitable site in Rotherham and modern equipment installed.

Hydrofluoric acid had previously been made by the company in Sheffield from fluorspar and sulfuric acid in a semi-continuous plant. A fully continuous rotary calcination process, partially developed at Regensburg in Germany in the 1940s, was acquired by

[1] For a summary of the early corporate history of Laporte, see ref. [1]. The company names and corporate structure have changed several times during the period under review. In this paper, for simplicity, the name 'Laporte' is used to mean any of Laporte Industries (Holdings) Ltd, Laporte Industries Ltd, Laporte Chemicals Ltd, Laporte Acids Ltd, and Laporte (Thailand) Ltd.

[2] For a brief history of James Wilkinson & Sons, see ref. [2].

Wilkinson after the war under the Allies' 'dismantling' scheme. Mr R. Y. Eagers[3], Chief Chemist at Wilkinsons, was one of the UK experts sent to investigate the achievements of the German fluorine and fluoride industry. He managed to acquire not only the technology but also the hardware: two large kilns were shipped from Regensberg, installed in the Rotherham works, and the process finally perfected. The state of the German fluorine industry in 1945 was described by Eagers and his fellow experts in BIOS Report 1595 [3].

The only fluorides made by Wilkinson in Sheffield had been sodium fluoride, sodium bifluoride, and ammonium bifluoride. After the move to Rotherham the product list was greatly expanded, and by 1962 it contained 50 different inorganic fluorides – simple salts, double salts, acid salts, fluoroborates, fluorosilicates, and the important potassium fluorotitanate used as a grain refiner in the aluminium industry. After its acquisition, Laporte rationalised the product range and by 1971 had reduced it to 28 products. Thereafter, Laporte followed the pattern of the British chemical industry in progressively abandoning the manufacture of commodity chemicals in the UK. By 1998 the only products still manufactured in Rotherham were potassium fluoride, an exceptionally pure grade of hydrofluoric acid for the electronics industry, and 4,4′-difluorobenzophenone, $(4\text{-}FC_6H_4)_2CO$ (see below). Contraction in the UK was compensated for by expansion overseas: the French fluoride manufacturer Soderec International S.A.R.L. was purchased in 1991; and a subsidiary company was established in Thailand in 1986, primarily to supply hydrofluoric acid to a local tantalum producer, but also for sale elsewhere in SE Asia (this company was disposed of in 1996).

The mining business

To supply its fluorspar needs, Wilkinsons had bought a derelict mine in the Derbyshire village of Eyam[4] in 1937, redeveloped it for its fluorspar, and built a minerals separation plant nearby. In 1945 the mine was registered as a separate company – Glebe Mines Ltd; other mine shafts were subsequently opened up[5]. Following its acquisition of Glebe Mines, Laporte built a minerals separation plant, the Cavendish Mill, on the open moor at Stoney Middleton. It produced barytes, a lead concentrate, and limestone, in addition to fluorspar; the fluorspar was sold to ICI and other companies, in addition to feeding the Rotherham plant. The mine was operated until September 1999, when it was closed because its fluorspar was no longer competetive with Chinese fluorspar on the world market. The Cavendish Mill continued to process fluorspar purchased from surface miners known locally as 'tributers'.

[3]R. Y. Eagers was later to write the classic book *Toxic Properties of Inorganic Fluorine Compounds*, published by Elsevier in 1969.

[4]Eyam (pronounced E'em) is famous for the tactic (voluntary isolation) adopted by its inhabitants to stop the spread of a 'plague' which struck the village in 1665, causing the deaths of about three quarters of the estimated population of 350.

[5]The early history of fluorspar mining in Derbyshire is recounted in a series of 'Special Reports on the Mineral Resources of Great Britain', published by the Geological Survey in 1916, 1917, 1933, and 1952. The last of these reports mentions that Wilkinsons had recently resunk the Ladywash mine and equipped the mill. The recent history of Laporte's Derbyshire mining activities in recounted by Dr J. V. Bramley, formerly General Manager of Laporte Minerals, in his paper published in *Mining Magazine*, *163*, No. 5 (1990), 328-333. The Laporte mines are on the famous Chatsworth Estate, described by the Duchess of Devonshire in her book, *The Estate: A View from Chatsworth*, published by Macmillan, London, 1990, pp. 196-200.

The organic fluorides business

Laporte in the 1980s acquired many small chemical companies. One of these was Wendstone Chemicals plc, Billingham, a producer of speciality organic chemicals. One of its fledgling products was 4,4′-difluorodiphenylmethane made from 4,4′-diaminodiphenylmethane by the Balz-Schiemann reaction on a 400-tpa scale [4, 5]. Oxdidation of the difluoro compound provides 4,4′-difluorobenzophenone (BDF), used by the ICI management buyout company Victrex as an intermediate in the manufacture of the engineering thermoplastic 'PEEK' (polyetheretherketone). The Balz-Schiemann reaction involves the synthesis and thermal decomposition of the bis(diazonium) tetrafluoroborate $(4\text{-}N_2^+C_6H_4)_2CH_2\ 2BF_4^-$, an inherently hazardous operation. Chemists at Wendstone, assisted by others in Laporte's Widnes laboratories and later by those in the Rotherham works, managed to tame this tiger and conduct the operation on an industrial scale in a controlled manner [4]. The bis(diazonium salt) is now manufactured and thermally decomposed at Rotherham, and the $(4\text{-}FC_6H_4)_2CH_2$ is oxidised to the corresponding diketone $(4\text{-}FC_6H_4)_2CO$ at Fine Organics on Teesside. The product is now firmly established on a scale of >1000 tonnes per year and production is expanding.

The Solvay connection

Laporte was a prime manufacturer of hydrogen peroxide from 1888 until 1992, when it sold its interest in the joint company Interox to its erstwhile partner Solvay. But Laporte still had a natural interest in chemical applications of hydrogen peroxide and its derivatives, and its research workers in Widnes continued to develop them. One of their discoveries concerns the oxidative one-pot trifluoromethylation of electron-rich arenes. This is normally done – with some difficulty – using bis(trifluoroacetyl) peroxide. The Widnes workers have shown [6] that the reaction can be effected safely with a mixture of commercial sodium percarbonate (sodium carbonate sesquiperhydrate) and trifluoroacetic anhydride.

Acknowledgements

This paper is based partly on an unpublished history of Laporte by Mr H. L. Salter, written for the company's centenary in 1988. I am grateful to Mr Ray Ward, Public Relations Manager of Laporte, for permission to use this material. I have also drawn on the aural histories of Mr D. J. Nichols (formerly Chief Accountant of Laporte Acids Ltd) and Mr R. Y. Eagers which were tape-recorded by Dr Eric Banks and Dr Kathleen Farrar (both from UMIST, Manchester) in the early 1980s. Other Laporte staff, past and present, have also assisted.

References

1 D. W. F. Hardie and J. D. Pratt (eds.), *A History of the Modern Chemical Industry*, Pergamon Press, Oxford, 1966, pp. 305-306.
2 P. J. T. Morris, C. A. Russell and J. G. Smith (eds.), *Archives of the British Chemical Industry*, British Society for the History of Science, Faringdon, 1988.

3 British Intelligence Objectives Sub-Committee (BIOS) Final report no. 1595, H.M. Stationery Office, London.
4 J. Regan, 'The Industrial Application of the Balz-Schiemann Reaction', in B. Pearson (ed.), *Speciality Chemicals: Innovations in Industrial Synthesis and Applications*, Elsevier Applied Science, London, 1991.
5 J. S. Moilliet, 'Industrial Routes to Ring-Fluorinated Aromatic Compounds', in R. E. Banks, B. E. Smart and J. C. Tatlow (eds.), *Organofluorine Chemistry: Principles and Commercial Applications*, Plenum Press, New York, 1994, pp. 195-219.
6 C. W. Jones, J. P. Sankey and W. R. Sanderson, 'The Oxidative Trimethylation of Arenes', in R. E. Banks (ed.), *Fluorine in Agriculture*, Paper 8, Chemserve (UMIST), Manchester, 1995. Used as the basis for *European Patent 700,885*.

BIOGRAPHIC NOTE

Alan Comyns

Alan Comyns has had an unusually varied career in academic, government, and industrial research laboratories. He graduated with first class honours in chemistry from the University of London in 1947 at the age of nineteen, and his subsequent PhD work, carried out in the Hughes-Ingold school of physical-organic chemistry at University College London, was followed by postdoctoral studies in the US at Caltech and the University of Wisconsin. He has worked at Harwell (UK), British Titan Products, Westinghouse Electric in Pittsburg, and National Lead in New Jersey. From 1974 to 1988 Alan was Product Research Manager, later Chief Scientist, at Laporte Industries in Widnes (UK); nowadays he is an independent consultant and author, specialising in market studies for inorganic chemicals and materials.

Alan's numerous publications include *Fluoride Glasses* [(ed.), John Wiley, Chichester, 1989 – still the only monograph in this field], *Dictionary of Named Processes in Chemical Technology* (Oxford University Press, Oxford, 1993; a key reference work which describes 3000 processes in 337 pages), and *Encyclopedic Dictionary of Named Processes in Chemical Technology* (CRC Press, Florida, 1999; an enlarged, updated version of the previous item). He edits (and largely writes)) *Focus on Catalysts*, a monthly newsletter published by the Royal Society of Chemistry.

Chapter 11

FLUORINE CHEMISTRY – A CHEMICAL GARDENER'S PARADISE[1]

DARRYL D. DESMARTEAU

Chemistry Department, Clemson University, Box 341905, Clemson, SC 29634-1905, USA

In the beginning

My researches in fluorine chemistry began in the laboratories of the late George H. Cady [1] in the fall of 1963. I had entered graduate school at the University of Washington in the spring of 1963 and selected Cady as my advisor, having first encountered him the fall of 1962 when he gave a lecture on sulfur oxyfluorides at Washington Sate University, where I was an undergraduate at the time. During that lecture he demonstrated the action of heat on a sample of the remarkable compound peroxodisulfuryl difluoride (FO_2SOOSO_2F) that had been made in his laboratory [2]. Taking a sealed glass ampoule containing colourless $S_2O_6F_2$ 'gas/liquid', he heated it gently with a Bunsen burner, generating an intense yellow gas; he then cooled the tube under tap water and the colour quickly disappeared. After repeating this sequence a few times, he explained that the colour was due to $FSO_3\cdot$ radicals arising from the reversible dissociation of the peroxidic $S_2O_6F_2$. I was absolutely fascinated by this demonstration and to hear how this peroxide was prepared:

$$2SO_3 + F_2 \xrightarrow[160\,^\circ C]{AgF_2} S_2O_6F_2$$

It was the first time I had met a person who had actually worked with fluorine. Later I learned that Cady was instrumental in the commercial production of fluorine, and that as a graduate student with Joel Hildebrand he made the first good measurements of the physical properties of elemental fluorine [3-5].

As an undergraduate I had carried out some research with H. H. Beatey, looking for the presence of sulfuryl iodide in a mixture of iodine and liquid sulfur dioxide. We never found sulfuryl iodide but I learned a lot about handling gases and other valuable techniques, and this experience convinced me that I wanted to go to graduate school. Since I thought I knew something about sulfur oxides and sulfur oxyhalides, Cady's research was very attractive. When I selected Cady as my advisor, he wanted me to work on the possible polymerization of SF_4 and on making the peroxide $P_2O_4F_4$, which would be isoelectronic with $S_2O_6F_2$. I did attempt to make $P_2O_4F_4$ by electrolysis of difluorophosphoric acid, but most of my time was spent on the chemistry of $S_2O_6F_2$.

---

[1] Dedication: In loving memory of my son Scott Warren DesMarteau (1966-1987).

The early 60s were exciting times in Cady's group as there was still a lot of interest in high-energy oxidizers related to the US space programme and military applications, and Cady had excellent research support from the Office of Naval Research. There were many excellent students and postdoctorals in Cady's group in 1963, and several of these individuals, like myself, continued work in fluorine chemistry long after leaving the University of Washington (*e.g.*, G. L. Gard, R. E. Noftle, F. Aubke). We had a lot of freedom to pursue our own ideas and my early ones were somewhat naïve. As soon as I began my studies in Cady's group I rapidly learned about many famous names in the field of fluorine chemistry, some of whom I would later meet (Haszeldine, Emeléus, Schmeisser, Knunyants, Miller, Ruff and Glemser, to mention but a few).

First new compound and xenon esters

The first new compound I synthesized was based on work by Hazeldine [6]. He had shown that iodine reacts with trifluoroacetic acid anhydride to give products based on cleavage of the C—O bond:

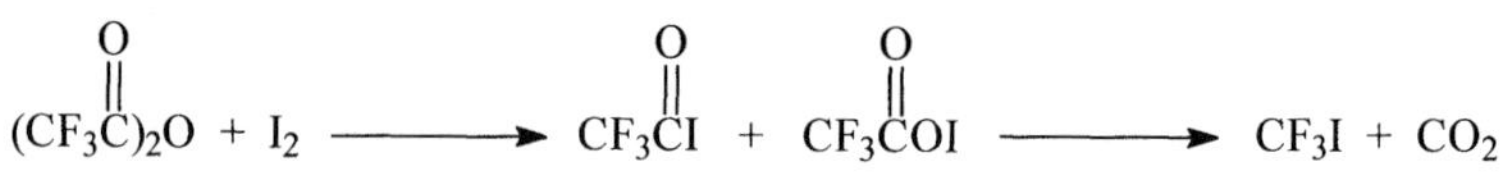

I reasoned that $S_2O_6F_2$ ought to react similarly, and to my delight it did:

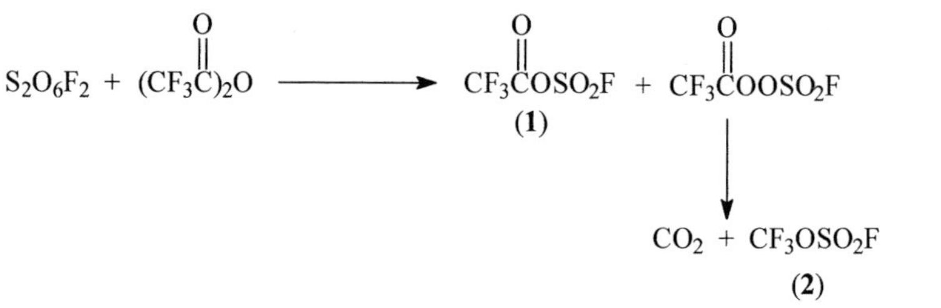

The acyl fluorosulfate $CF_3C(O)OSO_2F$ (**1**) was a new compound, and I took great pride in having made the first example of this class of compounds and proving that I actually had it [7]. (I can still recall realizing with wonderment how powerful ^{19}F NMR was as a structural tool; but Cady was a very classical chemist, so we also had to analyze our compounds by standard wet methods.) The perfluoroalkyl fluorosulfate CF_3OSO_2F (**2**) had been prepared earlier by other routes but mine was the best method for its synthesis. I went on to make a series of new acyl and alkyl fluorosulfates from all the fluorinated carboxylic acid anhydrides I could then get my hands on and became thoroughly hooked on synthesis: *it was exciting and fun*. I studied the reactions of $S_2O_6F_2$ with many other substrates and found a new way to prepare fluorodisulfate salts and also made the first phosphorus fluorosulfates, $OPF_{3-n}(OSO_2F)_n$ (n = 1-3) [8, 9].

When I began graduate work in 1963, Neil Bartlett had just reported the oxidation of Xe with PtF_6 and there was a lot of interest in this chemistry in Cady's group [10]. Xenon hexafluoride was prepared independently by Dudley, Gard and Cady while I was there [11], and had the nasty property of being rapidly hydrolysed to XeO_3 by ubiquitous water. This incredibly sensitive oxide once exploded in Gard's main vacuum line trap when he turned on the room lights!

I began to think about making a xenon fluorosulfate and first, by analogy with the photochemical synthesis of XeF_2 from xenon and fluorine, I tried the simple experiment of exposing a mixture of $S_2O_6F_2$ and Xe to sunlight on the roof of the chemistry building, Bagely Hall. From the instability of this compound, later prepared independently by Bartlett [12] and by me [13], this was doomed to fail from the outset, and next I tried the reaction of XeF_6 with $HOSO_2F$. My idea was that the formation of HF could lead to substitution of fluorine by the very electronegative fluorosulfate group. In practice, xenon hexafluoride readily dissolved in fluorosulfuric acid and $S_2O_6F_2$ slowly evolved. We assumed that an unstable xenon compound had formed and was decomposing to $S_2O_6F_2$, but, for reasons which will become apparent, we did not finish characterizing this reaction. If only we had run an NMR experiment on the fluorosulfuric acid containing the XeF_6, we would have easily seen that F_5XeOSO_2F was present, as we proved much later [14]. So it was that the first xenon ester to contain a polyatomic group bound to xenon and the first example of an Xe-O single bond had, in fact, already been achieved in 1965.

I was still trying to prepare the peroxide $P_2O_4F_4$ mentioned above and the result with fluorosulfuric acid prompted me to treat difluorophosphoric acid with XeF_6. Having developed a method for purifying $HOP(O)F_2$ [15], I carried out this reaction and to our surprise a dramatic formation of a white solid occurred. I remember how excited I was over this result, thinking that we had made a xenon difluorophosphate. Analysis of the volatile products however showed that all the phosphorus could be accounted for as OPF_3, hence we concluded that the white solid was XeO_3. Prior to this realization I had scaled up the reaction to give as much as 2 grams of this white solid; remarkably, it never exploded – except for the time I mistakenly added acetone instead of water to a flask containing it. Fortunately I was wearing leather gloves and a face shield when the flask disappeared! Fortuitously, as we published later, XeO_3 produced in this manner is stabilized somehow by an impurity, possibly $(HO)_xP(O)F_{3-x}$ [16], and the method used has remained the safest way to prepare XeO_3 right up to the present time.

A big explosion

In late 1965 in the midst of this xenon work, fellow graduate students from other research groups needed $S_2O_6F_2$ for IR and ESR studies of the fluorosulfate free radical [17, 18]. I also needed a large amount to finish work on phosphorus fluorosulfates and for other reactions I had in mind. The preparation of $S_2O_6F_2$ is relatively easy once one has a catalytic reactor set up, so I decided to make several hundred grams of the peroxide. We didn't have the best of temperature controllers on this reactor in Cady's group, and during the course of a production campaign lasting several days the temperature sometimes exceeded 165 °C; when this happens, a significant byproduct, fluorine fluorosulfate ($FOSO_2F$), is formed [19]. So in the process of collecting ~900 g of $S_2O_6F_2$ in a dry ice trap, I also collected ~200 g of $FOSO_2F$ in a large trap cooled in liquid oxygen. I didn't want this byproduct but it seemed like a real waste to let it go up the fume hood; hence we decided to save it for someone's future use by condensing it into a metal cylinder to store as a compressed gas. This proved to be a big mistake.

After vacuum transfer of $FOSO_2F$, the cylinder at −196 °C was placed inside a section of a 6-inch naval gun barrel in a hood next to my desk that we used for protection against potentially explosive reactions, for we knew that $FOSO_2F$ was a very reactive compound and so wanted to let it warm to room temperature with care. Shortly after the cylinder had been placed in the barricade, a friend who had just finished his first-quarter law school exams came by and we decided to go to my apartment and share a beer or two (5 actually!) to celebrate. After about 2 hours spent discussing 'world problems' we parted, and I returned to the laboratory where I reached inside the barricade to feel the cylinder of $FOSO_2F$; it was cold but had, I judged, warmed to above 0 °C. No problem, I thought, and donning my safety glasses *and* a face shield I removed the cylinder and laid it on its side in the hood. I then put a label on it, and was just reaching to pick it up when there was an incredible explosion. I was gravely injured, instantly losing my left hand and part of the forearm, part of my right hand and sustaining other injuries. My face shield and safety glasses were destroyed and my wristwatch was found embedded in the ceiling. The irony of all this is that if only I'd had another beer at my apartment, I could well have returned to the lab just later enough to find that the explosion had occurred in the barricade, injuring no one.

After 3 long months in hospital and another 3 with a monster cast on my right arm, then several follow-up surgeries, I was able to function on my own again – with a metal prosthetic device for a left hand that continues to be useful for handling hot objects, among other things. I have never looked back, although many people expected that I would give up chemistry. This never entered my mind, but I'd learned a very important lesson: it's what you don't know that will get you. We knew $FOSO_2F$ was very reactive, but since its discovery 9 years earlier there had been no evidence of its explosivity [20, 21]. In retrospect, it was a mistake to make and store such a large amount of a highly energetic compound which had not been thoroughly hazards tested.

I didn't do any more laboratory work for my PhD and wrote a thesis based on what I had done up to that time. It was certainly enough in terms of positive results even though several projects, including the characterization of the reaction of XeF_6 with $HOSO_2F$, were left unfinished.

On my own

After completing my PhD degree, I stayed at the University of Washington for 1 year as a visiting assistant professor, Cady letting me remain in his laboratory and continue research on my own. This gave me a chance to show that I could still be effective in the laboratory and provided the opportunity to look for an academic position. (I had to abandon job searching for a while after my accident due to the obvious uncertainties involved.) I went back immediately to working with reactive compounds and explored some reactions of bromine (I) fluorosulfate as a means of making covalent fluorosulfates [22]:

$$RX + BrOSO_2F \longrightarrow ROSO_2F + BrX \quad (R = \text{a variety of radicals; } X = Cl, Br)$$

Understandably, I had no desire at that time to work with anything that might lead to explosions, including XeF_6. But I hadn't given up the idea of attaching electronegative groups to xenon. After moving to Northeastern University as an assistant professor in 1967, I set up a laboratory and returned to xenon chemistry. I also headed off in a new research direction with fluorocarbon peroxygen compounds. There had been a lot of research on the latter in Cady's group, and also the work of W. B. Fox and his co-workers at Allied Chemical on $CF_3O_3CF_3$ had caught my interest [23]. I met Bill Fox at the University of Washington in 1966 when he was recruiting for Allied Chemical. It was after my accident, and I was demonstrating to him how well my prosthetic device worked by drinking coffee with it. The coffee cup slipped in my hook and the coffee doused Bill's suit pants. I was most embarrassed, but Bill and I became good friends. I also learned to accept the limitations of a functional hook as a substitute for a hand (over the years I've dropped and spilled a lot of things!) and it has not stopped me enjoying cars.

DesMarteau on cars

I became interested in cars when I bought my first automobile in 1956, a much-used 1949 Chevrolet. It wasn't a very good car, but I quickly learned to work on it (since I couldn't afford to pay others to do so) and carried out extensive modifications, turning it into a 'hot rod', which sometimes got me into trouble with the law. College, marriage, children and academic life took all of my time and money from the early 60s through to the 70s, and it was not until 1973, when I bought the 73 Datsun 240Z shown in the picture here, that I was able to renew my interest in cars. This was my first sports car and I loved it. In 1975 I entered my first racing event with this then stock car and became hooked on the sport. Soon I had modified the engine and suspension to be more competitive. When I moved to South Carolina in 1982, I

Plate 11.1.

pursued my hobby more vigorously and the Datsun was soon a full-blown race car, lightened by 700 lbs and the horsepower doubled from 150 to 300. I race the car in the Sports Car Club of America Solo I events, reaching speeds of up to 160 mph on road racing courses. I also run it in hill climbs, such as the Beech Mountain Hillclimb, where this picture was taken in April 1999. The car is registered for racing events under fluorine's atomic number, and it is painted gold and yellow/green, the latter representing the colour of fluorine gas. The lettering on the rear fender reads 'Powered by Fluorine': Fluorine Chemistry enables me to afford the car, so it is indeed powered by fluorine! Earlier the car was #19 for ^{19}F, but I changed to 9 when the system for allocating permanent numbers resulted in 19 being given to someone with no appreciation of its significance.

I have other sports cars that I am restoring – another 73 Datsun 240Z for my son Noel in California and a 71 Datsun 240Z for me. My street car on nice days is a beautiful Classic Roadsters 1965 racing Austin-Healey that I built during the period 1989-94; it is very authentic in outward appearance but otherwise contains almost nothing original, including the 5.0L Ford engine and drive train.

Xenon esters and fluorocarbon peroxygen compounds

My first research proposal to the NSF was on possible new xenon compounds and it received funding in 1968. Similarly the ACS-PRF gave us funding for fluorocarbon work in 1969. Soon we had evidence for various xenon(II) esters of CF_3CO_2H, $HOSO_2F$ and $HONO_3$ [13].[2] Others had also begun to think in these directions and soon xenon(II) esters were common [24]. We also pursued the XeF_6 chemistry with $HOSO_2F$ but our first publication on this turned out to be partially wrong [25]; I corrected this after Ron Gillespie challenged our results at the First ACS Winter Fluorine Conference in 1972 [14]. I had moved to Kansas State University in 1971, where facilities for synthetic chemistry were very much improved and we were able to pursue our work with fluorocarbon peroxygen compounds begun at Northeastern.

The fluorocarbon peroxide work began with a naïve idea. Based on some fluoride-based catalyst work done by Cady, $CF_3O_2CF_3$ was sometimes formed by the reaction of CF_3OF with COF_2. I reasoned that perhaps $CF_2(OF)_2$ might react similarly with COF_2, but the formation of $(CF_3OO)_2CF_2$ was never observed. However, I did encounter the unusual trioxide-peroxide $CF_3O_2CF_2O_3CF_3$, formation of which I still cannot explain. More importantly, however, under the right conditions this reaction also provided good yields of $CF_3OOC(O)F$ [26]:

$$CF_2(OF)_2 + xsCOF_2 \xrightarrow[-78\text{ to }-5\,^\circ\text{C}]{\text{CsF}} CF_3OF + CF_3OOC(O)F$$

Until recently, some 27 years after the event, the mechanism for this reaction was somewhat of a mystery; we now believe that it probably involves difluorodioxirane, a compound first isolated in 1993 through our Ausimont Connection (see later).

[2]There was to be a joint submission of our work with that of ref. [12], but this did not materialize. Thus our report appeared 6 months later.

The acyl peroxide $CF_3OOC(O)F$ had been shown previously by 3M scientists to hydrolyze to CF_3OOH, but no chemistry of this hydroperoxide had ever been carried out because methods then available for generating $CF_3OOC(O)F$ were not of preparative use [27]. Using our new method, we soon embarked on a study of the chemistry of CF_3OOH; and this was the beginning of a lot of interesting work on the synthesis of fluorocarbon peroxides.

Initial work focused on reactions of CF_3OOH as a weak acid in the presence of a base such as NaF [28]. A number of trifluoromethylperoxy esters were readily prepared:

$$RC(O)F + CF_3OOH \xrightarrow{NaF} RC(O)OOCF_3 + NaF \cdot HF$$

The acetyl ester ($R = CH_3$) was prone to explode on thermal shock, as I discovered to my cost one Saturday morning in 1970 when I was vacuum transferring a 1-gram sample of $CH_3C(O)OOCF_3$ from a storage tube to a vapour pressure apparatus and became impatient. So I picked up a 1-litre beaker of water and immersed the sample tube in it to warm it. A moment later the sample exploded, leaving me with a 1-inch piece of the beaker rim embedded in my right hand. For a brief period I could not look to see what had happened to my only hand for fear that more of it might be missing. Fortunately, the damage was easily repaired, but I vowed to be more careful in future.

I also prepared CF_3OOF by treating CF_3OOH with F_2 in the presence of CsF [29], and my co-worker Peter Bernstein prepared the first and (I think) still the only known fluorocarbon peroxide of phosphorus, $CF_3OOP(O)F_2$ [30]. When we moved to Kansas State University in 1971, my student Fred Hohorst moved with me and he used CF_3OOF and CF_3OOH to prepare the interesting peroxynitrate CF_3OONO_2 via reactions involving N_2O_4 and N_2O_5 [31]. We also began the first studies on the reactivity of the trioxide $CF_3O_3CF_3$, which readily added to various alkenes and other substrates [32, 33]:

$$CF_3O_3CF_3 + E \longrightarrow CF_3OEOCF_3 + CF_3OEOOCF_3$$

$(E = SO_2, SF_4, CO)$

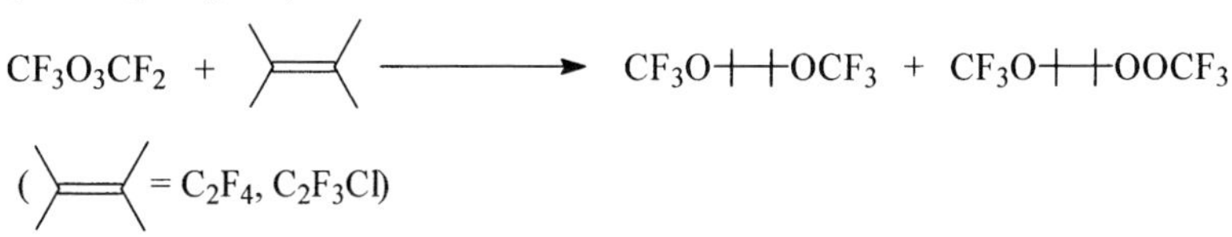

These were clearly radical reactions where $CF_3O^{\bullet}$ was the initiator andpropagation involved abstraction of CF_3O and CF_3OO units from $CF_3O_3CF_3$. Remarkably, the trioxide is kinetically very unreactive at temperatures below about 70 °C.

In the course of a few short years the number of known trifluoromethylperoxides had greatly expanded, and we went on to synthesise many other peroxides via reactions of CF_3OOCl [34] and CF_3OOF [35]. Subsequently we prepared $(CF_3)_3COOH$ [36-38] and SF_5OOH [39-42] and studied their chemistries, but hoped-for increases in the stabilities of R_FOOX compounds were not realized by using these new R_F groups. We also prepared SeF_5OOH and TeF_5OOH, but these compounds need to be more fully characterized [43]. Our ultimate goal was the preparation of a stable or readily isolable tetraoxide, such as $CF_3O_4CF_3$. To date we have not realized this goal, but I have not yet given up the quest.

The first xenon-nitrogen bond

After moving to Kansas we continued our research on xenon chemistry. Since essentially all reasonable strong oxy-acids had been explored as candidates for xenon esters, it was time to look for new directions. In a proposal to the NSF that was funded in 1968, I had suggested the possibility that a sufficiently electronegative nitrogen-centered group might participate in the formation of a xenon-nitrogen bond. This seemed rather far fetched at the time, but I included it in the proposal anyway. It ocurred to me that based on the work of John Ruff with $HN(SO_2F)_2$ [44] there might be a chance to make the first xenon-nitrogen bond, but we delayed starting this work until 1972 because the synthesis of $HN(SO_2F)_2$ was rather laborious, and it seemed like a long shot anyway. However, after trying several more accessible (and likely) nitrogen ligands without success, I had my student Robert LeBland make some $HN(SO_2F)_2$, which did indeed prove difficult to obtain pure. The very first reaction we ran with XeF_2 gave a white solid that we hoped was FXe-$N(SO_2F)_2$. It slowly decomposed at 22 °C to form Xe, XeF_2 and $[N(SO_2F)_2]_2$ as one would expect, but what if it was just some complex of XeF_2-$HN(SO_2F)_2$ and did not contain a real Xe−N bond?

I recall running the first Raman spectrum of our product in 1972 at about 2:00 am in the Physics Department at Kansas State University. We didn't yet have our own Raman instrument and the machine in the Physics Department wasn't really set up for typical 90 ° scattering, so it took us many hours to obtain the low-temperature spectrum. A very strong Xe−F stretch was observed at a position only 4 cm^{-1} different from that in the spectrum of XeF_2. This bothered us and we worried about it at length. Finally, we succeeded in obtaining an ^{19}F NMR spectrum of the compound in BrF_5 as a solvent and were able to see the $^3J_{Xe\text{-}F}$ coupling, the first example of long-range ^{129}Xe-F coupling. This convinced us that we had made $FXeN(SO_2F)_2$, so we sent a note to *Chemical Communications* in 1974 [45]. The review process was rather tough because everyone was skeptical and looked for alternative explanations for our data. A crystal structure would have solved the issue but we had no X-ray facilities at Kansas State, and the compound was insoluble or unstable in most every solvent except BrF_5. Trying to do a structure determination by long-distance collaboration seemed improbable. Finally, after three rounds of revision and replies to referees, the communication was accepted and I reported our discovery at the Fifth European Symposium on Fluorine Chemistry in Aviemore, Scotland, in September 1974 [46]. There remained many skeptics in the absence of an X-ray structure but we were quite confident that Xe−N bonds were real and ultimately made $Xe[N(SO_2F)_2]_2$ and $F[XeN(SO_2F)_2]_2^+AsF_6^-$ [47].

New nitrogen ligand for xenon

I was convinced in my mind that there must be other nitrogen ligands besides $N(SO_2F)_2$ capable of bonding to xenon. In 1976, I received a sample of $(C_4F_9SO_2)_2NH$ from chemists at Bayer who had developed such sulfonimides with apparent very high acidities [48]. Based on our experience with $HN(SO_2F)_2$, these compounds seemed to be good candidates for bonding to xenon, thus we were surprised when $(C_4F_9SO_2)_2NH$ failed to react with XeF_2. I then made one of the really good decisions in my career, and that was to repeat the reaction with $(CF_3SO_2)_2NH$ on the grounds that sometimes higher homologues of a given class of compound do not always behave as expected.

For reasons unknown to me, bis(trifluoromethylsulfonyl)imide had not been reported, so we obtained a sample of CF_3SO_2F from the 3M Company and set about preparing this N-H acid. With some difficulty my student Jerry Foropoulos succeeded in 1978 [49], and when we tried the reaction of his $(CF_3SO_2)_2NH$ with XeF_2 there was evidence of compound formation, but we could not isolate a pure product. At about this time we ran out of CF_3SO_2F and when I wrote to 3M telling them about our new acidic sulfonimide (with hindsight, this was a mistake!) and asking for more starting material, my request was ignored. Eventually, more than 6 months later and after several follow-up letters, they informed me that they could not supply us with more CF_3SO_2F. I make no public judgments on this matter but the reason turned out to be that 3M had independently prepared $(CF_3SO_2)_2NH$, which was subsequently patented by others, and is now sold as the lithium salt. In any event, Jerry and I pressed on and built our own Simons ECF setup to make CF_3SO_2F according to work published by Hazeldine [50]. After a lot of difficulty and one rather severe explosion (we guessed from $OF_2 + H_2$), we made several hundred grams of CF_3SO_2F and continued our work. Subsequently we found that $(CF_3SO_2)_2NSiMe_3$ would give a pure product on treatment with XeF_2 [51]:

$$XeF_2 + 2(CF_3SO_2)_2NSiMe_3 \longrightarrow Xe[N(SO_2CF_3)_2]_2 + 2MeSiF$$

We published this in 1982, relieved that the scientific community would now have to accept that Xe$-$N bonds were real. However, in the same year Gary Schrobilgen obtained a crystal structure of our original $FXeN(SO_2F)_2$ clearly showing the presence of a xenon-nitrogen bond, so it was no longer an issue [52]; I believe he was actually out to prove that the $N(SO_2F)_2$ group was actually O-bonded to xenon, *i.e.* FX$-$OSF(O)$=NSO_2F$.

Today Xe$-$N bonded compounds are commonplace [53] but it took 8 years to settle the issue of their existence. It was an interesting era, but when I moved to Clemson University in 1982 times were changing and funding agencies were becoming more concerned with applications and less concerned with curiosity-driven syntheses. It was difficult to build a case for applications of new xenon-element bonds, so I decided to end our studies in xenon chemistry.

Fluorocarbon peroxides, oxaziridines and *N*-halogen derivatives

The fluorocarbon peroxygen work that we were pursuing led to some major new developments in research directions. In 1976 we prepared the first perfluorocarbon oxaziridine by treating CF_3OOH with pentafluoroazapropene [54]:

(PFAPO)

We almost missed this interesting result because the ^{19}F NMR spectrum of the reaction mixture looked unintelligible. Fortunately my postdoctoral associate Bob Falardeau took a closer look, and we were led to conclude from IR and NMR analyses that we had produced the novel 3-membered ring compound. We were actually trying to make the unsaturated peroxide $CF_3N{=}CFOOCF_3$, which may indeed be an intermediate en route to the oxaziridine. Known to us as PFAPO (short for pentafluoroazapropene oxide), this unusual heterocycle directed us into some interesting fluorocarbon nitrogen chemistry.

The methods available for making perfluorocarbon imines were at best difficult and we needed a range of types to extend our oxaziridine studies. This didn't happen in fact for many years because the reaction used to make PFAPO was not general and worked only with 1,1-difluoro-2-aza-alkenes [55, 56]. In our quest for other imines we wanted to prepare the little known imine $CF_2{=}NF$, the plan being to prepare $ClCF_2NClF$ and dechlorinate it to give $CF_2{=}NF$ [57, 58]. The reaction of ClCN with ClF to give CF_2ClNCl_2 was well known, and my first Japanese postdoctoral associate, Akira Sekiya, and I reasoned that this must be a stepwise reaction so perhaps we could trap the intermediate with fluorine:

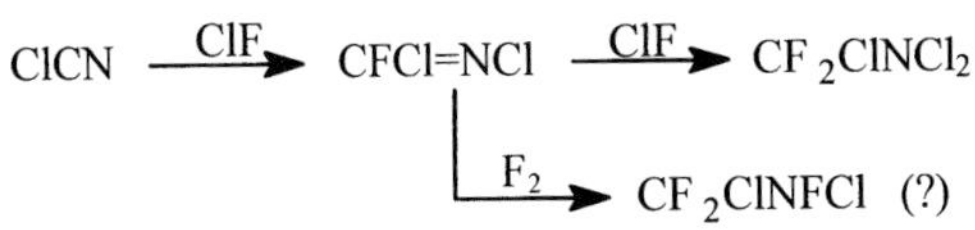

In fact, cyanogen chloride was found to react with a 1 : 1 mixture of $ClF{:}F_2$ to give the desired compound in high yield by the unanticipated route:

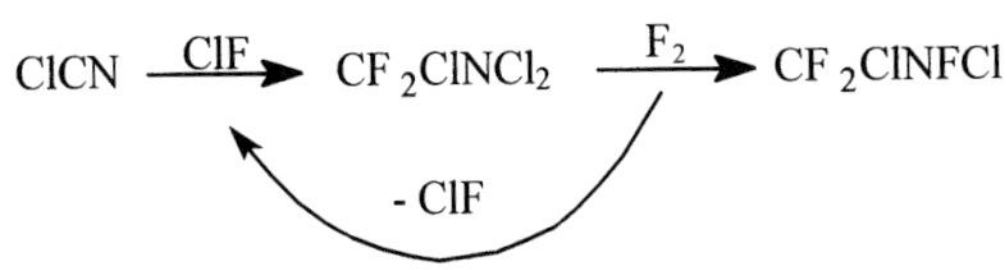

With this new synthesis to hand, we were on the way to $CF_2{=}NF$. However, the dechlorination of $CF_2ClNFCl$ with Hg was a miserable reaction compared to the well known analogous reaction of CF_2ClNCl_2 with Hg to form $CF_2{=}NCl$ [58]. We decided a polar solvent was needed and chose CF_3CO_2H; later trifluoroacetic acid anhydride proved superior, enabling high yields of pure $CF_2{=}NF$ to be achieved [59]. Despite this, conversion of $CF_2{=}NF$ to an oxaziridine was not realized, but some interesting work was done by trapping the nitranion CF_3NF^- generated by treating the imine with fluoride ion (Scheme 11.1) [60-64].

The chemistry of $CF_2{=}NF$ was so interesting that we sought related analogues and were able to prepare the novel imine $SF_4{=}NF$ [65]; this was even more reactive than $CF_2{=}NF$ and some very unusual sulfur-nitrogen derivatives were thereby obtained [66]. The idea of preparing $SF_4{=}NF$ was conceived while I was at the University of Heidelberg in 1979-80 as a Humboldt Research Fellow working in the laboratories of Professors Wolfgang Sundermeyer and Konrad Seppelt. I went to Germany to collaborate with Seppelt on xenon chemistry but became distracted by imine chemistry and didn't really do any serious work in that area. It was known that NSF_3 would react with ClF to give SF_5NCl_2, so I decided that NSF_3, being an analogue of ClCN, ought to undergo chlorofluorination to give SF_5NClF. It did, and the synthesis was completed as follows [67]:

$$NSF_3 \xrightarrow{F_2/ClF} SF_5NClF \xrightarrow{Hg/TF} SF_5NHF \xrightarrow{KF} SF_4{=}NF$$

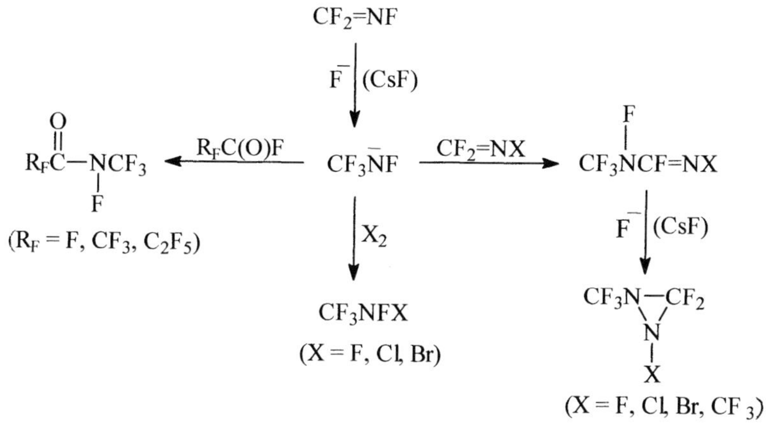

Scheme 11.1.

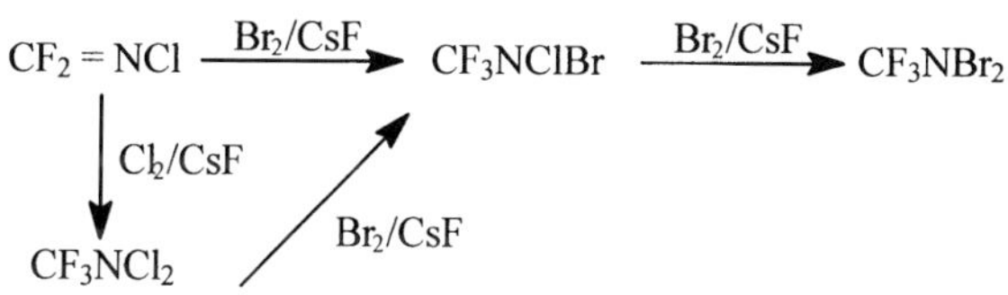

Scheme 11.2.

I can still recall the unexpectedly complex but beautiful ^{19}F NMR spectrum of $SF_4{=}NF$, which was analyzed by Seppelt as an A_2BCD spin system.

This imine work led us to explore fluoride-catalyzed reactions of other nitriles and we discovered that *N*-bromoimines could be readily made from halogenated nitriles [68].

$$RC{\equiv}N \xrightarrow{CsF/Br_2} RCF{=}NBr$$

$$(R = CF_3, C_2F_5, C_3F_7, CF_2Cl_2, CHF_2, CHCl_2, CCl_3)$$

These compounds were remarkably stable, contrary to the notion that *N*-bromo compounds are generally thermally unstable. In the course of this work we noted that CsF formed a complex with Br_2, and that this complex was important in the formation of the *N*-bromoamines. Subsequently, on another visit to Germany in 1989 with the help of a Humboldt-Preise and as a guest of Professor Konrad Seppelt (now at the Freie Universität, Berlin), we obtained an X-ray crystal structure of this complex. It turned out to be an unusual intercalation compound of bromine molecules in CsF [69], further use of which led to another interesting series of *N*-halogen compounds, including stable *N*,*N*-dibromo derivatives (*e.g.* Scheme 11.2) [70].

Cyanogen fluoride

The surprising stability of the N—Br bonds in the *N*-bromo compounds we had encountered led me to think about making $CF_2{=}NBr$. Initially, I could devise no route to this potentially interesting compound but finally the obvious occurred to me: based on the nitrile route to *N*-bromoimines of the type $R_FCF{=}NBr$, treatment of cyanogen fluoride with CsF/Br_2 ought to give $CF_2{=}NBr$. The problem was that although FCN might seem like a trivial starting material, known methods of making it were limited, and only chemists at DuPont had ever prepared a substantial amount of this simple compound [71]; also, their paper gave a discouraging report that condensed FCN was an energetic explosive. Fortunately, an unexpected source of small amounts of FCN soon became available: my postdoctoral worker Brian O'Brien found that flow pyrolysis of $CF_3CF_2CF{=}NBr$ at 450 °C gave good yields of FCN (+ C_2F_5Br) [72]. This enabled us to achieve our objective; 'overbromofluorination' gave *N*,*N*-dibromotrifluoromethanamine [73]:

$$\mathrm{FCN} \xrightarrow{\mathrm{CsF/Br_2}} \mathrm{CF_2{=}NBr} \xrightarrow{\mathrm{CsF/Br_2}} \mathrm{CF_3NBr_2}$$

Needing larger amounts of FCN, we decided to repeat the literature method involving pyrolysis of cyanuric fluoride at 1300 °C in a 'non-porous' carbon tube [71]. We immediately ran into a major problem: many laboratory furnaces will reach 1200 °C but higher temperatures are more expensive to achieve. After investing $6000 in a 1500 °C tube furnace, we discovered that any carbon tubing we could get our hands on was very porous. Finally, we invested all the scrap platinum in the Chemistry Department in the fabrication of a $3500 platinum pyrolysis tube. My thinking was that there were many examples in the literature of unusual compounds that could only be obtained using platinum equipment, hence the tube would come in useful for other things. The platinum tube indeed worked well, but since the optimum yields of FCN were obtained at about 1100 °C we could have avoided the rather expensive 1500 °C tube furnace [74]:

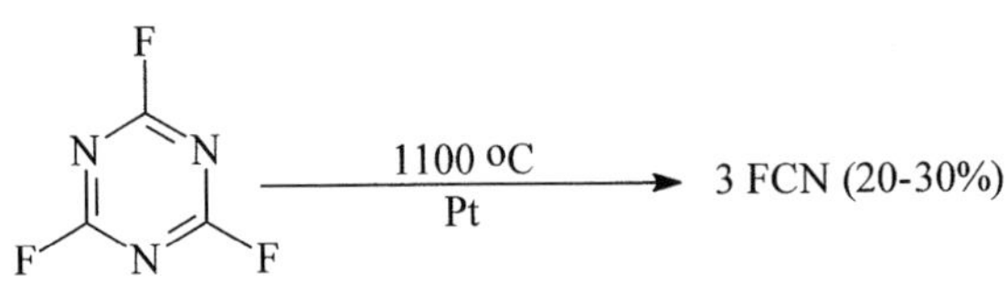

We found that FCN was actually well behaved and never observed any explosivity, although the literature is undoubtedly correct on this issue and it would be foolish to assume otherwise.

The availability of $CF_2{=}NBr$ and FCN led to a number of interesting new compounds with the N—Br bond being the most reactive as expected in the series $CF_2{=}NX$ (X = F, Cl, Br) [64]. Competitive reactions involving $CF_2{=}NF$, $CF_3N{=}CF_2$ and FCN and fluoride ion clearly established the relative reactivity of the anions implicated to be $CF_3NF^- > (CF_3)_2N^- > F_2CN^-$ [74]. In the course of this work, my student Bill Bauknight obtained the beautiful molecule 1,2-bis(trifluoromethyl)-3,3-difluorodiaziridine (Scheme 11.3).

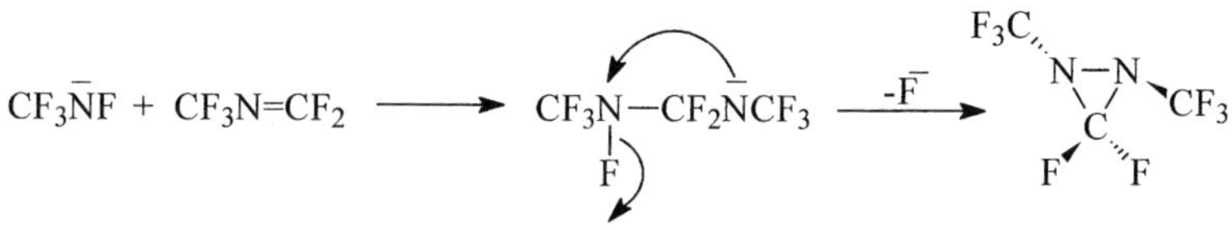

Scheme 11.3.

The Ausimont connection

Our work on oxaziridines and hypofluorites had attracted the interest of Ausimont, SpA (Montefluos), and in 1985 a collaboration continuing to the present was begun based on the company's interest in novel polymers and the use of hypofluorites to prepare perfluorovinyl ethers. We began a serious effort to find a better way to make PFAPO (pentafluoroazapropene oxide), which was of interest as a potential precursor of novel polymers. My first Italian coworker, Walter Navarrini, succeeded in making some PFAPO using concentrated H_2O_2 to oxidize $CF_3N{=}CF_2$, but the yields were very low; the less readily hydrolyzed imine $(CF_3)_2NCF{=}NCF_3$, however, gave the corresponding oxaziridine in better yield (40%), indicating that such perfluorinated heterocycles could be obtained using oxidants other than our original reagent, CF_3OOH [75]. Subsequently, use of CF_3OOH/KF was found to give good yields of the oxaziridine derived from the imine $(CF_3)_2C{=}NCF_3$, which was unreactive towards CF_3OOH alone [76]:

This suggested that CF_3OO^- was involved in this oxidation, and we soon demonstrated that the acylperoxy precursor of CF_3OOH could be used effectively as an *in situ* source of this peroxy anion [77], *e.g.*

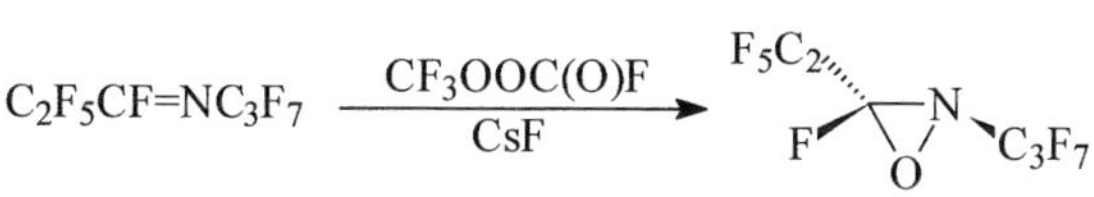

Then a breakthrough was made when use of the readily available peroxyacid *m*-$ClC_6H_4CO_3H$ (MCPBA), suitably dried and purified, was found to provide a general high-yield route to the oxaziridines needed [78]. We had actually tried MCPBA several years earlier without success, but an able postdoctoral worker, Viacheslav Petrov, found that purified MCPBA in an appropriate solvent worked well:

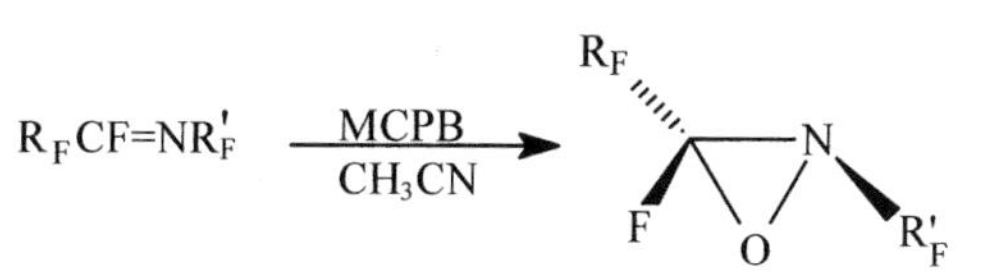

(R_F = *n*-C_3F_7, *i*-C_3F_7, *n*-C_4F_9; R'_F = C_2F_5, *n*-C_3F_7, CF_3)

This led to real advances in the chemistry of perfluorinated oxaziridines, including demonstrations of their utility as powerful and selective oxidants. Nearly all of our work emanating from the original PFAPO discovery in 1976 is summarized in an excellent review by my former coworkers Resnati and Petrov [79].

The quest for new nitrogen-containing polymers of the fluorocarbon class led us to think about the chemistry of well-known oxazetidines as possible precursors to amino-ether polymers not containing a weak N—O bond as in nitroso rubbers. Since PFAPO undergoes an interesting dimerization with SbF_5 [80], we tried this reaction with an oxazetidine, hoping to obtain a polymeric material. A beautiful crystalline trimer was formed with SbF_5 [81] and, if HF was present, another novel ring-opening reaction occurred [82]:

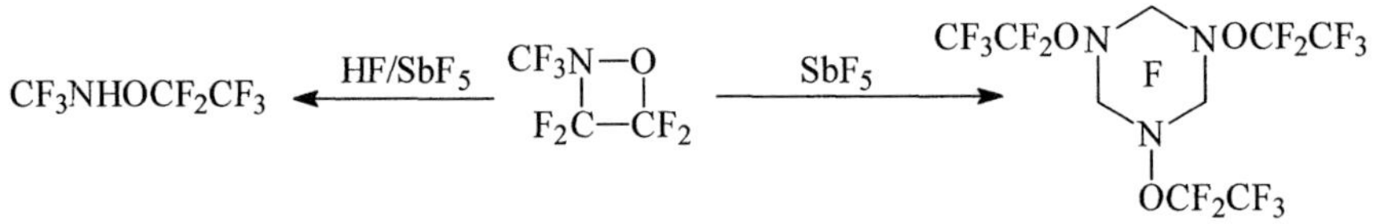

We speculated on the mechanism of the HF/SbF_5 reaction, but in retrospect the same intermediate, $CF_2{=}NOCF_2CF_3$, is probably involved in both reactions. Curiously, we never carried out the reaction of $CF_2{=}NOCF_2CF_3$ with SbF_5, despite the fact that this imine is readily prepared by dehydrofluorination of $CF_3NHOCF_2CF_3$ with KF. These results are partially described in some patents [83, 84], but much of this interesting chemistry has only been published in a thesis [85].[3]

Trifluoromethylations, perfluoroalcohols and perfluorodiketones

With the support of Ausimont to explore fluorocarbon nitrogen and oxygen derivatives, we were always on the lookout for novel routes to such compounds. When it became evident that Ruppert's reagent, CF_3SiMe_3,[4] was useful in nucleophilic trifluoromethylations [86, 87], we decided to study its reactions with perfluorinated systems, starting with acyl fluorides.

The trifluoromethylation of perfluoroacid fluorides with CF_3SiMe_3/MF (M = K, Cs) was found to proceed readily (Scheme 11.4), but was not catalytic in MF as in work carried out by Prakash *et al.* [88]. This work provided an easy route to a variety of tertiary monohydric alcohols and led us to explore the reaction with difunctional substrates. Naïvely we expected to make some interesting diols. What transpired was more interesting [89] (Scheme 11.4): diketones were produced which participated in facile cycloadditions with many polar multiple bonds, as illustrated by the case of acetone in Scheme 11.5, and reacted readily with water and related substrates to give diols [90]. The sodium salts of the diols can be used to form a variety of heterobicycles, as exemplified in Scheme 11.5 [91].

[3] At some point it will be described more fully in the regular chemical literature. I have learned over the years that letting students and postdoctoral fellows leave before delivering manuscripts on their work is a big mistake. Solemn promises to do this after leaving for a job are rarely kept.

[4] I had the pleasure of visiting Ruppert in Bonn in 1985, just before the 11th International Symposium on Fluorine Chemistry in East Berlin, and I well remember enjoying good beer as we sat overlooking the Rhine in Bonn, discussing chemistry and other topics. Ruppert's untimely death a few years later was a great loss to chemistry, and he never saw the fruits of his pioneering work with this remarkable reagent.

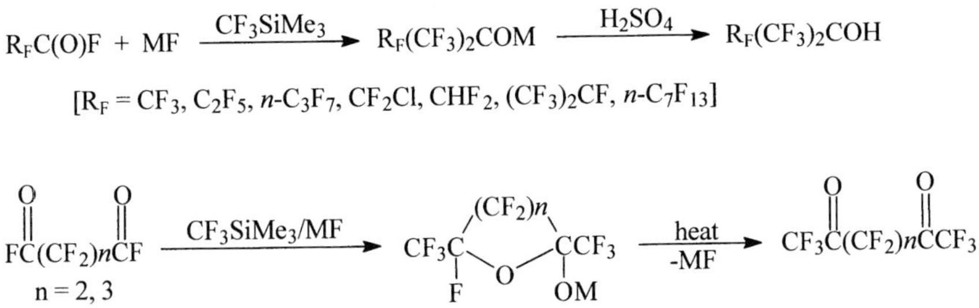

Scheme 11.4.

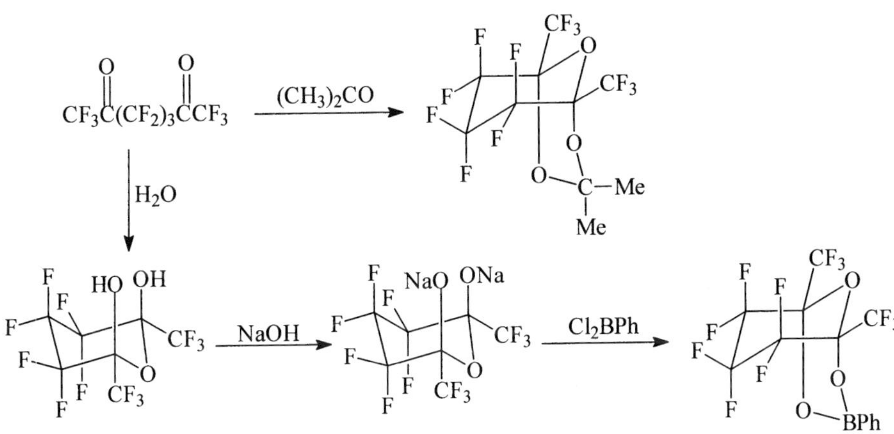

Scheme 11.5.

Hypohalites, new catalysts and dioxiranes

As mentioned earlier, Ausimont has a strong interest in fluorinated hypohalites, since these are the basis of commercial processes leading to various perfluorovinyl ethers. One of our goals was to make functionalized hypohalites such as $X(CF_2)_nOF$, where $X = C(O)F$, SO_2F, etc. One particularly interesting development in this research over many years was the search for the hypofluorite CF_3OCF_2OF, which is still unknown. After many abortive attempts to prepare this compound from the fluoroformate $CF_3OC(O)F$ and fluorine (Scheme 11.6), we decided that the problem lay with the starting material. Hoping that its higher homologues might behave differently, were set out to make some examples, having decided thatthe best route might be the addition of FOC(O)F to various fluoroalkenes.

Interestingly, FOC(O)F was first prepared by Cauble and Cady while I was a graduate student in Cady's group, and I well recall the tedious codistillations that Cauble had had to perform in order to isolate small amounts of this acyl hypofluorite following the photolysis of $FC(O)OOC(O)F\text{-}F_2$ mixtures, only to find that his product had limited stability [92]. This experience discouraged further investigations for 25 years until, undaunted, Antonio Russo and I set out to repeat Cauble's work because we were desperate for $R_FOC(O)F$ analogues of $CF_3OC(O)F$. We soon found that by changing the UV source to a low-wattage 254 nm

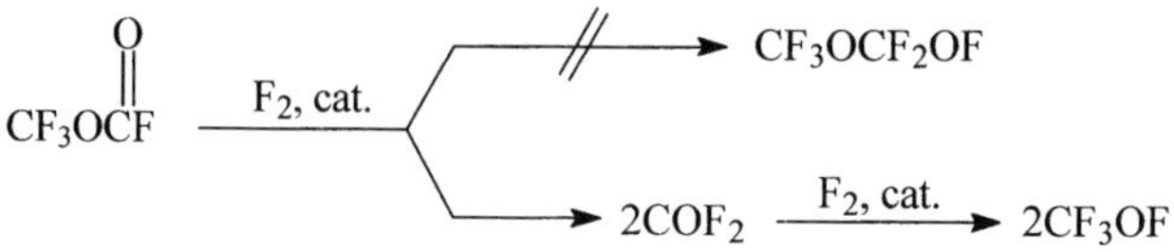

Scheme 11.6.

lamp considerably improved yields of FOC(O)F could be obtained, and the compound proved to be more thermally stabile than originally reported. We proceeded to make some $R_FOC(O)F$ samples for catalytic fluorination but found that a larger R_F group than CF_3 had no favorable effect, therefore we could not obtain higher homologues of CF_3OCF_2OF.

With a good source of FOC(O)F now available, we decided to look at other reactions. Many years earlier, when we were working in Kansas on new routes to bis(hypofluorites) of the type $R_FCF(OF)_2$ [93], I conceived the idea of trying to prepare a compound containing both an OCl and an OF function. In particular, I wanted to make $ClOCF_2OF$ because I thought it might possibly eliminate ClF under appropriate conditions to give the novel cyclic peroxide difluorodioxirane, a compound that had become of interest in work at Ausimont with $CF_2(OF)_2$ [94], *e.g.*

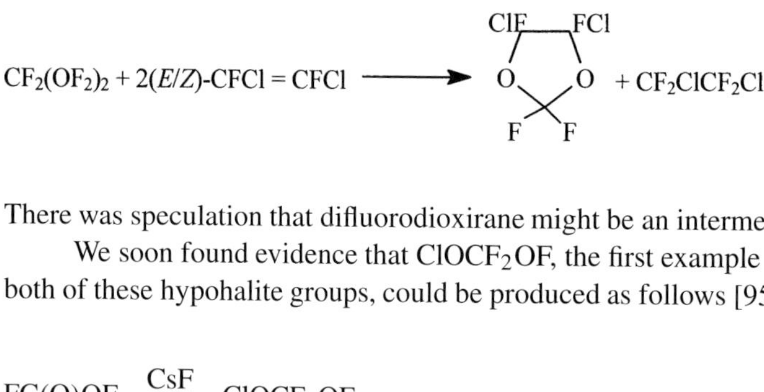

There was speculation that difluorodioxirane might be an intermediate in this reaction.

We soon found evidence that $ClOCF_2OF$, the first example of a molecule containing both of these hypohalite groups, could be produced as follows [95]:

$$FC(O)OF \xrightarrow[ClF]{CsF} ClOCF_2OF$$

Yields in this reaction were highly variable, and eventually we discovered that the metal fluoride needs to contain some water. Before this was recognized, we decided to try the reaction in a flow system to solve the yield problem. The infrared spectrum of the product could only be rationalized on the basis of the presence of a small high-symmetry ring compound, which we ultimately proved to be difluorodioxirane [96]. The structure of this beautiful molecule, thought to be produced as shown in Scheme 11.7, has been thoroughly investigated by both experimental and theoretical methods [97], and its chemistry is currently under active investigation in my laboratory. It is one of the joys of working in fluorine chemistry that in spite of the large number of compounds known, one can still discover new 5-atom varieties. Ironically, difluorodioxirane is not formed from $ClOCF_2OF$ and it is not the intermediate in the cycloaddition of $CF_2(OF)_2$ to alkenes.

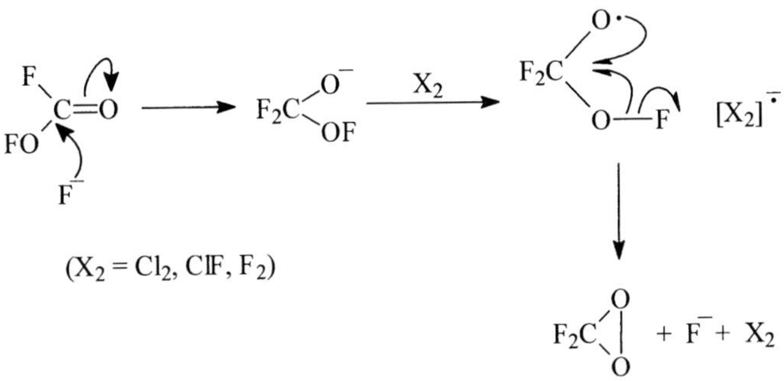

Scheme 11.7.

Superacids, electrophilic fluorination and fuel cells

After moving to Clemson University in 1982, we published the details of the synthesis and some properties of the sulfonimide $(CF_3SO_2)_2NH$ [98] mentioned earlier. This compound clearly possessed remarkable Bronsted acidity, and I assigned an able postdoctoral worker, Sukhjinder Singh, to explore further some of its chemistry and that of its analogues. In the course of this work we decided to see if we could prepare the *N*-fluoro derivative of $(CF_3SO_2)_2NH$. It was surprisingly easy and proceeded in high yield [99]:

$$(CF_3SO_2)_2NH + F_2 \xrightarrow{-196 \text{ to } 22\,^{\circ}C} (CF_3SO_2)_2NF + HF$$

At about this time a publication had appeared from a DuPont chemist on the use of *N*-fluoro-*N*-alkylsulfonamides to fluorinate carbanions [100], so we decided to test $(CF_3SO_2)_2NF$ in similar reactions, and, while using toluene as a solvent for a Grignard reaction, found that the toluene itself underwent electrophilic ring fluorination. We soon established that $(CF_3SO_2)_2NF$ is a very powerful and versatile selective fluorination reagent of the electrophilic class. This discovery led to a heightened interest in N-F fluorination and many new reagents were developed, but to date $(CF_3SO_2)_2NF$ is the most powerful one [101]. (An interesting aspect of this class of reagents is the rather easily predictable fluorinating power based on inductive effects of the substituents on nitrogen.) A number of my able co-workers have applied $(CF_3SO_2)_2NF$ to the selective fluorination of organic substrates [102], and the utility of this reagent is beautifully illustrated by the very recent successful synthesis of 14-fluorocamptothecin [103]. Camptothecin and some of its derivatives are important topoisomerase inhibitors and of strong interest in chemotherapy; the selective introduction of a single fluorine at the 14-position in this structurally complex molecule is quite remarkable.

In 1985, I received a call from Dan Scarpiello, a project manager at the Gas Research Institute in Chicago. He wanted to know if I would be interested in a programme they were promoting on fluorinated electrolytes for 'phosacid' (phosphoric acid) fuel cells. They were familiar with our work on $(CF_3SO_2)_2NH$, and when I heard of the possible budget level, I became very keen to collaborate with them! However, I was very busy being an administrator involved with the details of the construction of our new chemistry building at

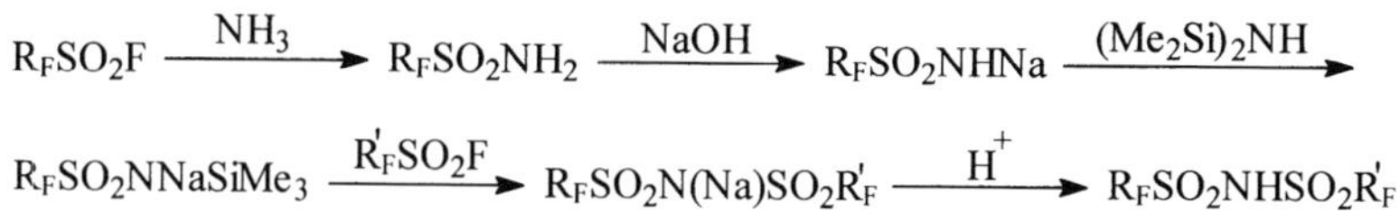

Scheme 11.8.

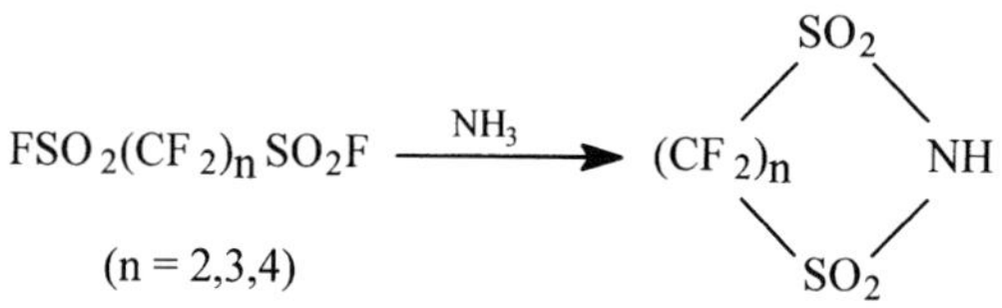

Scheme 11.9.

Clemson University and I didn't write a very good proposal, but it was sort of a 'done deal.' Thus we began a serious effort to prepare fluorinated superacids of carbon and nitrogen as potential electrolytes for fuel cells.

Early tests on $(CF_3SO_2)_2NH$ as an aqueous electrolyte in a test fuel cell at International Fuel Cells showed that the performance was superior to phosphoric acid under the same conditions. The excellent electrochemical properties of sulfonimides were confirmed by collaborators Professor E. Yeager and his co-workers at Case Western Reserve University [104]. This was exciting, and we set out to develop acids that could overcome the one shortcoming of $(CF_3SO_2)_2NH$: it is volatile at the practical operating temperature of a typical phosacid fuel cell. At first we attempted to solve this problem by replacing the CF_3 groups by large perfluorocarbon groups, using synthesis methodology evolving in our research [105] (Scheme 11.8).

Unfortunately, although the presence of large R_F groups did indeed result in lower volatilities, it also led to wetting of the Teflon® in the gas-permeable electrodes of the fuel cell, causing flooding, *i.e.* the fuel cell no longer worked.

The 3M Co. was invited by GRI to provide observers in this electrolyte programme, since the company was ultimately the most likely supplier of compounds and intermediates for the types of electrolytes under consideration. Fred Behr of 3M kindly supplied us with numerous sulfonyl fluorides, including perfluorinated 1,2-, 1,3- and 1,4-bis(fluorosulfonyl)alkanes. We utilized these only to rediscover the obvious fact that these compounds easily cyclize to 5-, 6- and 7-membered sulfonimides on treatment with ammonia (Scheme 11.9). These interesting cyclic compounds had actually been prepared already at 3M but we had forgotten this! In any event, they were too volatile and of little interest for the fluorinated electrolyte project. However, the geometry of these acids allows easy formation of large crystals, and the first single crystal X-ray structure determination of a compound containing a perfluorinated sulfonimide moiety on the silver salt of the acid where $n = 3$ [106]. The remarkable structures involved are lamellar in nature, with alternating infinite layers of ionic functions and fluorocarbon domains, and are a common feature of all bis(perfluoroalkylsulfonyl)imides and their carbon analogues [106-108]. Nowadays, other investigators rediscover this with increasing regularity.

The availability of 3M's perfluoro-α,ω-bis(fluorosulfonyl)alkanes led us to a very simple and successful idea. Any polyprotic strong acid is of low volatility because of the increased hydrogen bonding, *cf.* $HOP(O)F_2$-v-$(HO)_2P(O)F$-v-$(HO)_3PO$, so why not make difunctional sulfonimides? This again turned out to be quite easy, as demonstrated by post-doctoral Sharique Zuberi [104]:

$$2CF_3SO_2N(Na)SiMe_3 + FSO_2(CF_2)_nSO_2F \longrightarrow CF_3SO_2N(Na)SO_2(CF_2)_nSO_2N(Na)SO_2CF_3 \xrightarrow{H^+} CF_3NHSO_2(CF_2)_nSO_2NHSO_2CF_3$$

Alternatively, in the case of $n = 3$, 4 and higher, these novel compounds could be accessed via bis(silyl)amides $Me_3Si(Na)NSO_2(CF_2)_nSO_2N(Na)SiMe_3$ prepared from the corresponding bis(fluorosulfonyl)alkanes $FSO_2(CF_2)_nSO_2F$ using the functional group conversions shown in Scheme 11.8. Although these α,ω-bis(sulfonylimides) possessed electrochemical properties, the landscape was changing: GRI Advisory Board Members were beginning to raise the issue of cost of these new electrolytes and the estimates were clearly high (a 200 kW phosacid fuel cell utilizes *ca.* 2000 lbs of 85% phosphoric acid). At this point, an intriguing set of experiments carried out in Professor Yeager's group showed that the addition of small amounts of sulfonimides to phosphoric acid caused dramatic improvements in oxygen reduction during rotating disc voltametry on smooth platinum. This effect was confirmed by International Fuel Cells in a small test fuel cell [109], and the possibility of capturing this enhanced performance by incorporating a polymeric perfluorosulfonimide in the electrode assembly was raised. The amount of polymer required could be quite small, thus costs could be low, and long-term stability was envisaged. So in late 1989 we began a programme to prepare perfluorinated sulfonimide polymers. I had never really worked in polymer chemistry before, so here was a new adventure in fluorine chemistry.

The polymers envisaged were of two types: ionenes and ionomers. From the work on difunctional sulfonimides, it was clear that a linear polymer of the type $[-HNSO_2(CF_2)_nSO_2]_n-$ would be suitable since the acid functions would be well separated and successive ionization would not lead to decreased acidity as charge developed along the chain. Using difunctional monomers, we were successful in carrying out a step-reaction (condensation) polymerization to give novel ionenes [110]:

$$FSO_2(CF_2)_nSO_2F + Me_3Si(Na)NSO_2(CF_2)_nSO_2N(Na)SiMe_3 \longrightarrow [-N(Na)SO_2(CF_2)_nSO_2]_m- \quad (n = 4,\ 6,\ \text{etc.})$$

These polymers are under active investigation and provide remarkable Brønsted acids; however, they have become of more interest as battery electrolytes and of less interest for fuel cells because the polymers with average molecular weights of a few ten thousands are water soluble. The novelty of these new materials is clear when one tries to think of other examples of polymers of this type where the acid function is in the polymer main chain: these materials are unique.

The issue of preparing ionomers containing the sulfonimide function was a formidable one. Literature on the well-known perfluorinated sulfonic acid ionomers such as DuPont's Nafion® clearly suggested that one needed a perfluorovinyl ether monomer in order to carry out a successful copolymerization with tetrafluoroethylene (TFE). The preparation of such a monomer was quite a challenge, but eventually we succeeded in duplicating

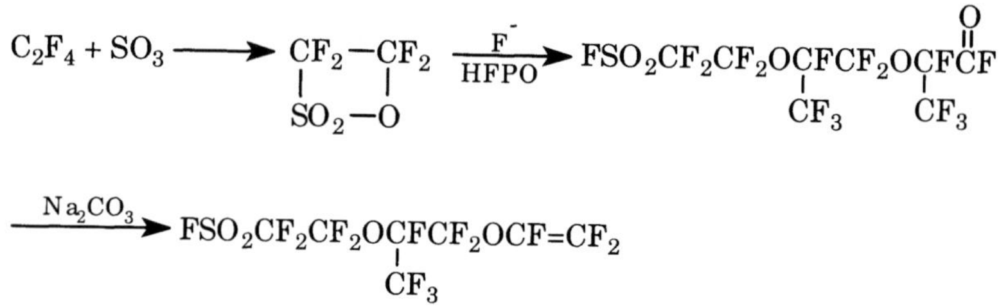

Scheme 11.10.

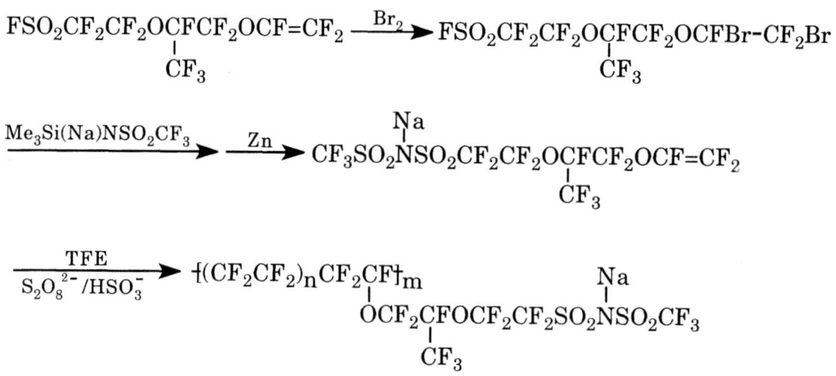

Scheme 11.11.

patent literature on routes to the critical starting materials (Scheme 11.10; HFPO = hexafluoropropene oxide). This may have been the first time that this chemistry was duplicated in an academic research laboratory.

With the fluorosulfonyl-substituted perfluorovinyl ether shown in Scheme 11.10 to hand, there remained two critical questions: (i) how could the SO_2F function be converted to $SO_2N(Na)SO_2R_F$ without affecting the perfluorovinyl group?; (ii) since the sulfonimide in acid form could not be used for copolymerization with TFE, would the water-soluble salt form undergo successful copolymerization? Conversion of the sulfonyl fluoride to the sulfonimide was achieved by extension of earlier methodology, with protection and deprotection of the double bond. This deceptively simply sequence required considerable effort to achieve in practice. Successful copolymerization of the vinyl ether with TFE was effected using a persulfate-based aqueous emulsion system [110] (Scheme 11.11) with the help of an able postdoctoral researcher, Jing-Ji Ma, who had had experience in China of preparing Nafion® copies.

High-molecular-weight copolymers of good quality were obtained by trial and error, as such polymerizations are exceedingly difficult to optimize [111, 112]. This work is ongoing and actual tests of our solution-cast perfluorosulfonimide as proton-exchange membranes in solid polymer electrolyte fuel cells have been very exciting. This has become a major research endeavor and the ability to vary the side chain in the ionomers in heretofore unknown ways has been equally exciting. A variety of monomers of the type below are under investigation.

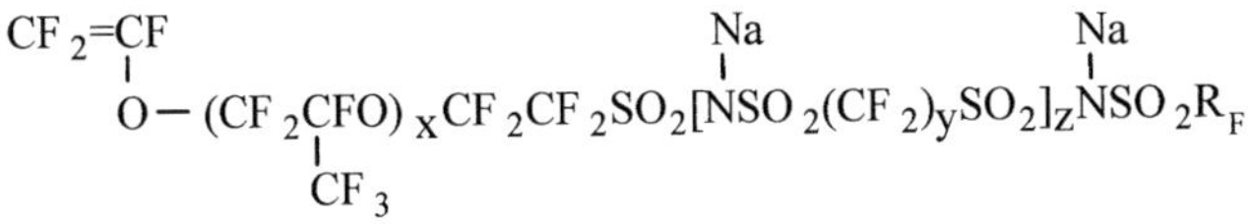

The successful arrival at new materials with strong potential for electrochemical and other applications was a particularly satisfying aspect of our fascination with fluorine. The trek from xenon-nitrogen compounds to practical polymers is an interesting example of how fundamental, curiosity-driven, basic research in academe often has unexpected outcomes. This well-known fact, repeated often by many investigators, should not have to be publicly emphasized. However, when it comes to funding basic research, decision makers often have lapses of memory. One of my favourite lectures I'm always happy to present to any willing audience is entitled 'From Noble Gas Compounds to Fuel Cells and Pharmaceuticals – An Example of the Benefits of Basic Research in Academe'.

Conclusion

From the above account of some of our research over the past 36 years, I hope it has become obvious to the reader why I chose the title 'Fluorine Chemistry – A Chemical Gardener's Paradise'. From the first personal discovery of a new fluorine compound in 1963 to exciting ongoing research on fuel cells, new materials and useful new reagents, it has been a rewarding and fun adventure. Just as the diversity in the plant kingdom is full of fascination, so too is the diversity with fluorine. I have never succeeded in creating a successful plant garden but I have had reasonable success in sowing the seeds of discovery in fluorine chemistry.

Acknowledgement

I wish to acknowledge the excellent work of many students and postdoctoral collaborators who contributed to the research described here. Due to space limitations not all of our interesting work carried out over the past 36 years could be included, and I apologize to those co-workers who are not mentioned in references or in the text. Similar remarks apply to numerous collaborations with many colleagues who have contributed to understanding the structure-property relationships of many of the fluorinated compounds and materials generated in this research, and who shared in my fascination with fluorine. My heartfelt thanks go to all the organizations who have supported my research in fluorine chemistry including the NSF, ACS-PRF, Research Corporation, Alfred P. Sloan Foundation, Alexander von Humboldt-Stiftung, ARO, DOE, ARPA, EPRI, EPA, GRI, Ausimont SpA, Dow Chemical Co., Central Glass Co., DuPont Company and the 3M Co.; and I acknowledge Dr Paul L. Tobey (1913-1996) whose generosity established the Tobey-Beaudrot Professorship in Chemistry at Clemson University, which I am privileged to hold. Lastly, I want to acknowledge the international community of fluorine chemists who have made my fascination with fluorine such an enjoyable experience.

References

1 D. D. DesMarteau, W. Navarrini, A. Zedda, A. Russo and V. Montanari, *J. Fluorine Chem.*, *71* (1995) 169.
2 F. B. Dudley and G. H. Cady, *J. Am. Chem. Soc.*, *79* (1957) 513; J. M. Shreeve and G. H. Cady, *Inorg. Synth.*, *7* (1963) 124.
3 G. H. Cady and J. H. Hildebrand, *J. Am. Chem. Soc.*, *52* (1930) 3839.
4 G. H. Cady, D. A. Rogers and C. A. Carlson, *Ind. Eng. Chem.*, *34* (1942) 443.
5 G. H. Cady, *J. Am. Chem. Soc.*, *56* (1934) 1432.
6 R. N. Haszeldine, *J. Chem. Soc.*, (1952) 4259.
7 D. D. DesMarteau and G. H. Cady, *Inorg. Chem.*, *5* (1966) 169.
8 D. D. DesMarteau and G. H. Cady, *Inorg. Chem.*, *6* (1967) 416.
9 D. D. DesMarteau and G. H. Cady, *Inorg. Chem.*, *4* (1966) 1829.
10 N. Bartlett, *Proc. Chem. Soc.*, (1962) 218.
11 F. B. Dudley, G. L. Gard and G. H. Cady, *Inorg. Chem.*, *2* (1963) 61.
12 N. Bartlett, M. Wechsberg, F. O. Sladky, P. A. Bulliner, G. R. Jones and R. D. Burbank, *Chem. Commun.*, (1969) 703.
13 M. Eisenberg and D. D. DesMarteau, *Inorg. Nucl. Chem. Letters*, *6* (1970) 29.
14 D. D. DesMarteau and M. Eisenberg, *Inorg. Chem.*, *11* (1972) 2641.
15 P. A. Bernstein, F. A. Hohorst, M. Eisenberg and D. D. DesMarteau, *Inorg. Chem.*, *10* (1971) 1549.
16 J. Foropoulos, Jr. and D. D. DesMarteau, *Inorg. Chem.*, *21* (1982) 2503.
17 P. M. Nutkowitz and G. Vincow, *J. Am. Chem. Soc.*, *91* (1969) 5956 and *J. Phys. Chem.*, *75* (1971) 712.
18 M. Parker, Ph.D. Thesis, University of Washington, 1967.
19 See ref. [2].
20 F. B. Dudley, G. H. Cady and D. F. Eggers, *J. Am. Chem. Soc.*, *78* (1956) 290.
21 G. H. Cady, *Inorg. Synth.*, *11* (1968) 155.
22 D. D. DesMarteau, *Inorg. Chem.*, *7* (1968) 434.
23 L. R. Anderson and W. B. Fox, *J. Am. Chem. Soc.*, *89* (1967) 4313.
24 K. Seppelt and D. Lentz, *Prog. Inorg. Chem.*, *29* (1982) 167.
25 M. Eisenberg and D. D. DesMarteau, *J. Am. Chem. Soc.*, *92* (1970) 4759.
26 D. D. DesMarteau, *Inorg. Chem.*, *9* (1970) 2179.
27 R. L. Talbott, *J. Org. Chem.*, *33* (1968) 2095.
28 P. A. Bernstein, F. A. Hohorst and D. D. DesMarteau, *J. Am. Chem. Soc.*, *93* (1971) 3882.
29 D. D. DesMarteau, *Inorg. Chem.*, *11* (1972) 193.
30 P. A. Bernstein and D. D. DesMarteau, *J. Fluor. Chem.*, *2* (1972/73) 315.
31 F. A. Hohorst and D. D. DesMarteau, *Inorg. Chem.*, *13* (1974) 715.
32 F. A. Hohorst, J. V. Paukstelis and D. D. DesMarteau, *J. Org. Chem.*, *39* (1974) 1289.
33 L. R. Anderson, D. E. Gould, W. B. Fox, F. A. Hohorst and D. D. DesMarteau, *J. Am. Chem. Soc.*, *95* (1973) 3866.
34 F. A. Hohorst and D. D. DesMarteau, *J. Inorg. Ncl. Chem., Supplement 1976*, p. 63.
35 N. Walker and D. D. DesMarteau, *J. Am. Chem. Soc.*, *97* (1975) 13.
36 S. L. Yu and D. D. DesMarteau, *Inorg. Chem.*, *17* (1978) 304.
37 S. L. Yu and D. D. DesMarteau, *J. Fluorine Chem.*, *12* (1978) 315.
38 S. L. Yu and D. D. DesMarteau, *Inorg. Chem.*, *17* (1978) 2484.
39 M. J. Hopkinson and D. D. DesMarteau, *J. Fluor. Chem.*, *7* (1976) 501.
40 M. J. Hopkinson, N. Walker and D. D. DesMarteau, *J. Org. Chem.*, *41* (1976) 1407.
41 D. D. DesMarteau and R. M. Hammaker, *Israel J. Chem.*, *17* (1978) 103.
42 D. D. DesMarteau, *J. Am. Chem. Soc.*, *94* (1972) 8933.
43 D. D. DesMarteau, unpublished results.
44 J. K. Ruff and M. Lustig, *Inorg. Synth.*, *11* (1968) 138.
45 R. D. LeBlond and D. D. DesMarteau, *Chem. Comm.*, (1974) 555.
46 D. D. DesMarteau and R. D. LeBlond, *Fifth European Symposium on Fluorine Chemistry*, Aviemore, Scotland, September, 1974, Abstract I-78.
47 D. D. DesMarteau, R. D. LeBlond, S. F. Hossain and D. Nothe, *J. Am. Chem. Soc.*, *103* (1981)7734.
48 J. N. Meusdorffer and H. Niederprum, *Chem. Ztg.*, *96* (1972) 582.

49 J. Foropoulos and D. D. DesMarteau, 2nd *Chemical Congress of the North American Continent*, Las Vegas, NV, August, 1980, Abstract Fluo 14.
50 T. Gramstad and R. N. Haszeldine, *J. Chem. Soc.*, (1956) 173.
51 J. Foropoulos and D. D. DesMarteau, *J. Am. Chem. Soc.*, *104* (1982) 4260.
52 J. F. Sawyer, G. J. Schrobilgen and S. J. Sutherland, *Inorg. Chem.*, *21*, (1982) 4064.
53 G. J. Schrobilgen, in G. A. Olah, R. D. Chambers and G. K. Surya Prakash, (eds.), *Synthetic Fluorine Chemistry*, Wiley, New York, 1992, p. 1.
54 E. R. Falardeau and D. D. DesMarteau, *J. Am. Chem. Soc.*, *98* (1976) 3529.
55 A. Sekiya and D. D. DesMarteau, *Inorg. Chem.*, *19* (1980) 1330.
56 Y. Y. Zheng and D. D. DesMarteau, *Inorg. Chem.*, *23* (1984) 644.
57 A. Sekiya and D. D. DesMarteau, *J. Am. Chem. Soc.*, *101* (1980) 7460.
58 A. Sekiya and D. D. DesMarteau, *Inorg. Chem.*, *20* (1981) 1.
59 A. Sekiya and D. D. DesMarteau, *J. Org. Chem.*, *46* (1981) 1277.
60 S.-C. Chang and D. D. DesMarteau, *Polyhedron*, *1* (1982) 129.
61 S.-C. Chang and D. D. DesMarteau, *J. Org. Chem.*, *48* (1983) 771.
62 S.-C. Chang and D. D. DesMarteau, *Inorg. Chem.*, *22* (1983) 805.
63 Y. Y. Zheng, C. W. Bauknight and D. D. DesMarteau, *J. Org. Chem.*, *49* (1984) 3590.
64 C. W. Bauknight and D. D. DesMarteau, *J. Org. Chem.*, *53* (1988) 4443.
65 D. D. DesMarteau and K. Seppelt, *Angew. Chemie*, *92* (1980) 659.
66 B. A. O'Brien and D. D. DesMarteau, *Inorg. Chem.*, *23* (1984) 2088.
67 D. D. DesMarteau, H. H. Eysle, H. H.Oberhammer and H. Gunther, *Inorg. Chem.*, *21* (1982) 1607.
68 B. A. O'Brien and D. D. DesMarteau, *J. Org. Chem.*, *49* (1984) 1467.
69 D. D. DesMarteau, T. Grelbig, S.-H. Hwang and K. Seppelt, *Angew. Chem. Int. Ed. Engl.*, *29* (1990) 1448.
70 Y.-Y. Zheng, Q. C. Mir, B. A. O'Brien and D. D. DesMarteau, *Inorg. Chem.*, *23* (1984) 518.
71 F. S. Fawcett and R. D. Lipscomb, *J. Am. Chem. Soc.*, *86* (1964) 2576.
72 B. A. O'Brien and D. D. DesMarteau, *Rev. Chem. Minerale*, *23* (1986) 621.
73 B. A. O'Brien, J. S. Thrasher, C. W. Bauknight, M. L. Robin and D. D. DesMarteau, *J. Am. Chem. Soc.*, *106* (1984) 4266.
74 C. W. Bauknight and D. D. DesMarteau, *J. Am. Chem. Soc.*, *112* (1990) 728.
75 W. Navarrini and D. D. DesMarteau, *U.S. Patent* 4874875 *C.A.*, *112* (1989) 159140y.
76 L. Brigante and D. D. DesMarteau, *J. Fluor. Chem.*, *53* (1991) 81.
77 V. A. Petrov and D. D. DesMarteau, *Mendeleev Comm.*, (1993) 87.
78 V. A. Petrov and D. D. DesMarteau, *J. Org. Chem.*, *58* (1993) 4754.
79 V. A. Petrov and G. Resnati, *Chem. Rev.*, *96* (1966) 1809.
80 W. L. Lam and D. D. DesMarteau, *J. Am. Chem. Soc.*, *104* (1982) 4034.
81 A. Malacrida and D. D. DesMarteau, *European Pat. Appl.; E.P.* 3533743 *C.A.*, *113* (1990) 23953d.
82 S. P. Kotun and D. D. DesMarteau, *Can. J. Chem.*, *67* (1989) 1724.
83 S. P. Kotun and D. D. DesMarteau, *European Pat. Appl.; E.P.* 353743 *C.A.*, *113* (1990) 39916e.
84 S. P. Kotun and D. D. DesMarteau, *European Pat. Appl.; E.P.* 353721 *C.A.*, *113* (1990) 58476c.
85 S. P. Kotun, Ph.D. Thesis, Clemson University, 1990.
86 I. Ruppert, K. Schlisk and W. Volbach, *Tetrahedron Lett.*, *25* (1984) 2195.
87 G. K. S. Prakash, R. Krishnamurti and G. A. Olah, *J. Am. Chem. Soc.*, *111* (1989) 393.
88 S. P. Kotun, J. D. O. Anderson and D. D. DesMarteau, *J. Org. Chem.*, *57* (1992) 1124.
89 J. D. O. Anderson, W. T. Pennington and D. D. DesMarteau, *Inorg. Chem.*, *32* (1993) 5079.
90 J. D. O. Anderson, Ph.D. Thesis, Clemson University, 1994.
91 J. D. O. Anderson, W. T. Pennington and D. D. DesMarteau, *Inorg. Chem.*, *35* (1996) 3188.
92 R. L. Cauble and G. H. Cady, *J. Am. Chem. Soc.*, *89* (1967) 5161.
93 A. Sekiya and D. D. DesMarteau, *Inorg. Chem.*, *19* (1980) 1328.
94 L. Brigante, S. Fontana, W. Navarrini, V. Tortelli and A. Zedda, *J. Fluorine Chem.*, *71* (1995) 111.
95 A. Russo and D. D. DesMarteau, *Inorg. Chem.*, *34* (1995) 6221.
96 A. Russo and D. D. DesMarteau, *Angew. Chem.*, *105* (1993) 956.
97 H. Burger, P. Weinrath, G. A. Arguello, B. Julicher, H. Willner, D. D. DesMarteau and A. Russo, *J. Mol. Spectroscopy*, *168* (1994) 607.
98 J. F. Foropoulos and D. D. DesMarteau, *Inorg. Chem.*, *23* (1984) 3720.
99 S. Singh, D. D. DesMarteau, S. S. Zuberi, M. Witz and H.-N. Hwang, *J. Am. Chem. Soc.*, *109* (1987) 7194.

100 W. E. Barelle, *J. Am. Chem. Soc.*, *106* (1984) 452.
101 G. S. Lal, G. P. Pez and R. G. Syvret, *Chem. Rev.*, *96* (1996) 1737.
102 W. Ying, D. D. DesMarteau and Y. Gotoh, *Tetrahedron*, *52* (1996) 15.
103 W. Ying, D. D. DesMarteau and G. B. Jones, *215th National ACS Meeting*, Dallas, TX, March 29-April 2, 1998, abstract ORGN 11.
104 M. Razaq, A. Razaq, E. Yeager, D. D. DesMarteau and S. Singh, *J. Applied Electrochem.*, *17* (1987) 1064.
105 D. D. DesMarteau, S. Zhu, W. T. Pennington, Y. Gotoh, M. Witz and S. Zuberi, *J. Fluorine Chem.*, *45* (1989) 24.
106 D. D. DesMarteau, S. S. Zuberi, W. T. Pennington and B. B. Randolph, *Europ. J. Solid State and Inorg. Chem.*, *58* (1992) 71.
107 D. D. DesMarteau, W. T. Pennington, K.-S. Sung, S.-Z. Zhu and R. Scott, *Europ. J. Solid State and Inorg. Chem.*, *28* (1991) 905.
108 L. Xue, D. D. DesMarteau and W. T. Pennington, *Angew. Chem.*, *36* (1997) 1333.
109 M. Razaq, A. Razaq, E. Yeager, D. D. DesMarteau and S. Singh, *J. Electrochem. Soc.*, *136* (1989) 385.
110 D. D. DesMarteau, *J. Fluorine Chem.*, *72* (1995) 203.
111 S. E. Creager, J. J. Sumner, J. J. Ma and D. D. DesMarteau, *J. Electrochem. Soc.*, *145* (1998) 107.
112 C. W. Bunker, B. Ma, K. J. Simmons, H. W. Rollins, J.-T. Liu, J.-J. Ma, C. W. Martin, D. D. DesMarteau and Y.-P. Sun, *J. Electroanal. Chem.*, *459* (1998) 15.

Chapter 12

PURSUING FLUORINE CHEMISTRY IN POLAND

WOJCIECH DMOWSKI

Laboratory of Fluoroorganic Compounds, Institute of Organic Chemistry, Polish Academy of Sciences, 01-224 Warsaw 42, Poland

Stepping into fluorine chemistry

Wojciech Dmowski

As often happens in human affairs, my entry into fluorine chemistry was just incidental and I progressed through a chain of fortunate events and the goodwill of a number of friendly persons. My story begins in 1967 when, as a young research assistant in the Institute of Organic Chemistry of the Polish Academy of Sciences in Warsaw,[1] I was asked to construct a fluorine cell for use in research on the fluorination of coal; this was the first time in my life that I had to learn more about fluorine than its position in the Periodic Table. Two years later, after combating numerous technical problems, a cell operating at 250 °C and 40-60 A capable of producing up to 40 g F_2 per hour, was ready. Subsequently, the cell was used to make several kilograms of bromine trifluoride ($Br_2 + 3F_2 \rightarrow 2BrF_3$) which was employed to fluorinate a variety of coals from Polish mines, affording highly-halogenated yellowish solids soluble in organic solvents [1].

During this period, an important and stimulating event took place in Warsaw: an International Seminar on Fluorine Chemistry (6-11 October, 1969) organized by our laboratory leader, Dr A. Ostaszyński. This seminar gave me the chance to become acquainted with several well known fluorine chemists, including D. W. A. Sharp (UK), H. Meinert (GDR) and G. G. Yakobson (USSR).[2] These contacts later smoothed my visits to fluorine laboratories at Humbold University in East Berlin (1971) and at the USSR Academy of Sciences in Novosibirsk and Moscow (1982).

The coal fluorination project terminated at the end of 1969 because Dr Ostaszyński emigrated to Sweden in 1971 and financial support from the Ministry of Mining was discontinued, but I was already infected incurably with 'a fluorine bacillus'. Fortunately, then,

---

[1] I was born in Warsaw on 17 March 1936 and studied chemistry at the city's Polytechnic. After graduation in 1958 (Dip. Eng. Chem.), I continued to study for a Masters degree (awarded in 1963) while also working as a research assistant at the Institute of General Chemistry of the Ministry of Chemical Industry.

[2] The late Professor Yakobson remains in my memory as a remarkable and extremely modest man. Despite his eminence he would insist on not crossing a threshold before a guest; since I was determined that he should take priority, entering a room or a laboratory in his company usually took several minutes!

when the fluorine group was incorporated into a large laboratory of macrocyclic compounds (lead by Dr R. Koliński) we were not forced to change direction. Looking for a new research topic in organofluorine chemistry, I became intrigued by the pioneering work of the DuPont group [2] and others on the fluorination of oxygen-containing functionalities with sulfur tetrafluoride. I was astonished by the versatility of this reagent and noted that the majority of papers reported rather random results, and that no systematic studies on structure-reactivity relationships had been carried out. Also, the reaction mechanisms proposed at the time seemed not very convincing to me. Therefore, I decided that there was plenty to play for in the area of deoxyfluorination with SF_4.

Doctoral work with sulfur tetrafluoride

Sulfur tetrafluoride is a toxic gas (b.p. −40.1 °C) which reacts readily with moisture, giving HF and SOF_2; hence its use requires great care, a high level of skill and the availability of metal equipment. Thanks to the assistance of two able technicians (P. Gwiazda and R. Woźniacki), autoclave facilities were established in Koliński's laboratory in the early 1970s and the production of SF_4 commenced on a 300-350 g scale from sulphur, chlorine and sodium fluoride, according to a DuPont method [3]. This enabled me to embark on a systematic investigation of the fluorination of alkane- and cycloalkane-carboxylic acids (more than 400 experiments were carried out). These studies led to the award of my first PhD degree (formally supervised by Dr R. Koliński and conferred by Professor P. Nantka-Namirski in 1974) and provided five publications [4-8].

Just in the early stages of the work we found that the formation of trifluoromethyl derivatives is not the only outcome of the reaction of SF_4 with carboxylic acids. Under correctly choosen conditions, alkanecarboxylic acids afforded, in addition to (trifluoromethyl)alkanes, considerable amounts of symmetrical bis(1,1-difluoroalkyl)ethers [4-6]; alkanedicarboxylic acids, particularly 1,2-dicarboxylic acids, gave high yields of cyclic tetrafluoroethers [7] (Scheme 12.1).

The formation of these tetrafluoroethers, coupled with a study on the rôle of HF as a catalyst, lead to the formulation of a mechanism for the reactions of SF_4 with carbonyl compounds which involves attack on carbonyl oxygen by SF_3^+ cations and RCF_2^+ cations as intermediates en route to both (trifluoromethyl)alkanes and tetrafluoroethers [8]

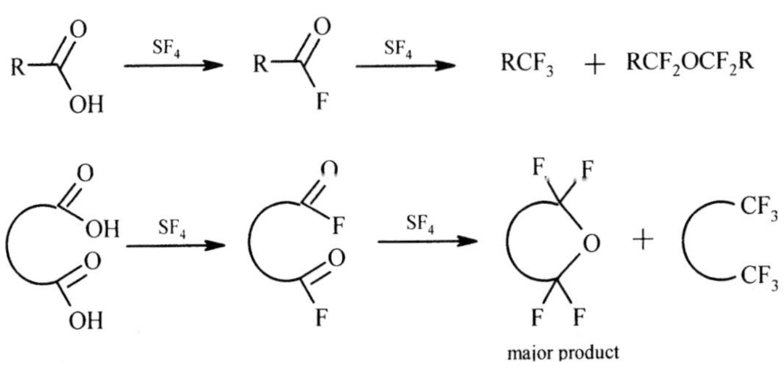

Scheme 12.1.

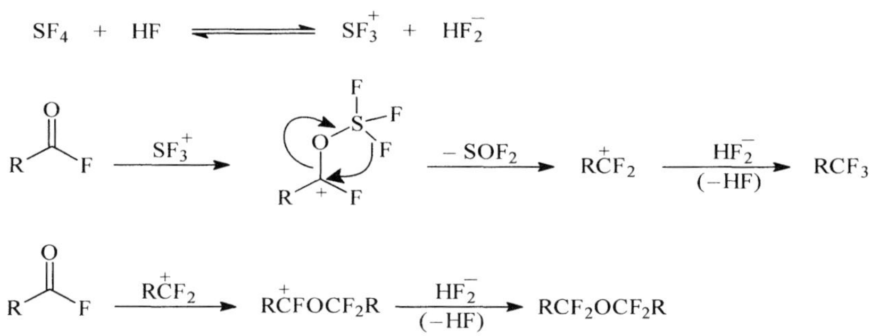

Scheme 12.2.

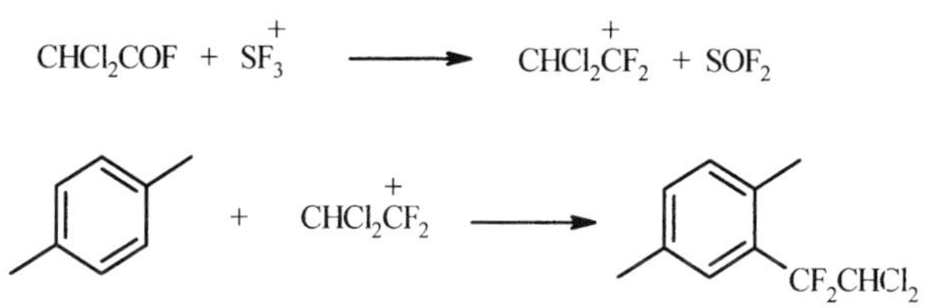

Scheme 12.3.

(Scheme 12.2); the participation of difluorocarbocations was later established experimentally in our laboratory by trapping them with aromatic hydrocarbons [9, 10] (Scheme 12.3). These fundamental studies on reactions of SF_4 with carboxylic acids, and later with aldehydes and amides (see below), brought international recognition to the fluorine group in Warsaw after being reviewed by Chia-Lin Wang for *Organic Reactions* [11] and presented in my plenary lecture at the 11th International Symposium on Fluorine Chemistry (Berlin, 1985).

Postdoctoral studies on fluoro-olefins

The next milestone on my 'fluorine way' was a two-year stint (1974-1976) in the UK at The University of Manchester Institute of Science and Technology (UMIST).

In the summer of 1974, the director of the F-team at UMIST, Professor R. N. Haszeldine, visited Poland and I was chosen to show him around the country. As we were driving together from Warsaw to Kraków and further to the Tatra mountains we talked about various matters – including chemistry, of course; and I soon realised that in a subtle manner I was being examined! Evidently I passed this examination because at the end of our journey I was invited to spend two years doing fluorine research at UMIST, commencing on October 1, after attending the upcoming 5th European Symposium on Fluorine Chemistry in Aviemore (Scotland). At this conference my respect for Professor Haszeldine increased still further, for he showed himself to be an invincible whisky drinker. After hosting a last-night party which went on until the early hours of the morning, he was absolutely fresh when we met again at 8 am, and later drove me south at speed in his Range Rover down Scottish serpentine roads. I would rather avoid describing my condition that morning!

Once safely at UMIST, I was appointed as a Senior Research Assistant, my task being to re-investigate oligomerization reactions of hexafluoropropene and tetrafluoroethylene under conditions of high fluoride ion concentration and in the presence of crown ethers. Crown ethers, I found, exerted a significant influence on the rate of hexafluoropropene oligomerization and on the yields of the already known thermodynamic dimer **1** and the trimers **2-4**, the latter being formed in yields up to 70%. No rate increase was observed in the oligomerization of tetrafluoroethylene, but new oligomers **5-9**, including perfluorodienes, which had not been previously reported, were isolated and their structures determined (note: in skeletal structures **1-9**, the convention [ ]$_F$ means that all bonds not shown are to fluorine).

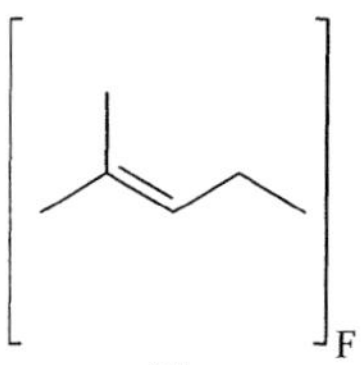

(1)

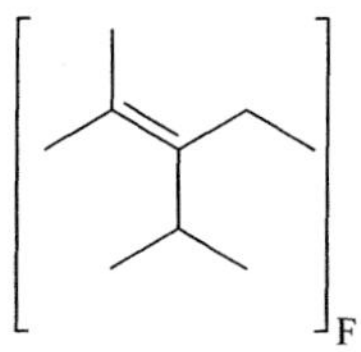

(2)

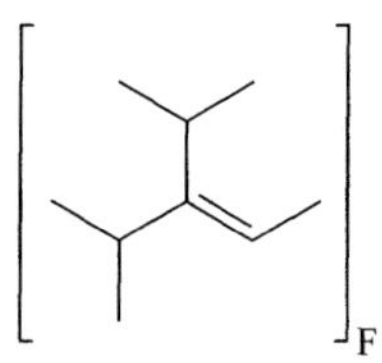

(3)

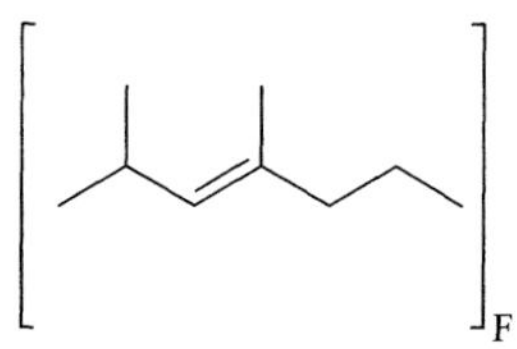

(4)

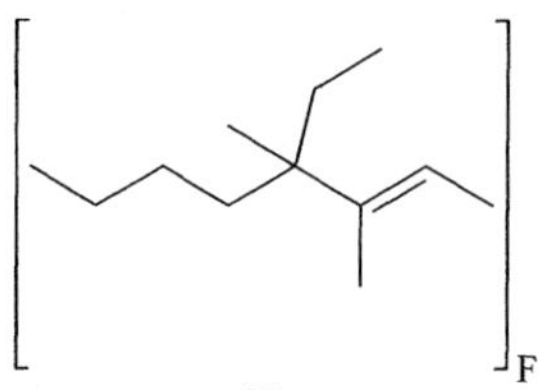

(5)

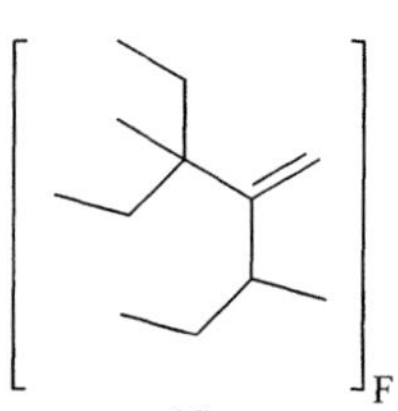

(6)

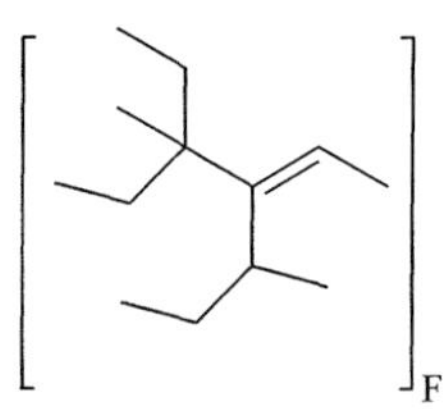

(7)

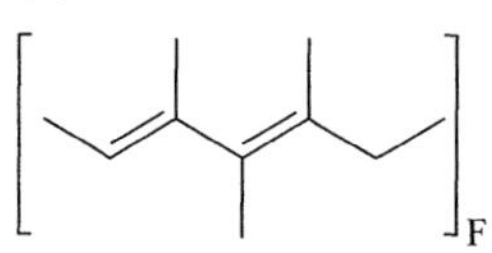

(8)

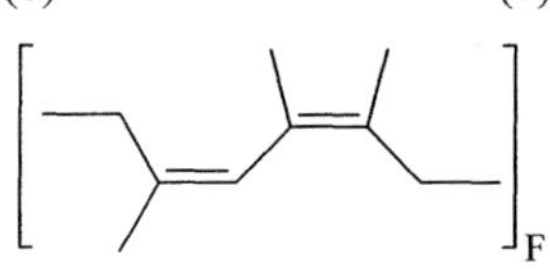

(9)

Co-oligomerization of tetrafluoroethylene and hexafluoropropene was achieved for the first time, giving ten new highly-branched perfluoro-olefins ranging from C_5F_{10} to $C_{11}F_{22}$. In addition, hexafluoropropene and tetrafluoroethylene were found to undergo fluoride-ion catalysed reactions with tetrahydrofuran and 2-methyltetrahydrofuran, affording 70-90% yields of products arising from insertion into C−H bonds; previously such reactions had been observed only under free-radical conditions.

My research at UMIST fulfilled the requirements for my second PhD degree, which was conferred on me by the Vice Chancellor of the Victoria University of Manchester in 1976. The thesis were prepared in just two months thanks to the assistance of my supervisor Dr W. T. Flowers and my wife Danuta, who was typing it as fast as I was able to write a manuscript. Unfortunately, only one preliminary paper based on my activities at UMIST has been published [12]. During my time there I learned how to manipulate and analyse gaseous perfluoro-olefins, hygroscopic fluoride ion sources and anhydrous solvents, and the whole experience greatly influenced my research activity for at least the next twelve years.

From the social and recreational point of view, the Manchester years were very happy ones. Danuta and I made a number of good friends, some of whom we are still in touch with (Eric Banks, Sean Duggan and Dick Krajewski), and we explored the beautiful countryside of the Peak District, the Lake District and North Wales; also, we spent wonderful holidays in Scotland and Cornwall. I also took the opportunity to visit fluorine laboratories in Durham and Birmingham where I had a chance to meet the important fluorine chemists Professors W. K. R. Musgrave, J. C. Tatlow and R. D. Chambers.

In the summer of 1976, we were invited by Professor Haszeldine to attend the EUCHEM Conference in Menton (France); this was an important event since it gave me a chance to mingle with the international fluorine society and to become acquainted with many European and American fluorine chemists. Amongst others of note, I met Jean'ne Shreeve, Bill Dolbier Jr., and Darryl DesMarteau (who later paved the way for me to visit America).

Back in Warsaw in the autumn of 1976, and now fascinated by fluoro-olefin chemistry, I undertook a study on nucleophilic reactions of substituted fluoroalkenes. 1-Tetrahydrofurylpentafluoropene and particularly 1-phenylpentafluoropenes (see Scheme 12.4), in which the electron density at the double bond is easily changed by attaching substituents (R) to the benzene ring, were choosen as model alkenes for this work. A number of structure-reactivity relationships were elucidated and a Hammett-type correlation established [13, 14]. The regio- and chemo-selectivities associated with these reactions are strongly influenced by the basicity and steric bulk of the attacking nucleophile and by the electronic character of the ring substituent R.

These studies in the field of fluoro-olefin chemistry fulfilled the requirements for a *habilitation,* and in 1985, I was awarded the degree of Doctor of Science by the Institute of Organic Chemistry of the Polish Academy of Sciences. By then, an important structural reorganisation of the Institute of Organic Chemistry had taken place and the new Director, Professor M. Mąkosza, conferred on the fluorine group the status of an independent laboratory, of which I was formally appointed Leader in 1981. The understanding and support of Professor Mąkosza, which allowed further successful development of our activities in the field of fluorine chemistry, cannot be overemphasized. Also of note is the rôle played by Professor Alois Haas of the Ruhr University, Bochum, in the promotion and knowledge of fluorine chemistry in Poland. In the early seventies this area of chemistry was not properly appreciated by higher authorities in the Academy. This situation changed dramatically when Professor Haas and a director of the Academy's Department of Mathematics, Physics

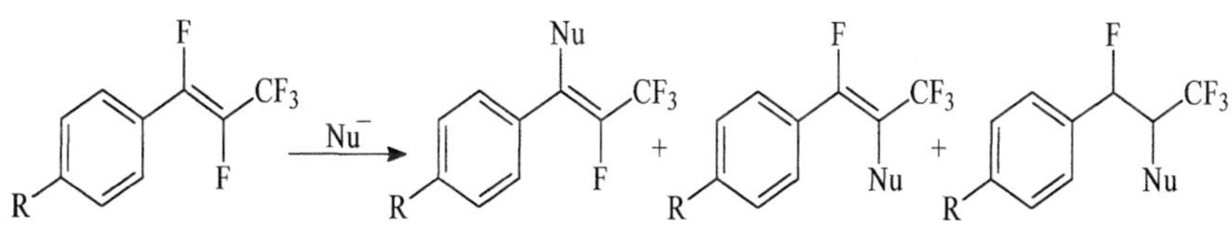

R = H, Me, MeO, Cl, CF_3; Nu^- = EtO^- / EtOH, R'Li / Et_2O (R' = Me, Et, *i*-Pr, *n*-Bu, *t*-Bu, *i*-Bu, Ph, Me_2N, Et_2N, $(CH_2)_4N$, $(CH_2)_5N$)

Scheme 12.4.

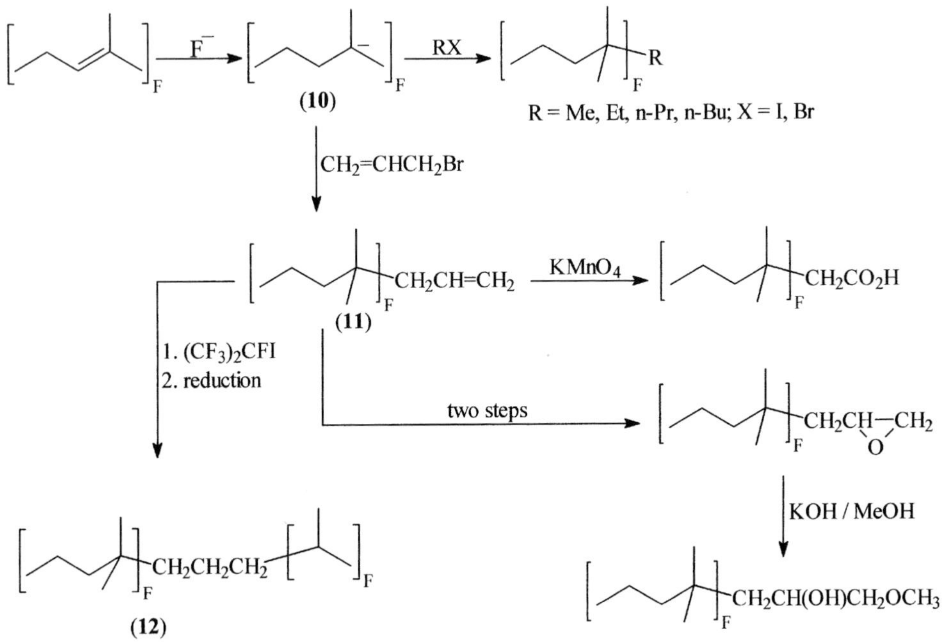

Scheme 12.5. All unmarked bonds are to fluorine.

and Chemistry journeyed together by train. By the end of their conversation the situation had changed in favour of fluorine chemistry[3].

Studies on fluoro-olefins were continued in Warsaw during the period 1986-1990. We investigated the generation of the perfluoro-(2-methyl-2-pentyl) carbanion **10** and its reactions with haloalkanes and haloalkenes (*e.g.*, see Scheme 12.5) [15]. Also, an expedient, large-scale laboratory synthesis of 5,5,6,6,7,7,7-heptafluoro-4,4-bis(trifluoromethyl)hept-1-ene (**11**) was developed and a number of its derivatives prepared (*e.g.*, see Scheme 12.5) [16-19]. Co-operative research involving the fluorosurfactants laboratory of the East Germany Academy of Sciences (Berlin-Adlershof), headed by Dr D. Prescher, led to preparation and investigation of some new fluorinated amphiphiles [18, 20]. The polyfluorinated alkane **12** ($C_{12}H_6F_{20}$, [21]) derived from **11** proved to be a perfect vitreous fluid with potential applications in ophthalmic surgery [22].

[3]Help given by Professor Haas to the Chemistry Department of the Jagiellonian University of Kraków during the hard times of marshall law (1981-1983) and following years cannot be overestimated; besides engineering vast support through the provision of chemicals, medicaments and laboratory equipment, he enabled numerous staff members of that Department (K. Bogdanowicz-Szwed, A. Kolasa, B. Zalewska and others) to spend time in Bochum, where they gained much expertise in the field of fluorine chemistry. In appreciation of his support, Professor Haas was honoured in 1989 by the Jagiellonian University through conferment of the degree of *Doctor Honoris Causa*.

Travelling on the fluorine circuit

Despite the political instability in Poland at the time, I engaged in some intensive 'fluorine travelling' during the years 1979-1984. In 1979, thanks to an invitation from Darryl DesMarteau and financial support provided by the American Chemical Society, I attended The Fourth Winter Fluorine Conference in Daytona Beach, Florida, and afterwards spent a happy weekend with Paul Tarrant and Bill Dolbier in Gainesville. Leaving Florida, I moved North to meet Bruce Smart at the DuPont Experimental Station in Wilmington (Delaware), the place when SF_4 chemistry was born, and thence to see the fluoro-olefins expert Heinz Koch in Ithaca, New York. An invitation from Jean'ne Shreeve enabled me to attend the Second Chemical Congress of the North American Continent, held in Las Vegas in 1980; together with my wife, I made a great American coast-to-coast tour by Greyhound buses, visiting the laboratories of famous fluorine chemists along the way [Miloš Hudlický in Blacksburg (West Virginia); Darryl DesMarteau in Manhattan (Kansas); Dayal Meshri in Tulsa (Oklahoma); Jean'ne Shreeve in Moscow (Idaho) and Donald Burton in Iowa City]. I went to the USA again in 1983, attending the Sixth Winter Fluorine Conference in Florida and calling in on Dick Lagow and Robert Soulen in Texas (Austin and Georgetown, respectively) and again on Dayal Meshri in Tulsa and Heinz Koch in Ithaca. Danuta and I twice toured Germany (in 1981 and 1984), visiting Reinhard Schmutzler (Braunschweig), Gerd Röschenthaler (Bremen), Herbert Roesky (Göttingen), Alois Haas (Bochum), Erich Klauke[4] (Bayer AG, Leverkusen), Günter Siegemund (Hoechst, Frankfurt) and others.

In 1982 and again in 1984, thanks to invitations from Professors G. G. Yakobson, V. E. Platonov and L. S. German, I visited fluorine laboratories in Novosibirsk and Moscow in the former USSR. During the second of these trips it was of great interest to observe the changes in the behaviour and outlook of Russian people which had taken place since my first visit. For example, the funeral of President Brezhniev was sheduled to take place during my flight from Moscow to Novosibirsk, and the authorities declared that at the exact time of the event all activities in the USSR must stop for five minutes. After landing, the young Russian who met me at the airport asked, with a smile, 'Did they stop the engines of your aircraft for five minutes?'; making such a joke would have been absolutely impossible two years earlier. I spent the next evening with Slava Platonov, and, over a couple of bottles of wine, we talked quite freely about worldwide political problems.

I made a prolonged visit to the Ruhr University, Bochum, in the mid-80s (mid-1985 till mid-1986) working in the Professor Haas's laboratory on the generation (from thiocarbonyl difluoride and fluoride ion in aprotic solvents: $CF_2{=}S + F^- \rightleftharpoons CF_3S^-$) and reactions of trifluoromethanethiolate ion [23]. The results are summarized in Scheme 12.6. Later, back in Warsaw, we generated the perfluoroisopropoxy anion [$(CF_3)_2CFO^-$] from hexafluoroacetone and fluoride ion and showed that it is not as stable as the CF_3S^- anion but can be trapped with tetrafluoropyridazine [24].

[4]Dr Klauke proved to be a real friend, helping us materially during the hard times we had in Poland. We shall always remember his generosity.

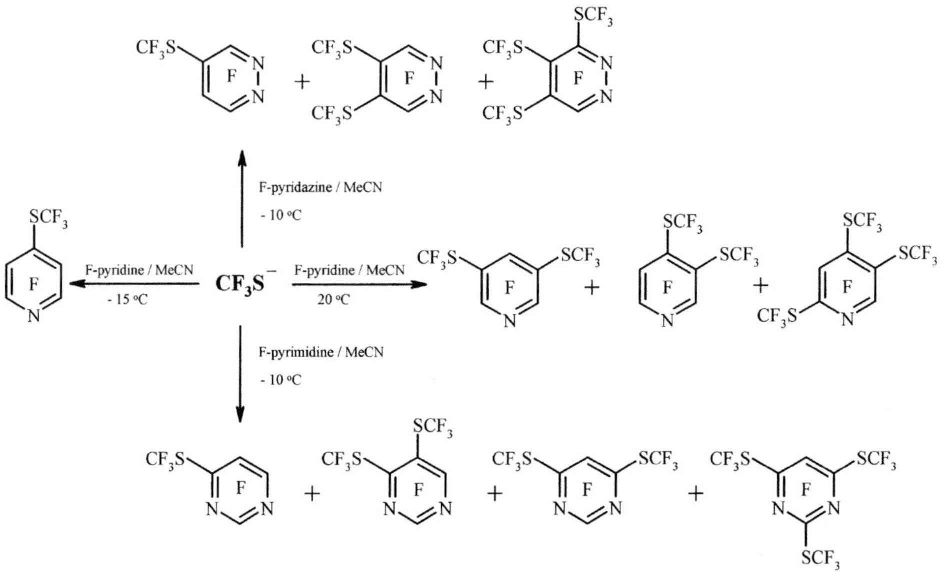

Scheme 12.6.

Back in Warsaw

My frequent absence from the laboratory in Warsaw did not impede its activities. Besides the work on fluoroalkene chemistry, described earlier, studies on the fluorination of carbonyl compounds with sulfur tetrafluoride were continued. Reactions with aliphatic aldehydes were shown to be more complex that thought previously, and under the proper conditions, beside the expected difluoromethyl derivatives, bis(1-fluoroalkyl) ethers and rearrangement products are formed in reasonable yields [25] (Scheme 12.7). This fascinating work, carried out as part of a PhD topic, was unfortunately not completed owing to the tragic death of the graduate student concerned. Another PhD topic, completed in 1981, led to the discovery of a direct and almost quantitative conversion of tertiary formamides into *N*,*N*-dialkyltrifluoromethylamines on treatment with SF_4 and anhydrous KF (acting as a scavenger for HF), *e.g.* $(C_2H_5)_2NCHO \rightarrow (C_2H_5)_2NCF_3$ [6, 27]. Like the popular SF_4 derivative $(C_2H_5)_2NSF_3$ known widely as DAST, these *N*-alkylated trifluoromethylamines will convert alcohols to monofluoroalkanes and carboxylic acids to acid fluorides [28].

Systematic investigations on the fluorocyclization of aromatic 1,2-dicarboxylic and tetracarboxylic acids *via* their reactions with sulfur tetrafluoride led to an evaluation of steric and electronic effects assiociated with the benzene ring substituents on the formation of 1,1,3,3-tetrafluoro-1,3-dihydroisobenzofurans *versus* the formation of trifluoromethylated benzenes [29]; some of the conversions achieved are shown in Scheme 12.8.

Since 1988, the activities of the Warsaw fluorine group have centred mainly on the synthesis and chemistry of trifluoromethyl derivatives using SF_4, of course, as the fluorinating agent. Some of our work was inspired by the report from Chambers' group [30] that 1,3,5-tris(trifluoromethyl)benzene readily forms the stable lithium derivative $1,3,5-(CF_3)_3C_6H_2Li$, which reacts conventionally with numerous electrophiles. Our at-

$$R'R''CHCHO \xrightarrow{SF_4} R'R''CHCHF_2 + R'R''CHFCH_2F + R'R''CHCHFOCHFCHR'R''$$

$R' = H, Me, Et, Pr, Bu$; $R'' = Me, Et, (CH_2)_4$, $(CH_2)_5$

$$(CH_3)_3CCHO \xrightarrow{SF_4} (CH_3)_2CFCHFCH_3 + (CH_3)_3CCHFOCHFC(CH_3)_3$$

Scheme 12.7.

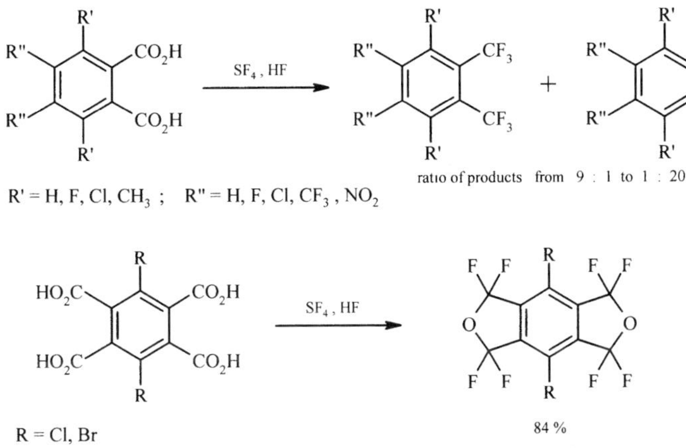

Scheme 12.8.

tempts to extend the lithiation reaction to tetrakis(trifluoromethyl)benzenes failed totally; instead, an unusual dearomatization leading to 2,5-cyclohexadiene-1-ylidene derivatives occurred [31, 32] (Scheme 12.9). Careful investigation of reactions of 1,2,4,5- and 1,2,3,5-tetrakis(trifluoromethyl)benzene and of 1,2,4-tris(trifluoromethyl)benzene with a number of alkyl- and aryl-lithium reagents, supported by crystallographic studies and semiempirical reaction modelling [33], were carried out as part of the programme; and, in cooperation with Professor Yurii L. Yagupolskii's group in Kiev, the novel stable pentakis(trifluoromethyl)phenyl carbanion was generated *via* treatment of 1,2,4,5-tetrakis(trifluoromethyl)benzene with a CF_3SiMe_3/TASF system [34].

Through the auspices of Professor Manfred Schlosser (University of Lausanne, Switzerland), we were invited at the end of 1993 to participate in a COST (European Cooperation in the Field of Scientific and Technical Research) programme dealing with CF_3 chemistry. Our remit was to investigate the regioselective functionalisation of 3,5-bis(trifluoromethyl)anisole and 1,3-bis(trifluoromethyl)benzene: this resulted in the synthesis of a number of bis(trifluoromethyl)aromatics, including 4,6-bis(trifluoromethyl)salicylic acid [35] and 4-methoxy-3-methyl-2,6-bis(trifluoromethyl)-benzaldehyde [36], an intermediate en route to trifluoromethylated analogues of aromatic retenoids. Subsequently, the 4th COST D2 workshop on CF_3 chemistry was held in Warsaw in May 1996, and this included a one-day symposium on fluoroorganic chemistry at which plenary lectures were delivered by internationally-recognized organofluorine specialists [Klaus Burger (Leipzig), Dick Chambers (Durham), Alois Haas (Bochum), Bernard Langlois (Lyon), Manfred Schlosser (COST D2 Principal Coordinator; Lausanne) and

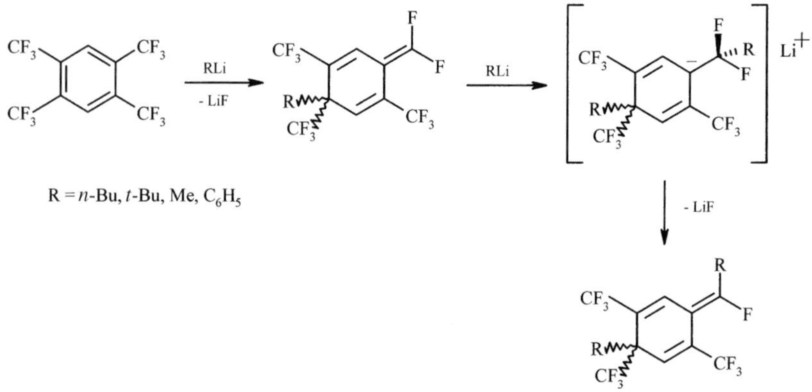

Scheme 12.9.

Claude Wakselman (Versailles)]. Amongst the 34 participants were chemists from the Polish universities of Łódź, Kraków, Poznań, Toruń, Warsaw, the Universities of Budapest (Hungary) and Zurich (Switzerland), and the Academy of Sciences of Ukraine (Kiev).

More recent work involving SF_4 as the fluorinating agent has included studies on SF_4 fluorination of bicyclo[2.2.2]oct-7-ene-*exo*-2,3,5,6-tetracarboxylic dianhydrides which led to the discovery of some unusual alkenes, *e.g. exo,exo*-4,8-etheno-1,1,3,3,5,5,7,7-octafluoro-2,6-dioxaperhydro-*s*-indacene (**13**) [37], a drive to synthesize enantiopure trifluoromethylated analogues of natural bioactive compounds from carboxylic acid precursors [practical molar-scale syntheses of (1*R*,3*S*)-(+)-3-(trifluoromethyl)camphonanic acid [38] and (1*R*,3*S*)-(−)-3-oxa-4-trifluoromethylcamphor [39], from commercially-available camphoric acid have been elaborated] and routes to ring-trifluoromethylated cyclopropane derivatives, potentially useful as precursors of pyrethroid pesticides.

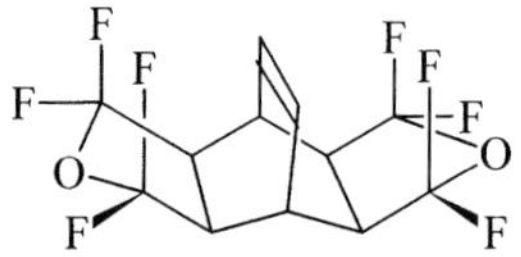

(13)

In another area, electrochemical fluorination techniques (learned in the spring of 1992 when I spent two months with Professor A. Laurent at the Université Claude Bernard in Lyon-Villeurbanne, France, as a visiting Professor[5]) have been used in studies involving the fluorination of cinnamates [40] (Scheme 12.10) and reactions of electrogenerated trifluoromethyl radicals with electron-deficient alkenes [41, 42] (Scheme 12.11).

[5] According to Polish tradition, only two groups of state officers are nominated by the President of the State: army generals and professors. The diplomas are handed personally by the President during a solemn celebration held in the presidential palace. This happened to me in May 1993 and I am proud to be the only person in my Institute who received the professorship from the first non-conmmunist president, Mr Lech Wałęsa.

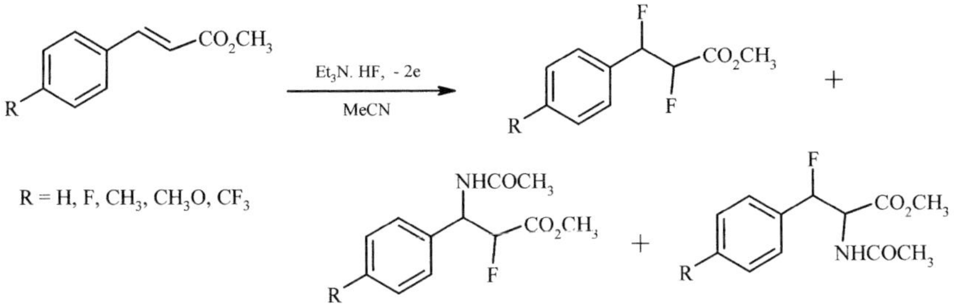

Scheme 12.10.

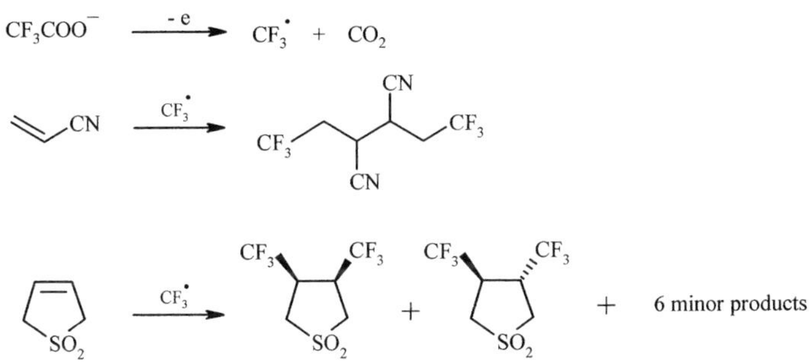

Scheme 12.11.

A way of life

Finally, I would like to say that working in the field of fluorine chemistry is for me much more than just an occupation or profession – it is the way to live. Nowadays, of course, all successful chemists need the help and co-operation of others, and I've been lucky in this respect. I've already mentioned a lot of persons who befriended me, most of them prominent fluorine chemists, and now I would like to pay tribute to those who by their skill and hard work incalculably contributed to the success of the fluorine chemistry group in Warsaw[6]. They are my present and former close friends and staff members: Dr Halina Plenkiewicz, Dr Jerzy Wielgat, Mrs Krystyna Piasecka-Maciejewska, BSc, the late Mr Ryszard Woźniacki, BSc, Mr Tomasz Kozłowski, BSc, the late Mrs Krystyna Usiekniewicz, BSc, Mrs Krystyna Tomaszewska, and my former PhD students Drs Maciej Kamiński, Jacek Porwisiak and Ireneusz Nowak.

References

1 A. Ostaszyński, A. Czerkawski and W. Dmowski, Pol. Pat., *80248* (1976).
2 W. R. Hasek, W. C. Smith and V. A. Engelhardt, *J. Am. Chem. Soc., 82* (1960) 543.

[6]Nowadays, fluorine chemistry is becoming increasingly popular in Poland; so besides the Warsaw team, a few other research groups are, to some extent, involved in this field, *e.g.* Professor H. Koroniak's group (University of Poznań), and those of Professor G. Młostoń (University of Łódź) and Dr A. Kolasa (University of Kraków).

3 W. C. Tullock, F. C. Fawcett, W. C. Smith and D. D. Coffman, *J. Am. Chem. Soc., 82* (1960) 539.
4 W. Dmowski and R. Koliński, *J. Fluorine Chem., 2* (1972/1973) 210.
5 W. Dmowski and R. Koliński, *Pol. J. Chem., 47* (1973) 1211.
6 W. Dmowski and R. Koliński, *Pol. J. Chem., 48* (1974) 1697.
7 W. Dmowski and R. Koliński, *Pol. J. Chem., 52* (1978) 71.
8 W. Dmowski and R. Koliński, *Pol. J. Chem., 52* (1978) 547.
9 J. Wielgat and Z. Domagała, *J. Fluorine Chem., 20* (1982) 785.
10 J. Wielgat, Z. Domagała and R. Koliński, *J. Fluorine Chem., 20* (1982) 785.
11 Chia-Lin Wang, *Organic Reactions, 34* (1985) 319.
12 W. Dmowski, W. T. Flowers and R. N. Haszeldine, *J. Fluorine Chem., 9* (1977) 94.
13 W. Dmowski, *J. Fluorine Chem., 15* (1980) 299; *21* (1982) 201; *26* (1984) 223; *26* (1984) 379; *29* (1985) 287.
14 W. Dmowski, *Pol. J. Chem., 60* (1986) 129.
15 W. Dmowski and R. Woźniacki, *J. Fluorine Chem., 36* (1987) 385.
16 W. Dmowski, H. Plenkiewicz and J. Porwisiak, *J. Fluorine Chem., 41* (1988) 191.
17 H. Plenkiewicz and W. Dmowski, *J. Fluorine Chem., 45* (1989) 389.
18 W. Dmowski, H. Plenkiewicz, K. Piasecka-Maciejewska, D. Prescher, J. Schulze and I. Endler, *J. Fluorine Chem., 48* (1990) 77.
19 H. Plenkiewicz and W. Dmowski, *J. Fluorine Chem., 51* (1991) 43.
20 D. Presher, J. Schulze, L. Richter, W. Dmowski and H. Plenkiewicz, *Tenside Surf. Det., 29* (1992) 337.
21 W. Dmowski, *J. Fluorine Chem., 50* (1990) 319.
22 J. Toczołowski, M. Gerkowicz and W. Dmowski, *Klinika Oczna, 93* (1991) 105 (in polish).
23 W. Dmowski and A. Haas, *J. Chem. Soc. Perkin Trans. I,* (1987) 2119; (1988) 1179.
24 K. E. Peterman and W. Dmowski, *Org. Prep. Proc. Int., 23* (1991) 760.
25 W. Dmowski, *J. Fluorine Chem., 32* (1986) 255.
26 W. Dmowski and M. Kamiński, *Pol. J. Chem., 56* (1982) 1369.
27 W. Dmowski and M. Kamiński, *J. Fluorine Chem., 23* (1983) 207.
28 W. Dmowski and M. Kamiński, *J. Fluorine Chem., 23* (1983) 219.
29 W. Dmowski, *J. Fluorine Chem., 65* (1993) 139.
30 G. E. Carr, R. D. Chambers, T. F. Holmes and D. G. Parker, *J. Organometal. Chem., 325* (1987) 13.
31 W. Dmowski and J. Porwisiak, *J. Fluorine Chem., 59* (1992) 321.
32 W. Dmowski, J. Porwisiak, J. Krajewski, A. Mishnyov and A. Kemme, *J. Fluorine Chem., 62* (1993) 15.
33 H. Koroniak, P. Fiedorow, W. Dmowski and J. Porwisiak, *J. Mol. Struct., 351* (1995) 187.
34 A. A. Kolomieitsev, V. N. Movchun, Y. L. Yagupolskii, J. Porwisiak and W. Dmowski, *Tetrahedron Lett., 41* (1992) 6191.
35 W. Dmowski and K. Piasecka-Maciejewska, *J. Fluorine Chem, 78* (1996) 59.
36 W. Dmowski and K. Piasecka-Maciejewska, *Tetrahedron, 54* (1998) 6781.
37 W. Dmowski, I. Nowak, P. Gluzinski and A. Kemme, *J. Org. Chem., 62* (1997) 1760.
38 W. Dmowski and K. Piasecka-Maciejewska, *Org. Prep. Proc. Int., 31* (1999) 207.
39 W. Dmowski and K. Piasecka-Maciejewska, *J. Fluorine Chem., 97* (1999) 97.
40 W. Dmowski and T. Kozłowski, *Electrochimica Acta, 42* (1997) 513.
41 W. Dmowski, A. Biernacki, T. Kozłowski, P. Gluzinski and Z. Urbanczyk-Lipkowska, *Tetrahedron, 53* (1997) 4437.
42 W. Dmowski and T. Kozłowski, *J. Fluorine Chem., 87* (1998) 179.

Chapter 13

BIOGRAPHICAL SKETCH OF PAUL TARRANT

WILLIAM R. DOLBIER, Jr.
Department of Chemistry, University of Florida, Gainesville, FL 32611-7200, USA

Formative years

Paul Tarrant

Paul Tarrant was born on All Saints' Day (November 1) 1914 in Birmingham, Alabama, USA. His full name then was Paul Tarrant Brittain, but his mother died soon after his birth, so he was raised by his maternal grandparents, James and Margie Tarrant, hence the formal change later. Paul had a happy childhood, playing baseball, swimming in the nearby creeks, and caddying golf when he got the chance, including once at the Southern Amateur Championship. He attended Robinson Grammar School, four blocks from his home, and Woodlawn High School, three miles away. During his high school and college years, he worked in his Uncle Paul's drug store in downtown Birmingham; his job as a soda jerk was really the first indication of intellectual potential and foreshadowed what he was to become.

After high school, Paul went to nearby Howard College (now called Samford University) in East Lake, which was within walking distance from home. His Uncle Felix helped pay his tuition for the first year, and Paul put in 36 hours per week (6:30-10:30 PM) at the drug store (and full time during the summers) while working on assistantships in the Physics and Chemistry Departments at Howard in order to cover his tuition fees and expenses. He graduated with a BS degree in chemistry in 1936 and entered Purdue University (Lafayette, Indiana) in the fall of that year.

At Purdue with McBee

During his last year at Howard, Paul met Viola Griffin, with whom he eloped in the summer of 1937, after one year on the Master's programme in chemistry at Purdue. Paul held an NYA (National Youth Administration) job at Purdue at first (paying a maximum of $22.50 monthly), but quickly also obtained a quarter-time assistantship; this gave him another $35 per month, for which he lectured a course in general chemistry. He chose

Professor Earl T. McBee as his research director. McBee had not yet started doing fluorine chemistry, but while Paul was in his group a fellow student visited Professor Albert L. Henne at Ohio State University to learn about current research in organofluorine chemistry. That was the first time that Paul became aware of the field of fluorine chemistry, which was in its infancy. Purdue's chemistry department was closely linked to many companies, such as Commercial Solvents, Dow, and Westinghouse, with research fellowships being endowed by these companies and research areas thus being dictated by them. For example, McBee's group was engaged in research on chlorination and nitration methodology, which was of interest to Commercial Solvents. One of Paul's projects while working with McBee was to try to convert 1,3-dichlorobutane to methylcyclopropane with zinc, in a manner analogous to the conversion of 1,3-dichloropropane to the parent cyclopropane, which at that time had gained some notoriety as an anaesthetic. So Paul's initiation into research was in the area of *chloro* organics. Later, when considering where to do his PhD degree, he was attracted to fluorine chemistry because of his experience with chlorine chemistry.

At Duke with Bigelow

Paul obtained his MS degree from Purdue in 1938 and returned home to teach high school physics and chemistry for two years in Birmingham (Ramsey Technical High School), at a salary of $1135 per year. Realizing that it was going to be difficult to support a family on that salary, he decided to seek a PhD degree, and travelled out to Stanford University (California) in 1940. Stanford's chemistry programme was somewhat in disarray at that time, since the Head of the Department for 25 years had just retired and a number of professors were not on campus when Paul arrived. As a result he decided not to enroll, and instead took a job at the Shell Development Company in Emoryville (CA), where he stayed for one year, working on plasticizers for rubber; he was earning $170 a month at the time he quit. Having time for a more considered decision about graduate schools, he thought long and hard about where he wanted to study for a PhD degree. He considered appying to schools at Duke, Indiana, Texas, and North Carolina but discarded the Texas option because it didn't have much money for assistantships, and eliminated Indiana because he thought the town was too small for Viola to find a good secretarial job. Using the College Blue Book, he learned that Professor Lucius A. Bigelow's group at Duke University (Durham, North Carolina) was active in the field of synthesis of organic fluorine compounds, so he applied there and was awarded an assistantship in 1941. Paul made his decision to work in the field of fluorine chemistry because he thought that 'fluorine chemistry would be something like chlorine chemistry'. Was that ever a wrong assumption!

When Bigelow was hired by Duke University in the late 1920s, he had been encouraged to do research in the field of fluorine chemistry by the Chairman of the Chemistry Department, Professor Paul M. Gross. Paul arrived at Duke holding a quarter-time assistantship, but after two months the department obtained a Navy contract that allowed him to move to a full-time research position at $150 a month while he worked on the development of inert fluorinated fluids needed for the Manhattan Project. Viola got a job next door in the Law School, and she and Paul would generally have lunch together in the Duke Gardens. It is interesting to note that all workers at Duke who were working with F_2 were required, as a preventative measure, to drink a quart (about a litre) of milk a day!

Bigelow's group generated their own F_2 and this made for some very interesting and exciting experiences during Paul's time at Duke. His work involved the direct fluorination of 'deactivated' aromatics, such as 4-chloro-1,3-bis(trifluoromethyl)benzene, although he also did a considerable amount of work on Cl-for-F exchange reactions using SbF_3 in HF. His PhD thesis was aptly entitled 'Fluorination of Organic Compounds,' and some of his results were published a few years later by Bigelow in papers entitled 'The Action of Elementary Fluorine upon Organic Compounds. XII. Vapor Phase Addition to Certain Deactivated or Condensed Aromatic Rings' [1] and 'Fluorine as a Halogen. Reaction with a Highly Deactivated Aromatic Nucleus' [2]. Paul graduated in 1944 and immediately went to work for the American Cyanamid Company in Stamford, Connecticut, at the recommendation of Professor Paul Gross, who was a consultant for the company. Through Gross, he knew that American Cyanamid was actively involved in the development of a fluorinated insecticide, fluoroacetamide.

At American Cyanamid

Paul's initial project at American Cyanamid was to develop a more practical route to fluoroacetamide. The earlier method, which introduced the fluorine via a diazonium route, had led to an explosion and the blinding of one researcher; so the company was looking for something better. Paul successfully developed an excellent Cl-for-F exchange process with chloroacetamide, using acetamide as the solvent. In the course of this work, he had reason to distil one kilogram of ethyl fluoroacetate in an open-bay lab! The entomologists who conducted the experiments on the effectiveness of ethyl fluoroacetate as a fumigant put a little in a Petri dish, and when it had evaporated, it was found that every insect in the huge greenhouse had been killed. It wasn't long thereafter when a worker mysteriously died, and testing led to the death of a few rabbits, that the clearly demonstrated great toxicity of monofluoroacetic acid derivatives led to the termination of the entire project; only later was it learned that the Allies (WWII) were aware of the extreme toxicity of this class of compounds, and had plans to make some on a large scale. Paul then started working on polymers derived from CTFE ($CF_2{=}CFCl$) and the preparation of trifluoroacetic acid from $CF_3CCl{=}CCl_2$ via a method suggested by Max Gergel. Incidentally the *first* paper with Paul's name on it derived from his work at American Cyanamid ('The Preparation of DDT using Hydrogen Fluoride as the Condensing Agent') [3].

American Cyanamid was a great place to work, and Paul indeed did work – 6 days a week from dawn to dark (in the winter). Saturday nights were generally spent playing cards (hearts) with Viola and her mother, Ava Griffin, who lived with them. In the end, however, Paul grew tired of the dreary Connecticut winters and sought to return to the South.

At UF

Paul applied for a few academic jobs at PhD granting institutions in the South and got an interview with the University of Florida (UF). Often in those days people were not brought in to be interviewed but generally seen at a central location, usually during an ACS meeting. Thus Paul was interviewed by Professor Jack Hawkins in Atlantic City, early in 1946. He was hired, along with about twelve other new faculty (hired to handle the expected increase in student numbers due to the return of the veterans to school), but only Paul, George Butler and T. W. Stearns ended up staying for more than a couple of years. Its easy to understand why they might have not stayed, given the conditions that prevailed in Gainesville and at the University in 1946. For a start, there was no decent housing for Paul and Viola. After staying in a rooming house for two months, they were moved to university-owned housing that was located at Stengel Field (a small airport located at the site of the current Butler Plaza on Archer Road), and for six weeks this accommodation had no hot water, and the wind, sand and mud (after rain) made life almost unbearable there. They survived somehow for a year, during which time Paul almost moved to Jefferson Chemical Company in Port Arthur, Texas. However, Paul and Vi decided that Port Arthur was even worse than Gainesville! To compound matters, the new faculty hires had been promised that there would be an extension built to the overcrowded Leigh Hall, which housed not only Chemistry but also the Pharmacy School. However, when the bids came in too high, the project was abandoned. Paul and George had been hired as Instructors, not Assistant Professors, and when they found out that they were the only Instructors on campus with PhD degrees, they protested to Dean Townes Leigh, who was also Chairman of the Department. The Departmental Policy and Procurement Committee, which was in charge of such decisions, refused to budge, but when Paul and George threatened to leave, Leigh gave in and they became Assistant Professors in 1947. This was not to be the only time that Paul and George had to stand up for themselves in those early years.

Before Paul and the others came to UF in 1946, the Chemistry faculty comprised organic chemists Leigh and Cash Pollard, physical chemist Jack Hawkins, water chemist A. P. Black, analytical chemist Fred Heath, and Vestus T. Jackson. It was an all-male school, except for the pharmacy and graduate programmes. Paul's teaching assignment during those early years included four sections of introductory organic chemistry for pharmacy students, which entailed two lectures and four discussion sessions per week. He also taught two 3-hour labs and a beginning course in physical organic chemistry (two lectures per week) to graduate students. There were about 35 students in the chemistry graduate programme at that time.

Paul's first graduate students were John Young and Henry Brown, with Dale Warner to follow shortly thereafter. The work of these students led to Paul's first papers in organofluorine chemistry at UF, *e.g.* 'The Preparation of Some Derivatives of Chlorofluoroacetic Acid' [4]. In those days many of the graduate students came from one's own undergraduate programme, and when Paul and George arrived on campus, they found that most of these undergraduates had already committed to join the research groups of one of the older faculty. Pollard had been the first professor at UF to direct PhD students, and he aggressively defended his turf as senior organic professor. Thus Paul had to work hard to hold his own and to build his research programme in those early years at UF.

Prior to WWII not many organic compounds containing fluorine were known; the aliphatic ones were most often reported by Frédéric Swarts, the Belgian chemist, and Al Henne at Ohio State, who hailed from Belgium. However, a great deal of interest in organic fluorine compounds derived from the Manhattan Project, and in about 1946 Hanford and others at DuPont published reports concerning reactions between fluoro-olefins and alcohols or amines. These results demonstrated that it was possible to prepare fluorinated aliphatic compounds without using HF, F_2 or SbF_3 and thus opened the field to more conventional organic chemists. Still, it was an adventure to make new fluorinated compounds, and even more so to *identify* them: one did not have the luxury of ^{19}F NMR!

The usual way to identify compounds in those days was by determination of molar refractions; of course one needed to have some idea of the molecular weight of a compound in order to calculate its 'theoretical' molar refraction. Paul gave me the following example of such difficulties: 'Once we treated $CF_2{=}CCl_2$ with phenylmagnesium bromide and obtained a product, $PhCF{=}CCl_2$, that we could not identify for several months (at the time, there was no precedent for such an addition-elimination reaction of a fluorochloro-olefin). When we finally guessed the correct structure and thus its molar weight, and calculated its expected molecular refraction, it hit the measured value right on the button'. Another reason why fluorine chemistry was so exciting in the '40s and '50s was that quite unexpected results were often obtained. Paul went on to say: 'We blew up several autoclaves and had many reaction products plastered on ceilings. Chemistry was more fun in those days. Today, those who carry out reactions in NMR tubes and identify products via the myriad of modern techniques available are definitely missing something'.

Paul, Vi and Mrs Griffin lived out at Stengel Field for a year, then moved to a duplex on the east side of town, at 1033 NE 8th Avenue. They lived in the duplex for one year, until construction of their home at 1723 NW 12th Road was completed. In the meantime Paul and Vi started their family, with daughter Linda being born in 1947.

The late 1940s were times of rapid growth and remarkable evolution with regard to the role of research at Universities in the United States, and young professors like Paul and George were to be the instruments of such change. In 1948, Lou Butz, a representative of the Office of Naval Research, came to the University to talk to people in the Chemistry Department about a new Federal initiative, that of *funding research*. (This small but successful initiative to fund university research, mostly in applied science, led shortly thereafter to the creation of the National Science Foundation.) Three chemistry faculty were awarded contracts by ONR as a result of the visit by Butz: Paul, George Butler, and Cash Pollard. Paul's was the first to come through, and in order to accept the $10 000 contract, he needed the signature of Dean Leigh, who told him that he'd first have to get it approved by the Dean of the Law School. Once the Dean had approved the contract, Leigh signed, which broke the ice with respect to all future research funding within the College of Arts and Sciences at the University of Florida. Paul's contract was to carry out research in the area of 'fluorine-containing olefin chemistry', and the contract lasted about three years. One published piece of work carried out under this contract involved a synthesis of ethyl difluoroacetate [5]:

$$CHF_2CF_2Cl + KOH/EtOH \rightarrow [CF_2{=}CF_2] \rightarrow CHF_2CF_2OEt$$
$$\rightarrow (\text{with } H_2SO_4)\ CHF_2CO_2H$$

Two years later, Quartermaster chemists from the Army's 'Arctic Rubber Program' came down to solicit help from Paul. They were interested in a much larger commitment from him – one which would include retaining him as a paid consultant. This was something new to UF. The authorities did not approve of any faculty member doing 'outside' work, and thus Paul's request was turned down, first by Dean Ralph Page and then by Vice President John Allen. Paul then went to see President J. Hillis Miller to discuss the matter. Miller, as is still the case with the majority of university presidents, found it hard to turn down any source of money; also, he understood the value of compromise and suggested that Paul go off State salary and 'do full-time research' for the duration of his contract. This Paul did for three years, until A. P. Black, the Chairman of the Department, insisted that he return to teaching. However, the Army contract rolled on for a total of 16 years. Initially they made monomers, including lots of fluorinated dienes, such as perfluoroisoprene, and studied their reactions. In the process, Paul's group carried out fundamental synthetic work, developing methodology based on free-radical reactions involving hydrocarbon or fluorinated alkenes, *e.g.* $CF_2BrCFClBr + CH_3CH{=}CH_2$ (at 80 °C, benzoyl peroxide) $\rightarrow$ $CF_2BrCFClCH_2CHBrCH_3$ $\rightarrow$ (with KOH in ethanol) $CF_2BrCFClCH{=}CHCH_3$ $\rightarrow$ (with Zn in isopropanol) $CF_2{=}CFCH{=}CHCH_3$ [6]. In 1956, Paul's group made 100 grams of $CF_3{-}N{=}O$ for the Army (using Hazeldine's photochemical $CF_3I + NO$ method) and he personally took it to Washington by air, carrying it in a sealed tube cooled in dry ice. What in the world would happen if he tried that today?

Paul's research programme evolved continuously, as reflected by the diversity of his publications during that 16-year Army contract period. As Paul said, 'We made a lot of compounds and educated a lot of students as a result of Army funding'. Among the 'students' he has mentioned to me were Al Lovelace, Bob Taylor, Marv Lilliquist, Mary Louise VanNatta and also a number of postdoctoral fellows, including Ron Richardson, Peter Johncock, David Sayers, Jim Heyes, Fred Drakesmith and Don Lomas, all of whom were trained in fluorine chemistry at either Durham or Birmingham University in the UK.

The mid-50s through the early '60s were the heydays for Paul's research group. Although never more than 8 strong, with 4 or 5 being supported by the Army, every day was fun, filled with exciting chemistry and stimulating discussions. At his farewell banquet at Wright-Patterson AFB in 1973, just before taking up his appointment as an Undersecretary of the Air Force, Al Lovelace commented, 'I have seen many other good research groups, but I don't believe that any one of them was as good as we were in those days at UF'. Al later went on to become Administrator of NASA.

The Army contract evolved into a subcontract with the Air Force, through TRW, and eventually to direct funding from Wright-Patterson AFB for another 3-4 years. Included among the work done with Air Force support were novel photochemical studies on cycloadditions of hexafluoroacetone to alkenes. During this time, Paul also had a Navy subcontract through MIT to make fluorinated fluids for gyroscopes, and a NASA subcontract through Cal Tech. In the early '70s he obtained an NSF grant, but by the end of that decade his funding had essentially dried up, consequently he did less research. He decided to retire in 1981 at age 66 so that he could give his full attention to teaching his second wife, Marian (a Yankee), how to become a rebel. His final paper was published in 1988 ('The Reaction of some 3- and 4-Fluorooxetanes with Acids' [7]).

Paul's professional achievements during his academic career at the University of Florida were considerable. He rose rapidly through the ranks at UF, attaining the rank of Professor in 1957. With a natural wit and an uncanny ability to defuse the most volatile of situations, he was the 'glue' of the Organic Division, serving as its Chairman for 15 years. In total, he directed the PhD degree work of twenty young men and women, as well as the MS studies of another twenty; also he served as mentor to about 25 postdoctoral fellows, being more like a surrogate father than a boss. (Paul still maintains close contacts with most of his former students and postdocs.) With 58 pioneering papers and 17 patents in the field of synthetic organofluorine chemistry to his credit, Paul certainly made significant contributions to the field of fluorine chemistry, and these have been acknowledged on numerous occasions. He received ACS recognition through the Florida Section Award in 1966, the Southern Chemist Award in 1963, and the Fluorine Division Award for Creative Work in Fluorine Chemistry in 1976.

Paul was very active in the Fluorine Division of the ACS (American Chemical Society), serving as secretary of the Fluorine Subsection of the Industrial Chemistry Section for several years, and, when the Fluorine Section was formed in 1960, becoming its first Chairman. He was instrumental in the continuation of the International Fluorine Symposia (the first was held in England at Birmingham University in 1959, with no plan then for a second), and helped to organize the second in the series at Estes Park, Colorado, in 1962. Estes Park proved so attractive and popular a site that the fourth meeting was also held there in 1968. By that time Paul had commenced his duties as Editor of *Fluorine Chemistry Reviews*, a Marcel Dekker series which ran to eight volumes during the period 1967-77.

PCR

During the early years at UF, Paul and George were always on the lookout for ways to make a little extra money. After all, their starting salaries were only $3200 per year, and raises rarely amounted to more than $100 per year. They came up with a number of abortive moneymaking projects before hitting on the idea of starting what was to become PCR.

Their first project was to make some super-high-energy racing fuel for a guy who came around looking for suckers, but although they made five gallons of nitropropane for this fellow, he never showed up to collect it and pay up (later, Paul and George found out that he was in jail for threatening an FBI agent). Next, they tried working with the Naval Stores Lab in Olustee where Professor Hawkins had been consulting. Olustee had discovered that the adduct derived by heating maleic anhydride with the resin acid abietic acid made a good paper sizer. So George and Paul obtained five gallons of pine tree drippings (collected for its turpentine) containing abietic acid, made 20 lbs of the adduct, and sent it off to the paper company for evaluation and, hopefully, purchase. Nothing came of that, so George, Paul and four others each put up $100 in partnership with Stan Wemberley (a friend and Associate Dean at UF) to buy fibreglass to build boats. After much frustration and many wasted fibreglass-making weekends, Stan returned $80 to each contributor and they called it quits.

Finally, in 1952, George and Paul decided to make *and sell* chemicals, specifically trifluoroacetone and various allyl compounds. They incorporated under the name of PenninsularChemResearch (PCR) with the purpose of 'doing research and making chemicals'! In choosing a name, they first tried combining their own names in some manner, but neither 'Tarbut' nor 'Buttar' sounded quite right, nor did 'Pine-Tree Chemicals', which was another candidate. Once they started the company, the most important thing they did was put an advert in C & E News. In addition to allowing them to sell some trifluoroacetone, their ad quickly led to a contract with American Viscose to make spinning machine lubricants. This was followed by two research contracts from the Air Force, the first to make antioxidants, and the second to synthesize fluoroalkylsilanes.

The first location of PCR was a shell of a building on NW 5th Avenue and 10th Street, near the water tower. Paul and George spent nights and weekends for 3-4 months building wooden benches, putting plumbing in etc. In those days, Paul and George ran the company by the seats of their pants, quoting to do research and make chemicals without really knowing whether they were making money or not. In 1957 they found a way to make $CF_3CH_2CH_2SiCl_2CH_3$, a precursor to a novel siloxane elastomer that had good low-temperature properties. Dow Corning was interested in the process and wanted to buy the patent rights. Their first offer was $25 000. George and Paul said that they would think about it, left the meeting, went to their hotel room, and burst out laughing: with assets of only $8000, Dow Corning could have had the *whole company* for considerably less than $25K! As it was, George and Paul took $30 000 for the patent rights, and this money allowed them to buy land and start construction, in 1958, of a modern facility at the current site of the company – the Airport Industrial Park. In addition to a small office building, they built many open-air structures, for safety reasons.

PCR grew quickly and prospered to such an extent that during the summer of 1960 Air Products came down to talk about the purchase of the company for $400 000! Paul was working in California that summer with Charlie Haber at the Naval Rocket Laboratory when he received word from Lee Gordon, then President of PCR, that the deal had fallen through.

At that time PCR was heavily involved in the preparation of high-energy propellants (NF compounds) for Allegeny Ballistics, ARPA and the Air Force. Also, George had discovered an allyl compound that proved to be a good flocculating agent, and which PCR developed and patented. This agent, called 'CATFLOC' was also of interest to Calgon, a water treatment company. In fact, Calgon had already built a plant to manufacture the material when it found out that PCR, 'some small company in Florida', held a Canadian patent on the process it was preparing to use. Thus Calgon offered to buy PCR, and the negotiations which followed led to the sale of PCR to Calgon in 1966 for ~40,000 shares in that company. Interestingly, the share price rose from $35 to $65 during the course of the negotiations, which lasted for 6 months. The number of shares was determined on the golf course, with Paul having to shoot an 8 on the last hole so as not to offend the Calgon representative.

After the Calgon purchase, George and Paul became 'consultants' to the company and so remained actively involved in the goings on. The Calgon executive in charge of PCR was Ralph Thompson, who, having seen a 1968 *Life Magazine* article in which 5-fluorouracil (5-FU) was being touted as a potential anti-cancer agent, challenged PCR to devise a good preparation of the drug. This Paul did, and as a result 5FU made millions for the company. Paul's method was quite simple: direct fluorination of uracil in water. To show how hit-and-miss Ralph was when using *Life Magazine* as a source of ideas, he later read that pest deer could be scared away by the odour of tiger dung and asked PCR to determine the identity of the 'active ingredient'. However, some halfhearted research did not discover the magic repellent.

Merck soon became interested in Calgon – not because of 5-FU, but because Calgon had acquired Pittsburg Activated Carbon Company, which made the charcoal-activated-carbon in cigarette filters. Merck acquired Calgon in 1968. George and Paul then became consultants for Merck, but Merck had not the slightest interested in PCR. In effect, they sat on the company, squelching business; there were to be no more government contracts or any business done with government agencies. Finally, in 1970, six junior executives at PCR decided they'd had enough of this backward movement and, with the backing of Dow, they purchased PCR back from Merck. These entrepeneurs were W. Arnold Dinkens, Gene Stump, Paul Shuman, Dale Warner, John Cochran and Max Petzold.

The 5-FU patent was finally issued in 1976 [8], and in 1977 PCR Puerto Rico was established to facilitate the compound's manufacture. PCR was acquired by the SCM Corporation in 1978, and, as time went by, Paul and George became increasingly separated from the company's activities. In 1986, SCM was acquired by Hanson plc, and just one year later PCR was re-acquired by Management, the Demetree family and Reichhold Chemical. Shortly afterwards, Hydrozo of Lincoln, Nebraska, acquired PCR and established vinylsilane manufacturing facilities. In 1994, PCR merged with Thoro (an ICI Americas, Inc., company), establishing Harris Speciality Chemicals, Inc., but was soon (1997) acquired by Lancaster Synthesis Ltd. (owned by British Tar) – a major catalogue company sited in NW England.

Family life and retirement

During those hectic years of the late '50s and '60s, Paul's personal life had its ups and downs. The family grew with the arrival of Paula and Sandy in 1953 and 1957, respectively, but Viola became ill in the '60s and eventually died of cancer in early 1971. Paul's grief was overwhelming, but the following year he was fortunate to meet, fall in love with, and marry Marian Christie, a widow from Chicago with two teenage daughters. Two years later they built their home at 2211 NW 26th Terrace, Gainesville, where they lived happily together until Marian's untimely death in 1997.

After retirement from UF, Paul continued his consulting not only with PCR, but also with Geigy in Tarrytown, NY (15 years); and during the 1980s he made a number of trips to Japan with Marian to consult for Daikin Industries. Nowadays, he continues to travel frequently with family and friends to exotic destinations around the world. He enjoys fishing on the Gulf with Gene Stump, keeping up his lake place, and tinkering with his computer *(tarrant@chem.ufl.edu)*, but most of all being with family and friends. Paul especially enjoys the frequent visits made by his former students and postdoctoral fellows; and every two years, on the weekend following the ACS Winter Fluorine Conference, he holds his traditional post-Conference party at his home in Gainesville.

References

1 R. Y. Tompson, P. Tarrant and L. A. Bigelow, *J. Am. Chem. Soc., 68* (1946) 2187.
2 L. A. Bigelow, R. Y. Tompson and P. Tarrant, *Ind. Eng. Chem., 39* (1947) 360.
3 J. H. Simons, J. C. Bacon, C. W. Bradley, J. T. Cassaday, E. I. Hoeberg and P. Tarrant, *J. Am. Chem. Soc., 68* (1946) 1613.
4 J. A. Young and P. Tarrant, *J. Am. Chem. Soc., 71* (1949) 2432.
5 J. A. Young and P. Tarrant, *J. Am. Chem. Soc., 72* (1950) 1860.
6 P. Tarrant and E. G. Gillman, *J. Am. Chem. Soc., 76* (1954) 5423.
7 P. Tarrant and R. N. Bull, *J. Fluorine Chem., 40* (1988) 201.
8 P. D. Schuman, P. Tarrant, D. A. Warner and G. Westmoreland, *US Patent 3954753* (1976) (to PCR, Inc.).

BIOGRAPHIC NOTE

William Dolbier

William Dolbier was born in New Jersey in 1939, and moved with his family to Haines City, Florida, in 1955, where he finished high school in 1957. He entered Stetson University as an engineering student, but his first experience with organic chemistry caused him to become a chemistry major, and he graduated with a BS degree in Chemistry in 1961. He immediately entered the PhD Programme at Cornell and joined the research group of Mel Goldstein, which set his future course as a physical organic chemist. Graduating with his PhD in 1965, he spent one and a half years expanding his physical organic horizons as a postdoc with Bill Doering at Yale University before accepting an Assistant Professor position at the University of Florida in the fall of 1966.

At UF he plied his trade as a physical organic chemist, using various kinetic and isotopic labelling techniques to study the mechanisms of cycloadditions and thermal rearrangements. Having an office adjacent to that of Paul Tarrant, he was inevitably exposed to the novel properties and reactivities of organofluorine compounds, and eventually he could resist the call no longer and began to apply his physical organic tools in the study of fluoro-organic reactivity. His initial ventures included work on the kinetic impact of fluorine substituents on cycloadditions of allenes and thermal rearrangements of cyclopropanes, which proved so exciting and productive that he never looked back and has devoted his talents almost exclusively to the field of fluorine chemistry ever since. In recent years, his primary research interest has been to determine the quantitative impact of fluorine substitution on radical reactivity. He also has been very active recently in the devlopment of commercial synthetic processes for the preparation of fluorinated [2.2]paracyclophanes for use by the semiconductor industry; this has led to a number of patents.

Bill moved through the ranks at the University of Florida, becoming Professor in 1975 and serving as Chairman of the Department from 1982 to 1987. He was an A. P. Sloan Fellow and a John Simon Guggenheim Fellow. Within the ACS Division of Fluorine Chemistry, he has served on the Executive Committee and as Chairman of the Division, and he acted as organizer of the 11th Winter Fluorine Conference. As of 1998, he'd published more than 175 papers, most of which involved synthetic and physical studies on fluoro-organic compounds. Currently he maintains a very active research group consisting of 5 PhD students and 5 postdoctoral fellows. Bill received the year 2000 ACS Award for Creative Work in Fluorine Chemistry.

Chapter 14

FLUOROPOLYMERS, STABLE NITROXIDES AND PERFLUOROALKYLATION

KALATHIL C. EAPEN

Senior Research Chemist, University of Dayton Research Institute, 300 College Park, Dayton, OH 45469-0168, USA

Kalathil C. Eapen

My introduction to fluorine chemistry began in the September of 1969, soon after I arrived in the UK from India to undertake a postdoctoral assignment at the University of Manchester Institute of Science and Technology (UMIST). I was in correspondence with Professor R. N. Haszeldine, FRS, a few months before my arrival in England, and was fortunate in obtaining financial support from the Commonwealth Foundation in the form of a scholarship. I was excited at the prospect of meeting Professor Haszeldine, whose fame exceeded national boundaries, and expected to see an old scientist with thick glasses and grey hair. I was in for quite a surprise: here was a distinguished, dynamic, middle-aged man who welcomed me to his lab and made me feel comfortable. He wanted me to meet Dr Eric Banks, who at the time was a Senior Lecturer at UMIST, to participate in further discussions so that my knowledge of organic chemistry could be assessed. I met Eric the next day without realizing that he was going to be my peer and friend for many years to come. We had a pleasant discussion and he asked me few questions on organic reaction mechanisms. I then returned to Professor Haszeldine's office to discuss possible research topics, details of which I cannot recall now, and I selected a difficult and high-risk project: the development of a 'self-curing' *Viton*®-type elastomer.

'Self-curing' *Viton*® *A* analogues

An important breakthrough in the search for heat and fluid resistant elastomers to meet the demands of the aerospace industry was the discovery of DuPont's *Viton*® *A* in the mid-1950s [1]. As is well known, *Viton*® *A* is a random copolymer of hexafluoropropene and vinylidene fluoride, $-[(CH_2CF_2)_xCF_2CF(CF_3)]_n-$, which in those days was crosslinked (cured; vulcanized) commercially by heating it with hydrocarbon diamines such as hexamethylenediamine or diamine carbamates [1, 2]. The thermal stability of amine-cured *Viton*® *A* is considerably lower than that of the raw elastomer owing to the

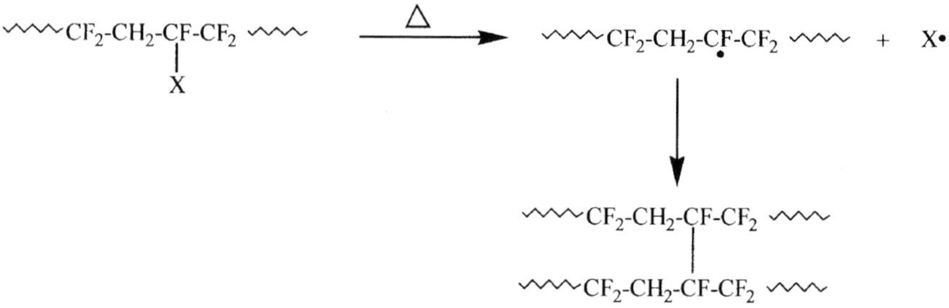

Scheme 14.1. Conceptual thermal self-cure mechanism for fluoroelastomers.

nature of the curing reactions and the relative instability of the hydrocarbon-type cross-links. Thus improved crosslinking techniques were required. One method under consideration at UMIST was to synthesize linear polymer chains incorporating units containing a pendant thermally-labile functional group (X), so that by simply heating the raw polymer, homolytic loss of the functional group would occur, enabling cross-links to form via coupling of macroradicals (Scheme 14.1).

Preliminary attempts to develop a 'self-curing' fluoroelastomer had been made at UMIST before I arrived, using SF_5 as the labile functional group X. In that work, modified *Viton* elastomers prepared by the emulsion copolymerization of vinylidene fluoride with hexafluoropropene in the presence of small amounts of the potential cure-site monomer perfluorovinyl sulfurpentafluoride ($CF_2{=}CFSF_5$) had been successfully cross-linked by heating [3]; however, the polymers prepared had lower molecular weights than commercial *Viton*® *A* and were less thermally stable [4]. Perfluorovinylsulfonyl chloride ($CF_2{=}CFSO_2Cl$) was also viewed as another promising cure-site monomer for preparing a 'self-curing' *Viton*, by virtue of the ease with which the SO_2Cl group undergoes thermal decomposition. (This monomer itself undergoes thermolysis at 150 °C giving chlorotrifluoroethylene and sulfur dioxide, presumably by a free-radical process.) My predecessor at UMIST, Dr Alan Jones, had studied the copolymerization of this monomer with other fluoro-olefins for his doctoral work [5]. Using a 1:5 monomer feed ratio of perfluorovinylsulfonyl chloride and vinylidene fluoride, and employing the hazardous bis(trichloroacetyl) peroxide as a low-temperature initiator, he had prepared a copolymer in about 15% yield as a viscous gum. It was a low-molecular-weight material, and my first task at UMIST was to attempt to obtain a higher-molecular-weight sample in reasonable yield. The first experiments involved getting familiar with an all-glass vacuum system and synthesizing perfluorovinylsulfonyl chloride via reported [5-7] methodology (Scheme 14.2).

My initial polymerization experiments were directed at preparing a copolymer of vinylidene fluoride and perfluorovinylsulfonyl chloride prior to making a terpolymer incorporating hexafluoropropene also. Trichloroacetyl peroxide was chosen as the low-temperature initiator so as to inhibit chain transfer leading to premature termination of growing radical chains, as well as to repeat Alan Jones' previous work [5]. As is well known, making and using trichloroacetyl peroxide [8] is a hazardous business because the isolated solid is liable to detonate at room temperature or below. Eric Banks advised me to exercise extreme caution when preparing this peroxide, and not to prepare more than

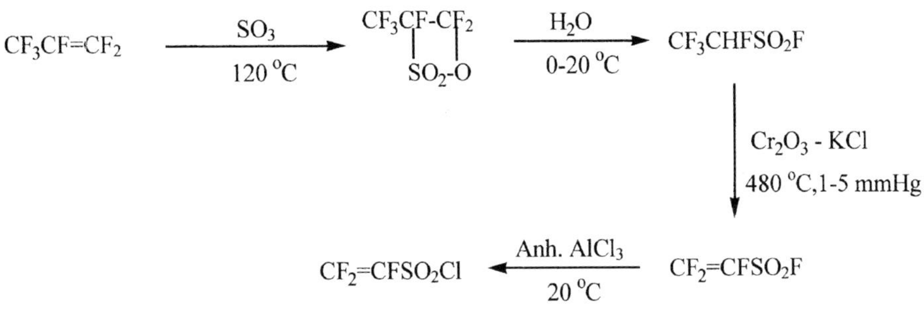

Scheme 14.2. Reaction steps used in the preparation of perfluorovinylsulfonyl chloride.

1.0 g at a time. To emphasize the danger, he told of a detonation resulting in personal injury to Dr Jones during preparation of the peroxide in the same laboratory. Thus, wearing head and face protection, leather lab coat and gloves, and safety glasses, I made less than 1 g of the peroxide at a time (as needed), handling it as a cold (below −20 °C) solution in CFC-113. Polymerizations initiated with trichloroacetyl peroxide, however, only confirmed the earlier work, *i.e.* formation of only a low-molecular-weight copolymer occurred, in low yield. Other peroxides such as bis(pentafluorobenzoyl) peroxide and di-*t*-butyl peroxide did not give promising results, and also attempts to produce terpolymers incorporating perfluorovinylsulfonyl chloride, vinylidene fluoride and hexafluoropropene proved abortive.

At this stage, it was increasingly becoming clear that the use of perfluorovinylsulfonyl chloride was not likely to provide a high-molecular-weight polymer. This conclusion was substantiated by the formation of small amounts of SO_2 in polymerizations at temperatures at which the monomer itself is stable, suggesting a competing chain-transfer process. Assuming that the initiating radical could also abstract chlorine from the monomer, the reaction products were carefully searched in a polymerization experiment with di-*t*-butyl peroxide. However, *t*-butyl chloride was not detected.

Since the copolymerization work with $CF_2{=}CFSO_2Cl$ had been unsuccessful, I decided to have a quick look at the copolymerization of perfluorovinylsulfonyl fluoride and vinylidene fluoride, even though the sulfonyl fluoride group was not viewed as thermally labile enough to be an X group (Scheme 14.1). Copolymerization using bis(pentafluorobenzoyl) peroxide gave only a 7% yield of a copolymer, so perfluoro-(2,4-dimethyl-3-oxa-2,4-diazapentane) was tried as the initiator. This NON compound had just been used successfully in the Banks-Haszeldine group by the then graduate student Tom Myerscough [9] to initiate homopolymerization of tetrafluoroethylene [possibly via slow homolytic decomposition at ambient temperature, $(CF_3)_2NON(CF_3)_2 \rightarrow (CF_3)_2NO\cdot + (CF_3)N\cdot$] and was available 'just along the bench' from me in lab F102. Copolymerization of perfluorovinylsulfonyl fluoride and vinylidene fluoride using Myerscough's NON initiator did give a white solid in about 35% yield, the IR spectrum of which showed the presence of pendant SO_2F groups, and I published a short note on this work a few years later [10].

Magic radicals

Since the project on the 'self-curing' fluoroelastomers did not produce the hoped-for results, it was decided to abandon it and start a new one. Discussions with Eric Banks and Professor Haszeldine led me to select a project on the first perfluorinated bis-nitroxide to be isolated.

The first perfluorinated nitroxide to be isolated was the mono-oxyl $(CF_3)_2NO\cdot$ (bistrifluoromethylamino-oxyl), known at UMIST as the 'magic radical' but first isolated elsewhere [11, 12], although it had earlier been postulated by Haszeldine as an intermediate in the photochemical dimerization of trifluoronitrosomethane [13]. Subsequently, other fluorinated nitroxides such as bis(2-chlorotetrafluoroethyl)amino-oxyl [14], difluoroamino-oxyl [15], bis(perfluoroethyl)amino-oxyl, bis(perfluoro-*t*-butyl)amino-oxyl, bis(perfluoroheptyl)amino-oxyl [16], perfluoropiperidino-*N*-oxyl [17], and perfluoromorpholino-*N*-oxyl [18] were prepared. (A comprehensive review of the chemistry of nitroxides is available [19].) During the period I was at UMIST, detailed studies on the chemistry of the 'magic radical' were being made by a number of graduate students, and I first encountered it during my work on Myerscough's NON initiator (see above). It was only natural that the next step would be to prepare a perfluorinated dinitroxide. The first attempt to do so was made by Myerscough [9] who succeded in preparing small amounts of perfluoro-2,5-diazahexane-2,5-dioxyl, $\cdot ON(CF_3)CF_2CF_2N(CF_3)O\cdot$, by oxidation of the hydrolysis product of an adduct, $2CF_3NO \cdot C_2F_4 \cdot PCl_3$, reported by Russian chemists [20, 21]. My new task was to prepare larger amounts of this persistent free radical and study its reactions in depth.

The reaction steps involved in the preparation of perfluoro-2,5-diazahexane-2,5-dioxyl which should perhaps, following the UMIST tradition, be referred to as the 'magic diradical', are shown in Scheme 14.3. It was not necessary to isolate the bis-hydroxylamine before oxidation, which was conducted by slow addition of the hydrolyzate to a hot (90 °C) solution of permanganate, with continuous removal of the volatile products at low pressure. Low-temperature fractionation of the volatiles gave the 'magic diradical' as a deep purple liquid (b.p. 55 °C at 769 mmHg) in 66% overall yield based on CF_3NO.

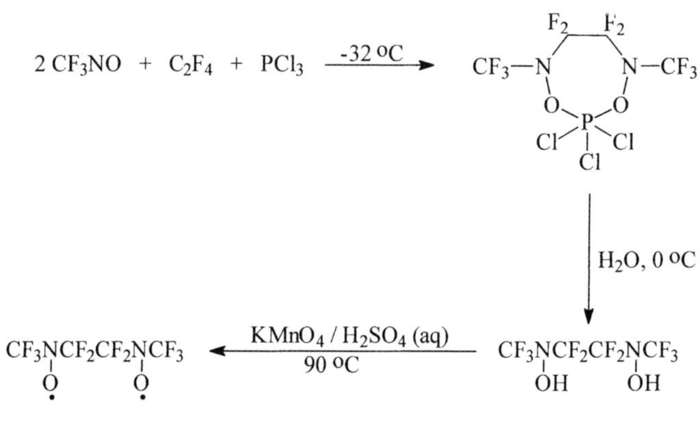

Scheme 14.3. Reaction scheme for the generation of the first stable perfluorodinitroxide, perfluoro-2,5-diazahexane-2,5-dioxyl [22, 23].

The first few oxidations were carried out on a 1 gram scale without any incidents, but on scale-up ($\times 7$) a yellowish orange material condensed along with the diradical in a $-96\,^{\circ}C$ trap. An attempt to separate this material (presumably an oxide of chlorine) from the diradical resulted in a violent explosion, which fortunately did not injure me, but left me shaken to the core. All further oxidations were therefore carried out only after removing chloride ion from the hydrolyzate with silver nitrate. No incidents occurred after implementation of this modification.

The rest of my work at UMIST involved studies on the reaction of the 'magic diradical' with HBr to obtain a pure sample of the bis-hydroxylamine, $CF_3N(OH)CF_2CF_2N(OH)CF_3$, and with various unsaturated halogeno-compounds. The reactions with $CF_2{=}CF_2$ and $CF_2{=}CFCF_3$, for example, gave 1:1 copolymers and novel eight-membered heterocyclic systems by 1:1 cycloaddition; condensed-phase reaction conditions favoured copolymer formation, while reactions in the gas phase at low pressure favoured cycloaddition. This work is documented in detail in two publications [22, 23] and in 'Chem. Soc.' *Specialist Periodical Reports* [24].

Return to India

By early 1971, my leave from the University of Calicut in Kerala, India, had come to an end, so I went home to resume my teaching career and soon slipped back into routine teaching. By that time, however, I was thoroughly 'under the spell' of organofluorine chemistry: I was fascinated by the novelty of the field, and by the unique chemistry of fluorinated compounds; added to this was the realization that most classes of hydrocarbon compounds can have fluorinated analogues, ranging from mono- to per-fluoro derivatives, thus providing for a virtually unlimited number of new and exciting compounds. However, I soon realized that conducting any long-term research work in organofluorine chemistry in India at that time was not a practical proposition: fluorinated compounds required had to be imported, and by local standards they were *very* expensive. Thus, I came to the conclusion that I would have to look outside India if I wanted to continue studies in fluorine chemistry. This time my efforts were directed at opportunities in the United States, and I started contacting prominent scientists in fluorine chemistry there. Success finally came in the mid-70s when I was awarded a National Research Council Senior Associateship to work with Dr Christ Tamborski at Wright-Patterson Air Force Base (WPAFB), Ohio.[1]

Christ Tamborski

[1] Dr Christ Tamboski, a graduate of the University of Buffalo (BA, 1948; PhD, 1953), was a Senior Scientist at the United State Air Force Materials Laboratory from 1955 until his retirement in 1986. A very well known postwar (WWII) pioneer of organofluorine chemistry, he has been a member of the Editorial Board of the *Journal of Fluorine Chemistry* from its inception until quite recently, and was involved in the formation of the American Chemical Society's Division of Fluorine Chemistry, which he has served well in numerous capacities (including that of Chairperson in 1971-72).

Fluorine research at WPAFB

When I first arrived in Dayton, Ohio, in September 1976 to start work on my new assignment, I had no idea whatsoever that this city would be home for me and my family for the rest of my career.[2] At WPAFB, the genuine interest and enthusiasm Dr Tamborski had in fluorine chemistry and organometallic chemistry was extremely stimulating to me, and our work on heterocyclic compounds containing perfluoroalkylether substituents was soon in full swing. Within a few months, Dr Loomis Chen joined us, followed by his wife Grace Chen, both as on-site contractors with the University of Dayton Research Institute (UDRI). Although differences of opinion have arisen at times, we have had long and productive careers together, lasting well after the retirement of Dr Tamborski. In September 1978, I joined UDRI and continued to work at WPAFB.

My work at WPAFB covered a variety of topics, including studies on both fluorinated and non-fluorinated materials; here, of course, I discuss only the former. Also included, for continuity, are details of work conducted solely by Loomis Chen and Grace Chen in collaboration with Christ Tamborski, as will be apparent from the references cited. Dr Tamborski, our mentor and advisor, was responsible for initiating most of the work until his retirement in 1986. One of the major objectives of the work was to generate soluble additives for perfluoropolyalkylether fluids, which we prefer to call PFPAE fluids [others refer to them as PFPE (perfluoropolyether) fluids]. The strategy was to select suitable substrates, taking leads from known additive materials used in mineral oils based on hydrocarbons, and introduce perfluoroalkyl or – better – perfluoroalkylether groups into those structures so as to make them soluble in the PFPAE fluids. This strategy worked in many cases though not with all substrates, as the powerful electron-withdrawing substituents altered the beneficial effects of some substrates too much to be of utility. Another requirement for the new additives was that they must possess high thermal and oxidative stability to ensure their survival at temperatures above 300 °C in the presence of air for extended periods of time. Obviously, the work involved significant amounts of trial and error.

Thus, the basic theme of work conducted over a period of two decades could be considered as 'Perfluoroalkylation and Perfluoro-oxa-alkylation' of a variety of substrates. Available fluorinated starting materials used in the effort were either perfluorinated iodides or acid fluorides. For convenience I shall discuss the work in separate sections according to the substrates studied or the intermediate compounds/products formed.

Heterocyclic compounds

Initially, I worked on benzothiazoles containing perfluoroalkylether substituents at the 2-position. Such compounds were already known, having been synthesized by imidate ester condensation [25] as well as lithiation of 2-aminobenzenethiol followed by reaction with acid fluorides [26]. We set out to synthesize them directly, utilizing readily available acid halides.

Interaction of equimolar amounts of 2-aminobenzenethiol and an acid fluoride at ambient temperature gave the *N*-acyl derivative rather than the *S*-acyl derivative, along

[2] I still live in Dayton with Susan, my wife of 33 years and a lady of great patience and tolerance. Our children, of whom we are very proud, have 'flown the nest' now: our daughter, Reenu, lives in Dallas (Texas) and is doing her fellowship in Pediatric Cardiology; and our son, Alex, is in Iowa City finishing his doctoral work in Pharmacology.

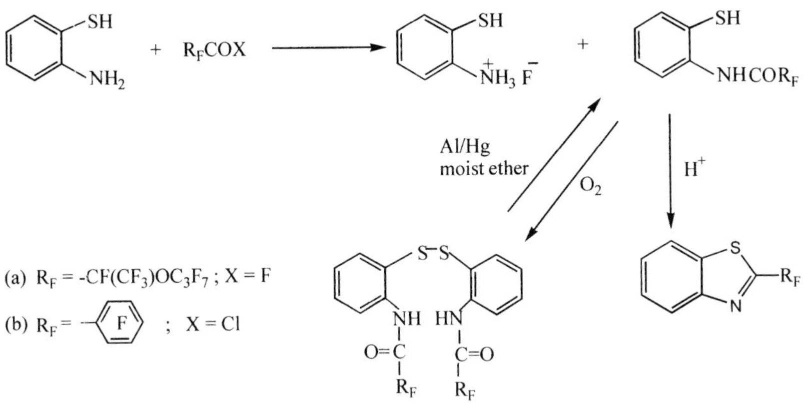

Scheme 14.4. Synthesis scheme for 2-substituted benzothiazoles [27, 28].

with an amine salt, as shown in Scheme 14.4. When R_F is a perfluoroalkylether group, the acyl derivative cyclizes slowly even at ambient temperature, perhaps due to the presence of traces of acid, to yield the benzothiazole. A side reaction that was observed was atmospheric oxidation of such *N*-acyl benzenethiols, giving the corresponding disulfides; reduction of these disulfides under neutral conditions using aluminum amalgam in moist ether gave pure *N*-acyl compounds that could be isolated and characterized without undergoing cyclization. When the R_F group was pentafluorophenyl, the *N*-acyl compound did not dehydrate readily at ambient temperature and could be isolated and characterized [27]. Using an essentially similar approach, additional 2-perfluoroalkylether benzothiazoles as well as bis-benzothiazoles (from diacid fluorides) were made in later years and found useful as soluble anti-oxidation-corrosion additives for PFPAE fluids [28].

An approach, similar to that used for the preparation of benzothiazoles (see above), was adopted initially for the synthesis of benzimidazoles containing 2-perfluoro substituents[3]. Acylation of *o*-phenylenediamine with a perfluoroalkylether acid fluoride at ambient temperature gave, in addition to the amine hydrofluoride, a monoamide, a benzimidazole and also a diamide. Here also, like the benzothiazole case, the monoamide was found to undergo slow dehydrative cyclization to the benzimidazole. Although it was possible to obtain up to 73% yields (based on GC analysis) of benzimidazoles by adjusting the molar ratio of the reactants as well as the experimental conditions, their isolation and purification proved tedious (column chromatography and repeated crystallizations). Fortunately, however, we were able to obtain excellent yields of the 2-(perfluoroalkylether)benzimidazoles by the method shown in Scheme 14.5. This involved fast monolithiation of *o*-phenylene diamine in diethyl ether with *n*-BuLi at 0 °C, followed by a lethargic reaction with fluorinated esters that took 5-7 days at ambient temperature for complete utilization of the ester. The adducts produced slowly eliminated LiOEt even under anhydrous conditions to

[3] A number of 2-perfluoroalkylbenzimidazoles had already been described in the literature when this work was started. The methods of preparation involved condensation of substituted *o*-phenylenediamines with perfluorinated carboxylic acids [29, 30], and reduction of *N*-(*o*-nitrophenyl) perfluoroalkylamides to the amine and concurrent cyclization [29].

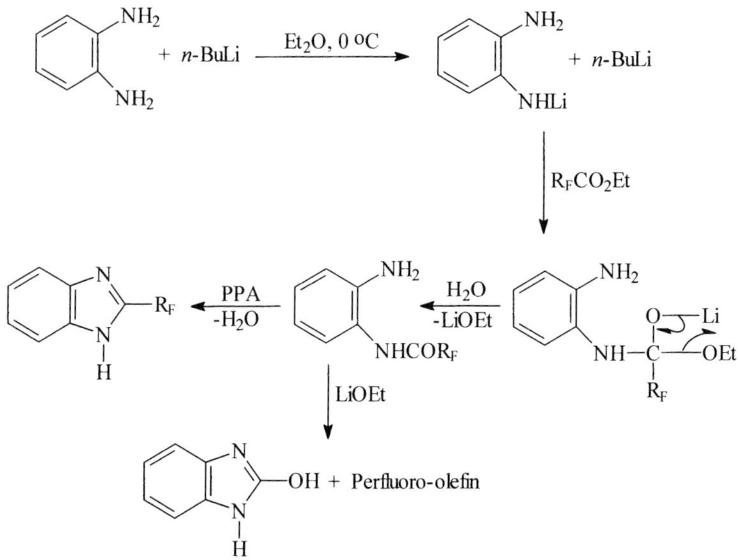

$R_F = CF(CF_3)[OCF_2CF(CF_3)]_nOC_3F_7$

Scheme 14.5. Preparation of 2-substituted benzimidazoles via amides [33].

yield the amides, a process that was obviously rapid when water was added at the end of the reaction. It was also observed that the amides formed underwent a side reaction with LiOEt in anhydrous ether giving rise to 2-hydroxybenzimidazole and fluorinated olefins; and the optimum yield of an amide was obtained about 24 hours after the addition of the ester. This decomposition of the amides by LiOEt is analogous to the decomposition of fluorinated esters by NaOR [31] and LiOEt [32]. It appears that this may be a general mode of decomposition exhibited by many different derivatives of perfluorocarboxylic acids in the presence of bases.

Amides synthesized as shown in Scheme 14.5 were readily cyclized to 2-perfluoroalkylether benzimidazoles by heating them with polyphosphoric acid at about 130 °C for 3 to 4 hours. 2-(Pentafluorophenyl)benzimidazole, was prepared simply by heating *o*-phenylene diamine with pentafluorobenzoic acid in polyphosphoric acid at 190-200 °C for 4 hours. The use of polyphosphoric acid was detrimental to acceptable yields in the synthesis of lower 2-perfluoroalkylbenzimidazoles (*e.g.* $R_F = C_2F_5$, C_3F_7). For instance, when *o*-phenylenediamine and *n*-heptafluorobutyric acid were heated together with excess polyphosphoric acid at 120 °C for 12 hours, less than 10% yield of 2-(heptafluoropropyl)benzimidazole was obtained. When polyphosphoric acid was omitted from this reaction, a 90% yield of the 2-(heptafluoropropyl)benzimidazole was obtained by heating the reactants at 115-120 °C for 4 hours. The same reaction conducted at 100-105 °C for about 6 hours gave only 15% yield of the benzimidazole, indicating significant effect of reaction temperature on product yield. Details of all this work, including thermal stablity data for some of the benzimidazoles prepared and information on their *N*-substituted derivatives, have been published [33].

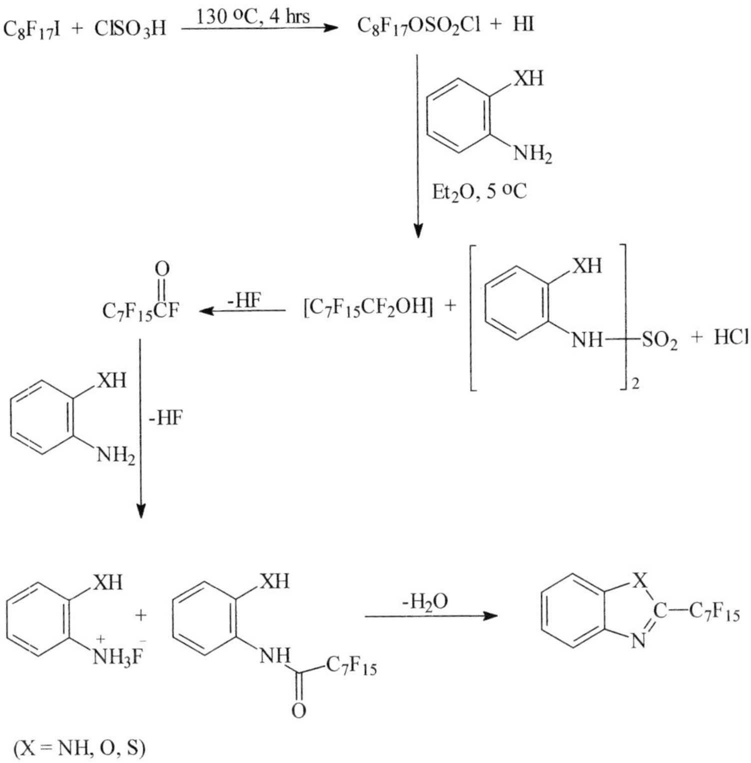

Scheme 14.6. Preparation of 2-substituted benzazoles via chlorosulphates [34].

A different method was developed later which could be applied to the preparation of 2-substituted benzoxazoles, benzothiazoles and benzimidazoles [34]. This involves slow addition of perfluoroalkyl or perfluoroalkylether chlorosulphates to *o*-substituted anilines in diethyl ether at about 5 °C. Products are mixtures of the corresponding heterocyclic compounds and precursor anilides except in the case of 2-hydroxyaniline, which gives only the anilides; however the latter can readily be converted to the benzoxazoles in fair to high yields (*e.g.* Scheme 14.6). No attempt was made to identify all the by-products formed. This method is particularly suitable when one has direct access to perfluorinated iodides since they are easily converted to chlorosulfates by treatment with chlorosulfonic acid.

A number of other heterocyclic systems containing perfluoroalkyl and perfluoroalkylether substituents were studied in our group, but I have included them later under 'perfluoroalkylations and perfluoro-oxa-alkylations' of aromatic systems because perfluorinated iodides feature as common starting materials. Also, a variety of heterocyclic compounds have been synthesized as shown in Scheme 14.7 in collaboration with Dr U. D. G. Prabhu, a visiting scientist (1981-1983) [35]. The 1,2-bis(trifluoroacetyl)benzene was prepared by thermal decomposition of 3-methoxy-1,3-bis(trifluoromethyl)-1-phthalanol which, in turn, was prepared from *o*-dibromobenzene via sequential lithium-bromine exchange and treatment with methyl trifluoroacetate [36].

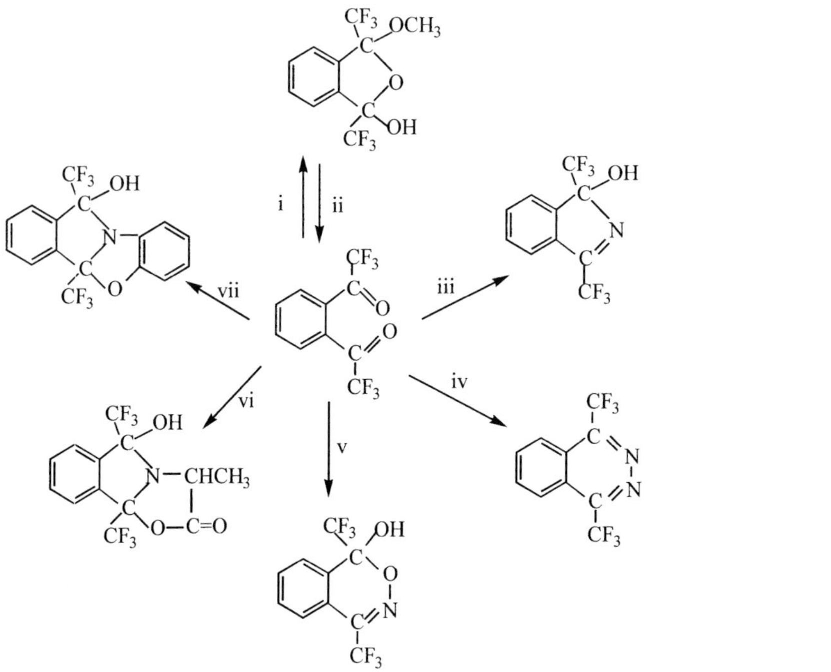

Reagents: i, Dry MeOH. ii, Heat with few drops of CF_3CO_2H at 280 °C, 5-6 h. iii, Dry NH_3. iv, $NH_2NH_2.2HCl$ in dry pyridine at 85-90 °C, 26 h. v, $NH_2OH.HCl$ in dry pyridine, reflux for 20 h. vi. *L*-Alanine in diglyme, 135 °C, 2 days. vii. *o*-Aminophenol, 130 °C, 4 h.

Scheme 14.7. Heterocyclic syntheses involving novel reactions of 1,2-bis(trifluoroacetyl)-benzene with nucleophiles [35].

Perfluoroalkylations and perfluoro-oxa-alkylations of aromatic systems

A series of publications based on perfluoroalkylations and, particularly, perfluoro-oxa-alkylations of a variety of aromatic and heterocyclic substrates came out of our work at WPAFB during the past ten years. As stated before, the work was undertaken to generate potential additive materials soluble in PFPAE fluids. The work was initiated by Dr Tamborski, continued even after his retirement, and provided useful additive materials.

One of the earliest methods of introducing a perfluoroalkyl group into an aromatic ring was reported by McLoughlin and Thrower in the late 1960s [37], namely cross-coupling of a perfluoroalkyl iodide and aromatic iodide with copper in an aprotic solvent; this procedure, which involves the generation of perfluoroalkylcopper species, is applicable to iodoaromatics containing functional groups such as OH, CO_2H, CO_2R, NH_2, NO_2 and OCH_3, and thus provides a convenient method of accessing a variety of perfluoroalkylated aromatics. Our objective was essentially to extend this reaction to include bromoaromatics and chlorodiazines as well as perfluoro-oxa-alkyl iodides [38-41].

Overall, a high degree of success was achieved [38-41]. Mono- and di-bromobenzenes, as well as derivatives containing a variety of functional groups were successfully utilized in cross-coupling reaction [38]; certain bromoarenes, however [*e.g.* o-$Br_2C_6H_4 + C_6F_{13}I \rightarrow C_6F_{13}C_6H_5 + o$-$C_6F_{13}C_6H_4Br + o$-, m- and p-$(C_6F_{13})_2C_6H_4 +$

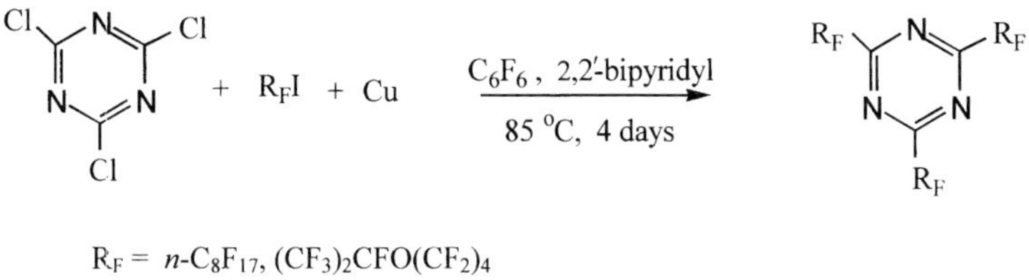

Scheme 14.8. Preparation of perfluorinated *sym*-triazines from cyanuric chloride.

$(C_6F_{13})_3C_6H_3$; m-$BrC_6H_4CO_2H + R_FOR'_FI \rightarrow C_6H_5CO_2H + m$-$R_FOR'_FC_6H_4CO_2H +$ unknowns], gave by-products, indicating the occurrence of competing reactions that may arise from the slower rate of reaction between R_FCu intermediates and aryl bromides than found with aryl iodides. Other bromoarenes successfully used in these cross-coupling include tribromobenzenes, bromodiphenylethers and a bromoterphenylether [39]. The beneficial catalytic effect of 2,2′-bipyridyl was also noticed in these reactions. Among the perfluoro-oxa-alkyl iodides used, those in which the ether oxygen is situated at position 5, *e.g.* $(CF_3)_2CFO(CF_2)_4I$, or higher, behave like perfluoroalkyl iodides and give excellent yields of substitution products; by contrast, ether iodides where the oxygen occupied the 3-position perform badly, giving only traces of the cross-coupled product. The reason for this unusual behavior is not clear at present, although our observations suggest that 3-oxa iodides do not form copper complexes as readily as their 'higher' homologues, and once a copper complex has formed, it is stabilized in some fashion, thus inhibiting reaction with the aromatic substrate. It should be noted that our earlier studies [42, 43] had shown that the 3-oxa lithium derivative $(CF_3)_2CFOCF_2CF_2Li$, generated from the corresponding iodide, is stable at −78 °C in diethyl ether for up to 24 hours, while n-$C_6F_{13}Li$ decomposes readily under the same conditions. From our observations on 3-oxa derivatives, as well as published information on a 2-oxa iodide [44], it seems that perfluoro-oxa-alkyl iodides having the general formula $R_FO(CF_2)_nI$, where $n = 1$ or 2, behave quite differently than those where $n = 4$ or 8. We were unable to obtain a suitable iodide having $n = 3$ in order to study its behaviour.

Secondary iodides such as $C_3F_7OCF_2CF(CF_3)OCF(CF)_3I$ were also used successfully in cross-coupling reactions with aryl iodides [40]; those with a pendant CF_3 group in between the oxygen and the iodine atom, did not react with aryl bromides. Although polar aprotic solvents such as DMSO or DMF are traditionally used in these cross-coupling reactions, better yields were obtained with secondary iodides when hexafluorobenzene containing small amounts of DMSO, DMF or DMAC was used, together with 2,2′-bipyridyl as a catalyst. Chlorinated diazines [41] and cyanuric chloride (Scheme 14.8) [45] were successfully utilized in cross-coupling reactions by Grace and Loomis Chen.

Fluoroketones and perfluorinated tertiary alcohols

Aryl perfluoroalkyl ketones and aryl perfluoroalkylether ketones were of interest to us since the carbonyl groups in these compounds could be converted to CF_2 groups by SF_4/HF fluorination, yielding stable materials for application as potentially useful fluids or model compounds. However, the development of the cross-coupling reaction (see above) made such compounds more readily available. Purely aliphatic perfluoroketones were used to prepare perfluorotertiary alcohols, some of which showed antiwear activity in PFPAE fluids [46]. One of the methods used to prepare the aryl ketones was Friedel-Crafts acylation. This method had been successfully employed long ago by others using acid halides of per- and poly-fluorinated aliphatic and aromatic acids, *e.g.* $C_5F_{11}C(O)Cl + C_6H_6 \rightarrow C_5F_{11}C(O)C_6H_5$ [47-49]. However, when we started work in this area, there were no reports of a Friedel-Crafts acylation having been performed with an acid halide containing a perfluoroalkylether function. Attempts to do so might have been discouraged by the reported substitution of fluorine α to the oxygen atom by chlorine when a perfluoroether is heated with anhydrous aluminum chloride at 150-200 °C [50]. Our studies established that such ether-containing acid halides can be successfully utilized in $AlCl_3$-promoted Friedel-Crafts reactions if the reaction temperature is maintained below 100 °C; under the conditions used, replacement of α-fluorine by chlorine was negligible [51], *e.g.*

$$C_3F_7OCF(CF_3)C(O)F + C_6H_6 \rightarrow C_3F_7OCF(CF_3)C(O)C_6H_5 \quad (57\% \text{ yield}).$$

Classical methods of synthesis of fluoroaliphatic ketones involve reactions between Grignard or organolithium reagents and fluorocarbon esters [52, 53]; product yields vary considerably, and the ketones are accompanied by variable yields of by-product secondary and tertiary alcohols in most instances. Tamborski *et al.* studied these reactions in great depth, and defined experimental conditions that would produce excellent yields of a number of ketones [54, 55], *e.g.* $CF_3C(O)OC_2H_5 + C_6H_5Li \rightarrow CF_3C(O)C_6H_5$; $C_2F_5O(CF_2CF_2O)_3C(O)OC_2H_5 + n\text{-}C_4H_9Li \rightarrow C_2F_5O(CF_2CF_2O)_3C(O)C_4H_9\text{-}n$. By extending the reaction to 1,1-dichlorobenzyllithium and perfluoroesters, perfluorotetraketones and their hydrates were procured for study [56], *e.g.* $C_6H_5CCl_2Li + EtO(O)C(CF_2)_3C(O)OEt \rightarrow C_6H_5CCl_2(O)C(CF_2)_3C(O)CCl_2C_6H_5 \rightarrow$ (via hydrolysis) $C_6H_5C(O)C(O)(CF_2)_3C(O)C(O)C_6H_5$. Reactions of perfluorinated Grignard and lithium reagents with diethyl carbonate or diethyl oxalate were exploited to synthesize symmetric or assymetric ketones, keto-esters and diketones [57], *e.g.* $(CF_3)_2CFO(CF_2)_2Li + EtOC(O)OEt \rightarrow (CF_3)_2CFO(CF_2)_2C(O)(CF_2)_2OCF(CF_3)_2$; $(CF_3)_2CFO(CF_2)_2MgBr + EtO(O)CC(O)OEt \rightarrow (CF_3)_2CFO(CF_2)_2C(O)C(O)OEt$; $(CF_3)_2CFO(CF_2)_2Li + EtO(O)CC(O)OEt \rightarrow (CF_3)_2CFO(CF_2)_2C(O)C(O)(CF_2)_2OCF(CF_3)_2$. Studies on the synthesis of substituted quinoxalines from fluorinated α, β-diketones and ketoesters via reactions with *o*-phenylenediamine followed [58], as did investigations on the formation of cyclic compounds from di- and tetra-ketones and their thermal behaviour [59].

These investigations conducted by Tamborski and his co-workers constitute perhaps the most detailed study made on fluorinated ketones to date. Note that hindered ketones in low to moderate yields were also made by refluxing α-substituted perfluoroacyl fluorides with alkali metal fluorides in anhydrous acetonitrile [60].

The above work made it possible to prepare a number of high-molecular-weight perfluorinated tertiary monohydric alcohols, ketoalcohols and diols, mostly via treatment of appropriate ketones with perfluoroalkylether lithium derivatives containing an ether oxygen α to the carbon atom attached to lithium [61, 62]. As mentioned earlier, such lithium reagents are unusually stable [42, 43] and can be prepared and used at −78 °C. Also used in the cases of a few high-molecular-weight perfluoro tertiary alcohols was Ruppert-reagent methodology, based on a recent publication by DesMarteau *et al.* [63] concerning the reaction of the 'CF_3^-' transfer agent $CF_3Si(CH_3)_3$ with perfluoroketones; of course, we used higher perfluoroalkyl and perfluoroalkylether trimethylsilanes [64]. Finally, it may be of interest to note that the lithium and sodium alkoxides of the perfluoroalkylether-type tertiary alcohol $(CF_3)_2CFOCF_2CF_2C(CF_3)_2OH$, reported by us years ago [42], were the first alkali metal alkoxides found to be liquids at ambient temperature.

Structure-property correlations

Our work on the synthesis of a variety of aromatic and heterocyclic compounds containing perfluoroalkyl and perfluoro-oxa-alkyl substituents enabled us to embark on a systematic study of their relative oxidative stabilities, using pressure differential scanning calorimetry (DSC) under oxygen. Unfortunately, due to changing priorities, this study could not be completed. However, some general trends emerged and proved useful in designing additives.

Another study that was conducted centered on the thermal stability of selected model perfluoroalkyl ethers. For this purpose a series of seven low-molecular-weight perfluoroalkyl ethers with different structural features were selected and synthesized, or obtained from Exfluor Corporation (prepared by direct fluorination of hydrocarbon analogues), and purified by spinning-band distillation or preparative gas chromatography. Their relative stabilities (semiquantitative) were determined using a flow pyrolyzer attached to a gas chromatograph, and the results ($T_{1/2}$ °C – the temperatures at which 50% of the compounds degraded under identical conditions) were compared with those of two perfluoroalkanes [65]. This study showed that perfluoroalkyl ethers in general are more stable than perfluorocarbons with comparable structures, *e.g.* $(CF_3)_2CFO(CF_2)_4OCF(CF_3)_2$, 827 °C; $(CF_3)_2CF(CF_2)_4CF(CF_3)_2$, 693 °C. It also showed that thermal stability decreased with increased branching as well as with increasing number of adjacent carbon atoms, consonant with the situation applying to perfluoroalkanes [66-68]. A branched PFPAE fluid was also investigated by studying its degradation in presence of anhydrous aluminum chloride [69].

Structure-property correlations were also made on chlorofluorocarbons while we were engaged in developing a chlorofluorocarbon-based non-flammable fluid, a project which was later abandoned on grounds of environmental safety. Several aliphatic chlorofluorocarbon model compounds were synthesized in which the number and position of the chlorine atoms varied, and their physical properties and thermal stabilities in the presence of selected metals were determined. One of the reactions that was studied in detail during the course of this study was the simple coupling of $CF_2ClCFClI$ with zinc [70]. Samples of this iodide are often contaminated with its isomer $CFCl_2CF_2I$, as was later pointed out to us by Eric Banks [71], and proved true for our sample; therefore an erratum was published [72] and zinc-coupling of the 2,2-dichloro isomer studied [73].

Our work on chlorofluorocarbons also led to the development of a novel general synthesis of 1,1,1-trihalopolyfluoroalkanes [74], namely heating a perfluoroalkyl iodide with an excess of anhydrous aluminum chloride or bromide, *e.g.* AlX_3 (X = Cl or Br) + $R_FCF_2I \rightarrow R_FCX_3$ ($R_F = n\text{-}C_7F_{15}$). The $AlCl_3$ reaction gives better yields when conducted in the absence of solvent in a stainless steel pressure vessel, but both reactions are exothermic, hence caution is advised. Other reactions used here to generate chlorofluoro model compounds include PCl_5-chlorination of perfluoroketones (*e.g.* $n\text{-}C_6F_{13}COCF_3 + PCl_5 \rightarrow n\text{-}C_6F_{13}CCl_2CF_3$), acids and acid derivatives (*e.g.* $n\text{-}C_7F_{15}CO_2H + PCl_5 \rightarrow n\text{-}C_7F_{15}CCl_3$) [75], and radical-induced addition of CCl_4 to terminal fluoro-olefins (*e.g.* n-$C_5F_{11}CF{=}CF_2 + CCl_4 +$ benzoyl peroxide $\rightarrow$ n-$C_5F_{11}CFCl$-CF_2CCl_3) [76].

Acknowledgements

In the course of my career in fluorine chemistry, I have had the great privilege to associate with some of the world-class experts in the field. If I have contributed anything to fluorine chemistry, it is largely due to this association, which initiated and sustained my interest in fluoro-organic compounds – a unique class of materials – and kept me fascinated by their chemistry. I am indeed most grateful for my association with Professors R. N. Haszeldine, FRS, and R. E. Banks, both now retired from UMIST (Manchester, UK), and Dr C. Tamborski (retired), a civilian scientist of the US Air Force. They provided me with both guidance and inspiration. I have enjoyed working with (among others) Dr K. J. Eisentraut (retired), Dr W. E. Ward, Dr H. L. Paige, Mr C. E. Snyder, Jr. and Ms L. J. Gschwender, all of the US Air Force, and with my colleagues at the University of Dayton Research Institute, particularly Dr K. A. Davis, Dr L. S. Chen (retired), and Mrs G. J. Chen (retired). I thank them all for their help and support during the course of my work.

References

1 S. Dixon, D. R. Rexford and J. S. Rugg, *Ind. Eng. Chem., 49* (1957) 1687.
2 *US Patent 3,039,992* (June 1962).
3 W. D. Morton, PhD Thesis, University of Manchester, UK, 1967.
4 P. Mitra (UMIST), private communication.
5 A. L. Jones, PhD Thesis, University of Manchester, UK, 1969.
6 R. N. Griffin and H. H. Gibbs, *US Patent 3,041,317* (1962).
7 R. E. Banks, G. M. Haslam, R. N. Haszeldine and A. Peppin, *J. Chem. Soc. (C)*, (1966) 1171.
8 W. T. Miller, A. L. Dittman and S. K. Reed, *US Patent 2,580,358* (1952).
9 T. Myerscough, PhD Thesis, University of Manchester, UK, 1970.
10 K. C. Eapen, *Current Science, 43* (1974) 179.
11 W. D. Blackley and R. R. Reinhard, *J. Am. Chem. Soc., 87* (1965) 802.
12 S. P. Makarov, A. Ya. Yakubovich, S. S. Dubov and A. N. Medvedev, *Doklady Chem., 160* (1965) 195.
13 R. N. Haszeldine and B. J. H. Mattinson, *J. Chem. Soc.*, (1957) 1741.
14 W. D. Blackley, *J. Am. Chem. Soc., 88* (1966) 480.
15 W. B. Fox *et al.*, *J. Am. Chem. Soc., 88* (1966) 2604.
16 W. D. Blackley, *US Patent 3,200,158* (1965).
17 R. E. Banks, K. Mullen and G. E. Williamson, *J. Chem. Soc. (C),* (1968) 2608.
18 G. F. Smith, PhD Thesis, University of Manchester, UK, 1970; R. E. Banks, A. J. Parker, M. J. Sharp and G. F. Smith, *J. Chem. Soc., Perkin Trans. 1*, (1973) 5.
19 A. R. Forrester, T. M. Hay and R. H. Thomson, *Organic Chemistry of Stable Free Radicals*, Academic Press, London and New York, 1986, p. 180.
20 A. Ya. Yakubovich, P. O. Gitel', Z. N. Lagutina and F. N. Chelobov, *Zhur. Obshchei Khim., 36* (1966) 163.
21 V. A. Ginsburg, L. L. Martynova, M. F. Lebedeva, S. S. Dubov, A. N. Medvedev and B. I. Tetel'baum, *Zhur. Obshchei Khim., 37* (1967) 1073.
22 R. E. Banks, K. C. Eapen, R. N. Haszeldine, P. Mitra, T. Myerscough and S. Smith, *J. Chem. Soc., Chem. Comm.,* (1972) 833.
23 R. E. Banks, K. C. Eapen, R. N. Haszeldine, A. V. Holt, T. Myerscough and S. Smith, *J. Chem Soc. Perkin Trans. 1,* (1974) 2532.

24 R. E. Banks and M. G. Barlow, 'Fluorocarbon and Related Chemistry', A Specialist Periodical Report, Chemical Society, London, Vol. 2, p. 233 (1974); Vol. 3, p. 235 (1976).
25 E. J. Soloski, G. J. Moore and C. Tamborski, *J. Fluorine Chem., 8* (1976) 295.
26 F. N. Jones and R. D. Richardson, *US Patent 3,666,769* (1972).
27 K. C. Eapen and C. Tamborski, *J. Fluorine Chem., 12* (1978) 271.
28 K. C. Eapen and L. S. Chen, *U.S. Statutory Invention Registration* H1537 (1996).
29 B. C. Bishop, A. S. Jones and J. C. Tatlow, *J. Chem. Soc.*, (1964) 3076.
30 Fisons Pest Control Ltd., *Belgian Patent 659,384* (1965) [C.A., *63* (1965) 18101h].
31 D. W. Wiley, *U.S.Patent 3,091,643* (1963) [C.A., *59* (1963) 11266e].
32 L. S. Chen and C. Tamborski, *J. Fluorine Chem., 19* (1981/82) 43.
33 K. C. Eapen and C. Tamborski, *J. Fluorine Chem., 18* (1981) 243.
34 L. S. Chen and K. C. Eapen, *J. Fluorine Chem., 49* (1990) 197.
35 C. Tamborski, U. D. G. Prabhu and K. C. Eapen, *J. Fluorine Chem., 28* (1985) 139.
36 U. D. G. Prabhu, K. C. Eapen and C. Tamborski, *J. Org. Chem., 49* (1984) 2792.
37 V. C. R. McLoughlin and J. Thrower, *Tetrahedron, 25* (1969) 5921.
38 G. J. Chen and C. Tamborski, *J. Fluorine Chem., 43* (1989) 207.
39 G. J. Chen, L. S. Chen and K. C. Eapen, *J. Fluorine Chem., 63* (1993) 113.
40 G. J. Chen, L. S. Chen and K. C. Eapen, *J. Fluorine Chem., 65* (1993) 59.
41 G. J. Chen and L. S. Chen, *J. Fluorine Chem., 73* (1995) 113.
42 K. K. Sun, C. Tamborski and K. C. Eapen, *J. Fluorine Chem., 17* (1981) 457.
43 L. S. Chen, G. J. Chen and C. Tamborski, *J. Fluorine Chem., 26* (1984) 341.
44 K. J. L. Paciorek, S. R. Masuda, J. G. Shih and J. H. Nakahara, *J. Fluorine Chem., 53* (1991) 233.
45 G. J. Chen and L. S. Chen, *J. Fluorine Chem., 89* (1998) 217.
46 C. E. Snyder, Jr., L. J. Gschwender, K. C. Eapen and G. J. Chen, *U.S. Patent 5,316,686* (1994).
47 J. H. Simons and E. O. Ramler, *J. Am. Chem. Soc., 65* (1943) 389.
48 J. H. Simons, W. T. Black and R. F. Clark, *J. Am. Chem. Soc., 75* (1953) 5621.
49 S. A. Anichkina, V. A. Barkhash and N. N. Vorozhtsov, Jr., *Zhur. Obshchei Khim., 38* (1968) 2493.
50 G. V. D. Tiers, *J. Am. Chem. Soc., 77* (1955) 4837, 6703 and 6704; *J. Org. Chem., 28* (1963) 1403.
51 K. C. Eapen, C. Tamborski and T. Psarras, *J. Fluorine Chem., 14* (1979) 243.
52 W. A. Sheppard and C. M. Sharts, *Organic Fluorine Chemistry*, W. A. Benjamin Inc., New York, 1969.
53 M. Hudlicky, *Chemistry of Organic Fluorine Compounds*, Horwood, Chichester (UK), 1976.
54 L. S. Chen, G. J. Chen and C. Tamborski, *J. Fluorine Chem., 18* (1981) 117.
55 L. S. Chen and C. Tamborski, *J. Fluorine Chem., 19* (1981/82) 43.
56 L. S. Chen and C. Tamborski, *J. Fluorine Chem., 26* (1984) 269.
57 L. S. Chen, G. J. Chen and C. Tamborski, *J. Fluorine Chem., 26* (1984) 341.
58 L. S. Chen, K. J. Eisentraut, C. S. Saba, M. T. Ryan and C. Tamborski, *J. Fluorine Chem., 30* (1986) 385.
59 L. S. Chen, A. V. Fratini and C. Tamborski, *J. Fluorine Chem., 31* (1986) 381.
60 L. S. Chen and K. C. Eapen, *J. Fluorine Chem., 55* (1991) 93.
61 G. J. Chen and L. S. Chen, *J. Fluorine Chem., 55* (1991) 119.
62 G. J. Chen and L. S. Chen, *J. Fluorine Chem., 59* (1992) 113.
63 S. P. Kotun, J. D. O. Anderson and D. D. DesMarteau, *J. Org. Chem., 57* (1992) 1124.
64 G. J. Chen, L. S. Chen, K. C. Eapen and W. E. Ward, *J. Fluorine Chem., 69* (1994) 61.
65 K. C. Eapen, L. S. Chen and G. J. Chen, *J. Fluorine Chem., 81* (1997) 143.
66 P. L. Coe, S. Sellers, J. C. Tatlow, H. C. Fielding and G. Whittaker, *J. Fluorine Chem., 18* (1981) 417.
67 R. E. Banks and J. C. Tatlow, *J. Fluorine Chem., 33* (1986) 227.
68 V. Tortelli, C. Tonelli and C. Corvaja, *J. Fluorine Chem., 60* (1993) 165.
69 K. C. Eapen, P. J. John and J. C. Liang, *Macromol. Chem. Phys., 195* (1994) 2887.
70 K. C. Eapen and C. Tamborski, *J. Fluorine Chem., 35* (1987) 421.
71 R. E. Banks, private communications (1987).
72 K. C. Eapen and C. Tamborski, *J. Fluorine Chem., 41* (1988) 443.
73 K. C. Eapen, *J. Fluorine Chem., 48* (1990) 17.
74 K. C. Eapen, K. J. Eisentraut, M. T. Ryan and C. Tamborski, *J. Fluorine Chem., 31* (1986) 405.
75 L. S. Chen and G. J. Chen, *J. Fluorine Chem., 42* (1989) 371.
76 L. S. Chen, *J. Fluorine Chem., 47* (1990) 261.

Chapter 15

FLUORINE CHEMISTRY IN ITALY

GIAMPAOLO P. GAMBARETTO[1]

Department of Chemical Process Engineering, University of Padua Via Marzolo 9, 35131 Padua (Italy)

Getting into fluorine chemistry

I first came into contact with fluorine chemistry back in 1957 after graduating in Industrial Chemistry from the University of Padua. Out of the various job offers I received, I chose a relatively new company called ICPM (Industrie Chimiche di Porto Marghera) which was developing alongside the Sicedison plant (subsequently Montedison) at Porto Marghera. This company had begun production of anhydrous HF and AlF_3 and was at the time planning the construction of a plant for manufacturing CFCs via a liquid-phase process. Fluorine chemistry immediately attracted me and was destined to become a life-long fascination. As I was the first chemist in the newly established research laboratory I was left free to decide on the projects to follow, an ideal situation for a young researcher.

During the eight years (1957-1965) I spent at ICPM (subsequently taken over in 1964 by Sicedison, which became Montedison in 1966), processes were developed and implemented in pilot plants for the production of halothane, benzotrifluoride and its derivatives, cryolite, vinyl fluoride and vinylidene fluoride, and Halon fire-fighting agents. Also, the company's first electrolytic cell for the production of elemental fluorine was installed and an experimental reactor set up for the production of sulfur hexafluoride. Halfway (1961) through this period I spent 6 months in Professor Colin Tatlow's research group at the University of Birmingham (UK) where I worked with Drs Colin Patrick and Jim Burdon on fluorination with CoF_3 and analytical problems involving HF.

Moving on to Padua

In 1965 I left industry and began my university career at the Institute of Industrial Chemistry at the University of Padua, where I was appointed by Professor Andrea Scipioni to teach 'Technology of Organic Reactions', then a new subject in Italy. At that time Padua was the first Italian university to implement the study of fluorine chemistry with the help of specific financial contributions from the CNR (Consiglio Nazionale delle Ricerche, National Council for Scientific Research). The first research undertaken involved fluorination processes with HF in the gaseous phase and the production and use of F_2. From that time onwards, contacts with industry and other Italian and foreign Universities increased.

---

[1] Professor Gambaretto occupies the Chair of Industrial Chemistry in the University of Padua.

Apart from the increasingly close links with Montedison (where I began my professional career) in the persons of Professor Dario Sianesi and his assistants (Drs Gherardo Caporiccio, Franco Gozzo, Martino Vecchio and others), new links were established with the other two companies working in the field of fluorine chemistry in Italy: Rimar (now Miteni), for electrofluorination processes, and Finchimica, for fluoro-aromatic products (trifluoralin and the like). Alongside these collaborations, an agreement was signed in 1969 between the Universities of Padua and Nice and the Polytechnic of Munster concerning collaboration and exchange of information in the field of fluorine chemistry. This agreement, which provided for two meetings per year between the researchers of the three institutes, including representatives of some industries by invitation only, led to a profitable collaboration lasting roughly ten years.

The photograph below (Plate 15.1) was taken on the day the agreement was signed.

Plate 15.1. Viewed left to right: Dr Eric Klauke, Prof. Giampaolo Gambaretto, Prof. Aimé Cambon, Prof. Claus Bliefert and Dr Guenter Siegemund.

The Italian fluorochemicals industry

Parallel to this surge of interest in fluorine chemistry at the University of Padua, which unfortunately had limited financial means, considerable expansion in research took place at Montedison through the opening of the company's two new research centres at Bollate and Linate at the beginning of the sixties. It was at these two centres that fluorine chemistry developed to the greatest extent, partly supported by work at the Universities of Padua and Milan.

Fluorine chemistry had begun in Italy at industrial level at the beginning of the fifties at Montecatini (a company which became Montedison in 1966), which bought PTFE tech-

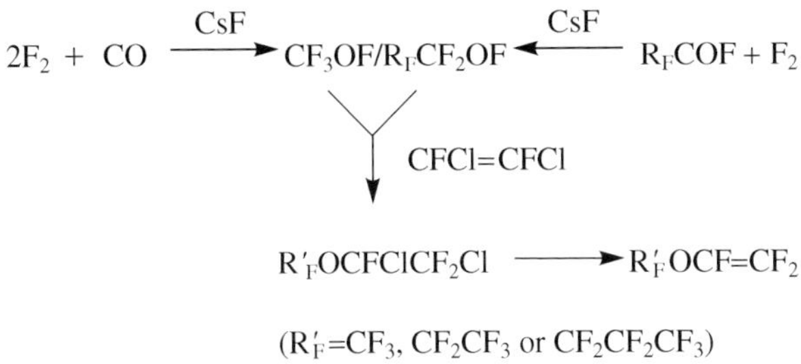

Scheme 15.1. Routes to perfluoro(alkyl vinyl ethers).

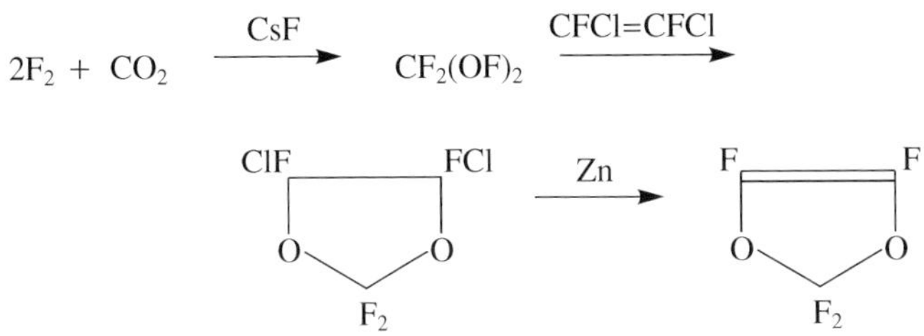

Scheme 15.2.

nology from the USA and applied it in its plants at Spinetta Marengo. Plant for the production of HF, cryolite and AlF_3 was built at Porto Marghera in 1957, followed in 1961 by the first plant for the production of CFCs (via a liquid-phase process, vapour-phase technology being implemented in 1966). In the meantime, proprietary technology for the production of F_2 was conceived at the Linate research centre, leading to important developments. The most spectacular and innovative of these was undoubtedly the low-temperature UV-assisted photopolymerization process for the synthesis of Fomblin® Y and Z perfluoropolyether fluids from perfluorinated olefins and oxygen. At that time the first Tecnoflon® proprietary fluoroelastomers were also developed, and the range of Algoflon® PTFE products was enlarged. In addition, industrial production of Esaflon® (SF_6) began.

After a period of relative stagnation in the seventies at Montedison, the eighties were marked by a huge commitment of resources to research and technological development. In the second half of the eighties, a new proprietary process based on fluorine and hypofluorite chemistry was estabilished for the production of sophisticated monomers such as the perfluoro(alkyl vinyl ethers) (Scheme 15.1). Thanks to the use of these monomers, new classes of fluorinated materials such as the PFA and MFA Hyflon® thermoprocessable perfluorinated polymers were developed, and the range of Tecnoflon® fluoroelastomers was enlarged with the introduction of peroxidically graftable points and improved properties at low temperatures.

The nineties have seen further significant developments in the chemistry of hypofluorites through work on a new and highly competitive process for the synthesis of polyfluorinated 1,3-dioxolanes (Scheme 15.2).

Monomers of this type represent the key to the latest generation of perfluorinated materials, amorphous perfluorinated polymers with a high glass transition temperature and high modulus. There are extremely interesting prospects for the use of these materials in advanced sectors such as the manufacture of polymeric optical fibres with low attenuation factors able to transmit data at speeds in excess of 3 Gbit/sec.

During the eighties Scientific co-operation between Ausimont and the University of Padua became stronger, and it is a pleasure for me to acknowledge the company's assistance with the organisation of the 7th and 10th European Symposia on Fluorine Chemistry, held in Venice in September 1980 and in Padua in September 1992, respectively.

Looking at the Italian fluorochemicals industry

Currently, the production plants operated by Ausimont, which is a sub-holding company of Montedison, are located at Porto Marghera and Spineta Marengo. At Porto Marghera, HF, the fundamental starting point for all organofluorine compounds, is produced by treating CaF_2 with sulfuric acid and used to manufacture chlorofluorocarbon (CFC) substitutes (HFCs and HCFCs) under the trade name Meforex® (M*). In 1994 a new plant was opened capable of producing derivatives of both the 120 [M*123 (CF_3CHCl_2); M*124(CF_3CHFCl); M*125(CF_3CHF_2)] and 130 series [M*134A(CF_3CH_2F)]. Spineta Marengo is Ausimont's most important fluorochemicals production site, manufacturing HCFC-22 (CHF_2Cl), SF_6 (using F_2), a range of fluorinated monomers (traditional and special), the perfluoropolyethers Fomblin® and Galden®, Algoflon® PTFE, Tecnoflon® fluorinated elastomers, and the thermoprocessable fluorinated copolymers Hyflon® PFA and Hyflon® MFA. Ausimont is not the only company associated with the field of fluorine chemistry in Italy. At Manerbio, in the province of Bergamo, Finchimica, produces fluoroaromatic intermediates and finished products, such as trifluralin and other weedkillers. This company, established in 1976, has a highly developed research/technology base and has reached the top of its sector in recent years, acquiring considerable market shares in the USA, Canada and Europe. At present it is the world leader for dinitroanilines derived from benzotrifluoride. Another Italian company of note in the fluorine field is Miteni. Established in the first half of the sixties under the name RIMAR, and producing oil-and-water repellent additives for fabrics, it specialised in electrofluorination routes to perfluorocarboxylic and perfluorosulfonic acids and perfluorinated amines. During the 70s, it became involved in the manufacture of benzotrifluoride derivatives and fluoroaromatic products as intermediates for fine chemicals. Production and also the number of plants expanded as the company increased its profile in the advanced intermediates sector associated with agrochemicals and pharmaceuticals (trifluralin and the antibacterial flucloxacillin being two of the most important), and in sectors associated with surfactants and perfluorinated fluids for industrial use. In consecutive phases during the 80s, aimed at increasing the company's presence in the world market, activities were transferred to two big international industrial holdings: MITsubishi Corporation and ENIchem Synthesis, hence the new company name MITENI, which today represents the continuation of RIMAR. In April 1996, Mitsubishi purchased the EniChem shareholding, thus acquiring full ownership of Miteni. Miteni products are widespread globally, and it sells over 90% of its products outside Italy (Europe, the Americas, Africa and Asia).

Finally, in the industrial sector, the big collection of companies involved in the fabrication of fluorinated polymers and elastomers and located in the valley between Bergamo and Brescia should not be forgotten. Approximately twenty companies operate in this area, the most important being Gapi, Guarniflon, Fluorseal, PMG and Fluorten; with an overall turnover of 300 billion lire in 1996, this group of small and medium-sized companies strongly geared to export produces O-rings, cables and other fluorinated rubber and plastic products to such high standards that they have attracted the attention of American companies. In the United States they call the location of these young yet already experienced enterprises (representing 60% of the entire Italian production in this business sector) 'Fluorine Valley' – perhaps even with a touch of envy. 'Fluorine valley' developed thanks to the spread, from the 50s onwards, of technology associated with Ausimont's Algoflon® and then Tecnoflon®. Today, 'Fluorine Valley' is characterised by companies with advanced transformation technologies competitive at world level.

Research & development

As regards current Italian research in fluorine chemistry, in the industrial sector Ausimont's basic research is concentrated at the Bollate laboratories because the Linate centre was closed between 1988 and 1990. Other research groups in specific sectors work at the two production plants at Spineta Marengo and Porto Marghera. Massimo Malavasi is the head of the research and development sector for the whole group. In addition to research into new products and technologies, the areas of greatest interest are: polymers (PTFE, Algoflon, Tecnoflon, Holar, Hyflon PFA and Hyflon MSA); fluorinated fluids (Fomblin and Galden); gaseous compounds (Meforex and Esaflon).

The Miteni plant at Trissino develops special technologies associated with electrofluorination and the amination and functionalization of fluoroaromatic products. Finchimica at Manerbio concentrates its studies on benzotrifluoride and its derivatives and fluorination processes involving HF. In particular, the aim is to promote synthetic methodology associated with plant protection products of the dinitroaniline family. Extremely important results have been obtained which impact process technologies and know-how for photochlorination, fluorination, nitration, hydrogenation and amination. Among the various fluoropolymer industries in the 'Fluorine Valley', considerable process technology research is being directed towards bringing increasingly sophisticated products to the marketplace.

Fluorine chemistry research at Italian universities is fairly limited at present owing to the lack of funds from the Ministry of Scientific Research and the CNR. Approximately 50-60% of the funds used in this sector come directly from industry and from contracts with the European Community. Almost all the research is carried out in the Department of Chemical Processes of Engineering (formerly the Institute of Industrial Chemistry) of the University of Padua and in the Department of Chemistry of the Polytechnic of Milan. I work with Professors Massimo Napoli and Lino Conte at Padua on electrofluorination of organic substances, direct fluorination methodology, the synthesis and characterisation of perfluorinated vinyl ethers, synthesis and physico-chemical properties of perfluorinated and semi-fluorinated (mixed R_F-R_H chains) compounds, and the reactivity of fluoroaromatic intermediates. Professors Pier Francesco Bravo and Giuseppe Resnati concentrate on research into the synthesis of fluorinated molecules of biological interest and fluorinated oxidizing agents.

As regards the near future of fluorine chemistry in Italy (and also abroad as far as I know), the switch of interest to fluorinated products that can be applied by means of 'soft' (low environmental impact) technologies and to fluorinated products designed to offer maximum performance as auxiliaries represents an opportunity for the scientific world and industrial R&D. Full exploitation of the idea of using fluorine only where and when strictly needed requires materials design abilities that presuppose scientific and technical mastery of how the fluorinated chain should be positioned and how it should interact in the final configuration, in addition to how it should be modified during processing to achieve the final configuration.

Soft application technologies could be widely implemented in the demand currently expressed across a wide range of industries. Super-super technical performance and chemical resistance is not always the only key to success. In particular, for technologies that presuppose application in a field such as high-performance anti-corrosion and cosmetic coatings, or where combination with technopolymers (nobilitation) offers top competitiveness, soft application technologies are the key to increasing these opportunities. The approach based on criteria that place fluorine on the surface in the required form or by surface treatments or via coating materials or by reactions with functionalised substrates will be successful. Fluorinated materials are already used to a limited extent, as components for surface treatments. Water-and-oil-repellent, stainproof, protective, drying and spreading treatments for application on many different surfaces (textiles, leather, paper, hard surfaces, plastics etc.) can, with the correct approach, employ fluorinated materials on a large scale in the future. Functionalisation of the fluorinated chains can open up an entirely new world of opportunities in the next few years provided that the big companies in fluorine chemistry and the researchers involved recognise that they have a rôle to play in this field.

Conclusions

As a whole the future of fluorine chemistry still offers many technological and scientific opportunities. As far as technology is concerned, the major players seem to have clearly identified their own routes. Specifically, in Italy Ausimont technology appears to be deeply routed in free-radical chemistry, as seen in the unique UV-assisted photocopolymerization process between fluoro-olefins and oxygen to give PFPEs (PerFluoroPolyEthers). Further development of this chemistry provided the hypofluorite route to perfluoro(alkyl vinyl ethers), and new sophisticated monomers like perfluorodioxoles and sulfonylated perfluorovinyl ethers. In this connection, there exists a clear diversification from the HFPO chemistry invented and brought to an industrial level by DuPont in the sixties. Accordingly, each industry seems to have clearly identified its own technology. The overall picture that emerges is less unitary than in the past, offering certainly more opportunities but, at the same time, involving greater risks.

Chapter 16

FLUORINE CHEMISTRY AT LEICESTER

JOHN H. HOLLOWAY and ERIC G. HOPE
Department of Chemistry, University of Leicester, University Road, Leicester LE1 7RH, UK

Background

Descriptions of the Manhattan Project – the name given to the United States Government's participation in the production of nuclear weapons towards the end of the Second World War – have generally celebrated the achievements of nuclear physics. The key role played by fluorine chemistry in the realization of nuclear weapons has received scant attention [1], and yet the separation of fissile U-235 on the requisite scale by gaseous diffusion of UF_6 was a key step, and its accomplishment an outstanding chemical and technical achievement [2, 3]. The British contribution to the war-time research programme which facilitated this was progressed in a number of laboratories including those of ICI General Chemicals. Significant work on the development of electrolytic fluorine generators at ICI was achieved by A. J. Rudge [4, 5] and the establishment of fluorine chemistry at Leicester can be traced back to the influence of this man. Between 1930 and 1933 he had been a research student with P. L. Robinson at the University of Durham's King's College, Newcastle upon Tyne. In September, 1948 he had been invited back to King's to talk about fluorine and Ray Peacock, who had graduated only that year, was appointed, as he put it, 'to be the dogsbody who did the experiments' [6]. These included demonstrating the spontaneous burning of asbestos by ClF_3 – a dangerous and frightening undertaking for a new graduate!

P. L. Robinson's own background was in Geology[1], but in the 1920s and 30s, he collaborated with H. V. A. Briscoe who had joined the department from Imperial College of Science and Technology in London. They started working on atomic weight determinations of boron to see whether this varied with source materials from different parts of the world. Later they carried out research on selenium, pseudo-halogens and rhenium. This was curtailed in 1936, but began again in 1946 after the Second World War. Briscoe and Robinson were among the first people in the UK to carry out preparative inorganic chemistry using vacuum lines; this technology was well established in the laboratory by the 1940s, and was helpful in developing the manipulation of fluorine. (Some of the inorganic techniques acquired or developed during this time were usefully described in an excellent book by Robinson and Dodd published in 1954 [7].)

Notwithstanding Ray Peacock's early and literal 'baptism of fire', he was clearly intrigued, and joined Robinson's research group. He was one of three students. The first

---

[1] P. L. Robinson was awarded a BSc degree at Armstrong College, Newcastle, in 1915 (R. D. Peacock believes this was in Geology). He was subsequently in receipt of an MSc from the same college in 1918 but the topic is unknown. His PhD thesis in 1926 was entitled 'A comparison of the Atomic Weights of Silicon from Different Sources', which was clearly chemical.

was involved in studies of the gas kinetics of N_2O; a second was working with pernitrous acid; and Ray commenced work on the chemistry of fluorine using a 10 ampere electrolytic cell for the production of elemental fluorine, loaned by ICI who were anxious to get some universities to carry out fluorine research [5].

Ray's first laboratory was housed in an elderly conservatory attached to an old laboratory. There was no fume cupboard of course, and unwanted fluorine was simply led out of the laboratory through a pipe passing through a hole Ray had drilled in the window frame. This was generally convenient for the occupants unless the wind was in the wrong direction! Work began with the preparation of ReF_6 and, quite soon, the new compounds $ReOF_5$ and ReO_2F_3 were also prepared. This gave him his first publication which was in the *Journal of the Chemical Society* [8]. The compounds prepared were characterized by chemical analysis and vapour density determination. There were no spectroscopic facilities of any kind but, in some later papers, X-ray powder methods were used to help with characterization. However, Ray *was* given considerable freedom to pursue his own interests and spent happy days picking chemicals off the shelves in the teaching laboratory and seeing how fluorine behaved towards them. Needless to say, more new compounds were discovered, including SeF_4 [9, 10]; and pure $SeOF_2$ [9] was prepared for the first time (this had first been made in an impure form by Prideaux and Cox in 1927 [11]).

Ray gained his PhD in 1951, Harry Emeléus (H. J. Emeléus) acting as his external examiner, and then spent a further two years at King's as a research assistant and one as a temporary lecturer before taking up an appointment at Imperial College, London. During this time, he carried out a significant amount of work on quadrivalent and quinquevalent fluororuthenates [12].

Ray Peacock's period in London was at an exciting time during which Jack Lewis (now Professor The Lord Lewis of Newnham) and the late Geoffrey Wilkinson (who later became Sir Geoffrey Wilkinson and Nobel Laureate for his work on sandwich compounds) established inorganic chemistry research laboratories at Imperial College, and Ron Nyholm (later Sir Ronald Nyholm) established himself at University College. It was also a fruitful period for Ray and papers on the trifluorides, MF_3 ($M = Ru, Ir, Rh$) [13], MoF_4 [14] and $[OsF_6]$ [15] along with more than a dozen papers and notes on complex metal fluorides, which are mostly summarized in a nice review he prepared for Professor F. Albert Cotton when he was editor of *Progress in Inorganic Chemistry* [16], were published. However, in 1958, he moved on to the University of Birmingham where he established an inorganic fluorine group alongside the already powerful organic fluorine team under Professors M. Stacey and J. C. Tatlow. Significant achievements during this period were the preparations of a number of novel binary fluorides, ReF_5, ReF_4 [17], OsF_5, OsF_4 [18], the confirmation of MoF_5 [19] and RuF_5 [20] and their characterization, the preparation of RuF_4 [21] and of the oxide fluorides $ReOF_4$ [17] and $RuOF_4$ [20].

Ray recognised the value of single-crystal X-ray work and set Tony Edwards (A. J. Edwards) a postgraduate student and, later, an ICI Post-doctoral Fellow, to learn the technique from the Birmingham crystallographer R. W. H. (Sam) Small. This was again productive, yielding the first structures of transition-metal pentafluorides (MoF_5 [19] and RuF_5 [22]) with their, until then, unknown fluorine-bridged bonds which have since emerged as an important component in the structures of many inorganic fluorine compounds. Other important achievements were the atmospheric-pressure synthesis of xenon tetrafluoride [23] and the preparation of the first xenon fluoride Lewis-acid pentafluoride complexes $XeF_2 \cdot 2SbF_5$ and $XeF_2 \cdot 2TaF_5$ [24] by reaction of what proved to be XeF_2/XeF_4 mixtures [25] with the appropriate pentafluoride. One of the postgraduate students contributing to this work was John Holloway who, after gaining his PhD in 1963, left to set up another inorganic fluorine group in the University of Aberdeen in Scotland. Around this time Ray was also beginning to look at the possibility of preparing mixed chloride fluorides of the transition metals. One of the last pieces of chemistry he accomplished in Birmingham with three of his postdoctoral people, Tony Edwards, Bernard Cohen and Mary Mercer[2], was the successful synthesis of WF_5Cl [26]. Soon afterwards, he was appointed to the Chair of Inorganic Chemistry at the University of Leicester and established a fluorine chemistry group here for the first time.

The Leicester Fluorine Group

Peacock leads the way

Noble-gas chemistry was continued by Ray himself on a visit to Argonne National Laboratory, where collaboration with Henry Selig and Irving Sheft led to the preparation and characterization of the $CsXeF_7$, $RbXeF_7$, Cs_2XeF_8 and Rb_2XeF_8 complex salts [27, 28], and with Henry Selig, produced the first evidence for the krypton fluoride complex, $KrF_2 \cdot 2SbF_5$ [29]. This field was also developed further in Leicester in the hands of a visiting Australian academic, Valda McRae. An important achievement was the preparation of crystals and the first structural determination of a xenon-difluoride complex, $XeF_2 \cdot 2SbF_5$ (or $[XeF]^+[Sb_2F_{11}]^-$) [30], carried out in collaboration with the crystallographer David Russell (D. R. Russell) in Leicester. David had studied for his PhD, along with Ray Kemmitt (R. D. W. Kemmitt), under Professor David Sharp (D. W. A. Sharp), initially at Imperial College, and subsequently at the University of Strathclyde in Glasgow, Scotland.

Professor Raymond D. Peacock

With Ray Kemmitt, also in Leicester, Ray Peacock began to try to make transition-metal carbonyl fluorides [31] and, again with the help of David Russell, solved the first crystal structure of such a species, $[Ru(CO)_3F_2]_4$ [32]. At the same time, interest in mixed chloride fluorides was developed with further work on tungsten chloride fluorides [33-35] and TeF_5Cl was also prepared and characterized [36]. Ray's move to Leicester brought him into contact with John Burgess whose formidable knowledge of the behaviour of ions in solution led to a long and productive joint research programme on measurement of the heats of hydrolysis of a wide range of fluorides and fluoride complexes. The first paper published was a measurement of the electron affinity of tungsten hexafluoride [37] and over the next 19 years up to Ray's retirement, around twenty papers providing information on both

[2] See Appendix 2, Chapter 30; Mary Mercer is now married to Professor David W. A. Sharp of the University of Glasgow.

electron affinities and fluoride-ion affinities, as well as details of the charge distributions in a range of hexafluorometallates were published, mostly in the *Journal of the Chemical Society* or the *Journal of Fluorine Chemistry*. The work also included collaborative studies with the research groups of Roland Bougon (France) and Karl Christe (K. O. Christe, USA) [38] and Ray's old friend Professor Rudolph Hoppe (Germany) [39]. Ray also engaged in a productive line of research involving reactions of MF_n species with Me_3SiX ($X = N_3$, NCO, NCS, CN) [40] and succeeded in the very difficult and hazardous synthesis of the fluoroazide $WF_5(N_3)$ with his exceedingly clever research technician John Fawcett, who not only prepared the compound but also determined the structure with David Russell [41] and the enthalpy of formation with John Burgess [42]. John also prepared and, with Ray, characterized, the nitrenes $ReF_5(NCl)$, $ReF_5(NF)$ [43] and $OsF_5(NCl)$ [44] and measured the enthalpy of formation of $ReF_5(NCl)$ [45].

Holloway comes to Leicester

In 1970, John Holloway was invited to apply for a lectureship in Leicester and to re-join Ray's group. When he arrived in January, 1971 he brought with him expertise on working with metal and Kel-F vacuum systems and carrying out fluorinations under pressure, and some of the special skills required to utilize mass spectrometry and vibrational spectroscopy for the characterization of highly reactive fluorides. These had been acquired largely under the tutelage of Henry Selig, John Malm (J. G. Malm), Howard Claassen (H. C. Claassen) and Larry Stein (L. Stein) at Argonne National Laboratory in the US, and with Jožef Slivnik at the Jožef Stefan Institute in Ljubljana (Slovenia) where John had spent every moment of leave he could muster whilst working in the University of Aberdeen. Thus, the interests and activities of the group broadened.

Professor John H. Holloway

Whilst at the University of Aberdeen, John had shown that pure XeF_2 could be prepared on a large scale by the photolysis of xenon/fluorine mixtures in Pyrex containers using sunlight [46, 47]. This was a rather slow process in the far northern latitude of Scotland at which Aberdeen is located[3], but this ready availability of good quality xenon difluoride enabled him to complete the characterization of XeF_2-metal pentafluoride adducts at first in Aberdeen [48, 49] and later in Leicester [50-53]. This showed that the solid-state structures can be interpreted in terms of ionic formulations involving $[Xe_2F_3]^+$ and $[XeF]^+$ cations and $[MF_6]^-$ and $[M_2F_{11}]^-$ anions but that the compounds contain weak interactions through fluorine bridging between the anions and cations. This work was accomplished with the help of Boris Frlec[4] who joined John from the Jožef Stefan Institute soon after he arrived in Leicester. John and Boris had become close friends while working in the group headed by

[3]Aberdeen is at lattitude 57 °10′ N, which is only a little further south than Stockholm in Sweden.

[4]After being in Leicester, Boris returned to become Assistant Director and Director of the Jožef Stefan Institute in Ljubljana. Subsequently he was Yugoslav Ambassador to Germany and Slovene Ambassador to Germany and is currently Foreign Minister for Slovenia.

Herb Hyman (H. H. Hyman) at Argonne National Laboratory in the late 1960s. The two of them were also highly successful in developing the related chemistry of krypton difluoride. They characterized the $[KrF]^+$ and $[Kr_2F_3]^+$ cations fully for the first time [54-56] and demonstrated the extraordinary oxidative fluorinating ability of KrF_2 by using it to prepare IF_7 and XeF_6 [54].

John was also fortunate soon afterwards in having Gary Schrobilgen join him as a post-doctoral fellow from Professor Ron Gillespie's (R. J. Gillespie) group in McMaster University in Canada[5], and so work on XeF_2 and KrF_2 complexes continued. John had commenced work with XeF_2 and stronger fluoride-ion bases such as WOF_4 just prior to Gary's arrival and Gary's considerable skill as a Raman and NMR spectroscopist moved the work on at a considerable pace. The complexes $XeF_2 \cdot WOF_4$ and $XeF_2 \cdot 2WOF_4$ were prepared [57-59] and it was shown that the ionic contribution to the bonding in both the solid state and in solution is reduced relative to that in the metal pentafluoride complexes, and that the structures are best formulated as covalent with fluorine bridges. Similar results were also obtained with $XeF_2 \cdot nMoOF_4$ ($n = 1$-4) species [59]. Extending the work to KrF_2 complexes revealed that, whereas the adducts with transition-metal oxide tetrafluorides could be obtained as crystalline solids with XeF_2, they can only be obtained in solution when KrF_2 is substituted [59]. This link with McMaster University, where Gary is now Professor of Chemistry, has been maintained and, more recently, John and Gary have shown that KrF_2 undergoes a phase transition at low temperature [60].

In Leicester, Gary and John also used the low-temperature oxidative fluorinating ability of KrF_2 to prepare (via its $[Kr_2F_3]^+$ and $[KrF]^+$ adducts) the pentafluoride of gold, AuF_5, for the first time [61]. This low-temperature fluorination required the most judicious manipulation of low-temperature baths to prevent the reaction getting out of hand, and on several occasions the FEP reactor tubes built up high pressures of krypton gas which resulted in them being detached from the valves and propelled across the laboratory like small rockets. On another occasion on a Saturday morning whilst they were trying to get a KrF_2/AuF_5 mixture (actually $KrF_2/[Kr_2F_3]^+[AuF_6]^-$ mixture) into solution in HF, the mixture incandesced in the tube and, as Gary threw it clear, it exploded leaving him (luckily) without fluorine burns but with a gold-plated hand! Efforts to prepare higher fluorides of chromium, osmium and neptunium were not conclusive.

Gary's early training in nuclear magnetic resonance spectroscopy and his subsequent work with Ron Gillespie provided the underpinning for the first comprehensive NMR study of a heavy nucleus, ^{129}Xe, in which Gary and John synthesised and contained some 25 compounds and flew them over to the laboratory of Pierre Granger in Nancy, France, where the spectra were run with the help of Christian Brevard of Brucker Spectrospin SA [62, 63]. The flights to Strasbourg were always interesting, involving making special arrangements with the airline and, usually, persuading the pilot that it was entirely reasonable and safe to carry samples of such things as XeF_6 and XeO_3 in Dewar vessels at dry ice temperature! Today, flying with such hazardous materials is simply not possible.

[5] Ron Gillespie and Gary Schrobilgen had independently also discovered the $[KrF]^+$ and $[Kr_2F_3]^+$ cations. Gary J. Schrobilgen is now Professor in McMaster University, Canada.

Hope arrives in Leicester

During this period, John had also commenced working on the fluorination of carbonyl fluorides which Ray Peacock had begun in Birmingham and pursued further in Leicester. Ray's successes are referred to earlier; what John brought to the problem were the techniques required to allow the reactions of noble-gas fluorides as fluorinating agents in solution to be carried out stoicheiometrically, the ability to monitor these spectrometrically and mass spectrometrically, and novel ways of growing single crystals. This quickly led to the characterization of mixed oxidation-state species such as $Re^{I}(CO)_5F \cdot Re^{V}F_5$ and $[Re(CO)_6]^+[Re_2F_{11}]^-$ [64, 65]. The presence of the differing oxidation states of the two metals, it was realised, permitted the powerfully oxidising ligand, fluoride, and the reducing ligand, CO, to co-exist in the same molecule. Further work with the ruthenium system also produced, once again, $[Ru(CO)_3F_2]_4$, which had been characterized earlier in Leicester [32] and put its preparation on a firm footing [66]. An understanding of the nature of these molecules via their X-ray structures [64, 67, 68] and vibrational, NMR and mass-spectral data [65-68] provided an excellent basis for predicting the nature of related compounds yet to be prepared [69]. However, progress slowed for a few years until developments in the NMR technique, and the arrival in 1987 of Eric Hope from the group of Bill Levason (W. Levason) in Southampton as an EPSRC Fellow with considerable NMR experience, permitted the group to revisit the field. Using mainly NMR methods, significant new advances have been made with a range of new iridium [70-72] and osmium and ruthenium compounds [72-78] being prepared. The work has demonstrated that immense control over the reactions is possible and has brought the attention of organometallic chemists to the value of fluoride as a ligand in organometallic chemistry. In addition, in 1997, Eric and John showed that further synthetic chemistry can be undertaken at the metal centre, the fluoride ligand, or at the other ligands in the complex (for a summary, see ref. [72]) which led to the preparation of the first air- and moisture-stable osmium and ruthenium fluorides [77] and the first triply-fluoride bridged late transition-metal dimers, *e.g.* $[Ru_2(\mu\text{-}F)_3(CO)_2(PPh_3)_4]^+$ [78]. The preparation of the moisture-stable $[MF_2(CO)_2L_2]$ complexes (L = phosphines, arsines or amines) [77] by the addition of the Lewis bases to $[MF_2(CO)_3]_4$ (M = Ru, Os) in CH_2Cl_2, acetone or THF, is of particular importance because the preparation of hydrolytically-stable species means that this type of chemistry is now within the reach of many inorganic chemists and is no longer the preserve of only the fluorine specialist.

Dr Eric G. Hope

Further spin-off from the carbonyl fluoride work was the discovery, in 1995, by Eric Hope that the fluorine ligand can be introduced into organometallic and coordination compounds by the reaction of anhydrous HF with methyl metal derivatives with elimination of methane [79]. More recently, it has been shown that a more widely applicable route is by reaction of metal hydrides, such as Ru^{II} or Os^{II} species, with the elimination of hydrogen [78]. However, reactions with five-coordinate Group 9 hydrides are not straightforward [80]. It is, of course, of considerable interest that the Leicester Fluorine Group, which at its origin was concerned with high-valent transition metal fluorides, oxide fluorides and their complexes, has pioneered and established the novel field of low-valent transition-metal fluorine chemistry and become significantly involved in the development of this area.

The use of xenon difluoride as a mild fluorinating agent in solution also gave rise to the preparation and characterization of a number of low-valent transition-metal species incorporating fluorine-containing ligands. The work began when John Holloway and Professor Evelyn Ebsworth (E. A. V. Ebsworth)[6] in Edinburgh University, Scotland, showed that xenon difluoride in solution reacts smoothly and in high yield with $[Ir(CO)Cl_2(PEt_3)_2(P'F_2)]$ to give $[Ir(CO)Cl_2(PEt_3)_2(P'F_4)]$, the first metal-$PF_4$ complex [81, 82]. Attempts to oxidatively fluorinate a number of other ligands attached to metal centres failed. However, realization that there were few established examples of transition-metal complexes containing small fluorinated ligands {known species included complexes containing carbon, $-CF_3$, $-C_2F_4$, phosphorus, $-PF_2$, $-PF_3$, $-PF_4$ [81, 82], and nitrogen, $-NF$ [43]}, coupled with the knowledge that insertion of fluorinated ligands at metal centres can impart dramatic changes to the chemistry at such centres, led them to make a variety of novel complexes with fluorine-containing ligands. Interesting syntheses in this area were carried out by two excellent postdoctoral workers, Russell Cockman (R. W. Cockman), who subsequently moved to BP Chemicals in Grangemouth, Scotland, and Paul Watson (P. G. Watson), now working with Professor Rüdiger Mews in Bremen, Germany, and the Edinburgh Group. These included the first syntheses of complexes containing $-NF_2$ [83], $-SF_3$, $-SOF$ [84], $-COF$ [85] and $-SeF_3$ [86] ligands. The last reaction was a particularly rare oxidative addition reaction of TeF_4 with *trans*-$[RhX(CO)(PEt_3)_2]$ to give unusual pentacoordinate monocationic complexes $[RhX(CO)(PEt_3)_2(TeF_3)]^+[TeF_5]^-$ [X = Cl, Br, NCS, NCO].

Whilst the primary interest of the Leicester group focused on metal-fluoride chemistry, in the late 1960s Ray Kemmitt's research in organometallic chemistry included work on fluorinated phosphine ligands such as perfluorotriphenylphosphine, $P(C_6F_5)_3$ [87]. As with the carbonyl fluoride project, interest waned in this area for a number of years until the arrival of Graham Saunders from the University of Oxford. Funded by BNFL Fluorochemicals Ltd. (now F2 Chemicals Ltd.), Graham's remit was to develop novel inorganic fluorine chemistry, and new work on the influence of fluorine in fluorinated ligands was initiated. Here, Graham observed the first examples of *cis*-puckered metallacyclic cores in Cp-Mo-imido dimers [88] and identified remarkable, quantitative, C−F bond activation reactions for his coordinated ligands [89] (see later) leading, most recently, to chiral-at-metal cationic complexes [90]. Building upon Graham's work in fluorinated ligand chemistry, two further post-doctoral appointments, Pravat Bhattacharyya from Imperial College and Alison Stuart from Paul Coe's group in Birmingham, have developed, with Eric Hope, the fluorous biphase approach to catalyst/product separation in homogeneous catalysis. This exploits the unique physical properties of perfluorocarbon solvents and requires the derivatization

[6] Evelyn A. V. Ebsworth was Professor of Chemistry at the University of Edinburgh, Scotland, and became Vice-Chancellor of Durham University, England, in 1990.

of metal-based catalysts with long perfluoroalkyl sidechains – ‘fluorous ponytails’. Alison’s approach has been a copper-mediated cross-coupling of an aryl iodide with a perfluoroalkyl iodide which has found such widespread application that the group is using about 1 kg of copper per month in the laboratory at the moment. The highlights of Alison’s work have been the first structural characterization of a perfluorocarbon-soluble metal complex (with David Russell) [91] and the first library of perfluoroalkyl-derivatised ligands [92]. Additionally, they have established the effectiveness of ‘spacer groups’ to minimise the electronic influence of the perfluoroalkyl substituents, carried out the first comprehensive evaluation of the coordination chemistry of the ligands and the influence of the perfluoroalkyl substituents on the properties of the complexes, and the criteria for preferential solubility of the complexes in perfluorocarbon solvents. Furthermore, their synthetic successes have led to collaborations with catalysis and synthesis groups in a number of universities and significant industrial interest. This has permitted the first direct comparison between reaction rates of a model catalytic system for fluorous and non-fluorous and non-fluorous ligands [93] and a patent application on the hydroformylation of long-chain and internal alkenes under fluorous biphase conditions with Professor David Cole-Hamilton (D. J. Cole-Hamilton) at St. Andrews University, Scotland [94]. However, in view of the costs associated with the perfluorocarbon solvents, further work is necessary to convince sceptical industrialists that this approach can be commercially viable. In August 1998 Eric was invited to present the Leicester work (alongside that of Horváth, Gladysz, Pozzi, Curran, Bergbreiter and DiMagno) at the first international symposium dedicated to Fluorous Biphase Systems at the 216th National American Chemical Society Meeting in Boston. In the following year, 1999, Alison started a prestigious Lloyd’s Tercentenary Fellowship at Leicester, investigating the extension of the fluorous approach to phase-transfer catalysis.

An adjunct to the fluorous biphase work has been the revelation that the ‘fluorous ponytail’ derivatisation of the metal catalyst renders them also soluble in supercritical CO_2. This has led Eric to collaborate with Dr J. Xiao at the University of Liverpool whilst, with Dr Andrew Abbott (A. P. Abbott) in Leicester, work on chemistry in supercritical hydrofluorocarbons [95] is in progress.

The extraordinary success of xenon difluoride as a mild fluorinating agent in opening up the field of transition-metal carbonyl fluoride chemistry was repeated again with the discovery of a second new class of compounds, the transition-metal chalcogenide fluorides. These were of interest because they are related both to transition-metal chalcogenide chlorides, bromides and iodides and to metal oxide-fluorides and fluorides. The synthetic challenge is that the chalcogenides themselves are so readily fluorinated that it is difficult to hold them bound to a metal in the presence of fluorine. The key was the preparation of WSF_4 from $WSCl_4$ by a postgraduate student Mal Atherton, now Dr M. J. Atherton working with BNFL, Preston, UK, who succeeded in fluorinating the $WSCl_4$ in solution with XeF_2 [96]. Eventually, it was shown that a whole class of compounds of this type could be made by interaction of a metal fluoride with the appropriate antimony chalcogenide (*e.g.* Sb_2S_3 or Sb_2Se_3). The compounds WSF_4 [97], $WSeF_4$ [98, 99], $MoSF_4$, $MoSeF_4$ [100], $ReSF_3$, $ReSF_4$ and $ReSF_5$ [97, 101] were all prepared. The first X-ray single crystal structure determinations on examples of this novel class of simple ternary compounds, WSF_4 and $ReSF_4$ [97, 102], which showed that they are closely related to the transition-metal fluorides, and the adduct $WSF_4 \cdot CH_3CN$ [102] were determined and the first electron diffraction studies on this type of compound, WSF_4 [103] and $WSeF_4$ [99] were completed in a collaboration with Professors David Rice (D. A. Rice) and Kenneth Hedberg at the Universities of Reading and Trondheim, Norway, respectively. The very difficult production of crystals and their study by X-ray techniques was accomplished by the Slovene chemist Dr Venčeslav Kaučič who now works at the National Institute of Chemistry and the University of Ljubljana, Slovenia, once again in collaboration with David Russell.

The several years around the appointment of Eric Hope (1987) saw immense change in the staffing of the Fluorine Group. Eric joined the Group as a Science and Engineering Research Council NATO Fellow. His arrival coincided with the promotion of John Holloway to a chair in inorganic chemistry and a rapid expansion of the Fluorine Group. In 1989, Eric was awarded an SERC Advanced Fellowship and followed this with a Royal Society University Research Fellowship in 1994. Ray Peacock, meantime, retired in 1991. Over this period, the skills base of the group broadened with the appointments of Wolfgang Dukat, from Dortmund University, and now with Höechst in Germany, Matthias Rieland, from Bochum University, now with Kalie-Chemie in Germany, Alan Brisdon, from Southampton University, now lecturing at the University of Manchester Institute of Science and Technology (UMIST), UK, and Paul Watson, Graham Saunders, Pravat Bhattacharyya and Alison Stuart, who have already been mentioned. This allowed considerable diversification, including work on the selective fluorination of organic compounds using metal fluorides and oxide fluorides [104-106] and spectroscopic investigation of fluorides. Spectroscopic aspects, which first interested John Holloway when working at Argonne National Laboratory, Chicago, in the late 1960s, have been developed by John, Gary Schrobilgen and Eric where NMR spectroscopy is concerned (see the noble gas and carbonyl fluoride work above), and by Eric, John, Alan Brisdon and the Southampton, UK, chemists Steve Ogden (J. S. Ogden) and Bill Levason (W. Levason) in the matrix isolation IR and uv-visible spectroscopic and EXAFS areas. In particular, manipulative methods for handling the more reactive fluorides have permitted matrix isolation spectroscopic studies of the hexafluorides [107, 108], while EXAFS studies on CrO_2F_2 and MnO_3F [109], hexafluorides and related fluoroanions [110] and osmium(VIII) oxide fluorides [111], have given valuable structural information not available by other means. Long, late-night vigils at the Daresbury Synchrotron radiation source, Cheshire, UK, punctuated only by visits to the 'Ring of Bells' (the nearest hostelry), were a feature of life for the Group throughout the 1990s.

The more recent work in Leicester on the preparation of novel organometallic compounds incorporating fluorinated ligands, carried out mostly by Graham Saunders, has given useful insights into C—F bond activation with concomitant C—C bond formation. The cleavage of the strong C—F bond is a challenge to the synthetic chemist and is receiving much attention at the present time and producing exciting results. Some fourteen papers have come out of the Leicester Group, the most interesting of which include the intriguing reaction between the metal complexes $[(\eta^5\text{-}C_5Me_4R)MCl(\mu\text{-}Cl)]_2$ (M = Rh, Ir; R = H, Me, Et) and dfppe $\{(C_6F_5)_2PCH_2CH_2P(C_6F_5)_2\}$ in ethanol or benzene. This involves the cleavage of two C—F bonds and two C—H bonds to give formation of two C—C bonds [89, 112]. The reaction exhibits complete regiospecificity and, in the rhodium case, remarkable regioselectivity. Similar reactions were also evident for iridium, but only in ethanol [90], showing that the solvent exerts a strong influence. As well as looking at the variation of the metal [112] and the solvent [90, 113] and the cyclopentadienyl ligand [90, 113] we have also reported on the effect of changing the halide ligand [114]. Graham took up a lectureship at Queen's University, Belfast in April 1996 and is continuing to develop C—F bond activation chemistry there.

In addition to the core chemistry carried out within the Fluorine Group at Leicester, there have been a number of very successful collaborations both within the UK and abroad. A long association with Boris Frlec and the late Darja Gantar from the Jožef Stefan Institute in Ljubljana, Slovenia, who were frequent visitors to Leicester, produced a wealth of metal difluoride adducts, described in some ten papers, including unusual polymeric cationic species such as $[(AgF)_n]^{n+}$ [115]. Extensive collaboration between Dr David Brown at the Atomic Energy Research Establishment, Harwell, UK, and with Dr Roland Bougon at the Centre D'Etudes Nucléaires de Saclay, France, has resulted in the preparation and study of the chemistry of actinide pentafluorides, the chemistry of uranium oxide fluorides, and the synthesis of new uranium chloride fluorides.

Neptunium pentafluoride was first prepared by John Holloway at Argonne National Laboratory in 1969 and was prepared again later in Harwell and shown to be isostructural with the high-temperature form of α-UF_5 [116]. Its chemistry was investigated and shown to differ somewhat from that of UF_5 and PaF_5 [116, 117]. In investigating these pentafluorides, considerable progress was made in achieving good synthetic routes and new preparative routes to other uranium pentahalides such as UBr_5 and UCl_5 [117-119]. Again, the success of the work was very much due to the patience and care of another very good postdoctoral worker, John Berry (J. A. Berry), who is now working at AERE. He held postdoctoral positions for several years to enable him to take long vacations to indulge his passion for photographing steam locomotives. Since the most interesting steam trains seem to exist in the former eastern block of Europe and South Africa, his appointment to the permanent staff in Harwell was somewhat delayed by a lengthy security clearance procedure!

The first evidence of Lewis-base character in UOF_4 and UO_2F_2 was provided by investigation of reactions of the oxide fluorides with a series of pentafluorides [120-124] and unusual, complex, ring and chain species have been characterized in the structures of $UOF_4 \cdot 2SbF_5$ [121] and $UO_2F_2 \cdot 3SbF_5$ [122]; the acetonitrile and triphenylphosphine oxide derivatives of these adducts and their UF_5 analogues have been shown to have monomeric structures [125].

Collaboration on the fluorination of fullerenes with the research group of Professor Harry Kroto, now Professor Sir Harry W. Kroto, Nobel Laureate, in the University of Sussex, UK, provided considerable excitement in Leicester. The possibility of being able to make a molecular-sized ballbearing, $C_{60}F_{60}$, was too intriguing to resist and John and Eric engaged with Roger Taylor in trying to understand and interpret the wealth of data that emerged. Perhaps the most exciting has been the initial identification of the first [60]- and [70]-fluorofullerenes [126], the generation of the chlorofluorofullerene $C_{60}Cl_{18}F_{14}$ [127] and the characterization of oxygenated derivatives from the reactions of the fluorofullerenes with water or aqueous methanol (*e.g.* [128-130]). These have yielded detailed information on the substitution and subsequent elimination processes that occur.

Finally, in collaboration with Professor Tony Legon (A. C. Legon) at Exeter University, UK, over a very short period (1995-1999) John has been involved in identifying for the first time a large series of Mulliken Inner and Outer complexes of fluorine and chlorine monofluoride with bases, including water and ammonia, which are snapshots of the first encounters between the two molecules *before* they react. Many of the mixtures studied explode violently under 'normal' conditions, but in the conditions of the experiment the chemical reagents are held together as weakly-bound bimolecular clusters in the low-temperature environment of a rapidly expanding gas jet. Out of an array of around thirty different complexes, the most interesting include the symmetric-top isotopomer of the $H_3N\cdots F_2$ adduct [131] and the planar $H_2O\cdots F_2$ adduct [132]. The excitement around this work was nicely summed up in an article by Michael Freeman in *Chemical and Engineering News* [133].

Into the future

The Leicester Fluorine Group has been fortunate in sustaining constant support from the British research councils and from a number of companies, including ICI Chemicals and Polymers, Johnson Matthey, BP Chemicals, various parts of British Nuclear Fuels, and F2 Chemicals Ltd. This has been much valued since it has not only provided challenging underpinning for much curiosity-driven research, but has provided chemical problems related to business needs and developments which have been exciting to tackle and have influenced the direction of research in the Group in interesting ways. The Group has also had the valuable input and enthusiasm of a number of gifted, enthusiastic and energetic postdoctoral workers, many whom are mentioned above, and a wealth of postgraduate students whom have worked energetically and with great good humour in a very collegiate atmosphere. We have all shared a fascination with fluorine which we hope will continue to sustain and strengthen the Group into the next millennium.

References

1 H. Goldwhite, 'The Manhattan Project', in R. E. Banks, D. W. A., Sharp and J. C. Tatlow (eds.), *Fluorine, the First Hundred Years (1886-1996)*, Elsevier Sequoia, Lausanne, 1986, Chap. 5, p. 109.
2 See the articles collected together in the March 1947 (Vol. 39, No. 3) issue of *Ind. Eng. Chem.*
3 C. Slesser and S.R. Schram (eds.), *Preparation, Properties and Technology of Fluorine and Organic Fluoro Compounds*, McGraw-Hill, New York, 1951.
4 For details of the ICI designs, see A. J. Rudge, in A. T. Kuhn (ed.), *Industrial Electrochemical Processes*, Elsevier, Amsterdam, 1971, p. 1 and A. J. Rudge and A. Davies, *Br. Pat. 861978* (1961) (to ICI).
5 A. J. Rudge, *The Manufacture and Use of Fluorine and Its Compounds*, Oxford Univ. Press, London, 1962.
6 R. D. Peacock, private communication.
7 R. E. Dodd and P. L. Robinson, *Experimental Inorganic Chemistry: a guide to laboratory practice*, Elsevier, Amsterdam, 1954.
8 E. E. Aynsley, R. D. Peacock and P. L. Robinson, *J. Chem. Soc.*, (1950) 1622.
9 E. E. Aynsley, R. D. Peacock and P. L. Robinson, *J. Chem. Soc.*, (1952) 1231.
10 R. D. Peacock, *J. Chem. Soc.*, (1953) 3617.
11 E. B. R. Prideaux and C. B. Cox, *J. Chem. Soc.*, (1927) 929.
12 M. A. Hepworth, R. D. Peacock and P. L. Robinson, *J. Chem. Soc.*, (1954) 1197.
13 M. A. Hepworth, K. H. Jack, R. D. Peacock and G. J. Westland, *Acta Crystallogr.*, (1957) 63.
14 R. D. Peacock, *Proc. Chem. Soc.*, (1957) 59.
15 G. B. Hargreaves and R. D. Peacock, *Proc. Chem. Soc.*, (1959) 85.
16 Much of this work is summarized in the review, R. D. Peacock, 'Fluorine Compounds of the Transition Metals', in F. A. Cotton (ed.), *Progress in Inorganic Chemistry*, Vol. II, Interscience, New York, 1960, p. 193.
17 G. B. Hargreaves and R. D. Peacock, *J. Chem. Soc.*, (1960) 1099.
18 G. B. Hargreaves and R. D. Peacock, *J. Chem. Soc.*, (1960) 2618.
19 A. J. Edwards, R. D. Peacock and R. W. H. Small, *J. Chem. Soc.*, (1962) 4486.
20 J. H. Holloway and R. D. Peacock, *J. Chem. Soc.*, (1963) 527.
21 J. H. Holloway and R. D. Peacock, *J. Chem. Soc.*, (1963) 3892.

22 J. H. Holloway, R. D. Peacock and R. W. H. Small, *J. Chem. Soc.*, (1964) 644.
23 J. H. Holloway and R. D. Peacock, *Proc. Chem. Soc.*, (1962) 389.
24 A. J. Edwards, J. H. Holloway and R. D. Peacock, *Proc. Chem. Soc.*, (1963) 275.
25 B. Cohen and R. D. Peacock, *J. Inorg. Nucl. Chem., 28* (1966) 3056.
26 B. Cohen, A. J. Edwards, M. Mercer and R. D. Peacock, *Chem. Commun.*, (1965) 14.
27 R. D. Peacock, H. Selig and I. Sheft, *Proc. Chem. Soc.*, (1964) 285.
28 R. D. Peacock, H. Selig and I. Sheft, *J. Inorg. Nucl. Chem., 28* (1966) 2561.
29 R. D. Peacock and H. Selig, *J. Am. Chem. Soc., 86* (1964) 3895.
30 V. M. McRae, R. D. Peacock and D. R. Russell, *Chem. Commun.*, (1969) 62.
31 R. D. W. Kemmitt, R. D. Peacock and I. L. Wilson, *Chem. Commun.*, (1968) 772.
32 R. D. Peacock, C. J. Marshall, D. R. Russell and I. L. Wilson, *Chem. Commun.*, (1970) 1643.
33 G. W. Fraser, M. Mercer and R. D. Peacock, *J. Chem. Soc.*, (1967) 1091.
34 D. M. Adams, G. W. Fraser, D. M. Morris and R. D. Peacock, *J. Chem. Soc.*, (1968) 1131.
35 G. W. Fraser, C. J. W. Gibbs and R. D. Peacock, *J. Chem. Soc.*, (1970) 1708.
36 G. W. Fraser, R. D. Peacock and P. M. Watkins, *Chem. Commun.*, (1968) 1257.
37 J. Burgess, I. Haigh and R. D. Peacock, *Chem. Commun.*, (1971) 977.
38 R. Bougon, T. Bui Huy, J. Burgess, K. O. Christe and R. D. Peacock, *J. Fluorine Chem., 19* (1982) 263.
39 J. Burgess, R. D. Peacock, R. Sherry, R. Hoppe and B. Müller, *J. Fluorine Chem., 16* (1980) 189.
40 These are mostly published in various issues of *J. Chem. Soc., Dalton Trans.*
41 J. Fawcett, R. D. Peacock and D. R. Russell, *J. Chem. Soc., Dalton Trans.*, (1980) 2294.
42 J. Burgess, J. Fawcett, R. D. Peacock and R. Sherry, *J. Fluorine Chem., 18* (1981) 173.
43 J. Fawcett, R. D. Peacock and D. R. Russell, *Chem. Commun.*, (1982) 958.
44 R. W. Cockman and R. D. Peacock, *J. Fluorine Chem., 30* (1986) 25.
45 J. Burgess, J. Fawcett and R. D. Peacock, *J. Fluorine Chem., 24* (1984) 341.
46 J. H. Holloway, *Chem. Comm., 22* (1966).
47 J. H. Holloway, *J. Chem. Educ., 43* (1966) 202.
48 J. H. Holloway and J. G. Knowles, *J. Chem. Soc. (A)*, (1969) 756.
49 J. C. Fuggle, D. A. Tong, D. W. A. Sharp, J. M. Winfield and J. H. Holloway, *J. Chem. Soc., Dalton Trans.*, (1974) 205.
50 J. Burgess, B. Frlec and J. H. Holloway, *J. Chem. Soc., Dalton Trans.*, (1974) 1740.
51 B. Frlec and J. H. Holloway, *J. Chem. Soc., Dalton Trans.*, (1975) 535.
52 B. Frlec and J. H. Holloway, *J. Inorg. Nucl. Chem., 38 (Supplement)* (1976) 167.
53 J. Fawcett, B. Frlec and J. H. Holloway, *J. Fluorine Chem., 8* (1976) 505.
54 B. Frlec and J. H. Holloway, *J. Chem. Soc., Chem. Commun.*, (1973) 370.
55 B. Frlec and J. H. Holloway, *J. Chem. Soc., Chem. Commun.*, (1974) 89.
56 B. Frlec and J. H. Holloway, *Inorg. Chem., 15* (1976) 1263.
57 J. H. Holloway, G. J. Schrobilgen and P. A. Taylor, *J. Chem. Soc., Chem. Commun.*, (1975) 40.
58 P. A. Tucker, P. A. Taylor, J. H. Holloway and D. R. Russell, *Acta Cryst., B31* (1975) 906.
59 J. H. Holloway and G. J. Schrobilgen, *Inorg. Chem., 19* (1980) 2632; J. H. Holloway and G. J. Schrobilgen, *Inorg. Chem., 20* (1981) 3363.
60 M. Al-Mukhtar, J. H. Holloway, E. G. Hope and G. J. Schrobilgen, *J. Chem. Soc., Dalton Trans.*, (1991) 2831.
61 J. H. Holloway and G. J. Schrobilgen, *J. Chem. Soc., Chem. Commun.*, (1975) 623.
62 J. H. Holloway, G. J. Schrobilgen, P. Granger and C. Brevard, *C.R. Acad. Sc. Paris, 282* (1976) C519.
63 G. J. Schrobilgen, J. H. Holloway, P. Granger and C. Brevard, *Inorg. Chem., 17* (1978) 980.
64 D. M. Bruce, J. H. Holloway and D. R. Russell, *J. Chem. Soc., Chem. Commun.*, (1973) 321.
65 D. M. Bruce, A. J. Hewitt, J. H. Holloway, R. D. Peacock and I. L. Wilson, *J. Chem. Soc., Dalton Trans.*, (1976) 2230.
66 A. J. Hewitt, J. H. Holloway, R. D. Peacock, J. B. Raynor and I. L. Wilson, *J. Chem. Soc., Dalton Trans.*, (1976) 579.
67 D. M. Bruce, J. H. Holloway and D. R. Russell, *J. Chem. Soc., Dalton Trans.*, (1978) 64.
68 D. M. Bruce, J. H. Holloway and D. R. Russell, *J. Chem. Soc., Dalton Trans.*, (1978) 1627.
69 D. M. Bruce and J. H. Holloway, Transition Metal Chem., 3 (1978) 217.
70 S. A. Brewer, J. H. Holloway, E. G. Hope and P. G. Watson, *J. Chem. Soc., Chem. Commun.*, (1992) 1577.
71 S. A. Brewer, A. K. Brisdon, J. H. Holloway, E. G. Hope, L. A. Peck and P. G. Watson, *J. Chem. Soc., Dalton Trans.*, (1995) 2945.

72 J. H. Holloway and E. G. Hope, *J. Fluorine Chem., 76* (1996) 209.
73 S. A. Brewer, J. H. Holloway and E. G. Hope, *J. Chem. Soc., Dalton Trans.*, (1994) 1067.
74 S. A. Brewer, J. H. Holloway and E. G. Hope, *J. Fluorine Chem., 70* (1995) 167.
75 S. A. Brewer, K. S. Coleman, J. Fawcett, J. H. Holloway, E. G. Hope, D. R. Russell and P. G. Watson, *J. Chem. Soc., Dalton Trans.*, (1995) 1073.
76 K. S. Coleman, J. H. Holloway and E. G. Hope, *J. Chem. Soc., Dalton Trans.*, (1997) 1713.
77 K. S. Coleman, J. Fawcett, J. H. Holloway, E. G. Hope and D. R. Russell, *J. Chem. Soc., Dalton Trans.*, (1997) 3557.
78 K. S. Coleman, J. H. Holloway, E. G. Hope and J. Langer, *J. Chem. Soc., Dalton Trans.*, (1997) 4555.
79 S. A. Brewer, L. A. Buggey, J. H. Holloway and E. G. Hope, *J. Chem. Soc., Dalton Trans.*, (1995) 2941.
80 H. C. S. Clark, K. S. Coleman, J. Fawcett, J. H. Holloway, E. G. Hope, J. Langer and I. M. Smith, *J. Fluorine Chem., 91* (1998) 207.
81 E. A. V. Ebsworth, J. H. Holloway, N. J. Pilkington and D. W. H. Rankin, *Angew. Chem., Int. Ed. Engl., 23* (1984) 630.
82 A. J. Blake, R. W. Cockman, E. A. V. Ebsworth, S. G. D. Henderson, J. H. Holloway, N. J. Pilkington and D. W. H. Rankin, *Phosphorus and Sulphur, 30* (1987) 143.
83 R. W. Cockman, E. A. V. Ebsworth and J. H. Holloway, *J. Chem. Soc., Chem. Commun.*, (1986) 1622.
84 R. W. Cockman, E. A. V. Ebsworth and J. H. Holloway, *J. Am. Chem. Soc., 109* (1987) 2194.
85 A. J. Blake, R. W. Cockman, E. A. V. Ebsworth and J. H. Holloway, *J. Chem. Soc., Chem. Commun.*, (1988) 529.
86 E. A. V. Ebsworth, J. H. Holloway and P. G. Watson, *J. Chem. Soc., Chem. Commun.*, (1991) 1443.
87 R. D. W. Kemmitt, D. I. Nichols and R. D. Peacock, *J. Chem. Soc. A*, (1968) 1898; 2149.
88 P. Bhattacharyya, J. Fawcett, J. H. Holloway, E. G. Hope and G. C. Saunders, *J. Chem. Soc., Dalton Trans.*, (1997) 199.
89 M. J. Atherton, J. Fawcett, J. H. Holloway, E. G. Hope, A. Karaçar, D. R. Russell and G. C. Saunders, *J. Chem. Soc., Chem. Commun.*, (1995) 191.
90 M. J. Atherton, J. H. Holloway, E. G. Hope and G. C. Saunders, *J. Organomet. Chem., 558* (1998) 209.
91 J. Fawcett, E. G. Hope, R. D. W. Kemmitt, D. R. Paige, D. R. Russell, A. M. Stuart, D. J. Cole-Hamilton and M. J. Payne, *J. Chem. Soc., Chem. Commun.*, (1997) 1127.
92 P. Bhattacharyya, D. Gudmunsen, E. G. Hope, R. D. W. Kemmitt, D. R. Paige and A. M. Stuart, *J. Chem. Soc., Perkin Trans. 1*, (1997) 3609.
93 E. G. Hope, R. D. W. Kemmitt, D. R. Paige and A. M. Stuart, *J. Fluorine Chem., 99* (1999) 197.
94 G. P. Schwarz, D. R. Paige, D. F. Foster, E. G. Hope and D. J. Cole-Hamilton, *UK Pat. Appl.*, (1998) 26619.
95 A. P. Abbott, C. A. Eardley, J. C. Harper and E. G. Hope, *J. Electroanal. Chem., 457* (1998) 1.
96 M. J. Atherton and J. H. Holloway, *J. Chem. Soc., Chem. Commun.*, (1977) 424.
97 V. Kaučič, J. H. Holloway and D. R. Russell, *J. Fluorine Chem., 23* (1983) 428.
98 M. J. Atherton and J. H. Holloway, *Inorg. Nucl. Chem. Letters, 14* (1978) 121.
99 K. Hagen, D. A. Rice, J. H. Holloway and V. Kaučič, *J. Chem. Soc., Dalton Trans.*, (1986) 1821.
100 J. H. Holloway and D. C. Puddick, *Inorg. Nucl. Chem. Letters, 15* (1979) 85.
101 J. H. Holloway, D. C. Puddick and G. M. Staunton, *Inorg. Chim. Acta, 64* (1982) L209.
102 J. H. Holloway, V. Kaučič and D. R. Russell, *J. Chem. Soc., Chem. Commun.*, (1983) 1079.
103 D. A. Rice, K. Hagen, L. Hedberg, K. Hedberg, G. M. Staunton and J. H. Holloway, *Inorg. Chem., 23* (1984) 1826.
104 W. W. Dukat, J. H. Holloway, E. G. Hope, M. Rieland, P. J. Townson and R. L. Powell, *European Patent, 92301531.7* (1992).
105 J. H. Holloway, E. G. Hope, P. J. Townson and R. L. Powell, *European Patent, 92301532.5* (1992).
106 W. W. Dukat, J. H. Holloway, E. G. Hope, M. Rieland, P. J. Townson and R. L. Powell, *J. Fluorine Chem., 54* (1991) 222.
107 J. H. Holloway, G. Stanger, E. G. Hope, W. Levason and J. S. Ogden, *J. Chem. Soc., Dalton Trans.*, (1988) 1341.
108 A. K. Brisdon, P. J. Jones, W. Levason, J. S. Ogden, J. H. Holloway, E. G. Hope and G. Stanger, *J. Chem. Soc., Dalton Trans.*, (1990) 715.
109 W. Levason, J. S. Ogden, A. K. Saad, N. A. Young, A. K. Brisdon, P. J. Holliman and E. G. Hope, *J. Fluorine Chem., 53* (1991) 43.

110 A. K. Brisdon, J. H. Holloway, E. G. Hope, W. Levason, J. S. Ogden and A. K. Saad, *J. Chem. Soc., Dalton Trans.*, (1992) 447.
111 S. A. Brewer, A. K. Brisdon, J. H. Holloway, E. G. Hope, W. Levason, J. S. Ogden and A. K. Saad, *J. Fluorine Chem., 60* (1993) 13.
112 M. J. Atherton, J. Fawcett, J. H. Holloway, E. G. Hope, A. Karaçar, D. R. Russell and G. C. Saunders, *J. Chem. Soc., Dalton Trans.*, (1996) 3215.
113 M. J. Atherton, J. Fawcett, J. H. Holloway, E. G. Hope, S. M. Martin, D. R. Russell and G. C. Saunders, *J. Organomet. Chem., 555* (1998) 67.
114 M. J. Atherton, J. Fawcett, J. H. Holloway, E. G. Hope, D. R. Russell and G. C. Saunders, *J. Organomet. Chem., 582* (1999) 163.
115 D. Gantar, B. Frlec, D. R. Russell and J. H. Holloway, *Acta Crystallogr., Sect. C, 43* (1987) 618.
116 D. Brown, B. Whittaker, J. A. Berry and J. H. Holloway, *J. Less Common Metals, 86* (1982) 75.
117 D. Brown, J. A. Berry and J. H. Holloway, U.K.A.E.A. Report, AERE-R 10425, 1982.
118 D. Brown, J. A. Berry and J. H. Holloway, *J. Chem. Soc., Dalton Trans.*, (1982) 1385.
119 D. Brown, J. A. Berry, J. H. Holloway and G. M. Staunton, *J. Less Common Metals, 92* (1983) 149.
120 R. Bougon, J. Fawcett, J. H. Holloway and D. R. Russell, *C.R. Acad. Sci. Paris, 287* (1978) C423.
121 R. Bougon, J. Fawcett, J. H. Holloway and D. R. Russell, *J. Chem. Soc., Dalton Trans.*, (1979) 1881.
122 J. Fawcett, J. H. Holloway, D. Laycock and D. R. Russell, *J. Chem. Soc., Dalton Trans.*, (1982) 1355.
123 J. H. Holloway, D. Laycock and R. Bougon, *J. Chem. Soc., Dalton Trans.*, (1982) 1635.
124 J. H. Holloway, D. Laycock and R. Bougon, *J. Chem. Soc., Dalton Trans.*, (1983) 2303.
125 J. H. Holloway, D. Laycock and R. Bougon, *J. Fluorine Chem., 26* (1984) 281.
126 J. H. Holloway, E. G. Hope, R. Taylor, G. J. Langley, A. G. Avent, T. J. Dennis, J. P. Hare, H. W. Kroto and D. R. M. Walton, *J. Chem. Soc., Chem. Commun.*, (1991) 966.
127 A. J. Adamson, J. H. Holloway, E. G. Hope and R. Taylor, *Fullerene Sci. and Tech., 5* (1997) 629.
128 R. Taylor, G. J. Langley, A. K. Brisdon, J. H. Holloway, E. G. Hope, H. W. Kroto and D. R. M. Walton, *J. Chem. Soc., Chem. Commun.*, (1993) 875.
129 R. Taylor, G. J. Langley, J. H. Holloway, E. G. Hope, A. K. Brisdon, H. W. Kroto and D. R. M. Walton, *J. Chem. Soc., Perkin Trans. 2*, (1995) 181.
130 O. Boltalina, J. H. Holloway, E. G. Hope, J. M. Street and R. Taylor, *J. Chem. Soc., Perkin Trans. 2*, (1998) 1845.
131 H. I. Bloemink, K. Hinds, J. H. Holloway and A. C. Legon, *Chem. Phys. Letts., 245* (1995) 598.
132 S. A. Cooke, G. Cotti, J. H. Holloway and A. C. Legon, *Angew. Chem., Int. Ed. Engl., 36* (1997) 129.
133 M. Freemantle, *Chem. Eng. News*, 14 October, (1996) 34.

Chapter 17

CHINESE RESEARCH IN ORGANOFLUORINE CHEMISTRY

CHANG-MING HU and WEI-YUAN HUANG

Shanghai Institute of Organic Chemistry, Chinese Academy of Sciences, 354 Fenglin Road, Shanghai 200032, P.R. China

Background

China is rich in fluorspar deposits, but only in the late fifties of the twentieth century did her fluorine chemistry and industry begin to develop. Before that, its fluorine industry was limited to the production of anhydrous HF and small amounts of simple fluorocarbons such as CF_2Cl_2, and CHF_2Cl for use as refrigerants, and only a few chemists were devoted to organofluorine chemistry. Things totally changed in the late fifties because after the withdrawal of the former USSR specialists the Chinese government determined to establish an atomic energy industry through its own efforts. This task was referred to then as the 111 (triple number one, *i.e.* uppermost, important and urgent) project, and large amounts of manpower and material resources were invested in this area.

Early in 1960, in order to fully utilize the country's limited resources in organofluorine chemistry – ranging from technical know-how to equipment and facilities – the Chinese Academy of Sciences decided that the main researchers in organofluorine chemistry should be centred on the Shanghai Institute of Organic Chemistry (SIOC). Thus a fluorine group at the Beijing Institute of Chemistry headed by Professor X. K. Jiang and Dr Q. Y. Chen (who received his PhD degree for work done in Knunyant's Laboratory in Moscow) moved to Shanghai to join the fluorine teams there headed by Professors W. Y. Huang and Y. Z. Huang, thereby founding SIOC's Department of Fluorine Chemistry. At the same time, the Institute of Synthetic Rubber (ISR) was set up in order to develop fluorinated monomers and polymers under the leadership of the Ministry of Chemical Industry. The title of this institute was changed to the Shanghai Institute of Organofluorine Materials (SIOFM) in 1980 and many industrial problems related to organofluorine chemistry continue to be investigated there.

Initially, there was no distinct division of research topics between the above two institutes. For example, the pyrolysis of CHF_2Cl to form tetrafluoroethylene was first studied at SIOC, while the polymerization and copolymerization of fluoromonomers was studied at SIOC, at the Changchun Institute of Applied Chemistry, and at the ISR.

As the stepwise development of the Chinese fluorochemicals industry progressed, however, a tremendous number of problems of a technical nature arose; and those, in turn stimulated action to enable prompt attention to be given to basic aspects of organofluorine chemistry. The need for basic research became especially acute in the late seventies, and as time progressed the situation moved to that which still exists, *i.e.* most of the basic research in fluoro-organic chemistry is done at SIOC, while industrial problems are tackled first at SIOFM and then at other industrial institutes.[1] The stepwise developments of the Chinese fluorochemicals industry referred to above included the production of numerous fluoromonomers (*e.g.* $CF_2{=}CF_2$, $CF_2{=}CFCl$, $CH_2{=}CF_2$, $CF_3CF{=}CF_2$) and derived polymers (fluids, elastomers and plastics) in increasing amounts, followed later by fluorinated surfactants, medicinals and agrochemicals; fluorination methodologies like electrochemical fluorination (Simons Process), direct fluorination with F_2 and exhaustive fluorination with high-valency metal fluorides were also developed simultaneously.

International connections

Around 1978, our country started implementing an 'open door' policy, and this enabled Chinese fluorine chemists to be sent abroad after being isolated from the outside world for more than 20 years. Between 1980 and 1983, therefore, G. Z. Ji, J. S. Ji and X. Y. Li went to West Germany; X. Y. Dai, C.M. Hu and J. A. Jiang went to the UK; and Y. C. Shen, Y. Y. Xu, J. X. Yao and C. X. Zhao were sent to the US to study organofluorine chemistry; all were staff members of SIOC, except for J. A. Jiang of SIOFM, and all of them assumed important or even leading roles once they returned home. Most of them became tutors to graduate students because by then the postgraduate education system had been re-established. Since many of the postgraduate students received PhD degrees and then went on to postdoctoral fellowships, a flourishing fluoro-organic community gradually came into being. Nowadays it is quite normal for Chinese students to go abroad to study fluorine chemistry, and there is a humorous saying that if you visit Professor D. J. Burton's laboratory in Iowa or Professor D. D. DesMarteau's in Clemson, South Carolina, you will find 'China towns' – 8 or even more Chinese students and postdocs working there at any one time. After an interruption of far too many years, a Chinese delegation headed by one of us (W.-Y. H.) was able to attend the ACS Winter Fluorine Conference in February 1979, and this provided the opportunity to visit numerous universities and industrial laboratories in the US after the meeting; later, in 1982, one of us (C.-M. H.), accompanied by J. A. Jiang (of SIOFM), became the first Chinese fluorine chemists to attend an ISFC (International Symposium on Fluorine Chemistry) – the 10th, held in Vancouver, Canada. Since then Chinese fluorine chemistry has always been represented at such meetings, and in the year 2003 China will be hosting the 17th ISFC in Shanghai.

The relationship between China and other countries has also been greatly enhanced through visits to Shanghai by overseas fluorine experts. Professor R. N. Haszeldine (UMIST, UK) came to SIOC in 1981 and was deeply impressed by our research activities; Professor N. Ishikawa (Tokyo Institute of Technolgy) and Professor N. Watanabe (Kyoto University) came in 1983; and in 1986 on the occasion of the 4th Chinese National Symposium on Fluorine Chemistry (a regular meeting, sponsored by the Chinese Chemical Society), Drs R. E. Banks (UMIST), J. Burdon and R. G. Plevey (Birmingham University), plus a delegation led by Professor Ishikawa, spent time with us. Through Professor Ishikawa, we were invited to participate in the 1987 Regular Meeting of Japanese-Soviet Fluorine Chemists; this was to be held in Tokyo but the delegates from the USSR failed to get visas

[1] Foreigners often have not been able to distinguish between the two Shanghai institutes – like Mr R. Iwa of Nippon Mektron in Japan who, in 1984, dropped into SIOC when his objective was to visit SIOFM.

from the Japanese Embassy, so that meeting became the 1st Bilateral Meeting for Japanese-Sino Fluorine Chemists – an event, mainly academic in nature, which is now held alternately in Japan and China every three years. To cater for industrial aspects of fluorine chemistry, a meeting sponsored mainly by SIOFM and Professor Ishikawa's F & F Research Centre (now F & F International, following Professor Ishikawa's death in 1991) was held in China in 1990; to date, four such Sino-Japanese Fluorine and Fine Chemicals meetings have been held, all in China. An international conference dealing with bromofluorocarbon (Halon) alternatives was held in Hang-Zhou City (Zhe-Jiang Province) in September 1990 (co-organized by the Zhe-Jiang Science and Technology Society, the Zhe-Jiang Institute of Chemical Engineering, the American Bureau of Environmental Protection, and the American Fire Fighting Society). A similar event dealing with CFC and Halon alternatives sponsored by the Chinese Chemical Society took place in Shanghai following the 14th ISFC in Yokohama, Japan. SIOFM and its subsidiary companies remain pre-eminent where the development of fluorine-containing materials and related technologies in China is concerned.

A necessarily brief survey of work carried out mainly at SIOC on academic aspects of organofluorine chemistry follows.

Sulfinato-dehalogenation

At the begining of the 1980s, W-Y. Huang *et al.* tried to convert the iodine-containing sulfonyl fluoride $ICF_2CF_2OCF_2CF_2SO_2F$ to the corresponding sulfinate, $ICF_2CF_2OCF_2CF_2SO_2Na$, by treating it with sodium sulfite in aqueous dioxan. To their surprise, the product was the disulfinate $NaO_2SCF_2CF_2OCF_2CF_2SO_2Na$, and this novel effective way of converting a CF_2I group into CF_2SO_2Na marked the discovery of the sulfinato-dehalogenation (halogen = I, Br, Cl) reaction in organofluorine chemistry [1]. Mechanistic studies having indicated that the original sulfinato-deiodination reaction observed involved single-electron oxidation of sulfite to $SO{\cdot}_2^-$, attention was turned to the use of sodium dithionite as an initiator of sulfinato-dehalogenation since the monomer – dimer equilibrium $S_2O_4^{2-} \rightleftharpoons 2SO{\cdot}_2^-$ exists in aqueous dithionite [2]. In order to improve the solubilty of the fluoro-organic substrates, co-solvents like aqueous acetonitrile were used, enabling conversions of the types $Cl(CF_2)_nI \rightarrow Cl(CF_2)_nSO_2Na$, $R_FBr \rightarrow R_FSO_2Na$, and $R_FCCl_3 \rightarrow R_FCCl_2SO_2Na$ to be achieved using $Na_2S_2O_4$ in aqueous nitrile [3-5]. The involvement of carbon-centred radicals in such conversions (*e.g.* $R_FI + SO{\cdot}_2^- \rightarrow R{\cdot}_F + I^- + SO_2$) is well supported by the results of experiments in which $Na_2S_2O_4$ has been used to initiate 1:1 addition reactions between polyfluorinated halides and unsaturated substrates (see Scheme 17.1 for examples [6, 7]). Some synthetic applications of this methodology are displayed in Scheme 17.2 [8-21].

Addition reactions initiated by conventional redox systems

C. M. Hu, F. L. Qing and others have studied in detail the applications of redox initiators in organofluorine chemistry, the following types having been included in this work: $(NH_4)_2S_2O_8/HCO_2NH_4$, $(NH_4)_2S_2O_8/HCO_2Na$ [24, 25]; $BrCo(dmgH)_2Py/Zn$ [26, 27]; Cp_2TiCl_2/Zn, Cp_2TiCl_2/Fe [28]; $CrCl_3 \cdot 6H_2O/Fe$ [29-31]; $NiCl_2 \cdot 6H_2O/Zn$, $NiCl_2 \cdot 6H_2O/Fe$ [32]; $PbBr_2/Al$ [33]; $SnCl_2/Zn$, $SnCl_2/Al$ [34]; $LnCl_3/Zn$ [35, 36]. In

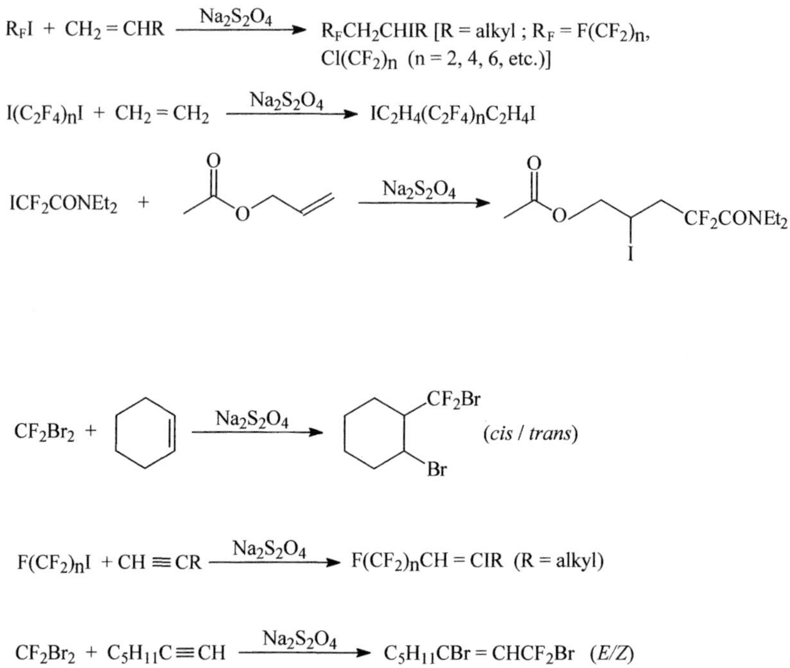

Scheme 17.1. Some exmples of SET-initiated 1:1 addition reactions between polyfluorinated halides and olefins or alkynes.

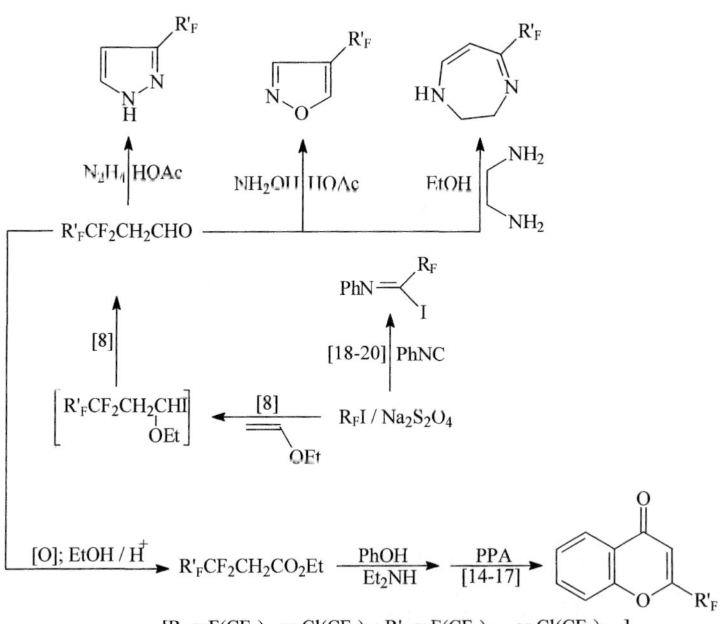

Scheme 17.2. Some of the wide-ranging synthetic applications of sulfinato-dehalogenation.

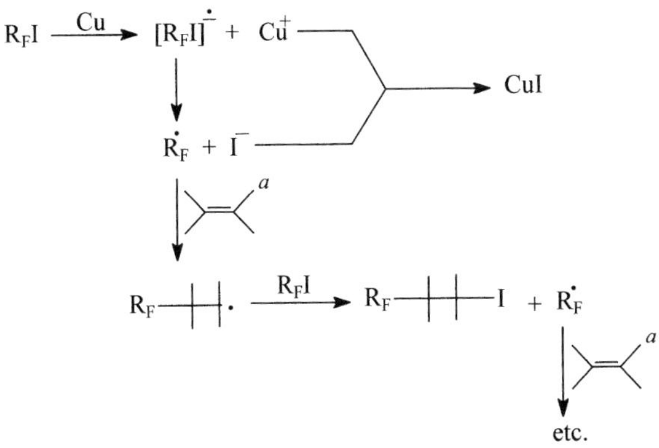

[a]General representation of an olefin

Scheme 17.3. Metal-based SET mechanism [for example, $R_FI = Cl(CF_2)_4I$, $FSO_2C_2F_4O(CF_2)_4I$].

general, the reactivity of the perhalogeno groups involved in the SET initiation steps proposed decreases in the order $CF_2I > CF_2Br \approx CF_2CCl_3 > CFCl_2 > CF_2Cl$, as established through the study of reactions such as:

$$CF_2ClCFCl_2 + CH_2{=}CHC_4H_9 \xrightarrow[HCO_2Na]{(NH_4)_2S_2O_8} CF_2ClCFClCH_2CHClC_4H_9$$

$$C_6F_{13}Br + CH_2{=}CHCO_2Et \xrightarrow{Co^{2+}/Zn} C_6F_{13}CH_2CHBrCO_2Et$$

Use of these initiators enables quite a lot of novel compounds to be synthesized which, in general, would be rather diffcult to make by conventional methods.

Metal-promoted addition reactions

Metal-promoted addition of R_FI to electron-rich alkenes and alkynes was established in 1980 by Q. Y. Chen *et al.* The metals investigated include Cu, Zn, Mg, Fe, Ni and Pd [37-43], and again such reactions are believed to occur via SET initiation (Scheme 17.3).

Carbenes and nitrenes

Work by X. K. Jiang in the 1950s on reactions between electrophiles and fluoroolefins included the synthesis of β-hydroxytetrafluoroethanesulfonic acid sultone from SO_3 and $CF_2{=}CF_2$ [44]. This sultone, now used extensively in the manufacture of perfluorinated ion-exchange membranes, has been shown by Q. Y. Chen *et al.* to be a useful source of difluorocarbene precursors [45-48]. This work has been extended to Burton-type trifluoromethylations utilizing $FO_2SCF_2CO_2Me$, CF_2ISO_2F, or CF_2XCO_2Me (X = Cl, Br, I) as sources of [$:CF_2$]-derived trifluoromethylcopper species [49-51]:

$$CF_2ISO_2F + RX \xrightarrow{Cu/DMF} RCF_3 + SO_2$$

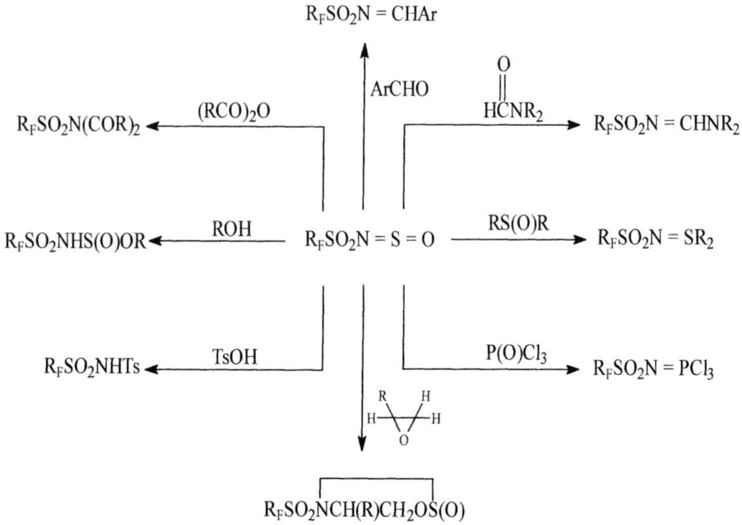

Scheme 17.4. Some reactions of *N*-sulfinylperfluoroalkanesulfonamides.

$$CF_2ClCO_2Me + RX \xrightarrow{KF/Cu/DMF} RCF_3 + CO_2 + MeX$$

$$FO_2SCF_2CO_2Me + RX \xrightarrow{Cu/DMF} RCF_3 + SO_2 + CO_2 + MeX$$

(R = aryl, benzyl, vinyl, allyl; X = Br, I)

F. L. Qing *et al.* have utilized this methodology to prepare trifluoromethylated analogues of retinoids [52].

Studies on bis(perfluoroalkanesulfonyl)carbenes, *e.g.* $(CF_3SO_2)_2C$:, generated by photochemical or thermal decomposition of the corresponding phenyliodonium methides have been pursued by Q. Y. Chen and S. Z. Zhu [53-58], and Zhu has studied nitrene formation from sulfonyl azides ($R_FSO_2N_3$) and *N,N*-dichlorosulfonamides ($R_FSO_2NCl_2$) [59-62].

Synthetic applications of *N*-sulfinylperfluoroalkanesulfonamides

N-Sulfinylperfluoroalkanesulfonamides, $R_FSO_2N{=}S{=}O$ [63], prepared by heating perfluoroalkanesulfonamides with an excess of $SOCl_2$, are versatile fluorine-containing building blocks, as illustrated by the syntheses of the types shown in Scheme 17.4 [64-69]

Phosphorus and arsenic ylides and onium salts

In the early 1960s, Y. Z. Huang and Y. C. Shen *et al.* embarked on a lengthy study of phosphorus ylides and phosphonium salts which has been extended to arsenic counterparts. Numerous fluorine-containing functionalized alkenes and alkynes have been synthesized

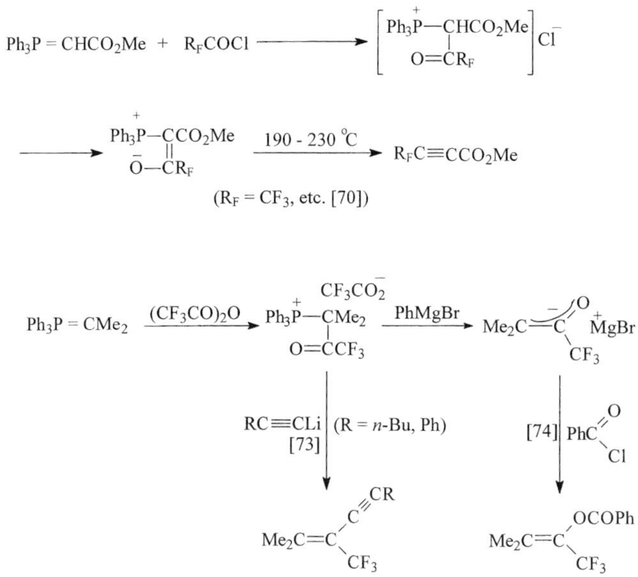

Scheme 17.5. Examples of syntheses achieved using phosphorus ylides.

which are not easy to prepare via conventional methods, and some of the achievements are indicated in Scheme 17.5 [70-77].

Spin-delocalization substituent constants

In nonradical chemistry the Hammett σ-ρ relationship has been applied successfully to explain structure-reactivity relationships mainly in terms of steric and polar effects. In free radical chemistry, however, it is complicated by a third major effect, namely spin-delocalization, *i.e.* a resonance effect involving the unpaired electron or spin. Quite a few radical chemists have dedicated their efforts to setting up a useful sigma-dot ($\sigma\cdot$) scale of spin-delocalization substituent constants, because a true $\sigma\cdot$ scale should be independent of steric and polar effects. Also it should be applicable to all kinds of carbon radicals in which spin delocalization can occur.

Studies begun in 1980 by X. K. Jiang *et al.* on the free-radical thermal cyclodimerization of *para*-substituted α,β,β-trifluorostyrenes have enabled a $\sigma_{JJ}\cdot$ scale to be established [78]. Using the $\sigma_{JJ}\cdot$ data found, and assuming that in the absence of measurable steric effects radical reactions might fall into one of four categories [(i) both polar and spin effects are important; (ii) polar effects dominate; (iii) spin-delocalization dominates; (iv) other complicating and interacting factors or effects are present], the long-standing puzzle concerning the failure of a delocalization effect to show up in the structure-reactivity correlationship analyses in many free radical reactions can now be explained successfully and explictly.

Other research

No account of our endeavours would be complete without mentioning Sun's work [79, 80] on the radiation of fluoropolymers, Zhang's study [81, 82] of fluorine-containing associating polymers, and Zhao's research on SET reactions of per- and poly-fluoroacyl peroxides [83, 84], all of which have attracted great attention from chemists worldwide.

References

1 W. Y. Huang, B. N. Huang and C. M. Hu, *Acta Chimica Sinica*, *39* (1981) 481.
2 E. G. Janzen, *J. Phys. Chem.*, *76* (1972) 157.
3 W. Y. Huang, B. N. Huang and W. Wang, *Acta Chimica Sinica* (Engl. Ed.), *43* (1985) 252.
4 W. Y. Huang, B. H. Wang and W. Wang, *Acta Chimica Sinica* (Engl. Ed.), *44* (1986) 68.
5 W. Y. Huang, B. N. Huang and J. L. Chen, *Acta Chimica Sinica*, *44* (1986) 45.
6 W. Y. Huang, W. Wang and B. N. Huang, *Acta Chimica Sinica* (Engl. Ed.), *44* (1986) 178.
7 W. Y. Huang and Y. M. Wu, *Faming Zhuanli Shenqing Gongkai Shuomingshu, CN 1,068,321*; *Chem. Abstr.*, *119* (1993) 160075g.
8 W. Y. Huang, L. Lu and Y. F. Zhang, *Chinese J. Chem.*, *8* (1990) 68.
9 X. Q. Tang and C. M. Hu, *J. Chem. Soc., Perkin Trans. 1*, (1994) 2161.
10 X. Q. Tang and C. M. Hu, *J. Chem. Soc., Chem. Commun.*, (1994) 631.
11 X. Q. Tang and C. M. Hu, *J. Fluorine Chem.*, *74* (1995) 9.
12 Q. F. Wang and C. M. Hu, *Tetrahedron lett.*, *39* (1998) 2377.
13 Q. F. Wang and C. M Hu, *J. Fluorine Chem.*, *94* (1999) 79.
14 W. Y. Huang, Y. S. Liu and L. Lu, *J. Fluorine Chem.*, *66* (1994) 263.
15 W. Y. Huang, Y. S. Liu and L. Lu, *J. Fluorine Chem.*, *66* (1994) 209.
16 W. Y. Huang and Y. S. Liu, *Heteroatom Chem.*, *6* (1995) 287.
17 Y. S. Liu and W. Y. Huang, *J. Chem. Soc., Perkin Trans. 1*, (1997) 981.
18 X. B. Yu and W. Y. Huang, *Tetrahedron Lett.*, *37* (1996) 7999.
19 X. B. Yu and W. Y. Huang, *J. Fluorine Chem.*, *84* (1997) 65.
20 X. B. Yu, Q. S. Zhang and W. Y. Huang, *Chinese J. Chem.*, *15* (1997) 278.
21 G. Zhao, J. Yang and W. Y. Huang, *J. Fluorine Chem.*, *86* (1997) 89.
22 D. J. Burton and L. J. Kehoe, *J. Org. Chem.*, *35* (1970) 3339.
23 A.Battats and B. Bontevin, *J. Fluorine Chem.*, *42* (1989) 215.
24 C. M. Hu and F. L. Qing, *J. Fluorine Chem.*, *49* (1990) 275.
25 C. M. Hu and F. L. Qing, *Tetrahedron Lett.*, *31* (1990) 1307.
26 C. M. Hu and Y. L. Qiu, *Tetrahedron Lett.*, *32* (1991) 4001.
27 C. M. Hu and Y. L. Qiu, *J. Org. Chem.*, *57* (1992) 3339.
28 C. M. Hu and Y. L. Qiu, *J. Fluorine Chem.*, *55* (1991) 113.
29 C. M. Hu and J. Chen, *J. Fluorine Chem.*, *69* (1994) 79.
30 C. M. Hu and J. Chen, *J. Chem. Soc., Chem. Commun.*, (1993) 75.
31 C. M. Hu and J. Chen, *Tetrahedron Lett.*, *34* (1993) 5957.
32 C. M. Hu and X. Q. Tang, *J. Fluorine Chem.*, *59* (1992) 401.
33 C. M. Hu and X. Q. Tang, *J. Fluorine Chem.*, *61* (1994) 217.
34 C. M. Hu and J. Chen, *J. Fluorine Chem.*, *67* (1994) 189.
35 D. Yu, G. Zhao and W. Y. Huang, *Tetrahedron Lett.*, *33* (1992) 8119.
36 D. Yu, G. Zhao and W. Y. Huang, *Tetrahedron Lett.*, *34* (1993) 1321.
37 Q. Y. Chen and Z. Y. Yang, *J. Fluorine Chem.*, *48* (1985) 399.
38 Q. Y. Chen, Z. Y. Yang and Z. M. Qiu, *KeXueTongBao*, *33* (1988) 1866.
39 Q. Y. Chen, Z. M. Qiu and Z. Y. Yang, *J. Fluorine Chem.*, *36* (1987) 149.
40 Q. Y. Chen, Y. B. He and Z. Y. Yang, *J. Fluorine Chem.*, *34* (1986) 255.
41 Q. Y. Chen and Z. Y. Yang, *J. Chem. Soc., Chem. Commun.* (1986) 498.
42 Q. Y. Chen, Z. Y. Yang, C. X. Zhao and Z. M. Qiu, *J. Chem. Soc., Perkin Trans. 1*, (1988) 563.
43 Q. Y. Chen and Z. Y. Yang, *Acta Chimica Sinica* (Eng. Ed.), *46* (1988) 155.

44 X. K. Jiang, *Acta Chimica Sinica*, *23* (1957) 330.
45 Q. Y. Chen and S. W. Wu, *J. Fluorine Chem.*, *47* (1990) 509.
46 Q. Y. Chen and S. W. Wu, *J. Chem. Soc., Perkin Trans I*, (1989) 2385.
47 D. B. Su, J. X. Duan, A. J. Yu and Q. Y. Chen, *J. Fluorine Chem.*, *65* (1993) 11.
48 D. B. Su, J. X. Duan and Q. Y. Chen, *J. Chem. Soc., Chem. Commun.*, (1992) 807.
49 Q. Y. Chen and S. W. Wu, *J. Chem. Soc., Chem. Commun.*, (1989) 705.
50 D. B. Su, J, X. Duan and Q. Y. Chen, *Tetrahedron Lett.*, *32* (1991) 7689.
51 J. X. Duan, D. B. Su and Q. Y. Chen, *J. Fluorine Chem.*, *61* (1993) 279.
52 F. L. Qing, J. Fan, H. Sun and X. Yue, *J. Chem. Soc., Perkin Trans 1*, *20* (1997) 3053.
53 S. Z. Zhu and Q. Y. Chen, *J. Chem. Soc., Chem. Commun.*, *20* (1990) 1459.
54 S. Z. Zhu, *Heteroatom Chemistry*, *5* (1994) 9.
55 S. Z. Zhu, *J. Fluorine Chem.*, *60* (1993) 289.
56 W. Guang and S. Z. Zhu, *Acta Cryst. Part C*, (1991) 1227.
57 S. Z. Zhu and A. W. Li, *J. Chinese Chem.*, *10* (1992) 458.
58 S. Z. Zhu, A. W. Li and K. Wu, *Chinese Chem. Lett.*, *3* (1992) 203.
59 S. Z. Zhu, *Tetrahedron. Lett.*, *33* (1992) 6503.
60 S. Z. Zhu, C. M. Zhou, A. W. Li and B. Xu, *J. Fluorine Chem.*, *67* (1994) 7.
61 S. Z. Zhu, *J. Chem. Soc., Perkin Trans. 1*, (1994) 2077.
62 S. Z. Zhu, B. Xu, C. Y. Qin and G. L. Xu, *Inorg. Chem.*, *36* (1997) 4909.
63 S. Z. Zhu, *J. Chem. Soc., Chem. Commun.*, *10* (1991) 723.
64 A. W. Li, B. Xu and S. Z. Zhu *et al.*, *J. Fluorine Chem.*, *68* (1994) 145.
65 A. W. Li, B. Xu, C. X. Wang and S. Z. Zhu, *J. Fluorine Chem.*, *69* (1994) 85.
66 S. Z. Zhu, J. Zhang and B. Xu, *J. Fluorine Chem.*, *71* (1995) 81.
67 S. Z. Zhu, B. Xu and J. Zhang, *J. Fluorine Chem.*, *74* (1995) 203.
68 S. Z. Zhu, C. Y. Qin and B. Xu, *J. Fluorine Chem.*, *79* (1996) 49.
69 S. Z. Zhu, B. Xu, J. Zhang and C. Y. Qin, *Phosphorus, Sulfur, and Silicon*, *112* (1996) 219.
70 Y. T. Huang, Y. C. Shen, K. T. Chen and C. C. Wang, *Acta Chimica Sinica*, *37* (1997) 47.
71 Y. T. Huang, Y. C. Shen, Y. K. Xin and Q. W. Wang, *Scientia Sinica*, (1981) 973; (Eng. Ed.), *25* (1982) 21.
72 Y. C. Shen and W. Qiu, *Tetrahedron Lett.*, *28* (1987) 449.
73 Y. C. Shen and W. Qiu, *J. Chem. Soc., Chem. Commun.*, (1987) 703.
74 Y. C. Shen and Y. Xiang, *J. Chem. Soc., Chem. Commun.*, (1991) 1384.
75 Y. C. Shen, Y. Xiang and W. Qiu, *Tetrahedron Lett.*, *32* (1991) 4953.
76 Y. C. Shen and Y. Zhou, *J. Fluorine Chem.*, *61* (1993) 247.
77 Y. C. Shen and S. Gao, *J. Org. Chem.*, *58* (1993) 4564.
78 X. K. Jiang, *Acc. Chem. Res.*, *30* (1997) 283.
79 J. Z. Sun, Y. F. Zhang, X. G. Zhong and W. X. Zhang, *Radiation Phys. Chem.*, *42* (1993) 139.
80 J. Z. Sun, Y. F. Zhang, X. G. Zhong and X. L. Zhu, *Radiation. Phys. Chem.*, *44* (1994) 665.
81 Y. X. Zhang, A. H. Da, G. B. Gutler and T. E. Hogen-Esch, *J. Polymer Sci., Part A: Polymer Chemistry*, *30* (1992) 1383.
82 Y. X. Zhang, Jin Yang, Ai-Hua Da and Yu-Qing Fu, *Polymers for Advanced Technologies*, *8* (1997) 169.
83 C. X. Zhao, F. T. Gamil and C. Walling, *J. Org. Chem.*, *48* (1983) 4908.
84 C. X. Zhao, X. K. Jiang and G. F. Chen, *J. Am. Chem. Soc.*, *108* (1986) 3132.

BIOGRAPHIC NOTES

Professor Chang-Ming Hu was born on Oct. 1, 1932 in Wu-Xi city, Jiang-Su Province, and graduated from Fu-Dan University, Shanghai, in 1953. He spent the next ten years as a Research Associate at the Chang-Chan Institute of Applied chemistry (Chinese Acadeny of Sciences) in Ji-Ling Province, moving in 1963 to the Shanghai Institute of Organic Chemistry where he is now Professor of organic chemistry. He spent the two-year period 1981-83 as a visiting scientist in Professor Tatlow's fluorine group at the University of Birmingham, UK. Professor Hu is a member of the Editorial Board of the *Journal of Fluorine Chemistry* and served on the steering committee of the 14th International Symposium on Fluorine Chemistry.

Chang-Ming Hu

Wei-Yuan Huang

Professor Wei-Yuan Huang was born in Putien, Fujian (China) on December 15, 1921. He graduated (BSc) from the Chemistry Department of Fu-Kien Christian university in 1943 and then received an MSc degree from Ling-Nan university in 1949, followed by a PhD degree from Harvard University, USA, in 1952. In his early research career he worced on steroids and other natural products, but changed his field of interest to organofluorine chemistry in 1960 and successfully developed a series of fluorine-containing materials in China. He discovered the sulfinato-dehalogenation reaction in 1981 and then proceeded study in detail the chemistry of pefluoroalkanesulfinates and perfluoroalkanesulfonyl halides. He was awarded a Moissan Medal in 1986 in Paris at the special international meeting *Centenary of the Discovery of Fluorine*. He has been a member of the Chinese Academy of Sciences since 1980.

Chapter 18

FLUORINE CHEMISTRY IN JAPAN

YOSHIRO KOBAYASHI and TAKEO TAGUCHI

Tokyo University of Pharmacy & Life Science, 1432-1 Horinouchi, Hacchioji, Tokyo 192-0392

and TAKASHI ABE

National Industrial Research Institute of Nagoya, Hirate-cho,1-1 Kita-Ku, Nagoya 462-8510, Japan

Preamble

The contributions of Japanese chemists and technologists to the development of fluorine chemistry since the early 1950s have been considerable, and it is not possible in the space available here to do justice to even the major research groups now active in Japan. In keeping with the spirit of this book, therefore, the discussion here is aimed mainly at providing a general picture of the organization of fluorine research in Japan and also some specific information relating to our personal experiences. A detailed account of Japanese contributions to fluorine chemistry is being prepared for publication in the *Journal of Fluorine Chemistry* under the guidance of Professor Tsuyoshi Nakajima (Regional Editor – Asia) of Kyoto University. A recent list of organofluorine compounds manufactured in Japan, together with details of the companies involved, can be found in ref. [1].

Introduction

It is generally accepted that the development of fluorine chemistry in the US and European countries owes much to its exploitation during World War II for the production of nuclear weapons [2]. Fluorine Chemistry in Japan has been developed without this influence, being associated simply with basic and general scientific technologies. Before World War II there was no remarkable academic activity in the field of fluorine chemistry. However, an episode known as 'the budding of Japan's fluorine chemicals industry' occurred in 1935 when the Daikin Company succeeded in producing CFC-12 (CF_2Cl_2); immediately, it was tested at the request of the Japanese Imperial Navy as a replacement refrigerant for ammonia in its new model submarine, 'I go 171'. During World War II some 120 tonnes of CFC-12 were manufactured [3].

After the war, it was realised in both academic and industrial circles that fluorine chemistry offered great potential for new discoveries and applications. This prompted the initiation of work on organofluorine compounds at the Government Industrial Research Institute of Nagoya (GIRIN) in 1952, and on inorganic fluorine chemistry at Kyoto University around 1960. As the numbers of researchers involved with fluorine chemistry grew, the pioneers at these centres (Professors Nobuatsu Watanabe and Teiichi Ando at Kyoto, Drs Kan Inukai, Shunji Nagase and Hiroshige Muramatsu at NIRIN) together with Professor Nobuo

Ishikawa[1] at the Tokyo Institute of Technology formed the Japanese Association of Fluorine Chemists, which held its first annual meeting (Fluorine Conference of Japan) in Tokyo in 1972. International recognition of Japanese Fluorine Chemists came in 1976, when the 8th International Symposium on Fluorine Chemistry (organised by Professors Watanabe and Ishikawa) was held in Kyoto. Further steps were taken by Professor Ishikawa to secure the internationalization of Japanese fluorine chemistry through arrangements for regular bilateral meetings held with Soviet and, later, Chinese fluorine chemists. The first biennial Soviet-Japanese meeting was held in February 1979. In 1990, through the efforts of Professor Watanabe, the Japan Society for the Promotion of Science constituted the 155th Committee on Fluorine Chemistry, and in 1994 the ISFC (14th meeting) returned to Japan (Yokohama).

Electrochemical fluorination as a locomotive for the development of fluorine chemistry at NIRIN, Nagoya
(by T. ABE)

Background

As mentioned earlier, systematic Japanese studies in organofluorine chemistry began mainly in the post World War II period. The opening moves were made in 1952 by scientists at the Government Industrial Research Institute of Nagoya (GIRIN), which later (1993) became the National Industrial Research Institute of Nagoya (NIRIN). The inception of the present Institute took place in the spring of 1952 after the integration of three National Institutes – two branch offices at Nagoya (Machinery Institute and Industrial Research Institute of Tokyo) and the National Porcelain Institute at Kyoto, which all belonged to the Ministry of International Trade and Industry (MITI). In those days, Japan was passing through a period of economic revival made necessary by the ruin associated with World War II, and the mission of MITI's National Institutes was not only to catch up with advanced countries generally but also to take the lead in industrial fields with prospects.

Takashi Abe

The development of a fluoropolymer industry was selected as one of the objectives after studying a 1945 US Department of Commerce report [5]. Basic studies associated with this aim were commenced at GIRIN in 1952 under the leadership of Dr Rimpei Kojima in the Inorganic Chemistry Section. His so-called fluorine chemistry group was soon reinforced by the arrival of new chemists, including Drs Nagase and Muramatsu, who are now

[1]The late Professor Nobuo Ishikawa (1926-1991), a particularly talented organiser and coordinator, was a leader among Japanese organofluorine chemists for many years. Following his retirement from the Tokyo Institute of Technology in 1987, he founded the Fluorine and Fine (F & F) Chemicals Research Centre in Tokyo to provide a worldwide information service for scientists and technolgists involved with fluorinated materials. For a brief résumé of his career, see ref. [4].

well known as pioneers of Japanese fluorine chemistry. At that time only a few chemical companies, notably Daikin, were producing fluorochemicals.

At GIRIN, the first work on fluoro-organic chemistry was conducted in parallel with experiments dealing with inorganic fluorides and centred on the preparation of chlorofluorocarbon precursors of fluoromonomers via Swarts-type reactions [6]. Alongside this, members of the Analytical Chemistry Section commenced work on methods of analysis of fluorine compounds. Some time later, chemists led by Dr Inukai of the Organic Chemistry Section modified their dyestuffs research to include fluorine-containing dyes, and eventually Inukai became leader of the fluorine chemistry group.

Improvements in Japan's economy from the 1960s onwards were reflected in the funding of GIRIN, and this enabled the Institute's budget to cover the installation, one by one, of modern analytical instruments (GC, IR, MS, NMR etc.). Progress in fluorine research done by Inukai, Nagase, Muramatsu, Teruo Ueda and me (Takashi Abe) in the Institute's Organic Chemistry Section (which became a Fluorine Chemistry Laboratory) increased accordingly; the major themes included: fluorination methods; fluorine-containing dyes; fluorinated monomers and polymers; bioactive fluoro-organic compounds; fluorinated semi-conductors; high-performance gas-permeable membranes; CFC and Halon alternatives. Throughout, Simons electrochemical fluorination (ECF) has played a key role in the promotion of our research projects, and continues to do so. Furthermore, the technology required for the safe and effective operation of ECF equipment has been transferred from the Institute to many fluorochemical industries in Japan. It is quite appropriate here, therefore, to concentrate on the history of the Institute's involvement with ECF.

Electrochemical fluorination (Simons Process [7, 8])

In principle, Simons ECF is very simple: very many organic compounds dissolve readily in anhydrous hydrogen fluoride (AHF), and the solutions formed conduct electricity. When current is passed, usually at 4. 5-6 V, free fluorine (nickel anodes are essential) is not evolved and the organic solute is generally perfluorinated (*i.e.*, completely fluorinated) [6, 7]. Invented by the American chemist Joseph H. Simons in 1941 (with publication delayed until 1949 [8]), this procedure is one of the few electrosynthetic techniques to have achieved commercialization.

After studying Simons' publications [9], Kojima's group at NIRIN commenced R & D studies on ECF in 1953. Experimental work was beset with difficulties at that time owing particularly to a shortage of equipment, tools, reagents, materials and coolants (AHF boils at 19. 5 °C). Thus, rods of silver solder for brazing purposes (*e.g.* the construction of packs of nickel electrodes) had to be made in-house, and the first cell body – fabricated by nickel casting – proved troublesome due to the presence of numerous small holes in the walls. The workers even had to face the hazardous challenge of making AHF from hydrofluoric acid (aqueous HF) which, happily, was available as a common reagent; in fact, AHF could be obtained through the courtesy of the Hashimoto Chemicals Company, but researchers had to collect the material and carry it with them on public transport from Osaka to Nagoya. After many setbacks, the group's efforts were crowned with success by the preparation of trifluoroacetyl fluoride from glacial acetic acid and its simple derivatives: $CH_3C(O)X \rightarrow CF_3C(O)F$ [X = OH, Cl, $OC(O)CH_3$], and the reaction conditions were investigated in some detail [10]. Since then, our ECF research has gone from strength to strength.

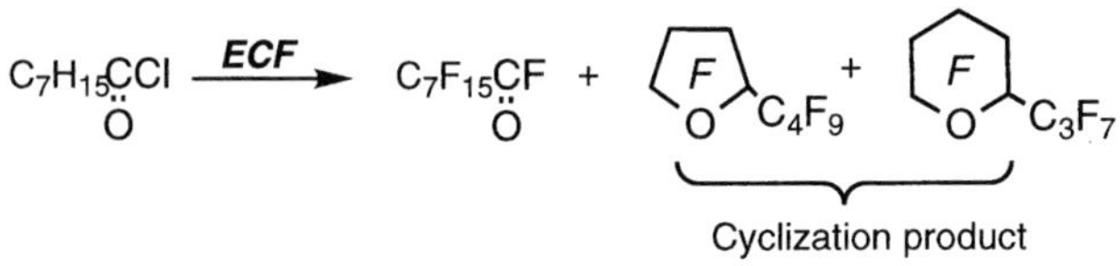

Scheme 18.1.

In 1959, Dr Nagase returned to the Institute after spending two years in the US working on direct fluorination (*i.e.*, using F_2) with the late Professor L. A. Bigelow at Duke University and took over the ECF research with Mr Hajime Baba. They conducted ECF experiments on several alcohols and carboxylic acid esters of short-chain alcohols, showing that perfluorocarboxylic acid fluorides, R_FCOF, could be obtained in reasonable yields [11]; this proved to be a fruitful line of investigation.

When I (TA) joined the Institute's fluorination group in 1964, the Japanese economy was expanding briskly and new work on preparative routes to fluorinated monomers was required [12]. In connection with this research theme, work was undertaken on the preparation of partially-fluorinated methanes and ethanes by ECF. At that time published information on the ECF of gaseous hydrocarbons – which are not really suitable for ECF owing to their very poor solubilities in AHF – was scarce [13]. However, a 'bubbler' was fabricated from a PTFE filter, making it possible to introduce gases as minute bubbles into a cell charged with AHF so that they rose through the pack of nickel electrodes. In this way, Nagase and his co-workers conducted ECF experiments on methane [14], ethane and ethylene and a variety of other gaseous substrates, *e.g.* CH_3Cl, $CHFCl_2$, $CH_2{=}CF_2$, CO, $COCl_2$, SO_2, thereby widening the scope of the Simons Process [7]. My interests centred on ECF-induced cyclizations, the best-known example of which is probably that shown in Scheme 18.1, which was discovered by Simons and commercialized by the 3M Company [8, 15]. Following a Government-sponsored spell of leave (1971-72) in Professor Jean'ne M. Shreeve's fluorine group in the US at the University of Idaho, I studied the synthesis of cyclic perfluoroethers via ECF of various types of carboxylic acids carrying α-alkyl substituents. This filled in quite a few blanks in the list of known cyclic perfluoroethers [16, 17], and also led to work on perfluorinated fluids for use in 'blood substitutes'.

The development of perfluorochemical-based oxygen carriers for use in 'artificial blood' was a hot topic in the 1970s, and the Green Cross Corporation (Osaka, Japan) had introduced an emulsified mixture of PFDC (perfluorodecalin) and PFTPA (perfluorotri-*n*-propylamine) as a blood substitute under the trade name *Fluosol-DA* [18]. However, use of this dual-PFC (perfluorochemical) product was viewed as a makeshift arrangement until a single PFC could be discovered which simultaneously provided stable emulsions and had a faster excretion rate from the human body. The possibility that the characteristics required might be found in a new perfluorinated heterobicylic compound led us to prepare a series of such bicyclic perfluoroethers via ECF (Scheme 18.2) [19].

Attempts to extend the work to the synthesis of perfluorinated bicycles containing both nitrogen and oxygen atoms were unsuccessful (*e.g.*, Scheme 18.3) [20].

Since the mid 1980s, we have used ECF techniques to prepare numerous nitrogen-containing perfluorocarboxylic acids and related perfluorinated *N*-vinylamines [21, 22] for

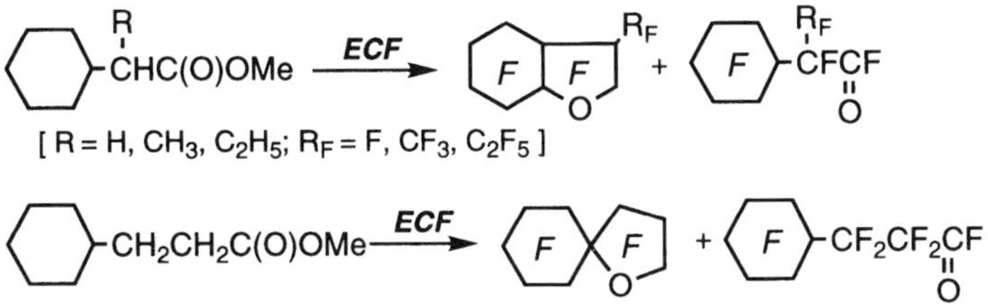

Scheme 18.2.

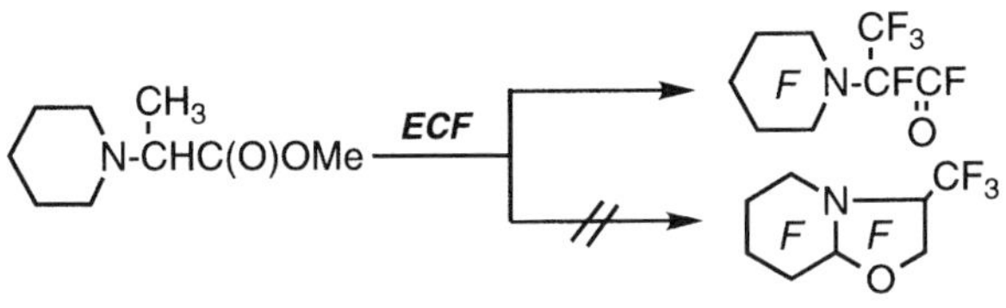

Scheme 18.3.

use in work on novel fluoropolymers and liquid crystals [23]. This research theme is still being pursued, and is related to the development of Halon alternatives containing $(CF_3)_2N$ groups which have excellent fire-extinguishing capabilities [24].

Fluorine in bio-organic chemistry: a 30-year memorandum
(by Y. KOBAYASHI and T. TAGUCHI)

After graduation (PhD, 1959) from the University of Tokyo (Professor Ochiai's group), one of us (YK) spent two years (1960 62) in the US doing postdoctoral research, first with Professor L. B. Clapp (Brown University) and then with Professor R. B. Woodward (Harvard University). On the way back to Japan, and armed with a letter of introduction from Professor Woodward, I visited DuPont's Central Research Department in Wilmington and found that more than half of the chemists there were working on fluorine topics. This was the first time that I had really taken note of fluorine chemistry, and it looked totally different from what I had ever seen. I was really impressed, and intuitively recognized the importance of fluorine chemistry to Japan and the need to accelerate its development there.

Y. Kobayashi

On my return to Japan, I worked in industry (Yawata Chemical Co.) for a time and did manage to initiate some fluorine research. Unfortunately, however, a general lack of understanding of this new and important work hindered progress, so I decided that the best way forward would be to aquire an academic post. In 1967, therefore, I joined the faculty of the Tokyo College of Pharmacy and remained there until retirement in 1989, when I moved to Daikin Industries to continue my researches in the area of bioactive organofluorine compounds.

My main co-workers during the past 30 years have been Dr Itsumaro Kumadaki (from 1968 until 1983, when he moved to a professorship at Setsunan University), my co-author here, Dr Takeo Taguchi (joined me in 1976 and was promoted to a professorship on my retirement), and Dr Katsuhiko Iseki of Daikin Industries (1989 onwards). Our major activities during this period are indicated below.

T. Taguchi

Initial studies (1968-1976): trifluoromethylation with 'CF_3Cu'

Our research up to the mid-70s centred on the synthesis of trifluoromethylated organic compounds and clarification of the reactivity of CF_3 groups attached to aromatic systems.

Our pioneering work on the trifluoromethylation of aryl halides with CF_3I in the presence of Cu in hot aprotic solvents such as $(Me_2N)_3PO$ (HMPA) (Scheme 18.4) was first announced in 1969 [25]. This method had some limitations owing to the reduction of the halides in certain cases. Later, it was found that the trifluoromethylcopper species involved could be isolated as an HMPA solution by filtering off undissolved Cu power; this solution reacted with a variety of organic halides under mild conditions [26]. An efficient synthesis of CF_3-substituted pyrimidine and purine nucleosides (*e.g.*, see Scheme 18.4) of interest as potential antiviral agents was developed using this modified procedure [27]. Following our work, several methods for the generation CF_3Cu species were reported, among which Burton's transmetallation of solvated trifluoromethylzinc reagents [formed from CF_2X_2 (X = Br, Cl) in DMF] by Cu(I) halides is noteworthy [28].

A trifluoromethyl attached to an aromatic ring was generally believed to be a very stable substituent, and little was known about its potential for chemical modification when we commenced our systematic study of trifluoromethylated aromatics with nucleophiles in 1968. We found that a CF_3 group showed a range of reactivity depending on its electronic environment and proposed mechanistic classifications [29]. Our findings related to an understanding of the biological response of some trifluoromethylated compounds; for example, Daniel Santi's proposed mechanism for the inhibition of thymidylate synthetase by 5-trifluoromethyl-2′-deoxyuridylic acid (Scheme 18.5) corresponds to the one we proposed for reactions of 3-(trifluoromethyl)quinoline [29].

Stabilization of strained molecules by CF_3 substituents (1973-1981)

The remarkable substituent effect whereby CF_3 groups stabilize highly-strained molecules was reported independently by Lemal and Haszeldine in 1969, and in 1973 we commenced our study on the synthesis of new strained-ring systems. At that time, the

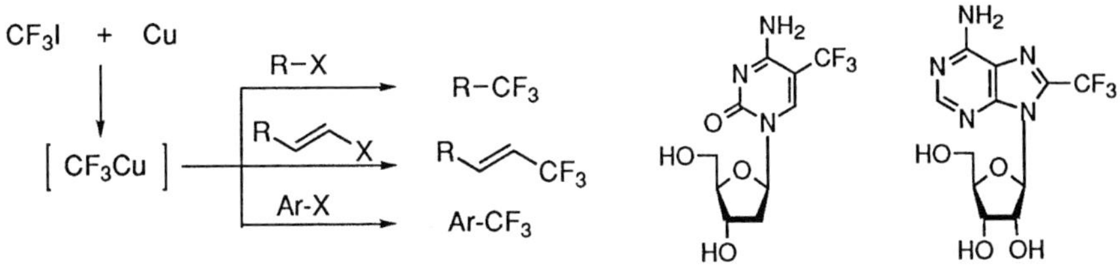

Scheme 18.4.

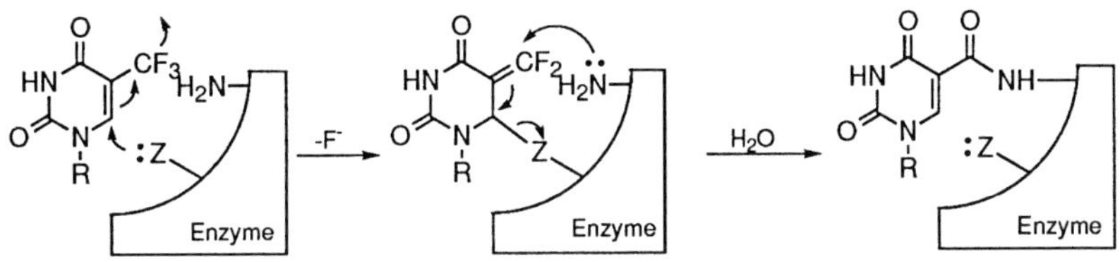

Scheme 18.5.

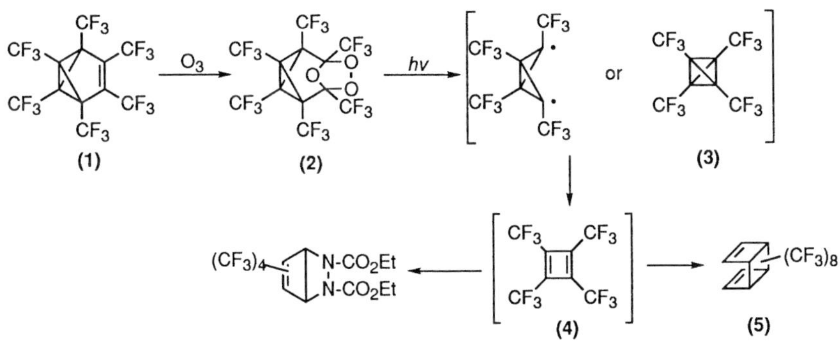

Scheme 18.6.

chemistry of such systems was attracting much attention on both theoretical and experimental fronts, so the competition to isolate and/or identify relevant compounds became quite severe.

Our first objective was to convert the highly stable $(CF_3)_6$-benzvalene **1** to $(CF_3)_4$-tetrahedrane **3,** as shown in Scheme 18.6. UV Irradiation of the matrix obtained by cooling a solution of the ozonide **2** in 3-methylpentane at −196 °C produced a yellow coloration, which persisted at the same temperature after the irradiation stopped. As the matrix thawed, however, the yellow colour disappeared and colourless crystals of the cyclobutadiene dimer **5** precipitated. The yellow-coloured substance was thought to be the cyclobutadiene monomer **4**. We tried several routes, including this ozonide route, but found no evidence for the formation of the $(CF_3)_4$-tetrahedrane [31, 32].

Our second objective was to synthesize valence-bond isomers of heteroaromatics. At first, we successfully determined the structure of Dewar thiophene **6,** obtained by irradiation of $(CF_3)_4$-thiophene. Following this work, Dewar pyrroles **7** (R = H, *t*-Bu, *c*-C_6H_{11}, Ph) and diphosphabenzene **8** and its benzvalene form **9** were prepared and their interesting properties investigated [31-33].

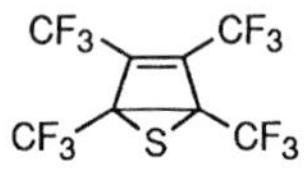

(6)

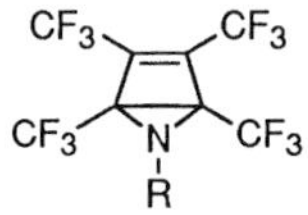

(7)

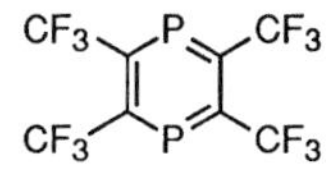

(8)

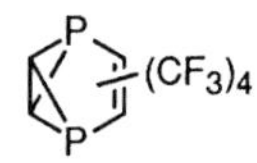

(9)

Synthesis of fluorinated bioactive compounds

Since we worked in a pharmaceutical department at a time when the importance of fluorine in medicinal chemistry and biochemistry was increasing dramatically, research projects associated with fluorinted bioactive compounds have featured prominently in our research programme since the early 1970s. CF_3-nucleoside chemistry gave us our first taste of this area, and following that we started a vitamin D project in 1977, then added several others notably those on retinals, arachidonic acid and its metabolites, sugars and amino acids. To promote these projects, extensive efforts were made to develop modern and sophisticated synthetic methods using fluorinated building blocks to achieve efficient preparations of structurally complicated or multi-functionalized bioactive molecules. During this period, one of us (YK) co-edited the well known books *Biomedicinal Aspects of Fluorine Chemistry* [with Professor Robert Filler (Illinois Institute of Technology, Chicago); 1982] and *Organofluorine Compounds in Medicinal Chemistry and Biomedical Applications* [with Professor Filler and Professor Lev Yagupolskii (Academy of the Ukraine, Kiev); 1993].

We started the vitamin D_3 project in 1977 as a collaborative work with Professor Ishikawa (Tokyo Institute of Technology), Professor DeLuca and Dr Yoko Tanaka (Wisconsin University at Madison). At that time, clarification of the physiological significance of the metabolism of vitamin D_3 attracted much attention, and the dihydroxylated metabolite $1,25(OH)_2D_3$ had come to be recognized as a hormonal active form, which might be an essential drug for bone diseases such as osteoporosis. To clarify this matter and, hopefully, to discover a therapeutically effective analogue, we designed and synthesized fluorinated compounds containing fluorine(s) at positions where metabolic hydroxylation occurred, thereby to block such metabolism. Of the analogues prepared, 26,26,26,27,27,27-hexafluoro-$1,25(OH)_2D_3$ **(10)** and 24,24-F_2-$1,25(OH)_2D_3$ **(11)** were highly potent and possessed long-lasting activity compared with that of $1,25(OH)_2D_3$ [34]. Moreover, we were able to propose a conclusive explanation for the physiological significance of metabolic hydroxylation on the side chain. Lately, at Daikin Industries Limited we have continued our second-stage study to design and synthesize new fluorinated analogues with significant anticancer activity separated from calcemic activity [35].

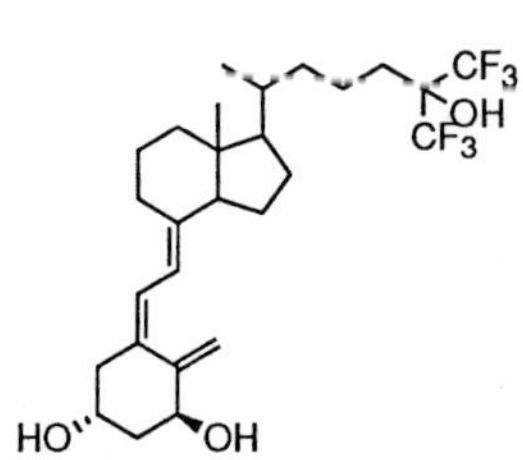

(10) 26,26,26,27,27,27-F_6-$1,25(OH)_2D_3$

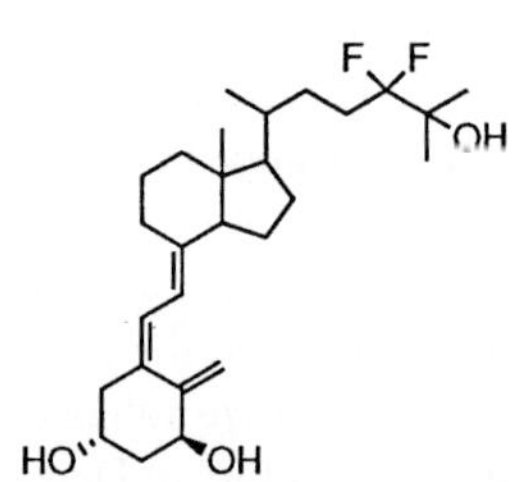

(11) 24,24-F_2-$1,25(OH)_2D_3$

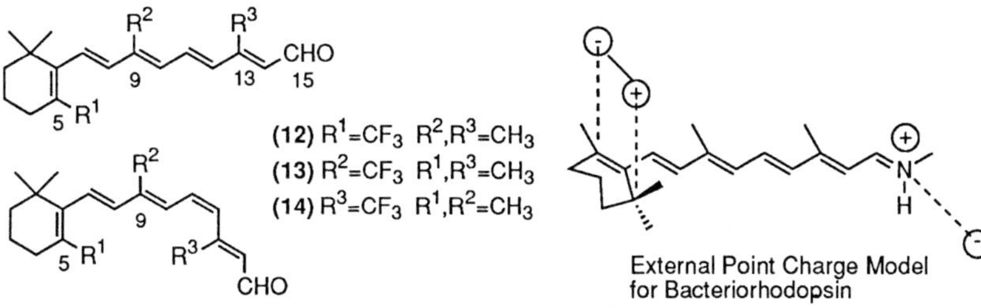

Scheme 18.7.

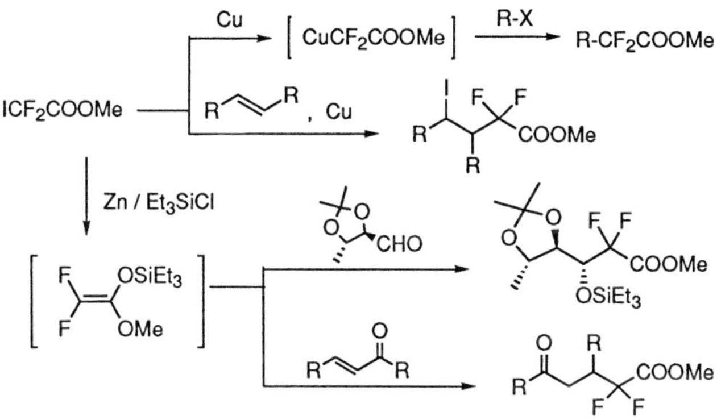

Scheme 18.8.

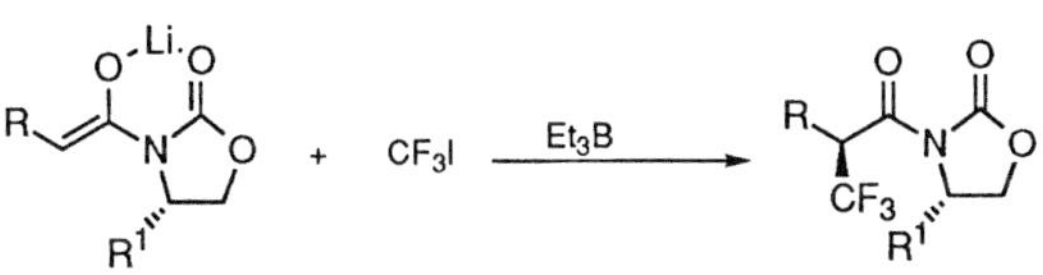

Scheme 18.9.

Collaborative work initiated in the early 80s with Professor Koji Nakanishi (Columbia University) and Professor Robert S. H. Liu (University of Hawaii) involved studies on the structure of photoreceptor proteins such as rhodopsine and bacteriorhodopsine, and the mechanism of the photo-activating process, using structurally modified retinals. The binding experiments of these retinals (**12**-**14**) with apoprotein, in particular that of the retinal having a CF_3 group on the ionone ring (**12**), provided supporting evidence for the external point-charge-model proposed by Nakanishi [36] (Scheme 18.7).

Developments in difluoroacetate chemistry (*e.g.*, Scheme 18.8) leading to efficient syntheses of difluoro analogues of sugars, amino acids and fatty acids of biological interest [37-39] exemplify our work on synthetic methodology during the period 1985-89.

Asymmetric synthesis of chiral fluoro-organic compounds

Since 1990 (at Daikin Industries Ltd.) and in addition to the second-stage study of vitamin D_3, mentioned above, development of efficient synthetic methodology for chiral fluoro-organic compounds has been our main research subject. A highly diastereoselective trifluoromethylation of the lithium enolate of chiral *N*-acyl oxazolidinones with CF_3I mediated by Et_3B as radical initiator was developed in 1993 [40] (Scheme 18.9). In one of our recent programmes it was found that difluoro- and bromofluoro-ketene silyl acetals show a unique temperature-dependent facial selectivity in a catalytic enantioselective aldol reaction, apparently a specific phenomenon associated with the fluorine-substituted ketene silyl acetals [41].

References

1 N. Ishikawa, in R. E. Banks, B. E. Smart and J. C. Tatlow (eds.), *Organofluorine Chemistry, Principles and Commercial Applications*, Plenum Press, New York, 1994, pp. 609-615.

2 H. Goldwhite, in R. E. Banks, D. W. A. Sharp and J. C. Tatlow (eds.), *Fluorine: The First Hundred Years (1886-1986)*, Elsevier Sequoia, Lausanne and New York, 1986, pp. 109-132.

3 H. Suzuki, *Kagaku (Chemistry)*, *48* (1993) 470.

4 Y. Kobayashi, *J. Fluorine Chem.*, *90* (1998) 175.

5 Bibliography of Scientific and Industrial Reports, distributed by the Office of the Publications Board, United States Department of Commerce (1945).

6 R. Kojima, M. Iwasaki and S. Nagase, *Nagoya Kogyou Gijyutu Sikensyo Hokoku (Report of Government Ind. Res. Inst., Nagoya)*, *3* (1954); R. J. Kojima, M. Iwasaki, S. Nagase and H. Baba, *ibid.*, *5* (1956) 225.

7 S. Nagase, *Fluorine Chem. Rev.*, *1* (1967) 77-106.

8 T. Abe and S. Nagase, in R. E. Banks (ed.), *Preparation, Properties and Industrial Applications of Organofluorine Compounds*, Horwood Ltd., Chichester (UK), 1982, pp. 19-43.

9 J. H. Simons, H. T. Francis, J. A. Hogg, W. J. Harland, W. H. Pearlson, T. J. Brice, W. A. Wilson and R. D. Dresdner, *J. Electrochem. Soc.*, *95* (1949) 47-64.

10 R. Kojima, T. Hayashi and S. Takagi, *Nagoya Kogyou Gijyutu Sikensyo Hokoku (Report of Govt. Ind. Res. Int., Nagoya)*, *9* (1960) 516.

11 S. Nagase and R. Kojima, *Bull. Chem. Soc. Jpn.*, *34* (1961) 1468; S. Nagase and R. Kojima, *Kogyo Kaguku Zasshi (J. Chem. Soc. Japan, Ind. Chem. Sec.)*, *64* (1961) 1397; S. Nagase, H. Baba and R. Kojima, *Bull. Chem. Soc. Jpn.*, *36* (1963) 29.

12 R. Kojima, S. Nagase, H. Muramatsu and H. Baba, *Kogyo Kagaku Zasshi (J. Chem. Soc. Japan, Ind. Chem. Sec.)*, *60* (1957) 499.

13 P. Sartori, *Angew. Chem.*, *75* (1963) 417.

14 S. Nagase, K. Tanaka and H. Baba, *Bull. Chem. Soc. Jpn.*, *38* (1965) 834.

15 A. J. Rudge, in A. T. Kuhn (ed.), *Industrial Electrochemical Processes*, Elsevier, Amsterdam, 1971, pp. 71-88.

16 T. Abe, K. Kodaira, H. Baba and S. Nagase, *J. Fluorine Chem.*, *12* (1978) 1.

17 T. Abe and S. Nagase, *J. Fluorine Chem.*, *13* (1979) 519; T. Abe, E. Hayashi, H. Baba, K. Kodaira and S. Nagase, *ibid.*, *15* (1980) 353.

18 M. Le Blanc and J. G. Riess, in R. E. Banks (ed.), *Preparation, Properties and Industrial Applications of Organofluorine Compounds*, Horwood Ltd., Chichester (UK), 1982, pp. 82-138.

19 T. Abe, H. Baba, E. Hayashi and S. Nagase, *J. Fluorine Chem.*, *23* (1983) 123.

20 T. Abe, E. Hayashi, H. Fukaya and H. Baba, *ibid.*, *50* (1990) 173.

21 T. Abe and E. Hayashi, *Chem. Lett.*, (1988) 1887; T. Abe, E. Hayashi and T. Shimizu, *ibid.*, (1989) 905.

22 A. Viji, R. L. Kirchmeier, J. M. Shreeve, T. Abe, H. Fukaya, E. Hayashi, Y. Hayakawa and T. Ono, *Inorg. Chem.*, *32* (1993) 5011.

23 Y. Hayakawa, H. Fukaya, E. Hayashi, M. Nishida, T. Abe, N. Nose, T. Shimizu and M. Tatemoto, *Polymer*, *36* (1995) 2807.

24 K. Takashashi, T. Inomata, H. Fukaya and T. Abe, in A. W. Miziolek and W. Tsang (eds.), *Halon Replacements – Technology and Science*, ACS Symposium Series 611, 1995, pp. 139-150.

25 Y. Kobayashi and I. Kumadaki, *Tetrahedron Lett.*, (1969) 4095.

26 Y. Kobayashi, K, Yamamoto and I. Kumadaki, *Tetrahedron Lett.*, (1979) 4071.

27 Y. Kobayashi, K. Yamamoto, T. Asai, M. Nakano and I. Kumadaki, *J. Chem. Soc., Perkin Trans. 1*, (1980) 2755.

28 D. M. Wiemers and D. J. Burton, *J. Am. Chem. Soc.*, *108* (1986) 832.
29 Y. Kobayashi and I. Kumadaki, *Acc. Chem. Res.*, *11* (1978) 197.
30 D. V. Santi and T. T. Sakai, *Biochemistry*, *10* (1971) 3598.
31 Y. Kobayashi and I. Kumadaki, *Acc. Chem. Res.*, *14* (1981) 76.
32 Y. Kobayashi and I. Kumadaki, *Top. Current Chem.*, *123* (1984) 103.
33 Y. Kobayashi and I. Kumadaki, in A. Katritzky (ed.), *Advances in Heterocyclic Chemistry*, Vol. 31, Academic Press, New York, 1982, p. 169.
34 Y. Kobayashi and T. Taguchi, in R. Filler and Y. Kobayashi (eds.), *Biomedicinal Aspects of Fluorine Chemistry*, Elsevier Biomedical Press and Kodansha, 1982, p. 30.
35 K. Iseki and Y. Kobayashi, in I. Ojima, J. R. McCarthy and J. T. Welch (eds.), *Biomedical Frontiers of Fluorine Chemistry*, ACS Symposium Series, No. 639, 1996, p. 124.
36 V. J. Rao, F. Derguini, K. Nakanishi, T. Taguchi, A. Hosoda, Y. Hanzawa, Y. Kobatashi, C. M. Pande and R. H. Callender, *J. Am. Chem. Soc.*, *108* (1986) 6077.
37 T. Taguchi, O. Kitagawa, T. Morikawa, T. Nishiwaki, H. Uehara, H. Endo and Y. Kobayashi, *Tetrahedron Lett.*, *27* (1986) 6103.
38 O. Kitagawa, A. Miura, Y. Kobayashi and T. Taguchi, *Chem. Lett.*, (1990) 1011.
39 O. Kitagawa, T. Taguchi and Y. Kobayashi, *Tetrahedron Lett.*, *29* (1988) 1803.
40 K. Iseki, T. Nagai and Y. Kobayashi, *Tetrahedron Asymmetry*, *45* (1994) 961.
41 K. Iseki, Y. Kuroki and Y. Kobayashi, *Synlett.*, (1988) 437.

Chapter 19

THE DISCOVERY OF SUCCESSFUL DIRECT FLUORINATION SYNTHESES: THREE ERAS OF ELEMENTAL FLUORINE REACTION CHEMISTRY

RICHARD J. LAGOW[1]

Department of Chemistry and Biochemistry, University of Texas at Austin, Austin, TX 78712-1167, USA

Eras defined

Our development of successful direct fluorination technology has passed through three distinct eras since 1963. The first era, which we shall call the *Inception*, comprises the years from 1964 through 1970, and was a period in which many novel organic and inorganic classes of fluorine compounds were prepared. However, the reactions proceeded on a long time scale (up to 7 days) and gave only two to five grams of product per run. Then came the *Interim* period in which great advances in capabilities and yields were made: it covered the years 1970 through 1992, and was an era when direct fluorination was done in ways that had much wider applications and capabilities. Reactions were performed using a number of different reactor styles and techniques. In this era, product yields normally fell in the 70-90% range, with product quantities per run ranging from five to forty grams.

The third era – *Renaissance* (1992 to the present day) – saw the range of fluorinated inorganic, organometallic, and organic compounds capable of being prepared by our techniques increase markedly. Product yields mostly exceeded 98%, and early in this era the first commercial-scale direct fluorination technology was invented at Exfluor Research Corporation in Austin, TX, by me and my colleagues, Tom Bierschenk, Tim Juhlke and Hajimu Kawa.

The Inception era

The strategy behind the early successful direct fluorination technology was generated by John L. Margrave and me at Rice University, Houston, Texas. Later, after I'd been appointed to the faculty of the Massachusetts Institute of Technology, we joined forces to describe that strategy in great detail in a 1979 review [1]. An important tactic was to heavily dilute fluorine with helium or nitrogen in the initial stages of a fluorination reaction, then gradually increase the fluorine concentration to bring the reaction to completion over several days. While this and other features of our work have been well documented [1-4], and now are widely understood, the circumstances surrounding the discovery of successful techniques for direct perfluorination of organic and organoelemental materials have not been discussed in detail previously. In fact, it all began with work on the fluorination of carbon itself.

[1] L. N. Vauquelin Regents Professor of Chemistry and Materials Science.

Graphite fluoride[2]

In the spring of 1963 through the fall of 1964, I was a young undergraduate student at Rice University who loved to play American football and became part of a superb research group built up by Professor John L. Margrave (Fig. 19.1). Ever since junior high school I'd wanted to be a high-temperature/high-energy chemist, but in some classroom lectures delivered by Margrave I also became fascinated with the possibilities of making fluorine compounds. Margrave graciously accepted me as a fledgling researcher in his laboratory and, quite significantly (in the story of direct fluorination) I focused first on the synthesis of graphite fluoride, working with two postdoctoral fellows, James Wood and Ram Badachappe, to learn how to handle fluorine.

Graphite fluoride had first been prepared in the 1930s by Ruff and later by the Rüdorffs and Palin and Wadsworth in the 1940s. At that time, perfluorinated (pure white) graphite fluoride had not been prepared. In Margrave's lab nearly white graphite fluoride had been obtained once or twice by fluorinating carbon at temperatures over 700 °C, but the success rate was low (about once in every 20 runs). By using electronic temperature controllers, I was able consistently to make twenty-gram quantities of snow-white superstoicheiometric graphite fluoride of composition $CF_{1.12}$ [1]. This impressed most of Margrave's research staff, and we sent samples to the National Aeronautic and Space Administration who reported that it was an excellent lubricant and stable at high temperatures. Later, at the US Army's Fort Monmouth research centre, Dr Herbert Hunger established that it was a terrific material for high-performance lithium batteries. This battery technology later spread to the Eagle Pitcher Battery Company and widely across Japan.

Margrave started MarChem Corporation and sold white $CF_{1.12}$ graphite fluoride samples under the tradename CFX® for ten US dollars per gram. I built a high-temperature reactor capable of preparing twenty grams of this material per day (Fig. 19.2) and between classes and football practice, was able to earn about a hundred dollars per day under an agreement whereby Margrave split the profits with me. This enabled me to buy a new Buick when I collected my PhD degree in 1969 and left Rice to join the faculty at MIT.

Why is this relevant? It so happens in the field of synthetic chemistry that John and I did not set out to solve the 'hundred-years-old' problem of how to control the reactions of elemental fluorine with organic compounds. Rather, there were questions about the structure of graphite fluoride, since it had proven impossible to get a single crystal structure of graphite fluoride itself. Proposals by several people, including the Rüdorffs and Lagow and Margrave, later on turned out to be correct.

The first successful direct fluorinations of organic compounds involved work that I judged might produce lower-molecular-weight structural analogues of 'white' graphite fluoride, the objective being to produce a small-scale version of graphite fluoride so that a crystal structure could be obtained. The problem was that fluorine was reported to react vigorously and often explosively at or below room temperature with all hydrocarbons whereas graphite only begins to react with fluorine at 400 °C. The subtrates chosen were

---

[2]Graphite fluoride is a solid, layered, nonstoicheiometric perfluorocarbon of empirical formula CF_x, where $0 < x \leqslant$ *ca.* 1.25, obtained by treating carbon with F_2; the value of x defines the grade of material under discussion [5]. The snow-white variety, composition $CF_{>1.0}$, is sometimes called superstoicheiometric poly(carbon monofluoride); this is generally believed to have a lamellar structure of weakly-coupled nongraphitic sheets constructed from an infinite array of *trans*-linked cyclohexane 'chairs', with each sp^2 carbon covalently bound to a fluorine atom [3, 5] and carbons on the edge of the sheet bound to two fluorine atoms.

Fig. 19.1. The author (right) and Professor John L. Margrave: No one could have had a better Research Advisor.

Fig. 19.2. Multi-tray high-temperature reactor for preparing graphite fluoride.

the polynuclear aromatic hydrocarbons coronene and ovalene. The literature contained no reports on reactions involving the interior carbons of such polynuclear hydrocarbons; electrophilic substitution of peripheral hydrogen was known, of course.

A 3/4″ thick glass-encased reactor with stout plates of nickel at the top and bottom was on hand in Margrave's lab to use as an observation vessel when fluorine bomb calorimetrists had trouble achieving the complete combustion of compounds in fluorine; this way they could observe what it took to achieve their objective.

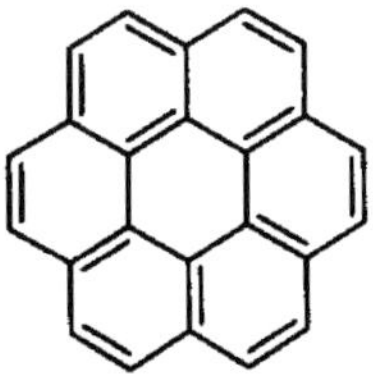

Coronene

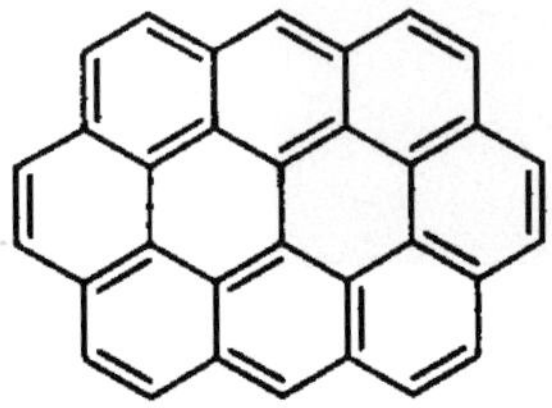

Ovalene

I crafted a 3″ by 3″ tray out of nickel foil, placed about one gram of finely powdered coronene on this handmade 'boat' and placed it in the observation reactor so that just what happened when elemental fluorine contacted the hydrocarbon could be clearly seen. (The manifold that we used to deliver and control the fluorine is shown in Fig. 19.2.) As I cracked open the heavy autoclave valves slightly on the fluorine line, a flame about an inch high was observed in the reaction vessel. This flame disappeared when the valve was turned off after a short delay. Incredibly, even with pure fluorine, if the pulses were short one could reignite the reaction for several short spurts of fluorination. The material in the centre of the sample of bright orange coronene powder charred and became black in colour, but surprisingly around the edges of the sample, there appeared to be a white ash. When I tried lower concentrations of fluorine in the ballast-tube fluorine source, the amount of white ash increased. Actually, I thought that the reaction flame was rather tame. Since I was looking for a white solid comparable to graphite fluoride, I carefully used a spatula to separate the 'ashes' from the obviously decomposed centre section of the coronene sample and measured their infrared spectrum. You can image my delight when a strong C-F stretching absorption appeared at 1200 cm^{-1}. Moving next to Margrave's time-of-flight mass spectrometer, the white product was found volatile enough to obtain a nice spectrum which contained a molecular ion corresponding to perfluoroperhydrocoronene as well as an $[M-F]^+$ peak. Later that day, I fluorinated ovalene (a burnt orange powder) and found again that all the aromatic systems were saturated with fluorine and all the hydrogen replaced, according to the results of MS and IR analyses.

From this work, I quickly concluded that if such a result could be obtained just by controlling the fluorine supply manually with a fairly coarse valve of the type used, then even greater success would surely be achieved if one could carefully measure and monitor as well as control the supply of elemental fluorine. At the time I had no real idea what the consequences would be; although later just what goes on in a simple elemental fluorine reaction was explained in detail [1, 2]. No careful calculations of the thermodynamics and kinetics of fluorine's reactions with hydrocarbons were available to guide me in those early days in Margrave's lab – just instinct and curiosity! Obviously, the central regions of the coronene and ovalene samples had become charred, crosslinked and decomposed. In retrospect, the fluorinated white powders obtained contained at least fifty percent of 'polymeric' (crosslinked) material as well as saturated perfluorinated analogues of the aromatic substrates.

My search for F_2 control devices led me to one of my colleagues, Romy Bautista who told me about new Monel mass flowmeters that worked electronically with a flow-cooling transducer. Margrave had one on order at the time, and I ordered a second one. Initially, though, I controlled fluorine flow with Monel needle valves and found that passage of mixtures of fluorine and helium for 24 hours over finely-powdered coronene or ovalene contained in a long nickel boat inside a nickel tube reactor (fitted with Teflon® O-rings) converted all of the so coloured hydrocarbons to snow-white fluorocarbon materials *with no charring whatsoever!*

My mind now turned to *fluorinations of possible commercial significance*. It did not take long to focus in that regard on the fluorination of finely-powdered polyethylene. In the same nickel apparatus, and using a continuous flow system starting with a 100% helium atmosphere then gradually increases the concentration of fluorine from 0 to 100%, I was able to prepare a perfluorinated polyethylene powder. When this new material was placed on a hot plate, I could clearly see that the high-temperature properties and inertness (to oxidation in air) were such that there was no doubt that is was perfluorocarbon in nature. At the time, the price of polytetrafluoroethylene was around \$6.50 a pound, so what a bonanza, I thought, if one could make that same material with fluorine at \$3.00 a pound and polyethylene at 30 cents a pound. (Later, it became clear that although the fluorinated polyethylene had all of the thermal and physical properties one associates with DuPont's Teflon®, it was crosslinked and possessed other disadvantages.)

I moved next to the fluorination of polyethylene bottles and vials. Surface fluorination gave them Teflon-like properties at a much lower price, and news of this was covered by *C&EN* in January 1970 [6]. Representatives of Air Products and Chemicals as well as Eddie Hedeya of Union Carbide really paid attention. The Rice University patent attorneys did not do a good job in protecting our intellectual property (they were oilfield attorneys), and both companies initiated work on the fluorination of plastic bottles. Air Products called their process 'Airopak', and Union Carbide labeled their technique the 'Linde Process' without licensing the Rice University patents. Margrave called the Rice University process the 'Fluorokote' technique. Margrave continued to suggest further research in the area, and I found samples (organic and inorganic) on my desk labelled 'Let's fluorinate this and see what happens. JLM'. So even though I continued to spend a lot of time playing football, I also kept four fluorination reactors running around the clock in my California-style hood, even on weekends. There were two things that I loved to do, one was synthetic chemistry and the other was making tackles and sacking quarterbacks (Fig. 19.3).

Although our direct fluorination advance first came to light worldwide in *C & E News* in 1970 [6], it had been reported verbally in the US at a major meeting. In the winter of 1969 Professor Margrave had announced that he had scheduled the first public lecture on this discovery at a fluorine conference at Marquette University, and wanted me, at 21 years of age, to stand in for him as he had a conflict. To help with this task, he gave me a copy of Harry Emeléus's new book *The Chemistry of Fluorine and Its Compounds* (Academic Press, 1969) to read on the planes connecting to Marquette. Earlier in the fall of 1969, I had been interviewed for a faculty opening at MIT and presented a lecture on our breakthrough

Fig. 19.3. Dick Lagow (left) and George Schulgen – outstanding Rice Owls players in 1966 (the Rice Owls lost to the US National Champions 27 to 24 when UCLA kicked a field goal in the last three seconds; Lagow made 22 tackles in the game).

in elemental fluorine chemistry. This was followed two weeks later by an interview at UC Berkeley where I met the distinguished inorganic fluorine chemist, Neil Bartlett. I was offered the post at Berkeley on the spot, and a week and a half later Al Cotton called to offer me an assistant professorship at MIT – which I accepted. Al Cotton and Neil Bartlett have become life-long friends of mine.

I remember meeting Darryl DesMarteau for the first time at Marquette. He was given a standing ovation before his talk as he was coming back from an accident in Cady's laboratory which had left him with only one natural arm. I have never seen anybody 'less handicapped' than Darryl. He is a strong fearless person with no mental scars and has become one of the world's best fluorine chemists. Also lauded with a standing ovation at the meeting was Jean'ne Shreeve, who had succeeded in being the first person to prepare tetrafluorourea, a synthetic target for many from 1965 through the mid-1970s, when defense programmes were in search of novel propellants and oxidizers.

I'd realized before going to Marquette that what had been tersely presented in *C&EN* was controversial and provided little proof that our discovery was indeed real. Margrave warned me not to discuss direct fluorination in too much detail or the patents which Rice University had applied for could be jeopardized. At the end of my lecture there was quite a lengthy question-and-answer session with members of the fluorocarbon hierarchy, who were unconvinced. That great gentleman and eminent fluorine chemist George Cady stood up and asked: 'Are you are saying that you can take naphthalene and convert it with elemental fluorine to perfluorodecalin and obtain a 60% yield?' 'Yes', I replied. 'Will you take your chalk then and show me on the board exactly how you do this?', he continued. Although Cady was his usual gracious self, he was obviously unconvinced when I gave more detail, while explaining that my instructions were not to give complete descriptions of the apparatus and techniques.

A year later, at the first Winter Fluorine Conference (St. Petersburg, Florida), I was invited to sit at Paul Tarrant's table, and so became aware of his great sense of humour (I already knew of Paul Tarrant's great chemistry). Midway into the meal, and after telling a lot of jokes and being extremely friendly, Paul leaned over and said, 'Organic compounds burn in elemental fluorine, don't they?' I thought carefully for several moments then answered, 'Paul, it really depends on how you do it' (which was a pretty good retort for a twenty-three year old). Seriously, fluorine chemists are some of the nicest people in the world, and over the years I have formed a great affection for these wonderful people.

As this era closed, we had proven that finely powdered hydrocarbon solids can be converted to fluorocarbon analogues using controlled elemental fluorine as a reagent [4, 7]. The reaction times were often long (sometimes five to seven days in length) and scales were small (usually 1-2 g), but products hardly dreamed of previously had been made and characterized. The game was on.

The Interim period

In 1969, I moved to MIT and rapidly built up a fluorine laboratory; this marked the beginning of the second era of direct fluorination synthesis with elemental fluorine in which the research goals changed. At this time, of course, it was necessary to raise research money, and in this regard, people in the Air Force funding system made major contributions to the work. First in line was the fluorine chemist, Dr Christ Tamborski, who strongly believed in direct fluorination and the potential of direct fluorination. The other key person was Dr Tony Matuszko of the Air Force Office of Scientific Research in Washington. At this stage, it was not in general possible to successfully fluorinate volatile samples or liquid materials. More importantly, the fluorination of volatile hydrocarbons, particularly those containing oxygen, had not been attempted except in a very cursory and exploratory way at Rice.

Tamborski invited me to visit Wright-Patterson Air Force Base and the US Materials Laboratory in Dayton, Ohio. He then asked if my group would focus on producing oxygen-containing compounds, particularly perfluoropolyethers. Tamborski wanted to know if it was possible for us to preserve carbon-oxygen bonds at all. I informed him that we did not know for certain but we thought we could succeed. We had some ideas that might facilitate that result, and we were certainly anxious to try. Tamborski arranged for a \$50 000 per year grant and told me that if we were able to convert oxygen-containing hydrocarbon compounds to perfluorocarbon analogues he would double this amount in the second year. In 1969 and 1970, this was a lot of research money.

Most of the oxygen-containing organic substrates in which the Air Force was immediately interested were volatile liquids. Obviously a new reactor system or set of systems would have to be designed. The way to proceed was with frozen liquids, but of course, even our freeze dried liquids in a powder form would have offered problems because only the surfaces would have been fluorinated.

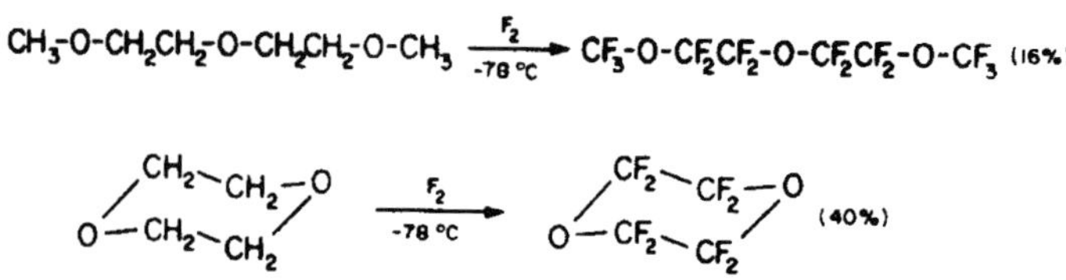

Scheme 19.1.

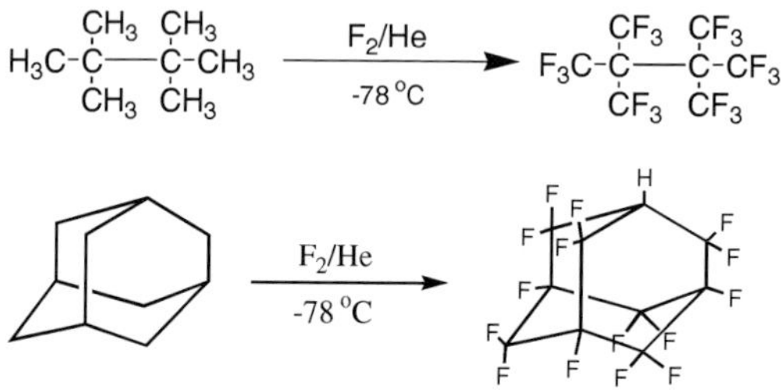

Scheme 19.2.

In the first days of the Lagow laboratory and throughout the next few years the key contributors were Dr Norma Maraschin, who was my first graduate student at MIT, Dr Ed Liu, and Dr James Adcock, who joined after receiving a PhD at the University of Texas at Austin. Various trials brought us the concept of the cryogenic 4- and 8-zone reactors [6] (Fig. 19.4) in which liquids and volatile materials were distilled slowly through low-temperature gradients until they became perfluorinated and then collected downstream. The Lagow research group with Jim Adcock at the bench was very successful in preserving carbon-oxygen bonds using this technique [8-10] (*e.g.* Scheme 19.1); he was such a talented chemist that I've always said 'having him on a project increased the yield by ten to fifteen percent'. Other new achievements included the synthesis of structurally unusual fluorocarbons by Norma Maraschin [11, 12] (*e.g.* Scheme 19.2) and the successful fluorination of metal alkyls by Ed Liu [13]. Products were still made on only a two gram scale but the work was noteworthy because the synthetic achievements were, in general, unprecedented.

Christ Tamborski was very proud of the success achieved with oxygen-containing organic compounds, particularly in the field of ethers. He wanted us to proceed with higher molecular weight materials, but at that time (1972) the basic research part of the military budget was cut so badly that the Wright-Patterson people decided to suspend support for academic chemistry. However, Tamborski was such a strong ally that he convinced Dr Anthony Matuszko to accept an AFOSR proposal on the same fluorination topic, so at least the Air Force Office was able to send me thirty-five thousand dollars. This was the first new grant that AFOSR had given to anyone for two years (since the end of the Sputnik funding era). Primary credit for the successes we enjoyed in the *Interim* period will always be due to Tony Matuszko and (with great appreciation) to Dr Don Ball, who was director of the Air Force Office of Scientific Research Chemistry Division. Tony was the best programme officer that my group has encountered in thirty years of chemistry. Matuszko took a lot of pride in what he supported, and he was a strong supporter of direct fluori-

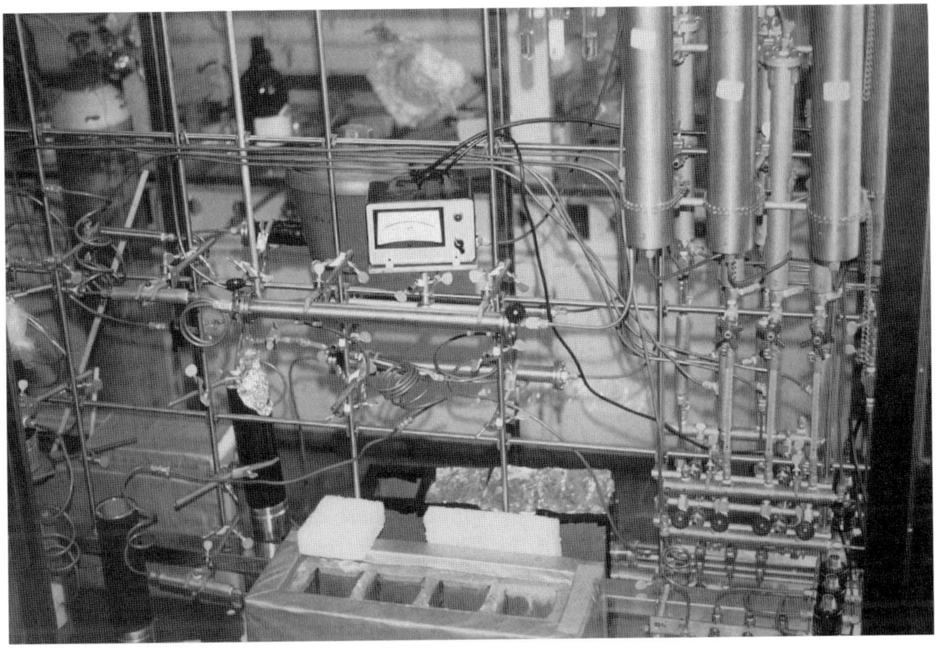

Fig. 19.4. The first 4-zone cryogenic fluorine reactors and reaction system.

nation. Throughout the years, Tamborski stayed a strong proponent of direct fluorination (he was right! – as revealed by progress during the early years of the Exfluor Research Corporation [14]).

During the first part of the *Interim* era and extending into the second part, there were still a good number of fluorine chemists who did not believe that our fluorination techniques worked. Some actually thought that we were faking the results and simply fabricating spectra of the unusual new compounds we reported. It is fun to be a source of controversy and have achievements doubted: one receives a lot of invitations to speak and is often challenged. I have been quoted as saying, 'If one makes controversial and substantial breakthroughs and is proven right, then there is a lot of fun to be had in the process'; however, I suspect that it might not be much fun at all to be in that position and be proven incorrect.

In the second phase of the *Interim* period, the leading researchers in my laboratory were PhD students, Hsu-Nan Huang, Todd Milsna, Win-Huey Lin and Robert Aikman. This was the era of the disk reactor, and amelioration of reactions with many new ether and polyether structures by scavenging HF with potassium fluoride or sodium fluoride to preserve these structures. The quantity of product was increased to the eight-to-ten grams range, and a very wide range of new classes of organometallic, inorganic and organofluorine compounds were prepared.

The Renaissance period

From the inception of the Lagow Laboratory in 1969 in the Department of Chemistry at the Massachusetts Institute of Technology, direct fluorinations were normally conducted on a five to twenty gram scale, and it was still not clear that it would be possible to scale up any of the reactions to commercial proportions. During this period over two thousand new organofluorine compounds and perfluorinated organometallic compounds were generated in the academic laboratories, first at MIT and then (post 1976) at The University of Texas at Austin. Leaders in the Lagow academic laboratory during the Renaissance phase were Dr Joel Kampa, Dr Han-Chao Wei, and Dr Tzuhn-Yaun Lin, and graduate students Ryan Callahan, Cameron Youngstrom, Dr Kuansen Sung and Dr Koichi Murata. They were major factors in our success. It is very significant, however, that one of the new concepts enabling high-volume production and scale-up of direct fluorination came from an experiment done 'just for fun' in our laboratory several years previously.

In 1982, my first University of Texas student, Robert Aikman, and I tried using a CFC solvent at −90 °C to moderate the direct fluorination of hexamethyltungsten, a compound first made in about 1974 by Geoff Wilkinson and his co-workers. Despite the weakness of the carbon-tungsten bonds in $W(CH_3)_6$ (probably in the 20-35 kcal mol^{-1} range), the yield of hexakis(trifluoromethyl)tungsten was about 50% (Scheme 19.3) when carried out in the reactor depicted in Fig. 19.5. (Note that this solution-phase reactor was very different from the reactors later developed by myself, Dr Thomas Bierschenk, Dr Timothy Juhlke and Dr Hajimu Kawa for use in the Lagow-Exfluor Process.) Wilkinson's new compound was more stable when complexed with triethylphosphine; Aikman's new perfluoroalkyltungsten counterpart was stable while uncomplexed in chlorofluorocarbon solvents but found to be even more stable when complexed with this phosphine.

Aikman and I found it amazing that 20-35 kcal C−W bonds were surviving collisions with F_2 to such an extent when the overall reaction leading to the conversion of just one C−H bond to C−F is more than 100 kcal exothermic. The intellectual picture that emerged from this finding is that the entire energy generated does not go into vibrational excitation of the tungsten-carbon bonds when a hydrogen attached to carbon in an entity RH is replaced in two steps by fluorine ($RH + F\cdot \rightarrow R\cdot + HF$; $R\cdot + F_2 \rightarrow RF + F\cdot$). Hence, about half of the substrate species survived the 36 step free-radical conversion $W(CH_3)_6 \rightarrow W(CF_3)_6$, in which each of the 18 two-part CH → CF stages generates something of the order of 103 kcal mol^{-1} of energy. Therefore the equilibrium constant (K) for this reaction is often greater than 10^{12}! Note that with the Lagow-Aikman reactor there was no continuous addition of substrate, the reactant was highly diluted, and the reaction was conducted on only about a half-gram scale. However, this experiment taught us that one needs to have rapid vibrational relaxation in reactions between F_2 and hydrocarbon moieties, and that cold 'inert' solvents efficiently promote such relaxation. Initiation of radical chain fluorination at low temperatures can be ascribed to molecule-induced homolysis of molecular fluorine, $RH + F_2 \rightarrow R\cdot + HF + F\cdot$, which is spontaneous at room temperature although the enthalpy is slightly positive (for $RH = CH_4$: $\Delta G_{298} = -3.8$ kcal mol^{-1} and $\Delta H = 3.9$ kcal mol^{-1}). Studies by researchers such as Dick Bernstein (UCLA), Doug McDonald (University of Illinois) and Dick Zare (Stanford) have revealed that the activation energy for abstraction of hydrogen atom from an alkane by molecular fluorine in the gas phase ranges from zero to 'as high as' 1 cal (*not kcal*) mol^{-1} [4]. This reaction is one of the fastest chemical reactions known.

Details of the design and operation of very successful lab-scale solution-phase fluorination reactors created in my academic laboratory at the University of Texas can be found

$$W(CH_3)_6 \xrightarrow[CFCl_3,\ -90\,^{\circ}C]{F_2/He} W(CF_3)_6 \sim 50\% \text{ yield} \xrightarrow[CFCl_3]{PEt_3} W(CF_3)_6 \cdot PEt_3$$

Scheme 19.3.

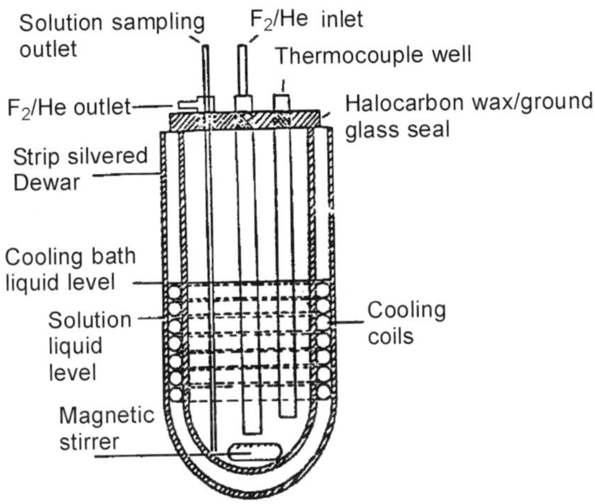

Fig. 19.5. Solution-phase reactor used to prepare $W(CF_3)_6$.

Fig. 19.6. Exfluor's talented staff – (left to right) Tim Juhlke, Tom Bierschenk, Hajimu Kawa and the author (far right) was very fortunate to have them.

in a recent review [4]. These have been used by Dr Han-Chao Wei to prepare, for example, the unique perfluorinated crown ethers **1-4** [4, 15].

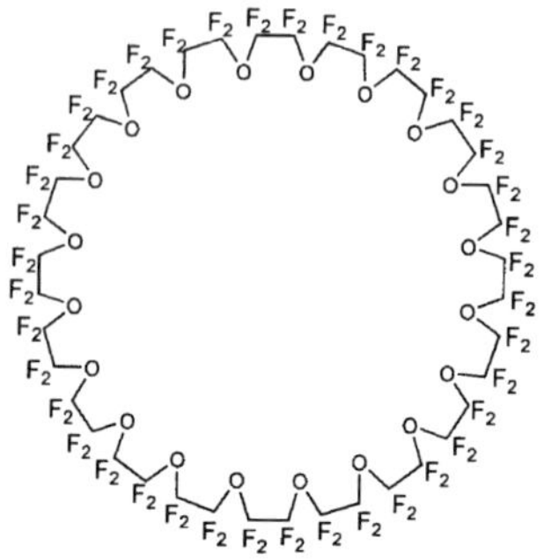

(1)

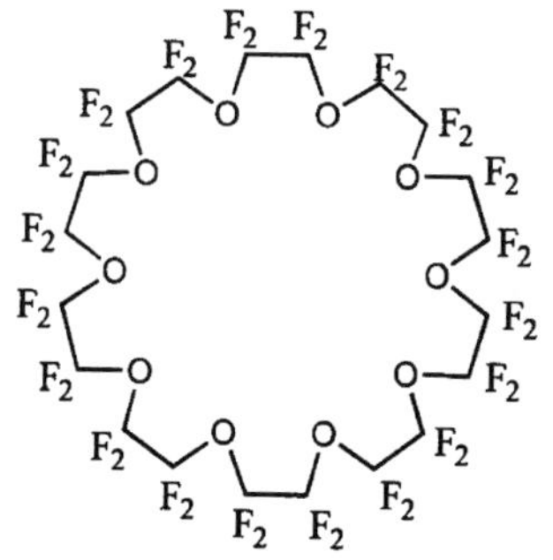

(2)

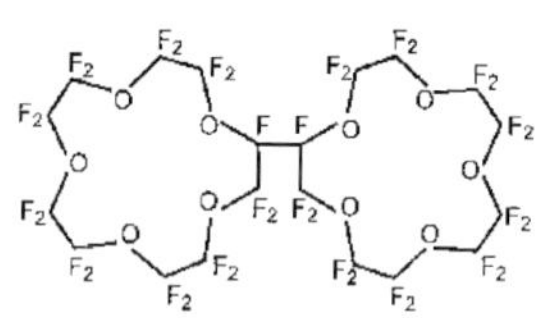

(3)

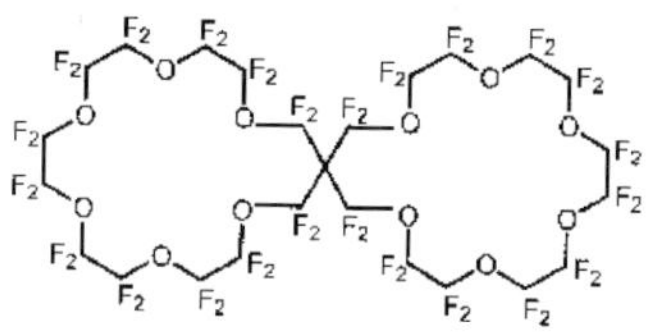

(4)

Liquid-phase direct fluorination technology has been developed further at the Exfluor research Corporation of Austin (Texas) in collaboration with two of my former graduate (PhD) students Dr Tom Bierschenk (University of Texas at Austin), Tim Juhlke (MIT) – both hired in 1982 – and Dr Hajimu Kawa (Tokyo Institute of Technology), who came in 1984. These talented people (Fig. 19.6) teamed up with me to generate the Lagow-Exfluor elemental fluorine process which is described in several US patents [16]. The procedure involved enabled extremely high yields of perfluorinated products to be achieved (often 95-99%) on a kilogram to multi-ton scale. Products sold by Exfluor include $CF_3(CF_2)_xCO_2H$ (x = 9, 10, 12, 14, 16), $HO_2C(CF_2)_xCO_2H$ (x = 4, 6, 8, 10), $CF_3(CF_2)_8Br$, and $Br(CF_2)_8Br$ [4]. For the records, I founded 'Exfluor' in 1987; it was funded primarily by Federal government research contracts and its goal was to develop direct fluorination technology.

Exfluor Research Corporation uses it commercial-scale fluorination reactors to produce extraordinarily pure fluorocarbons that are especially suitable as new biomedical and biomaterials since they are often orders of magnitude purer than those obtainable via other routes. Our procedures often produce only a single compound containing virtually no hydrogen and without complications arising from the occurrence of crosslinking during the fluorination. For example, some of the perfluoroethers synthesized contain residual hydrogen in concentrations below 3 parts per billion; to put that in perspective, in polytetrafluoroethylene made from tetrafluoroethylene monomer, the hydrogen content is several parts per million.

Epilogue

Way back in the late '60s, John Margrave and I realized immediately that we had made a substantial and perhaps very important discovery. In retrospect, however, I could not have predicted the breadth of the impact direct fluorination has had on synthetic capabilities in organofluorine and fluoro-organoelemental areas. I certainly learned a lot from John, and I have been very fortunate, first at MIT and then at the University of Texas and at Exfluor, to have been able to surround myself with excellent young chemists. I've always done my best to encourage them to think for themselves, to be innovative, and to try to do things that many would expect to fail. Even though direct fluorination isn't the answer to every synthetic challenge, it looks now that it will become the most broadly applicable general synthetic technique utilized by organofluorine chemists (both in the laboratory and on a commercial scale) during the next century.

Acknowledgement

Fluorine chemistry at the University of Texas is funded by the US National Science Foundation (CHE 9972888).

References

1 R. J. Lagow and J. L. Margrave, 'Direct Fluorination: A New Approach to Fluorine Chemistry', *Prog. Inorg. Chem.*, *26* (1979) 161.
2 R. J. Lagow, 'High-Yield Reactions of Elemental Fluorine', in R. E. Banks, D. W. A. Sharp and J. C. Tatlow (eds.), *Fluorine: The Fist Hundred Years (1886-1986)*, Elsevier Sequoia, Lausanne and New York, 1986, pp. 321-325.
3 R. J. Lagow, 'Direct Fluorination', in M. Howe-Grant (ed.), *Fluorine Chemistry: A Comprehensive Treatment*, John Wiley, New York, 1995, pp. 242-259. (Reprinted from the Kirk-Othmer Encyclopaedia of Chemical Technology.)
4 R. J. Lagow, in *Organofluorine Compounds (Methods of Organic Chemistry: Houben Weyl)*, Georg Thieme Verlag, Stuttgart, Vol. E10a, 1999, pp. 188-193 ('Reactions of Fluorine with Solids'), pp. 194-201 ('Reactions of Fluorine in the Presence of Solvents').
5 N. Watanabe and T. Nakajima, 'Graphite Fluoride', in R. E. Banks (ed.), *Preparation, Properties and Industrial Application of Organofluorine Compounds*, John Wiley, New York, 1982, pp. 297-322.
6 *Chemical & Engineering News*, 48, No. 2 (January 12) (1970) 41-42.
7 R. J. Lagow and J. L. Margrave, 'Direct Fluorination of Organic and Inorganic Substances', *Proc. Natl. Acad. Sci.*, *67* (1970) 4, 8A.
8 J. L. Adcock and R. J. Lagow, 'The Synthesis of the Perfluoroethers, Perfluoroglyme and Perfluorodiglyme by Direct Fluorination', *J. Org. Chem.*, *38* (1973) 3617.
9 J. L. Adcock and R. J. Lagow, 'The Synthesis of Perfluoro-1,4-dioxane, Perfluoro(ethyl acetate) and Perfluoropivaloyl Fluoride by Direct Fluorination', *J. Am. Chem. Soc.*, *96* (1974) 7588.
10 J. L. Adcock, R. A. Beh and R. J. Lagow, 'Successful Direct Fluorination of Oxygen-Containing Hydrocarbons', *J. Org. Chem.*, *40* (1975) 3271.
11 R. J. Lagow and N. J. Maraschin, 'The Successful Fluorination of Neopentane: A Challenge Met by Direct Fluorination', *Inorg. Chem.*, *12* (1973) 1459.
12 N. J. Maraschin, B. D. Catsikis, L. H. Davis, G. Jarvinen and R. J. Lagow, 'The Synthesis of Structurally Unusual Fluorocarbons by Direct Fluorination', *J. Am. Chem. Soc.*, *97* (1975) 513.
13 E. K. S. Liu and R. J. Lagow, *J. Organometal. Chem.*, *145* (1978) 161; *Chem. Commun.*, (1977) 450; *Inorg. Chem.*, (1978) 618 E.

14 R. J. Lagow, T. R. Bierschenk, T. J. Juhlke and H. Kawa, 'A New Synthesis Procedure for the Preparation and Manufacture of Perfluoropolyethers', in G. A. Olah, R. D. Chambers and G. K. S. Prakash (eds.), *Synthetic Fluorine Chemistry*, John Wiley, New York, 1992, pp. 97-126.
15 H. C. Wei, V. M. Lynch and R. J. Lagow, 'Synthesis of the First Perfluoro-spiro-bis-crown Ethers', *J. Org. Chem.*, *62* (1997) 1527.
16 *US Patent 5,093,432*; *5,461,117*; *5,332,790*; *5,322,904*; *5,322,903*.

Chapter 20

FLOGGING THE FLUOROCARBONS

DAVID M. LEMAL
Department of Chemistry, Dartmouth College, Hanover, NH 03755, USA

In the beginning

As a graduate student working with R.B. Woodward at Harvard on the configurations of desoxy sugars and a synthesis of yohimbine, I fancied myself a natural products chemist. During the seven years which followed as an instructor and assistant professor at the University of Wisconsin, however, my interests strayed in the direction of reactive intermediates and highly strained molecules. My students carried out the first synthesis of bicyclobutane [1] and, then at Dartmouth, obtained the first pure prismane, hexamethylprismane (**1**) [2].

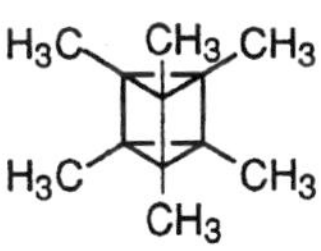

1

Our attempts to synthesize the parent prismane had failed, a feat later accomplished by Katz [3]. We considered our next best choice to be the synthesis of a prismane with the full D_{3h} symmetry of the parent, like **1**, but with strongly electron-withdrawing substituents for comparison with the electron-rich hexamethyl compound. Replacement of the 18 hydrogens of **1** by fluorine sounded ideal. This is how it happened that an undergraduate named Jim Staros, who has since had a distinguished career in biochemistry/molecular biology, made our first foray into organofluorine chemistry. Taking a long-shot approach to the problem, Staros prepared perfluorohexamethylbenzene (**2**) (weathering numerous explosions in the process) and irradiated it with 254 nm light in a flow system [4]. The result was a colourless liquid which contained in addition to starting material no fewer than three of its valence isomers: the benzvalene (**3**), the Dewar benzene (**4**) and the target prismane (**5**). In giving all three valence isomers, **2** is still the most "photoathletic" benzene known. Haszeldine's group independently discovered and thoroughly investigated the photo- and thermal transformations interconnecting this set of valence isomers [5, 6].

The contrast in properties between **3-5** and their methyl counterparts is stark indeed. Whereas hexamethylprismane (**1**) has been known to detonate violently and hexamethylbenzvalene isomerizes on a clean glass surface at room temperature [7], all three of the highly strained perfluorinated valence isomers are extremely stable. Study of the kinetics and thermodynamics of interconversion among the $(CF_3C)_6$ isomers led us to coin the term "perfluoroalkyl effect" to represent the composite of stabilizing influences (both steric and electronic) which perfluoroalkyl groups have on highly strained carbon frameworks [8]. The formation of isomers **3-5** and their remarkable properties, differing so dramatically from those of their hydrocarbon analogues, "hooked" me on fluorine chemistry. While my group continued to work in other areas of organic chemistry as well, organofluorine chemistry became an increasingly important focus of our research efforts.

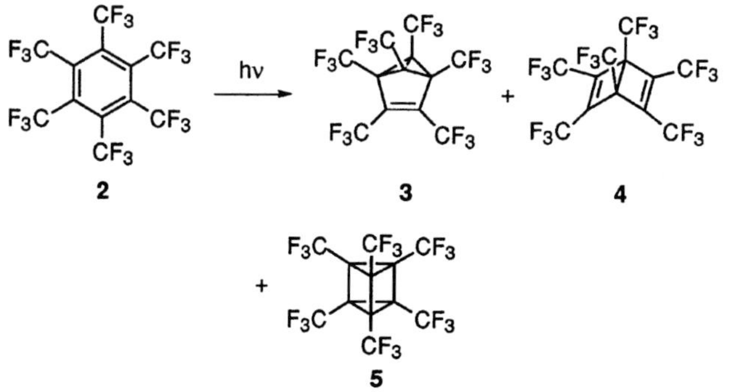

More perfluoroalkylated molecules

One valence isomer of perfluorohexamethylbenzene was still missing, the bicyclopropenyl **6**. To complete the first full set of benzene valence isomers, we synthesized **6** from hexafluoro-2-butyne and chlorotrifluorodiazirine [9]. Although the bicyclopropenyl is thermodynamically the least stable member of the set, it is also the most stable kinetically, aromatizing to **2** with a half-life of at least 2 hours at 360 °C.

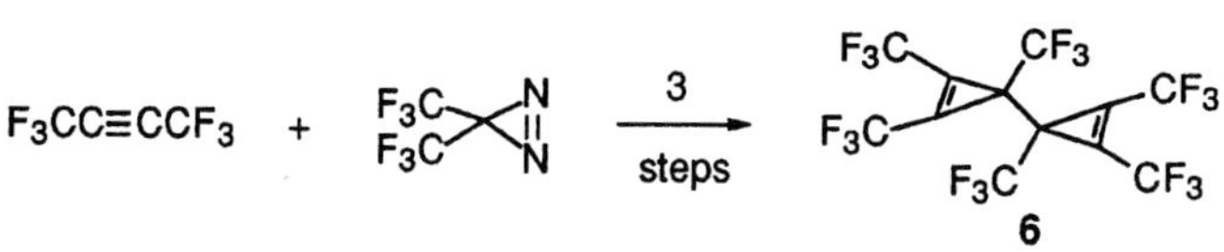

Regarding valence isomers of other aromatic systems, we synthesized Dewar furan **7** [10] and cyclopropenyl ketone **8** [11] from perfluoro-2-butyne, albeit via very different routes. The Dewar form isomerized to the ketone at about 100 °C, and the ketone, which was stable to 350 °C, aromatized almost quantitatively to the furan (**9**) at 250 °C with bromine as catalyst. Photochemically, the furan was found to interconvert with both of the other valence isomers [12].

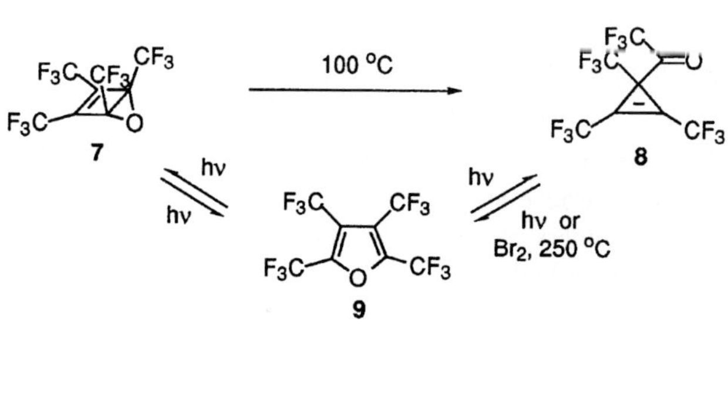

Heicklen [13] and Kobayashi [14] found that irradiation of thiophene **10** gives the Dewar isomer **11**, which very slowly reverts at 150 °C to **10**. We discovered that the labile, violet cyclopropenyl thioketone **12** is also formed as a minor isomer in the photolysis of **10** [15].

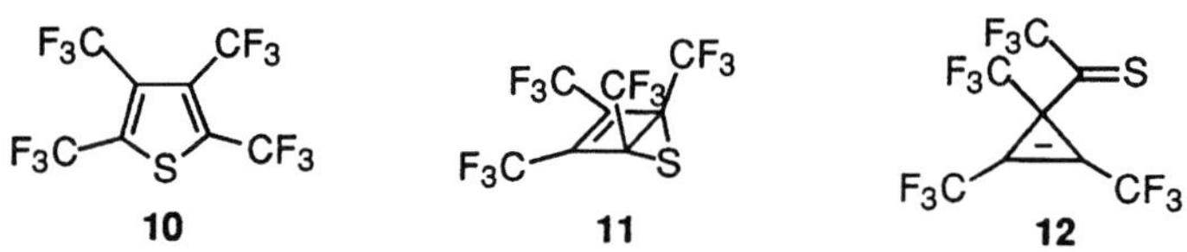

Oxidizing Dewar thiophene **11** with peroxytrifluoroacetic acid, we were surprised to obtain a crystalline compound whose ^{19}F NMR spectrum comprised a single line. Infrared and Raman spectra ruled out a structure with fourfold symmetry, however. Very low temperature NMR measurements by Bushweller confirmed that the molecule has the expected sulfoxide structure **13**, but undergoes a degenerate "walk" rearrangement (better described as a sprint) in which the sulfur migrates around the cyclobutene ring with inversion of configuration at each step ($\Delta H^{\ddagger} = 6.6 \pm 0.2$ kcal/mol, $\Delta S^{\ddagger} = -0.5 \pm 0.6$ e.u.) [16, 17].

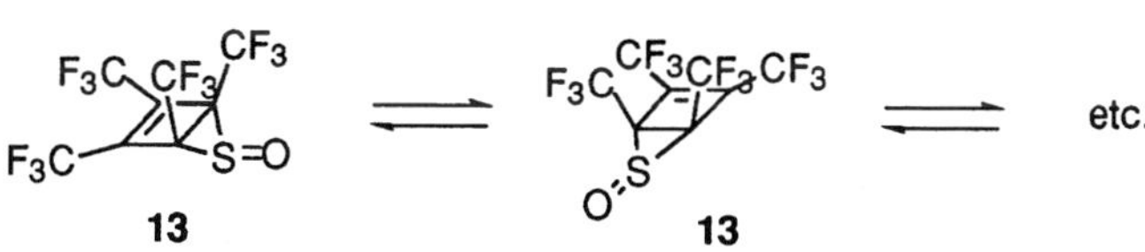

This finding led us to question whether **11** is also capable of degenerate rearrangement. High temperature NMR measurements gave an affirmative answer, but extrapolation of their rate constants showed that **13** rearranges 3×10^{10} times faster than **11** at 25 °C [17]!

The amazing facility of the sulfoxide rearrangement led us to recognize the existence of a class of reactions which we dubbed "pseudopericyclic" [16]. Because the geometry of **13** is not favorable for an allowed pericyclic process, we speculated that the lone pair on sulfur forms the new bond to carbon. We defined a pseudopericyclic reaction as "*a concerted transformation whose primary changes in bonding compass a cyclic array of atoms, at one (or more) of which nonbonding and bonding orbitals interchange roles.*" There is a striking distinction between pericyclic and pseudopericyclic reactions, for the latter *cannot be orbital symmetry forbidden*. Ironically, the rearrangement which led us to the concept of pseudopericyclic reactions is subject to other interpretation, and the whole idea lay fallow in the literature for many years [18]. Recently, Birney has revived it and, with a combination of high-level theory and experimental results, drawn attention to a series of reactions which are unequivocally pseudopericyclic [19]. Their very low activation energy barriers attest to the absence of orbital symmetry forbiddenness. A particularly simple example is the planar ring opening of lactone **14** to formyl ketene **15**, for which Ha *et al.* have calculated a barrier of just 3.5 kcal/mol [20]. The transformation entails motion of 4 electrons occupying

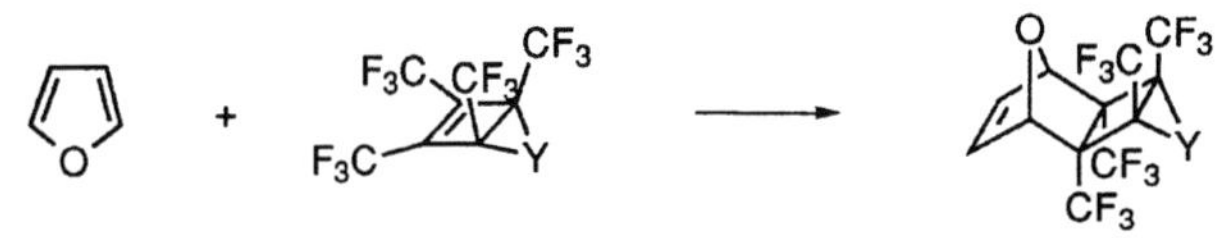

Y:	CH_2 (**16**)	S (**11**)	NH (**17**)	O (**7**)	F,F (**18**)
k (25 °C):	1.2	1.4	18	2200	$<3 \times 10^{-11}$

Fig. 20.1. Bicyclo[2.1.0] systems versus a cyclobutene: rates of Diels-Alder reaction with furan.

orbitals which are symmetric with respect to the plane, and 6 electrons in antisymmetric orbitals.

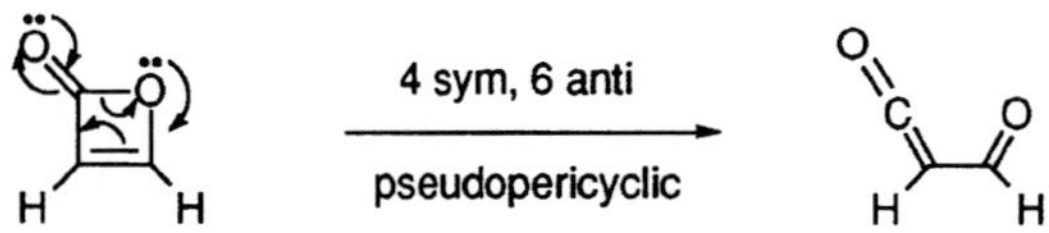

14 **15**

Before abandoning the Dewar isomers of 5-membered-ring heterocycles, we note that they can be regarded as cyclobutadiene derivatives and might therefore be antihomoaromatic. Though evidence for homoaromaticity in neutral species is unconvincing, in general [21], *anti*homoaromaticity can be a significant phenomenon even in neutral molecules. The unusual reactivity of bicyclo[2.1.0]pentene in cycloadditions is a case in point [22]. To learn whether this phenomenon plays a role in perfluoroalkyl-substituted systems, we compared a series of fluorinated bicyclo[2.1.0] systems with a model cyclobutene lacking the 3-membered ring by determining their rates of Diels-Alder addition to furan (Fig. 20.1) [12]. In the series Y = **16**, **11** < **17** < **7**, reaction rate parallels increase in the electronegativity of Y. These fast reaction rates stand in dramatic contrast to the inertness of **18**, for which the rate constant shown is a conservatively extrapolated upper limit. We regard these data as strong evidence that the perfluoroalkylated bicyclic systems are antihomoaromatic.

The electron-withdrawing ability of trifluoromethyl groups piqued our curiosity about the strength of 5*H*-perfluoropentamethylcyclopentadiene (**19**) as an acid. We synthesized the compound from Dewar thiophene **11** and discovered that its pK_a is $\leqslant -2$ [23]. Thus **19** is at least 18 orders of magnitude more acidic than the parent cyclopentadiene, and more acidic than nitric acid. We are not aware of any other carbon acid without conjugating substituents which approaches the acidity of this diene. Incidentally, Seppelt showed that 5*H*-pentafluorocyclopentadiene (**20**) does not differ greatly in acidity from the parent hydrocarbon (pK_a 13-15 vs 15.5, respectively) [24]. The contrast with **19** can be explained by lone pair-π repulsion in the conjugate base of **20** and negative hyperconjugative stabilization in that of **19**.

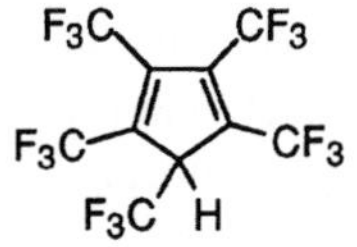

19

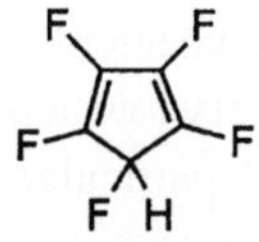

20

Perfluoroannulene chemistry

Clearly, fluoro and perfluoroalkyl substituents can influence the properties and chemistry of organic molecules very differently, for both steric and electronic reasons [25]. The balance of this account will relate some of our experiences with organic molecules bearing fluoro, as opposed to perfluoroalkyl, substituents. During the 1970's there was a great controversy over the nature of the ground state of cyclobutadiene, finally shown to be a rectangular singlet [26]. We became interested during that period in tetrafluorocyclobutadiene (**22**), and succeeded in generating it by photolyzing anhydride **21** in the vapour phase [27].

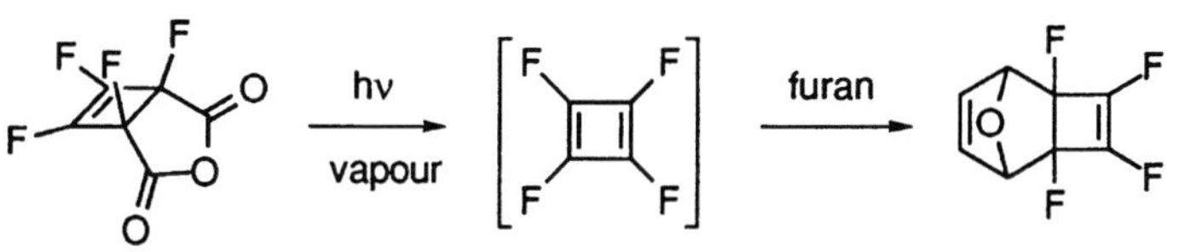

21 **22** **23**

Trapping it with furan as the adduct **23** revealed its presence as an intermediate in the photolysis. In the absence of furan, the product composition was strongly pressure-dependent. Perfluorocyclooctatetraene (**24**) was obtained at low pressures, but with 500 Torr of nitrogen present the product was tricyclooctadiene **25**, formally the Diels-Alder dimer of the cyclobutadiene.

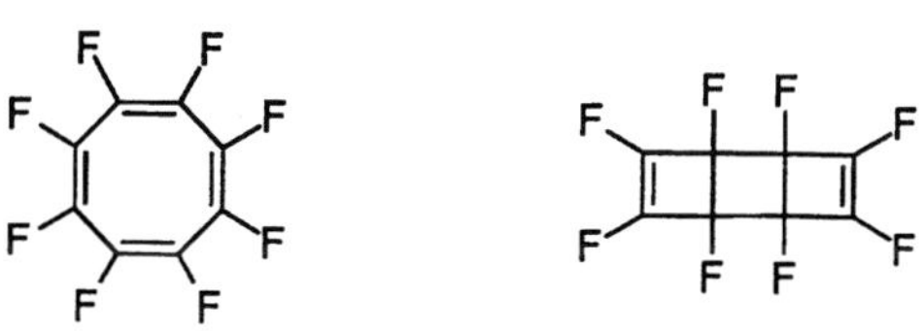

24 **25**

We surmised that the primary product in both cases was **25**, born very hot vibrationally by dimerization of the high-energy diene, and that the role of nitrogen was to preserve the strained tricyclic molecule by collisional cooling before it could rattle itself open to tetraene **24**.

Having published that interpretation, we were chagrined to learn that in the high pressure photolysis, **24** is present at short reaction times and the ratio of **24** to **25** rises steeply upon extrapolation back to time zero. Thus, the tetraene is the precursor of the tricyclic diene, not vice versa! It turns out that vapour phase irradiation of **24** yields *anti-* and *syn-***25** in the ratio 20:1, and that the isomerization is dramatically accelerated by nitrogen [28]. Presumably nitrogen does collisionally deactivate vibrationally hot **25** (or, less likely, an intermediate en route to **25**), thus preventing the very favorable *retro*-reaction.

24 *anti-***25** 20 : 1 *syn-***25**

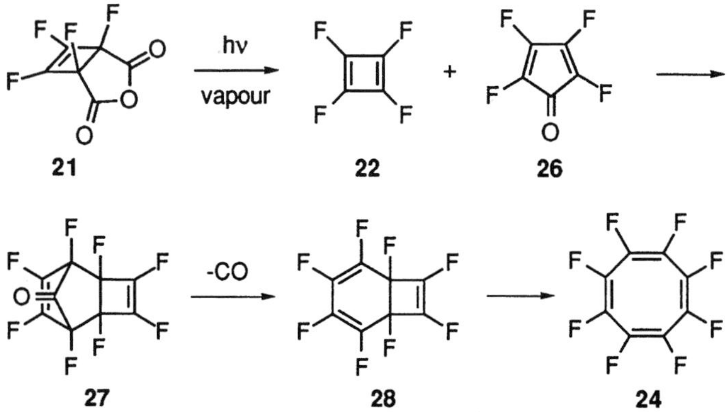

Scheme 20.1.

How is tetraene **24** formed if not in the obvious way, via **25**? We believe the answer is that cyclopentadienone **26** is also formed in the anhydride photolysis, and that cross dimerization occurs between diene **22** and dienone **26**. Extrusion of CO from the adduct (**27**) [29] yields bicyclooctatriene **28**, which spontaneously ring opens to tetraene **24** (Scheme 20.1). That final step is further discussed below. Maier has found that other cyclobutenedicarboxylic anhydrides yield both the cyclobutadiene and the cyclopentadienone upon photolysis [30].

Pyrolysis of the anhydride at temperatures above 500 °C gives dienone **26**, an orange compound which dimerizes rapidly below room temperature in the condensed phase but can be stored as a vapour at ~1 Torr for many hours at ambient temperatures [31]. Photolysis of the orange vapour quickly transforms it into tetraene **24**, almost certainly via decarbonylation of "housenone" **29** to cyclobutadiene **22**, which reacts with starting material as in Scheme 20.1. This finding therefore lends strong support to the mechanism depicted there.

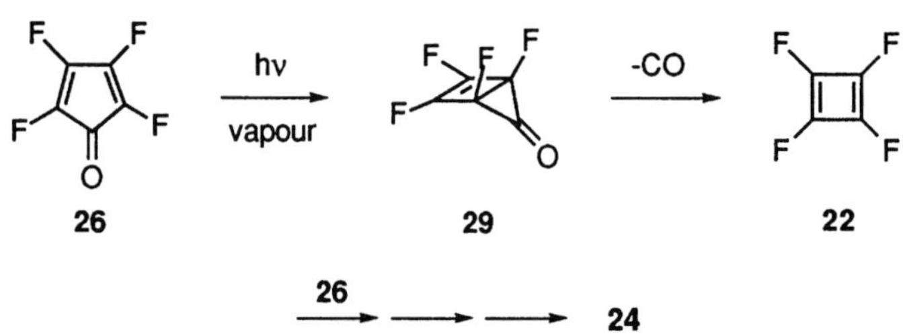

About 20 years after the completion of this work, we sent samples of anhydride **21** and *cis*- and *trans*-tetrafluoro-3,4-diiodocyclobutene to George Radziszewski at the NREL in Colorado. In an elegant study, he and Dartmouth undergraduate E. James Petersson obtained tetrafluorocyclobutadiene (**22**) from all three precursors by photolysis in argon matrices at 12 K [32]. Combining high level quantum mechanical calculations with linear dichroism measurements, they were able to assign its infrared spectrum and determine its structure convincingly. Remarkably, the molecule is nonplanar, with C_{2h} symmetry (Fig. 20.2).

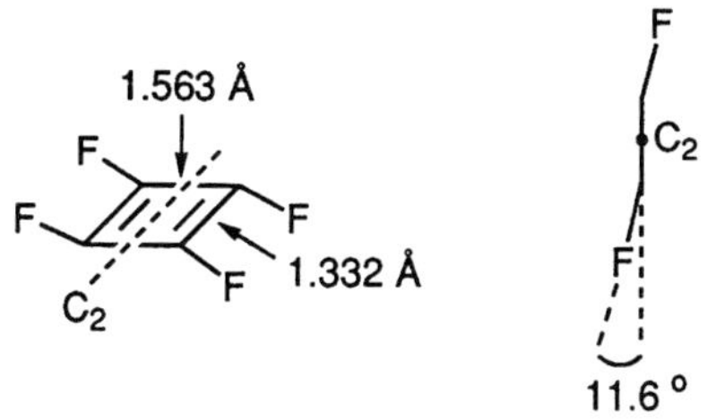

Fig. 20.2. The calculated structure of **22** (B3LYP/cc-pVDZ).

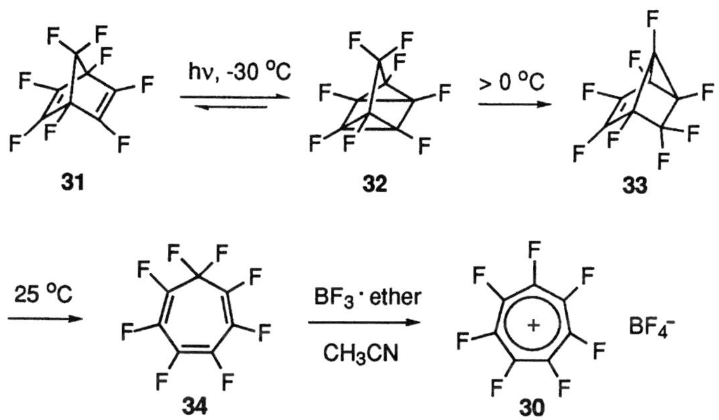

Scheme 20.2.

Nonplanarity is unprecedented among cyclobutadienes lacking bulky groups, and it is probably the result here of the confluence of three factors. Sigma electron withdrawal by the fluorines, lone pair-π repulsion and antiaromaticity all favour pyramidalization of the carbons.

Our interests included aromatic as well as antiaromatic systems, and were able to synthesize the perfluorotropylium ion (**30**) from hexafluorobenzene via perfluoronorbornadiene (**31**) and a remarkable series of rearrangements (Scheme 20.2) [33]. Irradiation of **31** at −30 °C with the full mercury arc established a photostationary state with perfluoroquadricyclane (**32**) in which the isomers are present in a 2:3 ratio, respectively. When the highly strained quadricyclane was allowed to warm above 0 °C in the dark, it rearranged spontaneously to tricycloheptene **33**. Standing at room temperature, that isomer suffered further rearrangement to perfluorotropilidene (**34**) [34]. Treatment with boron trifluoride etherate readily transformed **34** into perfluorotropylium tetrafluoroborate (**30**).

Interest in the chemistry of the new fluorocarbon perfluorocyclooctatetraene (**24**), obtained from anhydride **21** and dienone **26**, spurred us to find a more efficient synthetic route to the compound. Photocycloaddition of 1,2-dichlorodifluoroethylene to hexafluorobenzene provided the answer [35]. Dechlorination of the resulting stereoisomeric mixture of tricyclic adducts **35** with the help of ultrasound gave *anti*-**25**, which opened readily at 150 °C to **24** (Scheme 20.3) [36].

Reaction with diiron nonacarbonyl at room temperature transformed **24** into iron tetracarbonyl complex **36**, and at higher temperatures into the more interesting 1,2,3,6-η

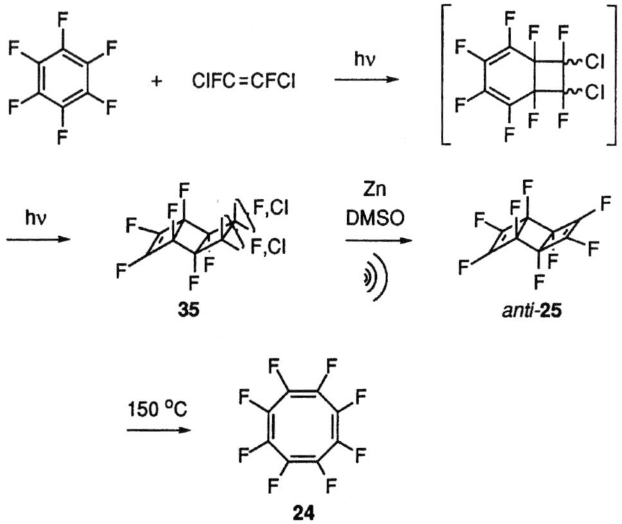

Scheme 20.3.

tricarbonyl complex **37** [37]. Here my colleague Russell Hughes took over the exploration of the organometallic chemistry of **24**; his students have prepared many transition metal complexes from it, representing an impressive variety of structural types [38].

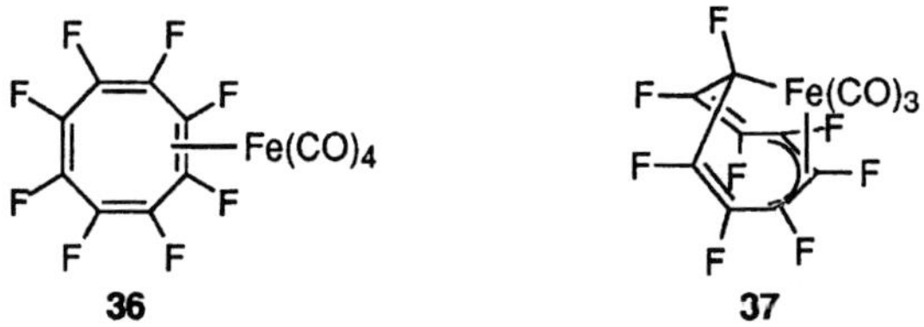

Bond-shift isomerization in **24** has a much higher barrier than in the parent hydrocarbon, as was revealed by ^{19}F NMR spin saturation transfer measurements with $7H$-perfluorocyclooctatetraene **38** [39]. $\Delta H^{\ddagger} = 21.1 \pm 0.9$ kcal/mol and $\Delta S^{\ddagger} = -10.2 \pm 2.5$ e.u. for **38** as compared with 10.0 kcal/mol and −9.7 e.u. for the hydrocarbon [40]. The difference is probably primarily steric in origin, as replacement of two adjacent fluorines in **24** with chlorines gives rise to stable bond-shift isomers (**39** and **40**) which interconvert only slowly at 150 °C [41].

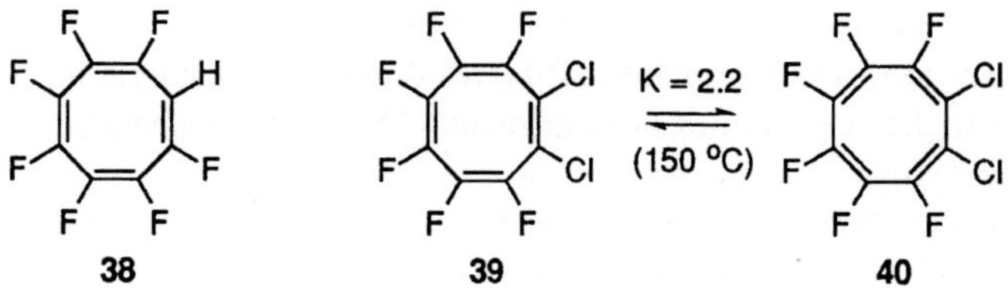

Regarding other $(CF)_8$ valence isomers, we synthesized bicyclic triene **28** and found that it has a half-life of only 14 minutes at 0 °C for ring opening to the tetraene [42]. In the equilibrium mixture of the two isomers at room temperature, **28** is present to the extent of just 0.2% [43, 44].

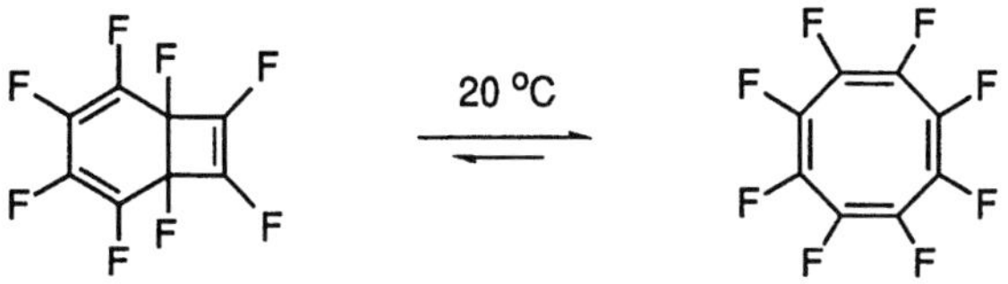

28 0.2% **24** 99.8%

Perfluorobarrelene (**44**) was also prepared in our laboratory in several steps from hexafluorobenzene via the benzene synthon **41** [45]. Diels-Alder reaction of **41** with bis(trimethylstannyl)acetylene [46] gave *endo*- and predominantly *exo*-**42**; reduction of the adducts with chromous chloride yielded **43**. Fluorinolysis of **43** with fluorine/helium in acetonitrile afforded the highly volatile, nicely crystalline barrelene **44**.

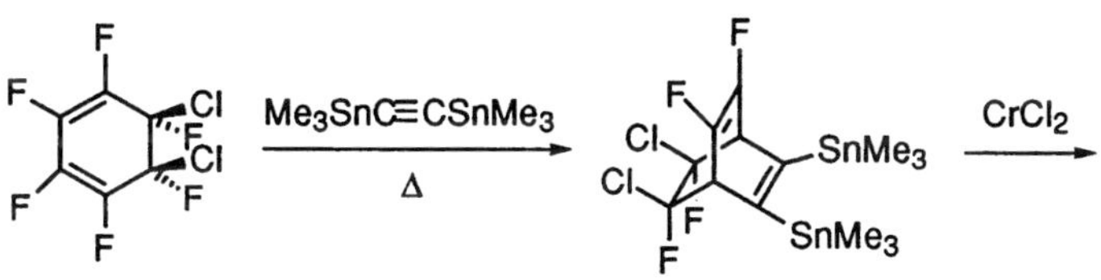

41 *exo*-**42** (+ *endo*-)

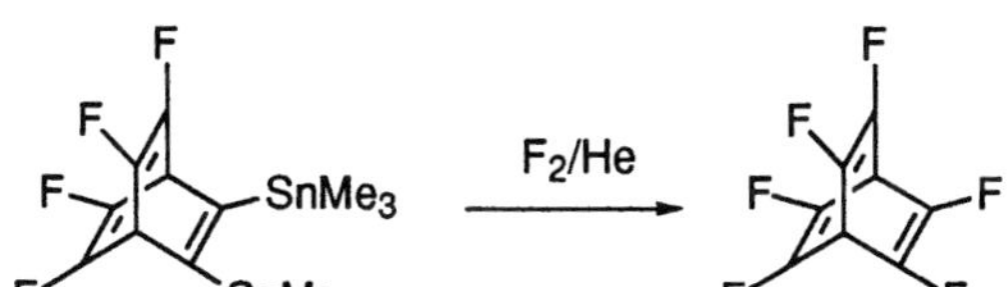

43 **44**

We hoped to rearrange **44** photochemically to perfluorosemibullvalene (**45**) in analogy to the behavior of the parent hydrocarbon [47], since we are interested in learning the effect of fluorine substitution on the barrier for degenerate Cope rearrangement in **45** [48]. As is typical, however, the fluorocarbon chose a different path from its hydrocarbon counterpart and isomerized with or without triplet sensitizer to the $(CF)_8$ thermodynamic sink, cyclooctatetraene **24**.

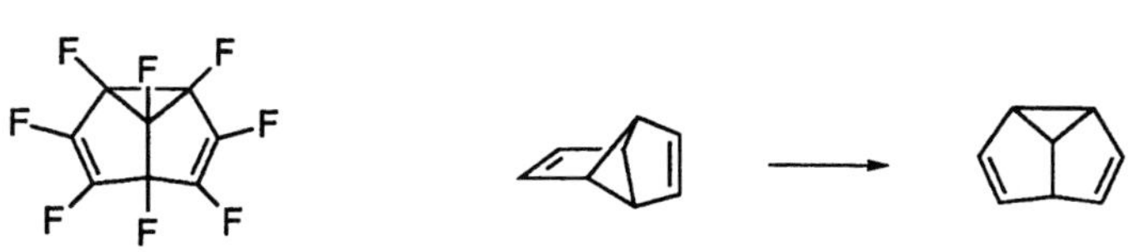

45 **46** **47**

Another approach to the semibullvalene was based on the very facile rearrangement of hydrocarbon diene **46** to the parent semibullvalene **47** [49], which occurs even at room temperature. Tetrachloride **48**, prepared by photochlorination of **35**, rearranged at 250 °C to **49** via the 1,5-cyclooctadiene [50, 51]. Dechlorination of **49** yielded dichloride **50**, but attempts to carry the reaction on to diene **51** invariably resulted in complete destruction.

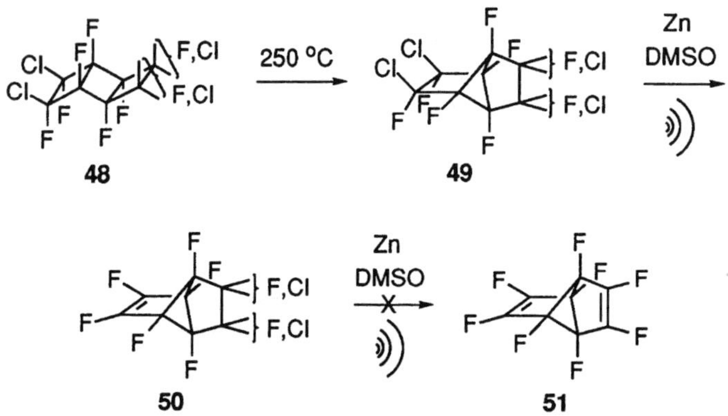

Dichloride **50** underwent both Diels-Alder additions and [2 + 2] cycloaddition with a triazolinedione. The adducts could be successfully dechlorinated, indicating that the problem lay in having both double bonds present in the twisted C_8 skeleton. The apparent lability of **51** can be understood by analogy to the parent hydrocarbon **46**, in which the double bonds interact with one another so strongly via the Walsh orbitals of the central 4-membered ring that the diene has significant UV absorption even at 300 nm ($\varepsilon \sim 190$) [52]. The interaction both raises the HOMO and lowers the LUMO, thus creating a highly reactive molecule [53].

Dechlorination of the exo pyrrole adduct of **50** led to **52**, which underwent allylic rearrangement at 80 °C to **53**. Since this compound is formally a Diels-Alder adduct of semibullvalene **45** with pyrrole, it was hoped that heating at a higher temperature in vacuum would afford the elusive **45**. At 130 °C only tetraene **24** was obtained, however, thus indicating that the semibullvalene is very fragile.

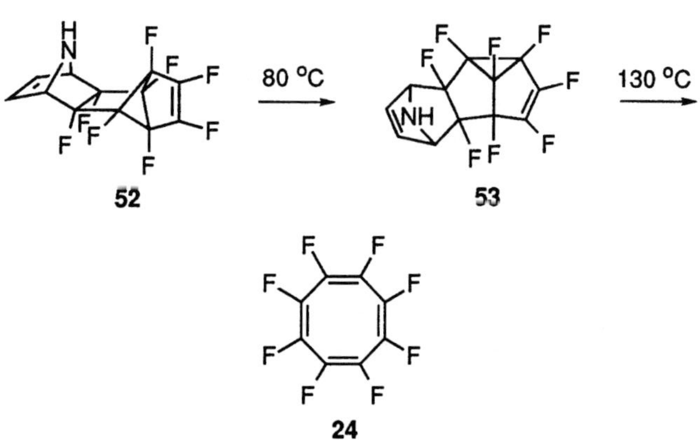

Chemistry of perfluorodienes and -trienes

We recognized that the considerable driving force for highly fluorinated double bonds to become saturated [54] has important implications for the reactivity of di- and polyunsaturated fluorocarbons, and decided to pursue them. The first diene to be examined was perfluoro-1,5-hexadiene (**54**), which was found to exist in closely balanced equilibrium with the strained bicyclo[2.2.0]hexane **55** at 250 °C and to yield the bicyclo[2.1.1]hexane **56** irreversibly at 300 °C [55].

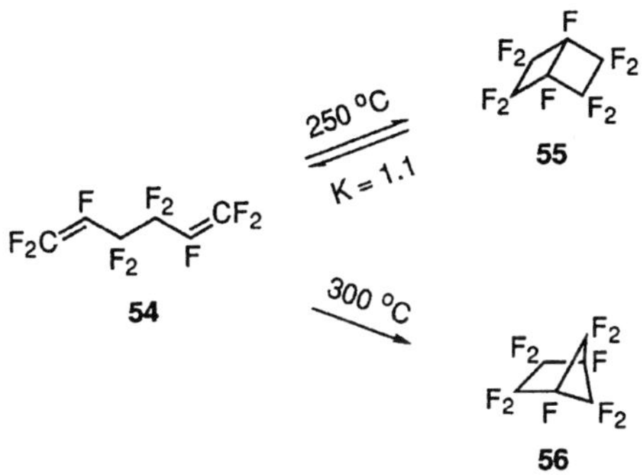

In contrast, the hydrocarbon parents of **55** and **56** ring open completely to hexadiene at elevated temperatures [56, 57]. Mercury-sensitized photolysis of **54** yields **56** and **55** in a ratio of 3-4:1.

At 300 °C perfluoro-1,6-heptadiene (**57**) cyclizes to give **58** and **59** in the ratio 9:1 [58].

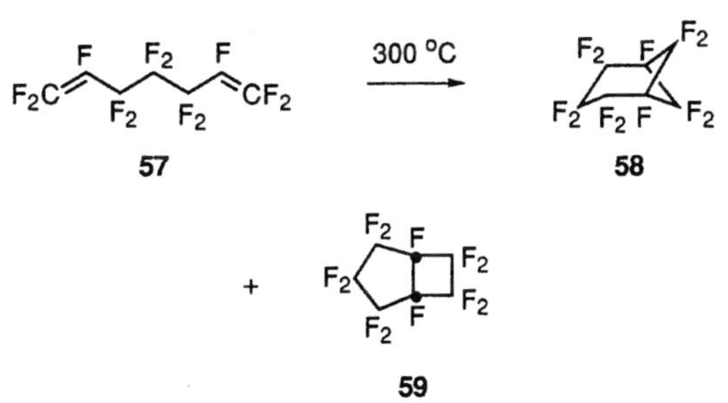

Again the same products result from mercury-sensitized photolysis, but now **59** predominates and some of its trans isomer is formed as well. Our finding that the favoured mode of internal cycloaddition of the double bonds, parallel versus crosswise, switches from one diene to the other and from thermal to photochemical reaction may be puzzling at first, but makes sense when it is recognized that the thermal cycloadditions prefer a 6-ring biradical intermediate and the photochemical reactions a 5-ring one (Table 20.1). Because of their smaller strain energies, 6-membered ring biradicals require less energy to form from the dienes than 5-membered; *i.e.* the late transition states reflect the relative energies of the biradical intermediates. The mercury-sensitized photoreactions, on the other hand,

TABLE 20.1
Preferred internal cyclization modes of perfluorodienes

Reaction type	1,5-Diene (**43**)	1,6-Diene (**46**)	Intermediate ring
Thermal	Parallel	Crosswise	6-Membered
Photochemical	Crosswise	Parallel	5-Membered

involve triplet states [59], which can be regarded as biradicals that possess much more energy than necessary for cyclization. Here the transition states come early, and reflect the fact that intramolecular radical attack on a double bond shows a strong preference for 5- over 6-membered ring formation [60, 61].

With a labelled perfluoro-1,5-hexadiene (**60**) we were able to show that Cope rearrangement to **62** takes place via a 1,4-biradical intermediate (**61**) [62], again in contrast to the concerted process characteristic of hydrocarbon dienes [63].

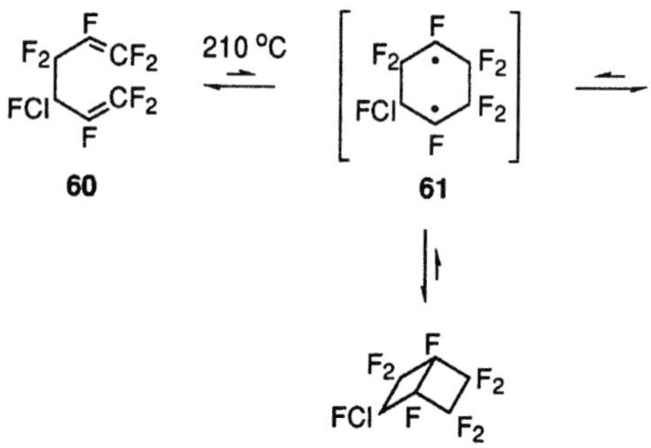

cis-Perfluoro-1,3,5-hexatriene (**63**) cyclizes reversibly to vinylcyclobutene **64** at 160 °C, and irreversibly to cyclohexadiene **65** at higher temperatures [64, 65]. Trans triene **66** cyclizes to **64** much more slowly than **63**, as required by the finding that ring opening of **64** gives the cis triene cleanly. The electrocyclic ring opening is thus highly torquoselective [66, 67], with the trifluorovinyl group rotating inward and the geminal fluorine outward in the transition state.

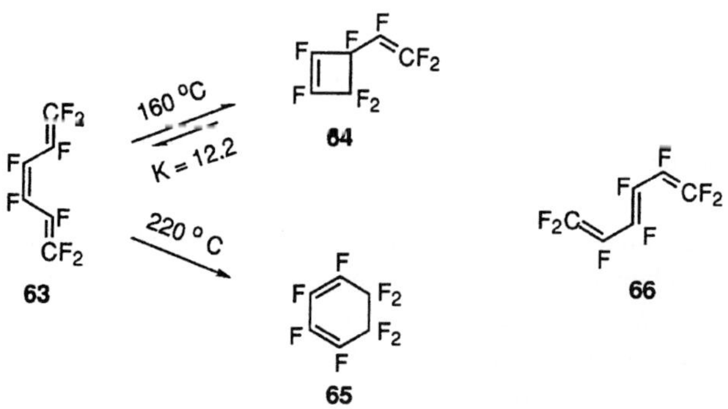

Thermal and photochemical internal cycloaddition of open-chain perfluorodi- and trienes to bicyclic systems proved to be quite general, and even highly strained tricyclics were produced. *cis*-Perfluoro-1,3,6-heptatriene (**67**), for example, gave **68** at 130 °C, **69** quantitatively at 250 °C, and upon mercury-sensitized photolysis gave **70** and **71** in low yield [58].

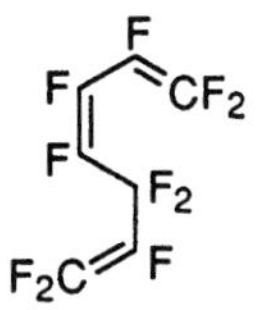

67

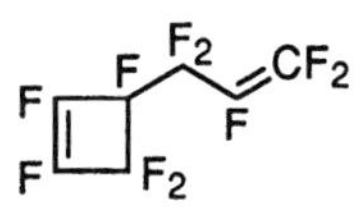

68

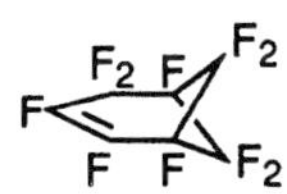

69

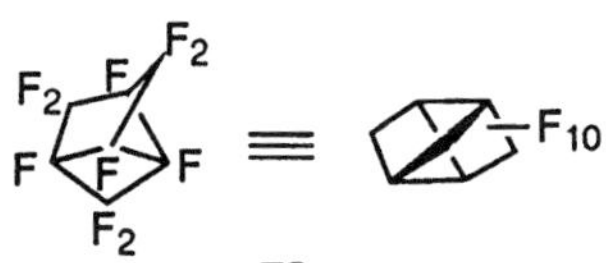

70

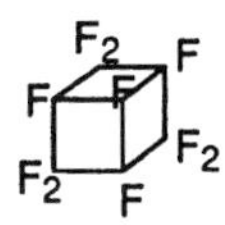

71

At 300 °C *cis*-Perfluoro-1,4,7-octatriene (**72**) isomerized to the far less reactive trans isomer (**73**) and cyclized to *trans*- and *cis*-**74** [68].

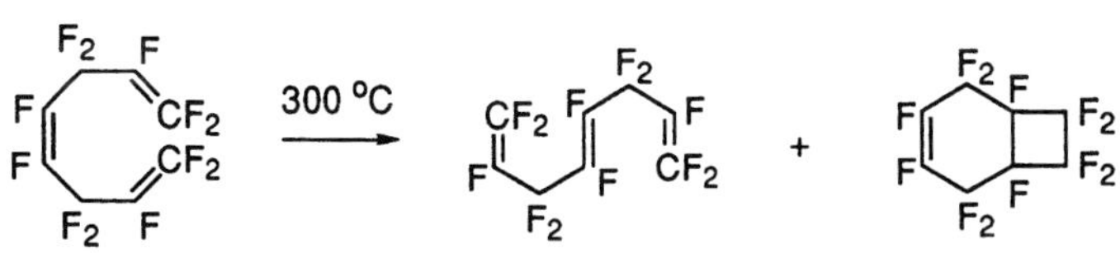

72 **73** *trans*- and *cis*-**74**
(1.6 : 1)

The behaviour of the corresponding 1,7-diene **75** was strikingly different. Unreactive at 300 °C, at 350 °C it gave exclusively fluorocarbons missing one or two CF_2 groups: **76**-**78** [68]. The mechanisms involved in the very surprising transformations of **75** remain a matter for speculation, but probably reflect the remarkable stability of difluorocarbene.

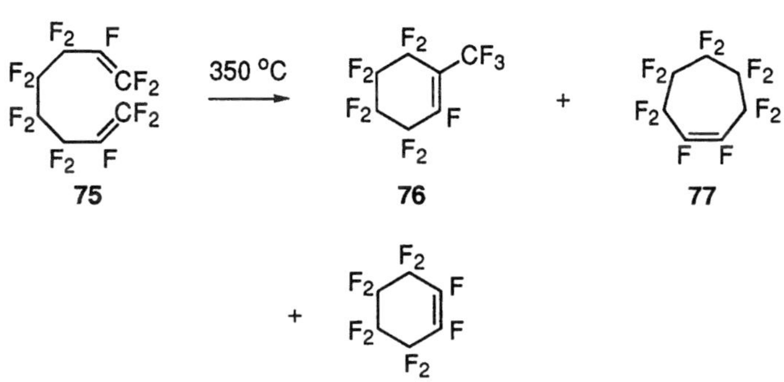

75 **76** **77** **78**

New highly strained molecules

Intrigued by the remarkable reactivity of Wiberg's very strained alkene bicyclo-[2.2.0]hex-1(4)-ene (**79**) [69], we set out to synthesize its perfluorinated counterpart (**80**) with confidence that it too would be very reactive in its own quite different way. A suitable precursor, dibromide **81**, was prepared from hexafluorobenzene. Treatment of **81** with an alkyllithium at low temperatures or with zinc and ultrasound at room temperature generated the alkene, as revealed by trapping experiments [70]. We discovered that a better precursor, diiodide **82**, can be made from **81** by what we call somewhat whimsically a "photo-Finkelstein reaction".

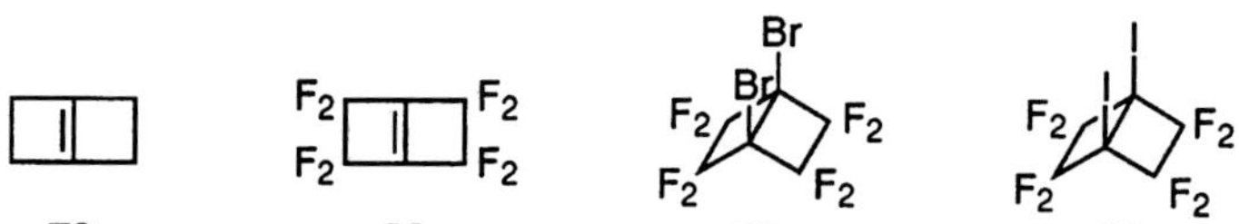

79 **80** **81** **82**

Like the Finkelstein reaction [71], this transformation replaces other halogens with iodine, but via electron transfer instead of the S_N2 mechanism. A solution of the dibromide in acetonitrile/ether containing potassium iodide is irradiated with ultraviolet light, which causes the iodide ion to undergo a CTTS (charge transfer to solvent) transition [72]. Solvated electrons cleave the C—Br bonds, allowing the introduction of iodines. Very mild conditions suffice to produce alkene **80** from diiodide **82**, *e.g.* mercury and ultrasound [70].

So generated, the alkene readily adds nucleophiles (*e.g.* water, methanol) and undergoes cycloadditions (Diels-Alder, [2 + 2], 1,3-dipolar). A third route to **80** has been found, namely, *retro*-Diels-Alder reaction of its adduct with *N*-benzylpyrrole (**83**) [73]. This method has made possible isolation and spectroscopic characterization of the labile alkene, which is caught in a cold trap and dissolved in a solvent before warming. Ab initio quantum mechanical calculations indicate that the carbon skeleton of the highly strained alkene is planar, a conclusion consonant with the absence of double bond stretching absorption in its infrared spectrum.

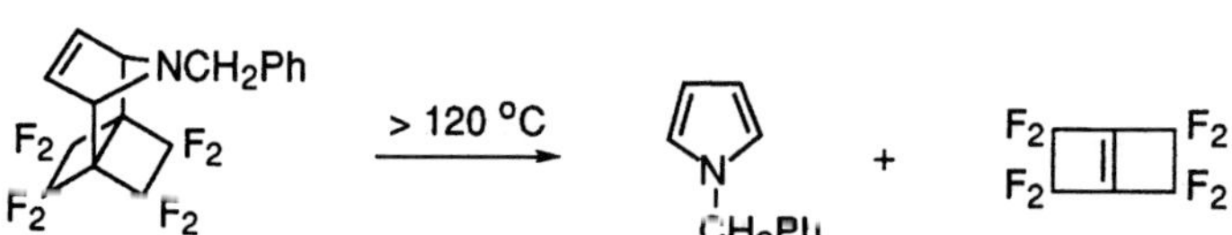

83 **80**

The product of trapping **80** with ethyl vinyl ether is the [2.2.2]propellane **84** [70], only the second [2.2.2]propellane to be isolated. The previously synthesized molecule with this extremely strained skeleton [74], Eaton's **85** [75], ring opened with a half-life of 28 minutes at 25 °C, but fortunately the half-life of **84** for decomposition to **86** and **87** is roughly 20 hours at room temperature.

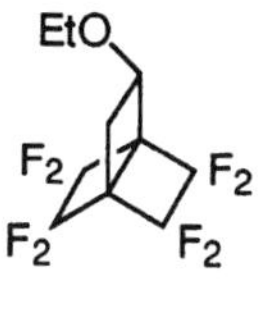

84

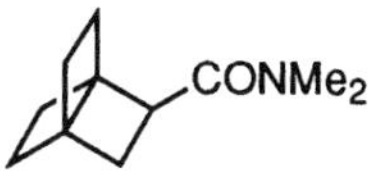

85

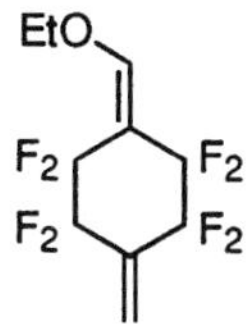

86

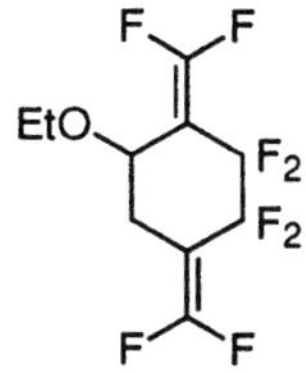

87

Since the strain in a [2.2.2]propellane is relieved almost completely upon cleavage of the central bond, that C—C bond is very fragile. Thus, bromine adds across the central bond of **85** instantaneously at $-70\ °C$ [75]. In contrast, propellane **84** stoutly resists electrophilic attack, for example by concentrated sulfuric acid in acetonitrile at ambient temperature, because of electron withdrawal by the fluorines. On the other hand, **84** is very susceptible to attack by free radicals or nucleophiles [76]. Halide ions readily cleave the central bond at room temperature, and in moist acetonitrile the net result is addition of HX to the bridgehead carbons. The addition is cleanly regioselective, giving only **88**. Presumably this is because the intermediate bridgehead anion **89** is lower in energy than the alternative anion, since electron withdrawal by the oxygen supplements that by the fluorines.

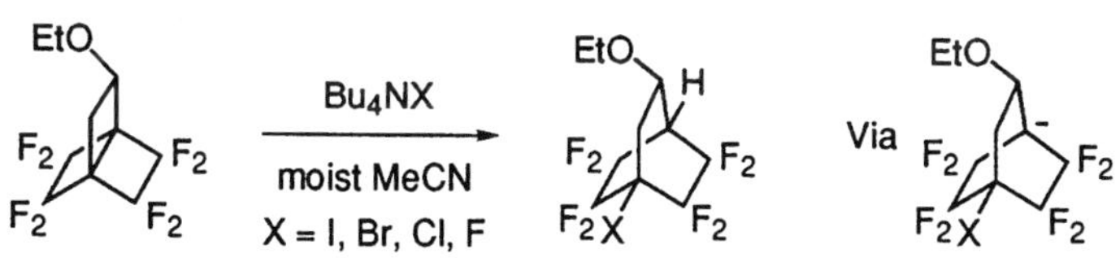

84 88 89

Perfluorinated keto-enol systems

In the course of a synthesis we had occasion to prepare bicyclic ketone **90**, and were surprised to find that a persistent "impurity" in the product was the corresponding enol (**91**) [77].

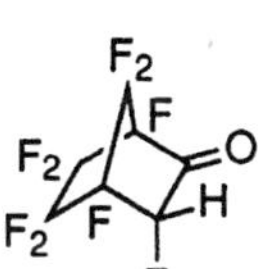

90

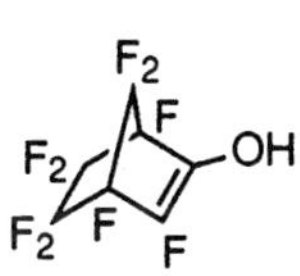

91

This led us to discover the intriguing work on perfluoroenols carried out by Bekker, Knunyants and their coworkers, mostly in the 1970s [78]. They showed that perfluoroenols are amazingly stable kinetically, presenting examples in which neither high temperatures nor powerful acids effected their ketonization. They found that, under the right conditions, pentafluoroacetone enol isomerized completely to the ketone [79], and thus made the assumption that the stability of perfluoroenols in general was only kinetic in nature. The equilibrium enol contents of simple hydrocarbon-derived ketones, of course, are miniscule; *e.g.* for acetone 4.7×10^{-9} and for cyclopentanone 1.1×10^{-8} (in water at 25 °C) [80].

We found that the equilibrium constant for enolization of **90** in carbon tetrachloride is 0.07, but in Lewis basic solvents such as ether, THF or acetonitrile the equilibrium lies too far to the right to detect ketone [81, 82]. This dramatic solvent dependence reflects the potent hydrogen bond donor ability of perfluoroenols. The Russian group had synthesized cyclobutanone **92** and its enol **93** independently and reported that the two could not be interconverted thermally or catalytically [83].

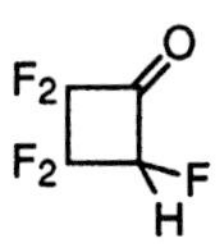

92

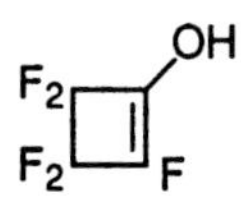

93

Strong acids left both unaffected, and bases rapidly destroyed them by elimination of hydrogen fluoride. We showed that the extremely weak base *N*-methylpyrrolidone was capable of equilibrating **92** and **93** without causing elimination, and found that the equilibrium constant for enolization of **92** was too large for us to measure (>250) even in carbon tetrachloride [78]! For cyclopentanone **94** [85], $K_{enol} = \sim 130$, and for cyclohexanone **95** ~ 0.33, both in carbon tetrachloride. On the other hand, in the case of acyclic perfluoroketones the enol content at equilibrium was negligible: $K_{enol} =< 0.005$ for butanone **96** and very small also for **97** [84].

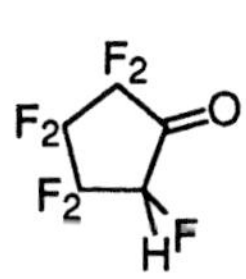

94

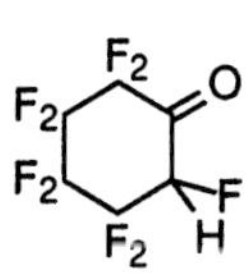

95

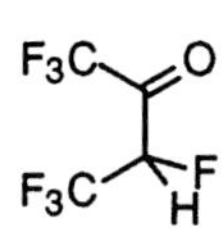

96

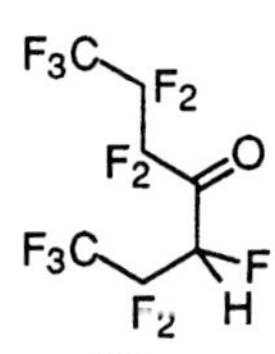

97

For these ketones, even strongly Lewis basic media failed to make any enol detectable. The >50 000-fold range of equilibrium constants represented here meant that we had discovered a gulf between acyclic and certain cyclic systems, a gulf that our ab initio quantum mechanical calculations had correctly predicted.

Calculations on families of isodesmic reactions revealed that the enols, not the ketones, are responsible for the cyclic/acyclic dichotomy, but its origin was still a mystery. Further experimental and especially theoretical work in our laboratory then established that the destabilization of acyclic enols is a consequence of severe steric repulsions that are present in highly fluorinated, vicinally substituted acyclic alkenes in general [86]. The well-known fact that *cis*- (**98**) and *trans*-perfluoro-2-butene (**99**), like their hydrocarbon parents, differ in energy by only ~1 kcal/mol [87] has misled chemists into thinking that steric interactions are slight in these molecules. In fact, steric repulsion is potent in both isomers; because buttressing extends it around the entire molecule, the trans form experiences almost as much strain as the cis.

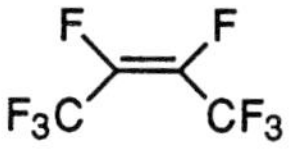

98

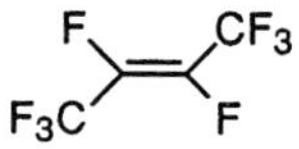

99

Returning to perfluorinated keto-enol systems, we note that the remarkable thermodynamic stability enjoyed by cyclic perfluoroenols relative to their ketones is attributable to destabilization of the keto form by fluorine substitution, an electronic effect [54, 81]. This phenomenon is present in acyclic perfluoroketones as well, but here the effect of higher ketone energy on keto-enol equilibria is negated to a large degree by the steric destabilization of their enols [86].

Functionalization of fluorocarbons

Much recent interest has focused on the challenge of functionalizing saturated fluorocarbons. The problem is both significant and thorny: significant because introduction of functional groups could considerably broaden the range of applications of these compounds [88, 89], and thorny because their inertness demands reaction conditions vigorous enough to bring about wholesale destruction. In other words, intermediate species along the reaction pathways are, in general, more reactive than the starting material [90, 91].

Attack on saturated fluorocarbons requires a potent electron donor, and the most vulnerable site for attack is a tertiary C—F bond. A probable sequence of events resulting from treatment of perfluoro(methylcyclohexane) (**100**) with an electron donor is depicted in Scheme 20.4. Crabtree has produced perfluoro(methylcyclohexene) (**103**) in this way at 11.9% conversion, but it was not separable from starting material and further reaction led to byproduct formation [92]. A similar reaction with perfluoro-(2-methylpentane) yielded the corresponding 2-methylpentene with a conversion of 1.4%.

We set out to intercept an early intermediate, radical **101** or anion **102**, in the reaction cascade shown in Scheme 20.4 in the hope of preventing further degradation. Attempts to trap **101** with the excellent hydrogen atom donor 1,4-cyclohexadiene were unavailing, as were our efforts to capture **102** with trimethylsilyl chloride [93]. We had been using powerful electron donors, however, and we recognized that slow delivery of electrons would allow more opportunity for trapping to compete with further reduction.

Accordingly, since light from an arc is effectively a dilute reagent, a photochemical reducing system was tried. Success with the "photo-Finkelstein" reaction described above led us to employ tetrabutylammonium iodide in tetrahydrofuran, irradiated with wavelengths above 280 nm [93, 94]. For the first time, our reaction mixtures gave well defined ^{19}F NMR spectra, the same spectra whether a trap was present or not. Calibration with an internal area standard showed disappointingly that the single mysterious product from perfluoromethylcyclohexane was present in yields of only 10-12%. When tetrabutylammonium iodide from a fresh bottle was used, though, the yield suddenly jumped to 40% and has since been increased to 60% [94]. This striking change, which resulted from a decrease in water content of the hygroscopic salt, provided the clue needed to understand

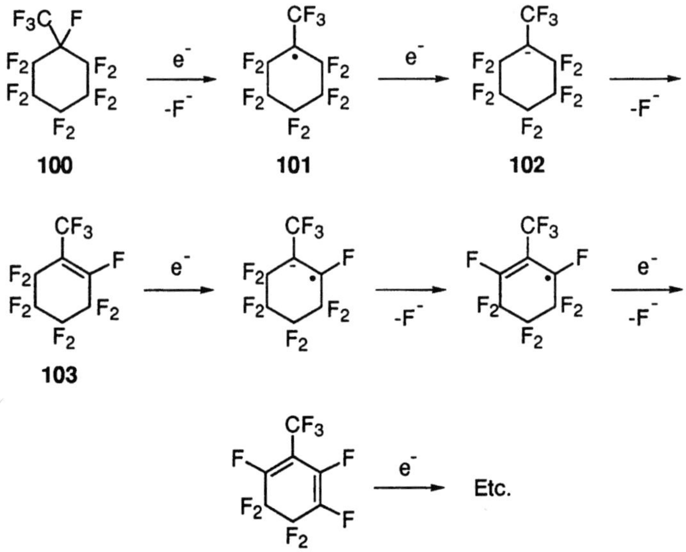

Scheme 20.4.

what was happening. We had assumed our reaction conditions were essentially anhydrous, but in fact water was playing a central role in the chemistry. The presence of some was essential, but too much, as in the early experiments, led to a disastrous reduction in yield.

The mysterious product was a tetrabutylammonium enolate **105**, apparently formed via interception of alkene **103** by hydroxide ion. Reduction of water to tetrabutylammonium hydroxide was to be expected under the reaction conditions. Deprotonation of intermediate enol **104** is very facile, as perfluoroenols are highly acidic [95].

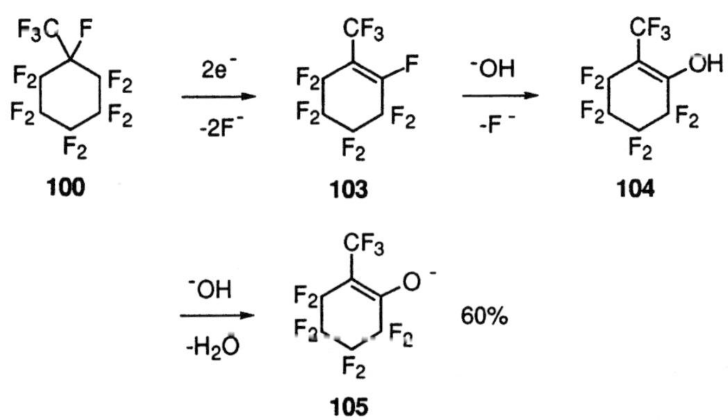

The good yield obtained of enolate **105** depends upon that fact, for *its negative charge protects it from accepting an electron and undergoing further loss of fluoride.* Enolate **105** can be transformed into a variety of derivatives, *e.g.* methyl ether **106** and trimethylsilyl ether **107** [94]. Moist air suffices to hydrolyze **105** to β-diketone enolate **108**.

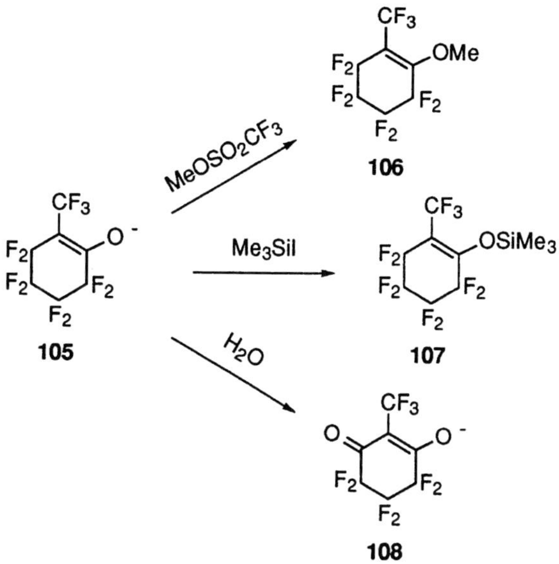

Other saturated fluorocarbons bearing tertiary fluorine, such as **109-111**, react similarly to **100**. The fact that **111** reacts at only one of the tertiary sites attests to the effectiveness with which a negative charge protects against further reduction, and the clean regioselectivity of the reaction is also noteworthy.

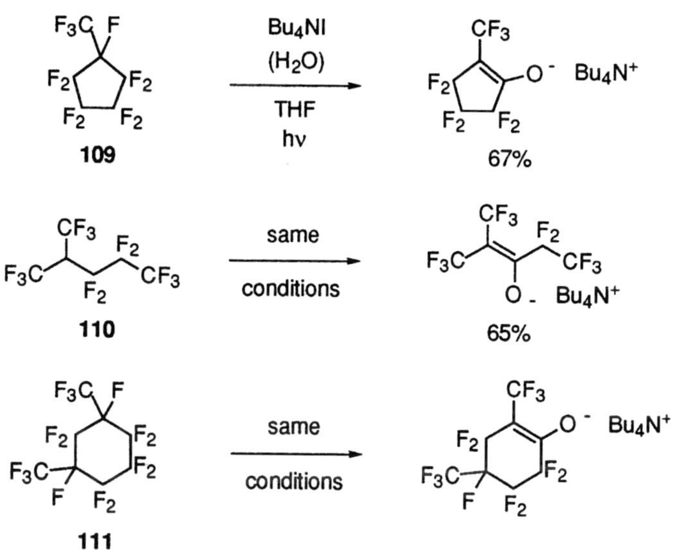

The behaviour of perfluorodecalin (**112**) is exceptional, but as such it provides evidence in support of the mechanism proposed above [94]. Since **112** has adjacent tertiary fluorines, the initially formed alkene, Δ^9-decalin **113**, is tetrasubstituted. Attack of hydroxide cannot result in addition-elimination as in the above cases, so S_N2' reaction takes place instead to give **114**. This trisubstituted alkene reacts analogously to **103**, proceeding all the way to the β-diketone enolate (**115**). Thus, the inability of the molecule to acquire a negative charge early in the reaction cascade results in the loss of many fluorines.

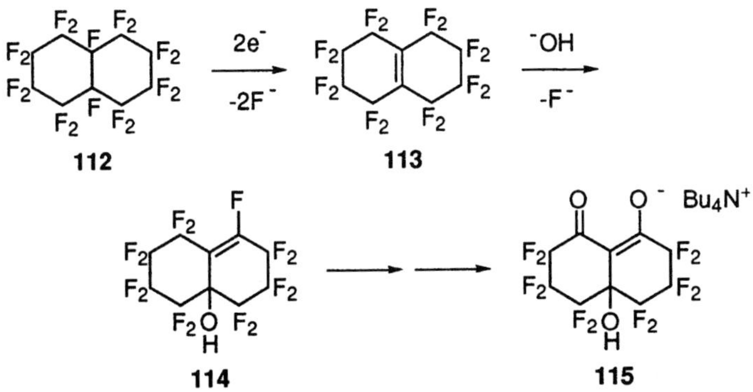

In summary, with the help of serendipity we have discovered the first general method for functionalization in good yield and without extensive fluoride loss of saturated fluorocarbons that possess a tertiary C—F bond. The method yields perfluoroenolates, useful synthetic intermediates for the preparation of a variety of other derivatives.

Outlook

Compared with the chemistry of hydrocarbons and their derivatives, organofluorine chemistry is still a youthful realm of science. The future holds a cornucopia of synthetic challenges, mechanistic puzzles and, of course, surprises. Because of its unique character the field has a special fascination for its aficionados, and there is much more fun to come.

Acknowledgement

I am very grateful for the diligent and skilful efforts of my coworkers whose accomplishments are cited in this account, and of the others too whose research in organofluorine chemistry went unmentioned for lack of space. Our work in this field has received generous support from the National Science Foundation, the Air Force Office of Scientific Research and the Petroleum Research Fund administered by the American Chemical Society.

References

1 D. M. Lemal, F. Menger and G. W. Clark, *J. Am. Chem. Soc.*, *85* (1963) 2529.
2 D. M. Lemal and J. P. Lokensgard, *J. Am. Chem. Soc.*, *88* (1966) 5934.
3 T. J. Katz and N. Acton, *J. Am. Chem. Soc.*, *95* (1973) 2738.
4 D. M. Lemal, J. M. Staros and V. Austel, *J. Am. Chem. Soc.*, *91* (1969) 3373.
5 M. G. Barlow, R. N. Haszeldine and R. Hubbard, *J. Chem. Soc., Chem. Commun.*, (1969) 202; *J. Chem. Soc. C*, (1970) 1232.
6 For a review on valence isomers of aromatic compounds, see Y. Kobayashi, *Topics in Current Chemistry*, *123* (1984) 103.

7 H. Ertl, J. P. Lokensgard and D. M. Lemal, unpublished results.
8 D. M. Lemal and L. H. Dunlap, Jr., *J. Am. Chem. Soc.*, *94* (1972) 6562. Others have argued that the perfluoroalkyl effect is entirely kinetic in nature. A. Greenberg, J. F. Liebman and D. Van Vechten, *Tetrahedron*, *36* (1980) 1161.
9 M. W. Grayston and D. M. Lemal, *J. Am. Chem. Soc.*, *98* (1976) 1278.
10 C. J. Boriack, E. D. Laganis and D. M. Lemal, *Tetrahedron Lett.*, (1978) 1015.
11 D. Wirth and D. M. Lemal, *J. Am. Chem. Soc.*, *104* (1982) 847.
12 D. Wirth and D. M. Lemal, unpublished results. Chambers' group reported photoisomerization of the furan **9** to cyclopropenyl ketone **8**. R. D. Chambers, A. A. Lindley and H. C. Fielding, *J. Fluorine Chem.*, *121* (1978) 337.
13 H. A. Wiebe, S. Braslavsky and J. Heicklen, *Can. J. Chem.*, *50* (1972) 271.
14 Y. Kobayashi, I. Kumadaki, A. Ohsawa and Y. Sekine, *Tetrahedron Lett.*, (1974) 2841.
15 M. W. Grayston and D. M. Lemal, unpublished observations.
16 J. A. Ross, R. P. Seiders and D. M. Lemal, *J. Am. Chem. Soc.*, *98* (1976) 4325.
17 C. H. Bushweller, J. A. Ross and D. M. Lemal, *J. Am. Chem. Soc.*, *99* (1977) 629.
18 J. P. Snyder and T. A. Halgren, *J. Am. Chem. Soc.*, *102* (1980) 2861. This group later identified another reaction which they did regard as pseudopericyclic. U. Henriksen, J. P. Snyder and T. A. Halgren, *J. Org. Chem.*, *46* (1981) 3767.
19 D. M. Birney and P. E. Wagenseller, *J. Am. Chem. Soc.*, *116* (1994) 6262; S. Ham and D. M. Birney, *Tetrahedron Lett.*, *35* (1994) 8113; P. E. Wagenseller, D. M. Birney and D. Roy, *J. Org. Chem.*, *60* (1995) 2853; D. M. Birney, *J. Org. Chem.*, *61* (1996) 243; S. Ham and D. M. Birney, *J. Org. Chem.*, *61* (1996) 3962; D. M. Birney, S. Ham and G. R. Unruh, *J. Am. Chem. Soc.*, *119* (1997) 4509; S. Ham and D. M. Birney, *Tetrahedron Lett.*, *38* (1997) 5925; D. M. Birney, X. Xu, S. Ham and X. Huang, *J. Org. Chem.*, *62* (1997) 7114.
20 M. T. Nguyen, T.-K. Ha and R. A. M. O'Ferrall, *J. Org. Chem.*, *55* (1990) 3251.
21 For an apparent exception, see J. F. Liebman, L. A. Paquette, J. R. Peterson and D. W. Rogers, *J. Am. Chem. Soc.*, *108* (1986) 8267.
22 W. R. Roth, F.-G. Klärner and H.-W. Lennartz, *Chem. Ber.*, *113* (1980) 1818; G. D. Andrews, J. E. Baldwin and K. E. Gilbert, *J. Org. Chem.*, *45* (1980) 1523.
23 E. D. Laganis and D. M. Lemal, *J. Am. Chem. Soc.*, *102* (1980) 6633.
24 G. Paprott and K. Seppelt, *J. Am. Chem. Soc.*, *106* (1984) 4060.
25 B. E. Smart, in R. E. Banks, B. E. Smart and J. C. Tatlow (eds.), *Organofluorine Chemistry, Principles and Commercial Applications*, Plenum, New York, 1994, p. 57.
26 G. Maier, *Angew. Chem. Int. Ed. Engl.*, *27* (1988) 309.
27 M. J. Gerace, D. M. Lemal and H. Ertl, *J. Am. Chem. Soc.*, *97* (1975) 5584.
28 A. C. Barefoot, III, W. D. Saunders, J. M. Buzby, M. W. Grayston and D. M. Lemal, *J. Org. Chem.*, *45* (1980) 4292; A. C. Barefoot, III, Masters Thesis, Dartmouth College, 1978.
29 M. W. Grayston, A. C. Barefoot, III, W. D. Saunders and D. M. Lemal, Abstracts of the Fourth Winter Fluorine Conference, Daytona Beach, Jan. 28-Feb. 2, 1979, p. 4; B. E. Smart, in S. Patai and Z. Rappoport (eds.), *The Chemistry of Functional Groups, Supplement D*, Wiley, New York, 1983, p. 603.
30 G. Maier, *Angew. Chem. Int. Ed. Engl.*, *13* (1974) 425.
31 M. W. Grayston, W. D. Saunders and D. M. Lemal, *J. Am. Chem. Soc.*, *102* (1980) 413.
32 E. J. Petersson, J. C. Fanuele, M. R. Nimlos, D. M. Lemal, G. B. Ellison and J. G. Radsiszewski, *J. Am. Chem. Soc.*, *119* (1997) 11122.
33 W. P. Dailey and D. M. Lemal, *J. Am. Chem. Soc.*, *106* (1984) 1169.
34 The triene had been synthesized previously by another route. D. J. Dodsworth, C. M. Jenkins, R. Stephens and J. C. Tatlow, *J. Chem. Soc., Chem. Commun.*, (1972) 803.
35 D. M. Lemal, J. Buzby, A. C. Barefoot, III, M. W. Grayston and E. D. Laganis, *J. Org. Chem.*, *45* (1980) 3118.
36 This sequence of steps gives yields superior to those of the published procedure.
37 A. C. Barefoot, III, E. W. Corcoran, Jr., R. P. Hughes, D. M. Lemal, W. D. Saunders, B. B. Laird and R. E. Davis, *J. Am. Chem. Soc.*, *103* (1981) 970.
38 For a review on organo-transition metal compounds with fluorinated ligands, see R. P. Hughes, *Adv. Organomet. Chem.*, *31* (1990) 183.
39 T. Spector, Ph.D. Dissertation, Dartmouth College, 1987.
40 R. Naor and Z. Luz, *J. Chem. Phys.*, *76* (1982) 5662.

41 G. D. Goldman, B. E. Roberts, T. D. Cohen and D. M. Lemal, *J. Org. Chem.*, *59* (1994) 7421.

42 R. F. Waldron, A. C. Barefoot, III and D. M. Lemal, *J. Am. Chem. Soc.*, *106* (1984) 8301.

43 M. M. Rahman, B. A. Secor, K. M. Morgan, P. R. Shafer and D. M. Lemal, *J. Am. Chem. Soc.*, *112* (1990) 5986.

44 In the case of the parent cyclooctatetraene, 0.01% of the bicyclic isomer is present at 100 °C. R. Huisgen, G. Boche, A. Dahmen and W. Hechtl, *Tetrahedron Lett.*, (1968) 5215; R. Huisgen and F. Mietzsch, *Angew. Chem. Int. Ed. Engl.*, *3* (1964) 83.

45 The barrelene was synthesized by P. Ralli and D. M. Lemal, unpublished work. For the benzene synthon, see W. P. Dailey, R. A. Correa, E. Harrison, III and D. M. Lemal, *J. Org. Chem.*, *54* (1989) 5511.

46 For a related example, see R. E. Banks, R. N. Haszeldine and A. Prodgers, *J. Chem. Soc., Perkin 1*, (1973) 596.

47 H. E. Zimmerman, R. N. Binkley, R. S. Givens, G. L. Grunewald and M. A. Sherwin, *J. Am. Chem. Soc.*, *91* (1969) 3316.

48 For semibullvalene itself the barrier is only 5.2 kcal/mol, and it may be lower yet for the perfluoro analog. D. Moskau, R. Aydin, W. Leber, H. Günther, H. Quast, H.-D. Martin, K. Hassenrück, L. S. Miller and K. Grohmann, *Chem. Ber.*, *122* (1989) 925.

49 J. Meinwald and D. Schmidt, *J. Am. Chem. Soc.*, *91* (1969) 5877; H. E. Zimmerman, J. D. Robbins and J. Schantl, *J. Am. Chem. Soc.*, *91* (1969) 5878.

50 For the cyclization of perfluoro-1,5-cyclooctadiene, see M. Prober and W. T. Miller, Jr., *J. Am. Chem. Soc.*, *71*(1949) 598; I. L. Karle, J. Karle, R. W. Broge, A. H. Fox and J. L. Hoard, *J. Am. Chem. Soc.*, *86* (1964) 2523.

51 M. A. Steffen, Senior Honors Thesis, Dartmouth College, 1985.

52 J. Meinwald and H. Tsuruta, *J. Am. Chem. Soc.*, *92* (1970) 2579.

53 P. Bischof, R. Gleiter and R. Haider, *J. Am. Chem. Soc.*, *100* (1978) 1036.

54 B. E. Smart, in J. F. Liebman and A. Greenberg (eds.), *Molecular Structure and Energetics*, VCH Publishers, Deerfield Beach, FL, 1986, Vol. 3, p. 141.

55 R. A. Correa, N. Jing and D. M. Lemal, *J. Org. Chem.*, *58* (1993) 6406.

56 C. Steel, R. Zand, P. Hurwitz and S. G. Cohen, *J. Am. Chem. Soc.*, *86* (1964) 679.

57 R. Srinivasan and A. A. Levi, *J. Am. Chem. Soc.*, *85* (1963) 3363.

58 N. Jing and D. M. Lemal, *J. Org. Chem.*, *60* (1995) 89.

59 J. G. Calvert and J. N. Pitts, Jr., *Photochemistry*, Wiley, New York, 1966, p. 68 *et seq.*

60 A. L. J. Beckwith and K. U. Ingold, in P. de Mayo (ed.), *Molecular Rearrangements*, Academic Press, New York, 1978, Vol. 2.

61 D. I. Schuster, G. Lem and N. A. Kaprinidis, *Chem. Rev.*, *93* (1993) 3, and references therein.

62 N. Jing and D. M. Lemal, *J. Am. Chem. Soc.*, *115* (1993) 8481.

63 K. N. Houk, S. M. Gustafson and K. A. Black, *J. Am. Chem. Soc.*, *114* (1992) 8565; M. Dupuis, C. Murray and E. R. Davidson, *J. Am. Chem. Soc.*, *113* (1991) 9756.

64 N. Jing and D. M. Lemal, *J. Org. Chem.*, *59* (1994) 1844.

65 The cyclization to **53** was also reported by R. Hrabal, Z. Chvátal and V. Dedek, *J. Fluorine Chem.*, *63* (1993) 185; Z. Chvátal, M. Brothankova, R. Hrabal and V. Dedek, *Czech. CS*, *241* (1988) 840.

66 W. R. Dolbier, Jr., H. Koroniak, D. J. Burton, A. R. Bailey, G. S. Shaw and S. W. Hansen, *J. Am. Chem. Soc.*, *106* (1984) 1871; W. R. Dolbier, Jr., T. A. Gray, J. J. Keaffaber, L. Celewicz and H. Koroniak, *J. Am. Chem. Soc.*, *112* (1990) 363.

67 W. Kirmse, N. G. Rondan and K. N. Houk, *J. Am. Chem. Soc.*, *106* (1984) 7989; N. G. Rondan and K. N. Houk, *J. Am. Chem. Soc.*, *107* (1985) 2099.

68 N. Jing and D. M. Lemal, Abstracts of the 206th National Meeting of the American Chemical Society, Chicago, IL, August 22-27, 1993, FLUO 13.

69 K. B. Wiberg, M. G. Matturro, P. J. Okarma, M. E. Jason, W. P. Dailey, G. J. Burgmaier, W. F. Bailey and P. Warner, *Tetrahedron*, *42* (1986) 1895; K. B. Wiberg, M. G. Matturro, P. J. Okarma and M. E. Jason, *J. Am. Chem. Soc.*, *106* (1984) 2194; K. B. Wiberg, W. F. Bailey and M. E. Jason, *J. Org. Chem.*, *39* (1974) 3803; J. Casanova and H. R. Rogers, *J. Org. Chem.*, *39* (1974) 3803; K. B. Wiberg, G. J. Burgmaier and P. Warner, *J. Am. Chem. Soc.*, *93* (1971) 246.

70 Y. Zhang, J. R. Smith and D. M. Lemal, *J. Am. Chem. Soc.*, *118* (1996) 9454.

71 H. Finkelstein, *Chem. Ber.*, *43* (1910) 1528.

72 D. C. Luehrs, R. E. Brown and K. A. Godbole, *J. Solution Chem.*, *18* (1989) 463; M.F. Fox and E. Hayon, *Trans. Faraday Soc. I*, *72* (1976) 1990.

73 C. J. Junk and D. M. Lemal, unpublished results.

74 For the parent [2.2.2]propellane, the strain energy has been calculated to be 97 kcal/mol. K. B. Wiberg, *J. Am. Chem. Soc.*, *105* (1983) 1227.

75 P. E. Eaton and G. E. Temme, III, *J. Am. Chem. Soc.*, *95* (1973) 7508.

76 J. R. Smith and D. M. Lemal, unpublished results.

77 R. A. C. Ceballos (R.A. Correa), Ph.D. Dissertation, Dartmouth College, 1990.

78 For a concise summary of the work, which encompasses 19 papers, see: H. Hart, Z. Rappoport and S. E. Biali, in Z. Rappoport (ed.), *The Chemistry of Enols*, Wiley, Chichester, 1990, p. 502.

79 R. A. Bekker, G. G. Melikyan, É. P. Lur'e, B. L. Dyatkin and I. L. Knunyants, *Akad. Nauk SSSR*, *217* (1974) 1320, Engl. transl., p. 572.

80 J. R. Keeffe, A. J. Kresge and N. P. Schepp, *J. Am. Chem. Soc.*, *112* (1990) 4862; J. Toulec, in V. Gold and D. Bethell (eds.), *Advances in Physical Organic Chemistry*, Academic Press, London, 1982, Vol. 18, p. 1; Y. Chiang, M. Hojatti, J. R. Keeffe, A. J. Kresge, N. P. Schepp and J. Wirz, *J. Am. Chem. Soc.*, *109* (1987) 4000.

81 R. A. Correa, P. E. Lindner and D. M. Lemal, *J. Am. Chem. Soc.*, *116* (1994) 10795.

82 P. E. Lindner, R. A. Correa, J. Gino and D. M. Lemal, *J. Am. Chem. Soc.*, *118* (1996) 2556.

83 R. A. Bekker, V. Ya. Popkova and I. L. Knunyants, *Dokl. Akad. Nauk SSSR*, *235* (1977) 103, Engl. transl., p. 370.

84 P. E. Lindner and D. M. Lemal, *J. Am. Chem. Soc.*, *119* (1997) 3259.

85 P. E. Lindner and D. M. Lemal, *J. Org. Chem.*, *61* (1996) 5109.

86 P. E. Lindner and D. M. Lemal, *J. Am. Chem. Soc.*, *119* (1997) 3267.

87 E. W. Schlag and E. W. Kaiser, *J. Am. Chem. Soc.*, *87* (1965) 1171.

88 J. Burdeniuc, B. Jedlicka and R. H. Crabtree, *Chem. Ber./Recueil*, *130* (1997) 145.

89 J. Burdeniuc, W. Chupka and R. H. Crabtree, *J. Am. Chem. Soc.*, *117* (1995) 10119.

90 With fluorocarbons containing 6-membered rings it is possible to stop the defluorination at the stage where the rings are aromatic. J. L. Kiplinger and T. G. Richmond, *J. Am. Chem. Soc.*, *118* (1996) 1805.

91 With perfluorodecalin, defluorination could be interrupted at the Δ^9-octalin stage, but this proved to be a special case because of the presence of vicinal tertiary fluorines. N. A. Kaprinidis and N. J. Turro, *Tetrahedron Lett.*, *37* (1996) 2373.

92 J. Burdeniuc and R. H. Crabtree, *J. Am. Chem. Soc.*, *118* (1996) 2525; G. Saunders, *Angew. Chem. Int. Ed. Engl.*, *35* (1996) 2615.

93 N. Ramchandani, Senior Thesis, Dartmouth College, 1996.

94 N. S. Stoyanov, Senior Honors Thesis, Dartmouth College, 1998.

95 R. A. Bekker, B. L. Dyatkin, N. M. Grushina, A. V. Ioganseen and G. A. Kurkchi, *Zh. Spectrosk.*, *28* (1978) 106, Engl. transl., p. 78.

BIOGRAPHIC NOTE

Professor David Lemal was born in Plainfield, New Jersey, on February 20, 1934. He received an AB degree from Amherst College, summa cum laude, in 1955 and a PhD from Harvard University in 1959. Following an Instructorship at the University of Wisconsin from 1958-1960, he was an Assistant Professor there until 1965. He joined the Dartmouth Chemistry Department that year as Associate Professor, became a full Professor in 1969 and served as Chairman during 1976-79. Since 1981 he has been the Albert W. Smith Professor of Chemistry.

David Lemal

Chapter 21

ADVENTURES OF A FLUORINE CHEMIST AT DuPONT

WILLIAM J. MIDDLETON
Chemistry Department, Ursinus College, Collegeville, PA 19426-1000, USA

Introduction

My first introduction to fluorine chemistry

I have been fascinated by chemistry ever since I was in grade school and discovered that vinegar reacts with soda, but my fascination with fluorine chemistry did not develop until I was assigned a project to prepare the then unknown difluoroacetylene while working at DuPont in their Central Research Department in Wilmington, Delaware. Actually, my first brief contact with fluorine chemistry (if you can call it that) was a total diaster. In fact, I didn't even know it was a contact with fluorine chemistry at the time. It occurred when I was taking my preliminary exams for my PhD degree at the University of Illinois. These 'prelims' were given to prospective PhD students before their final exam and thesis defence to determine if they had the necessary broad knowledge of chemistry for recipients of this degree. The question was, 'Explain why Teflon® is chemically inert'. I had to leave the answer blank since I didn't have any idea what Teflon® was. I couldn't even guess that it was a fluoropolymer. Not a very auspicious start for someone who one day would aspire to be a fluorine chemist!

Before DuPont

To start at the beginning, I was born in Amarillo, Texas on April 9, 1927. I spent the first twenty years or so of my life in Texas as a Texan, and the rest of my life trying to live up to (or maybe live down) the typical reputation that most Texans seem to have. I received my public school education in various towns in north and west Texas, ending up in Dallas. After a brief stint in the US Navy during World War II, I continued my education at North Texas State University[1] in Denton, receiving my BS degree in 1948 and MS degree a year later. More importantly, I met my wife Millie there, and we married in 1948. After more than fifty years of marriage, I still think Millie is the best thing that ever came out of the state of Oklahoma.

With the help of Millie's earning power, I continued my education at the University of Illinois in Urbana, Illinois, where I studied for my doctoral degree under Nelson Leonard (my thesis was on Reductive Cyclization). I was still a few years away from my first real contact with fluorine chemistry.

[1]When I enrolled, it was called North Texas State Teachers College; before I received my MS degree, the 'Teachers' part of the name was dropped; and now 'College' has been changed to 'University'.

On to DuPont

I was overjoyed to receive an invitation to work in DuPont's Central Research Department[2] at the Experimental Station in Wilmington, Delaware. At that time (1952), this department was regarded by many as the home of the best group of industrial chemists in the US doing fundamental research in organic chemistry, so I accepted this offer with great pleasure. Ironically, DuPont was the manufacturer of Teflon®, the material that had caused me grief during my 'prelims'.

My first project at DuPont had nothing to do with fluorine: shortly before I arrived at DuPont, a chemist named Dick Heckert had prepared tetracyanoethylene (TCNE), and when I turned up, both he and I were assigned the task of exploring the chemistry of this new, unique compound. Dick and I entered a friendly competition to see who could do the most interesting chemistry with TCNE, and I thought that I was doing very well on that score. Unfortunately, I can't feel too smug about that now, for Dick later became president of DuPont, and I remained at the bench for my entire career at the company.

Although the TCNE project was a scientific success (our work was even featured twice in *C&E News* stories, including once as the cover story [1]), it was a commercial flop, so the stage was set for my induction into the world of Fluorine Chemistry. In fact, I was thrust into the area almost against my will.

Fluoroacetylene

In 1955, a considerable amount of fluorine chemistry was being done at DuPont because of the commercial success associated with the company's Freons® and Teflon®; but in many ways we were very naive about some of the properties of fluorine and its compounds. There was a theory that the introduction of fluorine would always stabilize organic compounds owing to the extreme strengths of C—F bonds. So even though chloro- and dichloro-acetylene were known to be dangerously explosive, it was reasoned that fluoro- and difluoro-acetylene would be very stable and safe to handle. All this was theory, of course, because these compounds were unknown. In fact, there were no known compounds that contained fluorine bonded to an *sp*-hybridized carbon atom.

Although things were still going well for me in my work with TCNE, I was taken off that project and assigned a new task – to prepare fluoro- and difluoro-acetylene. This was quite a shock for me. In my previous project, I had been working with solids that in general were high-melting, crystalline compounds. Now I would be working with gases, a subject that was totally unfamiliar to me. But I received a much bigger shock a few weeks later when I was finally able to prepare the first sample of fluoroacetylene.

My first attempts to prepare the fluoroacetylenes by conventional dehalogenation and dehydrohalogenation reactions were unsuccessful, but a less conventional method yielded good results: quantitative yields of fluoroacetylene (**2**) were obtained by vacuum pyrolysis of fluoromaleic anhydride (**1**, Scheme 21.1) [2]. We believe that similar vacuum pyrolysis of difluoromaleic anhydride yielded difluoroacetylene, but this proved to be too unstable or reactive for us to isolate and conclusively identify.

[2]It was called the Chemical Department when I joined DuPont. Later the name was changed to the Central Research Department, which eventually merged with the Development Department.

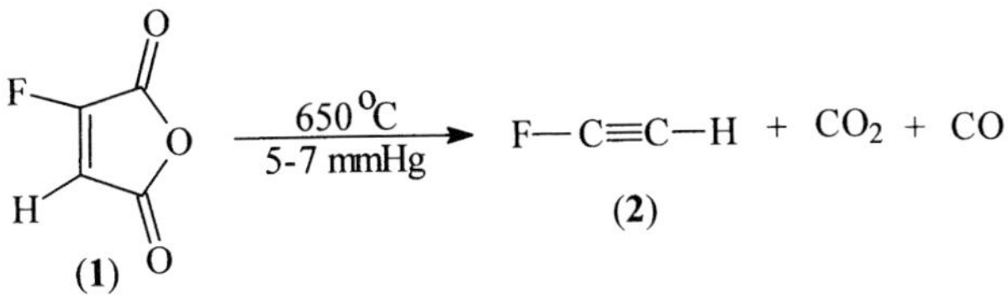

Scheme 21.1.

Just a few months earlier, we had acquired an exciting new analytical tool – our first nuclear magnetic resonance spectrometer and the only such instrument in the state of Delaware. It was sited in our building, so I thought this was a good opportunity to obtain my first fluorine ^{19}F NMR spectrum. I sealed a sample of fluoroacetylene in a tube and gave it to Harlen Foster, our spectroscopist. He inserted it into the probe and obtained a beautiful doublet on the oscilloscope, and then recorded it on pressure-sensitive tape, which was our recording device in those days. Before the sample could be removed from the probe, a deafening explosion occurred, and a black mushroom of smoke issued forth. The probe was completely gutted, but worse yet, the face of the heavy steel magnet was scored. To get the spectrometer back into operation, the probe had to be replaced and the face of the magnet reground. This required several weeks. You can imagine that I wasn't very popular with my co-workers for a while, having destroyed the only NMR spectrometer available to them.

The very next day, I had an explosion in my vacuum train while I was trying to purify another sample of fluoroacetylene. I wasn't seriously injured, but I did have over a hundred small puncture wounds in my left arm, which left me picking glass out of my arm for several weeks. These explosions so scared my supervisor, Bill Sharkey, that he immediately cancelled the project, and our safety engineer, who shall remain nameless, destroyed our remaining samples by shooting at them with a rifle in an open field. Such was my introduction to fluorine chemistry.

Perfluorothiocarbonyl compounds

I had no input into choosing my first two projects at DuPont – they were simply assigned to me; but in spite of the explosions, I found that I really liked organofluorine chemistry. I had no say in the choice of my next project either, which turned out to be one that kept me on the road to becoming a fluorine chemist. Subsequently, DuPont was very generous in letting *me* decide what to work on, and, in most cases, I continued to work in the field that I had learned to love – organofluorine chemistry – as one project seemed to flow naturally into another.

Thiocarbonyl fluoride

My next project was to investigate perfluoro polymers that contained heteroatoms in the chain or backbone. The hope was that we could prepare polymers with the chemical resistance and other good properties that Teflon® possessed, but would also be thermoplastic, so they could be moulded, or maybe even be elastomeric or rubber-like.

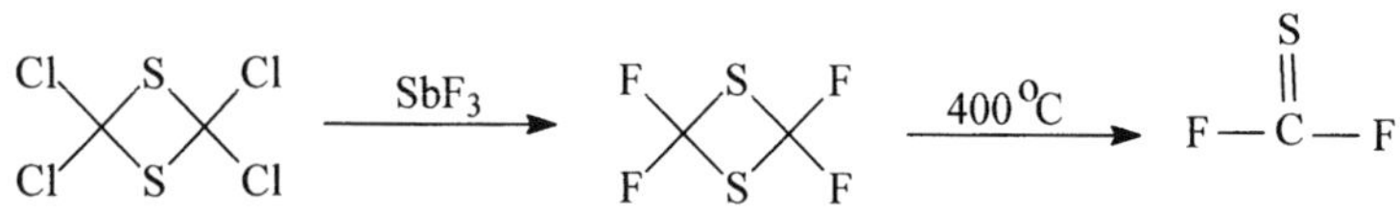

(3) (4) (5)

Scheme 21.2.

We were indeed able to prepare polymers of this type. The most interesting one was poly(thiocarbonyl fluoride) [3]. The monomer, thiocarbonyl fluoride (**5**), was prepared in high yield and high state of purity by the pyrolysis of 2,2,4,4-tetrafluoro-1,2-dithietane (**4**), which was in turn prepared by the fluorination of thiophosgene dimer (**3**; Scheme 21.2) [4]. Polymerization of thiocarbonyl fluoride by anionic initiation at low temperatures gave an elastomeric polymer which had alternating CF_2 groups and S atoms. Even before this polymer was cured, it was bouncier that the best natural rubbers after cure. My supervisor, Bill Sharkey, was so intrigued by its properties that he had a sample made into a golf ball. This ball had remarkable properties – it was sort of a super 'super ball'. If it was dropped onto a firm surface, it would rebound to more than 95% of the release height. If it was dropped from 5 feet, it would continue to bounce for a long time; and after 30 bounces, it would still be bouncing higher than a foot. Bill tried it out on a golf course, thinking that he might get a super-long-distance drive. He did, but not in the direction that he suspected. The problem was that the ball was very heavy – very dense – so it didn't travel far in the air; but, oh boy!, once it hit the ground it really travelled, bounding first in one crazy direction, then another.

Poly(thiocarbonyl fluoride) might have been commercial except for two things. It lacked insufficient stability toward non-aqueous bases to be a high value-in-use, inert, stable elastomer, and it was too expensive to produce to be a superior general-purpose elastomer. It did have one property that was useful, however: in its crystalline form, it had a low coefficient of friction and it was non-stick, like Teflon®. DuPont licensed the patent to a company interested in coating razor blades.

Hexafluorothioacetone

The work on thiocarbonyl fluoride led to our work on hexafluorothioacetone (HFTA) and other perfluorothiocarbonyl compounds. We first prepared HFTA by reacting bis(perfluoroisopropyl)mercury (**6**) (which could be prepared from HgF_2 and hexafluoropropylene) with molten sulfur [4]. HFTA is a deep blue liquid, b.p. 6 °C, that can be stored for extended periods of time at −78 °C, but slowly dimerizes to a dithietane (**7**) when stored at room temperature (Scheme 21.3).

Our original purpose was to form a polymer from HFTA, but this polymer turned out to be thermally unstable, probably because of steric strain. However, HFTA did have some remarkable properties. For example, it may be the most reactive dienophile and enophile now known that can be put in a bottle and stored. For example, it will react rapidly with the diene system in styrene, even at −78 °C, and with propylene, also at −78 °C, to give adducts [5]. The ene reaction with propylene is typical of many of HFTA's other reactions, in that it appears as though the sulfur atom is at the positive end of the C−S dipole, hence

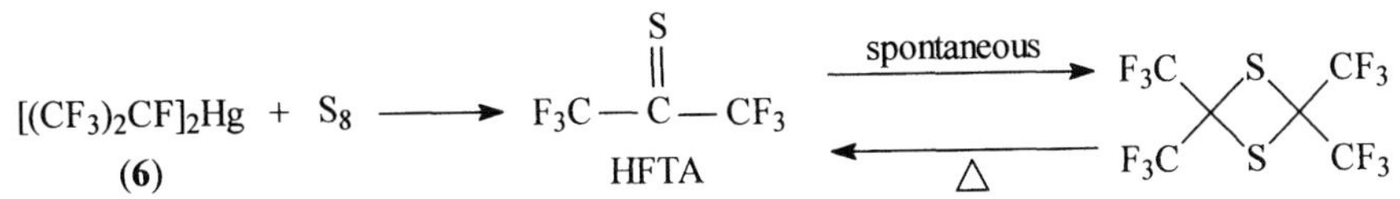

(6) (7)

Scheme 21.3.

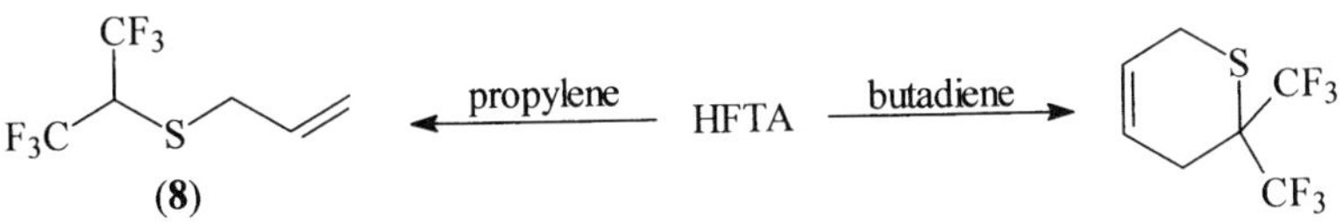

(8)

Scheme 21.4.

the sulfide **8** is produced instead of a thiol, as would be expected if it were to behave like hexafluoroacetone.

Hexafluoroacetone and its derivatives [6]

The work on HFTA led to our work with hexafluoroacetone (HFA) itself, including its many derivatives, which included (see Scheme 21.5): hexafluoroacetone imine (**9**) and 2,2-diaminohexafluoropropane (**10**) [7]; bis(trifluoromethyl)diazomethane (**11**) and the isomeric diazirine (**12**), and the carbene **13** derived from these compounds [8]; and 1,1-bis(trifluoromethyl)dicyanoethylene (**14**) [9]. We made many interesting discoveries while working with HFA, some by design, and some by accident. One example of an accidental discovery concerns midaflur.

Midaflur is 4-amino-2,2,5,5-tetrakis(trifluoromethyl)imidazoline (**15**), an extremely potent muscle relaxant [10]. This compound was first formed as an unexpected by-product when we were trying to make amino acids from HFA by the reaction of sodium cyanide with hexafluoroacetone imine (**9**; Scheme 21.6). My technician, Armand Bardales, noted a few crystals embedded in the pot residue after distillation. We fished these crystals out, identified them as the aminoimidazoline **15**, and then sent a new sample, made on purpose, to our routine biological screen. Much to our surprise, the compound was extremely potent as a muscle relaxant. It was given the generic name midaflur; and some have claimed that it was named after me, but I assure you that this was not the case.

Midaflur has some good properties. For example, it has very low toxicity. This was demonstrated when a massive dose was given to a beagle. The beagle went to sleep for a week, woke up, took a drink of water, and then went back to sleep for two days. When he finally awoke for good, he had suffered no ill effects except weight loss from not eating for several days, hence the facetious suggestion that midaflur might be the perfect diet drug. In addition to lack of toxicity, midaflur has many other good properties – for example, it does not depress heartbeat or breathing, as other muscle relaxants do; however, it failed in the clinic, partly because it was too potent, *i.e.* it relaxed all skeletal muscles, not just the spastic ones, and no way was found to terminate its effect prematurely once it was no longer needed, as in surgery.

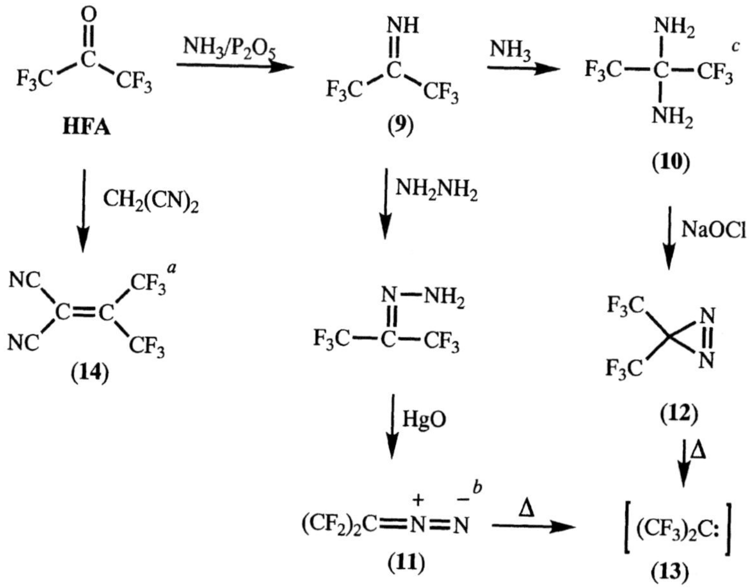

Scheme 21.5.

a An analogue of TCNE, undergoing many of the same reactions.

b A stable diazo compound.

c A stable *gem*-diamine, distillable at atmospheric pressure.

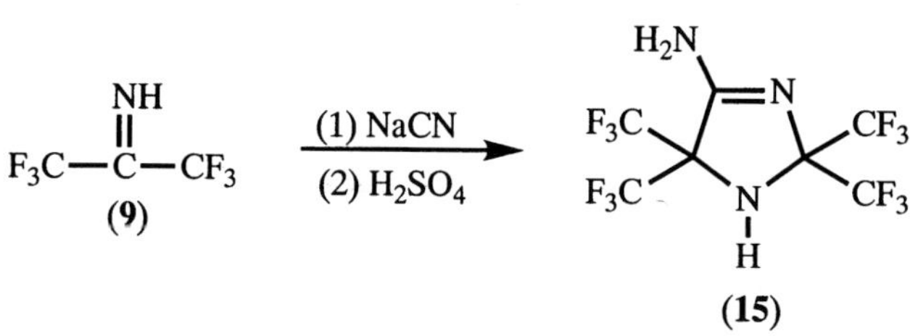

Scheme 21.6.

Fluoroalcohols

Another accidental discovery did lead to a commercial product; however, this discovery might be considered more serendipitous than accidental. We were trying to prepare some fluorothioketones through the corresponding fluoroalcohols, but the required alcohols were unknown, so first we had to prepare them. We reduced bis(perfluoroisopropyl) ketone with $LiAlH_4$ in diethyl ether to give the corresponding alcohol **16**, but found that we couldn't separate it from the ether solvent by distillation. Our new secondary alcohol had formed a very strong hydrogen bond with ether, and the resulting one-to-one complex could not be broken, either by distillation or recrystallization. This turned out to be a very general phenomenon: most highly-fluorinated secondary and tertiary alcohols form very strong hydrogen bonds with ether or other solvents or substrates that contain oxygen or nitrogen atoms [11].

$(CF_3)_2CFCH(OH)CF(CF_3)_2$ **(16)** $CF_3CH(OH)CF_3$ **HFIP**

It occurred to me that since these fluoroalcohols attached themselves so tenaciously to ethers, esters, amides, etc., perhaps they would also attach themselves to the surface of high-molecular-weight polymers; in other words, maybe they would form a fluorocarbon sheath around fibres of nylon or polyesters and render them soil and water repellant. It was an interesting notion, but it didn't work out that way: when I immersed pieces of nylon fabric or polyester fabric (Dacron®) in one of our new secondary fluoroalcohols, they dissolved like magic – just like sugar in water. Even polymers for which no good solvent was previously known dissolved. This was our serendipitous discovery – an excellent class of polymer solvents. DuPont marketed one of these new alcohols, hexafluoroisopropanol (HFIP)[3], for this use, for solvent welding of polyamides, polyesters and polyethers[4], and for analytical control in the manufacture of such polymers. Deuterated derivatives of these alcohols have also become useful NMR solvents.

Fluorinating reagents

Diethylaminosulfur trifluoride (DAST)

I feel that we were lucky to have made several accidental or serendipitous discoveries. However, most of our results have been the outcome of more directed research. An example is the development of DAST, or diethylaminosulfur trifluoride [12].

During the 1950s, William C. Smith and others at DuPont discovered that SF_4 could be used to replace the oxygen of carbonyl groups, and sometimes hydroxyl groups, with fluorine [13]. Unfortunately, SF_4, being a toxic gas, was difficult to handle and the harsh conditions often necessary for the fluorination reactions were unsuitable for sensitive organic compounds. We prepared several derivatives of SF_4 in the hope of finding a reagent that was easy to handle and could be used to replace hydroxyl groups with fluorine in sensitive molecules of biological interest. One of these derivatives, DAST (Scheme 21.7), seems to have fitted the bill fairly well judging from the number of investigators that used it in the decade after we first published our work [14]. We were able to obtain patents on DAST, but because the process of publication moves so slowly in an industrial company, another group of investigators beat us to the first journal publication, although they made no mention of the fluorination of alcohols, DAST's most useful fluorination reaction.

There are several reasons why it is of interest to insert fluorine into biological molecules. The reason I find most fascinating is the theory that fluorine can block sites of metabolism. After determining with model compounds that DAST did indeed replace OH groups with fluorine in sensitive molecules, we decided to test this theory. Our idea was to prepare a fluorine-containing derivative of the tranquilizer diazepam [Valium® (Roche)]

[3]US Patent 3,418,337 (1968).
[4]US Patent 3,245,944 (1966).

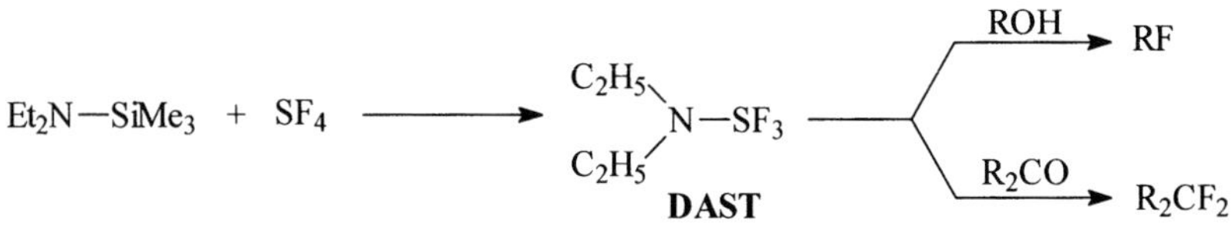

Scheme 21.7.

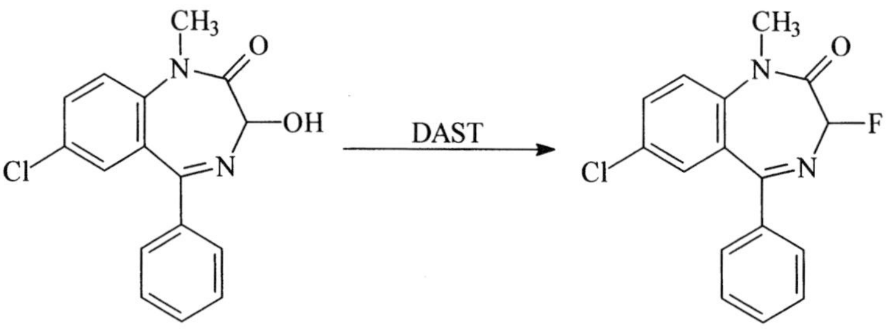

(17)

Scheme 21.8.

by replacing a hydrogen with a fluorine in the metabolically active 3-position. We hoped that this replacement would result in a compound with greater potency, since this derivative could not be metabolized by the normal pathway, and therefore would not be destroyed as rapidly. Using DAST, we were able to prepare 3-fluorodiazepam (**17**) in high yield from the corresponding hydroxy derivative (a metabolite of Valium®) [15] (Scheme 21.8), and found that it was indeed several times more potent in the model animals studied. I predict a bright future for pharmaceuticals and other biologically-active materials such as insecticides and herbicides that contain fluorine.

Tris(dimethylamino)sulfonium difluorotrimethylsilicate (TASF)

In trying to optimize the conditions for preparing DAST, we discovered that a sulfonium salt was formed instead if the SF_4 was added to the aminosilane (Scheme 21.9) instead of the other way around. One of the most interesting salts prepared this way, because it was highly crystalline and easy to isolate, was tris(dimethylamio)sulfonium difluorotrimethylsilicate (TASF) [16]. This salt is very soluble in organic solvents, and serves as a source of highly reactive fluoride ion due to the dissociation of the silicate anion and the large, charge-diffuse sulfonium cation that prevents significant ion-pairing. It can be used to replace various leaving groups (Cl, Br, I, TsO, CF_3SO_2O) with fluorine, even in non-polar solvents, to add fluoride ion to fluoro-olefins or fluorocarbonyl compounds to give stable isolable perfluorcarbanions (such as **19**) or perfluoroalkoxides (such as **18**), and to fulfil a variety of uses in general organic synthesis [17]. The alkoxide **18** is particularly noteworthy because the X-ray crystal structure shows that the C—F bonds are exceptionally long (1.390 and 1.397 Å) and the C—O bond is exceptionally short (1.227 Å), almost approaching the length of a double bond [18]. This appears to be one of the strongest bits of evidence now known for negative fluorine hyperconjugation (no-bond resonance).

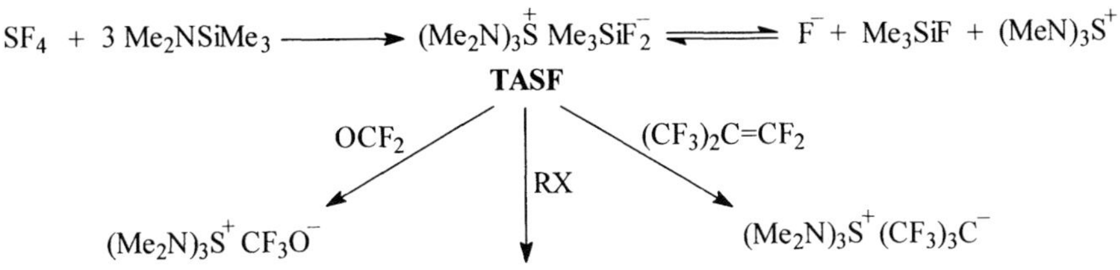

Scheme 21.9.

Fascinated by fluorine

Periodic table

A few years ago, I prepared a periodic chart of the elements (Fig. 21.1) to express how I feel about fluorine, and passed out copies to the audience when I gave my acceptance speech for the ACS Fluorine Award. One of the reasons why I am fascinated by fluorine is the extreme properties that fluorine and its compounds can exhibit. As a natural consequence of being the most electronegative element, some fluorine compounds have the distinction of being among the most reactive of all compounds, while others are found among the most inert of all compounds. Fluorine compounds also feature in lists of the least toxic of all compounds, and in those of the most toxic. Some of my own work illustrates this latter point.

Toxicity surprises

Freon® refrigerants and various fluorocarbon 'blood substitutes' are examples of some very non-toxic compounds. In our lab, however, we have prepared some very innocent-looking fluorine compounds that possess extreme toxicity.

One example is perfluoropinacol (**20**), which we prepared by a photolytic bimolecular reduction of HFA [11]. A single drop of **20** on the skin of a guinea pig is sufficient to kill it, whereas pinacol itself is practically non-toxic. Because it was an excellent solvent for polymers, we prepared over a pound of perfluoropinacol before we discovered its extreme toxicity. It is only by great good luck and good hygiene practices by almost all DuPont employees that no one was poisoned by this material before we destroyed it.

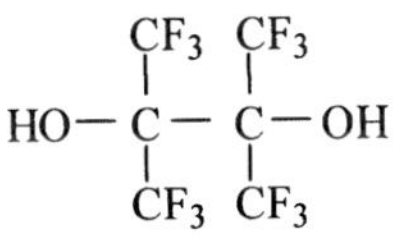

(20)

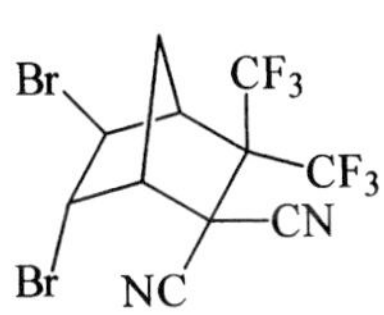

(21)

Another example is the norbornane **21**, derived from cyclopentadiene, bromine and 1,1-bis(trifluoromethyl)dicyanoethylene (**14**; Scheme 21.5) [19]. This norbornane is one of the most potent oral poisons known (oral LD_{50} 0.2 mg/kg), being more than four thousand times more toxic than lead arsenate. It was an extremely potent nematocide, but we aban-

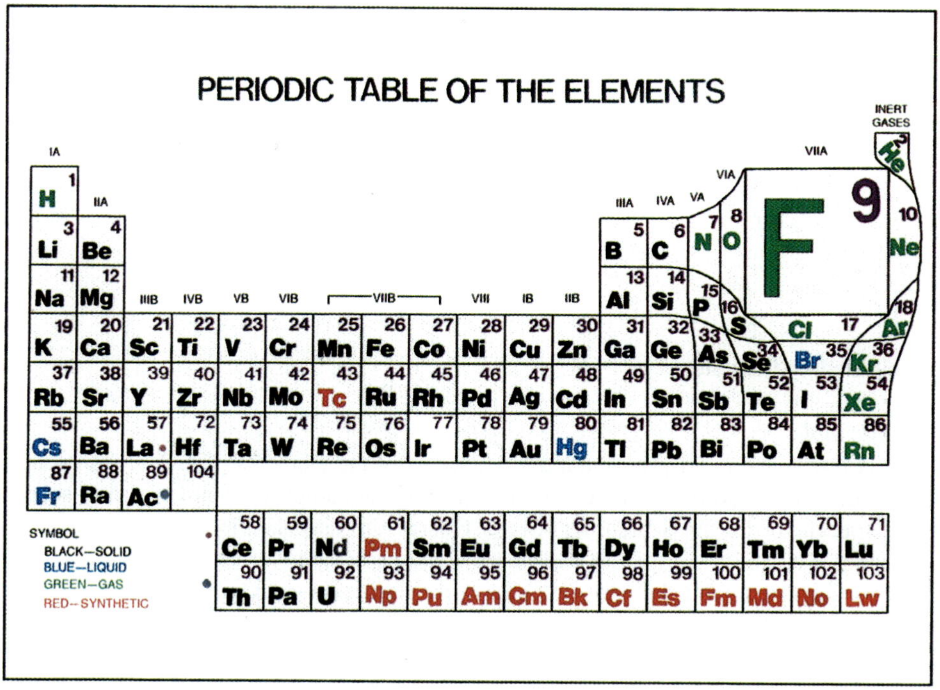

Fig. 21.1. A fluorine chemist's view of the Periodic Table.

doned all thoughts of field-testing it as soon as we discovered how toxic it was, and all existing samples were destroyed.

The perfluoroalkyl effect

Another fascinating aspect of fluorine chemistry is the remarkable tendency of perfluoroalkyl substituents to stabilize small-ring compounds. This perfluoroalkyl effect, which is thought to be a kinetic phenomenon, has enabled a number of compound types either unknown in hydrocarbon chemistry, or rarely isolable, to be prepared. In the course of our work we have found many examples of this effect, two of which I'll mention.

Ketones react with alkoxyacetylenes in the presence of BF_3 to give α,β-unsaturated esters in high yield (sometimes known as the Arens rearrangement.) The formation of an oxete as a cyclic intermediate has been postulated, but in general, oxetes are unstable and have not been isolated. However, we found that hexafluoroacetone reacts vigorously and exothermically with ethoxyacetylene without added catalyst to give the first example of an isolable oxete (**22**) that can be distilled and characterized [20]. It requires a mild pyrolysis or storage at room temperature for several days to be converted to the expected ester, **23**.

We observed other examples of the perfluoroalkyl effect when we combined our work with perfluorodiazo compounds and perfluorothiocarbonyl compounds. Usually, 1,3,4-thiadiazolines are too unstable to isolate, but a stable, isolable thiadiazoline (**25**) was formed when bis(trifluoromethyl)diazomethane was mixed with bis(trifluoromethyl)thio-

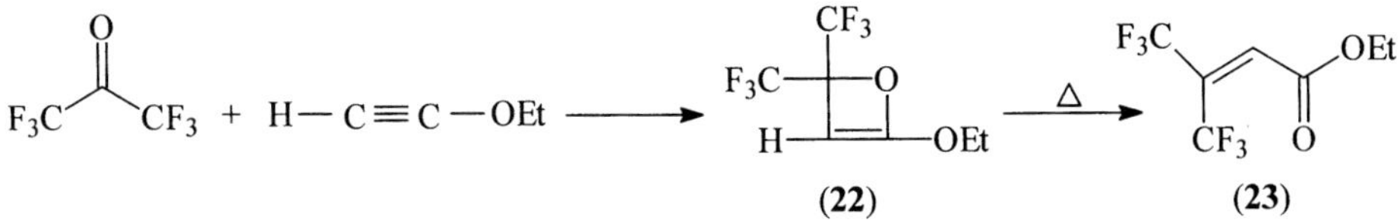

Scheme 21.10.

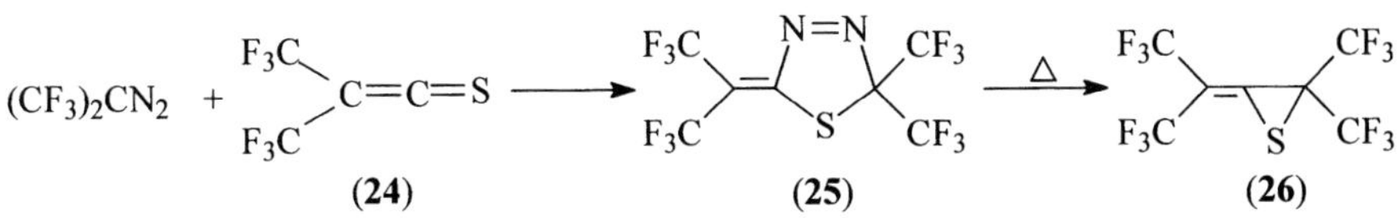

Scheme 21.11.

ketene (**24**) [21]. Mild pyrolysis of **25** caused extrusion of nitrogen to give the vinylepisulfide **26**, also stabilized by the trifluoromethyl groups.

A professional embarrassment

My several professional triumphs at DuPont were accompanied by a few professional embarrassments. One embarrassment in particular sticks in my mind. Our research had led us to prepare a number of CF_3-substituted phenylethylenes that possessed potent estrogenic activity [22], the most potent of which was an analogue (**27**) of diethylstilbestrol (DES), in which the ethyl groups had been replaced with trifluoromethyl groups. This compound had many times the potency of the very potent synthetic estrogen DES [23].

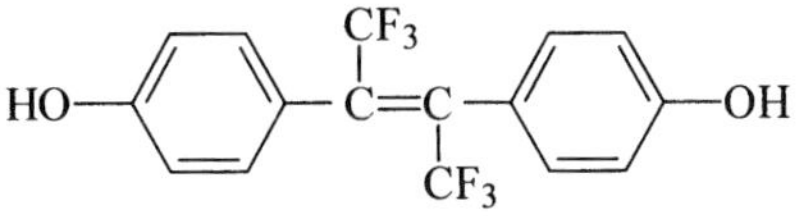

(27)

Using animal models, mainly rats, our biologist at DuPont, Jack Snyder, discovered that our compounds could be used as extremely effective post-coital antifertility agents, *i.e.* morning-after pills. But it was the interesting chemistry based on hexafluoroacetone or hexafluorothioacetone that was uppermost in our minds when we decided to give a paper on our research at the Washington DC ACS meeting in 1972. One of our PR men at DuPont, who was always looking for stories that would catch the public's eye (and who shall remain nameless), duly prepared a press release dealing with our work, but unfortunately the editors of the Wilmington papers wrote headlines for the story that really made me squirm. The headline in the *Wilmington Morning News* was 'DuPont Discovers Birth Control Pill for Rats', and the *Evening Journal*'s was even worse – 'Post Mating Pill for Rats is Developed'. Regrettably, my name was mentioned, and it was said that the compounds I had prepared were related to Teflon®, DuPont's fluorocarbon resin used in non-stick frying pans; the reports then went on to explain that the compounds could also be considered non-stick because they worked by preventing a fertilized ovum from sticking to the wall of the uterus. I received some very angry phone calls the next day, particularly from the people at DuPont who were involved with non-stick pans. They were concerned that people who read the article might believe that using non-stick cookware would make them impotent or infertile!

My DuPont colleagues

While at DuPont, I had the privilege of working and consulting with some of the best fluorine chemists anywhere in the world. There was always a free exchange of ideas, and we were always proud of each other's accomplishments. Many have considered the group of fluorine chemists assembled by the Central Research Department of DuPont during the last half of the 20th Century to be the greatest collection under one roof of industrial organic fluorine chemists anywhere in the world (the group included David C. England, Carl G. Krespan, William A. Sheppard, Bruce E. Smart, Maynard S. Raash, Charles W. Tullock, Frank S. Fawcett, William E. Burnette and William B. Farnham and myself, and for a brief time, Shlomo Rozen and Kirby Scherer). To add credence to this belief, three members of this group have been given the American Chemical Society Award for Creative Work in Fluorine Chemistry, the only such group to collect this many awards to date. Short biographies of the three award winners, of which I am one, are appended to this article.

After DuPont

I left DuPont in 1984 to teach chemistry at Ursinus College in Collegeville, Pennsylvania. I always wanted to try my hand at teaching in a small liberal arts college, and at Ursinus I finally got my chance. It was a delightful experience, but I found doing research there was a far different ball game than doing research at DuPont. For one thing, there was no fluorine NMR spectrometer available, and doing without ^{19}F NMR for an organofluorine chemist is almost like working blindfolded and handcuffed. I was totally dependent on friends in industry or research universities to run spectra for me. Research at a liberal arts undergraduate institution is also different from research at a university that has a graduate department, in that almost all of the research is done by junior or senior undergraduate students who can normally spend only a few hours each week in the lab, and by the time they finally become experienced, they graduate. Nonetheless, I believe that we were able to accomplish some useful research, and many of my students won awards at regional undergraduate research meetings.

One of our more significant discoveries was a new polarity scale for very non-polar substances [24]. This was a serendipitous discovery: one of my students, Beth Freed, noted that the fluorocarbon-soluble dye **28** she had prepared as a leak-detecting agent for CFCs had a distinctly different colour in every solvent that it was dissolved in, so we used its solvatochromic properties to devise a relative polarity scale.

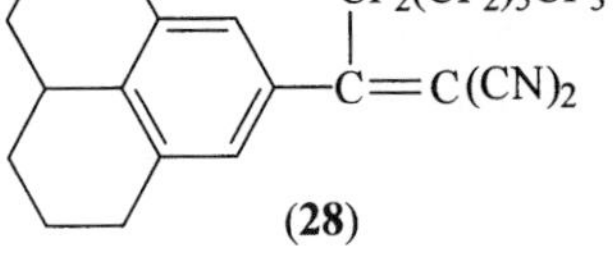

(28)

Another discovery [25] which may have utility concerns the selective monofluorination of benzylic compounds (*e.g.* $C_6H_5CH_2CN \rightarrow C_6H_5CHFCN$) via electrolysis (with alternating current) in pyridine/HF (Olah's Reagent) instead of the usual Simons electrolysis in straight HF with direct current, which results in polyfluorination.

Fluorine chemistry's future?

In the past few years, the concern has arisen that the number of investigators doing fluorine chemistry may drop because much government-supported research in the area has been terminated. This is particularly true in the US. However, I believe that this will be more than offset by the increased interest that chemists of all disciplines have in fluorine.

In the past, fluorine chemistry has been regarded by many as a specialized field – not in the main stream of chemical science. I think this was especially true for organic fluorine chemistry. However, that seems to be changing now since more and more chemists who do not regard themselves as fluorine chemists are utilizing fluorine in their research. This is certainly true in the biological and medicinal area, but it is also true in other areas as well. Consequently, there is a much broader interest in what the fluorine chemist is doing today and what he has done in the past. We seem to be entering the mainstream of chemistry now. I don't know what effect this will have on those of us who regard ourselves as fluorine chemists. I hope that we are up to the challenge.

Addendum

Chemistry, and particularly fluorine chemistry, has been my life's blood, but I also greatly enjoy another of my hobbies – writing poetry. On rare occasions the two mix, and I have actually had two of my poems published in scientific journals. The first one appeared in *C&E News* [26], and is a form of poetry called a double dactyl:

BURNER, BURNER,
BURNING BLUE

Flamity-blamity
Robert E. Bunsen, the
Chemist we all know of
Gas burner fame,

Worked to develop the
Gas and air mixer that
Characteristic'ly
Burns a blue flame.

The second poem was published in *Fluorine Technology Bulletin* [27]. The professor in this poem is fictional, but does bear some resemblance to me. It runs as follows:

PROFESSOR McGEE'S SOLUTION

Have you ever heard of a queer old bird named Thaddeus P. McGee?
He was of late a professor of State University.

Now Thaddeus was a teacher because he loved to teach chemistry,
But the real love he had was to work in the lab whenever his time was free.

He was renowned for once he had found a new solvent for stainless steel,
But, never shirking, he kept on working, for his solvent was far from ideal.

He said, to succeed, what he'd really need was a broth to dissolve everything,
And he dreamed of the day when the world would say, 'Thad, you're the chemistry king.'

Now it was no chore to dissolve the wood floor... a little strong acid would do,
And it was no task to dissolve a glass flask and a beaker or test tube or two.

And it's easy as pie to use a little lye to dissolve both flesh and hair,
And who couldn't use whatever acid they choose to dissolve rusty nails and such fair?

But the real challenge was to try to manage to find one solvent for all.
His students all say that he worked night and day to succeed in this task before fall.

When school started that fall, his students recall, they entered his lab through the door,
But they did not see Professor McGee. All they saw was a hole in the floor.

They looked all around. He was not to be found. All they saw was the hole in the floor.
They looked in the hole. Had McGee reached his goal? This hole was hard to ignore.

The hole was much more than a hole through the floor, for it was a bottomless pit.
His students surmised that McGee had devised a solvent that just wouldn't quit,

And his students all stated he miscalculated, for no flask could contain his new brew...
It dissolved the flask wall and that is not all, it dissolved the lab bench and ate through

The lab floor, and what's even more it dissolved a deep hole in the ground.
Some thought that McGee, who was nowhere to see, and was never again to be found,

Was now dissolved, too, into his new brew, and his solution flowed into the hole.
But don't shed a tear for Thaddeus, my dear, for in death he attained his life's goal.

References

1 *Chem. & Eng. News*, p. 50 (April 28, 1958); pp. 114-120 (April 11, 1960).
2 W. J. Middleton, *J. Am. Chem. Soc., 81* (1959) 803.

3 W. J. Middleton, H. W. Jacobson, R. E. Putnam, H. C. Walter, D. G. Pye and W. H. Sharkey, *J. Polymer Sci.: Part A, 3* (1965) 4115.
4 W. J. Middleton, E. G. Howard and W. H. Sharkey, *J. Am. Chem. Soc., 83* (1961) 2589.
5 W. J. Middleton, *J. Org. Chem., 30* (1965) 1390, 1395.
6 C. G. Krespan and W. J. Middleton, *Fluorine Chemistry Reviews, 1* (1967) 145-196.
7 W. J. Middleton and C. G. Krespan, *J. Org. Chem., 30* (1965) 1398.
8 W. J. Middleton, D. M. Gale and C. G. Krespan, *J. Am. Chem. Soc., 88* (1966) 3617.
9 W. J. Middleton, *J. Org. Chem., 30* (1965) 1402.
10 W. J. Middleton and C. G. Krespan, *J. Org. Chem., 35* (1970) 1480.
11 W. J. Middleton and R. V. Lindsey, *J. Am. Chem. Soc., 86* (1964) 4948.
12 W. J. Middleton, *J. Org. Chem., 40* (1975) 574.
13 W. C. Smith, C. W. Tollock, E. L. Muetterties, W. R. Hasek, F. S. Fawcett, V. A. Englhardt and D. D. Coffman, *J. Am. Chem. Soc., 81* (1959) 3165-3166.
14 M. Hudlicky, 'Fluorinations with Diethylaminosulfur Trifluoride and Related Aminosulfurans,' in *Organic Reactions*, Vol. 35, John Wiley & Sons, Inc., New York, 1988, pp. 513-637.
15 W. J. Middleton, E. M. Bingham and D. H. Smith, *J. Fluorine Chem., 23* (1983) 557-572.
16 W. J. Middleton, *Organic Synthesis, 57* (1985) 221-225.
17 O. W. Webster, W. R. Hertler, D. Y. Sogah, W. B. Farnham and T. V. RajanBabu, *J. Am. Chem. Soc., 105* (1983) 5706; R. Noyori, I. Nishida and J. Sakata, *J. Am. Chem. Soc., 105* (1983) 1598; T. V. RajanBabu and T. Fukunaga, *J. Org. Chem., 49* (1984) 4571; M. Fujita and T. Hiyama, *J. Am. Chem. Soc., 107* (1985) 4085.
18 W. B. Farnham, B. E. Smart, W. J. Middleton and D. A. Dixon, *J. Am. Chem. Soc., 107* (1985) 4565.
19 W. J. Middleton and E. M. Bingham, *J. Fluorine Chem., 20* (1982) 297.
20 W. J. Middleton, *J. Org. Chem., 30* (1965) 1307.
21 W. J. Middleton, *J. Org. Chem., 34* (1969) 3201.
22 W. J. Middleton, D. Metzger and J. A. Snyder, *J. Med. Chem., 14* (1971) 1193.
23 W. J. Middleton, *U.S. Patent 3,683,009* (1972); Ger. Equivalent, *C.A. 73*, 14425m.
24 B. K. Freed, J. Biesecker and W. J. Middleton, *J. Fluorine Chem., 48* (1990) 63-75.
25 S. M. Lee, J. M. Roseman, C. B. Blair, E. P. Morrison, S. J. Harrison, C. A Stankiewicz and W. J. Middleton, *J. Fluorine Chem., 77* (1996) 65-70.
26 K. M. Reese, *Chem. & Eng. News*, p. 66 (Nov. 8, 1993).
27 W. J. Middleton, *Fluorine Technology Bulletin*, No. 19 (1995) 25.

BIOGRAPHIC NOTES

William J. Middleton

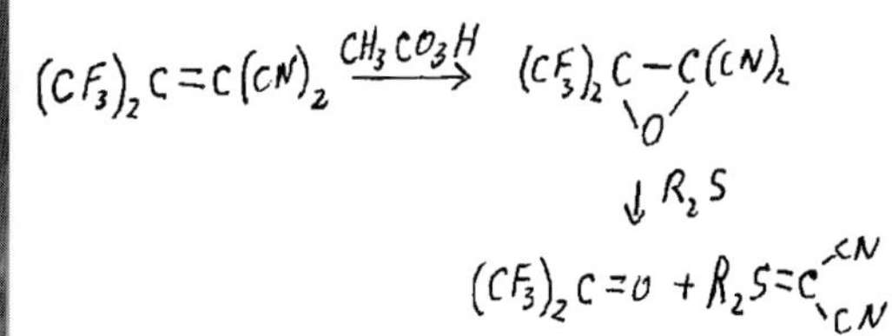

William J. Middleton (Bill) was the recipient of the 1982 ACS Award for Creative Work in Fluorine Chemistry. Born in Amarillo, Texas, in 1927, he holds BS and MS Degrees in chemistry from North Texas State University and a PhD in organic chemistry from the University of Illinois. He joined DuPont's Central Research & Development Department in 1952 as a research chemist and for the next 32 years carried out research, primarily in the field of organofluorine chemistry; his work resulted in 107 US patents on fluorine-containing pharmaceuticals, agricultural chemicals, polymers, solvents and reagents. Bill left DuPont in 1984 and joined the chemistry staff at Ursinus College in Collegeville, Pennsylvania, where he taught organic chemistry and continued his research over the next eleven years, advising 36 junior and senior undergraduates in their research projects and theses. Currently he is emeritus professor of research at Ursinus College. During his career, Bill has published more than 80 scientific papers and has been an invited speaker at numerous domestic and foreign universities and scientific meetings. In 1995, he received the Laughlin Professional Achievement Award for distinguished service to Ursinus College. Among his non-chemistry related honours, he was chosen to be the Poet of the Year, 1996, by the *Feelings Poetry Journal*.

David C. England received the American Chemical Society Award for Creative Work in Fluorine Chemistry in 1985. He was born in Portland, Oregon, USA, graduated from Oregon State College in 1940 and received a PhD degree in organic chemistry from the University of Wisconsin three years later. He then joined DuPont's Chemical Department (now Central Research & Development Department) as a research chemist and is proud of the fact that he spent almost 40 years there 'working at the bench' on interesting projects. Dave retired in 1982, but not before he had published 30 scientific papers and received

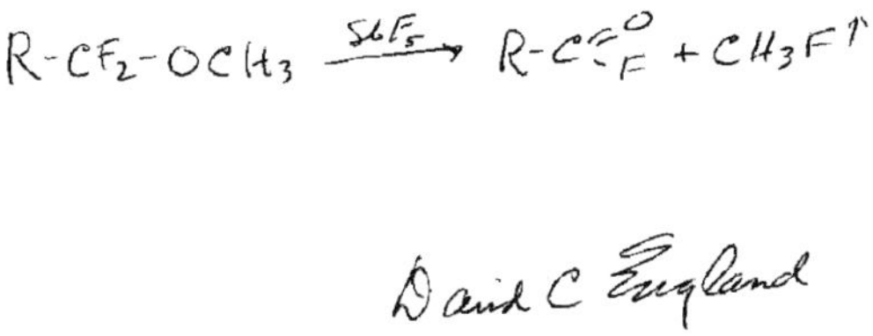

David C. England

61 US patents. Some of the more interesting compounds he has synthesised and elucidated the chemistry of include the extremely reactive perfluorocyclobutanone, the deep blue perfluorocyclobuta-1,2-dione and the perfluoro-β-sultone derived from tetrafluoroethylene and sulfur trioxide, which is the starting material for commercial fluorinated ion-exchange polymers.

In addition to the ACS Fluorine Award, Dave's achievements have been recognized by the presentation of the ACS Delaware Section Award (1984), DuPont's Pederson Award for Technical Excellence (1995) and DuPont's Lavoisier Medal for Technical Achievement (1997).

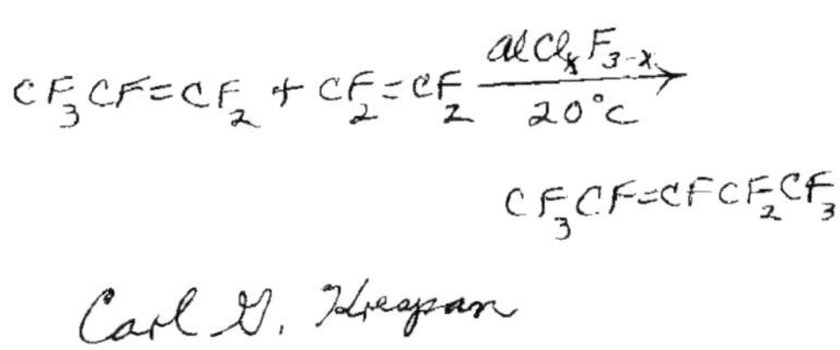

Carl G. Krespan

Carl G. Krespan received the American Chemical Society Award for Creative Work in Fluorine Chemistry in 1987. He was born in 1926 and received his BS Degree in chemistry from the University of Rochester in 1948 and his PhD in organic chemistry from the University of Minnesota in 1952. Carl joined DuPont's Central Research & Development Department in 1952 and carried out much significant research in organofluorine chemistry during his 41-year career there. His broad interests in organofluorine chemistry include new synthetic methods and novel structures, anionic, cationic and free- radical reactions of fluoro-olefins and fluoroketones and the chemistries of fluorinated sulfates, aminoimidazolines, diazo compounds, diazirines, thiiranes and cyclic polysulfides. Among his many achievements is a technology he developed which gave rise to the high-performance Viton® GLT class of fluoroelastomers.

Carl's involvement with the ACS Division of Fluorine Chemistry has included two terms as chairman and the organization (with Professor Alan Clifford) of the First Winter Fluorine Conference. He is an author of more than 60 scientific papers and has received 71 US patents. In addition to winning the 1987 ACS Fluorine Award, he was presented with DuPont's Pederson Award for Technical Excellence in 1995.

Chapter 22

FLUORINE CHEMISTRY: THE ICI LEGACY

RICHARD L. POWELL (with contributions by D. W. Bonniface, R. D. Bowden, P. Edwards, H. C. Fielding, P. Gamlen, J. Hutchinson, S. Korn, S. Lee, A. McCulloch, J. S. Moilliet, D. Moody, T. A. Ryan, R. Salmon, L. Shipp and N. Winterton)

Department of Chemistry, UMIST, Manchester, M60 1QD, UK

Introduction

Despite its profound restructuring over the past decade, ICI, in 1999, remains a primary producer of fluorine chemicals, a field which it helped to pioneer in the 1930s and 40s. Some of the Company's businesses which nutured fluorine chemistry have been inherited by other companies, hence the title of this chapter. Originally, Eric Banks formally invited only myself (RLP) to write an article about ICI's involvement in fluorine chemistry; however, he gave me carte blanche to involve other ICI fluorine chemists should that be necessary in view of the diversity of the subject to be covered. I have taken full advantage of that offer. Since fluorine chemists are likely to form the predominant readership of *Fascinated by Fluorine*, the content of this chapter has been biased towards their interests. Significant engineering developments, such as CFC plants, have been mentioned only briefly, while topics which foreshadowed developments elsewhere, such as the fluorination of polyethylene, have been given more prominence even though they provided ICI with little or no commercial return.

Until 1943, ICI's fluorine chemistry was essentially limited to a single research group in General Chemicals Division, which was based in NW England at Runcorn and Widnes, so there is no difficulty in presenting a coherent narrative up to that point. Problems then arise through the setting up of a second research group in 1944, followed by further fragmentation from the 1950s onwards as new Divisions, such as Pharmaceuticals and Plastics, and subsidiaries such as Plant Protection Ltd., were established, each with an interest in fluorine-based products. The ultimate split occurred in 1993 with the demerger of Zeneca, which took the speciality and bioactive fluorochemical businesses, while ICI retained the fluorinated refrigerants and fluoropolymers. Since this chapter is focused on the technical development of fluorine chemistry within ICI, organisational changes which affected the subject have been mentioned but briefly or simply ignored where they are judged to be of little interest to the reader. In the interests of clarity, post-1945 each development has generally been followed through to its present day position... or to its demise. A very significant portion of the fluorine chemistry inherited by Zeneca pre-dates demerger and therefore has been included in the sections provided by Zeneca contributors.

For the early history, which is now beyond living memory, reports indexed in the ICI report system have been my major source of information; although these are written in a matter-of-fact style typical of both scientific papers and company literature, the fascination that those pioneering industrial fluorine chemists had for their subject is still very apparent.

Regarding events from the 1960s onwards, many of the chemists actually responsible for the work have contributed to this chapter or have been contacted for information.

In the beginning

Imperial Chemical Industries Ltd., generally known as ICI[1], was founded in 1926 by the merger of Brunner Mond, a producer of soda ash, Nobel Industries, British Dyestuffs Corporation and United Alkali Company [1] to meet the challenges of large chemical combines elsewhere in the world, notably DuPont and IG Farben. In the re-organisation following merger, the electrolytic caustic/chlorine manufacturing facility at Runcorn, originally belonging to Brunner Mond, was linked with sulfur and other miscellaneous chemicals from United Alkali based at Widnes (on the opposite bank of the River Mersey from Runcorn) to form General Chemicals Division. United Alkali's ammonia soda ash facility at Hillhouse on the coast 50 miles north of Runcorn combined with Brunner Mond's soda ash manufacturing units to form Alkali Division; this was headquartered at Winnington, 11 miles south of Runcorn. These sites were destined to play important roles in ICI fluorochemical research and production.

ICI's new manufacturing Divisions each developed their own styles. Some wag commented that, 'in Alkali Division (which was staffed by graduates largely drawn from Oxford and Cambridge) gentlemen pretended to be chemists, whereas in General Chemicals Division (where graduates largely hailed from the newer, "redbrick" universities) chemists pretended to be gentlemen'. Rivalry certainly existed between the two organisations, and in the 1950s each had quite separate fluorine chemistry research programmes; these were ultimately brought together in 1964 by uniting the two Divisions into a single grouping called Mond Division, which became the Company's centre of fluorine expertise.

The genesis of ICI fluorine chemistry

The earliest report in the ICI archives on organofluorine compounds appears to be a 1932 'paper' assessment of the emerging CFCs and HCFCs as potential diluents for fumigants, notably ethylene oxide and ethylene dichloride, to provide nonflammable formulations. CFC-11 ($CFCl_3$) was identified as the preferred candidate, but '*an instruction was received from the Research Manager to discontinue investigation of this subject owing to the patent position.*' By 1934, General Chemicals Division had successfully operated a concentrated aqueous HF process at the semi-technical scale. A report by D. W. F. Hardie, reviewing the literature up to 1933, clearly indicates the Division's interest in exploring opportunities offered by organofluorine compounds, both aliphatic and aromatic. Hardie was well aware of the development of the Freon™ products as refrigerants, solvents and firefighting agents, and he also noted that chlorofluoroethylenes and '*fluorine derivatives of chloroform*' had been tested as anaesthetics. He speculated that, although inorganic fluorine compounds found widespread use as insecticides, nothing was known about the applications of organofluorine compounds for this purpose, an area which was ultimately to prove very successful for ICI (and subsequently Zeneca) some 50 years later.

[1] Imperial Chemical Industries plc formally changed its name to ICI plc in 1999.

ICI's work on organofluorine chemistry commenced in the 1930s with studies on routes to CFCs, notably CFC-11 and CFC-12, by F. D. Leicester. He investigated the reaction of HF and CCl_4 in the vapour-phase over various solid catalysts, finding that CrF_3 on coke was most effective, – an intriguing observation considering ICI's later success in developing highly-active chromia catalysts. Leicester's vapour-phase route to CFC-12 was patented in 1937 [2] after a review of the patent literature failed to find any relevant prior art. The review alluded to further experimental work by Leicester using a catalyst *'comprising a mixture of Cr_2O_3 and CrF_3'*, which produced good yields of CF_2Cl_2 and $CFCl_3$ at 450 °C; this is the first reference to a fluorinated chromia catalyst in the ICI archives. In 1935, Leicester designed a semi-technical plant capable of making '2 *cwt*' (*i.e.* 2 hundred weight or 112 lb, equivalent to ~51 kg) per week, and a 1000 tpa production plant whose capital cost he calculated to be £7470 based on a detailed evaluation. The semi-technical plant was operating by May 1937. The reactor section consisted of two electrically-heated tubes each containing 4 kg of catalyst. The latter was initially prepared by coating 0.125-0.25 inch coke, previously desilicized with 10% hydrofluoric acid, with a paste of fluorinated chromia. Although designed for continuous operation, the plant was first operated only on a daily basis since catalyst degradation resulted in blocked reactor tubes. A successful catalyst was prepared by soaking 0.25 inch desilicized coke in an aqueous chromium fluoride solution to give a 5% Cr loading, and the plant was operated continuously using this catalyst, each campaign lasting about 1 week (actual running time). The reaction temperature appears to have been around 450 °C. The second continuous run produced 16 kg of CFC-11 and 30 kg of CFC-12, which were separated using a batch pressure still. Looking back with the hindsight of 60 years, the causes of problems they experienced are easy to understand; even today corrosion and blockages are not unknown! The people who ran this semi-technical plant were the pioneers of vapour-phase chromia catalyst fluorination technology, a field in which ICI was to establish a technical lead with its KLEA™ 134a (CF_3CH_2F) plants of the 1990s.

In the mid 1930s ICI also carried out laboratory studies on the synthesis and properties of organofluorine compounds. Various other chlorofluoroethanes, including CFC-113 ($CF_2ClCFCl_2$), were prepared and their effects on dyed silk assessed; presumably they were being tested as dry cleaning solvents? Chlorofluoroethenes were also synthesised by Zn/ethanol dehalogenation or by HCl elimination, but only $CF_2{=}CFCl$ was reported to polymerize readily. The synthesis of vinyl fluoride by the addition of HF to acetylene was also investigated and ethyl fluorosulfonate was prepared by the action of fluorosulfonic acid on ether or ethylene. In 1936 a variety of fluoroaromatic compounds were prepared. Fluorobenzene was synthesised by *in-situ* decomposition of benzenediazonium fluoride while α-fluoronaphthalene, *o*- and *p*-fluorochlorobenzene, 4,4′-difluorobiphenyl and fluorobenzene were obtained via the decomposition of the corresponding diazonium tetrafluoroborates. Nitration/reduction of fluorobenzene gave *o*-aminofluorobenzene, which was converted to 1,2-difluorobenzene. α-Fluoronaphthalene and various chlorofluorobenzenes were reported to be bacteriocides when emulsified in water with Turkey red oil, the first indication from ICI work that fluoro-organics could have useful bioactivity. By the end of the 1930s, however, ICI's investigations into CFCs and fluoroaromatics appear to have ceased, and were not resumed until 1944.

ICI was also noted for its early development of fluorine cell technology. According to an internal review written by J. W. Thompson in 1939 an investigation into the generation of fluorine started in May 1934 at the Runcorn Lab of General Chemicals Division, but the name of the research chemist is not recorded. The justification for the work appears to have been rather speculative... *'Fluorine is the most active chemical known and some of its compounds possess unique properties. It was therefore proposed to study fluorine compounds produced by direct fluorination...'*. The programme does not appear to have been given a high priority since the cell and fume cupboard were not ready until October 1935 and electrolysis was not started until December. The high-temperature cell installed consisted of a 6-inch diameter copper pot (which also served as the cathode), fitted with a 1 inch graphite rod anode (surrounded by a 4-inch diameter copper diaphragm) and filled with molten KHF_2 as electrolyte. Unfortunately it produced no fluorine, primarily because the electrical insulation failed.

Influenced by the publication of DuPont patents [3] in May 1936, a low-temperature cell was constructed whose nickel-lined mild steel body was also the anode. The cathode was an iron rod surrounded by a nickel diaphragm. Although the cell was ready in May 1937, the first fluorine was generated only in late 1938 when Thompson took over the programme. Unfortunately, rapid corrosion of the diaphragm resulted in the mixing of cathodic hydrogen with the fluorine, causing an explosion outside the cell. Despite various modifications the design proved unreliable. A third cell consisting of a copper pot, 4.5 inches deep and 4 inches in diameter, fitted with tubular nickel electrodes operated successfully with an 'efficiency' of about 60% at a current of 5 amps. A fourth larger cell was ordered with a design based on the second cell but equipped with a monel diaphragm; this appears to have been satisfactory.

Despite this work the future of fluorine chemistry in the Company was not assured. General Chemicals Division started supplying 60% aqueous HF to Thorium Metals in 1937/38 from a 200 tpa plant at Widnes. However, this venture was not profitable and there was discussion about ceasing production and purchasing HF for resale, or even exiting the HF market completely. Perhaps if this had happened there would have been no ICI fluorine chemistry, and this chapter would not have been written.

The war years (1939-45)

The Company's ability to generate and handle elemental fluorine was crucial to the UK's contribution to the development of the atomic bomb. ICI's involvement in this programme is well described by Reader [1]. The UF_6 development was devolved to a separate company with the cover name 'Tube Alloys' whose records are not part of the ICI report archive. Although ICI fluorine chemists undoubtedly contributed to the success of UF_6 production by developing reliable fluorine cells and producing highly-fluorinated fluids, few explicit links with Tube Alloys are obvious from the ICI reports, presumably reflecting the intense security surrounding the development of the atomic bomb.

In 1939, Dr A. J. (Jack) Rudge, well known outside ICI for his monograph on fluorine technology [4], became involved in the development of fluorine cells and continued the work of J. W. Thompson. Low-temperature (0 to 20 °C) cells were still favoured, but high anode corrosion rates, low current densities and short cell lives were major problems. In an attempt to overcome these problems, a flowing mercury cathode cell – similar to that used for chlorine manufacture – was tested with a nickel anode and an electrolyte containing 30 g of $KF \cdot HF$ and 60 g of HF. Unfortunately a mixture of hydrogen and fluorine was generated by the cell and no potassium could be detected in the mercury; not surprisingly therefore, this approach was dropped. In 1941 H. H. Skinner noted that the lowest rate of nickel loss observed by Rudge was 1 g per 2 g of fluorine produced and suggested on the basis of a literature review that graphite or carbon anodes such as those advocated by Fredenhagen and Henne looked promising.

By September 1942 it was clear that low-temperature cells were not suitable for large-scale fluorine manufacture. In an influential paper, Cady, Rogers and Carlson [5] confirmed the ICI experience, and in May 1942 C. C. Moss started work on a medium-temperature (~80 °C) cell based on Cady's design which retained a nickel anode with an electrolyte composition of KF/1.8HF. Although the new cell design was still considered 'troublesome', between October 1942 and May 1944 more than 500 kAh were passed through such cells and 250 kg of fluorine produced to meet British Government requirements.

In 1944, H. (Harry) R. Leech, who had extensive experience in electrochemical processes, became manager of the fluorine cell group [6]. Together with Jack Rudge, who now worked for him, Leech was to play an influential role in the development of ICI's F_2-based chemistry. In the early days of medium-temperature cell development carbon and graphite electrodes were tried with little success, and work did not resume until 1944 when Rudge commenced a research programme into these materials. All graphites examined suffered either disintegration or polarization. Although many carbons also failed, a highly porous, granular, filter carbon manufactured by the National Carbon Co. of America showed promise and by 1945 the essential features of the ICI type fluorine cell were in place. Rudge reported in 1946 that although anode breakage was a problem after prolonged operation, current efficiencies were >90% and fluorine gas purity was 98-99%. The anode was free from the chemical attack and sludge formation experienced with nickel electrodes, and the frequent explosions caused in previous designs by the mixing of H_2 and F_2 were now rare. Among the other reported advantages of the medium-temperature cell was that '*electrolyte burns not very serious*', while for the low-temperature cell (which contained more HF) electrolyte burns were noted to be '*very serious*'. Clearly, safety perceptions have changed much over the past 55 years. Leech and Rudge concluded in 1946 that '*Researches in the General Chemicals Division laboratories over the past ten years have shown that the principal technical difficulties in fluorine production have been overcome.*'

Starting in 1941 Rudge also investigated processes for manufacturing chemically inert, perhalocarbon fluids. Liquid-phase fluorinations of partially-chlorinated paraffin wax (ICI's Cerechlor^{TM}), chlorinated kerosene and medicinal paraffin were unsuccessful, but vapour-phase fluorinations, especially using cobalt trifluoride, proved to be effective. The equipment consisted of a pair of electrically-heated reactors 3 inches in diameter and 2 feet 6 inches long, each containing 800-900 g of CoF_2. While the CoF_3 in one tube was being used to fluorinate an organic vapour, the CoF_2 in the other was refluorinated to CoF_3 at temperatures up to 400 °C. The feeds included pure *n*-heptane, toluene, and 2,3-dimethylpentane, but the best results were obtained with benzotrifluoride, efficiencies at fluorination temperatures of 350-400 °C reaching as high as 80% based on carbon. Even in the early 1940s Rudge clearly recognised that the incorporation of some fluorine into a substrate before perfluorination was advantageous, a ploy which has been developed extensively over the past 20 years.

Visions for the future

In 1943, C. C. Tanner produced a techno-commercial proposal for a future ICI fluorine chemicals business. He commented that 15 years previously the only fluorine chemicals on the market were 60% aqueous hydrofluoric acid, sodium fluoride and sodium silicofluoride. Since that date '*astonishing progress*' had been made in development of CFCs as safe refrigerants in the US, with production also starting in Germany just before the outbreak of WWII. Tanner anticipated that in the '*near future*' the UK would require 1000 to 2000 tpa of '*Freon*^{TM}' type refrigerants. He also suggested that if PTFE had been available at that time it would have been in great demand for pipe gasketing and for replacing rubber in chemical plants; and he foresaw that, because of its very specific properties, it would complement rather than replace existing polymers. He also speculated that fluorine compounds would find applications in the electrical and dyestuffs industries. Although not indicated in the available documents, this report presumably influenced the setting up of a research group to investigate CFCs and related compounds under J. H. Brown; this group was quite separate from Leech's elemental fluorination team.

Tanner pointed out that the existence of British patents (presumably filed by DuPont?) had inhibited the development of a UK fluorine products industry, and opined that this was '*a scarcely adequate excuse for inaction*'. He concluded, '*In a nutshell the problem is to learn by 1946, the date of expiry of the patents we should infringe by present day manufacture, the art that has been steadily built up abroad in the last 15 years.*'

Tanner proposed a remarkably detailed research, development and construction programme to achieve his vision. He recommended that further surveys be carried out to appraise the postwar outlets for fluorine chemicals and hence extend his own assessment of the CFC refrigerant market. His plan included the construction of a 485 tpa anhydrous HF plant, a 1000 tpa '*Freon*^{TM}' 12 plant with provision for doubling its capacity, and provided detailed flow sheets, costings and profitability forecasts. He also proposed that a semi-technical plant be built primarily for the production of '*tetrafluorodichloroethane for the production of TFE*' but which could also be used for making small quantities of other fluorinated derivatives and for gaining experience in the technique of fluorination. The ultimate tribute to Tanner is that by the mid-1950s his vision had largely been realised, although with some modifications, *e.g.* the manufacture of tetrafluoroethylene (TFE, $CF_2{=}CF_2$) from HCFC-22 (CHF_2Cl) not CFC-114 (CF_2ClCF_2Cl).

In a series of four reports written in 1946, Rudge and Leech not only reviewed fluorine production technology and the chemistry of fluorine compounds, but also speculated about potential commercial applications. They clearly preferred medium-temperature (70-120 °C) fluorine cells with carbon anodes, which they anticipated being built in 1000 amp units each capable of generating 5.6 tpa of F_2. They also recommended that small cells be developed for laboratory use to encourage research into fluorine chemistry (particularly in academia) and that neat fluorine be offered for sale compressed to 200 psi in steel cylinders. They wrote, '*It cannot be doubted that there is a new field of chemistry to be explored, as vast as the classical field based on hydrocarbon skeletons. And it can hardly be doubted that discoveries of considerable industrial importance are likely to be made in this field. These discoveries are not exactly on the doorstep at this moment. Much has to be done in the development of new techniques, new procedures new materials of construction, new instruments and methods of measurement; much has to be done in the preparation of certain of the basic materials, and in the discovery of "opening up" reactions and a few reactive groups to be inserted by such reactions.'* They were clearly fascinated by the prospects for fluorine chemistry.

Essentially the Rudge-Leech vision was the programme, academic and industrial, pursued in the UK and elsewhere during the decades following the end of the WWII. In 1948 J. H. Brown and H. R. Leech, presumably as the respective managers of ICI's CFC and elemental fluorination research groups, jointly reviewed the then current fluorochemical knowledge and suggested a future programme for the Company which, in many respects, reiterated Rudge and Leech's suggestions of 2 years earlier. Interestingly, one of the synthetic targets proposed was pentafluoropyridine, a compound first made by UK academic fluorine chemists some 12 years later.

External contacts

The war years inhibited international scientific exchange, but immediately hostilities ceased General Chemicals Division made a determined effort to catch up on developments elsewhere. The Company had access to captured IG Farbenindustrie reports (including studies at Hoechst) dealing with of CF_3CH_2Cl (HCFC-133a), chlorofluoropropanes, -butanes and -olefins. J. H. Brown visited DuPont in mid-1946 to discuss, *inter alia,* progress made in organofluorine chemistry during the war years. His report concentrates mainly on the detailed manufacture, handling and polymerization of TFE, noting that '*The plant and process remain substantially as described by C. G. Harris after his visit in 1944*'. Interestingly, concern was expressed over unknown toxic by-products present in the $H(CF_2)_nCl$ ($n = 2, 3, 4 \ldots$) 'heavies' from TFE distillation; of course we now know that the culprit is perfluoroisobutene. Also discussed was '*the excessive weakness, tiredness, nausea and sore throat*' experienced by people who had inhaled fumes from heated PTFE, symptoms which have subsequently been known collectively as 'polymer fume fever'. The stabilization of TFE was also covered in the meetings: DuPont reported that terpenes would stabilize liquified TFE and α-pinene became ICI's preferred ICI stabilizer until the Company ceased transporting liquid TFE in the early 1990s. R. Joyce of DuPont '*expressed the opinion that monomer stored under pressures too low to permit the formation of a liquid phase would be more stable than when a liquid phase was present*'. Interestingly, this approach has recently been adopted for the safe storage and transportation of trifluoroethylene monomer, which has also proved to be potentially explosive. Various methods for fabricating PTFE were discussed, and, in modified forms, these are still used today. Perhaps the most interesting observation was that DuPont had discovered that fibrous PTFE could be produced by heavy rolling of PTFE polymer to produce a thin transparent sheet, followed by breaking it up in a blade mill. The porous material obtained, which more resilient than solid PTFE, was being used for gaskets and appears to be an ancestor of Gore-Tex^{TM}. Co-polymers of TFE, vinyl fluoride (VF, $CH_2{=}CHF$) and chlorotrifluoroethylene (CTFE, $CF_2{=}CFCl$) were also discussed.

In October 1946, A. F. Benning of DuPont visited ICI's Widnes Laboratories to discuss fluorine production with Leech, Rudge and C. G. Harris, the General Chemicals Divsion Research Director. The discussions were centred on published information, particularly that presented at the September 1946 fluorine chemistry symposium held in Chicago. Like ICI, DuPont had concentrated initially on Ni anode cells and had been producing perfluorinated fluids using CoF_3. Benning alerted the ICI chemists to the forthcoming publication in 1947 of fluorine chemistry developed during the Manhattan project[2].

In 1948 a further visit was made to DuPont by Leech, W. Bridge, and R. B. Mooney whose report about the generation of elemental fluorine and its applications makes fascinating reading... at least for a fluorine chemist! It seems that while ICI was significantly ahead in fluorine cell technology, DuPont's cobalt trifluoride reactor at Jackson Laboratory equipped with a mild-steel paddle stirrer was superior to the static-bed reactor unit used at Runcorn. At the Central Experimental Station in Wilmington the ICI people met R. S. Scheiber and C. Tullock who had been using elemental fluorine to synthesise inorganic fluorine compounds, including IF_5, MoF_6, SO_2F_2, and POF_3. Although they had intended to generate their own fluorine using a cell supplied by DuPont's Jackson Lab, Schreiber and Tullock decided to buy a cylinder of fluorine from the Pennsylvannia Salt Manufacturing Corp. since it appeared to involve less hassle... It seems that the choice between cell and cylinder fluorine for research was just as pertinent 50 years ago as it is today.

The ICI people record that Du Pont's Benning had delivered a lecture before the Scientific Forum of the New York Electrical Society on 19 November 1947. Although they did not attend it, they appear to be have been given a detailed preview since they noted '*that in lieu of lecture demonstrations, which in this field were considered possibly too risky, Dr Benning had had prepared a coloured, talking film of the experiments which he wished to demonstrate*'. It ran for 15 minutes and showed now well-known experiments such as a pre-heated iron rod burning in fluorine and chlorine trifluoride inflaming a piece of cotton wool. The ICI group felt this was '*not good publicity for fluorine or for the reactive products which can be made from it*'. They were concerned that such demonstrations would scare potential research workers away from this field; Leech and his colleagues would have preferred '*to tempt people to work in this relatively new and unexplored field by demonstrating the ease, simplicity and safety of the techniques required*'... techniques which ICI had done much to develop in the UK during WWII.

Despite their disapproval of the demonstration of the more spectacular effects of elemental fluorine, the ICI people attended a lecture by J. F. Gall, Pennsalt's Assistant Research Supervisor, entitled 'Fluorine and its Compounds', given before the Philadelphia Section of the ACS in the Franklin Institute on 20 November 1947. They commented that it was as much a lively social event as a scientific meeting, attended by several hundred members and far different in style from equivalent meetings in England. To quote from the ICI report, '*A whole series of lecture demonstrations was carried out, not all successfully. The fume cupboard capacity in the lecture theatre was unsatisfactory, and the smell of fluorine soon reached the back of the theatre. The lecturer explained that so long as the smell was bearable, the amount of fluorine was well below the toxic limit for short exposures.*' Clearly Pennsalt and DuPont had very different corporate cultures when it came to demonstrations for live audiences. Gall's lecture included an intriguing experiment in which he ground up black fluorspar and showed that evolved gases darkened KI paper, which he claimed was evidence for naturally-occurring free fluorine.

The ICI group also attended the 21st Exposition of the Chemical Industries, held in New York on 1-6 December 1947, where the American Chemical Society exhibit was based on the theme '*The Chemical Industry of today and tomorrow is built on yesterday's*

[2]Presumably the famous March 1947 issue of Industrial and Engineering Chemistry [*39* (1947) 236-434].

fundamental research.' This was exemplified by tracing the evolution of two modern developments of the chemical industry, fluorine chemistry and the silicones. A 50 amp Harshaw fluorine cell and cans of reactive fluorides such as CoF_3, MnF_3 and AgF_2 were featured, but the ICI people felt that the '*present state of* *commercial* *development in these fields seems a trifle exaggerated.*' Surprisingly, since he was a key figure in the development of ICI fluorine cells and the fluorination of organic compounds, the circulation list for the report on which the immediately preceding paragraphs are based does not include Rudge.

Elemental fluorine: the postwar period

After WWII, ICI improved the reliability of industrial scale, medium-temperature F_2 cell technology and introduced standard 10 amp and 60 amp cells suitable for lab use to encourage wider use of fluorine in research, especially in academia. During the period 1941-45 the Company supplied a number of nickel anode cells to Haworth and Stacey's group at Birmingham University (and to other universities), including both low and medium temperature types, which were followed in 1949 by a 10 amp cell and a 60 amp cell on free loan. It is a tribute to the basic design of these units that several are still operating in the UK and are only now being surpassed by sealed cells from Fluorogas Ltd. which are based on a BNFL design. Cobalt fluoride reactors were still being operated by ICI to produce perfluorocarbons, but appear to have been closed down in the early 1950s when contracts for perfluorinated fluids had been completed.

ICI identified sulfur hexafluoride (SF_6) as a product which would exploit the Company's competence in fluorine generation and would be in demand as the need for electric power grew. Following a research programme in the late 1940s, an SF_6 plant was operating by 1952. By modern standards the basic reactor was very crude: essentially a jet of fluorine gas was directed downwards through a tube situated just above the surface of a pool of molten sulfur; the SF_6 thus generated flowed outwards to the walls of the reactor where it was isolated from the feed fluorine by the inlet tube. ICI manufactured SF_6 until 1974 when it exited the business and mothballed its fluorine production cells; the latter were never operated again and were finally dismantled in 1978.

Having developed an industrial process to fluorine, ICI wanted to be able to supply it to potential customers as well as use it in-house, but the problems of handling such a reactive material as a compressed gas or liquid were well appreciated. Indeed, during his 1946 visit to ICI, DuPont's Benning is reported as saying that he would only use compressed fluorine behind a 5-inch brick wall. Chlorine trifluoride (ClF_3), a liquid boiling at 11 °C and briefly investigated by Skinner and Mooney in 1941, appeared to be a more practical fluorinating agent than F_2 and was studied by W. Hook and Rudge between 1946 and 1951. Useful information on the properties of ClF_3 was obtained from German sources via the Chemical Defence Experimental Station at Porton Down, which presumably had acquired it at the end of the War. In his well-known contribution to 'Mellor' [7], Leech reported that the Germans were thought to have produced about 1000 t of ClF_3, codenamed N-Stoff, for use in incendiary devices and flame throwers rather than as a fluorinating agent. Following experience gained on a small unit, a 60 tpa plant was constructed in 1951 based on a design developed by Rudge.

Chlorine trifluoride was used by R. Le C. Burnett and J. (John) Muray in 1950-1951 to produce Florube™ fluids by the chlorofluorination of chloroaromatics, notably hexachlorobenzene, octachloronaphthalene (containing up to 50% of hexachlorobenzene) and decachlorobiphenyl (containing 50-75% of hexachlorobenzene). ICI's interest in highly-fluorinated or chlorofluorinated hydrocarbons as chlorine-resistant lubricants dates back to 1941, presumably related to the activities of Tube Alloys. Reactions were carried out at 100-150 °C by passing ClF_3 through stirred suspensions of the chloroaromatics in previously-made samples of liquid products. As the reactions proceeded the solid substrates dissolved and more were added as required. The Florube™ products were used for sealing and lubricant duties in severe chemical environments, notably on chlorine and fluorine plants. They were ultimately superseded by the more stable fluids derived from the polymerization of CTFE.

A contributor to the ClF_3 programme was A. (Arthur) A. Banks, who had studied interhalogen compounds under Emeléus at Cambridge just after the War. Hook and Banks[3] also tested chlorine trifluoride/hydrogen flames for cutting and welding copper, which they hoped would be easier to control than the fluorine/hydrogen flames being tested in the US. Chlorine trifluoride was supplied to university departments considered capable of handling it in order to promote interest in fluorine chemistry; for example, Musgrave at Durham received a 10 lb cylinder which the editor of this book remembers using during his PhD studies. According to Leech [6], the major external customer for the product was BNFL for use in the production of uranium hexafluoride. However, fluorine proved to be better than chlorine trifluoride, so ICI supplied liquid fluorine in tankers to the BNFL Springfield site 50 miles north of Runcorn [6]. In 1959, ICI quoted BNFL for the design and construction of a 250 tpa fluorine plant, and by the mid 1960s BNFL became self-sufficient in fluorine based on ICI technology. In 1967, ICI sold its small-scale (10 and 60 amp) fluorine cell business to Sherman Chemicals (Birmingham, UK). The Company continued to generate fluorine until it withdrew from the sulfur hexafluoride market in 1974, thereby ending its forty-year involvement with this amazingly reactive element.

Rudge's postwar work included the fluorination of polyethylene (PE) [8], a material which had been discovered in the Winnington High Pressure Laboratory in 1935 by ICI's Alkali Division. His interest in this topic undoubtedly stemmed from his study of *in-situ* fluorination of PE gaskets for fluorine cells in 1941. PE was exposed to a stream of fluorine for 10.5 hours, and since the material showed little effect he installed PE gaskets in a cell. After 20 days operation they had suffered only slight surface attack in the cell, but only to a depth of 0.03 of an inch; the affected layer could be removed by scratching with a knife. In contrast, Neoprene™ discs became hard and brittle. Following further work at Winnington by L. (Leslie) Seed in 1943-46 which produced materials containing 46.5% fluorine, Rudge and Warren in 1952-54 produced essentially perfluorinated (*i.e.* 76% by weight fluorine) PE by sandwiching a thin (0.003 inch) amorphous film between metal gauze to dissipate the heat of reaction and fluorinating it for 39 hours at temperatures up to 80 °C. The initial fluorination was carried out at room temperature using F_2 diluted with nitrogen; then gradually the concentration of fluorine and the temperature were increased until in the final stage neat F_2 at 80 °C was being used. The product, a highly crossed-linked polymer with a molecular weight lower than that of typical PTFE grades, could not compete with PTFE so the project was abandoned. With hindsight perhaps Rudge and

[3]No relation to Eric Banks; but the two were occasionally confused, Eric sometimes being asked about his PhD studies with Emeléus!

Warren's most important observation was that partially-fluorinated PE had a perfluorinated surface with an unfluorinated core and was resistant to boiling xylene for short periods, clearly foreshadowing the modern process for the surface fluorination of plastic containers to render them impervious to organic solvents.

Although Rudge continued in fluorine chemistry until the end of his career in 1967, publishing his monograph in 1962, his role in the development of ICI fluorine chemistry after the mid-1950s appears to have been limited. His strong espousal of elemental fluorine as a key way of accessing new fluorine-containing products based on his successes in the 1940s was probably unattractive to ICI's General Chemicals Division, which was finding the rapidly-expanding CFC/HCFC market a much more attractive commercial proposition than speciality fluorochemicals. Even his fluorination of polyethylene, with its emphasis on essentially complete fluorination, appears to have been an attempt to develop an alternative route to a perfluoropolymer based on elemental fluorine which could compete with TFE polymerization.

The growth of the CFCs

(by R. L. Powell and A. McCulloch[4])

In 1944 a second organofluorine research group was set up under J. H. Brown which focused on the development of CFCs, HCFCs (W. B. Whalley, J. H. Richards and W. M. Owens) and fluorinated monomers (J. Chapman and C. W. Suckling). As well as the development of CFCs, high priority was given to Chapman's work on the synthesis of TFE (via pyrolysis of HCFC-22) and its polymerization. A report by Brown reviewing ICI's knowledge about TFE clearly shows that valuable information was gained from the US patent applications provided to the Company under the ICI-DuPont Agreement and from a liason visit to DuPont in 1943 by R. Hill of Dyestuffs Division. He reported that DuPont were storing liquid TFE in cylinders cooled to -80 °C as a safety precaution, although 'terpene B' was being advocated as a stabilizer. Suckling worked on (trifluoromethyl)styrenes for low refractive index polymers intended for '*making up achromatic lenses with other more highly refractive resins*', an intriguing concept which foreshadowed the idea of fluorocoated polymer optical fibres 40 years later. In summarizing the work of his group in early 1945 Brown acknowledged that cobalt trifluoride '*...has lately become available...but its relatively difficult preparation makes it of little value unless some specific use is found for it.*' This comment was surprising considering that it was already being used to manufacture pure perfluoro(methylcyclohexane) which had been used in the uranium hexafluoride process; perhaps this is an early manifestation of the uneasy relationship between speciality and bulk fluorine chemicals, a dilemma which ICI General Chemicals Division and its successors never properly resolved.

---

[4]Archie McCulloch graduated from UMIST in 1963 then joined ICI's General Chemicals Division. In 1977 he became a member of the team developing CFC replacements, subsequently becoming the senior chemist responsible for existing CFC and HCFC technology. In 1987 he moved to the Environmental Science Group, developing ICI's technical view on the environmental impact of CFCs, HCFCs and HFCs; he represented ICI on the science committee of AFEAS. Archie retired from ICI in November 1999.

The erection of a liquid-phase semi-technical plant to produce CFCs 11 ($CFCl_3$) and 12 (CF_2Cl_2) was started in 1944 and production commenced in 1946. Although internal reports often used 'Freon^{TM}' to refer to the CFCs and HCFCs, ICI needed its own trademark and two candidates, 'Diclon' and 'Arcton', were submitted to the Company's trademark unit. Arcton^{TM} was preferred, although ICI's Commercial Director is quoted as saying that it was not '*especially euphonious*', a comment which seems strange to those of us who have used it regularly for many years.

In late 1945, through a commercial agreement with Kinetic Chemicals, the DuPont subsidiary responsible for Freon^{TM} manufacture, ICI gained access to the design and operation of DuPont's low-pressure 11/12 plants. ICI sanctioned its own 1000 tpa 11/12 plant in 1946 to be built at Rocksavage Works[5], Runcorn, which was commissioned in June 1949. Significant modifications and operating experience were required before the plant achieved its flowsheet production of 20 t/week; by 1958, however, following further extensions the plant was capable of producing 120 t/week. In 1955, ICI sanctioned the construction of a high pressure 11/12 plant based on its own design; this commenced operation at the beginning of 1958. Although originally designed for an output of 7200 tpa (5500 t CFC-12 and 1700 t CFC-11), its initial capacity in fact appears to have been 8500 tpa since all the equipment was oversized. The nameplate capacity had been raised by 1964 to 13 500 tpa with relatively minor modifications and in fact proved operationally capable of 18 000 tpa. This plant was progressively extended in line with the rapidly increasing demand for CFCs, so by the early 1970s it was capable of producing 80 000 tpa.

The development of an HCFC-22 (CHF_2Cl) process was a target of J. H. Brown's original 1944 research programme, in particular to provide feedstock for PTFE manufacture. A flowsheet and preliminary design for a 500 tpa plant was drafted in 1951 based on a semi-technical plant already operating at Widnes. Provision was also made to recover pure HCFC-21 ($CHFCl_2$), the intermediate in the conversion of chloroform to HCFC-22. Although the process was largely based on ICI R and D it was influenced by DuPont's experience. In contrast to ICI's first 11/12 plant, which operated at near atmospheric pressure, the 22 plant was designed to run under a pressure of 120 psi with a reactor temperature of 70 °C. The unit was operational by 1954 and by 1958 had a capacity of 1000 tpa. Initially, the major application for 22 was captive PTFE production, but this was overtaken in the 1960s by external sales to the air conditioning and refrigeration markets. Subsequent uprates increased the plant capacity to its current value of 30 000 tpa.

ICI's most significant contribution to the technology of fluorinated fluids was its development of an efficient chromia catalyst for vapour-phase chlorine/fluorine exchange. Although ICI pioneered this approach in the late 1930s (see above) and some further research was carried out by E. Tittensor and R. L. McGinty in 1945/46, the development of modern catalysts was initiated following a literature review by P. G. N. Leonard in 1959. He concluded that vapour-phase fluorination offered promising routes to two-carbon CFCs, notably 114 (CF_2ClCF_2Cl). An experimental programme resulted in the patenting by J. Chapman and D. C. Horner of 'black chromia' catalyst for the manufacture of CFCs and was exemplified for CFCs 114, 114a (CF_3CFCl_2) and 115 (CF_3CF_2Cl) [9]. McGinty discussed

[5]Rocksavage Works, which became ICI's main site for the production of CFCs and HCFCs, was named after the Rocksavage Family which originally owned the local hall and farmed the land on which the site was built.

the topic with Emeléus during a consultancy visit in 1961. Despite this programme, ICI decided in 1961 to license a liquid-phase plant design from the Allied Chemical Corporation to meet the rapidly growing market for CFCs 113 and 114. The plant, commissioned in mid-1964, had a nameplate capacity of 550 tpa of 113 and 150 tpa of 114. Although a severe corrosion problem initially made it difficult to operate the plant at its 114 design rate this was ultimately overcome in 1966. However, the sales of this product were predicted to grow more rapidly than originally anticipated, and CFC-115 also emerged as a potential product since it was required for formulating with HCFC-22 to produce the azeotropic refrigerant 502. ICI decided to develop a vapour-phase process to fluorinate perchloroethylene to 114 and 115. Chapman and Horner's work on the ICI chromia catalyst had been continued by D. (David) J. Viney and R. (Robin) A. Woolhouse, who developed a vapour-phase route to 114 from 113 to the semi-technical scale using a pelleted chromia produced by ICI's catalyst group. Woolhouse also developed a 114/115 process which he took from lab scale to semi-technical scale, demonstrating that it could be operated at pressures in the range 140-190 psi with a catalyst life time of 700 hours at temperatures typically in the range 350 to 500 °C: $CCl_2{=}CCl_2 + Cl_2 + HF \rightarrow CF_3CF_2Cl$ (115) + CF_2ClCF_2Cl(114) + CF_3CFCl_2 (114a). In June 1969, ICI sanctioned a plant to produce 2000 tpa of 114 and 1000 tpa of 115; this was commissioned in 1972.

The standard method for preparing the catalyst used in the plant was developed by R. (Ray) G. Thorp using a pragmatic experimental approach. Detailed kinetic and mechanistic studies were carried out by R. J. Perry, J. (John) H. Forster and especially K. (Ken) Waugh (now Professor of Physical Chemistry at UMIST) at ICI's Corporate Research Laboratories at Bozedown. Waugh determined the Arrhenius parameters for Cl/F exchange in CCl_4 and C_2Cl_6 on prefluorinated chromia, using temperature-programmed desorption.

Over the years many chemists have contributed to the evolution of ICI's CFC and HCFC technology, but those particularly associated with it have been R. (Bob) L. McGinty, D. (Don) Lomas and A. (Archie) McCulloch. Although Bob McGinty was little known outside the Company, he worked on fluorine chemistry from joining the Company in the mid-1940s through to his retirement in the mid-1970s. Initially he worked on a range of projects in J. H. Brown's group, including PTFE process routes. In the early 1950s he established himself as the expert on the chemistry of the Arcton™ liquid-phase processes and has a lasting claim to fame as the first person to synthesize the inhalation anaesthetic 'Fluothane™'. Bob passed on his knowledge to Don Lomas, who had studied fluorine chemistry through PhD studies with Ken Musgrave at Durham University and postdoctoral work with Paul Tarrant in Florida before joining ICI. Along with troubleshooting on the liquid-phase Arcton™ processes that, by the late 1970s, made CFCs 11, 12, 113 and 114 and HCFCs 22 and 133a (CF_3CH_2Cl), Don continued the development of the chromia catalyst, concentrating on improvements to the production process that led to better catalyst life. He moved on to a senior planning role in the Arcton Business before his untimely death in the late 1980s.

Archie McCulloch succeeded Don Lomas as senior chemist responsible for ICI's Arcton™ technology and continued in the same vein, developing the chromia catalyst with incremental improvements to its durability so that it had to be replaced only at intervals of several months rather than weeks. In the late 1970s, collaboration started with John Winfield and Geoff Webb at the University of Glasgow on the mechanism of chromia catalysis [10, 11]. Through funded research and consultancy this has continued to the present day, and the understanding of the behaviour of chromia in real process operation which resulted from the combination of these mechanistic studies and plant observations was invaluable in the development of processes to CFC alternatives.

Fluon^{TM} PTFE polymer

(by Laurie Shipp[6])

ICI's interest in TFE and PTFE appears to date from 1943 when R. Hill visited DuPont under the ICI/DuPont technology sharing agreement. In 1944, Chapman, working in J. H. Brown's newly formed research group, started to investigate the cracking of HCFC-22 and, by the late 1940s, ICI's General Chemicals Division was operating a semi-technical unit at Widnes; between April 1947 and August 1948 this produced 4.6 t of Fluon^{TM}-brand. The embryonic business suffered a serious set-back in 1950 when a TFE facility on the Widnes site exploded. Responsibility for PTFE was transferred to the newly-formed Plastics Division, and by 1953 a 20 tpa pilot plant was operating at ICI's Hillhouse facility near Blackpool, followed in 1956 by a 200 tpa commercial plant on the same site.

Undoubtedly ICI's greatest contribution to the fluoropolymers area has been the HCFC-22 (CHF_2Cl) steam-cracker process for TFE production which contributed significantly to reducing monomer costs. The original TFE process, developed by DuPont, involved the pyrolysis of HCFC-22 in an electrically-heated platinum-lined tube, a crude method producing a host of by-products including low-molecular-weight polymer. In the ICI process, developed in the early 1960s, HCFC-22 vapour was injected into super-heated steam at 1000+ °C, which greatly decreased the number and quantities of by-products. This not only increased feedstock conversion efficiency but also simplified the distillation, since the fewer by-products meant less azeoptropes requiring separation from the TFE product and the unchanged CHF_2Cl. For this advance, together with the improvement in free-flowing granular powders for fabrication, ICI's Plastics Division was awarded a Queens award to Industry for technical achievement in 1967. The steam cracker technology was incorporated in ICI's American plant, sited at Bayonne New Jersey and commissioned in 1970, and in the joint ICI-Asahi Glass plant at Chiba, Japan, commissioned in 1983.

ICI researchers made significant technical contributions to the understanding of TFE and PTFE, identifying the factors which could induce TFE to explode, both in the presence and absence of air, and carried out important studies into on the molecular and crystal structure of PTFE. ICI's Plastic Division Research Department pioneered the application of crystallography (1950s), electron microscopy (1960s) and differential scanning calorimetry (DSC) (1970s) to the study of PTFE. Using DSC, another transition was discovered at 339 °C, above the generally accepted transition at 327 °C, which led to a quicker and more accurate method for comparing molecular weights of variously prepared PTFE batches. This advance in analysis was instrumental in improving the polymerization technique for coagulated dispersion polymers used in expanded PTFE tape and sheets. Although ICI

[6]Laurie Shipp joined ICI directly from school in 1946. He first worked on fluoropolymers part-time in 1951 at ICI's Welywn garden City site north of London where he investigated PCTFE (*poly*chlorotrifluoroethylene) before being switched full-time to PTFE in 1953. He was involved with technical development of ICI's Fluon^{TM} for the rest of his career, becoming recognized, both inside and outside the Company, as ICI's leading expert in this area. He retired in 1983.

produced grades of PTFE containing trace co-monomers, it did not diversify into other fluoropolymers. Several technical assessments were carried out over the years, but none came to commercial fruition.

In 1998 ICI sold its interest in the Chiba (Japan) facility to Asahi Glass; this was followed in September 1999 by the sale of the rest of the Fluoropolymers Business to the the same company.

BCF^{TM} firefighting agent

(by Neil Winterton[7])

ICI's interest in brominated firefighting agents can be traced back to 1950 when a literature survey was carried out to find alternative firefighting agents to toxic methyl bromide. Although bromochloromethane (CH_2BrCl) emerged as the preferred material it was noted that a number of bromofluoro-methanes and -ethanes could be potential agents, including bromochlorodifluoromethane (BCF^{TM}, CF_2ClBr). In 1954, ICI prepared experimental quantities of BCF by the gas-phase bromination of HCFC-22 at 500 °C ($CHF_2Cl + Br_2 \rightarrow CF_2BrCl + HBr$), and, after successful firefighting trials, 500 lbs were produced in 1955. While this work was underway, however, Firestone was granted a patent [12] to this route which ICI licensed in 1957. ICI's attempts to sell BCF^{TM} to fire equipment manufacturers met with a poor response until 1960, when the Wright Air Development Center (USA) published a report into firefighting agents which suddenly increased interest in BCF. In the early 1960s, with BCF^{TM} sales growing rapidly, a small plant previously used to manufacture the inhalation anaesthetic Fluothane^{TM} ($CF_3CHBrCl$; see later) was modified to produce BCF at a rate of 120 tpa during the period 1963-1965. Sales were boosted considerably by the British Defence Ministry's award of a contract to Pyrene Ltd. to replace all its carbon tetrachloride (CTC) extinguishers by units containing BCF. In the late 1960s, UK sales were further helped by the banning of CTC in fire extinguishers on toxicity grounds.

The first ICI plant specifically designed for BCF^{TM} manufacture was commissioned in 1967 with a capacity of 750 tpa, which, by 1973 had been extended to over 3000 tpa. In 1972 G. (Graeme) S. Milne had developed a kinetic model of the plant based on 10 fundamental equations; a further 13 equations were included to model the effect of chlorine present in the bromine feed. By-product CF_2Br_2 was shown to arise via difluorocarbene: $CHF_2Cl \rightarrow CF_2 + HCl$; $CF_2 + Br_2 \rightarrow CF_2Br_2$. BCF^{TM} capacity was subsequently extended to 5500 tpa, but the plant was finally closed in 1993 in accordance with the requirements of the Montreal Protocol. ICI considered developing a replacement for BCF^{TM} and indeed assessed CF_3CHFBr, synthesized by the bromination of CF_3CH_2F (HFC-134a) under conditions similar to those used for converting HCFC-22 to BCF. This two-carbon compound proved to have a fire-fighting performance comparable to that of BCF, but was

[7] Neil Winterton graduated from University College London (BSc, 1965; PhD, 1968). After holding a post-doctoral fellowship at the University of North Carolina, Chapel Hill (1968-1970), he became a temporary lecturer at the University of Warwick (1970-71) and then Research Fellow (1971-1973). He joined ICI Corporate laboratory in 1973, moving later to Mond Division where he became responsible for research into the chemistry and technology of chlorine and chlorinated products, including the development of substitutes for materials phased out under the Montreal Protocol. He was a visiting Industrial Professor at the Queen's University, Belfast, in 1994/5. Shortly after retiring from ICI at the end of 1998 he joined the Leverhulme Centre for Innovative Catalysis, University of Liverpool.

rejected because of its ozone depletion potential. Since no agent was identified which simultaneously satisfied ICI's principal criteria, *i.e.* environmentally acceptable, adequate performance and a fit with the Company's expertise in halocarbon technology, it was decided to withdraw from the fire extinguishing agent market completely.

The Winnington contribution

(by J. Hutchinson[8], with assistance from J. R. Case, H. C. Fielding and H. L. Roberts)

While the first fluorine chemistry in ICI started in the General Chemicals Division at Runcorn/Widnes, another line of fluorine chemistry started in the Alkali Division laboratories at Winnington near Northwich, Cheshire. Arguably, this second strand was a result of PE having been discovered in these laboratories by Gibson and Fawcett in the 1930s. Research and development on PE continued at Winnington until 1957, when it was transferred to ICI Plastics Division at Welwyn Garden City. However, a year before this transfer took place, the decision was made to set up an 'Exploratory Group' under G. (George) Manning with the aim of identifying the next big product – 'the next Polythene™'.

It was recognised that there was going to be an increasing demand, mainly from the aerospace industry, for oils and greases having high thermal and chemical stability. With this in mind, various projects aimed at making inorganic polymers were proposed, including one by H. (Hugh) L. Roberts who had just been recruited from the University of Oxford (1956). Given the stability of sulphur hexafluoride, Roberts proposed the synthesis of polymers having the same relationship to sulphur hexafluoride as PTFE had to carbon tetrafluoride, *i.e.* the synthesis of polymers with the general structure $SF_5(SF_4)_nSF_5$. Attempts to synthesize SF_4Cl_2, a potential monomer for the required polymer, were unsuccessful. However, much novel chemistry of sulphur-fluorine compounds was established [13], including a high-yield synthesis of the new hexahalide SF_5Cl [14] and the subsequent reaction of this compound with tetrafluoroethylene to give a series of telomers, $SF_5(CF_2CF_2)_xCl$ ($x = 1, 2, 3$) [15, 16]. Uses for these novel telomers, which were eventually prepared on a multi-kilo scale, were explored but the materials lacked the required thermal stability. Furthermore, the chlorine atom was insufficiently reactive for them to be useful for the synthesis of fluorochemical surfactants or surface treatments corresponding to those derived from fluorocarbon iodides and acids.

Early in the 1960s, because of its acknowledged expertise in 'hot tube' chemistry, the Winnington Group was approached by ICI Pharmaceuticals Division with a request to synthesise a sample of hexafluorobenzene. It was envisaged that this would be made by the original Désirant method of pyrolysing $CFBr_3$. Some hexafluorobenzene was made by this process but it was also decided to explore the halex route from hexachlorobenzene ($C_6Cl_6 + KF \rightarrow C_6F_6$). This task fell to H. (Harold) C. Fielding – his first foray into fluorine chemistry. Having developed the technology for preparing hexafluorobenzene

[8]Following his graduation (BSc) from Durham University in 1959, John Hutchinson studied under Ken Musgrave for his PhD, which he gained in 1962. After 3 years at Durham as a postdoctoral research assistant investigating highly-fluorinated pyridines, he joined ICI Mond Division and worked on TFE oligomers from 1965 to 1973. He resumed his interest in fluorine chemistry in 1982 and worked on a variety of topics, including CFC replacements, until he retired from ICI in 1992. From 1992 to 1998 he was the (BNFL) F2 Chemicals Ltd. senior research fellow at Durham University, investigating uses of F_2 in organic synthesis.

by this route [17, 18], he used it to prepare pentafluoropyridine from pentachloropyridine [19], which was available by the vapour-phase chlorination of pyridine from the laboratories of ICI General Chemicals Division, Runcorn. (This process was being carried out on a significant scale, and pentachloropyridine was donated to the Musgrave-Chambers group at Durham University where the difficult pyridine/PCl_5 reaction was being used to procure C_5Cl_5N *en route* to C_5F_5N.) The chemistry of hexafluorobenzene was explored at Winnington and a route to 1,4-dichlorotetrafluorobenzene, a potential starting material for perfluoropolyphenylenes, was developed. Also, many compounds derived from hexafluorobenzene in the Winnington laboratories were evaluated as plant-protection chemicals at the ICI Experimental Station, Jealotts Hill in Southern England. Similarly, the chemistry of pentafluoropyridine was explored, mainly by a team at Runcorn led by Maurice Green[9], and again, many derivatives were evaluated as plant-protection chemicals (some of this work is described in a later section).

Interestingly three teams in the UK were independently investigating the halex route to pentafluoropyridine at the same time, namely the Musgrave-Chambers (Durham), Banks-Haszeldine (Manchester) and ICI (Winnington) groups. Amongst the reactions of hexafluorobenzene and pentafluoropyridine which were investigated by Fielding at ICI were those with tetrafluoroethylene and fluoride ion (*i.e.* an in-situ source of $C_2F_5^-$) in dipolar aprotic solvents [20, 21]. Not only was it found that all ring fluorines in the substrates could be replaced sequentially by C_2F_5 but that the reaction products contained some non-aromatic perfluorinated liquids. These liquids turned out to be branched oligomers of tetrafluoroethylene – mainly compounds **1-3** [22]; clearly these unexpected products were worthy of further study.

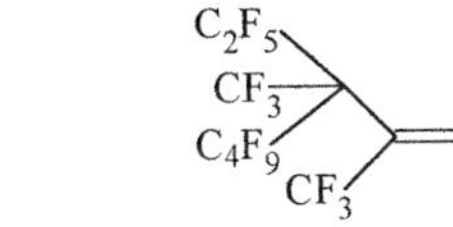

(1) 'Tetramer' - *cis* and *trans* isomers in approximately equal quantities

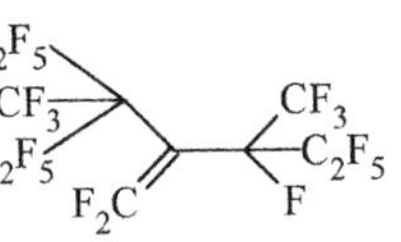

(2) 'Pentamer'

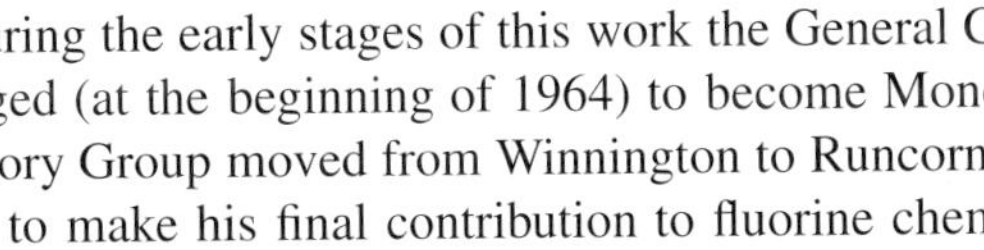

(3) Major components of 'Hexamer'

During the early stages of this work the General Chemicals and Alkali Divisions of ICI merged (at the beginning of 1964) to become Mond Division. Most members of the Exploratory Group moved from Winnington to Runcorn, with Jack Rudge joining Harold Fielding to make his final contribution to fluorine chemistry before retiring. The unique properties of fluorochemical surfactants and surface coatings were well known, and it was believed that the oligomers of TFE could be exploited to synthesize such materials (Scheme 22.1). The team was necessarily expanded to develop the chemistry involved,

[9] Sadly, Maurice Green died in 1998 while this chapter was being researched.

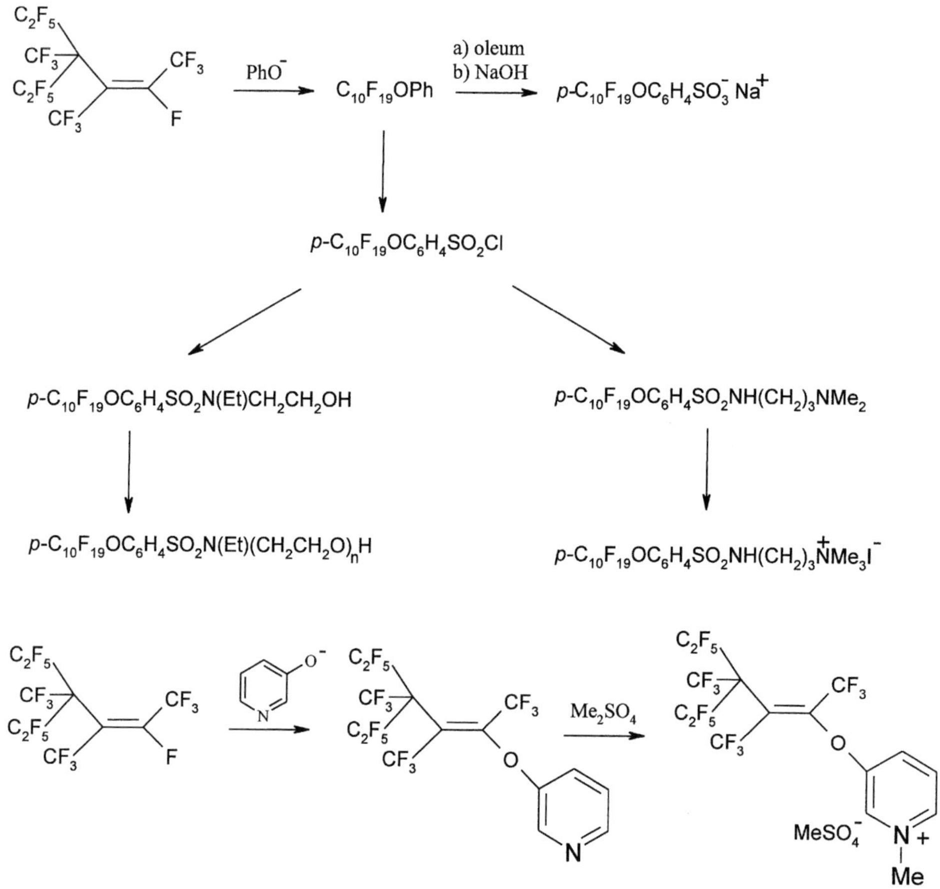

Scheme 22.1. Examples of surfactants derived from TFE oligomers.

and recruits included Roy Deem and John Hutchinson[8] from the Manchester and Durham fluorine schools respectively; later (1970) R. (Dick) L. Powell joined the team and was introduced to fluorine chemistry for the first time.

The chemistry of TFE pentamer (**2**), the most abundant product in the reaction mixture, was mainly developed within ICI, but in collaborative programmes with Durham and Birmingham Universities (Dick Chambers [23] and Paul Coe [24, 25]) the chemistries of the tetramer (**1**) and hexamer (**3**) were also explored. Textile treatments derived from TFE oligomers did not deliver adequate performance due to the branched structure of the fluorinated group, whereas surfactants [26], which were given the trademark Monflor™, proved to be successful in a range of niche applications, *e.g.* in self-levelling floor polishes and as crude oil and gas recovery enhancers. The most important single product was the anionic sulfonate surfactant, MD313, sold as an additive for protein-based foams typically used to fight aircraft and fuel fires. The fluorochemical surfactant stabilized the foam against hydrocarbons, preventing its premature collapse. In 1976 the Monflor™ products and MD 313 moved from the New Ventures Group, where they had been nurtured in the early 1970s, to General Chemicals Business Group.

The development of TFE oligomerization and the oligomer based products was an undoubted technical and commercial success. However, when market conditions in the chemical industry became more difficult in 1980, the anomalies of operating an isolated speciality business supplying niche applications alongside a high-tonnage CFC/HCFC business became apparent. As a result, General Chemicals Group decided to withdraw from the fluorochemicals surfactants business in mid-1981.

Environmental concerns – tackling ozone depletion

(by A. McCulloch[4])

Following their introduction in the early 1930s, CFCs came to be used in many different industries because their properties matched Midgley's original requirements almost exactly – they were essentially nontoxic, nonflammable and chemically stable [27]. From refrigeration to solvent cleaning, from polyurethane foam manufacture to aerosols, CFCs came to play an important role in enhancing the quality of peoples' lives throughout the world. The requirement for primary safety to users was met so well that venting to atmosphere became a principal means of disposal of unwanted refrigerants. Thus, when demand grew dramatically postwar, the consequent growth in CFC emissions also became problematic, and by the early 1970s it became clear that a significant fraction of each year's production was being released into the environment.

Using his newly-invented electron-capture detector, Jim Lovelock was able to show by 1970 that CFCs 11 and 12 were ubiquitous trace constituents of the atmosphere [28]. In 1972, ICI representatives, together with those from other major CFC manufacturers, attended a meeting initiated by DuPont to review the possible environmental fate of CFCs. Their objective was well expressed by Ray McCarthy of DuPont: '*Fluorocarbons are intentionally or accidentally vented to the atmosphere world-wide at a rate approaching one billion pounds per year. These compounds may be either accumulating in the atmosphere or returning to the surface, land or sea, in pure form or as decomposition products. Under any of these alternatives it is prudent that we investigate any effects which the compounds may produce on plants or animals now or in the future.*'

This prompted the CFC manufacturers to investigate the possible environmental effects of CFCs, and the Fluorocarbon Panel was set up under the auspices of the Manufacturing Chemists Association (now the Chemical Manufacturers Association, CMA) in Washington DC. The original aim was to assess the smog-forming potential of CFCs (since found to be insignificant); however, the course of the programme was altered in 1974 when, based on Lovelock's work, Mario Molina and F. Sherwood Rowland propounded their hypothesis of ozone depletion by CFCs, challenging the science community to find evidence of ozone depletion [29].

From the beginning, the Fluorochemical Industry responded responsibly to the implied environmental threat. Through the auspices of the Fluorocarbon Panel, ICI and other major CFC producers funded extensive research programmes to test the validity of the hypothesis using independent external experts in gas-phase reactions and atmospheric chemistry. ICI, like other manufacturers, needed to understand properly the extent of any problem and had to base that understanding on high-quality scientific research which would ultimately be published as peer reviewed papers in reputable journals.

The replacement of CFCs: 1974 to 1980
(by A. McCulloch and R. L. Powell)

In parallel with the atmospheric chemistry studies, ICI ran a research programme to find alternatives to the CFCs. The primary target was to replace the major products, namely CFC-12 (CF_2Cl_2), which was used as a refrigerant and as the propellant for most aerosols, and CFC-11 ($CFCl_3$), the main aerosol vapour-pressure depressant and PU foam-blowing agent. The customers still preferred the properties of low toxicity, nonflammability and good thermal and chemical stability for which the CFCs had originally be chosen as refrigerants. Additionally, customers wanted fluids which would also match the physical properties of the CFCs for which their current products were designed. Only fluorocarbons could provide this combination of desirable features, and Don Lomas, J. (Jock) S. Moilliet *et al.* drew up a list of a potential candidates of the HFC and HCFC classes, including 32 (CH_2F_2), 125 (CF_3CF_2H), 143a (CF_3CH_3), 21 ($CHFCl_2$), 31 (CH_2FCl), 134a (CF_3CH_2F), 124 (CF_3CHFCl), 123 (CF_3CHCl_2) and 133a (CF_3CH_2Cl). HCFCs were considered acceptable because they were destroyed in the troposphere by reaction with hydroxyl radicals and were seen as very much part of the solution. Subsequently they have proved to be a stop-gap solution only: they are now included in the Montreal Protocol, but, in view of the fact that they have much less effect on stratospheric ozone, their phase-out is delayed to year 2030. The European Union, however, has opted for a total HCFC phase-out by year 2015.

Compounds such as 21 and 133a (possible replacements for CFC-11) were already available from ICI's plants. The other compounds were produced in the laboratory through a programme run initially in 1974 by Jock Moilliet and, from the beginning of 1975, by Dick Powell. From the beginning, ICI's chromia catalyst was a key factor in the Company's development of CFC replacements. Lab samples of the candidates were obtained by HF fluorination of suitable chlorinated feedstocks over chromia catalyst in 3 foot long × 1 inch diameter lab reactors operating at atmospheric pressure which were capable of producing up to 10 kg of material per week. An antimony halide-catalyst liquid-phase reactor operating under pressure was used in an attempt to produce 31 and 32, but it proved more difficult to control and was less productive than a chromia-catalysed, gas-phase flow reactor.

The samples obtained were used to determine toxicity, physical properties and rate of destruction by hydroxyl radical. ICI's Industrial Hygiene Research Laboratory (subsequently renamed Central Toxicology Laboratory, CTL) provided the Arcton™ Business with key toxicological data, as well as being a major contributor to the industry's co-operative programme. As the data accumulated during 1975 and 1976, it became apparent that the only sensible candidate for the replacement of CFC-12 (CF_2Cl_2, b.p. -29 °C) was HFC-134a (CF_3CH_2F, b.p. -27 °C), which had a reasonably similar vapour pressure, no adverse toxicity indications in the preliminary screen, and, containing no chlorine, could not contribute to stratospheric ozone destruction as postulated by Rowland and Molina. HCFCs 21, 31, and 133a proved to be unacceptably toxic for use either as refrigerants or in aerosols.

As 134a emerged as a leading CFC replacement, effort was focused on the process routes. The original test samples produced by Moilliet and subsequently by Powell had been generated by the fluorination of 133a with HF over a chromia catalyst, but conversions were limited to 10-20%. At first this was attributed to catalyst activity limitations but an experiment by Powell in which a mixture of 133a and 134a was passed over chromia with HCl resulted in a 10-fold reduction in the concentration of the 134a. Clearly the reaction was equilibrium limited. It would be pleasing to claim that this valuable insight resulted from an experiment designed specifically to test the idea, but in fact a different, now forgotten, hypothesis was being tested. The implications for basing a 134a process on an equilibrium-limited 133a vapour-phase fluorination were profound... an external recycle of 85 to 90% of feeds would be required, greatly increasing the physical size and therefore capital cost of the plant for a given output compared to previous CFC and HCFC processes where equilibrium limitations were not a problem. A further complication was the formation of the toxic by-product 1122 ($CF_2{=}CHCl$) through the elimination of HF from 133a, especially when the operating temperature of the catalyst was increased to counteract decreasing activity. Powell recognised that, like 134a formation, this could be an equilibrium process, hence by passing the reaction mixture over a catalyst at a lower temperature this might be reversed. He demonstrated the principle by passing a 133a/134a/1122 mixture recovered from a previous fluorination with HF back over the catalyst at temperatures around 200 °C; as anticipated, the 1122 content was greatly decreased. S. (Stephen) Potter showed that by linking a laboratory chromia fluorination reactor directly to a low-temperature 1122-removal reactor, the process could be operated to produce 134a free from 1122 [30]. The process was scaled up on a new, computer-controlled, semi-technical plant at Widnes Experimental Site that contributed extensively to catalyst developments.

No problems had appeared in the initial toxicity tests on 134a, so the decision was made to undertake 90-day animal inhalation exposure trials which would require half a tonne of high-purity 134a. This material was made by the addition of HF to trifluoroethene over chromia catalyst, using laboratory reactors operated around the clock by a shift team. Although the reaction required heating for initiation, it was then sufficiently exothermic to run without the further application of heat. Indeed the reaction was controlled by blowing air over the reactor tube. Although this was an exceedingly expensive method of producing 134a, it provided very pure material requiring only a single distillation.

A range of speculative routes to 134a were considered, the most promising of which was invented by J. (John) I. Darragh [31]. In contrast to CFC-114 (CF_2ClCF_2Cl) produced by liquid-phase fluorination of $CCl_2{=}CCl_2$, ICI's Arcton^{TM} 114 from its vapour-phase chromia process was in fact a mixture of 114 and its isomer 114a (CF_3CFCl_2), the quantity of the latter varying depending upon the operating conditions. Darragh demonstrated that if a mixture of 114 and 114a was passed over a Pd catalyst supported on alumina or charcoal, the 114a was preferentially hydrogenated to 134a, and under properly selected conditions the 114 was largely unaffected. In principle, recovered 114 could be isomerized to 114a. The route was further studied by McCulloch and developed to the semi-technical scale. Hydrogenation was not in the mainstream of Mond Division technology, so the semi-technical plant was designed and constructed initially at Ardeer in Scotland with the help of O. (Owen) Lambert. From the outset it was designed to run unattended to avoid the shift operation that had proved so hugely expensive, and this semi-technical plant and its subsequent re-incarnation at Widnes Experimental Site went on to make many hundreds of kilogrammes of 134a at a reasonable cost.

However, based on its accumulated laboratory and semi-technical experience and on the perceptions of integration into manufacturing sites, ICI selected the chromia-catalysed HF fluorination of 133a (CF_3CH_2Cl) as the basis of its preferred process, and in 1979 plans were being drafted to build the first production plant in the US. The urgency to replace the CFCs appeared to be reduced when the best available computer models available at the time indicated that the rate of ozone depletion was about 3% a year; also, it was felt that while the CFCs would have to be replaced, there was sufficient time to achieve this by a phased introduction of replacements. Consequently, the potential market which ICI had anticipated for 134a disappeared, so the Company shelved its development programme, but had sufficient prescience to retain its semi-technical facility for producing 134a and to store a few tens of kilos of the compound itself.

Time of transition: 1980-1985

From ICI's earliest interest in fluorine-containing compounds in the 1930s until 1980 the Company's fluorine chemistry expertise was centred in the Runcorn-Widnes area. In the late 1970s a growing interest in ring-fluorinated aromatics emerged; the challenge to manufacture these compounds was taken up the the newly-formed Fine Chemical Manufacturing Organisation (FCMO) of Organic Chemicals Division at Blackley, north Manchester, which had access to considerable expertise in aromatic chemistry derived from dyestuffs and related fine chemical products. Jock Moilliet was recruited from Runcorn to provide fluorochemical expertise, especially where the handling of liquid HF was involved; thus a second centre of fluorochemical expertise emerged whose development is described in a later section.

At Runcorn the interest in CFC replacements was quiescent, but Dick Powell, working for the Arcton™ refrigerant business, investigated fluorinated compounds boiling between room temperature and about 50 °C as potential working fluids for vapour-recompression heat pumps, the intended market being industrial converting 'waste' heat in the range 60 to 100 °C to process steam at 150 °C. The relatively high fuel prices at the time made this a potentially attractive proposition. The work was part of a collaborative programme with F. (Fred) Drakesmith and M. (Mark) Bertinat of the Electricity Council R & D Centre (ERDC), now E A Technology, at Capenhurst near Chester, and included a project with Dick Chambers at the University of Durham to investigate the synthesis of perfluorocarbon ethers by the CoF_3 fluorination of partially-fluorinated ethers. A conscious decision was made to avoid chlorine-containing compounds and to concentrate on various hydrofluorocarbons, fluoroethers and amines. In terms of thermodynamic performance, the most attractive fluorinated compound was 2,2,2-trifluoroethylamine ($CF_3CH_2NH_2$, b.p. 35 °C), which also had the advantage of nonflammability.

A second programme with other collaborators explored the potential for fluorocarbon fluids in absorption heat pumps which involved the solution of the working fluid in a high-boiling solvent. The programme was specifically aimed at higher-efficiency gas 'boilers' for British circulating water central heating systems which operate at 80 °C. For low-pressure absorption heat pumps, trifluoroethylamine/*m*-chlorophenol and 1,1,2,2,3,4-hexafluorocyclobutane/*NN'*-dimethylimidazolidone emerged as the most promising combinations, but neither could deliver the desired performance. For high-pressure systems, HFC-134a and HCFC-124 (CF_3CHFCl) with various amide and ether solvents were evaluated as working fluids; 124 was the best, but again it was not possible to achieve the required efficiency with the high output temperature typical of British central heating systems. Although these heat pump fluid projects were unsuccessful, methods for measuring the physical properties of the fluorocarbon fluids and their interactions with solvents developed in ICI by P. (Peter) Rathbone and R. (Bob) Wheelhouse provided a basic competence for assessing HFCs and lubricants when the CFC replacement programme was subsequently resumed.

In the mid 1980s a New Fluorochemicals R&D Group under J. H. (Hugo) Steven was set up to investigate the potential for entering the fine chemicals fluorine market, notably perfluorinated inert fluids and trifluoroacetaldehyde (fluoral), the latter being produced by the fluorination of trichloroacetaldehyde over ICI's chromia catalyst. In addition to being a product in its own right, fluoral was seen as an excellent intermediate for trifluoroethanol, trifluoroacetic acid and trifluoroethylamine. In 1985, a small-scale production plant was built and operated by R. (Roy) D. Bowden which supplied fluoral to ICI's Pharmaceutical Business. The project was effectively halted when increasing effort had to be devoted to the development of CFC replacements and the fluoral plant was adapted to the production of 134a.

From the late 1970s to the early 1990s, J. (John) Beacham was the senior manager whose remit included fluorochemicals research at Runcorn. He provided encouragement and support for scientists in the area, and in particular ensured that in the period 1980 to 1986 a nucleus of fluorochemical expertise remained available to the Arcton^{TM} Business which could be expanded when the need arose for the development of the CFC replacements.

The replacement of CFCs: 1985-1998

(by A. McCulloch and R. L. Powell)

Measurements carried out by the British Antarctic Survey in 1985 showed major, temporary losses of ozone in Antarctica each Spring – the 'Antarctic Ozone Hole' as this popularly became known [32]. This galvanised the regulators into action and caused all of the industrial programmes for alternatives to CFCs to be re-appraised rapidly. In 1987, the world's governments signed the first international agreement to protect the global environment – the *Montreal Protocol*. Whilst the original Protocol mandated only a 50% reduction in CFC production and consumption by 1st July 1999, subsequent amendments in 1990 and 1992 introduced phaseout schedules for all substances that were implicated in the depletion of the ozone layer, including CFCs and HCFCs [33] (see Table 22.1).

The adoption of the Montreal Protocol led to a significant resurgence, by both producers and users of CFCs, in activities that had lain dormant since the late 1970s. The fluorocarbon producers invested hundreds of millions of pounds sterling in firstly identifying, then researching and developing, and finally manufacturing alternatives to CFCs. Toxicological testing of these alternatives was carried out in parallel with both their development and the construction of the production facilities [34]. This unprecedented step, which carried a high financial risk (normally toxicological work is carried out prior to investment in manufacture), was a clear response to what was seen as a priority issue. Toxicological testing was further accelerated by the formation of a consortium, the *Programme for Alternative Fluorocarbon Toxicity Testing* (PAFT) which carried out the testing for a group of 12 companies (in the case of HFC-134a), thereby avoiding the need for each company to carry out its own tests and to pool the results of tests that had been carried out in-house, in ICI's case at CTL.

TABLE 22.1
Environmental concentrations and properties of the principal CFCs and alternative fluorocarbons

Compound	Formula	Current atmospheric concentration (ppt[a])	Atmospheric lifetime (years[c])	Ozone depletion potential[d], relative to CFC-11 = 1	Global warming potential over 100 years[e], relative to $CO_2 = 1$
CFC-11	$CFCl_3$	264	45	1	4000
CFC-12	CF_2Cl_2	530	100	1	8500
CFC-113	$CFCl_2CF_2Cl$	83	85	0.8	5000
CFC-114	CF_2ClCF_2Cl	15	300	1	9300
CFC-115	CF_2ClCF_3	5	1700	0.6	9300
HCFC-22	CHF_2Cl	125	11.8	0.06	1700
HCFC-123	$CHCl_2CF_3$	no data	1.4	0.02	93
HCFC-124	$CHFClCF_3$	no data	6.1	0.02	480
HCFC-141b	CH_3CFCl_2	7	9.2	0.11	630
HCFC-142b	CH_3CF_2Cl	8	18.5	0.07	2000
HFC-125	CHF_2CF_3	1[b]	32.6	0	2800[f]
HFC-134a	CH_2FCF_3	7[b]	13.6	0	1300[f]
HFC-152a	CH_3CHF_2	1[b]	1.5	0	140[f]

[a] Parts per trillion (1 in 10^{12}). Data from *Scientific Assessment of Ozone Depletion:* 1998, World Meteorological Organisation Global Ozone research and Monitoring Project report No. 44, WMO Geneva, 1999, except where indicated.
[b] Personal communication: Georgina Sturrock, CSIRO, 1999.
[c] Data from *Scientific Assessment of Ozone Depletion*: 1998.
[d] Data from *Handbook for the International Treaties for the Protection of the Ozone Layer: the Vienna Convention* (1985), *the Montreal Protocol* (1987), United Nations Environment Programme, Nairobi, 1996.
[e] Data from *Climate Change* 1994: *Radiative Forcing of Climate Change*, Intergovernmental Panel on Climate Change, Cambridge University Press, Cambridge, 1995, except where indicated.
[f] Data from *Climate Change* 1995: *the Science of Climate Change*, Intergovernmental Panel on Climate Change, Cambridge University Press, Cambridge, 1996.

With the CFCs being phased out due to their impact on the ozone layer, the fluorocarbon producers needed to assure both themselves and the public that they were not producing products which would also have an adverse environmental impact. This led to the formation of a second consortium, the *Alternative Fluorocarbons Environmental Acceptability Study* (AFEAS) whose role was to test, through independent experts in academic and government institutions, the environmental impact of the alternatives and their breakdown products. The work of both these panels is now complete: toxicological testing has progressed to the point where sufficient is known about the compounds [35], and their environmental behaviour and that of their breakdown products has been completely characterized [36]. In an interesting side-issue of this work it was found that trifluoroacetic acid, a minor breakdown product of HFC-134a, is present in the contemporary environment in quantities far greater than could have come from the products of the fluorocarbon industry. Its source is an environmental puzzle that has yet to be solved [37].

When concern over the possible depletion of the ozone layer resurfaced in the 1985 the remit of the New Fluorochemicals R&D Group at Runcorn led by Hugo Steven was expanded to include the development of new CFC replacements, and within a year this had become its principal task. The first major target was a re-assessment of 134a, the accepted replacement for CFC-12, the largest tonnage refrigerant used in automobile air-conditioning, domestic refrigerators and industrial refrigeration. Brain-storming sessions were used to generate a large number ideas, which, after paper evaluation, were reduced to six for experimental study that would provide data for the economic evaluation. The hydrogenation of CFC-114a and the fluorination of 133a, two routes extensively studied in the 1970s, were included in the short list and the latter emerged as the most attractive approach, not least because of confidence in the effectiveness of ICI's chromia catalyst. In 1988, the semi-technical unit, built 20 years earlier to develop the chromia catalyst, and the recently commissioned fluoral production plant were modified for 134a production, allowing ICI to provide development quantities of the refrigerant to customers and for toxicity testing, as well as giving valuable insights into catalyst performance.

Although the decision that the ICI Fluorochemicals Business would become a major manufacturer of 134a with a large plant in the US was made in 1987, management knew that this would take several years to realize. The demand for 134a was expected to grow from 1991 onwards, so it was decided to build a smaller manufacturing facility in the UK which would be operating by 1990 while simultaneous developing the American plant. This ambitious target represented a 'fast track' approach, and to ensure success senior management promoted close collaboration between the chemists generating the laboratory results and the engineers designing the plant. Multidisciplinary joint development teams were formed with chemists and engineers working alongside each other in the same office. It was accepted that 'fast track' would mean that some details of the design work would need to be modified when more precise lab data became available, but the success of this approach to technical management was demonstrated when the world's first full-scale HFC-134a plant at Runcorn commenced beneficial production on 18 October 1990 with a nameplate capacity of 1200 tpa. Learning from the operation of the UK plant was incorporated into the design of the much bigger 10 000 tpa capacity unit built in the US (the world's biggest market for 134a because of automobile air-conditioning) and commissioned on time in mid-1993. A third plant of 5000 tpa capacity was planned for Japan to serve the Asia Pacific region; production commenced in 1994.

Although many breakthroughs were made during the programme, two stood out as being of fundamental importance. The first was an improved catalyst [38] and the second was the concept of the 'reverse series reactor' system [39]. The fluorination of trichloroethylene is usually written as two steps: $CCl_2{=}CHCl + 3HF \rightarrow CF_3CH_2Cl + 2HCl$; $CF_3CH_2Cl + HF \rightarrow CF_3CH_2F + HCl$. To convert this chemistry into a manufacturing process two fundamental problems must be addressed. The first reaction is thermodynamically very favorable but is highly exothemic, so if the heat of reaction is not dissipated efficiently the catalyst can become deactivated by overheating. In contrast, the second reaction is equilibrium limited and requires a higher operating temperature than the first reaction, but is only slightly exothermic. In the earlier designs (Fig. 22.1a) of the 134a plant, trichloroethylene was fluorinated in reactor R1 to 133a which was then separated from the HCl co-product by distillation and fed to the second reactor, R2, where it was converted to 134a. This reactor sequence is designated a 'forward series' plant configura-

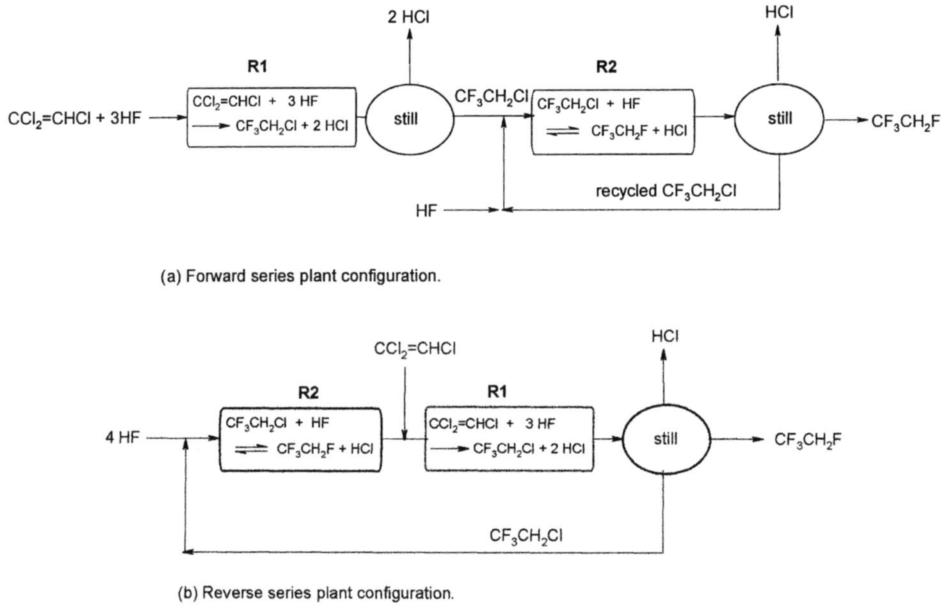

Fig. 22.1.

tion. In the reverse series configuration, shown in Fig. 22.1b, trichloroethylene is converted to 133a in reactor R1, which now immediately precedes the product (and only) still and the excess of 133a is recycled to reactor R2, which is operated at a higher temperature. So why is this better than the more conventional arrangement? The HF required for the fluorination of trichloroethylene to 133a is present during the 133a-to-134a reaction, helping to drive the equilibrium to the right-hand side. The mixture of HF, 134a, HCl and unreacted 133a is cooled to approximately the temperature of reactor R1 and the trichloroethylene added. Although the temperature is adquate to convert trichloroethylene to 133a, it is too low to for the back reaction of 134a to 133a, despite the extra HCl generated. The combined heat capacities of HF, 134a, HCl and excess 133a from R1 acts a heat reservoir, preventing the temperature of the catalyst from rising to an excessive level. Additionally, an HCl still is eliminated.

The second major breakthrough was the invention by J. (John) Scott of a modified chromia catalyst which was more active than the conventional catalyst, especially for the conversion of 133a to 134a. This enabled R2 to be operated at lower temperatures, significantly increasing catalyst life between reactivation or replacement and reducing the quantities of by-products such as 125 and 143a.

These improvements reduced by-product formation by more than 90% and made ICI's 134a process technology a world leader. In 1994, for their contributions to the development of the 134a process, catalyst chemist John Scott and engineers F. (Frank) Maslen, D. (Denis) Henderson and R. (Rachel) Steven were presented with the McRobert Award, the most prestigious engineering prize in UK for which all branches of the discipline compete. The prize money from the Award was donated to the Engineering Education Scheme, which is dedicated to improving the awareness of engineering in schools. Although individuals had to be nominated for the Award it was recognised as a team effort; so all those who had contributed to the success of the 134a process and their families were thanked by invitations to a day-long trip on a special train named *The Klea Adventurer* which travelled through some of the most magnificent scenery in the North of England.

The Fluorochemicals Business decided that its new HFC product range should be given its own tradename to distinguish it from its chlorine-containing Arcton™ CFC and HCFC range. A competition, with bottles of champagne as the prize, was held for present and retired employees in the North West of England. The prize was won by the wife of a retired publicity officer, Laurie Allen, who suggested 'Clea' because this was the name of her new granddaughter. Since an existing word could not be used as a trademark, the spelling was changed to 'KLEA' and was registered for worldwide use. Somewhat whimiscally, the attraction of 'KLEA' was further enhanced on learning that 'Clea' was related to the name 'Chloe' which is derived from the Greek word 'khloros', meaning green or verdant. (Not surprisingly 'khloros' is also the origin of the name 'chlorine', the only green element.) Subsequently the refrigerants business was officially named the 'Klea' Business, the lower case letters distinguishing it from the product trademark.

Preliminary work in the 1970s had indicated that 134a had only low solubility in the mineral oil lubricants normally used with the CFCs. Equipment manufacturers generally insisted upon high solubility to ensure that oil blown into the refrigeration circuit with the refrigerant vapour was washed back to the compressor. From the first contacts with the automobile companies and other refrigeration equipment manufacturers, it was clear that any supplier of 134a would also be expected to provide a compatible lubricant, at least for development; this was a significant departure from previous practice where the major lubricant companies had supplied oils quite independently of the refrigerant manufacturers. Like other companies, ICI realised that more polar synthetic lubricants were required, but in contrast to other refrigerant manufacturers was fortunate in having its own speciality oil business which could supply potential polyether or possibly ester lubricants candidates. Initial work concentrated on the polyethers, but these were shown to be generally incompatible with aluminium-based alloys commonly used in compressors. Through an extensive programme of work by S. (Stuart) Corr of the Klea Business Applications team and the ICI Lubricants Business, a range of polyol esters were developed for use with 134a and other HFCs. These are marketed under the trademark Emkarate™.

Although the first major target of the Klea Business was the development of 134a as a CFC-12 replacement, the increasing environmental concern in the late 80s over the effects of all chlorine-containing refrigerants meant that replacements would be required for the HCFCs, fluids which had originally been considered as part of the solution to ozone depletion. Early in 1990 Dick Powell was asked to prepare a critical review of likely candidates. It was clear that the properties of the low-boiling refrigerants HCFC-22 (CHF_2Cl) and azeotrope R502 ($CHF_2Cl + CF_3CF_2Cl$) could not be readily matched by single fluids or azeotropes, and that various blends of 125, 32, 134a and perhaps 143a would be necessary to meet customers' requirements. A number of people contributed to the subsequent development of blends, notably F. T. (Tom) Murphy, J. D. (Dave) Morrison, A. A. (Andy) Lindley and Stuart Corr. ICI's first success in introducing blends was the use of KLEA™ 407C (rather than R 22) by the major UK confectionary producer Cadbury in the air-conditioning system of their huge chocolate products warehouse, a building about a third of a kilometre long.

Although the major thrust of Klea's programme was the development of long-term replacements for the major CFC and HCFC refrigerants, the Business responded to an approach in 1990 by K. (Kasuo) Takamasa of Sanyo to develop CFC-free replacements for 502 and 503 in biomedical freezers to help his company meet its policy of being CFC-free by 1995. Dick Powell selected an azeotrope of 22 and 218 as a replacement for 502, and an azeotrope of 23 (CHF_3) and 116 (CF_3CF_3) for azeotrope R503 ($CHF_3 + CF_3Cl$). Takamasa and his team found that they were near 'drop-in' replacements, and within 2 years Sanyo was selling equipment containing these new refrigerant mixtures, which were given the product names TP5R2 and TP5R3.

Refrigerant KLEA^{TM} 32

(by Tony Ryan[10])

In 1992 the Klea Business decided that in addition to 134a, HFC-32 (CH_2F_2) would also be an important long-term refrigerant. Material for development and initial sales was produced by utilising the hydrogenation of CFC-12 over a supported Pd catalyst, a route not viewed as necessarily suitable for bulk production. A total of 85 potential synthesis routes were considered as candidates for the large-scale manufacture of HFC-32. Chlorine/fluorine exchange in dichloromethane ($CH_2Cl_2 + 2HF \rightarrow CH_2F_2 + 2HCl$) and the hydrodechlorination of HCFC-22 ($CHF_2Cl + H_2 \rightarrow CH_2F_2 + HCl$) were considered as the promising options; but it became clear that a process based upon the fluorodeoxygenation of formaldehyde might offer the lowest-cost and most environmentally-friendly option.

The hypothetical reaction of formaldehyde with hydrogen fluoride ($CH_2O + 2HF \rightarrow CH_2F_2 + H_2O$) is thermodynamically limited at all reasonable temperatures and pressures. Nevertheless, a process based upon this net stoicheiometry was selected as the preferred manufacturing route to HFC 32 in ICI, the thermodynamic restrictions being circumvented by operating in two distinct stages. It was known from the literature [40] that paraformaldehyde ($(CH_2O)_n$) dissolves into liquid anhydrous HF at room temperature. Although no attempts were made to isolate the compound, the product obtained when fluoroolefins reacted with this solution were suggested to be consistent with the formation of the 'hypothetical' bis(fluoromethyl) ether. It was found that an HF/$(CH_2O)_n$ equilibrium mixture contained bis(fluoromethyl) ether (CH_2FOCH_2F, BFME), poly(formaldehyde) and water [$2(CH_2O)_n + 2n HF \rightarrow n CH_2FOCH_2F + n H_2O$] from which BFME could be isolated by distillation. Passage of vaporized BFME or BFME/HF mixtures over a heated catalyst [41] resulted in the decomposition of BFME into difluoromethane and formaldehyde ($CH_2FOCH_2F \rightarrow CH_2F_2 + CH_2O$). A variety of Lewis acid type catalysts were shown to be effective.

[10]Tony Ryan graduated (BSc) in 1974 from Liverpool University, where he also gained his PhD (1977). He then joined the Solvents and Monomers Business of ICI's Mond Division where he studied $COCl_2$ and COF_2 as feedstocks for the production of chlorocarbons and fluorocarbons respectively. From 1987 to 1996 he was a member of the team investigating routes to HFCs. He left ICI in 1999 and is now senior research manager at Waterlink Sutcliffe Speakman Carbons Ltd, UK.

HFCs – more than just refrigerants

Like the CFCs, HFC-134a, originally developed as refrigerant, is finding applications in other areas. The Klea Business has developed propellant grades suitable for metered-dose inhalers used to dispense anti-asthma drugs. Leader of the technical support team for this application is an ex-Birmingham fluorine chemist, P. (Peter) P. Clayton, and the technical service manager is ex-UMIST fluorine chemist, T. (Tim) Noakes. The Klea Business has also explored other applications for 134a, *e.g.* as an extraction solvent for materials from biomass, especially flavours, fragrances and bioactives, an application first described in a patent assigned to Advanced Phytonics Ltd [42]. This work was extended in a patent assigned to ICI to blends of 134a with co-boiling additives such as dimethyl ether, hydrocarbons and methyl chloride, which modify the solvent properties of 134a [43]. HFC-134a also shows promise as an HPLC solvent [44]. Although ICI has become a major supplier of HFCs especially into the refrigerant area, it decided not to develop fluorinated replacements for its Arklone™ 113 solvent. Partially-fluorinated ethers were considered as potential 113 replacements [45], but, for commercial reasons, it was decided not to develop new fluorinated solvents and to concentrate on the environmentally-acceptable chlorinated solvents $CCl_2{=}CCl_2$, $CCl_2{=}CHCl$ and CH_2Cl_2.

Over the years CFCs and HCFCs have proved to be excellent feedstocks for a range of other fluoro-organic compounds. Could 134a also generate a whole new range of fluorinated products? Following initial work by Paul Coe on the thermal decomposition of 134a, it was shown that it can be readily converted over a chromia catalyst or in a steam cracker to trifluoroethylene ($CF_2{=}CHF$, TrFE; $CF_3CH_2F \rightarrow CF_2{=}CHF + HF$). The Klea Business has supported several programmes at UK universities to further the chemistry of 134a. Dick Chamber's group (Durham University) has investigated free radical, electrophilic and nucleophilic addition reactions of TrFE, which had previously been little explored compared to those of TFE or hexafluoropropene. Paul Coe and Jim Burdon (Birmingham University) generated $CF_2{=}CFLi$ from the reaction of 134a with BuLi and showed that it is an excellent reagent for the introduction of the $CF_2{=}CF$ group via attack on electrophilic species [46]. Jonathon Percy, also at Birmingham, has developed the chemistry of the analogous chlorotrifluoro-lithium ($CF_2{=}CClLi$) obtained by treating the 134a intermediate CF_3CH_2Cl with BuLi. Worthy of mention, although not funded by ICI, is the work of Alan Brisdon at UMIST: from $CF_2{=}CXLi$ ($X = F$, Cl) he has generated an extensive range of new compounds containing $CF_2{=}CX$ groups linked to a variety of metallic and nonmetallic elements.

The development of Klea™ 134a and the other HFCs within the time-scale set by the requirements of the Montreal Protocol was a tremendous technical challenge to the chemists and engineers involved over the past 12 years. Despite the problems inevitably thrown up by such a demanding programme the goals were met. From its humble beginnings in Widnes, ICI's Klea Business has grown to the point where there are plants in the UK, America and Japan. In mid-1999 ICI showed its confidence in future of Klea™ 134a by announcing capital investment to increase its world-wide capacity to 65 000 tpa.

Biologically-active fluorine compounds

Although fluoronitroaromatics were investigated as bacteriocides by ICI in the late 1930s, the first sustained investigation into the biocidal properties of fluorochemicals appears to have been undertaken by C. (Charles) W. Suckling in General Chemicals Division in collaboration with J. L. Charlton of Hawthorndale Laboratories, an early predecessor of ICI's (and thus Zeneca's) agrochemical business. In the late 1940s and early 1950s, Suckling prepared the halonitroaromatics **4-8** which Charlton assessed for fungicidal activity. The fluorinated compounds showed significantly greater activity than their chlorine, bromine or iodine analogues, with structures having F *ortho* to NO_2 being the most active.

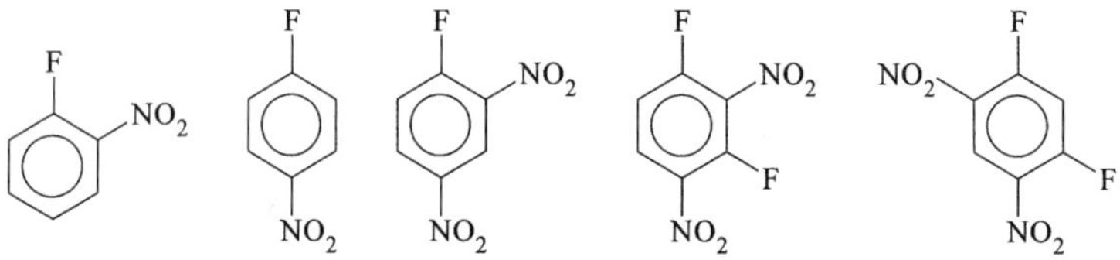

(4) **(5)** **(6)** **(7)** **(8)**

Suckling clearly established himself as ICI's first expert on the biological effects of fluorinated compounds and in 1956 produced a comprehensive literature review on the biological effects of fluorinated compounds which was circulated to the newly established 'ICI Pharmaceutical Division' and to 'ICI Plant Protection Ltd.'

Inhalation anaesthetics
(by J. S. Moilliet[11])

In 1950, apparently at the request of ICI's Dyestuffs Division at Blackley, Manchester, Suckling was given the task of searching for a new fluorine-containing inhalation anaesthetic. J. Raventos, a pharmacologist employed by Pharmaceuticals Division which was set up shortly after the programme started, was responsible for testing the candidates.

Suckling was strongly influenced by two previous pieces of work. The first was Ferguson's ideas on chemical potential as a general measure of biological potency, which Suckling was already using to rationalize the fungicidal activity of fluoronitro-aromatics. The second was a paper by Robbins [47], who had tested 42 'alkyl fluorides' as potential anaesthetics, compounds not synthesised specifically for this application but merely because they were available. By applying 'Ferguson's Principle' to Robbin's experimental results, Suckling concluded that '*...alkyl fluorides were structurally non-specific... The major differences in anaesthetic power remarked by Robbins such as the superiority of bromo over chloro compounds and of dihalo over monohalo compounds are clearly due to differences in vapour pressures and not to intrinsic anaesthetic properties.*' This was Suckling's key insight into the selection of candidate inhalation anaesthetics, *i.e.* specific chemical structure is not paramount.

Although he also considered the possibility of linking to chemical structure the adverse effects reported by Robbins and others, such as a tendency to cause convulsions, Suckling felt that there was insufficient evidence to draw any strong conclusions. Nor could he confidently predict the rates of induction and recovery. He relied on structure correlations only to select potentially stable compounds, favouring those containing CF_3 and CF_2

[11] After graduating from Birmingham University and carrying out research under Colin Tatlow, Jock Moilliet joined Mond Division in 1969 and worked on anaesthetics, fluoro-monomers and CFC replacements. He transferred in 1982 to ICI Organics Division where he worked on fluoroaromatics. In 1993 he retired from ICI and moved to his current post at (BNFL) F2 Chemicals Ltd to develop the use of F_2 in organic synthesis.

groups, but rejecting those containing iodine. Thus right from the start it was impossible to predict the structural requirements of a good anaesthetic, and throughout the research programme it was clear that the whole process was very much a matter of empiricism, even for such relatively simple molecules. Based on Robbins results, Suckling constructed a short list of potential candidates which included $CF_3CHClBr$, perhaps because it was structurally close to CF_3CHBr_2, which was considered by Robbins as worthy of further testing, and to CF_3CHCl_2, which Suckling considered Robbins had rejected prematurely. Bob McGinty, the synthetic chemist working for Suckling, produced $CF_3CHClBr$ relatively early in the synthetic programme and the compound proved to have most of the properties desired in a good anaesthetic. As expected, it was nonflammable and stable to soda-lime, which is used to remove exhaled carbon dioxide in the closed-cycle anaesthetic equipment. Anaesthesia was smooth, with rapid induction, and side effects were minimal. It was easy and cheap to produce and, most important of all, could be manufactured with a total impurity level of less than 20 ppm. Not surprisingly it quickly moved into clinical trials, but, surprisingly in comparison with modern practice, it was it only patented at this stage.

Sold by ICI under the trade mark 'Fluothane', $CF_3CHBrCl$ was so superior to previous anaesthetics that by the early sixties it had become the leading product in this area, and there was little incentive to continue searching for further agents. However, by the early 1970s competitive fluorinated inhalation anaesthetics had emerged, notably enflurane ($CHFClCF_2OCHF_2$), introduced by Airco (now Anaquest) in 1972. Fluothane^{TM} had been used successfully in many millions of operations but was not without side effects. For example, it had the property of lowering the blood pressure, which could be turned to advantage to reduce bleeding, but for a patient in shock could be dangerous. As a result, Pharmaceuticals Division restarted the inhalation anaesthetic programme in the 1970s after the lapse of more than a decade to find a successor to Fluothane^{TM}. The research group, based at Alderley Park, Cheshire, was led by W. G. (Glynne) M. Jones in a joint programme with the research chemists of General Chemicals Division, Runcorn, including the author of this section.

The new investigations concentrated more on fluorinated ethers than alkanes, partly because fewer of these been made and tested, and partly because it was hoped they would have less side effects than alkanes. It was by then a different world. New techniques meant that only one or two grams of material were needed for the preliminary screening, allowing wide use of preparative GC; and the availability of analytical GC meant that there was much more confidence in the purity of the test samples. Working with Colin Tatlow's Group at Birmingham University, extensive use was made of high-valency metal fluorides to make a whole range of fluorinated model compounds. This enabled many more compounds to be tested but some, including 2*H*/3*H*-hexafluoro-1,4-dioxan, which was a near miss, would have been a real challenge to make by any sensible route. Two compounds emerged as the most promising: $CHF_2CHFOCHF_2$ which was originally synthesised by CoF_3 fluorination of $C_2H_5OCH_3$ and its monochloride $CF_2ClCHFOCHF_2$. Neither was superior to Fluothane^{TM}, enflurane ($CHFClCF_2OCHF_2$) or another candidate, isoflurane ($CF_3CHClOCHF_2$), being developed by Airco. The field of inhalation anaesthetics was obviously far more competitive in the seventies than twenty years previously.

By the late 1970s ICI's inhalation anaesthetic programme was competing with a parallel programme to develop an injectable anaesthetic. It was clear that the Company would not introduce two competing products simultaneously. The decision was made to develop the non-fluorinated injectable agent 2,6-diisopropylphenol, which was introduced under the trademark Diprivan^{TM}, and to terminate the inhalation research programme. Unlike many drugs, effective anaesthetics have a very long commercial life. Fluothane^{TM} is still going strong after 45 years and shares the market with enflurane and isoflurane, and more recently with desflurane ($CF_3CHFOCHF_2$) and sevoflurane [$(CF_3)_2CHOCH_2F$]. Partly this arises from anaethetists staying with products which they thoroughly understand from experience. But also the manufacturing process for Fluothane^{TM} is less complex than those associated with its competitors, so it is able to compete effectively on price, an especially important consideration in the developing world.

Fluorine-containing pharmaceuticals
(by S. A. Lee[12] and P. N. Edwards[13] of AstraZeneca)

The replacement of hydrogen, or occasionally oxygen, by fluorine has been one of the most successful modifications used to produce 'designer molecules' in many spheres of activity. This has been used with enormous success in the design of pharmacologically-active molecules in essentially all therapeutic areas. The not-too-dissimilar molecular size of fluorine to covalently bound hydrogen, and near identity with oxygen, combined with chemical inertness and very powerful electron-withdrawing properties gives rise to a unique combination of very desirable properties in drug design. The following examples show how fluorine has been used in drug candidates by ICI and Zeneca Pharmaceuticals. In common with most pharmaceutical companies, many of these compounds did not reach the sales range – falling at one of the several hurdles that must be overcome before launch, such as adverse toxicological findings in animals and inadequate therapeutic efficacy in man. Nevertheless, a large amount of interesting and valuable chemistry has been investigated and some of it developed for use on a large scale. It is impossible to cover all the ICI/Zeneca compounds in this brief account, but it is hoped that the diversity of the challenges posed by the introduction of fluorine, and mostly overcome, will be conveyed by the following examples.

Histamine H2 blockers. – This class of compound has proved to be of great value in the treatment of duodenal ulcers and other gastric disorders which are exacerbated by an acidic environment. An ICI candidate H2 blocker was ICI 162846 (**9**), which, at the time it was put into development, was by far the most potent compound to reach that status – and we believe that still to be true today.

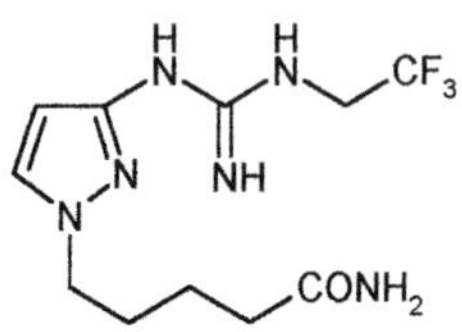

(**9**) ICI 162846

[12]Stan Lee graduated from Oxford University (1966) where he also gained his DPhil (1968). He then joined the Process Development Department of ICI's (now AstraZeneca's) Pharmaceuticals Business where he has remained throughout his career. He is now Senior Team Manager responsible for the development of manufacturing syntheses of pharmaceutical compounds. His interest in fluorine chemistry dates from 1981 when he investigated routes to a fluorine-containing β-blocker.

[13]Dr Phil Edwards was a Senior Research Associate (now retired) and worked in medicinal chemistry within Zeneca/ICI Pharmaceuticals throughout the whole of his career.

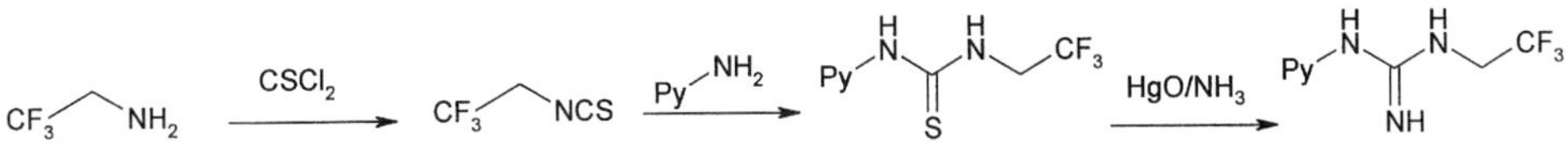

Py = 1-(4-carboxamidobutyl)pyrazol-3-yl

Scheme 22.2.

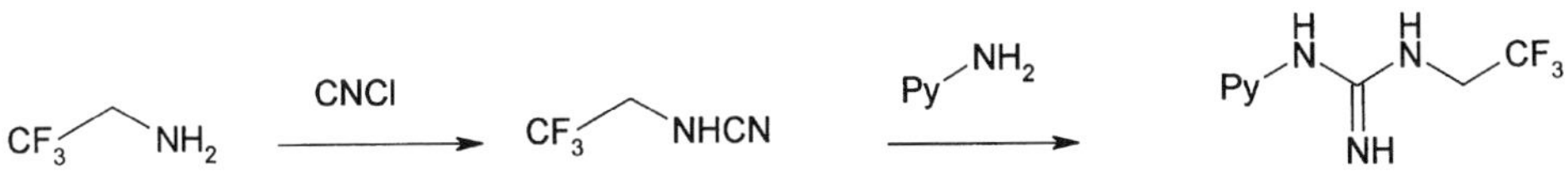

ICI 162846 (**9**)

Py = 1-(4-carboxamidobutyl)pyrazol-3-yl

Scheme 22.3.

Ts = *p*-toluenesulphonyl

Scheme 22.4.

The introduction of the 2,2,2-trifluoroethyl group into the guanidine moiety had many beneficial pharmacological and pharmokinetic advantages, but it gave rise to considerable large-scale production issues. Many fluorine-containing intermediates are only available in gram quantities from the chemical supply houses. The projected demands of ICI 162846 indicated that tonne quantities of intermediates would be required. A short-term solution was required since the large capital expense of a special large-scale plant could not be justified for the early development phase of a speculative pharmaceutical. ICI 162846 was originally synthesized in the research laboratory using trifluoroethyl isothiocyanate (derived from trifluoroethylamine and thiophosgene) (Scheme 22.2).

This synthesis was not suitable for large-scale manufacture due to environmental and toxicity issues and availability of the reagents used. The alternative synthesis which was devised (Scheme 22.3) took advantage of the expertise of manufacture and handling of cyanogen chloride on the ICI Grangemouth site, near Edinburgh. However, trifluoroethylamine was only available in laboratory quantities and initial experiments at ICI Runcorn showed that although it was possible to displace the chlorine in trifluoroethyl chloride (CF_3CH_2Cl, HCFC-133a) with ammonia, yields were modest and very high pressures were needed. Proceeding via this route would therefore have incurred the considerable delay involved in the construction of a specialized large-scale plant; so a viable synthesis from commercially- available trifluoroethanol (Scheme 22.4) was developed and used to produce trifluoroethylamine in bulk at ICI's Stevenston site in Scotland. ICI 162846 is an excellent example of the inter-divisional co-operation that was one of the strengths of ICI, and over 800 kg of pure ICI 162846 were produced before the development was terminated.

$$CCl_3CHO \xrightarrow{HF/Cr_2O_3\ \text{catalyst}} CF_3CHO$$

Scheme 22.5.

Thromboxane antagonists. – This series of 1,3-dioxans is based on the structure of naturally occurring thromboxane A2. By blocking the site of action of thromboxane A2 (**10**), a possible treatment for asthma, premature blood clotting and other cardiovascular disorders is potentially available. The dioxan ring is critical to the pharmacology: all activity is lost once the ring is opened. The stereospecific synthesis of this ring proved to be a major challenge, but was successfully achieved. Unfortunately the lead compound ICI 159651 (**11**; made from acetone) was shown to have a very short half-life of less than one minute (and hence activity) at pH 2. Introduction of the electron withdrawing CF_3 group at the labile position of the dioxan ring increased the half-life in acid to over 7 years. Synthesis of ICI 185282 (**12**) required the corresponding aldehyde trifluoroacetaldehyde (fluoral), which was only available in small quantities at the time. ICI Mond Division developed the vapour-phase fluorination of readily available chloral hydrate, shown in Scheme 22.5, and this was run in a continuously-operated small plant at Widnes to produce multi-kilogram quantities of fluoral hydrate.

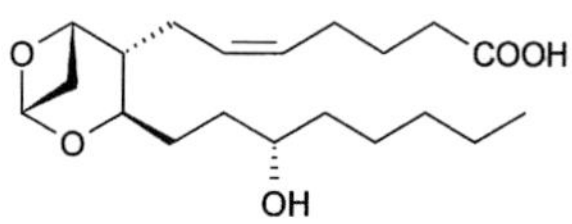

(10) **Thromboxane A2**

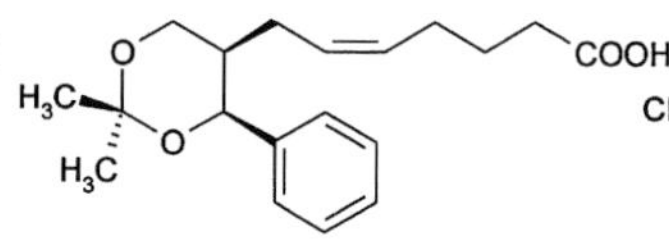

(11) **ICI 159651**

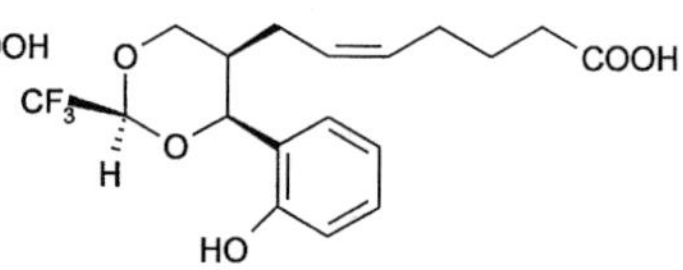

(12) **ICI 185282**

Elastase inhibitors. – ICI 200880 (**13**) was the first of a series of compounds which were being developed as potential treatments for bronchitis, emphysema and other related pulmonary disorders. A key part of the molecule is the trifluoromethylketone moiety which, like the related group in **12**, was derived from fluoral.

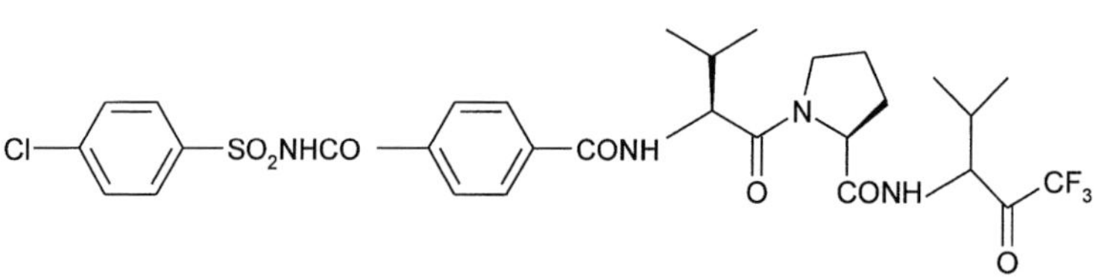

(13) ICI 200880

Since the Widnes fluoral plant was no longer available, however, a 'short-term' synthesis of fluoral hydrate was developed which involved the reduction of ethyl trifluoroacetate with aqueous sodium borohydride. Because this reduction is performed in an aqueous medium, the fluoral immediately forms the stable hydrate, thus preventing over-reduction to trifluoroethanol. This process does not require any special plant and has been successfully operated on the 100 kg scale.

Anti-androgens. – Prostate cancer is very prevalent in older men. One of the first forms of treatment was surgical removal of the testes. Use of anti-androgens can achieve beneficial effects without the need for surgery. One such compound is ICI 176884 (**14**) which is now marketed as the successful drug Casodex^{TM}. A key intermediate for the synthesis is 4-fluorothiophenol, made by Grangemouth Works starting from fluorobenzene, which is manufactured there on the kilotonne scale by diazotisation of aniline in liquid hydrogen fluoride.

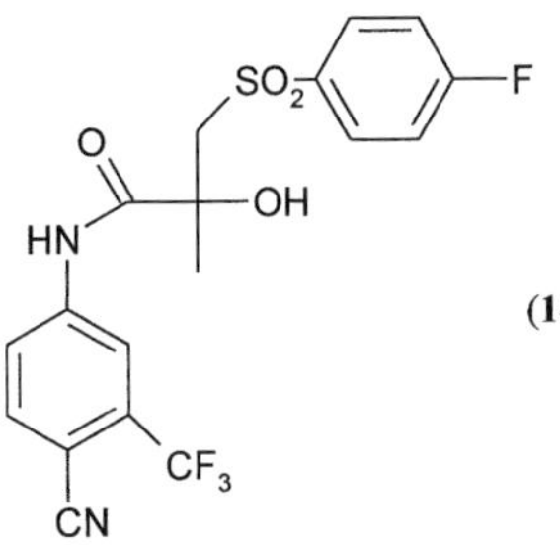

(**14**) ICI 176884

ICI Agrochemicals

Fluorine makes its mark (by R. Salmon[14], AstraZeneca Agrochemicals). – In the late sixties researchers in ICI Plant Protection (later to become ICI Agrochemicals) began to explore the chemistry of ring-fluorinated pyridines with the aim of finding agriculturally useful compounds. Its interest continued into the seventies when the herbicidal properties of 3,5-dichloro-2,6-difluoro-4-hydroxypyridine (**15**; Haloxydine), were discovered. Various fluorinated pyridines, made at ICI Mond Division, were dispatched to Jealott's Hill for further elaboration within Clive Tomlin's group, as well as to Roy Bowden's team at Runcorn. David Cartwright joined this project in the early seventies, and he used his knowledge of fluoropyridines in several of his later projects. In those early years the chemists at Jealott's Hill had to send samples to Runcorn for ^{19}F NMR analysis. It took about two weeks from sending off the compounds to getting the spectra back! Unfortunately, none of the compounds made at the time were suitable for commercialization.

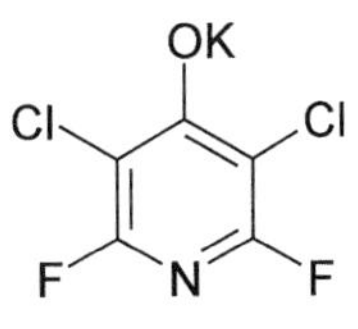

(**15**) Haloxydine

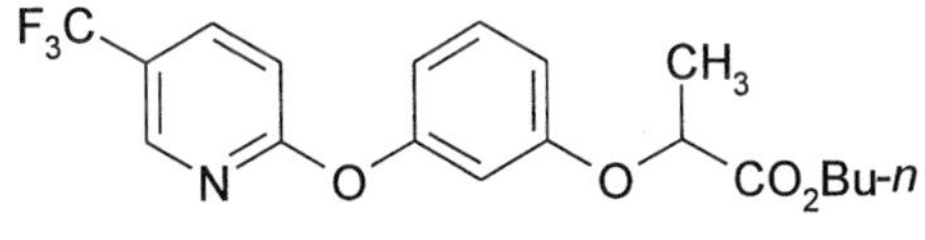

(**16**) Fluazifop-butyl

[14]After graduating from Bath University in 1970, Roger Salmon joined ICI Plant Protection Ltd at Jealott's Hill. He has worked on insecticide and herbicide discovery thoughout his career, specializing in agents containing fluorine.

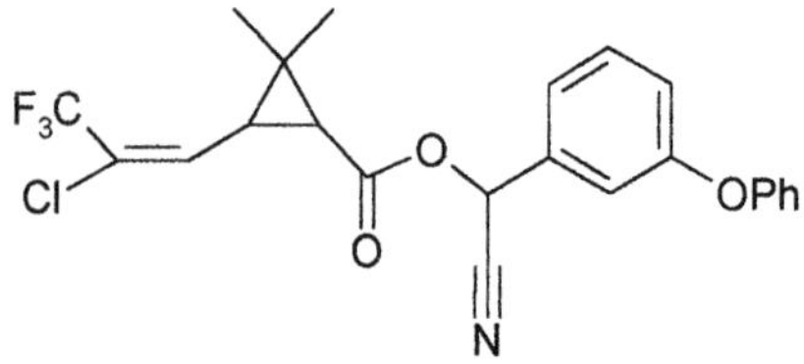

(**18**) Lambda-cyhalothrin

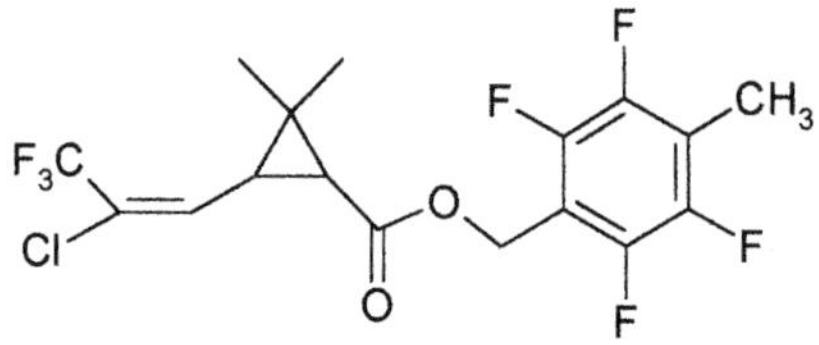

(**19**) Tefluthrin

However, the concept of incorporating fluorine into novel pesticides continued in the design of other pharmacophores within ICI Agrochemicals. An example is the development of the herbicide fluazifop-butyl, sold by under the tradename FUSILADE^{TM} (**16**), which contains a (trifluoromethyl)pyridine moiety. Some of the initial research work, undertaken by Dave Cartwright and his team, included synthesizing a range of substituted (trifluoromethyl)pyridine analogues for screening. Two of the initial routes used to make the (trifluoromethyl)pyridines involved treating picolinic acid derivatives with sulfur tetrafluoride and heating (trichloromethyl)pyridines with hydrogen fluoride in pressure reactors. Starting materials made at Jealott's Hill were transported to the pressure laboratories at Winnington for A. (Alan) G. Breeze to fluorinate, the products being returned to Jealott's Hill for subsequent elaboration.

Meanwhile, further fluorine chemistry was underway at Jealott's Hill in Roger Huff's team orientated towards making some highly-potent, fluorine-containing pyrethroid type insecticides that included lambda-cyhalothrin (**18**). Synthesis and screening of the various isomers soon identified a commercially-viable product and subsequent work in Mike Robson's team led to the product being marketed as KARATE^{TM}. Continued research in this area by Nazim Punja's team led to the discovery of the first soil-acting pyrethroid, tefluthrin (**19**). Work on new pyrethroid insecticides continued up to the nineties; final projects in this area included the work of Mike Bushell's group on insectides for use in rice and that of Alan Whittle, Roger Salmon and their teams, on acaricidal pyrethroids. Ted McDonald, who was the insecticide chemistry manager for this period, provided much support and encouragement for the development of fluorine-containing agents. In the late eighties and early nineties the considerable commitment to the development of fluorine-containing compounds led to the construction of a high-pressure lab at Jealott's Hill operated by Alf Williams, and to the installation of facilities for producing fluorinated intermediates on the kilogram scale managed by Peter Cleare in the process development group.

Over the years, the chemistry department at Jealott's Hill has grown in size and many newcomers – too numerous to mention – have been involved in some aspect of fluorine chemistry within their research responsibilities. Through its collaborations with universities, via chemists such as Patrick Crowley, the Company has continued to encourage research into the development of new methods for incorporating fluorine into molecules.

Dave Cartwright has highlighted the major new structural classes of pesticides containing fluorine that emerged worldwide during the period 1978-93 in a review [49] which included the major contributions made by the AstraZeneca Agrochemical Business (formerly ICI Agrochemicals before the demerger of Zeneca). There remains a strong interest in using fluorine to optimize the properties of biologically-active molecules, as evident from the patent applications filed by ICI (now Zeneca) Agrochemicals.

Haloxydine: a candidate selective herbicide which nearly made it (by R. D. Bowden[15]). – Haloxydine (**15**) was first synthesized by one of Musgrave's postdoctoral fellows at Durham University, J. (John) Hutchinson[8], who subsequently joined ICI Mond Division.[16] Haloxydine was a 'natural' for ICI Mond Division: the chemists at Runcorn had 80 years of experience in chlorination, which was highly relevant to the vapour-phase chlorination of pyridine to pentachloropyridine, a key intermediate (Scheme 22.6). Mond Division already intended to become a pyridine manufacturer to meet the growing demand for ICI's existing herbicide Paraquat^{TM}, so a second in-house requirement would have been welcome, but in fact the proposed plant was never built. Unlike Paraquat^{TM}, which is a total herbicide and is virtually un-translocated, Haloxydine showed species selectivity and movement within the plant. It was effective at a rate as low as 60 g per ha and was rapidly biodegraded in the soil. Development was clearly attractive.

Plant Protection Ltd, an ICI subsidiary which had no manufacturing arm, and Mond Division, where Paraquat^{TM} was already being made, were keen to become the producers. Teams of biologists at Jealott's Hill and of chemists and engineers at Runcorn were built up; while the biologists tried to understand the huge discrepancies between their greenhouse and early field trial results, the chemists were attempting to make them enough haloxydine. In 1969, laboratory reactors became bigger and bigger, culminating in a chlorination unit built in a walk-in fume cupboard and notionally capable of producing 15 tpa. By the end of 1969 a properly engineered 20 tpa unit incorporating a 3-stage process (Scheme 22.6) had been built at Widnes. At last enough material was being supplied to Plant Protection for them to carry out field trials around the world and to eliminate the uncertainties.

By Feb 1970 plans were being drawn for a 100 tpa haloxydine plant due to become operational in time for the 1972 growing season. The inevitable scale-up problems were solved, not least in the final step where reaction using alkali (KOH) also generated 20% of the product resulting from nucleophilic hydroxylation at the 2-position; use of KOAc solved the problem, giving the 4-isomer with 99% selectivity and generating volatile AcF as the by-product. Amusingly, the American patent examiner initially refused the application on the grounds that the reaction was 'unprecedented in organic chemistry'!

But it was too late. Haloxydine was being targeted as a pre-emergent herbicide in cereals, but with a wide range of other applications being explored. The margin between application rate and the rate at which it caused crop damage was always narrow, but eventually the real problem emerged: haloxydine was simply too water-soluble. Its effectiveness

[15]Roy Bowden graduated from Imperial College, London, and after completing his PhD at Queen Mary College, London, joined Mond Division in 1966. Most of his career was spent as a development chemist, and many of the products which he was involved with included fluorine. He retired from ICI in 1993 and is now R&D manager for (BNFL) F_2 Chemicals Ltd.

[16]Through the Department of Scientific and Industrial Research (DSIR), a British Government Agency which funded the fellowship, the route to compound **15** was patented and the compound offered to industry for testing. The ICI subsidiary Plant Protection Ltd. found it sufficiently promising to take out a licence.

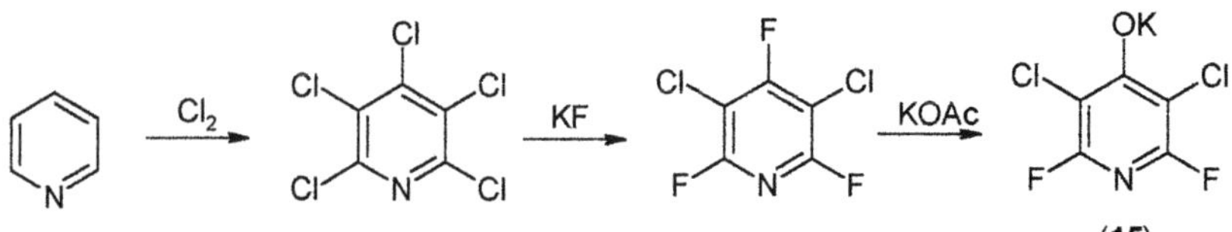

(15)

Scheme 22.6.

depended on the rainfall pattern after application, so in a dry climate it gave impressive results; but in areas of variable rainfall, it could be washed down to the same level at which the crop had been drilled. Australian farmers liked it, but British farmers didn't; not surprisingly, senior management decided to abandon the project despite the considerable investment which had been made. But at least somebody made money out of haloxydine: John Hutchinson, Ken Musgrave and the University of Durham received some royalty payments from ICI for several years.

2-Chloro-5-(trifluoromethyl)pyridine: a triumph for chlorofluorination (by D. Boniface[17]). – Commercialization of FUSILADE™ (**16**) required the development of a process to 2-chloro-5-(trifluoromethyl)pyridine (CTF; **20**), a formidable challenge which was undertaken by General Chemicals Division Research Group, Runcorn. In an early route to CTF, devised by Reilly Tar & Chemical Corporation, 5-carboxy-2-pyridone was treated sequentially with SF_4 and PCl_3 [50] – a selective, high-yield route, but not viable for large-scale manufacture.

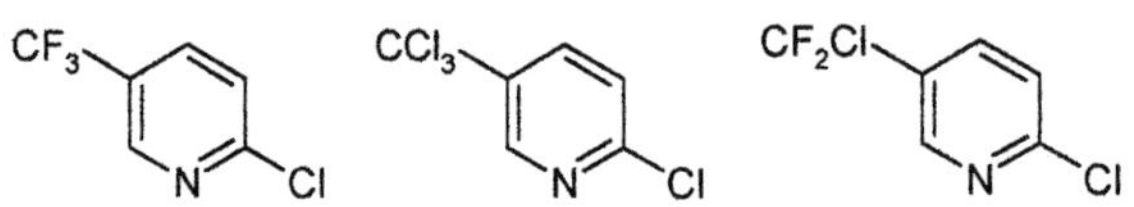

(**20**) CTF (**21**) CCMP (**22**) CDF

2-Chloro-5-(trichlororomethyl)pyridine (CCMP; **21**) was the obvious precursor to CTF; but it was a difficult compound to synthesise cleanly from readily available feedstocks, although it could be converted to CTF by a various fluorinating agents, such as KF/sulpholane [51], SbF_3, and $SbCl_5$/HF. For a viable process it was clearly necessary to start from commercially-available 3-picoline (Scheme 22.7). Ishihara Sangyo Kaisha Ltd patented a liquid-phase, UV-initiated chlorination of 3-picoline in refluxing carbon tetrachloride, to produce CCMP, followed by fluorination with antimony pentachloride in liquid hydrogen fluoride to produce CTF [52]. However, a combination of vapour-phase chlorination and liquid-phase fluorination seemed to offer a possible approach. Chlorination of 3-picoline was highly exothermic and vapour-phase chlorination was difficult to tame, but control of temperature and the chlorine/3-picoline feed ratio proved to be the key; CCMP

[17]Dave Bonniface graduated (BSc) in 1964 from the University of Nottingham, where he also gained his PhD in 1967 before taking up a postdoctoral fellowship in the US at Michigan State University. He joined ICI Dyestuffs Division in 1968. He was involved in the preliminary techno-commercial evaluation of routes to CFC replacements in 1976 and in 1982 became senior chemist for 2-chloro-5-(trifluoromethyl)pyridine development. Subsequently he joined the group developing CFC replacements in the late 80s and became team leader for 134a process development. He retired from ICI at the end of 1998.

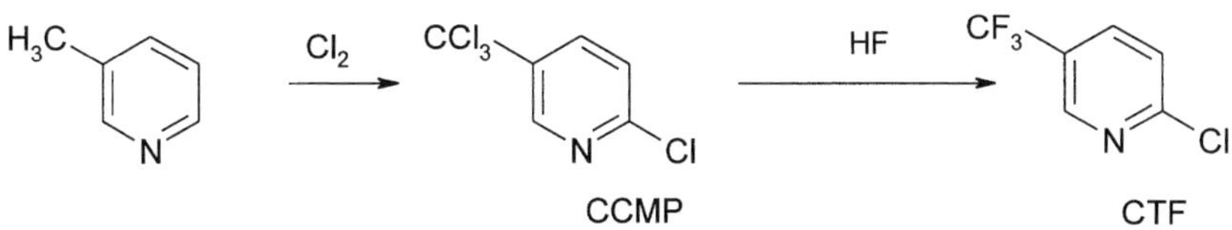

Scheme 22.7.

yields approaching 50% were ultimately achieved, although purification of the fairly involatile product accounted for significant further loss. Hydrofluorination of CCMP with HF in the liquid phase proved facile and was operated on a modest scale to produce material for regulatory and field trial requirements. Unfortunately, the process was perceived to have too high a capital cost to operate at full manufacturing scale; additionally, the efficiency of the chlorination stage was barely satisfactory and the product (CCMP) was found to be an extremely strong skin-irritant.

These considerations focused development on novel combined 'chlorofluorination' processes for the direct conversion of 3-picoline to CTF (**20**); both liquid- and vapour-phase systems were investigated. G. (Graham) Whittaker and his team at Runcorn reacted 3-picoline with chlorine in liquid anhydrous HF, but obtained poor yields of CTF, the major product being 2-chloro-5-(chlorodifluoromethyl)pyridine (CDF; **22**). Concerns about materials corrosion rates in this reaction mixture prevented more forcing conditions being applied [53]. Vapour-phase reactions were investigated by ICI's Japanese partner, Ishihara Sangyo Kaisha, who carried out simultaneous chlorination and fluorination of 3-picoline to CTF in a fluidized bed of aluminium fluoride promoted with transition metals [54]. In the early 1980s, ICI built a full-scale plant based on this technology which is still operating in the UK.

Scouting work on a number of other substrates (*e.g.* methylthiophenes, lutidines, xylenes) was carried out which demonstrated the applicability of this chlorofluorination technology to a range of trifluoromethyl-substituted aromatics. The real achievement of the CTF process was the successful commercial application of chlorofluorination to the direct conversion of hydrocarbons to fluorocarbons. This idea had previously been claimed in the patent literature, but this was first time it had been operated on a production scale.

Aromatic fluorine chemistry[18]

(by Stewart Korn[19] and Dave Moody[20])

From the 1930s onwards ICI periodically assessed the fluoroaromatics area, looking for potential products, but its commercial entry into this area only occurred in the early 1980s, via the Fine Chemicals Manufacturing Organisation (FCMO) based at Black-

[18]In July 1999 AstraZeneca sold its fine chemicals business, including fluoroaromatics, to Avecia.

[19]Stewart Korn graduated from Glasgow University in 1969 and gained his PhD from McMaster University, Canada, in 1973. After a postdoctoral year at Ohio State University, he joined ICI's Organics Division, Blackley (Manchester), in 1974.

[20]Dave Moody graduated from the University of St Andrews, where he also gained his PhD before joining ISC (Avonmouth, Bristol) in 1988. After 3 years he moved to ICI where he developed expertise in the production of fluoroaromatics, an area in which he continues to work.

ley, Manchester. The first compound to be manufactured was 2-fluorobenzoyl chloride for use by Plant Protection Division. The process, developed by Jock Moilliet and Mike Howarth, involved the handling and reaction of hydrogen fluoride for which a new facility was established at Grangemouth Works in Scotland by Terry Hollis and Stuart Strathdee. The demand for fluoroaromatics, especially at the tonne scale, was increasing both inside and outside the Company. ICI's Advanced Materials Group, Wilton, were forecasting large tonnage requirement for BDF ('benzophenone difluoride' or more correctly 4,4′-difluorobenzophenone) the key intermediate for its PEEK (*poly*-ether-ether-ketone) high-performance speciality polymer. Initially the Company considered investing significantly in the manufacture of 4,4′-difluorodiphenylmethane (DFDPM) via HF-diazotisation of the corresponding and readily-available diamino compound, and the process was developed to the 1 tpa pilot scale at Huddersfield works by Howarth, Moilliet and R. (Roger) Findon. Subsequently, however, DFDPM was replaced by fluorobenzene as the main target to meet demands both inside and outside ICI, so a project was initiated to develop a 1500 tpa fluorobenzene plant based on continuous diazotisation of aniline in liquid HF and immediate decomposition of the diazonium fluoride salt. The kinetic data for this difficult system, determined with great skill by J. (John) Atherton and J. (John) Umbers, was used to design a semi-technical unit based on a cascade of 1 litre pots, run by Kevin Williams. The full size fluorobenzene plant, engineered by Keith Carpenter, opened in summer 1989 at Grangemouth.

ICI's Pharmaceutical and Agrochemical Businesses were generating an ever-increasing number of fluorine-containing bioactive products, which provided potential opportunities for the Fluoroaromatics Business. Some grew into major products, *e.g.* 2-chloro-4′-fluoroacetophenone (CFAP), developed by Ian Hodgkinson and manufactured at Huddersfield Works on the hundreds of tonnes scale from 1992. The fluorobenzene plant was extended to 2000 tpa in the early 1990s. To realize the full potential of ICI's fluorobenzene technology, a study was undertaken by David Moody to assess whether the plant could be adapted to produce other fluoroaromatics. As a result the plant was successfully modified in 1995 at low capital cost to accommodate a process for 3-fluorotoluene, which is now a major product (300 tpa). As the result of the last 20 years efforts, Avecia (ex-AstraZeneca) is now one of the world's leading producers of fluoroaromatics.

Project 'MERLIN'

(by Phil Gamlen[21] and Harold Fielding[22])

ICI Mond Division mounted a major programme, codenamed 'MERLIN', in 1984 aimed at finding a way to break the DuPont and Asahi Chemical stranglehold on the supply

[21]Phil Gamlen gained his DPhil from the University of Oxford in 1975 and joined ICI Mond Division, where he worked initially on electrochemical synthesis of organic compounds, including some fluorocarbons. He has had a variety of research management, product development and business roles. His present post is Head of ICI's Science and Technology Policy and Strategy Unit.

[22]Harold Fielding graduated from the University of Liverpool in 1954 where he also gained his PhD (1957) before joining ICI Alkali Division at Winnington. He is particularly noted for his contributions to the chemistry of TFE oligomers. The development of fluorinated membranes was his final research project in fluorine chemistry before his retirement from ICI in 1991. He is still actively involved in chemistry through membership of local committees of The Royal Society of Chemistry and The Society of Chemical Industry.

of perfluoro-ionomer membranes that lie at the heart of the modern chloralkali electrolyser. A smaller programme, mounted some some 4 or 5 years before in conjunction with ICI Australia, had explored radiation grafting of fluorinated monomers on to fluoropolymers as a possible technology for membrane manufacture, but had not been successful. By contrast, the MERLIN project started from the premise that success would require advances in several fields in parallel. Accordingly, a team was assembled by the project manager, Phil Gamlen, with specialists to investigate the underlying mechanism of ion transport in perfluoro-ionomer membranes, the manufacture of perfluoropolymer films and their conversion into ion-selective membranes. A key aspect of the work was synthesis of new perfluoromonomers that fell outside the scope of existing patents. Two important contributors to this group were Ian Shirley and Harold Fielding.

In order to get the various elements of the project up and running in parallel, much work was done in synthesising and processing the existing known monomers and polymers. This ensured that the necessary polymerization and membrane fabrication expertise was ready when the synthesis programme started to deliver sufficient quantities of novel monomers. An important breakthrough was the realization that incorporation of rings (perfluorocarbocyclic or cyclic perfluoroether) in the co-monomer would give a less-mobile side chain, would provide scope for incorporating multiple acid groupings, and would have a variable geometry depending on the substitution pattern of the precursor. Best of all, the materials would be patentable [55].

The established chemistry for the manufacture of the monomer for copolymerization with TFE *en route* to a perfluorinated chloralkali membrane, such as Nafion^{TM}, is shown in Scheme 22.8.

The target for the ICI programme was to produce substituted perfluorocyclohexane monomers such as **24** and **25**. A number of routes were explored and one of the most productive is exemplified in Scheme 22.9. The monomers produced were successfully copolymerized with TFE; indeed terpolymers with TFE and the more usual PVES (**23**) co-monomer were also made and found to afford much better control of the molecular and equivalent weights. Polymers were processed into membranes and tested not only in brine electrolysis but also in fuel cell applications. Interestingly the novel materials were surprisingly effective when used in fuel cells. The physical and chemical properties of the new materials were extensively characterized using the battery of measurement skills that had been developed.

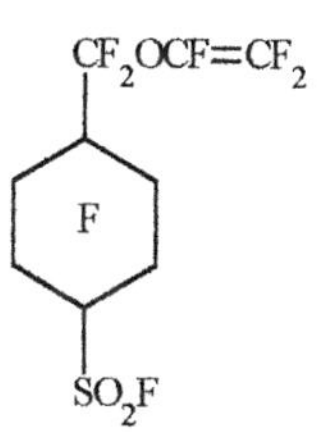

(24)

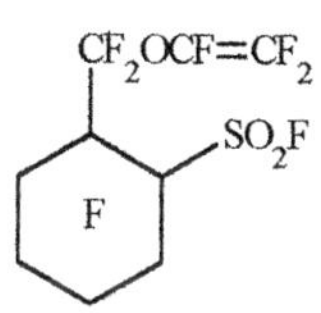

(25)

In the end, however, Mond Division did not build their own perfluoro-ionomer membrane plant, but the demonstration of a patent-protected capability enabled a much better commercial deal to be negotiated.

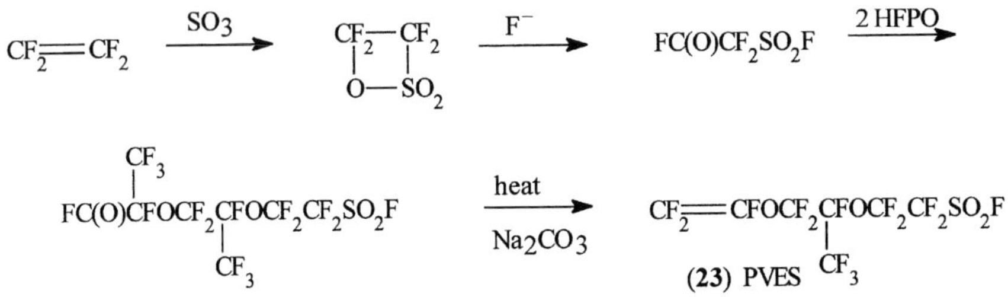

HFPO = hexafluoropropene oxide

PVES = 'perfluorovinylethersulfonyl fluoride'

Scheme 22.8.

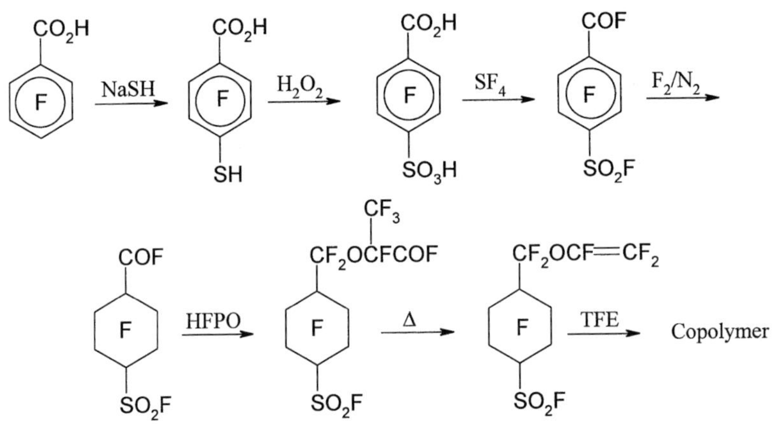

Scheme 22.9.

UK university contacts

ICI has been fortunate to draw upon the extensive UK academic expertise in fluorine chemistry over the past 50 years, both through invited lectures, research projects and consultancies. The first consultants in the 1940s were Harry Emeléus, who served regularly until the late 1960s, and Maurice Stacey, about whom Leech had misgivings because he also advised ICI's UK competitor, Imperial Smelting Corp. Bob Haszeldine consulted occasionally in the 1950s, while David Sharp and Ron Nyholm were regular consultants from the mid-1960s to the mid-70s. Over the past 25 years the Company has appreciated the continuing input of Dick Chambers, John Holloway, Colin Tatlow, John Winfield, Paul Coe, Jim Burdon, Ray Plevey, James Clark and Eric Banks. Eric's contribution to the debate about the impact of fluorinated fluids on the environment has certainly made him known, at least by reputation, even to the senior managers of the Klea Business. The contributions of Mike Dove and Ray Peacock are also acknowledged.

The growing influence of the younger generation of academic fluorine chemists, notably Eric Hope (Leicester), Jonathon Percy (Birmingham), David O'Hagan (Durham), Alan Brisdon (UMIST) and Graham Sandford (Durham), has also been appreciated.

Postscript

The past achievements of ICI fluorine chemists will continue to contribute to future wealth generation, if not in ICI itself then in other organisations. BNFL has continued the development of the original ICI F_2 production technology whose genesis can be traced back as far as the mid-1930s. Through their subsidiary F2 Chemicals Ltd., BNFL have applied F_2 to the production of fluoro-organic compounds – a development which Leech and Rudge would have surely approved, or perhaps envied. Four ex-ICI fluorine chemists, who took early retirement, have contributed significantly to that work. On demerger from ICI in 1993, Zeneca directly inherited fluoro-aromatics, fluorine-containing drugs (including Fluothane[23]), and agrochemicals, which continue to be profitable. ICI's achievement in developing highly-efficient chromia catalysts for Cl/F exchange, probably its greatest contribution to fluorination chemistry, is a key part of the Klea Business. Likewise HCFC-22 steam cracking technology developed by ICI is still very important in TFE production.

As ICI has clearly demonstrated over the past 7 years, large companies change their identities by buying or selling businesses, or merging or demerging, in response to commercial pressures. There can be no sentiment about such changes – the only relevant criterion is whether such changes enhance the wealth-creating ability of the Company. The 'past' ICI in which fluorine chemistry flourished is very different from the 'present' ICI. Although to talk about the value of ICI's – or any other company's – contribution to fluorine chemistry is a useful convenience in organising a book, the tribute really belongs to the individuals who carried out the work and made the inventions, and to those managers who championed projects and created the conditions under which technical progress could be made.

Acknowledgements

I wish to thank ICI plc and Zeneca plc for permission to publish the material in this chapter. I appreciated having access to the ICI report archive which provided much of the material for the early advances. I acknowledge the help of friends from ICI Halochemical, Fluoropolymer and Klea Businesses, and I would particularly like to thank my co-authors who I badgered into producing contributions. A special word of thanks is due to Lorraine Ellams and Sharon Jones who had the dusty task of finding old ICI reports and files.

References

1 W. J. Reader, *Imperial Chemical Industries: A History: The First Quarter Century, 1926-52*, Oxford University Press, London, 1975.
2 F. D. Leicester, *Br. Pat. 468,447* (1937) (to ICI).
3 A. F. Benning and W. S. Calcott, *US Pat. 2034458* (1936) (to DuPont).
4 A. J. Rudge, *The Manufacture and Uses of Fluorine and its Compounds*, Oxford University Press, London, 1962.
5 G. H. Cady, D. A. Rogers and C. Carlson, *J. Ind. Eng. Chem.*, *34* (1942) 443-8.

[23]Although Fluothane™ is a Zeneca product, its manufacture is contracted to the Klea Business.

6 H. R. Leech, reminiscences taped by R. E. Banks and K. R. Farrar, (UMIST) on 8th January 1980.

7 H. R. Leech, in *Mellor's Comprehensive Treatise on Inorganic and Theoretical Chemistry*, Suppl. I, Vol. II, Longmans Green, London, 1956, 15 et seq.

8 A. J. Rudge, *Brit. 710523* (1954) (to ICI).

9 J. Chapman and D. C. Horner, *Br. Pat. 976,883* (1964) (to ICI).

10 L. Rowley, G. Webb, J. M. Winfield and A. McCulloch, *Applied Catalysis*, *52* (1989) 69-80.

11 L. Rowley, J. Thomson, G. Webb, J. M. Winfield and A. McCulloch, *Applied Catalysis A: General*, *79* (1991) 89-103.

12 *Br. Pat. 713522* (1954) (to the Firestone Tire and Rubber Company).

13 H. L. Roberts, *Quart. Revs.*, *15* (1961) 30.

14 H. L. Roberts, *Br. Pat. 875209* (Appln.1959) (to ICI).

15 H. L. Roberts, *Br. Pat. 877961* (Appln. 1958) (to ICI).

16 J. R. Case, *Br. Pat. 931600* (Appln. 1960) (to ICI).

17 H. C. Fielding, *Br. Pat.* 1071323 (Appln. 1965) (to ICI).

18 H. C. Fielding, L. P. Gallimore, H. L. Roberts and B. Tittle, *J. Chem. Soc. C.*, (1966) 2142.

19 H. C. Fielding, *Br. Pat. 1198476* (Appln. 1967) (to ICI).

20 H. C. Fielding, *Br. Pat. 1076357* (Appln. 1965) (to ICI).

21 H. C. Fielding, *Br. Pat. 1133492* (Appln. 1966) (to ICI).

22 H. C. Fielding and A. J. Rudge, *Br. Pat. 1082127* (Appln. 1965) (to ICI).

23 S. Bartlett, R. D. Chambers, J. R. Kirk, A. E. Lindley, H. C. Fielding and R. L. Powell, *J. Chem. Soc. Perkin Trans. 1*, (1983) 1235.

24 P. L. Coe, A. Sellars, J. C. Tatlow, H. C. Fielding and G. Whittaker, *J. Fluorine Chem.*, *32* (1986) 151.

25 P. L. Coe and N. C. Ray, *J. Fluorine Chem.*, *53* (1991) 15.

26 J. Hutchinson, *Fette, Seifen, Anstrichmittel*, *76* (1974) 158.

27 T. Midgley, Jr., *Ind. Eng. Chem.*, *29* (1937) 239-244.

28 J. E. Lovelock, *Nature*, *230* (1971) 379.

29 M. J. Molina and F. S. Rowland, *Nature*, *249* (1974) 810-812.

30 S. E. Potter, *US Pat. 4158675* (1979) (to ICI).

31 J. I. Darragh, *Brit. Pat.* 1578933 (1980) (to ICI).

32 J. C. Farman, B. G. Gardiner and J. D. Shanklin, *Nature*, *315* (1985) 207-210.

33 *Montreal Protocol on Substances that Deplete the Ozone Layer*, United Nations Environment Programme, Nairobi, Kenya, 1997 (complete current text at *www.unep.ch/ozone/home/html*).

34 J. H. Steven and A. McCulloch, *IChemE. Env. Prot. Bull.*, *3* (1989) 3-7.

35 *Testing to Extremes*, PAFT, Washington, 1997.

36 *Proceedings of Workshop on Atmospheric Degradation of HCFCs and HFCs*, NASA, NOAA & AFEAS, Washington, 1995.

37 J.-C. Boutonnet *et al.*, *Human and Ecological Risk Assessment*, *5* (1999) 59.

38 J. D. Scott and R. A. Steven, *US Pat. 5243107* (1993); *US Pat. 5382772* (1995); *US Pat. 5395996* (1995); *US Pat. 5744658* (1998) (all to ICI).

39 J. D. Scott and M. J. Watson, *US Pat. 5281568* (1993) (to ICI).

40 V. Weinmayr, *J. Org. Chem.*, *28* (1963) 492.

41 L. Burgess, J. L. Butcher and T. A. Ryan, *WO 9312057* (1991) (to ICI).

42 P. Wilde, *US Pat.* 5512285 (1996) (to Advanced Phytonics).

43 T. N. Noakes, R. L. Powell and P. Wilde, *Brit. Pat. 2288552* (1995) (to ICI).

44 A. J. Handley, R. D. Clarke and R. L. Powell, *US Pat. 5824225* (1995) (to ICI).

45 N. Winterton and D. McBeth, *Eur. Pat 0450855* (1991) (to ICI).

46 J. Burdon, P. L. Coe, I. B. Haslock and R. L. Powell, *Chem. Comm.*, (1996) 49-50; *J. Fluorine Chem.*, *85* (1996) 151-153.

47 B. H. Robbins, *J. Pharmacol.*, *86* (1946) 197.

48 C. W. Suckling and J. Raventos, *Brit. Pat. 767779* (1957) (to ICI).

49 D. Cartwright, in R. E. Banks, B. E. Smart and J. C. Tatlow (eds.), *Organofluorine Chemistry: Principles and Commercial Applications*, Plenum Press, New York, 1994, 237-262.

50 *US Pat. 4,249,009* (1981) (to Reilly Tar).

51 *US Pat. 4,745,193* (1988) (to ICI).

52 R. Nishima, K. Fujikawa, T. Haga and K. Hase, *Jap. Pat. 80 22617* (1980) (to Ishihara Sangyo Kaisha Ltd.).

53 G. Whittaker, *Eur. Pat. Appln. 14033* (1980) (to ICI).
54 R. Nishiyama, K. Kujikawa, I. Yokomichi, Y. Tsuju and S. Nishimura, *Ger. Pat. 3008081* (1980) (to Ishihara Sangyo Kaisha Ltd.).
55 H. C. Fielding, P. H. Gamlen and I. M. Shirley, *EP 0331321* (6 September 1989) (to ICI).

BIOGRAPHIC NOTE

Dick Powell graduated from King's College, University of London, in 1965 and carried out research into organophosphorus chemistry at the same college, gaining a PhD in 1968. After 2 years postdoctoral work in the US at Rutgers University, New Jersey, where he also studied phosphorus chemistry, he joined Harold Fielding's section in ICI's Mond Division in 1970. He worked on fluorine-based projects throughout his career in ICI and was especially concerned with the development of CFC replacements. In 1985 he was appointed a Division Research Associate with special responsibility for fluorine chemistry, a senior scientific ladder post. He retired from ICI at the end of 1998 and is now a visiting professor at UMIST.

Dick Powell

Chapter 23

FLUOROCARBON EMULSIONS – DESIGNING AN EFFICIENT SHUTTLE SERVICE FOR THE RESPIRATORY GASES – THE SO-CALLED 'BLOOD SUBSTITUTES'

JEAN G. RIESS[1]

MRI Institute, Medical Center, University of California at San Diego, 410 Dickinson St, San Diego, CA 92103-1990, USA

Tu es mon sang, ma vie et ma lumière.
Pierre de Ronsard (1584)

Introduction

A fluorocarbon[2] emulsion, *Oxygent*™ (Alliance Pharmaceutical Corp., San Diego, CA), is now in the final stages of clinical evaluation for use as a therapeutic oxygen delivery system – a so-called 'blood substitute'. Thirty odd years have passed since Leland Clark dropped his mouse in a beaker of an oxygenated liquid fluorocarbon and Bob Geyer for the first time infused a fluorocarbon emulsion into an animal. Thirty years of research and development, which have brought us from a position of total ignorance about the physiological effects of these compounds to one of sufficient mastery to accept the wild idea that fluorocarbons could flow through our veins. As is so often the case, product development was much more complicated than anticipated, given the need for a solid basis of fundamental knowledge about the 'biology' of fluorocarbons, the need for a proper understanding and accurate definition of the product's utility, and the difficulties in getting a new concept and unconventional type of drug through all the usual and non-usual clinical and regulatory hurdles. Add to this some psychological barriers, perhaps, to accepting such a non-natural means of delivering oxygen to tissues, some cultural reluctance to having the all-sacred red and emblematic symbol of life, love and death replaced with some milky-white synthetic stuff[3].

Man has been in pursuit of a surrogate for blood for at least a century. In the last twenty years, the AIDS crisis emerged as a new driving force behind this effort. It then became more generally appreciated that blood transfusion was not without risks and side

[1]The author is also a member of the Board of Directors of Alliance Pharmaceutical Corp. in San Diego; however, the opinions expressed here are solely his and not necessarily those of the Corporation.

[2]Nowadays the term fluorocarbon, once reserved for compounds containing only carbon and fluorine, is commonly used when discussing any type of *highly* fluorinated organic compound. This has led to the introduction of the specific term perfluorocompound (PFC) to describe *fully* fluorinated organic hydrocarbons. The fluorocarbons discussed here are PFCs and derivatives (group descriptor perfluorochemicals) rationally derived from them by limited substitution of fluorine or skeletal modification (*e.g.* insertion of a nitrogen atom). The IUPAC-authorized prefixal symbol *F* used when naming compounds conveys the sense of perfluoro, *e.g.* *F*-alkyl ≡ perfluoroalkyl = C_nF_{2n+1}.

[3]Can one imagine toddler Dracula with milk dripping from his toothy-pegs?

effects. More recently, new reasons for developing a blood substitute came from the projected shortage in the developed countries.[4] Baby boomers have now reached an age where the likelihood of their needing blood increases while their tendency or ability to donate blood decreases. Also recent is the important realization that banked blood can be significantly less effective than fresh blood. Oxygen carriers may offer a unique chance for devising strategies that will allow optimal use of one's own fresh blood and mitigating the anticipated blood shortage. For the developing countries, the availability of such a blood substitute may be the best hope to meet future health care needs.

To think of using fluorocarbon emulsions as oxygen carriers, one must think of oxygen as a drug. To be useful to the body, oxygen must be transported to the tissues, where it provides energy for metabolism. In Nature, hemoglobin is the oxygen transporter. In the approach discussed here, fluorocarbons assume that rôle. They increase the blood's capacity to carry and deliver oxygen, which diffuses into the tissues as carbon dioxide diffuses out of them. By boosting the oxygen-delivery capacity of the blood, use of a fluorocarbon emulsion is expected to provide the surgeon or critical care physician an alternative to blood transfusion to cope with an acute oxygen deficit. This, therefore, would buy time, increase the margin of safety and the physician's confidence in delaying or avoiding transfusion of donor blood. It would help promote transfusion avoidance strategies during surgery and could profoundly impact medical practice. Fluorocarbon emulsions will also be invaluable for delivering oxygen in situations when blood is unavailable or ineffective.

This chapter reflects the author's personal experience with, and perceptions of, the history of the development of fluorocarbons as oxygen carriers, with his share of intuition, both right and wrong, good and bad fortune, enjoyment and frustration, detours and encounters, etc., as science actually goes. It by no means pretends to be exhaustive. The limited selection of references definitely reflects the author's point of view and biases. The reader will find numerous additional references in recent review papers [2-6] and proceedings of meetings on blood substitutes [7-9].

Entry into the field: How a coordination chemist got involved in 'blood substitutes'

Of chance and grants as motivators

Even when a particular subject has driven a substantial portion of one's professional life, there obviously was a point in time when one did not have the faintest notion of even the very existence of this particular subject. It is fascinating to try to backtrack to the original instant and circumstances that led to one's involvement and the reasons behind it, as well as to identify the major go/no-go decisions that were confronted along the way.

Let's be candid: the onset of our 'blood substitute' programme had little to do with any fascination with blood or the noble quest for new ways of saving the lives of our contemporaries. My first fuzzy inkling about getting involved in using fluorocarbon emulsions as *in vivo* oxygen carriers was the consequence of a meeting in the early 70s at Péchiney Ugine Kuhlmann in France (PUK, an ancestor of Elf-Atochem) where I was a consultant. The meeting was about finding new applications for a series of *F*-alkyl iodides that the

[4]Deficit in blood donation is chronic in most parts of the world. In the United States the number of transfusions has increased by 3.7% between 1994 and 1997, while blood collection decreased by 5.5%. Extrapolation of the present trends led to projecting a *ca.* 250 000 units deficit as soon as year 2000 [1].

company had recently developed. It so happened that I had read a short paragraph in some professional newspaper about rats who had survived for several hours with a fluorocarbon emulsion in their veins in lieu of blood, enough to get the writer of the article excited and the reader intrigued about the prospect of developing an 'artificial blood'. 'Why not develop some artificial blood?' I suggested, mentioning this paragraph, half-jokingly, so as not to be taken for a naive fool who had swallowed some science fiction hook. 'No, this was not a joke, and it could even be something big', intervened Raymond Hamelin, who was part of the panel and had heard rumors that other companies, including Allied, DuPont and 3M, were interested in the matter... 'and if we, at the University of Nice, were to submit a sensible research project on that theme, PUK might very well want to sponsor it'.

Subsequent discussions in Nice with biochemist and physiologist friends were, however, all but encouraging: how could one expect that such crude, foreign preparations could substitute for the subtle and bewilderingly complex, adaptive haemoglobin/red blood cell machinery involved in delivering oxygen and retrieving carbon dioxide in our body? How could such simplistic concoctions respect the numerous vital equilibrium and rate constants involved in these processes? Not to mention the likelihood that toxic effects would accompany the infusion of such large doses of these little known and unnatural products. Everything about a fluorocarbon emulsion was so different from the real thing: the nature of the carrier, of course, and the mechanism by which it takes up O_2 and CO_2; but also the availability of oxygen to the tissues, the carrier's packaging system (surfactant-coated emulsion droplets versus red blood cells), not to speak of the colour. And blood has so many other functions... No, no way could this work.

The Pioneers

Results of our literature search were, on the other hand, provocative, though rather scarce. First, there was the famous, now classic experiment of Clark and Gollan in 1966, in which they showed that a mouse could survive while breathing an oxygen-saturated liquid fluorocarbon [10]. The ability of fluorocarbons to dissolve large amounts of gases had been known for quite a while [11, 12]. What Clark and Gollan demonstrated in a most dramatic fashion, was that animals could survive such a treatment, *i.e.* that the fluorocarbon supported respiration and caused the animal no harm. This experiment stimulated imaginations and officially marked the entry of fluorocarbons into the biomedical field[5].

The following year, Sloviter and Kamimoto prepared a crude *emulsion* of FX-80 [a 3M Company product consisting primarily of F-(2-n-butyltetrahydrofuran) **1**, Fig. 23.1] which they perfused into isolated rat brains (the most sensitive of organs with respect to oxygen supply). They demonstrated that the emulsion was as effective as, if not superior to, a suspension of red blood cells in a buffered electrolyte solution for maintaining the spontaneous electrical activity and metabolic function of these brains [16].

Then, in 1968, Geyer *et al.* published the truly incredible[6] rat blood substitution experiment that had piqued my interest. These authors were able to replace virtually all

---

[5] Fluorocarbons had actually already been used in cellular microsurgery [13] and in a blood oxygenator [14]; a technique using Freon® 113 ($CF_2ClCFCl_2$) had been developed to isolate viruses [15].

[6] Now we have become blasé about it! Bloodless rats have been used as part of emulsion screening and testing protocols; see for example [17, 18]. But think about it: rats running around literally bloodless (with no complaints about any loss of identity)!

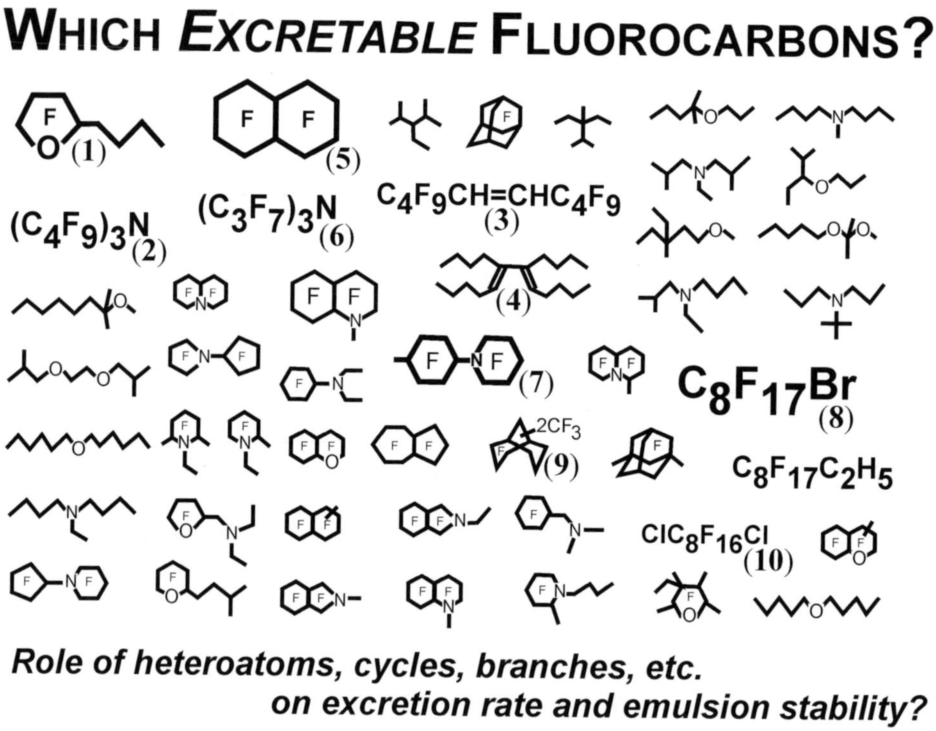

Fig. 23.1. Which fluorocarbon for intravascular use? Here are some of those that were investigated (those with numbers are referred to in the text); note the diversity of their structures. All of the compounds in this chart are fully fluorinated unless definitely indicated otherwise, skeletal ('stick') structures having been drawn for clarity.

of the red blood cells of rats with a poloxamer-stabilized emulsion of *F*-tributylamine (2), and to keep the animals alive in an atmosphere of pure oxygen for about eight hours [19]. During this time, while their haematocrit (vol. % of red blood cells in 'blood') was essentially zero, these rats displayed normal activity. This was the first blood substitution experiment in live animals in which a fluorocarbon assumed the oxygen transport function of blood.

Eventually, we got our hands on the proceedings of the first symposium that had been held on 'Inert Organic Liquids for Biological Oxygen Transport', which took place in Atlantic City in 1969 [20].[7] These proceedings offered new data, concepts and perspectives. Improved emulsions, with lower particle sizes, complemented by serum, had allowed *continuous* survival of rats with haematocrits of 3 or 4 (*i.e.* better than 90% red blood cell replacement) [21]. A greater than threefold increase in oxygen cathode current had been measured from polarographic electrodes implanted into the cortex of oxygen-breathing cats perfused with an FC-43 (mainly *F*-tributylamine)/*Pluronic*® F-68 emulsion, indicating a significant increase in cerebral oxygen tension [22]. The possibility of preserving isolated

[7]Note the pertinence of the title; it is only later, in 1974, that the more alluring, but misleading title 'Blood Substitute' was used; 'Artificial Blood', the press' favorite, is even one step further from reality.

heart, kidney, lung and liver had been explored. For example, in an isolated rat liver perfusion experiment, it had been shown that an FX-80/albumin emulsion delivered oxygen to the organ in greater amount than an erythrocyte suspension [23]. Further papers explored the use of fluorocarbons in liquid breathing and blood oxygenators. By that time, emulsified fluorocarbons had already been injected intravenously in hamsters, rabbits, cats, dogs, chickens, mice and frogs. As one would expect, contradictory results, side effects, and a host of unexplained observations were also reported. Hindsight tells us that the emulsions utilized were often too crude, and the fluorocarbon, when FX-80 was used, too volatile, to allow satisfactory results. The sentence with which Clark closed the symposium captured all our attention: 'Perhaps a science will develop for the purpose of tailoring surfactants and oxygen-transport liquids purely for biological work'.

Of fluorocarbons as haemoglobin?

Two radically distinct options are available to the chemist who wants to design an oxygen carrier, depending on whether the dioxygen molecule is chemically bound to, or simply dissolved, in the carrier [3, 24]. The first option implies coordinating the O_2 molecule to a metal chelate; the second will seek the most effective and most inert solvents for gases, namely fluorocarbons. In the former case, stoichiometric binding is involved; in the latter, transport capacity is simply dependent on the gas' partial pressure, according to Henry's law. The first approach is definitely biomimetic; the second utilizes a class of compounds that have never been observed in nature.

One may wonder why a coordination chemist – which I was at that time – would choose the fluorocarbon route. Besides the above-mentioned, circumstantial opportunity, which no doubt played a major role in our involvement, it must be said that the prospect of mimicking haemoglobin seemed somewhat out of reach. As it turned out, the attempts that were being made at reproducing haemoglobin's oxygen-transport function using metal chelates did not yield very encouraging results. Basolo [25], Collman [26], Baldwin [27], Traylor [28] and Tsuchida [29] and their coworkers treated us with elegant chemistry, but their picket-fenced, bridged, double-bridged, capped or strapped, or polymer-bound porphyrins did not function properly in biological environments and conditions.[8] The use of native, cell-free haemoglobin had been investigated for over a century,[9] but proved toxic and ineffective [34]. Numerous groups were working at modifying cell-free haemoglobin so as to restore its oxygen delivering capacity, prevent its immediate clearance through the kidneys, and reduce its side effects, the most potent of which being renal toxicity and vasoconstriction, but the results were not yet convincing [24]. Mimicking Dame Nature is seldom easy!

Let's just say that fluorocarbon chemistry seemed much simpler than biomimetic coordination chemistry and that it appeared as though it would be fun to explore such an alien route. Some motivation may even have come from the preposterous temptation of

[8]More recently, Tsuchida prepared even more sophisticated iron picket-fence porphyrin complexes with pickets consisting of fatty chains terminated by phosphocholine; lauryl imidazole was also coordinated to the metal. This construct was then incorporated into a phospholipid bilayer in liposomes, or into a phospholipid film in a fat emulsion. The product now transported oxygen *in vivo*, but biocompatibility was still not achieved [30]. More recently, albumin-conjugated metal chelates have been investigated [31].

[9]See, for example, a disclosure to the French Academy of Sciences dated 1886 (Fig. 23.2) [32]. Early studies of administration of haemoglobin to humans were reviewed in 1916 [33].

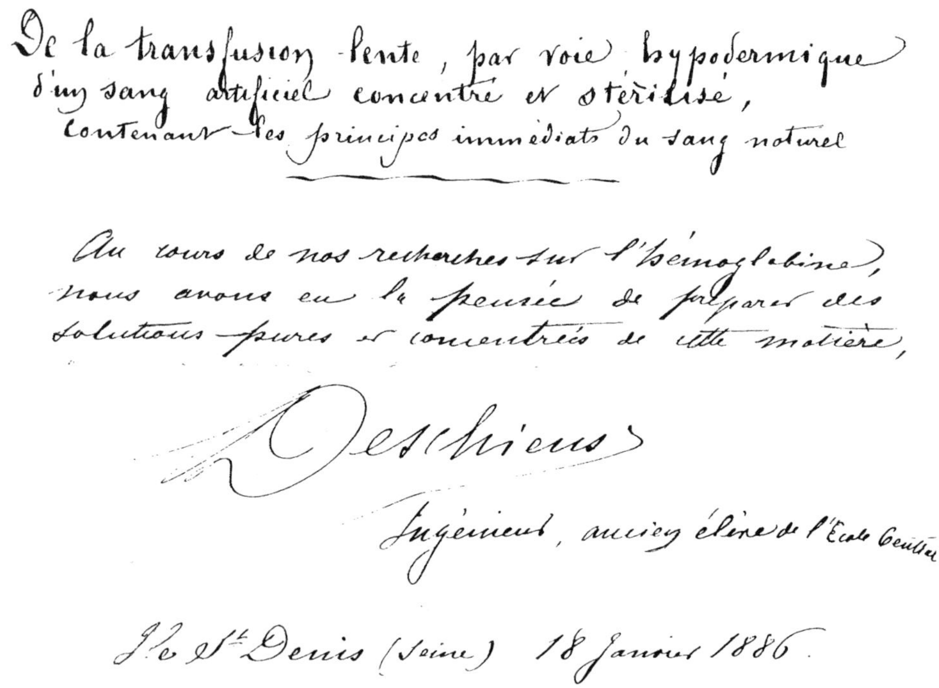
De la transfusion lente, par voie hypodermique d'un sang artificiel concentré et stérilisé, contenant les principes immédiats du sang naturel

Au cours de nos recherches sur l'hémoglobine, nous avons eu la pensée de préparer des solutions pures et concentrées de cette matière,

Deschiens

Ingénieur, ancien élève de l'Ecole Centrale

Ile St Denis (Seine) 18 janvier 1886.

Fig. 23.2. Excerpts from a disclosure to the French Académie des Sciences, dated 1886, about using a solution of haemoglobin from lysed human red blood cells as an 'artificial blood' [31].

beating Nature with a simpler, man-made *solution* (Fig. 23.3). Actually, as we found out later, although the synthesis of the fluorocarbons was indeed rather simple, their colloid chemistry was far less obvious... but so fascinating.

Liquid fluorocarbons are, as we know, the best possible solvents of gases, due to very weak intermolecular interactions [11, 35]. A number of factors also make them extremely inert. These factors include the high strength of the C−F bond, the reinforcement of the backbone C−C bonds when the carbons are fluorinated, the dense packing of the fluorine atoms around the backbone, and the repulsion exercised by the densely electron-coated fluorines on their environment. As we also know, fluorocarbons are extremely hydrophobic and essentially immiscible with water or serum. This is certainly an advantage from the biological inertness standpoint, but it requires that PFCs be dispersed in an aqueous phase, in the form of an emulsion or other colloidal system, prior to infusion in the vasculature.

In parallel, several coordinated-oxygen carrying products are presently being developed, all based on haemoglobin. This haemoglobin comes either from outdated human blood, from bovine blood, or is genetically engineered. It is either chemically or genetically modified,[10] or encapsulated into liposomes [24, 34, 37-39, 39a]. Several of these products eventually went into Phase II or III clinical trials for diverse indications.

[10]One company used transgenic swine to produce human haemoglobin [36]; an additional, heart-breaking issue in this case is that you cannot roast and eat the flesh of these animals: it would be anthropophagism.

Fig. 23.3. Two options have been created for *in vivo* oxygen transport and delivery.

First steps: New fluorocarbons

The data provided by the proceedings of the Atlantic City meeting did not exhaust our skepticism, and raised innumerable new questions, especially about the behavior of fluorocarbons and fluorocarbon emulsions *in vivo*. Had it not been for the perspective of securing some badly needed funding from a prestigious partner, we might not have strayed out of our way and into fluorocarbon (and 'artificial blood') chemistry. Or perhaps it was the strangeness of the approach and the leap into the unknown that appealed to us? Who knows?

Analysis of the existing papers indicated clearly that chemists could contribute significantly to clarifying some fundamental questions by providing the appropriate fluorocarbons. Essentially all the work published so far had indeed been done with some of the perfluorochemicals that were most available at that time (principally FX-80 and FC-43 from the 3M company), which often had poor definition and purity. It is also important to remember that, at that point in time, there was no clue whatsoever about the structural and physicochemical characteristics required for the fluorocarbons to fit the 'blood substitute' application.

The F-nn'E series

The plan we submitted to PUK was, that by taking advantage of the highly pure starting materials they produce by their tetrafluoroethylene telomerization technology, we would synthesize new families of perfluorochemicals that would be well defined and pure; we felt indeed that compounds that were destined to be injected into people in large amounts needed to be rather pure! It would be inappropriate to speak of tailor-made products, since there were no patterns or measurements provided. Using *F*-alkyl iodides as starting materials (a building block approach), rather than introducing fluorine 'directly' via electrochemical or CoF_3-based synthesis, was expected to go a long way towards ensuring the product's purity. Moreover, each family was to consist of a homologous series of compounds, so as to allow incremental variations in properties, and the determination of badly needed structure/property relationships.

We got funded and settled down to work. None of us had done any perfluorocarbon chemistry before. My own experience of fluorine chemistry was limited to the synthesis of a few dexamethasone derivatives, the investigation of some boron trifluoride/phosphorus trioxide reactions, a study of the redistribution of fluorine between phosphorus and silicon and other elements, a study of the molecular dynamics of tungsten oxyfluoride complexes, some fluorophosphorane chemistry, and the assessment of phenyltetrafluorophosphorane as a fluorinating agent for alcohols [40].

The doctoral thesis work of Maurice Le Blanc, Georges Santini and François Jeanneaux (our own private pioneers) produced a range of novel perfluorochemicals, among which were the bis(*F*-alkyl)ethenes **3** [41]. Our principal contacts at Produits Chimiques Ugine Kuhlmann (a branch of PUK) were René Lichtenberger, who has been very effective in promoting fluorine chemistry in our country, and later our long-time friend André Lantz.

The 1,2-bis(*F*-alkyl)ethenes **3** (subsequently code-named F-nn′E, where n and n′ are the numbers of carbons in the *F*-alkyl groups and E indicates the olefinic nature of the compound), were obtained from *F*-alkyl iodides in a two-step process that could easily be scaled up for industrial production [41]. The materials isolated were essentially *trans* isomers. Initially we did not intend to keep the unsaturation, but the internal double bond turned out to be so inert that we left it there.[11]

Other series of fluorocarbons that were synthesized at that time include the higher molecular weight (too high as it turned out) tetrakis(*F*-alkyl)butadienes **4**, which were conceived to provide bundles of covalently interconnected *F*-alkyl chains [44].[12]

We proudly published our work in the chemical literature, displayed the new compounds – clear and pure, and listed some of their physical characteristics, but no one in the medical community expressed interest; nor did Pharmuka, the pharmaceutical branch of PUK. Funding dried up and we decided to quit the field: end of our first fluorocarbon period.

Second wind

Combining fluorocarbon and colloid chemistries

A review paper in *Angewandte Chemie* that I wrote with Maurice Le Blanc in 1978 was intended to wrap up our results and thoughts on fluorocarbon-based blood substitutes –

[11] F-44E and F-66E, for example, remain unaffected when heated at 130 °C with diethylamine or bromine for several weeks, by *meta*-chloroperbenzoic acid, or by various biomimetic oxidation systems [42, 43]. Degradation can be achieved by heating **3** with concentrated alcoholic potassium hydroxide, *i.e.* under conditions where the glass container itself is attacked.

[12] Further papers on *F*-alkyl derivatives concerned the preparation and reactivity of *F*-alkylated dienes [45], alkynes [46], enamines and enaminoketones [47, 48], enimines and ethenylamines [49], isoxazoles and isoxazolines [50], and explored the unusual reactivity of *F*-alkyl copper, calcium and magnesium compounds [40, 51-53].

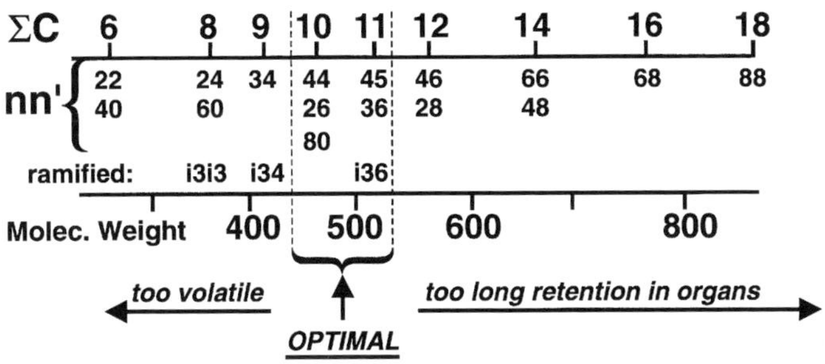

Fig. 23.4. The bis(*F*-alkyl)ethene (F-nn′E) family [41–43]. The bracket delimitates the (narrow) window of molecular weights acceptable for i.v. use.

a sort of testament [54]. Ironically, this review raised enough interest, requests for samples and reprints (several hundreds), invitations to lecture on the topic, and proposals for collaboration, that it brought us back into the game. By then, however, we had learned our lesson: if we wanted to contribute to the field in any significant way, we had to move beyond synthesizing chemicals. We had to learn how to formulate, prepare and characterize fluorocarbon-in-water emulsions, *i.e.*, get acquainted with yet another new branch of unfamiliar chemistry. Moreover, since these emulsions needed to be injectable *in vivo*, some biocompatibility aspects had to be taken into account; hence also the absolute necessity to establish a dialogue and collaborate with physiologist and physician colleagues (strange fellows – lots of (mutual) education needed), who would test our emulsions and experiment with them. At this point I would like to mention the early and effective collaboration we then initiated with Dr René Follana, the then-director of our local blood bank (Alpes Maritimes), and his group, at a time when the notion of 'Blood Substitutes' was far from popular among the average blood banker.[13]

So we wrote new project proposals; Atochem provided us with a new grant; and further help came from various governmental institutions. We were under way again! We started by synthesizing new fluorocarbons, including branched ones[14], and extended our F-nn′E series, of which eighteen members were synthesized over the years (Fig. 23.4) [41-43]. We now tested the new compounds, along with the earlier ones, on cell cultures for purity and biocompatibility. Oxygen and CO_2 solubilities were measured (Fig. 23.5) [41], as well as critical solution temperatures and Hildebrand's solubility parameters [57]. Evidence for metabolism or enzymatic degradation was sought, but never found. The fluorocarbons extracted after several weeks from the liver of rats were chromatographically and spectroscopically identical to the product administered. The higher gas-dissolving capacities and compressibility (the two are related) of F-44E with respect to its saturated analogue was attributed to the 'notch' introduced in the structure by the centrally located double bond, which is thought to facilitate the formation of the cavities that host the gas

[13] Subsequent to the HIV-contaminated blood scandal that shook most countries' blood banking system [55], I have often been asked how come we had not yet been able to provide a substitute that would relieve some of the pressure.

[14] There was a claim that branching accelerated excretion [56], which, unfortunately, was not substantiated by our results.

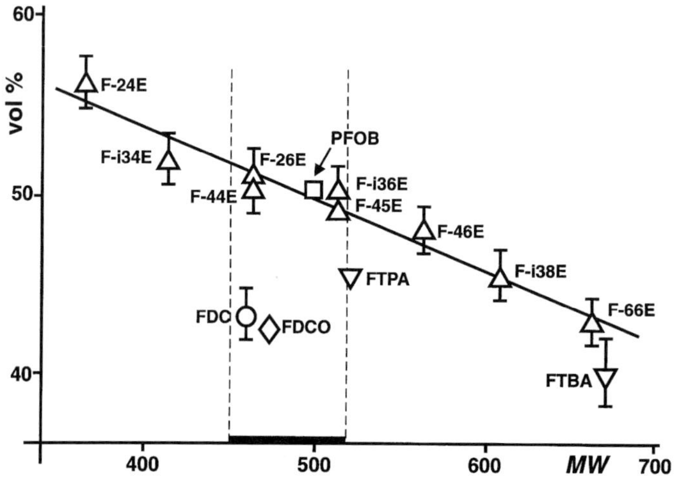

Fig. 23.5. O_2-solubility as a function of molecular weight within a homologous series of fluorocarbons, the bis(*F*-alkyl)ethenes (F-nn′E); for comparison, data are also provided for *F*-decalin (FDC), *F*-tripropylamine (FTPA), *F*- tributylamine (FTBA), *F*-octyl bromide (PFOB or perflubron), and *F*-α,β-dichlorooctane (FDCO) [5, 42, 87].

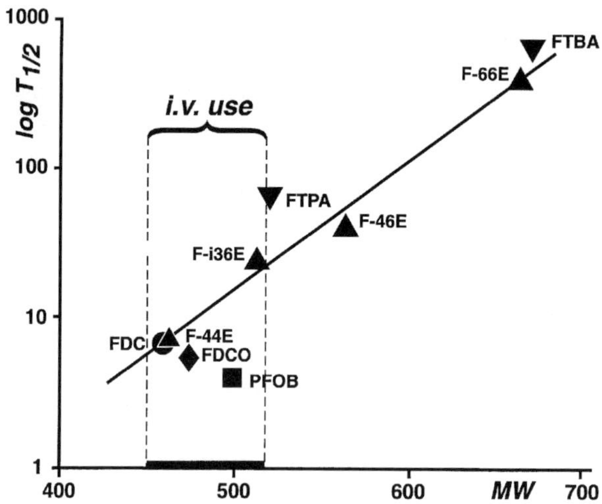

Fig. 23.6. Organ retention (on a log scale) versus molecular weight for several members of the F-nn′E fluorocarbons [42, 43]; for comparison organ retention of FTBA, FDC, FTPA, PFOB and FDCO are also plotted.

molecules [58]. Organ retention half-times were measured [42, 43] for four of the compounds (Fig. 23.6), leading to the selection of F-44E[15] for intravascular use and F-66E for non-i.v. applications and exploratory work.

[15] 1,2-Bis(*F*-butyl)ethene (F-44E) was eventually selected by DuPont for the development of *Therox*®, a concentrated fluorocarbon emulsion that was available for several years for experimental use. An F-44E emulsion is presently being developed by Neuron Therapeutics (Malvern, PA) for treatment of stroke.

We bought a sonicator, a used Manton-Gaulin high-pressure homogenizer, threw together a make-shift agitating sterilizer, acquired a particle size analyzer, and started formulating and turning out emulsions. We immersed ourselves in colloid chemistry and started learning the hard way to juggle with multi-parameter formulation and process optimization (see, for example, [5, 59, 60]). We became accustomed to some queer new equations.[16] Geyer's rat exchange-perfusion model [61] was set up at the Blood Bank in Nice (with the help of Tom Goodin who was with Bob Geyer at Harvard at that time), for the purpose of testing these emulsions. It was quite a sight watching the animals' eyes and ears turn white as their blood was being drained and replaced by the emulsion (Fig. 23.7). Definitive survival of over 60% of the animals with about 95% of their blood replaced meant that the emulsion was 'good', *i.e.* well tolerated (the emulsion's O_2-delivering ability was no longer in question). Most amazing was that these essentially bloodless rats behaved, fed, groomed, urinated and responded to audible and visual stimuli as if nothing had happened!

Where fluorocarbon physiology becomes the name of the game

Our first participation in meetings on blood substitutes in Las Vegas [62] and San Francisco [63], in March and September 1982, introduced us to a new, highly competitive scientific community. That's when I met in person the historical figures, Leland Clark, Henri Sloviter and Bob Geyer; a few chemists (not many, actually), notably Kirby Scherer and Dick Lagow, who were both pioneering the use of elemental fluorine for synthesizing fluorochemicals; Kazumasa Yokoyama, Kouichi Yamanouchi and many others from the teams of the Green Cross Corp. (Osaka, Japan) and their US subsidiary, Alpha Therapeutics (Los Angeles, CA), who at that time were the leaders in the field; David Long who was investigating with Bob Mattrey (both physicians from the University of California, San Diego) radiopaque fluorocarbons for use as diagnostic agents; and many others. Further exchanges and collaborations were initiated. In San Francisco, we became acquainted with the haemoglobin competition, which appeared to struggle with even bigger problems than we did. We also held a small 'European' meeting in Nice with some anaesthesiologist and physiologist friends, Simon Faithfull (then at the Univ. of Rotterdam, The Nederlands), Ken Lowe (Univ. of Nottingham, UK), Joachim Lutz (Univ. of Würzburg, Germany) and others, to try to share our still meagre science and vast ignorance.

Within the next few years a considerable body of new findings and data was reported [63-65]. Numerous new fluorochemicals were synthesized and evaluated [42, 56, 66, 67]. Their pharmacokinetics were investigated. Excretion data were accumulated. The realization that some fluorocarbons, and in particular F-decalin (**5**), were excreted faster than others removed a major roadblock, making possible the development of the first commercial emulsion, *Fluosol®*, by the Green Cross Corporation [68, 69].

By then, some interest had started to rise among French companies. This interest was, however, tempered by the cost (actually grossly underestimated) of the projected development effort, and was soon killed off by the news in 1983 that the US Food and Drug Administration (FDA) had turned down the new drug application (NDA) filed by the Green Cross Corporation for *Fluosol.* I argued – unsuccessfully – that the shortcomings of *Fluosol* were

[16] As an example: $\dot{D}O_2 = CaO_2 \times CO$ (oxygen delivery is equal to oxygen concentration in arterial blood times cardiac output).

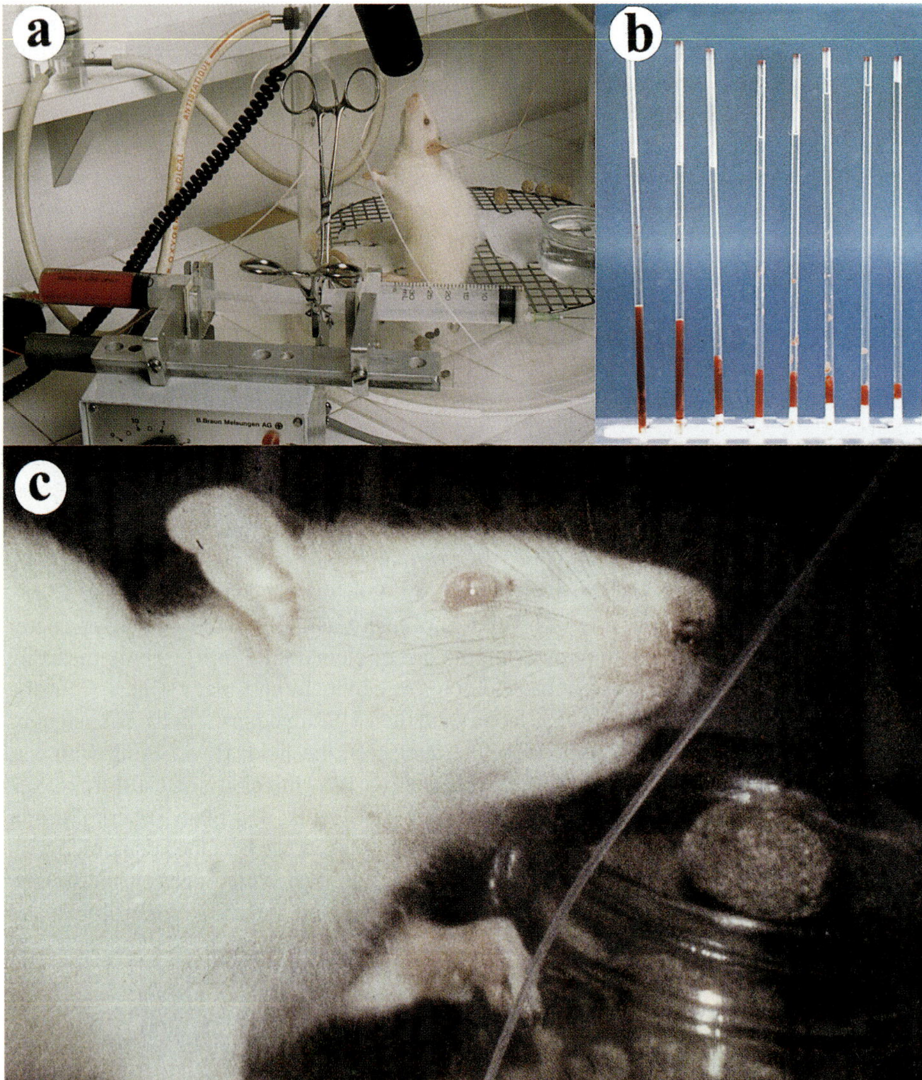

Fig. 23.7. 'Making' bloodless rats: a) the awake animal has a double-lumen catheter inserted in its heart's left atrium; its blood (in the syringe at the left) is being drained through one channel of the catheter while the emulsion (syringe at the right) is infused in the other; being deprived of its blood does not refrain the rat's interest for the oxygen probe that has intruded its domain; b) as the exchange transfusion proceeds the animal has less and less haemoglobin (red) and more and more fluorocarbon droplets (white) in its blood (as seen here after centrifugation); c) after the operation is completed the animal's initially red albino eyes have become translucent and its ears quite white (courtesy of Dr R. Follana, Institut A. Tzank, St-Laurent-du-Var).

inherent to that particular preparation and not to fluorocarbon emulsions in general; that we were in the position of developing a better product with a better fluorocarbon; and that the difficulties met by the competition would help us make up for a later start. Resources dried up again and our efforts had to slow down. By then, however, we had acquired some skills in surfactant and colloid chemistry and emulsion technology. Also, we had started to gain a clearer perception of the requirements that needed to be fulfilled for fluorocarbons and emulsions to be acceptable for intravascular use [70].

What was wrong with Fluosol?

Fluosol was a rather dilute (10% by volume) emulsion of a 7:3 wt/wt mixture of *F*-decalin (**5**) and *F*-tripropylamine (**6**) with a surfactant system consisting of *Pluronic*® F-68 and small amounts of egg yolk phospholipids and potassium oleate, complemented by various salts, hydroxyethylstarch, glycerol and glucose [68, 71]. The shortcomings of *Fluosol* have been reviewed to satiety [5, 70, 72-74]: prolonged organ retention of *F*-tripropylamine; complement activation attributed to *Pluronic*; excessive dilution; limited intravascular persistence; insufficient stability; and lack of user friendliness. Poor stability meant that the emulsion had to be frozen for shipment and storage. The product actually came as three separate preparations: the frozen stem emulsion and two annex salt solutions, which had to be admixed sequentially to the carefully thawed stem emulsion; and the reconstituted product had to be used within eight hours.

The principal reason why *Fluosol* did not gain approval in 1983 was, however, that the proposed indication, anaemia, and the strategy for its use were incompatible with the product's short circulation life. Stabilization of highly anaemic surgical patients upon administration of the drug was demonstrated, but the effect, being short-lived, did not change the outcome for these patients who, being Jehovah Witnesses, refused blood transfusion [75-77]. *Fluosol*'s rejection unfortunately cast unjustified – but persistent – doubt on the entire fluorocarbon approach to O_2 delivery. Approval of *Fluosol* was subsequently obtained in 1989 for a more realistic indication: use in conjunction with percutaneous transluminal coronary angioplasty [78, 79].[17] However, lack of user-friendliness due to the lengthy and cumbersome multistep reconstitution procedure that needed to be performed prior to administration, and the development of autoperfusion catheters, impeded the product's commercial success.

But one must render unto Caesar that which is Caesar's: the development of this early fluorocarbon emulsion and its availability to many researchers allowed it to be firmly established that such a preparation could indeed carry, and did deliver, the expected amount of oxygen to tissues. It also helped establish that large doses of such products could be administered to humans (several thousand patients have received *Fluosol*) and animals without significant side effects. The availability of *Fluosol* also allowed collection of a considerable amount of valuable new information. Last, but not least, the very analysis of its deficiencies played a major role in formulating the criteria and setting the standards for the next generation of injectable fluorocarbon emulsions, and strongly contributed to triggering new research to this end.

Subsequently, an emulsion in many respects close to *Fluosol* (10% fluorocarbon by volume; also based on *F*-decalin, but incorporating a different higher-molecular-weight tertiary *F*-amine, cyclic this time, namely *F*-[1-(4-methylcyclohexyl)piperidine]

[17] In this procedure, a catheter fitted with a balloon is introduced into a blood vessel that is partially obstructed by an atheroma plaque. The balloon is inflated to crush the plaque and reopen the artery. During the inflation period the myocardium is, however, deprived of oxygen. That's where oxygenated *Fluosol*, when administered distal to the balloon, was shown to be beneficial.

(**7**); again a poloxamer was used as the emulsifier), was developed in Russia under the tradename *Perftoran* (Russian Academy of Sciences/Perftoran Company, Pushchino, Russia) [80]. The Russian health authorities approved *Perftoran* in 1997 for a wide range of indications.[18,19]

Defining the criteria for intravascular fluorocarbon use

Excretion rate versus emulsion stability – a major dilemma

Surprisingly, at the time *Fluosol* was initially rejected by the FDA, there was still no common understanding of the criteria fluorocarbons and their emulsions had to fulfil in order to qualify for i.v. use, and what their structural characteristics should be (Fig. 23.1). To preserve their funding, scientists tend to (or, maybe, are coerced to) defend their personal views and pet compounds, at the risk of losing vision. Thus, because *F*-decalin (**5**) was the first perfluorinated compound shown to be eliminated from the body within a reasonable timescale [83, 84], some swore only by cyclic and polycyclic compounds.[20] Others, because very stable emulsions had been obtained with *F*-tributylamine, felt that heteroatoms were needed to provide emulsion stability.

Being a chemist and therefore a somewhat peripheral (or 'out of the box') observer, probably made it easier for me to analyze the available data objectively. The 1984 paper 'Reassessment of criteria...' [70] showed clearly that neither cyclization nor the presence of heteroatoms *per se* had any significant impact on excretion rate. For over fifty compounds for which data were available, organ retention half times were shown to be, within experimental error, an exponential function of the fluorocarbon's molecular weight.[21] The range of molecular weights acceptable for i.v. administration was established to be 460-520, which is a very narrow range: barely more than one CF_2 group and less than a CF_3!

This analysis also identified a 'puzzling deviant point' for which excretion was faster than would have been predicted on the sole basis of molecular weight: this point was *F*-octyl bromide (perflubron[22], **8**), whose half-life in the body had been determined by Long *et al.* to be only about 4 days [86], in spite of a molecular weight of 499 (*cf.* *F*-decalin: 7 days for a MW of 462). The faster excretion of perflubron was subsequently attributed to the lipophilic character introduced by the terminal bromine atom, which facilitates uptake of the fluorocarbon by circulating lipids and transit through the organism [87, 88].[23]

---

[18] *Perftoran* has even been used to sober up dead drunk patients! [81]. Accelerated oxidative metabolism of ethanol may be the basis for the cure.

[19] The Russian blood-substitute effort was tarnished by a sad event: Professor Felix Beloyarzev, who was in charge of the program [82], was challenged by competitors apparently better connected politically; he was demoted, and eventually committed suicide.

[20] It is interesting to note that in certain patents on fluorocarbon O_2-carriers, the inventors restrict themselves voluntarily – but unnecessarily – to cyclic compounds.

[21] The belief that cyclization accelerated excretion probably originated from graphs in which organ retention was plotted against *number* of carbon atoms in the molecule [85], overlooking the fact that each cyclization supposes the ejection of two atoms, in our case two fluorine atoms, hence a non-negligible 28 atomic mass units.

[22] An USAN (United States Adopted Name) designation that sounds rather bizarre to a chemist's ear! Likewise, believe it or not, *F*-pentane (C_5F_{12}) has become 'perflenapent'!

[23] As so often, innovation lies in deviant data (provided they are not artifactual).

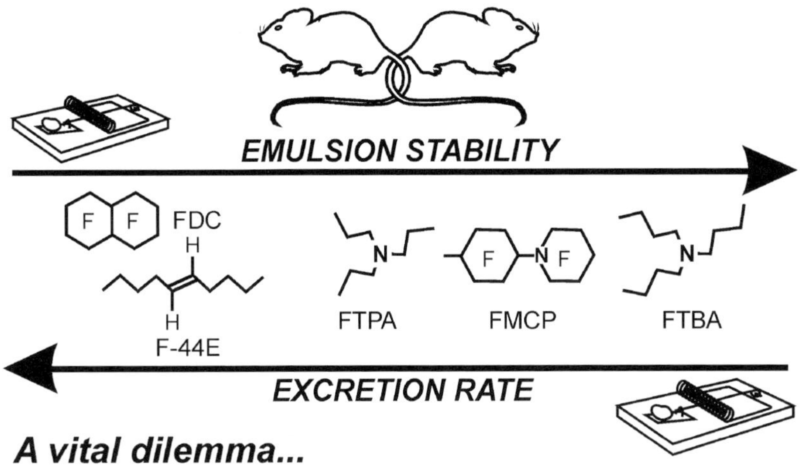

Fig. 23.8. Fluorocarbons with fast excretion rates usually don't render stable emulsions; fluorocarbons that render stable emulsions are usually retained in the body for an excessive length of time: this raises a cardinal dilemma [70].

Another vital condition for fluorocarbon emulsions to be of practical use is that they be stable. The principal mechanism by which particle size increases in such emulsions was determined to be molecular diffusion (Ostwald ripening) [89, 90], even in highly concentrated emulsions, where coalescence could have been favoured [91, 92]. During Ostwald ripening, individual molecules leave the smaller droplets, where the chemical potential is higher as a result of the Kelvin effect, and join larger ones. The rate of diffusion is, according to the Lifshitz-Slezov equation, proportional to the solubility of the dispersed phase in the continuous phase [90, 93-95]. Ostwald ripening will become slower when the fluorocarbon is less soluble and diffusible in water, hence is heavier. Unfortunately, organ retention increases exponentially with molecular weight. This situation is well illustrated by the observation that *F*-decalin (MW 462), which is excreted with a half-life of *ca.* 7 days, tends to give poorly stable emulsions, while *F*-tributylamine (MW 671), which gives very stable emulsions, is retained in the body for several years. A major dilemma (Fig. 23.8) was thus identified [70], for which several solutions were subsequently devised (see below).

Further criteria for selecting fluorocarbons and designing emulsions were discussed. The candidate fluorocarbon needed to be well defined and pure, which would seem quite obvious for a product destined to be injected by hundreds of grams in our veins; but this was far from being the case at that time.[24] It also needed to be industrially feasible and

[24]In 1982, in the abstracts of an ACS symposium on Organofluorine Compounds in Medicine and Biology [62] one could read the following statements concerning the different fluorocarbon products investigated: 'All were mixtures... composed of a primary PFC and related isomers and perfluorinated products that co-distill'; 'All contained partially fluorinated (toxic) compounds which were removed by exhaustive treatment (5 days to 3 weeks) with $KOH/HNEt_2$'; 'Not a single chemical compound'; 'They are multiple component mixtures'; 'Analysis and purity is still a complex issue'; 'About 10 peaks in vapor phase chromatography'; 'We still have to learn how to make them'.

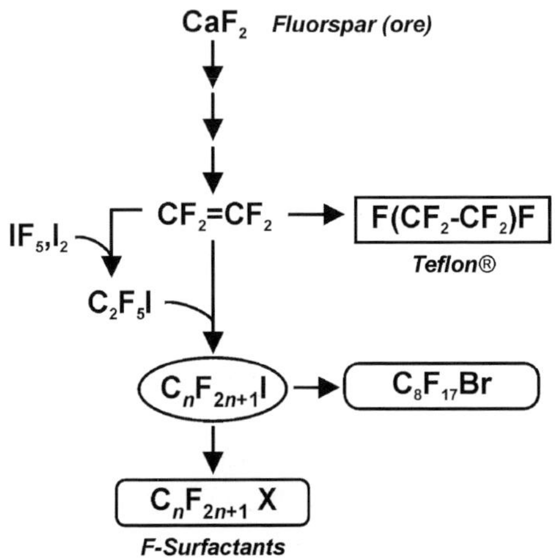

Fig. 23.9. Commercial access to F-octyl bromide (perflubron): just one single step from a pivotal industrial fluorochemical intermediate on the route to large tonnage fluorosurfactants.

cost-effective. Where the O_2-dissolving capacity was concerned, we have shown that linear fluorocarbons had a definite advantage over cyclic ones [70, 87]. It was also deemed necessary to increase the emulsion's fluorocarbon concentration. The guidelines for selecting fluorocarbons[25], surfactants, emulsion formulation, emulsification process, and product characterization have been further refined as experience was gained (see, for periodical reviews, refs. [4, 5, 72, 96].

Perfluorooctyl bromide

F-Octyl bromide (**8**) (the aberrant point on the excretion rate versus molecular weight graph) combines a number of advantages that make it unique among candidate fluorocarbons for therapeutic use. In addition to its exceptionally fast excretion, it can easily be produced in 99.9% purity using the tetrafluoroethylene telomerization route[26]; it is just one step from the industrially well-developed perfluoroalkyl halides (Fig. 23.9); it is amongst the fluorocarbons that have the highest O_2 and CO_2 solubilities (Fig. 23.5); its emulsions show improved stability when phospholipids are the emulsifier; finally, it is radiopaque, thus allowing its use as a contrast agent in radiography. Despite of its bromine atom, F-octyl bromide was demonstrated to be inert in the conditions of processing, storage and use relevant to its applications for O_2 delivery.

[25] Choosing an inappropriate fluorocarbon can be fatal. One clear example is that of Adamantech (Marcus Hook, PA, USA), a subsidiary of Sun Oil, who locked onto 'F-(dimethylbicyclononane)' (**9**) (actually a highly complex mixture of mono- and poly-cyclic fluorocarbons obtained by CoF_3 fluorination of methyladamantane [67], again in the belief that cyclization would help excretion), and failed to produce an acceptable emulsion, in spite of spending big money on the project.

[26] Prof. Clay Sharts from San Diego State University recently disclosed that Bill Sheppard (from Sheppard and Sharts [97]) had convinced David Long to switch from electrochemically produced F-octyl bromide (82% n-$C_8F_{17}Br$, 17% *iso*-$C_8F_{17}Br$ and 17 trace products) to very pure F-octyl bromide produced by DuPont by telomerization [98].

Subsequently, another company, HemaGen-PFC (St-Louis, MO), also adopted a lipophilic fluorocarbon, F-α,ω-dichlorooctane (**10**), for its emulsion [99], in spite of lesser O_2 solubility.

Fluorine is good for you

The fact that fluorocarbons are synthetic materials presents obvious advantages from the standpoints of supply, safety and cost-effectiveness. As compared to haemoglobin-derived products, fluorocarbon emulsions have the enormous advantage of being independent from the collection of blood and the cumbersome and costly extraction and purification of haemoglobin, whether from human, bovine or recombinant (*E. coli* or other) sources. Fluorocarbons do not interfere with nitric oxide (the endothelium-derived relaxing factor) as most haemoglobin products do, which results in vasoconstriction. There are no side effects related to product oxidation, degradation or reactivity – certainly another advantage of fluorocarbons over the multifunctional, sensitive and fragile haemoglobin protein. In contrast to blood or red blood cells, fluorocarbon emulsions carry no risk of transmission of diseases, no requirement for cross-matching, no concern over shortage of supply, have a much longer shelf life, and demand much less stringent storage conditions. Parenteral emulsion technology is well established as a result of routine use of fat emulsions for nutrition. The present fluorocarbon emulsions are terminally heat-sterilized under standard conditions, which is currently impossible with the haemoglobin products. Finally, the cost of production of the emulsions is several times lower than that of any of the haemoglobin products.[27]

Although blood is now safer than ever, the fear created by an AIDS epidemic is still omnipresent.[28] The so-called blood substitutes are being developed for times of shortage, when blood is not available, for bridging the time to transfusion, for improving patient safety and reducing exposure to allogeneic (donor) blood, in particular during surgery, or for oxygen delivery applications where blood is not effective, such as improving neurobehavioural outcome after cardiopulmonary bypass, treatment of stroke or myocardial infarction, for organ preservation, etc.

Further peripeteia – the birth of a dedicated Company

After all industrial development plans had been abandoned in France, our lab was again in a low-tide-mode where fluorocarbons were concerned. My 1984 paper on the criteria for selecting fluorocarbons for i.v. use [70] attracted some attention, however, and I was approached by several companies and venture capitalists.[29] Eventually, it was Pierre Janin, a French entrepreneur, who offered sufficient real support for us to launch a new program of synthesis and evaluation of fluorosurfactants and of design of new emulsion

[27] Investment in a commercial-scale plant is probably one full order of magnitude lower.

[28] In a recent poll in Canada and the US, over three-quarters of the people questioned expressed concern with the safety of blood supply in these countries; many stated a preference to receive a blood substitute and were prepared to pay more for such a substitute [100].

[29] I have wonderful memories, among others, of the time we spent together, Leland Clark, Bob Geyer, Pierre Bothorel (a microemulsion expert from Bordeaux) and me in palaces in New York and London, while Robert Shaw, who subsequently founded HemaGen-PFC, tried to entice us to work for him.

systems. Janin had a small company, Otisville Biopharm (Otisville, NY), which raised horses, produced cell culture media, and had some interest in blood substitutes, including a patent on fluorocarbon microemulsions [101].

It was Otisville's newly hired President, Duane Roth, who was instrumental in bringing together the people and elements needed to take up the challenge of developing a therapeutic oxygen carrier seriously. Duane is a strong willed, dedicated and creative individual, and an outstanding salesman. One of my first goals was to convince him that the best candidate fluorocarbon identified so far was *F*-octyl bromide[30] in view of its fast excretion rate and of the stability of its phospholipid-based emulsions, two of the weakest points of *Fluosol*. The most extensive experience with *F*-octyl bromide was at that time in the hands of Fluoromed, a San Diego-based start-up company founded by David Long. Fluoromed also held patents on concentrated emulsions, which would address another of the perceived weaknesses of *Fluosol*. In addition, *F*-octyl bromide had potential in radiology [102]. I brought the two companies together and, in 1989, after much negotiating, Otisville and Fluoromed decided to merge. The newly born company was baptized Alliance Pharmaceutical Corp. The early times were heroic.[31] I was fortunate (and innocent) enough to accept spending a sabbatical in San Diego around that time and was entrusted with organizing the company's Research and Development effort. Quite an experience for an academic used to small budgets, sparse recruiting possibilities and faint accountability.[32]

Maximizing emulsion stability and performances

An essential lesson from the *Fluosol* experience was that the new emulsion had to be ready for use, hence significantly more stable. Frozen storage, complex reconstitution, and a short window of time for use were simply not acceptable. The target was at least one year of shelf stability under standard refrigeration storage conditions for the actual injectable preparation. It was also imperative to achieve small particle sizes and to keep them small during the sterilization procedure and throughout the shelf life of the product. Small particle sizes, in the 0.1-0.2 μm range, translate into longer intravascular persistence and reduced side effects, as these features depend on phagocytosis and macrophage activation [84, 103, 104]. In addition, it was deemed necessary that the emulsion should be relatively concentrated, yet fluid. The manufacturing process needed to be as 'forgiving' as possible, *i.e.* insensitive to small changes in processing conditions, so bottle-to-bottle and lot-to-lot consistency could be easily achieved[33].

The result of intensive team effort, insight, planning, experimentation and optimization was *Oxygent*™, a ready-for-use 60% w/v, osmotically- and pH-balanced emulsion of *F*-octyl bromide (Fig. 23.10).

[30] And not our own F-44E, although I still believe that it is superior in many respects to *F*-decalin and other cyclic fluorocarbons.

[31] The company started in a few offices and labs rented here and there. We manufactured the first pharmaceutical-grade emulsions in Ireland, which was probably also Alliance's first teamwork experience, and definitely my first effort at conducting preparative colloid chemistry under pharmaceutical GMP conditions. We started worrying about timelines, mourning over dead pigs, and doing essentially everything over again for lack of appropriate-quality supplies or adequate documentation.

[32] Being grilled by industrial colleagues, and potential future partners, during the course of their 'due-diligence' processes, was another interesting new experience.

[33] This is where considerable know-how comes in!

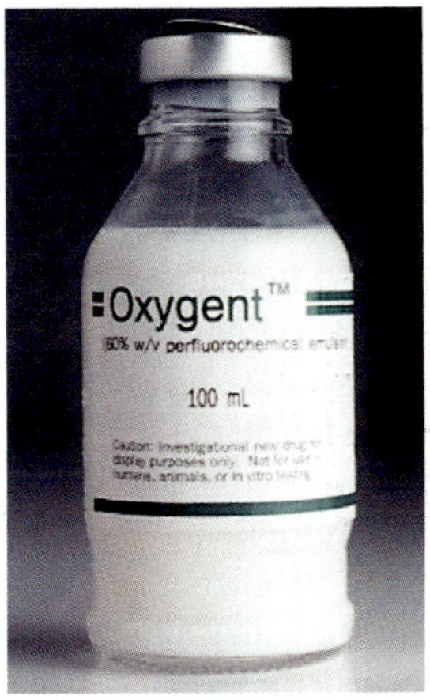

Fig. 23.10. A dose of *Oxygent* (AF0144), as used in the ongoing clinical trials (courtesy of Alliance Pharmaceutical Corp.).

Fig. 23.14. (See text on p. 409.) View of the manufacturing plant, with its battery of electronically controlled high-pressure homogenizers, utilized for producing *Oxygent* (courtesy of Alliance Pharmaceutical Corp.).

Heavy fluorocarbons as stabilizers

As previously noted, molecular diffusion, the principal cause of particle size increase in submicronic fluorocarbon emulsions, can be counteracted by adding a higher-molecular-weight ('heavier') fluorocarbon. This principle had actually been used in *Fluosol*, where *F*-tripropylamine is the heavy fluorocarbon, and in *Perftoran*, which utilizes *F*-[1-(4-methylcyclohexyl)piperidine] (**7**) (30%) as the stabilizer; however, these two fluorocarbons have organ half-lives of about 65 and at least 90 days, respectively. Selecting a *lipophilic* 'heavy' fluorocarbon, *F*-decyl bromide (n-$C_{10}F_{21}Br$), whose bromine atom provides the desired lipophilic touch, allowed us to minimize the increase in organ retention [105].

The present *Oxygent* formulation (AF0144) consists of 60% fluorocarbon by weight – primarily *F*-octyl bromide (perflubron; supplied by Elf Atochem, Pierre Benite, France) and a small amount of *F*-decyl bromide, a higher homologue, as the stabilizing additive. Egg yolk phospholipids is used as the emulsifier. Osmolarity is adjusted with sodium chloride, and pH with a phosphate buffer. Minute amounts of α-D-tocopherol and EDTA are added to protect the phospholipids against oxidation. Average droplet size, after heat sterilization, is about 0.16 μm and viscosity is around 5 cPs (extrapolated at zero shear rate). The product has a shelf life of 18 months when stored at 5-10 °C.

Fluorosurfactants

The synthesis of specially made fluorosurfactants was also advocated as a means of stabilizing fluorocarbon emulsions [106-108]. Interfacial tension is indeed another term of the Lifshitz-Slezov equation to which the fluorocarbon's diffusion rate is proportional, and fluorosurfactants are capable of reducing fluorocarbon/water tensions to very low values.[34] Commercial fluorosurfactants are usually complex mixtures, obviously not destined and not suited for biomedical uses [111, 112]. There was a clear desire for fluorosurfactants acceptable for use in the medical and pharmaceutical areas in general, as well as for research in physical chemistry and colloid chemistry (see below).

According to the fluorosurfactant programme that I launched in the mid-80s, the new compounds were to be well defined and pure; would be neutral, anionic, cationic or zwitterionic, single- or double-tailed, with or without identical tail groups; and would have a large diversity of polar head groups. A modular molecular design was elected that allowed stepwise variation in hydrophilic, lipophilic and fluorophilic characters, size and shape, chemical functions available for further derivatization, etc. (Fig. 23.11; see also examples in Fig. 23.12) [112-114]. The polar heads were generally derived from natural products and included various types of polyols, anhydropolyols, mono- and disaccharides, amino acids, amine oxides, phosphoramides, phosphocholine, phosphatidylcholine, and other phosphatides and lipids, as well as telomers with a variable number of tris(hydroxymethyl)aminomethyl groups. Diverse junction units between head and tail(s) were utilized, such as ester, ether, thioether and amide groups. Nearly two hundred new molecules were synthesized (for reviews and references see [112-115]). Purity and definition being prerequisites for fluorosurfactants to be used in biomedical research and pharmaceuticals, synthetic strategies and laboratory practices were selected that integrated these requirements from the onset.

[34]Microemulsions, *i.e.* thermodynamically stable dispersions, of fluorocarbons involving fluorosurfactants have been reported (see, for example, [101, 107, 109, 110]), but biocompatibility was not achieved.

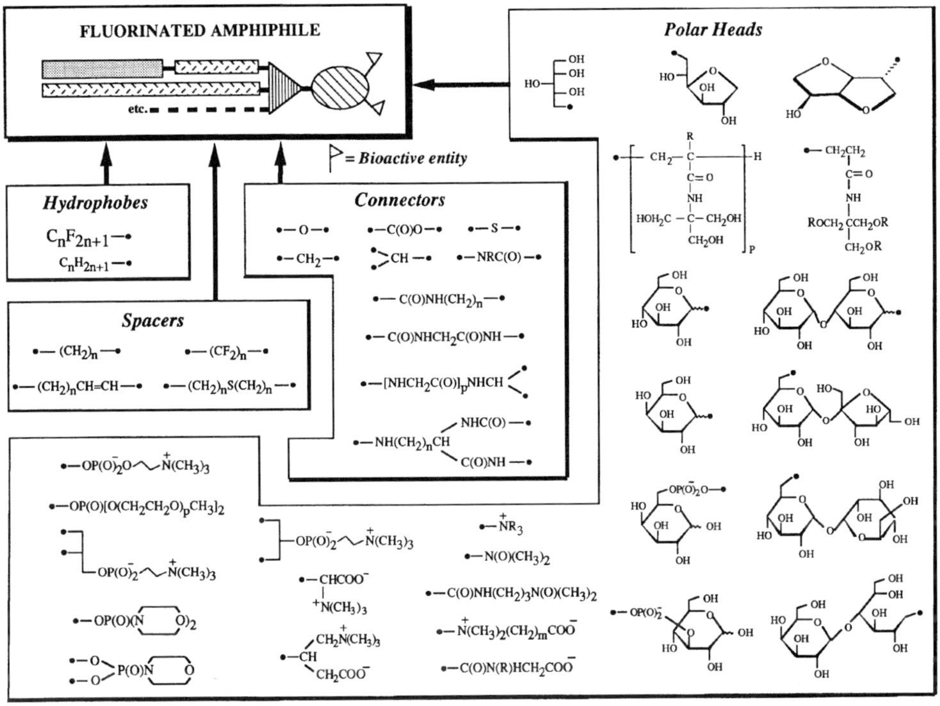

Fig. 23.11. A versatile modular design for fluorosurfactants; the building blocks represented are among those present in fluorosurfactants synthesized and investigated by, or in collaboration with the Unité de Chimie Moléculaire [112, 113, 234] (from [233], with permission).

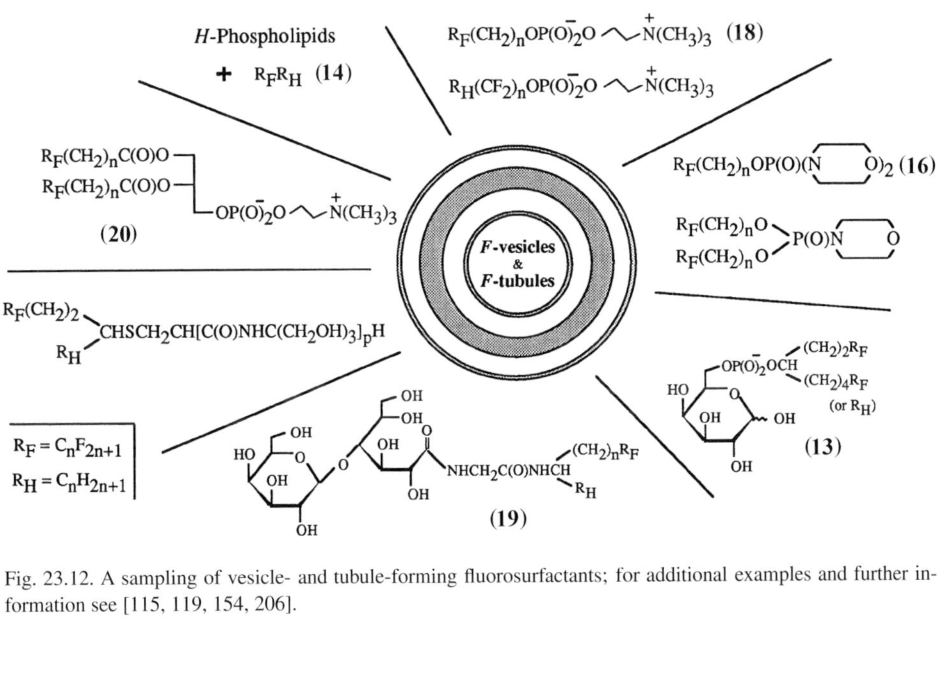

Fig. 23.12. A sampling of vesicle- and tubule-forming fluorosurfactants; for additional examples and further information see [115, 119, 154, 206].

Our surfactant programme also comprised a physicochemical arm, with the measurement of their surface activity and emulsion stabilization capability, and the exploration of their aggregation behavior; and a biological arm with the determination of hemolytic activity, impact on the growth and viability of cell cultures [116], and estimation of acute toxicity in mice [117]. An interesting finding was that fluorosurfactants usually had very low or undetectable hemolytic activity, in spite of high surface activity [118, 119].

Fluorosurfactants proved, as predicted, to be extremely effective in stabilizing fluorocarbon emulsions. Not all of them, though. In some cases, small amounts of one single fluorosurfactant did the job. In other cases, the fluorosurfactant needed to be used in conjunction with another surfactant. For example, some closely related derivatives of trehalose **11** and maltose **12** have dramatically different behaviours: the former gave highly stable emulsions, while the second did not even allow the preparation of an emulsion, probably reflecting differences in polar head conformation and hydration [120, 121].

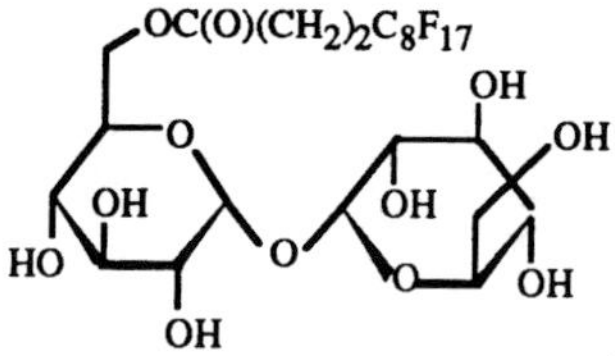

(11)

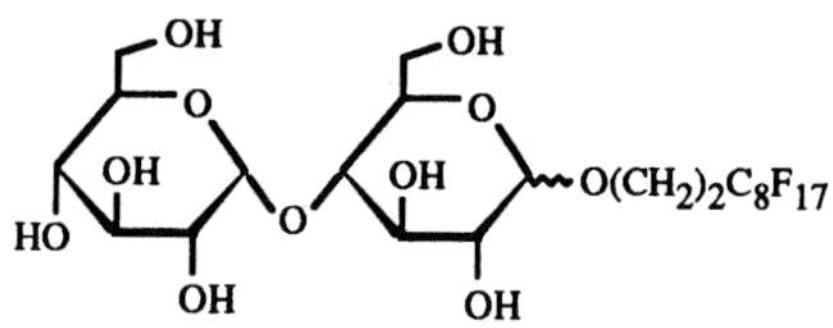

(12)

On the other hand, strong synergistic stabilizing effects were found between the latter maltose-derived surfactant and a poloxamer, indicating hydrogen-bond interactions between the two species [122]. In some cases there was no significant change in particle size distribution for the six years during which the emulsion was monitored [5]. Certain anionic sugar derivatives such as **13** (Fig. 23.12), which were capable of giving stable emulsions by themselves, were synergistic with phospholipids but not with poloxamers [121]. These differences in behaviour are still unpredictable at the present stage of our knowledge. As we will see later, fluorosurfactants also allowed us to prepare and stabilize a range of other emulsions, gels, films, vesicles, suspensions, and further novel colloidal systems with unique structures, properties and potential.

Fluorosurfactants thus achieved more than expected from the physico-chemical standpoint. Disappointment was therefore all the greater when we realized how strongly the uncertainties that exist about the pharmacology of these new materials would hinder their development. The cost and duration of their evaluation are indeed considerable, and the fact that the surfactant is not the active component, but only part of a delivery system, is a serious handicap to spending such time and money. The recently disclosed use of a fluorosurfactant as the stabilizer of an *F*-pentane-based diagnostic contrast agent [123] that has been licensed in Europe (which supposes that inocuity has been established for this fluorosurfactant) is encouraging and hopefully will break this reluctance.

Fluorocarbon-hydrocarbon diblocks as molecular dowels

Since egg yolk phospholipids already provided rather stable fluorocarbon emulsions, and sensing that getting fluorosurfactants tested would not be a small affair, we devised yet another means of increasing the stability of fluorocarbon emulsions. It consisted of

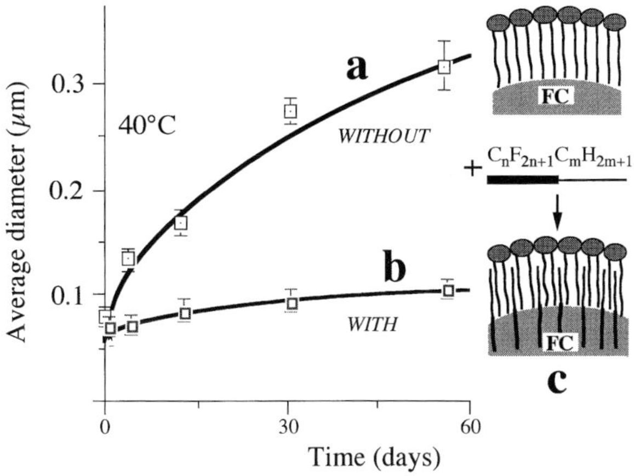

Fig. 23.13. Particle size increase over time at 40 °C in a concentrated *F*-octyl bromide/egg yolk phospholipids (EYP) emulsion prepared (a) with the phospholipids alone and (b) with an equimolar mixture of EYP and $C_6F_{13}C_{10}H_{21}$ diblock; (c) hypothetical stabilizing 'dowel effect' of the diblock compound at the fluorocarbon droplet/phospholipid film interface (adapted from ref. [5], with permission).

supplementing standard phospholipids with mixed fluorocarbon-hydrocarbon diblock compounds, such as $C_nF_{2n+1}C_mH_{2m+1}$ (FnHm, **14**) or $C_nF_{2n+1}CH{=}CHC_mH_{2m+1}$ (FnHmE, **15**) [124]. The idea was to improve the adhesion of the fatty acid tails of the phospholipids onto the fluorocarbon droplet. The diblocks were expected indeed to concentrate preferentially at the interface between the fluorocarbon droplets and the surfactant film that surrounds them, and to behave as molecular 'dowels' at this interface (Fig. 23.13). Extremely stable emulsions were obtained [124, 125] that will perhaps provide the basis for a future generation of products. Although the exact mechanism by which stabilization is achieved has not yet been fully elucidated, it has been shown that the diblocks were more effective than a 'heavy' fluorocarbon additive of similar molecular weight [5].[35] and also that their presence had a definite impact on the characteristics of the phospholipid film [127]. Use of fluorosurfactants or co-surfactants or fluorocarbon-hydrocarbon diblock co-surfactants are probably the only way there is to effectively stabilize emulsions of volatile fluorocarbons such as *F*-pentane (b.p. 29.5 °C).

Gaining further indispensable fundamental knowledge

Development of a novel therapeutic approach cannot be seriously considered without a sound, cutting-edge scientific basis. The availability of *Fluosol* for research, and later of *Oxygent*, and the efforts that accompanied the development of these products, produced a wealth of results and considerably increased our knowledge of the *in vivo* behaviour of fluorocarbons. By small brush strokes a comprehensive picture was progressively formed. Significant progress was made in our understanding of how oxygen was delivered to tissues and in which circumstances fluorocarbon emulsions would be most effective [2]. Both

[35] Pluronic/diblock combinations were also investigated, but proved less effective [126].

a convective and a diffusive mechanism were proposed [128].[36] The effects of the emulsion on the reticuloendothelial system (RES; an essential part of our defense system) were investigated [129]. The mechanism by which fluorocarbons are eliminated from the body was unraveled [130-133]: the fluorocarbon droplets infused in the vasculature are foreign particles and are handled as such by the body. They are progressively phagocytized and removed from circulation by macrophages, the larger droplets being removed first. The fluorochemical is temporarily stored in the liver, spleen and other organs of the RES. It is then recirculated by lipid carriers and is eventually excreted through the lungs in the expired air. The advantages of using lipophilic fluorocarbons as the gas carrier and phospholipids as the emulsifier were confirmed. No evidence for metabolism has ever been reported for any of the fluorocarbons investigated. The mechanism of the side effects that can accompany the clearance of particulates from the blood stream was extensively investigated [103]. It was shown that these effects are the natural consequence of macrophage activation during the phagocytosis step. Other side effects that were attributed to the fluorocarbon in some early emulsions were subsequently traced to the surfactant or to some improper characteristic of the emulsion, such as too large particles or inadequate osmotic pressure [65]. The anaphylactoid reactions observed with some patients in the case of *Fluosol* were attributed to the poloxamer, *Pluronic* F-68, used as the principal emulsifier in this product, and were no longer seen when phospholipids were used [134]. Finally, the efforts directed at injectable emulsions certainly contributed in a non-negligible way to promoting the development of the chemistry of fluorinated colloids.

Further potential uses of fluorocarbon emulsions were explored, including salvage of ischaemic tissues resulting from myocardial infarction or stroke [135-138][37] use in cardiopulmonary bypass surgery [139, 140]; reduction of air microemboli associated with cardiopulmonary bypass [141, 142]; use to increase the sensitivity of tumour cells to radiotherapy and chemotherapy [143-146]; preservation of semen, tissues, severed limbs and isolated organs [147-150]; improvement of cell cultures [151]; use as diagnostic contrast agents [102, 152, 153]; use for drug delivery and as new tools in biomedical research [154].

We were lucky enough to be associated with the exploration of some of these applications, including use in and effect on cell cultures [155-157], neuroprotection against air emboli [158], synthesis and evaluation of new contrast agents [159], diagnosis of liver tumors using computed tomography [160], the staging of cancer by lymphography [161], cancer radiotherapy in nude mice grafted with human tumour lines [162, 163], organ preservation [149, 164], and stabilization of isolated organ models for physiological research [165].[38]

---

[36] It was realized on this occasion that our knowledge of *in vivo* oxygen transport and delivery was far less advanced than we would have liked, which amply justifies the active annual gathering of an ISOTT (International Society on Oxygen Transport to Tissues).

[37] One of the most intriguing applications investigated is a treatment of stroke by which Osterholm *et al.* inject the oxygenated emulsion directly into the subarachnoid spaces of the brain, thus bypassing the vascular system; results on cats with experimentally induced stroke indicated spectacular restoration of electrical brain activity and substantial reduction in infarct size [138].

[38] Eventually, I found myself in charge of organizing, along with Thomas Chang and Bob Winslow for the haemoglobin branch, the 5th International Symposium on Blood Substitutes (San Diego, March 1993). Thanks to contributions from most of the prominent scientists in the field, the Proceedings of the Symposium became a valuable tool for all those working in the area [7, 8]. My gratitude still goes to Gwen Rosenberg, who did not spare her efforts to make sure that everything would go according to plan.

Collaborative studies with Alliance's researchers involved the development, characterization and evaluation of various highly concentrated fluorocarbon emulsions [166 - 169]. Further studies explored the structure of fluorocarbon/phospholipid emulsions [170], their rheology [168], the effect of filtration as compared to heat sterilization [171], the interaction of PFCs with liver cytochromes [172]; the absence of protein-solubilizing properties of some fluorosurfactants [173], and various Langmuir film [174, 175] and bilayer membrane [176] structure studies.

Maturation

Meanwhile, Alliance pursued its development. New faces and new talents joined the team[39]. Experience was gained in emulsion technology and many other vital areas. A biological research facility was built. Analytical and microbiological departments, biostatistics, engineering, bioengineering, validation and calibration, and quality assurance groups, a library and information services were formed. Clinical and regulatory departments, a patent service and a documentation system, so essential in the pharmaceutical industry, were established. Pilot plants, and eventually a complete, fully automated manufacturing plant, capable of producing about 750,000 doses of *Oxygent* per year, were built (see Fig. 23.14, p. 403). During that period, I was exposed to such vital, yet non-chemical, considerations as business development, financing, marketing, pharmacoeconomics, clinical protocol design and regulatory strategies, matrix management and the theory of constraints, about which I had not the faintest idea.[40]

As it grew, the company added new products to its development pipeline, each one with its specific requirements and development strategy. It also added numerous new patents to its patent portfolio. Partnerships were signed with big corporations,... which usually slowed down the pace of product development immediately.[41] Creative strategies for product use, clinical protocols and regulatory approaches were devised. The company continued to strive for excellence in its core technologies: basic and applied fluorocarbon, surfactant and colloid chemistry, parenteral emulsion technology, understanding of tissue oxygen needs and delivery mechanism (understanding of the product was far from immediate; fine tuning continued as clinical experience is gained). Strong bonds with the best experts and investigators from academia in both the United States and Europe were

---

[39]Listing by name all those who contributed to this development effort is no longer possible (Alliance employs now over 250 people); I would just like to mention – and thank – a few fellow chemists from the early days, mainly fluorocarbon colloid chemists: David Klein, Leo Trevino, and Jeff Weers and his merry companions.

[40]How to entertain an annual burn-rate of some 60 million dollars per year without having a single product on the market still remains a process at which I marvel.

[41]Getting a large corporation with deep pockets involved seems indispensable at a given point in the development of a pharmaceutical product. While the smaller company appears most adept at turning a discovery into a properly defined product, the implication of a larger partner appears inevitable when it comes to pursuing large-scale Phase III clinical trials for multiple applications, full-scale manufacturing, and world-wide distribution. Partnering does not go without frustrations. Differences in cultures, priorities and pace, perhaps also in the individual sense of ownership and urgency, sometimes a surprisingly fast turnover rate among team members, a tendency to reinvent the wheel and start all over again, and internal constraints and politics in both organizations, can seriously rattle a partnership.

established through research grants, consultant agreements, and clinical testing arrangements.

The road was bumpy, the target moving, problems kept surfacing and were solved as we went, pitfalls were numerous,[42] and the outer world both supportive and merciless. Some wrong paths were explored. Excitement and doubt alternated. The company got through growth crises and adjustments inevitable in any fast-growing living community. Did we invest enough efforts into educating our people, our partners and the authorities about our goals, concepts and drugs?

Product efficacy

No, fluorocarbons are by no means a lesser equivalent of haemoglobin or red blood cells. These products are fundamentally different. Oxygen uptake by fluorocarbons is not limited by saturation of some coordination sites; oxygen is more readily available to tissues; the proportion of oxygen unloaded is much higher; emulsion droplets are much smaller and many times more numerous than red cells; etc., etc. And advantage can be taken from these differences.

The high pO_2 under which fluorocarbons are used (the patients usually breathe pure oxygen in order to maximize efficacy) provides a strong driving force for diffusion of oxygen into tissues. Extended multiparameter simulations of the effect of a given dose of a given emulsion in a given clinical situation, on tissue oxygenation, oxygen consumption, delay to transfusion, etc. were conducted [128, 177]. Such simulations are precious for identifying the actual situations and patient populations for which the product should be most effective, and for optimizing its use. This research also developed into RODA™ (Real-time Oxygen Dynamics Analyzer), a minimally invasive, on-line computerized device destined to monitor in real-time the oxygenation and haemodynamic status of surgical and intensive care patients [178].

Extensive animal experimentation with *Oxygent* established that even small doses of fluorocarbon could contribute significantly to tissue oxygen consumption, especially in the case of low haemoglobin levels and elevated cardiac output. For example, canine models that mimic normovolemic haemodilution and surgical bleeding were used. The dogs treated with the emulsion (2.7 g of fluorocarbon per kg of body weight) could lose 70 mL of blood per kg (about two-thirds of their blood), compared to 10 mL/kg only for controls, before the partial pressure of oxygen in the mixed venous blood fell below the initial 100%-O_2-breathing baseline [179, 180]. Tissue pO_2 of skeletal muscle, gut and brain was seen to increase significantly in response to the administration of the emulsion (see also Fig. 23.15). A further study of this type concluded that the emulsion was as effective as fresh autologous red blood cell transfusion in maintaining tissue oxygenation in such a model [181].

[42] They include, among others, loss of focus, insufficient planning, meeting mania, inadequate supervision or excessive micromanagement, excessive endogenous bureaucracy, loss of corporate memory, use of preconceived notions, hindsight biases, incorrect correlations, overconfidence and inordinate contentment, as well as defeatism, insufficient trust among team members, finger-pointing and butt-covering behaviour, lack of realism and wishful thinking, and so on and so forth. Unemotional analysis of any mistake or failure (including those of competition) is mandatory and perhaps the only way to progress.

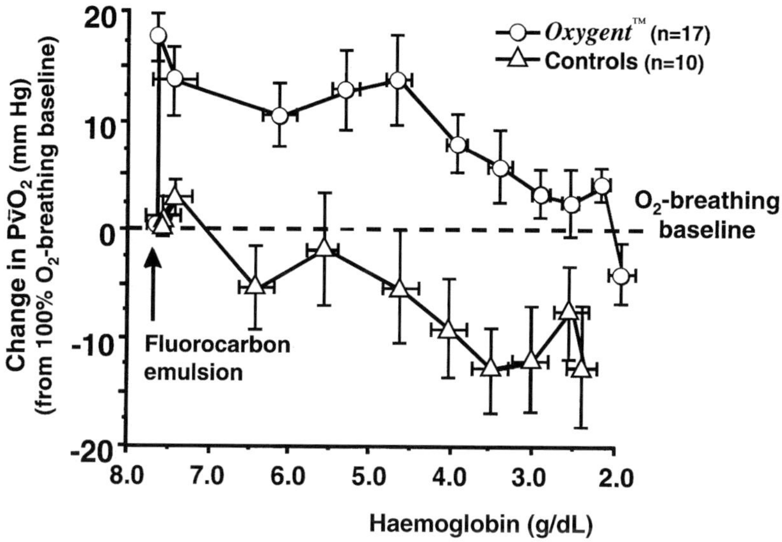

Fig. 23.15. Efficacy of a fluorocarbon emulsion, *Oxygent*, in a canine model mimicking surgical blood loss after acute normovolemic haemodilution. Both the treatment group and the control group animals are haemodiluted (from haemoglobin about 14 to about 8 g/dL), breathe oxygen, and lose blood (hence haemoglobin; x-axis) in a controlled manner. Mixed venous oxygen tension ($P\bar{v}O_2$, the oxygen tension in blood after tissues and organs have been irrigated; y-axis) reflects the adequacy of tissue oxygenation. The difference in tissue oxygenation in the two groups of animals is significant; the fluorocarbon-treated dogs (1.35 g fluorocarbon per kg body weight in this example) still enjoy adequate tissue oxygenation at haemoglobin levels one fifth of normal (courtesy Dr P. Keipert, Alliance Pharmaceutical Corp.).

Physiological efficacy of fluorocarbon emulsions for delivering O_2 *in vivo* had already been amply demonstrated with *Fluosol* [74, 77, 182, 183]. It should be reemphasized that *Fluosol* had failed, not for lack of efficacy in delivering oxygen or because of side effects, but because of poor physico-chemical characteristics and inappropriate strategy of use.

Using *Oxygent*, efficacy was further demonstrated in canine models of cardiopulmonary bypass [139]; preservation of brain stem function following experimental brain stem ischaemia in dogs [184]; increase in retinal oxygenation in cats [185]; restoration of cerebral oxygen delivery in severely haemorrhaged rats [186]; improvement of systolic function and reduction in edema when blood cardioplegia was supplemented with fluorocarbons in dogs with experimentally induced myocardial ischaemia [140]. Drug approval requires, however, that not only physiological efficacy be proved, but also that clinical benefit be established. This is difficult for blood substitutes because of the complexity and diversity of the actual clinical situations encountered in which the product could be useful.

Extensive preclinical and clinical investigations were obviously also devoted to establishing the product's safety. No side effects were found, other than those expected from the progressive phagocytosis of the particles [103, 187]. These effects consist of 'flu-like symptoms' such as headache, fever, chills and nausea. They were all transient and fully reversible within 12-24 h. They were significantly reduced, both in terms of intensity and frequency, through formulation and process optimization [5, 187]. An increase in pulmonary residual volume, which had been observed with F-decalin in certain animal species [188], was subsequently shown to result from stabilization and retention of air bubbles in the alveoli [189]. This phenomenon was shown not to be relevant to humans, who have much wider airways [190]. After having established the nature of this phenomenon, Ernie Schutt and co-workers turned it into a novel principle of stabilizing micron-sized air bubbles, which was subsequently developed into a new contrast agent for ultrasound imaging [153, 191].

Therapeutic indications and clinical evaluation

Let us remember again that fluorocarbon emulsions are essentially a drug delivery system, the actual drug being oxygen. And what better antihypoxic drug than oxygen could there be when there is a need to correct an oxygen deficit and salvage hypoxic tissues? Fluorocarbon emulsions provide a simple and elegant means of delivering this drug rapidly when the need is identified. Our oxygen carrier is expected to provide the physician with an alternative to blood transfusion for temporarily alleviating tissue hypoxia and its consequences in the surgical and critical care patient. Moreover, such a product could be used to *prevent* the occurrence of ischaemic events in a patient at risk of inadequate tissue oxygenation during certain surgical operations, and could therefore revolutionize standard patient care. One of the target indications for fluorocarbon-based therapeutic oxygen carriers, which takes into account their aptitude for delivering oxygen *and* the fluorocarbon's relatively short intravascular half-life, is their use during surgery in combination with acute normovolemic haemodilution (ANH) [3, 187, 192, 193].[43] The principle of this method, called augmented-ANH^{SM} (A-ANH), is schematically depicted in Fig. 23.16: when during the operation the physician would normally give a transfusion to the haemodiluted patient, he gives the emulsion instead; the patient's tissue oxygenation thus remains in safe control and he now loses something far less precious than his blood; if his haemoglobin concentration reaches a level that is deemed too low, or at the end of the intervention, the patient gets his own blood back. A-ANH is expected to result in increased patient safety and in reduced need for donor blood transfusion. Because the emulsion enables the anaesthesiologist to perform ANH in a safer and more effective way it should help ANH to become available to a broad population of surgical patients[44].

Phase II clinical trials involving patients undergoing elective surgery with ANH determined that *Oxygent* was significantly more effective than *fresh* autologous blood[45] at reversing physiological transfusion triggers, and significantly delayed the need for a subsequent blood transfusion [187, 193]. No safety issues were raised. These results confirm the extensive experimental animal work and clinical studies that had been conducted over the years. They establish that fluorocarbon emulsions do deliver oxygen in a physiologically

[43] A practice in which 2-4 units of blood are withdrawn from the patient immediately before surgery, replaced by a volume expander, and returned to her/him when a transfusion becomes necessary.

[44] Over 60% of all blood transfusions are given in conjunction with a surgical procedure.

[45] And fresh blood is known to be more immediately effective than stored blood; stored red blood cells take 6-12 h to reach full oxygen delivering capacity, due to so-called storage lesions, including loss of the allosteric effector 2,3-diphosphoglycerate.

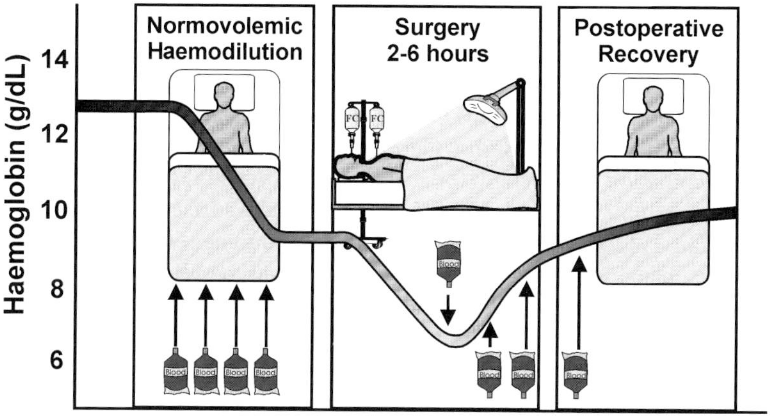

Fig. 23.16. The augmented acute normovolemic haemodilution (A-ANH) method: Three to four units of blood are withdrawn from the patient just before he or she undergoes surgery, and are replaced by a volume-expanding saline or colloidal solution. When, during surgery, the need for transfusion is determined, a fluorocarbon emulsion is administered in lieu of blood and prevents tissue hypoxia from occurring. When the haemoglobin concentration reaches a level that is considered unsafe, or at the end of the surgical procedure, the patient receives his/her own blood back. The A-ANH method is expected to provide an increased safety margin and to avoid or reduce the patient's exposure to allogeneic blood (courtesy of Alliance Pharmaceutical Corp.).

significant way. Is it necessary to prove that oxygen is beneficial to hypoxic tissues? Well controlled, multicenter Phase III trials are currently under way in Europe and the United States for this indication. *Oxygent* will be manufactured and distributed together with Baxter Healthcare Corp. (Deerfield, IL).[46]

Trauma is another critical situation where the emulsions should prove beneficial, especially during the prehospital 'golden hour' period, which largely determines the outcome for the patient. During this period, blood is usually not available and transfused banked blood is not effective yet. Fluorocarbon emulsions could provide a unique means of stabilizing the patient waiting for a transfusion or other intervention, and should therefore find their place in any ambulance or rescue vehicle.

Fluorocarbon emulsions could also be useful in cardiopulmonary bypass surgery [139, 141, 142]. Added to the priming solution in the extracorporeal circuit, they could potentially reduce the need for allogeneic red blood cell transfusion and increase oxygen supply. In addition, they would dissolve the tiny air bubbles that may be introduced in the circuit, which can cause microemboli and can lead to serious post-surgical neurological dysfunction [194, 195]. The ongoing clinical trials are encouraging [196]. Additional possible uses for fluorocarbon-based oxygen delivery systems under investigation have been listed earlier.

Fluorocarbon emulsions should also play a significant rôle in allowing developing countries, especially those without an established blood banking system, to benefit from an injectable oxygen carrier and thus from improved patient care.

[46]It is interesting that after two attempts with haemoglobin products, one cross-linked chemically and the other by genetic engineering, Baxter eventually turned to a fluorocarbon emulsion-based oxygen carrier product.

Other fluorocarbon products in an advanced stage of clinical evaluation by Alliance are *LiquiVent*®, which consists of neat *F*-octyl bromide to be used for treating acute respiratory failure by liquid ventilation [197-200], and *Imagent*®, a preparation of micron-size air bubbles osmotically stabilized by *F*-hexane vapour, to be used as an ultrasound contrast agent (co-developed with Schering AG, Berlin, Germany) [153]. *LiquiVent* is intended for use with mechanical ventilation as a therapy for acute lung injury and respiratory distress syndrome resulting, for example, in infants, from deficiency in lung surfactant, or in adults from trauma or other causes [200, 201]. The product is expected to help recruit collapsed alveoli, facilitate the exchange of the respiratory gases, promote the removal of foreign material and debris from the lung, and reduce the patient's exposure to harmful mechanical ventilation (baro or volutrauma). *Imagent* is destined for assessment of cardiac function and diagnosis of perfusion defects and tumors in organs, such as liver, kidney, breast and prostate [191, 202]. Clinical evaluation for improvement of endocardial border delineation has now been successfully completed and the product is awaiting licensure by the FDA. Still another product, *PulmoSpheres*™, is a novel type of drug-delivering perforated microshell with wide-range potential, that can be suspended in virtually any nonaqueous solvent or propellant, or used in dry powder form, thus providing an entirely new strategy for formulating asthma and other drugs. Such products could be used, among others, to deliver bronchodilators or steroids to the lung. This delivery system could, in particular, solve some of the problems encountered with metered-dose inhaler products when switching from CFCs to HFCs in compliance with the Montreal protocol.

Further colloidal systems with a fluorocarbon phase

Introducing fluorine into a molecule, product or scientific domain virtually always results in novel behaviour or unmatched performances [203, 204]. Colloid chemistry is no exception. The synthesis of homologous series of pure fluorosurfactants gave us the opportunity to engineer a variety of novel colloidal systems with challenging structures and/or unique attributes (Fig. 23.17), and thus to contribute to the development of a fertile branch of chemistry that, surprisingly, is still somewhat neglected by fluorine chemists, the chemistry of fluorinated colloids [115, 154, 205-207].

Reverse emulsions, multiple emulsions and gels

Stable reverse (*i.e.* water-in-fluorocarbon) emulsions were produced. The challenge consisted of stabilizing a dispersion of fine droplets of water in one of the most water-repellent media possible. Compared to direct emulsions, molecular diffusion of water through the fluorocarbon phase, which leads to Ostwald ripening, is facilitated by the extremely low intermolecular cohesion of liquid fluorocarbons. Stable reverse emulsions were nevertheless obtained when using highly fluorophilic surfactants, such as the (*F*-alkyl)alkyldimorpholinophosphate **16** [208]. Such emulsions have potential for the delivery of drugs, lung surfactant, possibly vaccines or genes, and other bioactive material through the pulmonary route [209, 210].

Hydrocarbon-in-fluorocarbon emulsions, *i.e.* totally waterless dispersed systems, were devised. Stability was achieved using appropriate fluorocarbon-hydrocarbon diblocks of type **14** [211]. These systems could find uses for the delivery and controlled release of lipidic material, including moisture-sensitive agents. Various kinds of multiple emulsions

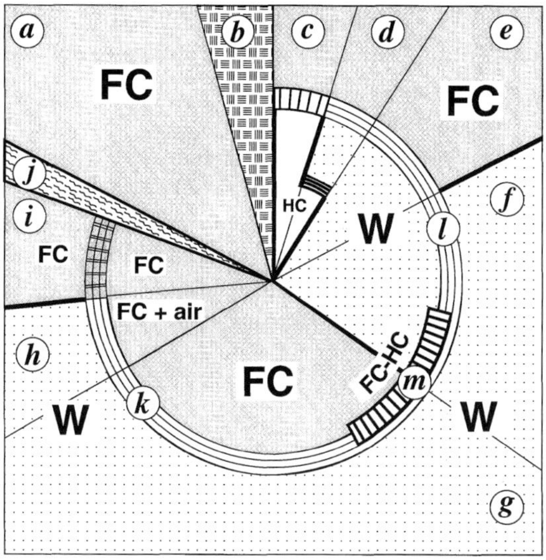

Fig. 23.17. Schematic representation of fluorocarbon-based colloidal systems presently under investigation (examples of potential biomedical applications): (a) neat fluorocarbon (liquid ventilation, diagnosis, ocular tamponade, cell culture); (b) suspension of nanocrystals in a fluorocarbon (drug delivery); (c) apolar hydrocarbon-in-fluorocarbon emulsion (delivery of lipophilic drugs, especially to the lung); (d) example of a multiple emulsion with non-miscible fluorocarbon, hydrocarbon oil and water microcompartments (multidrug delivery); (e) water-in-fluorocarbon reverse emulsion (delivery of hydrophilic drugs through the pulmonary route); (f) fluorinated liposomes, tubules and other self-assemblies of fluorosurfactants in water (controlled delivery); (g) fluorocarbon-in-water emulsion (injectable oxygen-carriers) and high internal phase ratio gel-emulsion (topical use); (h) gaseous fluorocarbon-stabilized air bubbles (ultrasound contrast agents); (i) porous drug-loaded microspheres suspended in a fluorocarbon (delivery to and through the lung); (j) gel with a continuous fluorocarbon phase (skin protection); (k) Langmuir film comprising fluorinated amphiphilic components; (l) black lipid membrane comprising fluorinated components; (m) molecular dowel-reinforced film or membrane (FC = fluorocarbon, HC = hydrocarbon, W = water, FC-HC = fluorocarbon-hydrocarbon diblock; from ref. [233], with permission).

were also obtained, including some novel combinations that simultaneously involve three distinct, non-miscible phases: a fluorocarbon, a hydrocarbon and water [211]. An internal lipidic phase can, for example, be separated from the external continuous fluorocarbon phase by an intermediate aqueous layer. Both lipophilic and hydrophilic agents can be loaded in such multicompartment systems.

Gelifying fluorocarbons could not be an easy task in view of the fluorocarbon's extremely weak cohesive forces and its inability to dissolve the standard gelation agents. Several types of gels have nevertheless been reported [212]. Some are very rich in water and consist of a dispersion of water droplets in an external phase made of a water-in-fluorocarbon microemulsion; the surfactants used were F-alkylated polyethoxylated alcohols [213]. Other gels contain up to 99% of fluorocarbon and have a compartmentalized structure, with polyhedral fluorocarbon domains (polyaphrons) separated by a thin reverse film of hydrated surfactant; the surfactant was the F-alkylated amine oxide $C_7F_{15}C(O)NH(CH_2)_3N(CH_3)_2O$ **17** [214]. Still another type of fluorocarbon-rich gel was obtained by dispersing fluorocarbon/hydrocarbon diblocks together with phospholipids in the fluorocarbon; gelation occurred when adding a small amount of water, which probably triggers the formation of long entangled micelles of hydrated surfactant [215]. Ongoing research concerns the structure and dynamics of the fluorocarbon/water and fluorocarbon-hydrocarbon interfacial films with diverse surfactant systems.

Self-organization of fluorosurfactants – fluorinated supramolecular constructs

Exploring the self-aggregation behaviour of fluorosurfactants proved to be extremely rewarding. Due to their extreme hydrophobicity, fluorinated tails confer to surfactants an extremely powerful driving force for them to collect and organize at interfaces. As compared to non-fluorinated analogs, fluorosurfactants also have a much stronger tendency to self-aggregate into a rich variety of discrete molecular assemblies when dispersed in water and other solvents (Figs. 23.17 and 23.18) [115, 154, 205-207].

Among the findings we enjoyed most were the formation of highly stable vesicles, even from very short *single*-chain fluorinated amphiphiles, such as **18** ($R_F = C_8F_{17}, n = 2$), without any need for supplementary associative forces [216]; the achievement of sturdy microtubules from *non*-chiral, non-hydrogen bonding, single-chain fluorosurfactants such as **16** [217-219], thus questioning the currently accepted understanding of tubule formation; and the design of a variety of fluorinated films, membranes and vesicles, including those made from combinations of standard phospholipids with fluorocarbon-hydrocarbon diblocks (Fig. 23.17) [119, 220, 221]. The latter combinations can result in considerably increased vesicle stability, reduced membrane permeability and lower rate of vesicle fusion [154, 220, 222, 223]. Rather amazing were the Langmuir films made of phospholipids and fluorocarbon-hydrocarbon diblocks which, depending on pressure, consist of an homogenous mixed amphiphile film, or phase-separated domains, or a two-storied construct with the diblock jumping on top of the phospholipid film (Fig. 23.19b) [224]. The effect of a hydrocarbon spacer on micellization and adsorption of fluorosurfactants was investigated, leading to questioning the 1 $CF_2 \approx 1.5$ CH_2 equivalency rule [225]. The lyotropic and thermotropic behaviours of fluorinated phospholipids were studied in detail [206, 226]. Vesicles made from fluorinated amphiphiles have an internal fluorinated film within their bilayer membranes (Figs. 23.12 and 23.19), which usually provides higher stability and lower permeability [119, 206, 227]. It was also observed that this internal fluorinated film can have a significant repercussion on a particle's behaviour in a biological milieu. For example, it can affect its *in vitro* or *in vivo* recognition [228, 229] or the enzymatic hydrolysis of phospholipid components [230] (Fig. 23.20). Nanotubules were obtained from anionic glucophospholipids [231] and their structure and reversible conversion into vesicles were examined [176, 232]. Fluorinated bilayers were used as templates for polymerization [232a]. We continue to explore how in such systems the interplay between hydrophobic, lipophobic and fluorophobic interactions can be exploited maximally for the purpose of developing new supramolecular assemblies of amphiphiles and controlling their properties [154, 233].[47] This effort will hopefully contribute providing new tools for research and the basis for next generation and novel fluorocarbon-based products with therapeutic value.

[47] The fluorocarbon emulsion and fluorosurfactant aggregation efforts initiated in Nice have now essentially moved to the Institut Charles Sadron in Strasbourg, where Marie-Pierre Krafft has built a clean room dedicated to the preparation of fluorinated colloidal systems, including emulsions, gels, Langmuir films, nanotubules, etc.

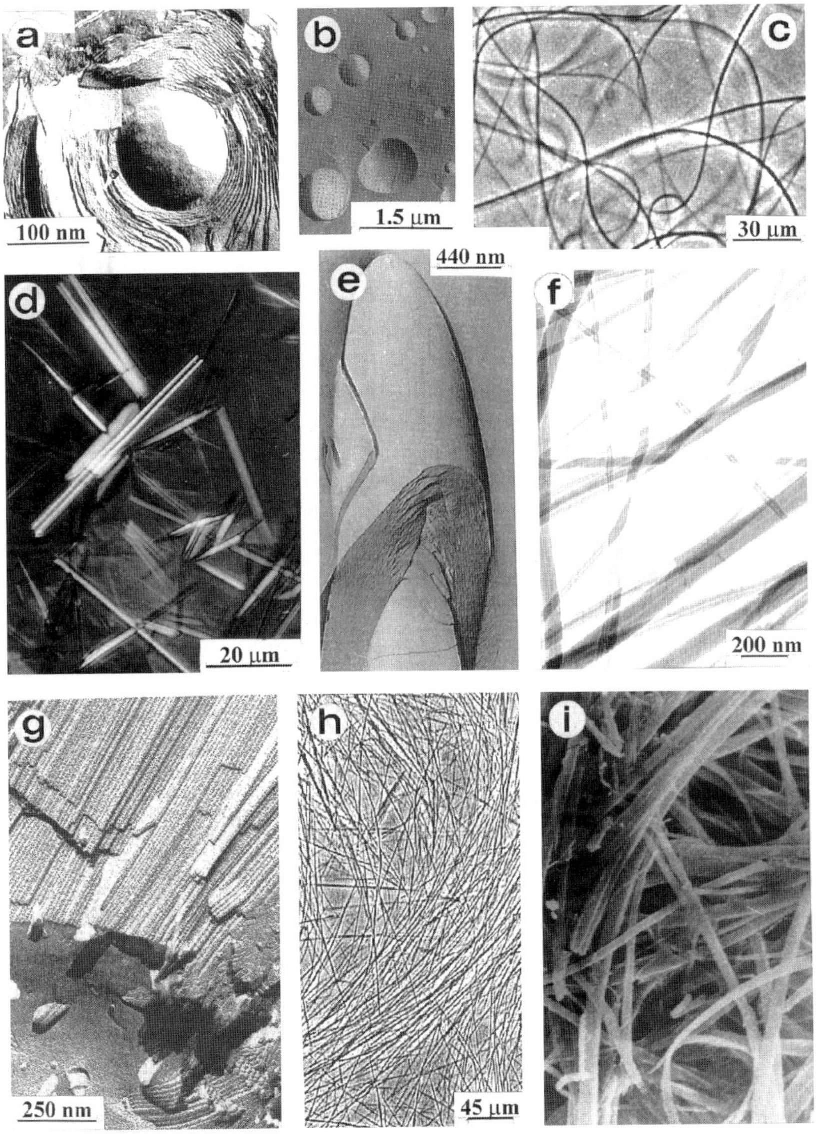

Fig. 23.18. From the family album: (a) multilamellar vesicle made from an (F-alkyl)alkyldimorpholinophosphate **16** ($R_F = C_8F_{17}$, $n = 5$) in water at 35 °C (freeze fracture EM; private collection) [217]; (b) small unilamellar vesicles made from glycolipid **20** (freeze fracture electron micrograph (EM)) [236]; (c) long flexible tubules made of non-chiral **16** ($R_F = C_8F_{17}$, $n = 5$) in water at low temperature (optical microscopy) [217]; (d) tubules from non-chiral **16** ($R_F = C_{10}F_{21}$, $n = 2$) in an ethanol/water mixture (polarized light) [219]; (e) rolled-up bilayer membranes, as found in tubules made of a 1:1 mixture of **16** ($R_F = C_8F_{17}$, $n = 5$) and **16** ($R_F = C_{10}F_{21}$, $n = 2$) in water [219]; (f) tubules made from the anionic glucose phosphate **13** (cryotransmission EM) [176]; (g) non-lamellar, ribbon-like phase made from fluorinated phospholipid **19** ($R_F = C_6F_{13}$, $n = 10$; freeze fracture EM) [226]; (h) metallized fibers made from phosphocholine derivative **18** ($R_F = C_8F_{17}$, $n = 2$); fiber formation was induced by Nd^{3+} in water (optical microscopy; courtesy Dr F. Giulieri, Univ. of Nice) [237]; (i) fibers formed by a fluorinated alcohol, $C_{10}F_{21}C_7H_{14}OH$, in methanol (scanning EM; courtesy Dr F. Giulieri) [237]. See Fig. 23.12 (p. 405) for structures **13**, **16**, **18** and **19**.

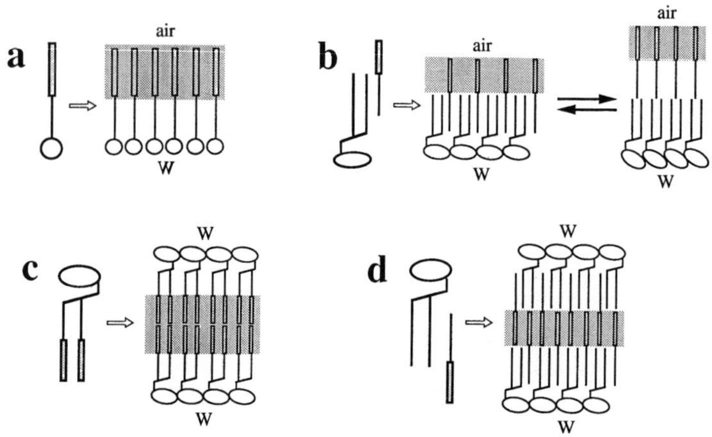

Fig. 23.19. Organizing space at the molecular level: (a) Langmuir film made from a fluorinated amphiphile; (b) fluorinated Langmuir film made of a combination of a standard phospholipid such as dipalmitoylphosphatidylcholine DPPC and a fluorocarbon-hydrocarbon diblock such as F8H16; upon compression the latter jumps out and on top of the phospholipid film [224]; (c) fluorinated bilayer membrane made of fluorinated phospholipid **19** [120] (d) bilayer membrane made from a combination of a standard (hydrocarbon) phospholipid and fluorocarbon-hydrocarbon diblock **14** or **15** [221]; see also Fig. 23.13 for the use of FnHm diblocks at a fluorocarbon/water interface. The shaded zones are fluorinated.

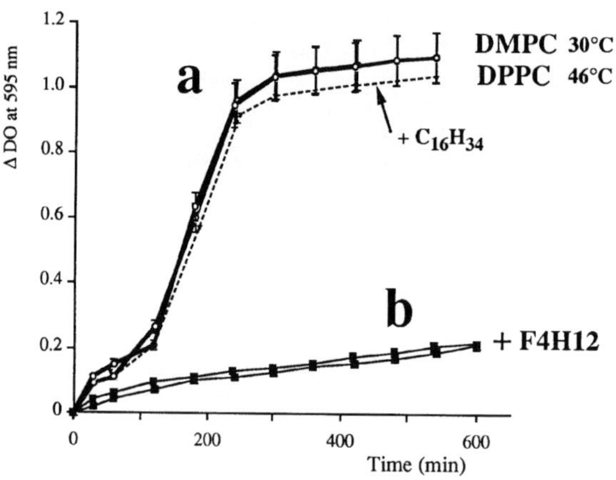

Fig. 23.20. Hindering the enzymatic hydrolysis of phospholipids in liposomes by incorporation of FnHm diblocks **14** into the bilayer membrane: action of porcine pancreatic phospholipase A_2 (as measured by monitoring the release of the fatty acid with a pH indicator), (a) on vesicles made of dimyristoylphosphatidylcholine (DMPC) or dipalmitoylphosphatidylcholine (DPPC) alone, and (b) after incorporation of an equimolar amount of $C_4F_9C_{12}H_{25}$ (F4H12); dotted line is with $C_{16}H_{34}$ instead of F4H12 (adapted from ref. [230], with permission).

In retrospect

With now three large market-size innovative products in late Phase III clinical trials, and more products in its development pipeline, Alliance is faced with unique expectations, challenges and the workload to match.

Alliance's lead product, *Oxygent*, a fluorocarbon-based blood substitute, should play a significant role in helping to relieve the projected blood supply shortage. It should also allow the development of strategies that maximize the use of the most precious of resources, one's own fresh blood. Finally, it could provide the developing countries with an effective alternative to human blood banking.

Being part of this story, and following it throughout its research and development phases, was for the author a most enthralling and rewarding, though often stressful experience. It was an unforeseen chance to add the zest of a pharmaceutical industrial experience to an essentially academic career, a unique opportunity to widen his horizons and circle of friends and relations.

Pharmaceutical development is characterized by a level of complexity unfamiliar to the average chemist. The involvement of patients of widely different health status and suffering a diversity of conditions, and of physicians and other caregivers, and their interactions and personal biases, introduces hundreds of (interrelated) variables in clinical evaluation, and an enormous amount of 'noise' in clinical data. Top that with regulatory, manufacturing, marketing and other constraints and you will get a hint of the complexity of the situations that need to be dealt with. This complexity cannot be ignored. Simplification (or ingenuity) is a prerequisite for enterprise. Oversimplification, on the other hand, increases the risk of failure dramatically. Management of complexity is the nemesis and responsibility of present-day leaders. Our leaders are no longer expected to provide solutions to problems, but rather to catalyze the emergence of solutions through mobilization of the complementary collective understanding, intelligence and creativity of a group of (talented and experienced) individuals.

At Alliance, I got a hands-on notion of how complex the formula and conditions for a successful injectable oxygen carrier needs to be: the right fluorocarbon;[48] extensive surfactant and colloid chemistry; a strong patent position; a bushel of naiveté; the appropriate blend and interplay of scientific, medical, engineering, and non-technical skills;[49] strong ties with both academia and the investment community; solid teamwork and creative thinking; clear consensual product development strategies; judicious clinical, regulatory and marketing approaches; a stable, yet flexible organization with exquisite and objective communication; effective decision making, prioritizing and negotiation-based resource-deconflicting procedures; courageous leaders, dedicated champions and realistic managers; a clear-minded Board of Directors; operational excellence and everyone's commitment to 'do it right'; add carloads of DMFs, CMCs, SOPs, ICHs, cGMPs, BPRs, DRs, INDs, MAAs, ERs, OOSs, OOTs, cCRPs, NPVs, 510(k)s, 10Qs, CRTs, CROs, SAEs, LSDs,[50] etc., and more paper work; substantial financial resources, stable supportive market conditions, and ardent, loyal partners; sweat, time, focus, focus again and time again, some luck, and out-and-out perseverance. Have we been successful (and fortunate enough)

[48] Does it not sound paradoxical that, although the fluorocarbon 'only' plays the role of an inert delivery fluid, so few of them actually qualify for this use?

[49] 'Biodiversity' among personnel is a definite asset, probably a must, for an R & D-oriented company to be prolific.

[50] You know: Drug Master File; Chemistry, Manufacturing and Controls documents; Standard Operating Procedures; International Conference on Harmonization guidelines, current Good Manufacturing Practices; Batch Production Records; Discrepancy Reports, Investigational New Drug applications; etc., etc.

in properly identifying, collecting and combining all these indispensable ingredients? Did we do it right? We shall soon know.

Acknowledgements

I wish to thank my former students, technicians and coworkers of the Unité de Chimie Moléculaire in Nice (and more particularly the heroic SOFFT group), colleagues and friends at Alliance and elsewhere (as well as competitors), who over all these years helped us stay on line against all the odds. Grateful thanks to Gwen, Jolene, Marie-Pierre and Eric B. for having nit-picked at this manuscript with reckless abandon and Lora Reece for help with the artwork. I am also indebted for the support provided by numerous institutions and industries, particularly the Centre National de la Recherche Scientifique, Atochem and Alliance Pharmaceutical Corp. Finally, I owe a lot to Professor Guy Ourisson for having, during my thesis years in Strasbourg, given me some lateral vision and indulged my inclination for exploring side lanes.

References

1 M. Sullivan, Conference of the United States Public Health Service, Advisory Committee on Blood Safety and Availability (Bethesda, MA, April 1999).
2 N. S. Faithfull, 'The role of perfluorochemicals in surgery and the ITU', in J. L. Vincent (ed.), *Yearbook of Intensive Care and Emergency Medicine*, Springer-Verlag, Brussels, 1994, pp. 237-251.
3 T. F. Zuck and J. G. Riess, 'Current status of injectable oxygen carriers', *Crit. Rev. Clin. Lab. Sci., 31* (1994) 295-324 .
4 J. G. Riess, 'Fluorocarbon-based oxygen delivery: basic principles and product development', in T. M. S. Chang (ed.), *Blood Substitutes: Methods, Products and Clinical Trials*, Karger Landes Systems, New York, 1998, pp. 101-126.
5 M. P. Krafft, J. G. Riess and J. G. Weers, 'The design and engineering of oxygen-delivering fluorocarbon emulsions', in S. Benita (ed.), *Submicronic Emulsions in Drug Targeting and Delivery*, Harwood Academic Publ., Amsterdam, 1998, Chap. 10, pp. 235-333.
6 P. Keipert, 'Perfluorochemical emulsions: Future alternatives to transfusion', in T. M. S. Chang (ed.), *Blood Substitutes: Principles, Methods, Products and Clinical Trials*, Karger Landes Systems, New York, 1998, pp. 127-156.
7 T. M. S. Chang, J. G. Riess and R. M. Winslow (eds.), *Blood Substitutes, General. Proc. Vth Intern. Symp. on Blood Substitutes, Vol. 1*, in *Art. Cells, Blood Subst., Immob. Biotech., 22* (1994) 123-360.
8 J. G. Riess (ed.), *Blood Substitutes, the Fluorocarbon Approach, Proc. Vth Intern. Symp. on Blood Substitutes, Vol. 3*, in *Art. Cells, Blood Subst., Immob. Biotech., 22* (1994) 945-1543.
9 E. Tsuchida (ed.), *Artificial Red Cells*, John Wiley & Sons, New York, 1995; E. Tsuchida (ed.), *Blood Substitutes – Present and Future Perspectives*, Elsevier Science Publ., Amsterdam, 1998.
10 L. C. Clark and F. Gollan, 'Survival of mammals breathing organic liquids equilibrated with oxygen at atmospheric pressure', *Science, 152* (1966) 1755-1756.
11 J. Gjaldbaek and J. H. Hildebrand, 'The solubility of nitrogen in carbon disulfide, benzene, normal- and cyclo-hexane, and in three fluorocarbons', *J. Am. Chem. Soc., 71* (1949) 3147-3150.
12 J. Gjaldbaek, 'The solubility of hydrogen, oxygen, and carbon monoxide in some non-polar solvents', *Acta Chem. Scand., 6* (1952) 623-633.
13 M. J. Kopac, 'Recent developments in cellular micrurgy', *Trans. New York Acad. Sci., 17* (1955) 257-265.
14 S. Howlett, D. Dundas and D. C. Sabiston, 'Fluid fluorocarbon as oxygenator in experimental extracorporeal circulation', *Arch. Surg., 91* (1965) 643-645.
15 A. E. Gessler, C. E. Bender and M. C. Parkinson, *Trans. New York Acad. Sci., 18* (1956) 701-703.
16 H. A. Sloviter and T. Kamimoto, 'Erythrocyte substitute for perfusion of brain', *Nature, 216* (1967) 458-460.

17 R. P. Geyer, 'Perfluorochemical blood replacement preparations', in *Proceed. IVth Int. Symp. Perfluorochemical Blood Substitutes*, Excerpta Medica, Amsterdam, 1979, pp. 3-32.
18 R. Naito and K. Yokoyama, *Perfluorochemical Blood Substitutes Fluosol-43, Fluosol-DA, 20%, and 35% for preclinical studies as a candidate for erythrocyte substitution*, Green Cross Corp., Osaka, Japan (Tech. Info. Ser. n° 5&7, 1978, 1981).
19 R. P. Geyer, R. G. Monroe and K. Taylor, 'Survival of rats totally perfused with a fluorocarbon-detergent preparation', in J. C. Norman (ed.), *Organ Perfusion and Preservation*, Appleton-Century-Crofts, New York, 1968, pp. 85-96.
20 Proceed. Symposium on Inert Organic Liquids for Biological Oxygen Transport (Atlantic City, 1969), in *Fed. Proceed., 29* (1970) 1695-1820.
21 R. P. Geyer, 'Whole animal perfusion with fluorocarbon dispersions', *Fed. Proceed., 29* (1970) 1758-1763.
22 L. C. Clark, S. Kaplan, F. Becattini and G. Benzing, 'Perfusion of whole animals with perfluorinated liquid emulsions using the Clark bubble-defoam heart-lung machine', *Fed. Proceed., 29* (1970) 1764-1770.
23 L. Triner, M. Verosky, D. V. Habif and G. G. Nahas, 'Perfusion of isolated liver with fluorocarbon emulsions', *Fed. Proceed., 29* (1970) 1778-1781.
24 E. Dellacherie, P. Labrude, C. Vigneron and J. G. Riess, 'Synthetic carriers of oxygen', *Crit. Rev. Therap. Drug Carrier Syst., 3* (1987) 41-94.
25 R. D. Jones, D. A. Summerville and F. Basolo, 'Synthetic oxygen carriers related to biological systems', *Chem. Rev., 79* (1979) 139-179.
26 J. P. Collman, 'Synthetic models for oxygen-binding hemoproteins', *Acc. Chem. Res., 10* (1977) 265-272.
27 J. E. Baldwin and P. Perlmutter, 'Bridged, capped and fenced porphyrins', *Host Guest Complex Chemistry III*, (1984) 181-220.
28 T. G. Traylor, 'Synthetic model compounds for hemoproteins', *Acc. Chem. Res., 14* (1981) 102-109.
29 E. Tsuchida, K. Konda and H. Sata, 'Formation and decomposition of the oxygen complex of ferroprotoporphyrin bound to the water-soluble polymer ligand in aqueous solution', *Inorg. Chem., 15* (1976) 352-357.
30 K. Kobayashi, E. Tsuchida and H. Nishide, 'Totally synthetic hemes: their characteristics and oxygen-carrying capacity in dogs', in E. Tsuchida (ed.), *Artificial Red Cells*, John Wiley & Sons, New York, 1995, pp. 93-116.
31 E. Tsuchida, T. Komatsu, K. Hamamatsu, Y. Matsukawa, A. Tajima, A. Yoshizu, Y. Izumi and K. Kobayashi, 'Exchange transfusion with albumin-heme as an artificial O_2-infusion into anesthetized rats: physiological responses, O_2-delivery and reduction of the oxidized hemin sites by red blood cells', *Bioconj. Chem., 11* (2000) 46-50.
32 Deschiens. De la transfusion lente, par voie hypodermique d'un sang artificiel concentré et stérile contenant les principes immédiats du sang naturel, *Disclosure to the French Academy of Sciences* (1886).
33 A. W. Sellards and G. R. Minot, 'Injection of hemoglobin in man and its relation to blood destruction, with special reference to the animals', *J. Med. Res., 34* (1916) 469-493.
34 R. M. Winslow, *Hemoglobin-based Red Cell Substitutes*, The Johns Hopkins University Press, Baltimore, 1992.
35 J. G. Riess and M. Le Blanc, 'Solubility and transport phenomena in perfluorochemicals relevant to blood substitution and other biomedical applications', *Pure Appl. Chem., 54* (1982) 2383-2406.
36 R. Kumar and B. N. Manjula, 'Human hemoglobin derived from transgenic swine: a starting material for 'blood substitute' formulations', in A. S. Rudolph, R. Rabinovici and G. Z. Feuerstein (eds.), *Red Blood Cell Substitutes*, Dekker, New York, 1998, pp. 309-324.
37 T. M. S. Chang (ed.), *Blood Substitutes: Methods, Products and Clinical Trials*, Karger Landes Systems, New York, 1998.
38 R. N. Winslow, K. D. Vandegriff and R. Intaglietta (eds.), *Advances in Blood Substitutes. Industrial Opportunities and Medical Challenges*, Birkhauser, Boston, 1997.
39 A. S. Rudolph, R. Rabinovici and G. Z. Feuerstein (eds.), *Red Blood Cell Substitutes: Basic Principles and Clinical Applications*, Dekker, New York, 1998.
39a J. G. Riess, 'Les transporteurs d'oxygène (substituts du sang) et leurs composants', in D. Muster (ed.), *Biomatériaux*, Elsevier, Paris, 2001.
40 J. G. Riess, 'Organometallic chemistry, coordination chemistry, main group chemistry. What frontiers nowadays?', *J. Organometal. Chem., 281* (1985) 1-20.
41 G. Santini, M. Le Blanc and J. G. Riess, 'The reaction of perfluoroalkylcopper compounds with 1-bromo-2-perfluoroalkylethylenes', *Tetrahedron, 29* (1973) 2411-2414; M. Le Blanc and J. G. Riess, 'A strategy

for the synthesis of pure, inert perfluoroalkylated derivatives designed for blood substitution', in R. Frey, H. Beisbarth and K. Stosseck (eds.), *Oxygen Carrying Colloidal Blood Substitutes*, Zuckschwerdt Verlag, Munchen, 1982, pp. 43-49.

42 F. Jeanneaux, M. Le Blanc, J. G. Riess and K. Yokoyama, 'Fluorocarbons as gas-carriers for biomedical applications: 1,2-bis(*F*-butyl)ethene as a candidate O_2/CO_2 carrier for second generation blood substitutes', *Nouv. J. Chim., 8* (1984) 251-257.

43 C. Arlen, Y. Gauffreteau, F. Jeanneaux, M. Le Blanc and J. G. Riess, 'Le *F*-isopropyl-1, *F*-hexyl-2 éthène comme transporteur des gaz respiratoires à usage intravasculaire', *Bull. Soc. Chim. Fr.*, (1985) 562-567.

44 F. Jeanneaux, G. Santini, M. Le Blanc, A. Cambon and J. G. Riess, 'Synthèse de tétrakis(perfluoroalkyl)-1,2,3,4 butadiènes-1,3 par duplication des bis(perfluoroalkyl)-1,2 iodoéthènes en présence de cuivre: une nouvelle série de transporteurs de gaz dissous pour usages biologiques', *Tetrahedron, 30* (1974) 4197-4200.

45 M. Le Blanc, G. Santini, J. Guion and J. G. Riess, 'Action des perfluoroalkylcuivre(I) sur les perfluoroalkyléthylènes', *Tetrahedron, 29* (1973) 3195-3201.

46 M. Le Blanc, G. Santini, F. Jeanneaux and J. G. Riess, 'The preparation of 1-hydroperfluorohexyne, octyne and decyne', *J. Fluorine Chem., 7* (1976) 525-530.

47 M. Le Blanc, G. Santini and J. G. Riess, 'Enamines et énaminocétones perfluoroalkylées', *Tetrahedron Lett.*, (1975) 4151-4152.

48 M. Le Blanc, G. Santini, J. Gallucci and J. G. Riess, 'The preparation and spectral characterization of a series of perfluoroalkylenamines and perfluoroalkylenaminoketones', *Tetrahedron, 33* (1977) 1453-1956.

49 J. Gallucci, M. Le Blanc and J. G. Riess, 'Addition of primary amines to perfluoroalkylethynes, Formation of perfluoroalkylenimines, perfluoroalkylenaminoketones, and N,N-bisperfluoroacylethenylamines', *J. Chem. Research* (1978) 5131-5150.

50 J. Gallucci, M. Le Blanc and J. G. Riess, 'Cycloaddition of nitrile oxides to perfluoroalkyl ethenes and ethynes. Synthesis of perfluoroalkylated isoxazoles and isoxazolines', *J. Chem. Research*, (1978) 2529-2544.

51 G. Santini, M. Le Blanc and J. G. Riess, 'Perfluoroalkylcalcium derivatives: reactions on carbonyl compounds', *J. Chem. Soc. Chem. Commun.*, (1975) 678-679.

52 G. Santini, M. Le Blanc and J. G. Riess, 'Some unexpected reactions of perfluoroalkynylmagnesium halides', *J. Organometal. Chem., 102* (1975) C21-24.

53 G. Santini, M. Le Blanc and J. G. Riess, 'Reactions of perfluoroalkycalcium derivatives with ketones and aldehydes', *J. Organometal. Chem., 140* (1977) 1-9.

54 J. G. Riess and M. Le Blanc, 'Perfluorocompounds as blood substitutes', *Angew. Chem. Int. Ed. Eng., 17* (1978) 621-634.

55 R. Madhok, C. D. Forbes and B. L. Evatt (eds.), *Blood, Blood Products and HIV*, Chapman & Hall Medical, London, 1994.

56 K. Yokoyama, R. Naito, Y. Tsuda, C. Fukaya, M. Watanabe, S. Hanada and T. Suyama, 'Selection of 53 PFC substances for better stability of emulsion and improved artificial blood substitutes', *Prog. Clin. Biol. Res., 122* (1983) 189-196.

57 J.-J. Grec, J. G. Riess and B. Devallez, 'Etude de solvants perfluoroalkylés à usage biomédical: températures critiques supérieures de solubilité de bis(*F*-alkyl)éthènes dans l'hexane et vitesses d'excrétion: paramètres de solubilité, grandeurs d'excès de mélanges d'acides carboxyliques et de composés perfluoroalkylés', *Nouv. J. Chimie, 9* (1985) 637-643.

58 M. A. Hamza, G. Serratrice, M. J. Stébé and J. J. Delpuech, 'Solute-solvent interactions in perfluorocarbon solutions of oxygen, An NMR study', *J. Am. Chem. Soc., 103* (1981) 3733-3738.

59 M. P. Krafft, J.-P. Rolland and J. G. Riess, 'Detrimental effect of excess lecithin on the stability of fluorocarbon/lecithin emulsions', *J. Phys. Chem., 95* (1991) 5673-5676.

60 C. Cornélus, F. Giulieri, M. P. Krafft and J. G. Riess, 'Impact of the structure of phospholipid dispersions on the stability of fluorocarbon/phospholipid emulsions for biomedical uses', *Colloid Surf., 70* (1993) 233-238.

61 R. P. Geyer, "Bloodless' rats through the use of artificial blood substitutes', *Fed. Proceed., 34* (1975) 1499-1505.

62 Amer. Chem. Soc., Division of Fluorine Chemistry, Symposium on Organofluorine Compounds in Medicine and Biology (Las Vegas, 1982), abstracts.

63 R. B. Bolin, R. P. Geyer and G. J. Nemo (eds.), *Advances in Blood Substitute Research*, in *Prog. Clin. Biol. Res., 122* (1983).

64 K.K. Tremper (ed.), *Perfluorochemical Oxygen Transport*, in *Int. Anesth. Clin., 23* (1985).

65 J. G. Riess, 'Hemocompatible fluorocarbon emulsions', in C. P. Sharma and M. Szycher (eds.), *Blood Compatible Materials and Devices: Prospectives Towards the 21st Century*, Technomic Publ. Co., Lancaster, 1991, Chap.14, pp. 237-270.

66 K. Yokoyama, C. Fukaya, Y. Tsuda, T. Suyama, R. Naito, K. Yamanouchi and K. Scherer, 'Screening of new perfluorochemicals (PFC's) as candidates for use in fluorochemical emulsion blood substitutes – structures and biological properties' (abstract), in *Amer. Chem. Soc. Natl. Meeting Abstracts*, Las Vegas, 1982.

67 R. E. Moore and L. C. Clark, 'Synthesis and physical properties of perfluorocompounds useful as synthetic blood candidates', in R. Frey, H. Beisbarth and K. Stosseck (eds.), *Oxygen Carrying Colloidal Blood Substitutes*, Zuckschwerdt Verlag, Mainz, 1982, pp. 50-60.

68 K. Yokoyama, K. Yamanouchi and T. Suyama, 'Recent advances in a perfluorochemical blood substitute and its biomedical application', *Life Chem. Reports, 2* (1983) 35-92.

69 K. Yokoyama, T. Suyama and R. Naito, 'Development of perfluorochemical (PFC) emulsion as an artificial blood substitute', in R. Filler and Y. Kobayashi (eds.), *Biomedical Aspects of Fluorine Chemistry*, Elsevier Biomedical, Amsterdam, 1982, pp. 191-212.

70 J. G. Riess, 'Reassessment of criteria for the selection of perfluorochemicals for second-generation blood substitutes: Analysis of structure/property relationships', *Artif. Organs, 8* (1984) 44-56.

71 R. Naito and K. Yokoyama, 'An improved perfluorodecalin emulsion', in G.A. Jamieson and T.J. Greenwalt (eds.), *Blood Substitutes and Plasma Expanders*, A. Liss, New York, 1978, pp. 81-89.

72 J. G. Riess, 'Post-Fluosol progress in fluorocarbon emulsions for in vivo oxygen delivery', *La Trasf. del Sangue, 32* (1987) 316-334.

73 H. Ohyanagi, Y. Saitoh, T. Mitsuno, M. Watanabe, K. Yamaouchi and K. Yokoyama, 'A new perfluorochemical emulsion: An overview', *Art. Organs, 14* (1990) 199-200.

74 J. G. Riess, 'Fluorocarbon-based in vivo oxygen transport and delivery systems', *Vox Sang., 61* (1991) 225-239.

75 S. A. Gould, A. L. Rosen, L. R. Sehgal, H. L. Sehgal, L. A. Langdale, L. M. Krause, C. L. Rice, W. H. Chamberlain and G. S. Moss, 'Fluosol-DA as a red-cell substitute in acute anemia', *New England J. Med., 314* (1986) 1653-1656; see also the correspondance this paper aroused in the *New England J. Med., 315* (1987) 1677-1678.

76 K. K. Tremper, E. M. Levine and K. Waxman, 'Clinical experience with Fluosol-DA (20%) in the United States', *Int. Anesthesiol. Clin., 23* (1985) 185-197.

77 R. K. Spence, S. McCoy, J. Costabile, E. D. Norcross, M. J. Pello, J. B. Alexander, C. Wisdom and R. C. Camishion, 'Fluosol DA-20 in the treatment of severe anemia: Randomized, controlled study of 46 patients', *Crit. Care Med., 18* (1990) 1127-1230.

78 K. M. Kent, M. W. Cleman, M. J. Cowley, M. Forman, C. C. Jaffe, M. Kaplan, S. B. King, M. Krucoff, T. Lassar, B. McAuley, R. Smith, C. Wisdom and D. Wohlgelernter, 'Reduction of myocardial ischemia during percutaneous transluminal coronary angioplasty with oxygenated Fluosol', *Am. J. Cardiol., 66* (1990) 279-284.

79 M. B. Forman, J. M. Perry, B. H. Wilson, M. S. Verani, P. R. Kaplan, F. A. Shawl and G. C. Friesinger, 'Demonstration of myocardial reperfusion injury in humans: Results of a pilot study utilizing acute coronary angioplasty with perfluorochemical in anterior myocardial infarction', *J. Am. Coll. Cardiol., 18* (1991) 911-918.

80 G. R. Ivanitsky and S. I. Vorobyev, *Perftoran Blood Substitute with Gas-Transporting Function*, Perftoran Co., Pushchino, Russia, 1997.

81 K. Vaiciulis, Z. Dobrovolskiene, N. Petrauskiene and G. Zukauskas, 'Perfluorocarbon emulsion as a new medication in intensive care unit', *Acta Med. Lituanica*, (1995) 3-8.

82 F. F. Beloyarstev, E. I. Mayevsky and B. I. Islamov, *Ftorosan-oxygen carrying perfluorochemical plasma substitute*, Acad. Sci. USSR, Pushchino, 1983.

83 L. C. Clark, F. Becattini, S. Kaplan, V. Obrock, D. Cohen and C. Becker, 'Perfluorocarbons having a short dwell time in the liver', *Science, 181* (1973) 680-682.

84 T. Fujita, T. Sumaya and K. Yokoyama, 'Fluorocarbon emulsion as a candidate for artificial blood', *Eur. Surg. Res., 3* (1973) 436-453.

85 R. E. Moore and L. C. Clark, 'Chemistry of fluorocarbons in biomedical use', *Int. Anesth. Clin., 23* (1985) 11-24.

86 R. M. Mitten, A. R. Burgan, A. Hamblin, G. Yee, D. C. Long, D. M. Long and R. F. Mattrey, 'Dose related biodistribution and elimination of 100% PFOB emulsion', *Biomat., Artif. Cells, Artif. Organs, 16* (1988) 683-684.

87 J. G. Riess and M. Le Blanc, 'Preparation of perfluorochemical emulsions for biomedical use: principles, materials and methods', in K. C. Lowe (ed.), *Blood Substitutes: Preparation, Physiology, and Medical Applications*, Ellis Horwood Ltd., Chichester, 1988, pp. 94-129.

88 J. G. Weers, 'A physicochemical evaluation of perfluorochemicals for oxygen transport applications', *J. Fluorine. Chem., 64* (1993) 73-93.

89 S. S. Davis, H. P. Round and T. S. Purewal, 'Ostwald ripening and the stability of emulsion systems: an explanation for the effect of an added third component', *J. Colloid Interf. Sci., 80* (1981) 508-511.

90 A. S. Kabalnov and E. D. Shchukin, 'Ostwald ripening theory: applications to fluorocarbon emulsion stability', *Adv. Colloid Interf. Sci., 38* (1992) 69-97.

91 L. Trevino, L. Solé-Violan, P. Daumur, B. Devallez, M. Postel and J. G. Riess, 'Molecular diffusion in concentrated fluorocarbon emulsions and its effect on emulsion stability', *New J. Chem., 17* (1993) 275-278.

92 J. G. Weers and R. A. Arlauskas, 'Sedimentation field-flow fractionation studies of Ostwald ripening in fluorocarbon emulsions containing two disperse phase components', *Langmuir, 11* (1995) 474-477.

93 I. M. Lifshitz and V. V. Slezov, 'Kinetics of diffuse decomposition of supersaturated solid solutions', *Soviet Physics Jept., 35* (1959) 331-339.

94 W. I. Higuchi and J. Misra, 'Physical degradation of emulsions via the molecular diffusion route and the possible prevention thereof', *J. Pharm. Sci., 51* (1962) 459-466.

95 M. Postel, J. G. Riess and J. G. Weers, 'Fluorocarbon emulsions: the stability issue', in *Ref. 8*, pp. 991-1005 (1994).

96 J. G. Riess and M. P. Krafft, 'Elaboration of fluorocarbon emulsions with improved oxygen-carrying capabilities', *Adv. Exp. Med. Biol., 317* (1992) 465-472.

97 W. A. Sheppard and C. M. Sharts, *Organic Fluorine Chemistry*, Benjamin, New York, 1969.

98 C. M. Sharts, 'William A. Sheppard', *J. Fluorine Chem., 90* (1998) 197-199.

99 R. J. Kaufman, 'Clinical development of perfluorocarbon-based emulsions as red cell substitutes', in R. M. Winslow, K. D. Vandegriff and M. Intagietta (eds.), *Blood Substitutes: Physiological Basis of Efficacy*, Birkhauser, Boston, 1995, pp. 53-75.

100 'Hemosol poll indicates future public demand for blood substitute products in US and Canada', *Canada NewsWire,* (April 30, 1998).

101 H. L. Rosano and W. E. Gerbacia, 'Fluorocarbon microemulsions', *US Patent 3,778,381* (1973).

102 D. M. Long, C. B. Higgins, R. F. Mattrey, R. M. Mitten and F. K. Multer, 'Is there a time and place for radiopaque fluorocarbons?', in R. E. Banks (ed.), *Preparation, Properties, and Industrial Applications of Organofluorine Compounds*, Ellis Horwood Ltd, New York, 1982, pp. 139-156.

103 S. F. Flaim, 'Pharmacokinetics and side effects of perfluorocarbon-based blood substitutes', in *Ref. 8,* pp. 1043-1054 (1994).

104 P. E. Keipert, S. Otto, S. F. Flaim, J. G. Weers, E. A. Schutt, T. J. Pelura, D. H. Klein and T. L. Yaksh, 'Influence of perflubron emulsion particle size on blood half-life and febrile response in rats', in *Ref. 8,* pp. 1169-1174 (1994).

105 J. G. Weers, J. Liu, T. Fields, P. Resch, J. Cavin and R. A. Arlauskas, 'Room temperature stable perfluorocarbon emulsions with acceptable half-lives in the reticuloendothelial system', in *Ref. 8,* pp. 1175-1182 (1994).

106 L. C. Clark, E. W. Clark, R. E. Moore, D. G. Kinnett and E. I. Inscho, 'Room temperature-stable biocompatible fluorocarbon emulsions', *Prog. Clin. Biol. Res., 122* (1983) 169-180.

107 G. Mathis, P. Leempoel, J. C. Ravey, C. Selve and J. J. Delpuech, 'A novel class of nonionic microemulsions: Fluorocarbons in aqueous solutions of fluorinated poly(oxyethylene) surfactants', *J. Am. Chem. Soc., 106* (1984) 6162-6171.

108 J. G. Riess, 'Blood substitutes: Where do we stand with the fluorocarbon approach?', *Current Surg., 45* (1988) 365-370.

109 P. Chabert, L. Foulletier and A. Lantz, 'Procédé de préparation de liquides à applications biologiques et transporteurs d'oxygène', *Fr. Patent 2,452,513* (1975).

110 K.-V. Schubert and E. W. Kaler, 'Microemulsifying fluorinated oils with mixtures of fluorinated and hydrogenated surfactants', *Colloids Surf., 84* (1994) 97-106.

111 E. Kissa, *Fluorinated Surfactants, Synthesis, Properties, Applications*, Marcel Dekker, New York, 1994.

112 J. G. Riess and M. P. Krafft, 'Fluorinated materials for in vivo oxygen transport (blood substitutes), diagnosis and drug delivery', *Biomaterials, 19* (1998) 1529-1539.

113 J. G. Riess, C. Arlen, J. Greiner, M. Le Blanc, A. Manfredi, S. Pace, C. Varescon and L. Zarif, 'Design, synthesis, and evaluation of fluorocarbons and surfactants for in vivo applications: New perfluoroalkylated polyhydroxylated surfactants', *Biomat., Artif. Cells, Artif. Organs, 16* (1988) 421-430.

114 J. Greiner, J. G. Riess and P. Vierling, 'Fluorinated surfactants intended for biomedical uses', in R. Filler, Y. Kobayashi and L. Yagupolskii (eds.), *Organofluorine Compounds in Medicinal Chemistry and Biomedical Applications*, Elsevier, New York, 1993, pp. 339-380.

115 J. G. Riess, 'Highly fluorinated systems for oxygen transport, diagnosis and drug delivery', *Colloids Surf., 84* (1994) 33-48.

116 M. Le Blanc, J. G. Riess, D. Poggi and R. Follana, 'Use of lymphoblastoid Namalva cell cultures in a toxicity test. Application to the monitoring of detoxification procedures for fluorocarbons to be used as intravascular oxygen-carriers', *Pharm. Res.*, (1985) 246-248.

117 A. M. Mahé, J. Manoux, A. Valla, R. Follana, L. Zarif, J. Greiner, P. Vierling and J. G. Riess, 'Perfluoroalkylated surfactants: Relationships between structure and acute toxicity in mice', *Biomat., Art. Cells., Immob. Biotech., 20* (1992) 1025-1028.

118 J. G. Riess, S. Pace and L. Zarif, 'Highly effective surfactants with low hemolytic activity', *Adv. Mater., 3* (1991) 249-251.

119 J. G. Riess, 'Fluorinated vesicles', *J. Drug Targeting, 2* (1994) 455-468.

120 A. Milius, J. Greiner and J. G. Riess, 'Fundamental differences of behaviour between some neutral and anionic sugar-derived perfluoroalkylated surfactants', *New J. Chem., 16* (1992) 771-773.

121 J. G. Riess, J. Greiner, S. Abouhilale and A. Milius, 'Stabilization of fluorocarbon emulsions by sugar-derived perfluoroalkylated surfactants and co-surfactants', *Prog. Coll. Polym. Sci., 88* (1992) 123-130.

122 L. Zarif, A. Manfredi, C. Varescon, M. Le Blanc and J. G. Riess, 'Synergistic stabilization of perfluorocarbon-Pluronic F-68 emulsions by perfluoroalkylated polyhydroxylated surfactants', *J. Amer. Oil Chem. Soc., 66* (1989) 1515-1523.

123 D. M. Worah, D. R. Kessler, A. R. Meuter, M. Huang, J.-M. Correas and S. C. Quay, 'Perflenapent emulsion. Ultrasound contrast media', *Drug Future, 22* (1997) 378-385.

124 J. G. Riess, L. Solé-Violan and M. Postel, 'A new concept in the stabilization of injectable fluorocarbon emulsions: the use of mixed fluorocarbon-hydrocarbon dowels', *J. Disp. Sci. Technol., 13* (1992) 349-355.

125 J. G. Riess, C. Cornélus, R. Follana, M. P. Krafft, A. M. Mahé, M. Postel and L. Zarif, 'Novel fluorocarbon-based injectable oxygen-carrying formulations with long-term room-temperature storage stability', *Adv. Exp. Biol. Med., 345* (1994) 227-234.

126 H. 'Meinert and A. Knoblich, 'The use of semifluorinated alkanes in blood substitutes', *Biomat., Artif. Cells, Immob. Biotech., 21* (1993) 583-595.

127 C. Cornélus, M. P. Krafft and J. G. Riess, 'About the mechanism of stabilization of fluorocarbon emulsions by mixed fluorocarbon/hydrocarbon additives', *J. Colloid Interf. Sci., 163* (1994) 391-394.

128 N. S. Faithfull, 'Mechanisms and efficacy of fluorochemical oxygen transport and delivery', in *Ref. 7*, pp. 181-197 (1994).

129 J. Lutz, 'Effect of perfluorochemicals on host defense, especially on the reticuloendothelial system', *Int. Anesth. Clin., 23* (1985) 63-93.

130 Y. Tsuda, K. Yamanouchi, K. Yokoyama and T. Suyama, 'Discussion and considerations for the excretion mechanism of perfluorochemical emulsion', *Biomat., Artif. Cells, Artif. Organs, 16* (1988) 473-484.

131 W. R. Ravis, J. F. Hoke and D. L. Parsons, 'Perfluorochemical erythrocyte substitutes: disposition and effects on drug distribution and elimination', *Drug Metabolism Rev., 23* (1991) 375-411.

132 V. V. Obraztsov, A. S. Kabalnov, A. N. Sklifas and K. N. Makarov, 'A novel model describing excretion of perfluorochemicals from living organisms – dissolution of fluorocarbons in lipid components of blood', *Biofizika, 37* (1992) 379-383.

133 Y. Ni, D. H. Klein and D. Song, 'Recent developments in pharmacokinetic modeling of perfluorocarbon emulsion', *Art. Cells, Blood Subst., Immob. Biotech., 24* (1996) 81-90.

134 D. E. Hammerschmidt and G. M. Vercellotti, 'Limitation of complement activation by perfluorocarbon emulsions: Superiority of lecithin-emulsified preparations', *Biomat, Artif. Cells, Artif. Organs, 16* (1988) 431-438.

135 W. L. Holman, D. C. McGiffin, W. V. A. Vicente, R. D. Spruell and A. D. Pacifico, 'Use of current generation perfluorocarbon emulsions in cardiac surgery', in *Ref. 8,* pp. 979-990 (1994).

136 R. A. Kloner and S. Hale, 'Cardiovascular applications of fluorocarbons in regional ischemia/reperfusion', in *Ref. 8,* pp. 1069-1081 (1994).

137 J. D. Ogilby, 'Cardiovascular applications of fluorocarbons: current status and future direction. A critical clinical appraisal', in *Ref. 8,* pp. 1083-96 (1994).

138 B. Bose, J. L. Osterholm and A. Triolo, 'Focal cerebral ischemia: reduction in size of infarcts by ventriculo-subarachnoid perfusion with fluorocarbon emulsion', *Brain Research, 328* (1985) 223-231.

139 W. L. Holman, R. D. Spruell, E. R. Ferguson, J. J. Clymer, W. V. A. Vicente, C. P. Murrah and A. D. Pacifico, 'Tissue oxygenation with graded dissolved oxygen delivery during cardiopulmonary bypass', *J. Thorac. Cardiovasc. Surg., 119* (1995) 774-785.

140 R. S. Mosca, T. J. Rohs, R. R. Waterford, K. F. Childs, L. A. Brunsting and S. F. Bolling, 'Perfluorocarbon supplementation and postischaemic cardiac function', *Surgery, 120* (1996) 197-204.

141 P. Ménasché, E. Pinard, A. Desroches, J. Seylaz, P. Laget, R. P. Geyer and A. Piwnica, 'Fluorocarbons: a potential treatment of cerebral air embolism in open-heart surgery', *Ann. Thorac. Surg., 40* (1985) 494-497.

142 B. D. Spiess, B. Braverman, A. W. Woronowicz and A. D. Ivankovich, 'Protection from cerebral air emboli with perfluorocarbons in rabbits', *Stroke, 17* (1986) 1146-1149.

143 R. G. Evans, B. F. Kimler, R. A. Morantz and S. Batnitzky, 'Lack of complications in long-term survivors after treatment with Fluosol and oxygen as an adjuvant to radiation therapy for high-grade brain tumors', *Int. J. Radiat. Oncol. Biol. Phys., 26* (1993) 649-652.

144 S. Rockwell, 'Perfluorochemical emulsions and radiation therapy', in *Ref. 8,* pp. 1097-1108 (1994).

145 B. A. Teicher, 'An overview on oxygen carriers in cancer therapy', *Art. Cells, Blood Subst., Immob. Biotech., 23* (1995) 395-405.

146 S. Stern and M. Guichard, 'Efficacy of agents counteracting hypoxia in fractionated radiation regimes', *Radiother. Oncol., 41* (1996) 143-149.

147 A. R. Smith, W. Van Alphen, N. S. Faithfull and M. Fennema, 'Limb preservation in replantation surgery', *J. Plast. Reconstr. Surg., 75* (1985) 227-237.

148 L. Brazile, J. Clarke, E. P. Green and C. Haisch, *Transplant. Proceed., 28* (1996) 349-351.

149 E. J. Voiglio, L. Zarif, F. C. Gorry, M. P. Krafft, J. Margonari, X. Martin, J. Riess and J. M. Dubernard, 'Aerobic preservation of organs using a new perflubron/lecithin emulsion stabilized by molecular dowels', *J. Surg. Res., 63* (1996) 439-446.

150 R. J. Thurston, M. S. Rogoff, T. R. Scott and N. Korn, 'Effects of perfluorochemical diluent additives on fertilizing capacity of turkey semen', *Poultry Science, 72* (1993) 598-602.

151 P. Anthony, K. C. Lowe, M. R. Davey and J. B. Power, 'Strategies for promoting division of cultured plant protoplasts: beneficial effects of oxygenated perfluorocarbon', *Biotech. Techniques, 9* (1995) 777-782.

152 R. F. Mattrey, 'The potential role of perfluorochemicals (PFCS) in diagnostic imaging', in *Ref. 7,* pp. 295-313 (1994).

153 E. G. Schutt, T. J. Pelura and R. M. Hopkins, 'Osmotically stabilized microbubbles sonographic contrast agents', *Acad. Radiol., 35* (1996) 188-190; E. G. Schutt, J. G. Riess, L. Dellamare, D. Klein and L. Trevino, 'Perfluorohexane as an osmotic stabiliser of a microbubble ultrasound contrast agent', *16th Int. Symp. Fluorine Chemistry*, Durham, UK, July 2000, Abstract 1P-121.

154 M. P. Krafft and J. G. Riess, 'Highly fluorinated amphiphiles and colloidal systems, and their applications in the biomedical field – A contribution', *Biochimie, 80* (1998) 489-514.

155 J. J. Grec, B. Devallez, H. Marcovich and J. G. Riess, 'Emploi de dérivés perfluorés pour le contrôle des cultures cellulaires', in *Colloque National de Génie Biologique et Médical*, Toulouse, 1982.

156 M. Mathy-Hartert, M. P. Krafft, C. Deby, G. Deby-Dupont, M. Meurisse, M. Lamy and J. G. Riess, 'Effects of perfluorocarbon emulsions on cultured human endothelial cells', *Art. Cells, Blood. Subst., Immob. Biotech., 25* (1997) 563-575.

157 M. Mathy-Hartert, C. Deby, M. P. Krafft, G. Deby, J. J. Bradfer, A. DeRoover and M. Lamy, 'Absence of effects on human endothelial cells after prolonged contact with a fluorocarbon emulsion stabilized with molecular dowels, as compared to Eurocollins or University of Wisconsin solutions', *Transplantation* (in press).

158 E. Pinard, A. M. Desroches, M. Le Blanc, R. Charbonne and J. G. Riess, 'Influence des fluorocarbures sur l'oxygénation et la circulation sanguine du cerveau', in *Proc. Colloque Le Globule Rouge et autres Transporteurs d'Oxygène*, Lyon, France, 1987.

159 V. Sanchez, L. Zarif, J. Greiner, J. G. Riess, S. Cippolini and J. N. Bruneton, 'Novel injectable fluorinated contrast agents with enhanced radiopacity', in *Ref. 8*, pp. 1421-1428 (1994).

160 J. N. Bruneton, M. N. Falewee, E. François, P. Cambon, J. G. Riess, C. Balu-Maestro and A. Rogopoulos, 'Preliminary clinical results using perfluorooctylbromide (PFOB) for CT imaging of the liver, spleen and vessels', *Radiology, 170* (1989) 179-183.

161 G. L. Wolf, D. M. Long and J. G. Riess, 'Percutaneous lymphography with PFOB emulsions', *Radiology, 177* (1990) 366.

162 E. Lartigau, C. Thomas, M. Le Blanc, J. G. Riess, D. Long, C. Long, E. P. Malaise and M. Guichard, 'New high concentration O_2 carrying perfluorochemical emulsions: Toxicity, radiosensitivity of GM-CFC and development of metastases in mice', *Int. J. Radiat. Oncol. Biol. Phys., 16* (1989) 1153-1156.

163 C. Thomas, J. G. Riess and M. Guichard, 'Influence of the 100% w/v perfluorooctylbromide (PFOB) emulsion dose on tumour radiosensitivity', *Int. J. Radiat. Biol., 59* (1991) 433-445.

164 S. Khalfoun, P. Janin, M. C. Machet, B. Arbelle, M. Lacord, A. Locatelli, H. Salmon, J. G. Riess, Y. Gruel, H. Nivet, P. Bardos and Y. Lebranchu, 'Xenogenic cellular interaction in an ex vivo model of pig kidney perfused with human lymphocytes', *Transplant. Proceed., 27* (1995) 2210-2211.

165 L. Bouley, M. P. Krafft, P. Dutoit, P. Bercik, J. G. Riess and P. Kucera, 'Viability of the rat ileum perfused with various oxygen carriers', *VIII Int. Symp. Blood Substitutes*, Tokyo, 1997.

166 D. C. Long, D. M. Long, J. G. Riess, R. Follana, A. Burgan and R. F. Mattrey, 'Preparation and application of highly concentrated perfluoroocyl bromide fluorocarbon emulsions', in T. M. S. Chang and R. P. Geyer (eds.), *Blood Substitutes*, Dekker, New York, 1989, pp. 441-443.

167 M. P. Krafft, M. Postel, J. G. Riess, Y. Ni, T. J. Pelura, G. K. Hanna and D. Song, 'Drop size stability assessment of fluorocarbon emulsions', *Biomat., Artif. Cells, Immob. Biotech., 20* (1992) 865-868.

168 J. G. Riess, J. L. Dalfors, G. K. Hanna, D. H. Klein, M. P. Krafft, T. J. Pelura and E. G. Schutt, 'Development of highly fluid, concentrated and stable fluorocarbon emulsions for diagnosis and therapy', *Biomat., Artif. Cells, Artif. Organs, 20* (1992) 839-842.

169 J. G. Riess, S. F. Flaim, D. H. Klein and J. G. Weers, 'The relative physicochemical and biological attributes of perflubron emulsions', in S. I. Vorobyev and G. R. Ivanitsky (eds.), *Proc. Int. Symp. Blood Substitutes*, Russian Academy of Sciences, Pushchino, 1996, pp. 73-90.

170 M. Postel, P. Chang, J. P. Rolland, M. P. Krafft and J. G. Riess, 'Fluorocarbon/lecithin emulsions: identification of EYP-coated fluorocarbon droplets and fluorocarbon-empty vesicles by freeze-facture electron microscopy', *Biochim. Biophys. Acta., 1086* (1991) 95-98.

171 C. Cornélus, M. P. Krafft and J. G. Riess, 'Filtration as an alternative to heat-sterilization for injectable fluorocarbon emulsions', in *Proc. 1st. World Congress on Emulsion*, Paris, 1993.

172 J. Lutz and M. P. Krafft, 'Longitudinal studies on the interaction of perfluorochemicals with liver cytochromes P-450 by means of testing the route of detoxification of pentobarbital', *Adv. Exp. Med. Biol., 411* (1997) 391-395.

173 C. Der Mardirossian, M. P. Krafft, T. Gulik-Krzywicki, M. Le Maire and F. Lederer, 'On the lack of protein-solubilization properties of two perfluoroalkylated detergents, as tested with neutrophil plasma membranes', *Biochimie, 80* (1998) 531-541.

174 M. P. Krafft, F. Jeanneaux, F. Giulieri and M. Goldmann, 'Monolayers made from double-chain fluorosurfactants derived from monomorpholinophosphate' (in preparation).

175 Y. Ferro, F. Giulieri and M. P. Krafft, 'Disorganizing effect of the hydrocarbon spacer in monolayer behavior of perfluoroalkyl surfactants' (in preparation).

176 T. Imae, K. Funayama, M. P. Krafft, F. Giulieri, T. Tada and M. Matsumoto, 'Small-angle scattering and electron microscopy investigation of nanotubules made from a perfluoroalkylated glucophospholipid', *J. Colloid Interf. Sci., 212* (1999) 330-337.

177 N. S. Faithfull, G. E. Rhoades, P. E. Keipert, A. S. Ringle and A. Trouwborst, *Adv. Exp. Med. Biol., 361* (1994) 139.

178 N. S. Faithfull and G. E. Roades, 'Systems for measuring blood oxygen levels', *U.S. Patent 5,634,461* (1997).

179 P. E. Keipert, N. S. Faithfull, J. D. Bradley, D. Y. Hazard, J. Hogan, M. S. Levisetti and R. M. Peters, 'Oxygen delivery augmentation by low-dose perfluorochemical emulsion during profound normovolemic hemodilution', *Adv. Exp. Med. Biol., 345* (1994) 197-204.

180 S. F. Flaim, 'Perflubron-based emulsion: efficacy as a temporary oxygen carrier', in R. M. Winslow, K. D. Vandegriff and M. Intaglietta (eds.), *Blood Substitutes: New Frontiers*, Birkhauser, Boston, 1997, pp. 91-132.
181 O. P. Habler, M. S. Kleen, J. W. Hutter, A. H. Podtschaske, M. Tiede, G. I. Kemming, M. V. Welte, C. O. Corso, S. Batra, P. E. Keipert, N. S. Faithfull and K. F. W. Messmer, 'Hemodilution and intravenous perflubron emulsion as an alternative to blood transfusion: effects on tissue oxygenation during profound hemodilution in anesthetized dogs', *Transfusion, 38* (1998) 145-155.
182 K. K. Tremper, A. E. Friedman, E. M. Levine, R. Lapin and D. Camarillo, 'The preoperative treatment of severely anemic patients with a perfluorochemical oxygen-transport fluid, Fluosol-DA', *New Engl. J. Med., 307* (1982) 277-283.
183 T. Mitsuno and H. Ohyanagi, 'Present status of clinical studies of Fluosol-DA (20%) in Japan', *Int. Anesth. Clin., 23* (1985) 169-183.
184 J. Guo, J. White and H. Batjer, 'Intravenous perflubron emulsion administration improves the recovery of auditory evoked potentials after temporary brain stem ischemia in dogs', *Neurosurgery, 36* (1995) 350-357.
185 R. D. Braun, R. A. Linsenmeier and T. K. Goldstick, 'New perfluorocarbon emulsion improves tissue oxygenation in cat retina', *J. Appl. Physiol., 72* (1992) 1960-68.
186 K. F. Waschke, M. Riedel, D. M. Albrecht, K. van Ackern and W. Kuschinsky, 'Effects of a perfluorocarbon emulsion on regional cerebral blood flow and metabolism after fluid resuscitation from hemorrhage in conscious rats', *Anesth. Analg., 79* (1994) 874-882.
187 J. G. Riess and P. E. Keipert, 'Update on perfluorocarbon-based oxygen delivery systems', in E. Tsuchida (ed.), *Blood Substitutes – Present and Future Perspectives*, Elsevier Science Publ., Amsterdam, 1998, Chap. 7, pp. 91-101.
188 L. C. Clark, R. E. Hoffmann and S. L. Davis, 'Response of the rabbit lung as a criterion of safety for fluorocarbon breathing and blood substitutes', *Biomat., Artif. Cells, Immob. Biotech., 20* (1992) 1085-1099.
189 E. Schutt, P. Barber, T. Fields, S. Flaim, J. Horodniak, P. Keipert, R. Kinner, L. Kornbrust, T. Leakakos, T. Pelura, J. Weers, R. Houmes and B. Lachmann, 'Proposed mechanism of pulmonary gas trapping (PGT) following intravenous perfluorocarbon emulsion administration', in *Ref. 8,* pp. 1205-1214 (1994).
190 T. Leakakos, E. G. Schutt, J. C. Cavin, D. Smith, J. D. Bradley, C. A. Strnat, U. Del Balzo, D. Y. Hazard, S. Otto, T. K. Fields, P. E. Keipert, D. H. Klein and S. F. Flaim, 'Pulmonary gas trapping differences among animal species in response to intravenous infusion of perfluorocarbon emulsions', in *Ref. 8,* pp. 1199-1204 (1994).
191 S. L. Mulvagh, D. A. Foley, B. C. Aeschbacher, K. K. Klarich and J. B. Seward, 'Second harmonic imaging of an intravenously administered echocardiographic contrast agent', *J. Am. Coll. Cardiol., 27* (1996) 1519-1525.
192 P. E. Keipert, N. S. Faithfull, D. J. Roth, J. D. Bradley, S. Batra, P. Jochelson and K. E. Flaim, 'Supporting tissue oxygenation during acute surgical bleeding using a perfluorochemical-based oxygen carrier', *Adv. Exp. Med. Biol., 388* (1996) 603-609.
193 D. R. Spahn, R. van Brempt, G. Theilmeier, J.-P. Reibold, M. Welte, H. Heinzerling, K. M. Birck, P. E. Keipert, K. Messmer, 'Perflubron emulsion delays blood transfusions in orthopedic surgery', *Anesthesiology, 91* (1999) 1195-1208; D. R. Spahn, 'Current status of artificial oxygen carriers', *Adv. Drug Del. Res., 40* (2000) 143-151.
194 P. Ménasché, E. Pinard, A. Desroches, J. Seylaz, P. Laget, R. P. Geyer and A. Piwnica, 'Fluorocarbons: A potential treatment of cerebral air embolism in open-heart surgery', *Ann. Thorac. Surg., 54* (1992) 392-3.
195 B. D. Spiess and R. P. Cochran, 'Perfluorocarbon emulsions and cardiopulmonary bypass: a technique for the future', *J. Cardiothorac. Vasc. Surg., 10* (1996) 83-90.
196 D. W. Amory, B. Leone, N. D. Croughwell, W. D. White, K. M. B. Richardson, B., C. F. Osgood, L. Gerstle and M. F. Newman, 'Perflubron emulsion administration during cardiopulmonary bypass – preliminary results of increased cerebral blood flow' (abstract), *Anesth. Analg., 86* SCA6 (1998).
197 K. M. Sekins, L. Nugent, M. Mazzoni, C. Flanagan, L. Neer, A. Rozenberg and J. Hoffman, 'Recent innovations in total liquid ventilation system and component design', *Biomed. Instru. Technol., 33* (1999) 277-284.
198 R. Hirschl, T. Pranikoff, P. Gauger, R. Schreiner, R. Dechert and R. Bartlett, 'Liquid ventilation in adults, children, and full-term neonates', *Lancet, 346* (1995) 1201-1202.
199 D. M. Steinhorn, M. C. Papo, A. T. Rotta, A. Aljada, B. P. Fuhrman and P. Dandora, 'Liquid ventilation attenuates pulmonary oxidative damage', *J. Crit. Care, 14* (1999) 20-28.

200 C. L. Leach, J. S. Greenspan, D. Rubenstein, T. H. Shaffer, M. R. Wolfson, J. C. Jackson, R. DeLemos and B. P. Fuhrman, 'Partial liquid ventilation with perflubron in premature infants with severe respiratory distress syndrome', *New Engl. J. Med., 335* (1996) 761-766.

201 J. S. Greenspan, M. R. Wolfson and T. H. Shaffer, 'Liquid ventilation: clinical experience', *Biomed. Instru. Technol., 33* (1999) 253-259.

202 L. Galiuto, A. N. DeMaria, K. May-Neuman, U. Del Balzo, K. Ohmori, V. Bargawa, S. F. Flaim and S. Iliceto, 'Evaluation of dynamic changes in microvascular flow during ischemia-reperfusion by myocardial contrast echocardiography', *J. Am. Coll. Cardiol., 32* (1998) 1096-1101.

203 R. E. Banks, B. E. Smart and J. C. Tatlow, *Organofluorine Chemistry, Principles and Commercial Applications*, Plenum Press, New York, 1994.

204 R. E. Banks, 'Fluorine the enabler', in R. E. Banks and K. C. Lowe (eds.), *Fluorine in Medicine in the 21st Century*, Rapra Technol, Paper 1, 1994, pp. 1-17.

205 J. G. Riess and M. P. Krafft, 'Fluorinated phosphocholine-based amphiphiles as components of fluorocarbon emulsions and fluorinated vesicles', *Chem. Phys. Lipids, 75* (1995) 1-14.

206 J. G. Riess, F. Frézard, J. Greiner, M. P. Krafft, C. Santaella, P. Vierling and L. Zarif, 'Membranes, vesicles, and other supramolecular systems made from fluorinated amphiphiles', in Y. Barenholz and D. D. Lasic (eds.), *Handbook of Nonmedical Applications of Liposomes*, CRC Press, Boca Raton, Vol. III, 1996, Chapter 8, pp. 97-141.

207 J. G. Riess, 'Du fluor dans nos artères', *New J. Chem., 19* (1995) 891-909.

208 V. M. Sadtler, M. P. Krafft and J. G. Riess, 'Achieving stable, reverse water-in-fluorocarbon emulsions', *Angew. Chem., Int. Ed. Engl., 35* (1996) 1976-1978.

209 M. P. Krafft, V. Sadtler and J. G. Riess, 'Water-in-fluorocarbon emulsions for pulmonary drug delivery', *Proceed. Int. Symp. Control. Rel. Bioact. Mater., 22* (1995) 464-465.

210 J. G. Riess and J. G. Weers, 'Emulsions for biomedical uses', *Curr. Opinion Colloid Interf. Sci., 1* (1996) 652-659.

211 M. P. Krafft, V. M. Sadtler and J. G. Riess, 'Multiple emulsions with a fluorocarbon continuous phase', *Intl. Symp. Fluorine Chemistry*, Vancouver, 1997.

212 M. P. Krafft, 'Fluorocarbon gels', in S. Magdassi and E. Touitou (eds.), *Novel Cosmetic Delivery Systems*, Dekker, New York, 1998, Chap. 10, pp. 195-219.

213 J. C. Ravey and M. J. Stébé, 'Water-in-fluorocarbon gel emulsions; structures and rheology', *Colloids Surf., 91* (1990) 237-257.

214 M. P. Krafft and J. G. Riess, 'Stable highly concentrated fluorocarbon gels', *Angew. Chem., Int. Ed. Engl., 33* (1994) 1100-1101.

215 M. P. Krafft and J. G. Riess, 'Gels inverses avec une phase fluorocarbure continue', *Fr. Pat. 2,737,135* (1995).

216 M. P. Krafft, F. Giulieri and J. G. Riess, 'Can single-chain perfluoroalkylated amphiphiles alone form vesicles and other organized supramolecular systems', *Angew. Chem., Int. Ed. Engl., 32* (1993) 741-743.

217 F. Giulieri, M. P. Krafft and J. G. Riess, 'Stable flexible fibers and rigid tubules made from single-chain perfluoroalkylated amphiphiles', *Angew. Chem., Int. Ed. Engl., 33* (1994) 1514-1515.

218 F. Giulieri and M. P. Krafft, 'Self-organization of single-chain fluorinated amphiphiles with fluorinated alcohols', *Thin Solid Films, 284-285* (1996) 195-199.

219 F. Giulieri and M. P. Krafft, 'Non-chiral amphiphiles that give stable microtubules' (in preparation).

220 L. Trevino, F. Frézard, M. Postel and J. G. Riess, 'Incorporation of a perfluoroalkylalkane (R_FR_H) into the phospholipid bilayer of DMPC liposomes results in greater encapsulation stability', *J. Liposome Res., 4* (1994) 1017-1028.

221 L. Trevino, F. Frézard, J. P. Rolland, M. Postel and J. G. Riess, 'Novel liposome systems based on the incorporation of (perfluoroalkyl)alkenes (FmHnE) into the bilayer of phospholipid liposomes', *Colloids Surf., 88* (1994) 223-233.

222 L. Trevino, M. P. Krafft, F. Frézard, F. Giulieri and J. G. Riess, 'Fluorinated liposomes – use of (*F*-alkyl)alkyl components to impart impermeability to the liposomal membrane', *Proceed. Intern. Symp. Control. Rel. Bioact. Mater., 21* (1994) 606-607.

223 M. P. Krafft and Y. Ferro, 'Semi-fluorinated alkanes as components and stabilizers of fluorinated colloids, Impact on the fusion of vesicles', *Polymer Preprints, 39* (1998) 938-939.

224 M. P. Krafft, Q. Huo, S. Vidon, R. Leblanc and F. Giulieri, 'Monolayers made from a combination of DPPC with a mixed fluorocarbon/hydrocarbon amphiphile', in *Am. Chem. Soc. Natl. Meeting*, Boston, 1998.

225 V. M. Sadtler, F. Giulieri, M. P. Krafft and J. G. Riess, 'Micellization and adsorption of fluorinated amphiphiles: Questioning the $1CF_2 \sim 1.5CH_2$ rule', *Chem. Eur. J., 10* (1998) 1952-56.

226 C. Santaella, P. Vierling, J. G. Riess, T. Gulik-Krzywicki, A. Gulik and B. Monasse, 'Polymorphic phase behaviour of perfluoroalkylated phosphatidylcholines', *Biochim. Biophys. Acta, 1190* (1994) 25-39.

227 F. Frézard, C. Santaella, P. Vierling and J. G. Riess, 'Permeability and stability in buffer and in human serum of fluorinated phospholipid-based liposomes', *Biochim. Biophys. Acta, 1192* (1994) 61-70.

228 N. Privitera, R. Naon and J. G. Riess, 'Phagocytic uptake by mouse peritoneal macrophages of microspheres coated with phosphocholine or polyethylene glycol phosphate-derived perfluoroalkylated surfactants', *Int. J. Pharm., 120* (1995) 73-82.

229 C. Santaella, F. Frézard, P. Vierling and J. G. Riess, 'Extended in vivo blood circulation time of fluorinated liposomes', *FEBS Lett., 336* (1993) 481-484.

230 N. Privitera, R. Naon and J. G. Riess, 'Hydrolysis of DMPC or DPPC by pancreatic phospholipase A_2 is slowed down when (perfluoroalkyl)alkanes are incorporated into the liposomal membrane', *Biochim. Biophys. Acta, 1254* (1995) 1-6.

231 F. Giulieri, F. Guillod, J. Greiner, M. P. Krafft and J. G. Riess, 'Anionic glucophospholipids – A new family of tubule-forming amphiphiles', *Chem. Eur. J., 2* (1996) 1335-1339.

232 T. Imae, M. P. Krafft, F. Giulieri, T. Matsumoto and T. Tada, 'Fibril-vesicle transition and their structure. Investigation by microscopy and small angle scattering', *Prog. Colloid Polym. Sci., 106* (1997) 52-56.

232a M. P. Krafft, L. Schieldknecht, F. Giulieri, M. Schmutz, E. Nacache and N. Poulain, 'Polymerization of isodecylacrylate in the bilayer of fluorinated vesicles', *ACS Natl. Meeting*, San Francisco, March 2000; S. P. Wang, R. Lunn, M. P. Krafft and R. M. Leblanc, 'One and a half layers? Mixed Langmuir monolayer of 10,12-pentacosadiynoic acid and a semifluorinated tetracosane', *Langmuir, 16* (2000) 2882-2886.

233 J. G. Riess and M. P. Krafft, 'Fluorocarbons and fluorosurfactants for in vivo oxygen transport (blood substitutes), imaging and drug delivery', *Mat. Res. Soc. Bull., 24* (1999) 42-48.

234 J. G. Riess and J. Greiner, 'Perfluoroalkylated sugar derivatives as potent surfactants and co-surfactants for biomedical uses', in G. Descotes (ed.), *Carbohydrates as Organic Raw Materials*, VCH, Weinheim, 1993, pp. 209-259; updated in *Carbohydrate Res.*, (2000).

235 P. E. Keipert, J. D. Bradley, N. S. Faithfull, D. Y. Hazard, M. Spooner, K. M. Mackley, P. D. Rusheen, S. Batra and S. F. Flaim, 'Effects of *Oxygent* (perflubron emulsion) on O_2 transport in a dog model of hemodilution and surgical bleeding', *FASEB J., 8* A901 (1994).

236 C. Guedj, B. Pucci, L. Zarif, C. Coulomb, J. G. Riess and A. A. Pavia, 'Vesicles and other supramolecular systems from biocompatible synthetic glycolipids with hydrocarbon and/or fluorocarbon chains', *Chem. Phys. Lipids, 72* (1994) 153-173.

237 M. P. Krafft and F. Giulieri (unpublished).

BIOGRAPHIC NOTE

Jean G. Riess

Jean G. Riess received his chemical engineer and doctorat-ès-sciences degrees (the latter with Professor Guy Ourisson in 1963) from the University of Strasbourg. He then spent two years with Professor John Van Wazer at Monsanto's Central Research Dept. in St. Louis, Mo, learning some phosphorus and transition metal chemistry. In 1968 he became a professor at the University of Nice, France, where he founded, directed and eventually became the honorary director of the Unité de Chimie Moléculaire, which was associated with the Centre National de la Recherche Scientifique. Professor Riess's research successively involved organic and inorganic phosphorus chemistry, transition metal complexes, organometallics, and eventually perfluorochemicals. His main present interests are in fluorocarbons, fluorinated amphiphiles, and their colloid chemistry, including injectable fluorocarbon emulsions for *in vivo* oxygen delivery, fluorinated self-assemblies and various types of drug delivery systems. Professor Riess has published about 360 papers, holds some 25 patents, has served on numerous councils and committees, and has chaired or co-chaired international conferences on phosphorus chemistry and on blood substitutes. He has won various honours, including awards from the French Academy of Sciences, French Chemical Society, Alexander von Humboldt Forschung, City of Nice and Controlled Release Society, as well as Alliance's first Distinguished Contribution Award. He also holds an honorary Research Associate position at the University of California, San Diego, and sits on the Board of Directors of Alliance Pharmaceutical Corp.

Mailing address: J.G.R., Les Giaines, 06950, Falicon, France; e-mail: jgr@allp.com

Chapter 24

SOME ASPECTS OF FLUORINE CHEMISTRY IN GÖTTINGEN

HERBERT W. ROESKY

Institut für Anorganische Chemie, Universität Göttingen, Tammannstraße 4, D-37077 Göttingen, Germany

In his widely read book *Le Fluor et ses Composés*, published in 1900, Henri Moissan described how he produced fluorine for the first time in 1886 and demonstrated that it is an element with an extraordinary high reactivity. It was clear from the outset therefore that this element challenges chemists to work with it. This is one of the two reasons why I started to practise fluorine chemistry. The other stems from the long tradition of fluorine chemistry in Göttingen.

Due to the rapid development of science in the early 1900s, the University of Göttingen founded a Faculty of Mathematics and Science in 1922. This incorporated the Chemistry Department, which up to that time had been part of the Faculty of Medicine. *Otto Wallach (1889-1915)*, *Adolf Windaus (1915-1944)*, and *Hans Brockmann (1945-1972)* were the outstanding organic chemists of the time. *Gustav Tammann (1908-1929), Arnold Eucken (1929-1950)* and *Wilhelm Jost (1953-1971)* became the pioneers of an excellent school of physical chemistry. The chair of inorganic chemistry was held by *Richard Zsigmondy (1908-1929), Hans von Wartenberg (1933-1948)* and, later on, by *Oskar Glemser (1952-1979)*. Fluorine chemistry in Göttingen was started by *von Wartenberg*, promoted strongly by *Glemser*, and is now my field of endeavour.

The von Wartenberg era

Hans von Wartenberg [1] was a chemist blessed with excellent experimental skills, who openly declared that he liked particularly tricky experimental problems. This special mentality strongly influenced his choice of research topics. His research publications (more than 150 papers) reveal that he did not specialize in only one field but was an 'universal chemist', working in physical, organic and inorganic areas of his subject. His PhD thesis, completed in Berlin in 1902, describes research on oxyhalides of mercury; he then became a co-worker of Walter Nernst, who at that time was professor of physical chemistry in Göttingen.

Hans von Wartenberg

His stay in Göttingen at that time was not very long because *Nernst* left the university in 1905 and went to Berlin taking his young co-worker with him.

During his time with Nernst, von Wartenberg designed and constructed electrical equipment containing platinum and porcelain or iridium components for use in thermochemical studies on inorganic compounds at temperatures up to 1300 °C or 2000 °C, respectively. His work in this area continued throughout his research career, the final piece involving determination in 1959 of the heats of formation of La_2O_3 and $La(OH)_3$.

In 1913 von Wartenberg moved from Berlin to the University of Danzig, where he first became a full professor of physical chemistry, and later (1916) full professor of inorganic chemistry. It was in Danzig that he commenced his pioneering researches on fluorine and its compounds, his major interest being in the thermochemistry of inorganic fluorides; and by 1921 he had conducted his first experiments using elemental fluorine, generated electrolytically using a cell based on the technical procedure developed by Moissan and Ruff.

In 1932 von Wartenberg was offered the chair of inorganic chemistry at the University of Göttingen; he accepted, and thus became the successor to Tammann and Zsigmondy. With the financial support of the Rockefeller foundation, the then-existing Institute of Inorganic Chemistry was enlarged and renovated, providing him with enough space to expand his research interests. Unfortunately his work was interrupted in 1936 when, due to the political situation, he was forced to retire and leave the Institute. At the same time the chair he held was abolished, hence for about ten years inorganic chemistry was no longer viewed as an important discipline in the Chemistry Department. However, von Wartenberg's work was not stopped completely but only interrupted, because his colleague R. W. Pohl offered him some rooms in the Institute of Physics where he was able continue to do some experiments. Inevitably, his research output decreased, but not its quality. During the years of his forced retirement he confirmed that $KF{\cdot}3HF$, which had been proposed earlier (1925-27) by Lebeau and Damiens as an electrolyte in fluorine cells, was a definite compound. Moreover, for the first time he obtained pure white CuF_2 and used it in calorimetric studies on copper fluorides, observing that this compound dissociates at its melting point, giving CuF and F_2;on cooling, the CuF disproportionated into CuF_2 and Cu. Furthermore, he measured the heat of formation of AgF_2 and succeeded in preparing colourless crystals of PbF_4, CeF_4 and BiF_5. Finally, he produced CrF_3, CrF_4 and CrF_5 from chromium metal and fluorine.

After the end of World War II (1945) von Wartenberg was reinstated to his former position at Göttingen, and retired normally in 1948. Even after his retirement he worked regularly in his laboratory and always came up with new and often surprising results. Like many great scientists von Wartenberg was quite a character. For example, he liked to sit in a rocking chair and move to and fro while speaking to his co-workers and colleagues. Since he always spoke with a very low voice, those listening to him had to move with him, back and forth. So people coming out of his laboratory after sessions of some twenty or thirty minutes were always sick!

The successor to the chair of inorganic chemistry was *Oskar Glemser*, who came to Göttingen in 1951.

Oskar Glemser's reign

Oskar Glemser graduated at the University of Stuttgart and then moved to the Technische Hochschule in Aachen in 1939. There he discovered new metal oxides and hydroxides of nickel, molybdenum and tungsten. In Göttingen, influenced by von Wartenberg, he changed his research field and began to work with inorganic fluorides – sulfur-nitrogen-fluorine compounds and various non-metal and metal fluorides. Since that time, fluorine chemistry remained the major research preoccupation for Glemser and many of his co-workers.

Oskar Glemser

Shortly after Glemser's arrival in Göttingen von Wartenberg showed him a fluorine generator which he had constructed, and said that he could use it to produce F_2 for experiments such as the direct fluorination of S_4N_4(tetrasulfur tetranitride, an orange-yellow solid)[1]. However this nitride reacted explosively with pure F_2 or even F_2 diluted with N_2. So a procedure involving AgF_2 (made using F_2) as the fluorinating agent was developed. Thiazyl fluoride (N≡SF) and a number of other sulfur-nitrogen-fluorine compounds were obtained. This was the beginning of a long research period for Glemser in the field of fluorine chemistry, with NSF and NSF_3 proving to be key substances in the promotion of his group's researches. The highlights of Glemser's sulfur-nitrogen-fluorine chemistry are compiled in Scheme 24.1. The exceptional properties of fluorinated compounds in comparison to those of other halides, and the idea of applying them as insecticides or pesticides, prompted many chemists to begin to work in this field [2].

Oskar Glemser, now in his 89th year, has received many honours (Member of the Akademie der Wissenschaften in Göttingen, the Österreichische Akademie der Wissenschaften in Vienna, the Deutsche Akademie der Naturforscher 'Leopoldina' in Halle, the Centro di Logica e Scienze Comparate in Bologna, the New York Academy of Science, and the American Association for the Advancement of Science; Associate Member of the Jozef Stefan Institute in Ljubljana). He retired in 1979 but is still working with a small group of researchers. One of the most important events in Glemser's career as a fluorine chemist was the 2nd European Symposium on Fluorine Chemistry, which was held in the 100-year-old building at the University of Göttingen in 1968 (the Department of Chemistry did not move to a new modern building outside the old city until 1972); some 200 fluorine chemists from all over the world attended. In 1970 he received the Liebig Denkmünze and was the President of the German Chemical Society in 1976-77 and President of the inorganic division of IUPAC (1969-73). Moreover, he was honoured with the fluorine award of the American Chemical Society (1994).

[1] Working with elemental fluorine in those early days was a very dangerous task, made worse by the fact that at the time the Institute was more than 100 years old and had wooden floors and partly old equipment. The only more or less safe place was a solid stone building where *Friedrich Wöhler* accomplished his famous chemistry in the last century (it was said that even Wöhler did experiments with hydrogen fluoride). Later on, in the fluorine chemistry period, this building was called the 'witches kitchen'. The 'witches' (young enthusiastic students) more than once produced big explosions, which resulted in fires, injuries and cauterisations, and it is said that one doctor close by earned a fortune during the 'elemental fluorine period'.

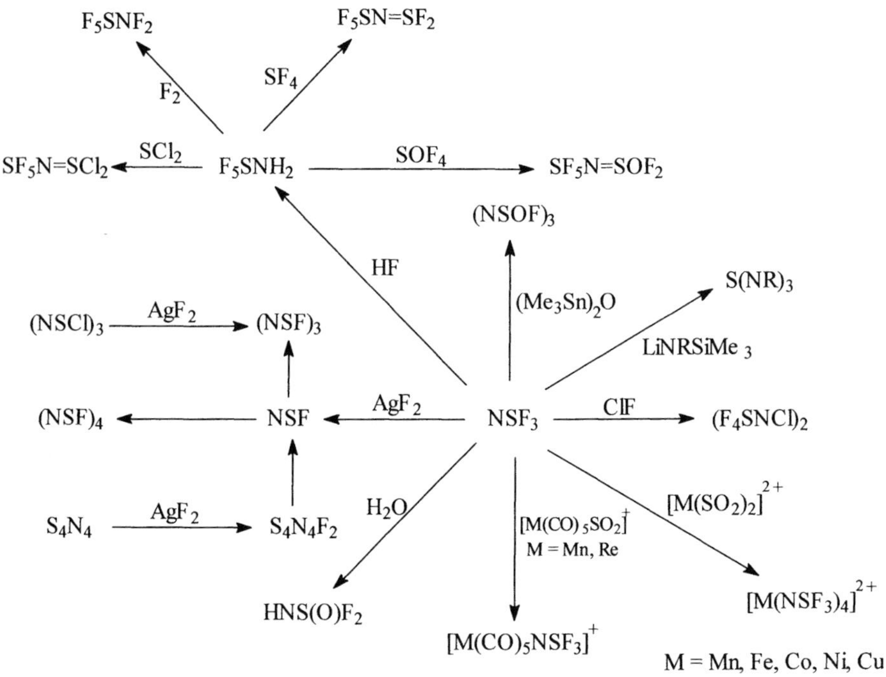

Scheme 24.1. Highlights of Glemser's N-S-F research.

Accomplishments of Herbert Roesky's group

Herbert W. Roesky

After receiving a Diplom in chemistry at Göttingen in 1961, I started working for a PhD in Glemser's group, using elemental fluorine and metals such as chromium, manganese, lead, bismuth and rhenium in an autoclave or in a fluidized bed to generate the corresponding fluorides in high oxidation states [3, 4]. Since fluorine was not commercially available in Germany then, we had to prepare it using the well-known electrochemical process. I still remember the frequent explosions we had when the cell was first turned on due to the water content of the HF we used.

After finishing my PhD work I took up a postdoctoral position (1963-1966) in Göttingen and at DuPont's Experimental Station in Wilmington, where, under the supervision of Earl Muetterties, I was asked to work in aluminium chemistry. However, I had enough freedom to develop my own chemistry in the phosphorus-fluorine field, and prepared a new class of thiosphosphates from alkali-metal fluorides and P_4S_{10}

which contained anions **1-3** [5].

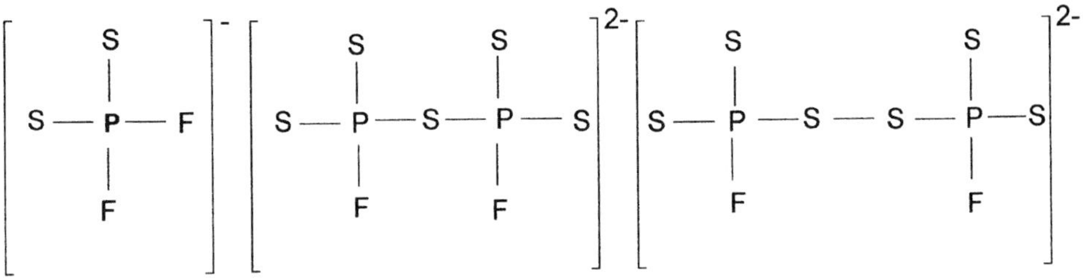

(1) **(2)** **(3)**

Caesium difluorothiphosphate can also be prepared in almost quantitative yield via the reaction $2SPF_3 + 2CsF \longrightarrow CsS_2PF_2 + CsPF_6$ using acetonitrile as a solvent. Studies on $(S_2PF_2)_2$, prepared from $CsS_2PF_2$and bromine, and the corresponding acid $HSP(S)F_2$ have revealed the existence of a large class of metal chelates whose distinguishing features are relatively high volatility and unusual reactivity. Established examples of this class include mono-chelates of Cu(I), Ag(I), bis-chelates of divalent Mn, Fe,Co, Ni, Pd, Zn, Cd, and Hg and tris-chelates of trivalent Cr and Co [6]. Volatile derivatives of Ni, Co, Mn, Cr, Zn, and Cd have been prepared using R(F)P(S)SH (R = Me, Et) and the metals as starting materials [7].

The thiophosphoryl halides SPF_3 and $SPFCl_2$ can be converted to the corresponding amides $S{=}PF_2NH_2$ and $S{=}PFClNH_2$ [8, 9], which have a broad chemistry; *e.g.* SPF_2NH_2 has been used to obtain $SPF_2N{=}PF_3$ and numerous analogues ($SPF_2N{=}PF_2X$, where X = Br, NH_2, or OH), $SPF_2N{=}PCl_3$, $SPF_2{=}PF_2N{=}PCl_3$, $SPF_2N{=}PCl_2N{=}PCl_3$, $SPF_2N{=}PF_2N{=}C{=}NSiMe_3$) [10-12]. In this connection it is worth mentioning that the first $-N{=}PBr_3$ compound was obtained from FSO_2NSO and PBr_5; FSO_2NSO, prepared from FSO_2NH_2 and $SOCl_2$, turned out to be a versatile compound for preparing FSO_2N-derivatives (Scheme 24.2) [13-16]. Furthermore, the availability of the tetraza compound $FSO_2N{=}S(NMe_2)_2{=}NSO_2F$ enabled work on the chemistry of aza analogues of sulfate to be initiated.

Subsequently these types of reaction were extended to CF_3SO_2NSO, prepared from $CF_3SO_2NH_2$ and $SOCl_2$ [17, 18]. The most prominent derivatives obtained were $CF_3SO_2N{=}SF_2{=}NSO_2CF_3$, $CF_3SO_2N{=}PCl_3$, $CF_3SO_2N{=}SF_2O$, $CF_3SO_2NHSO_2F$ and $CF_3SO_2N{=}PBr_3$.

In 1970 I moved with 10 very good students from Göttingen to Frankfurt, where I held a chair in inorganic chemistry for nearly ten years. In the following years we demonstrated the use of tin-nitrogen compounds for the preparation of nitrogen-containing derivatives based on the facile formation and cleavage of Sn−N bonds [19]. (This type of cleavage reaction was later frequently copied by other groups.) For example CF_3SO_2NSO was found to react with $(Me_3Sn)_3N$ to give $CF_3SO_2N(SnMe_3)_2 + Me_3SnNSO$ [20], while the cyclotriphosphazene $P_3N_3F_6$ and $(Me_3Sn)_3N$ gave $P_3N_3F_5N(SnMe_3)_2$, accompanied by elimination of Me_3SnF [21, 22]. From there, $P_3N_3F_5N{=}S{=}O$, $P_4N_4F_6(NSO)_2$ and $P_3N_3F_5NCO$ were synthesised via treatment of the corresponding P-N-F compounds with

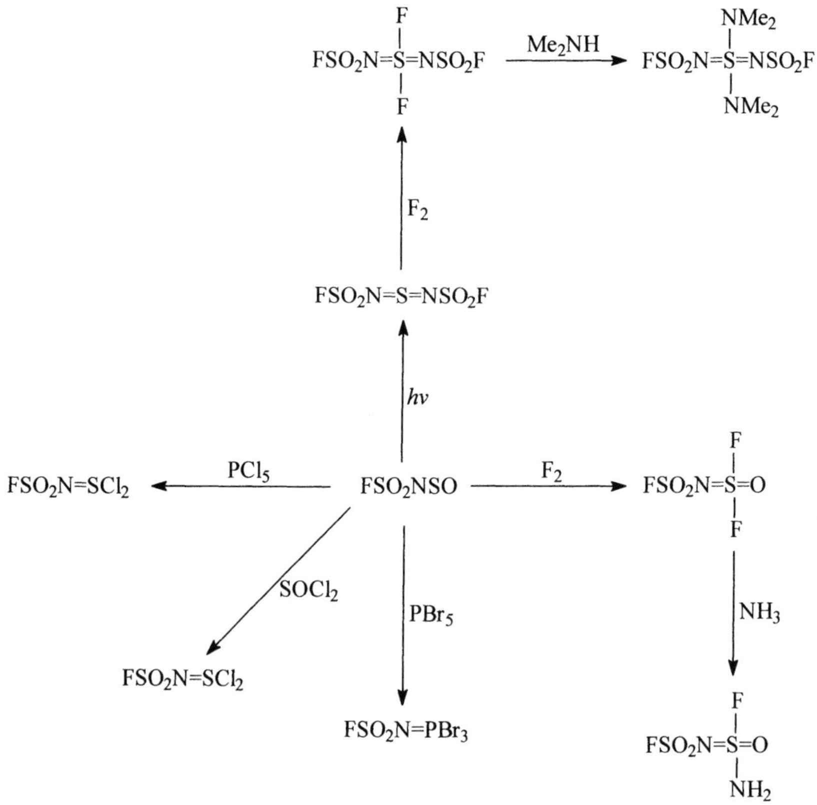

Scheme 24.2. Syntheses based on FSO_2NSO.

$SOCl_2$, and used as starting materials for the preparation of various derivatives, such as $P_3N_3F_5NSNP_3N_3F_5$ [23], $P_4N_4F_6N_2S$ [24, 25], and $P_3N_3F_5N_3S_3$ [26].

In Frankfurt, at the beginning of the 1970s, we isolated the perfluoroalkanesulfinic acids $CF_3S(O)OH$ and n-$C_4F_9S(O)OH$, the first members of their class to be prepared. They were obtained as stable (in glass at room temperature) colourless liquids following treatment of their hydrazinium salts (from $2R_FSO_2F + 5N_2H_4 \longrightarrow 2R_FSO_2^-[N_2H_5]^+ + N_2 + 2[N_2H_5]^+F^-$; $R_F = CF_3$, n-C_4F_9) with concentrated sulfuric acid [27]. Also during that decade we showed that $(FSO_2NSO_2)_2$ and $(CF_3SO_2NSO_2)_2$ form the 1:1 adducts $FSO_2N{=}SO_2S_4N_4$ and $CF_3SO_2N{=}SO_2py$ with S_4N_4 and pyridine, respectively [28, 29]. Comparable adducts were obtained with FSO_2NCO [30]. A ten- and a twelve-membered sulfur-nitrogen ring was isolated using the CF_3SO_2 group as an electron-withdrawing substituent, namely $(CF_3SO_2NS)_4S_2$ and $(CF_3SO_2NS_2)_4$, respectively [31, 32]. Smaller S-N rings such as $FSO_2N_3S_3$ and $CF_3SO_2N_3S_3$ were prepared from S_3N_2O and FSO_2NSO and CF_3SO_2NSO, respectively, via elimination of SO_2 [33]. The analogous phosphorus-containing compound $P_3N_3F_5N_3S_3$ was prepared in high yield from $P_3N_3F_5N(SnMe_3)_2$ and $S_3N_2Cl_2$ [34, 35].

In the 1980s we developed some extensive co-ordination chemistry using AsF_6^- salts, preferentially of Ag^+, Zn^{2+} and Cd^{2+}. These systems are soluble in liquid SO_2 and eas-

TABLE 24.1
Hexafluoroarsenate complexes [36-53]

$[Zn(OS_3N_2)_6][AsF_6]_2$	$[Ag][Ag_2(CH_2O)_6][AsF_6]_3$
$[Cd(OS_3N_2)_6][AsF_6]_2$	$[Ag(OCH_2CH_2)_8][AsF_6]$
$[Ag(OS_3N_2)_6][AsF_6]$	$[Ag_2(Ph_2S_2)_4][AsF_6]_2$
$[Zn(S_2N_2CO)_2][AsF_6]_2$	$[Ag_2(Ph_2Se)_4][AsF_6]_2$
$[Zn\{S(NSO)_2\}_2][AsF_6]_2$	$[Ag(1,3\text{-}C_4H_8O_2)_3][AsF_6]$
*$[Ag(S_8)_2]AsF_6$	$[Ag\{(CN)_2\}_2]_n[AsF_6]_n$
*$[Ag(S_4N_4O_2)_4]AsF_6$	$[PhPH_2Ag\{\mu\text{-}(PhPH)_2\}]_2[AsF_6]_2$
$[Ag_4\{S(NSO)_2\}_9][AsF_6]_4$	*$[Ag\{S_n(CN)_2\}][AsF_6](n = 3, 4)$
$[Zn\{P(O{=}P(OC_2H_5)_2)_3\}_2][AsF_6]_2$	*$[Ag(NCS)_2]_n[AsF_6]_n$
$[Ag_2\{(CH_2S)_3\}_5][AsF_6]_2$	

* = highlights of the series.

ily form adducts with ligands having soft donors according to Pearson's principle. The complexes isolated are listed in Table 24.1 [36-53], the highlights of this series being the silver complexes of S_8, cyanogen and thiocyanogen. The cyanogen and thiocyanogen complexes form two- and one-dimensional polymers, which are the prototypes of recent materials research using CN-containing starting materials. The geometry around the silver cation was found to be linear, square planar and tetrahedral, respectively. The S_8 complex $[Ag(S_8)_2][AsF_6]$ is the only co-ordination compound of S_8 reported so far.

Studies on hexafluoroacetone and tetrafluoro-1,2-ethanedisulfenyl dichloride

My researches on carbon-fluorine systems have focused on reactions involving hexafluoroacetone (HFA), tetrafluoro-1,2-ethanedisulfenyl dichloride, or 1,3,5-tris(trifluoromethyl)benzene. Results of work on reactions between hexafluoroacetone and CN- or SCN-compounds of main group and transition elements are summarised in Scheme 24.3; cyclization reactions involving the CN-moiety are catalysed by base, preferentially triethylamine. Of particular interest are the adducts of HFA with $(SCN)_2$ and $Hg(SCN)_2$, which are precursors for various new compounds containing six-membered rings [54-61]. Moreover, it was found that the dimer of hexafluorothioacetone reacts with nitriles of germanium, phosphorus, and arsenic to yield comparable insertion products [62, 63] (Scheme 24.3).

Tetrafluoro-1,2-ethanedisulfenyl dichloride was used extensively for the preparation of fluorine-containing heterocycles (Scheme 24.4) [64-66]. Of special interest are the reactions with cyclic ketones which result in the formation of spirocyclic ring systems via substitution of the activated hydrogens alpha to the carbonyl group. Furthermore, cleavage reactions involving reagents containing Si—N, As—Si and P—Si-bonds leads to various heterocycles (Scheme 24.4).

A definite highlight from our development of sulfur-nitrogen chemistry was the synthesis of stable thione *S*-imides containing trifluoromethyl groups (**4**, R = 1-adamantyl or 1,1,3,3,-tetramethylbutyl). Prepared from $(CF_3)_2CClSCl$ and $Me_3SiN(H)R$, with subsequent metallation and elimination of the metal chloride, these thione *S*-imides undergo [2+3]-cycloaddition reactions with norbornene [72], and combine with dichlorogermylene at room temperature to yield [3+1] cycloadducts (*e.g.* **5**, R = 1-adamantyl) [73] in which the bond angles of the four-membered CSNGe ring are all close to 90°.

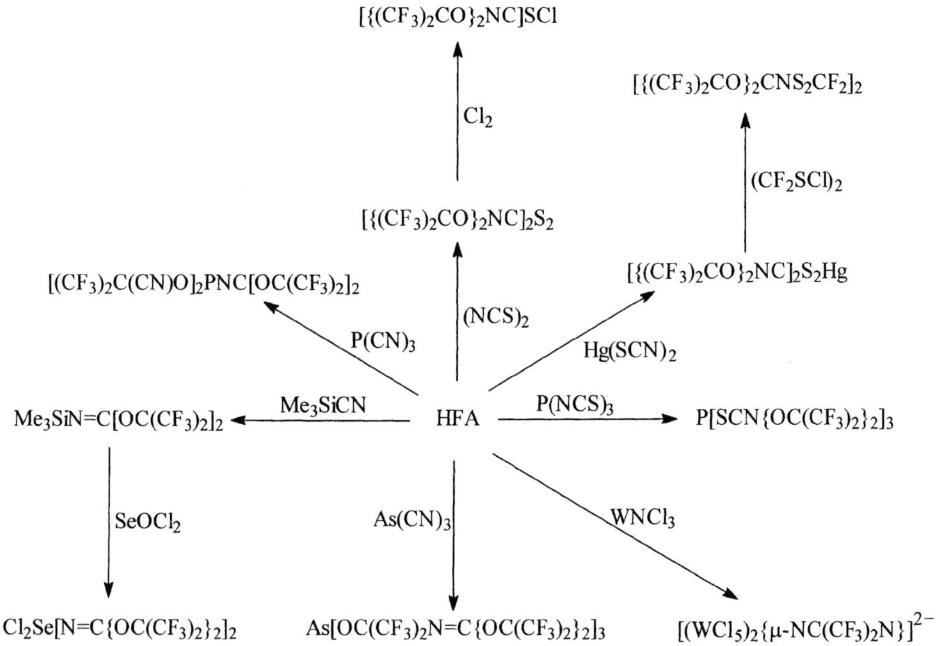

Scheme 24.3. Reactions of hexafluoroacetone (HFA).

$(CF_3)_2C{=}S{=}NR$

(4)

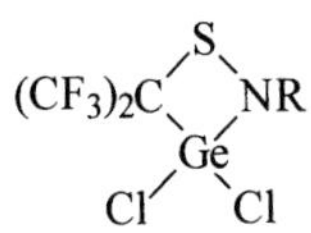

(R = 1-adamantyl)

(5)

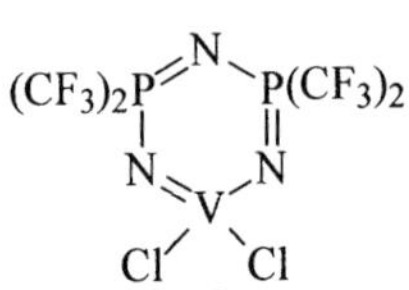

(6)

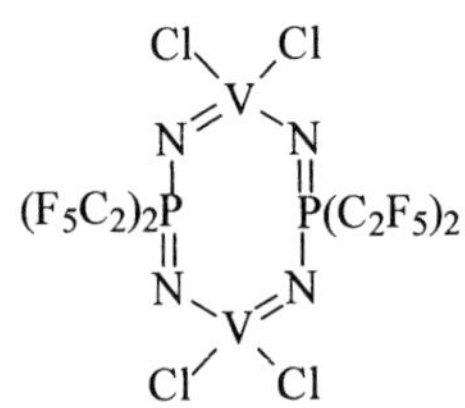

(8)

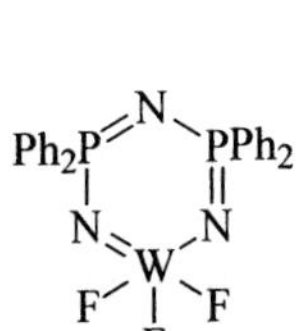

(7)

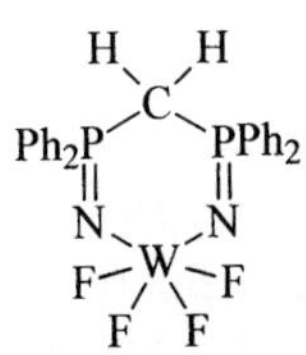

(9)

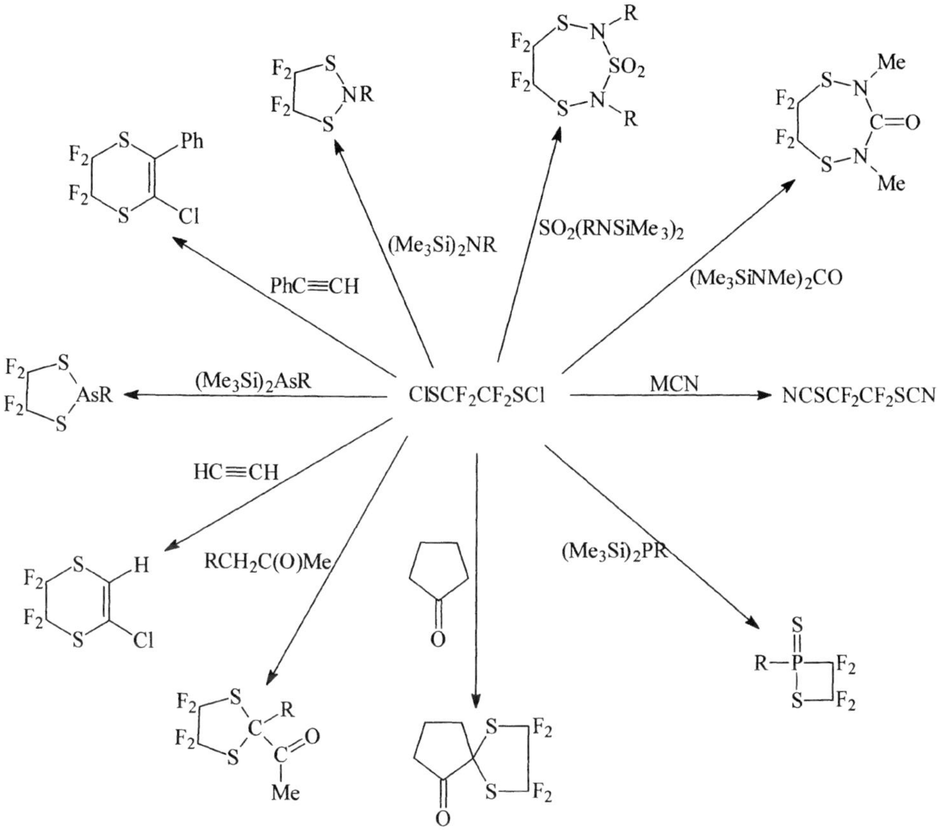

Scheme 24.4. Reactions of tetrafluoro-1,2-ethanedisulfenyl dichloride.

Chemistry without borders between main group and transition metal chemistry

This concept was used in a particularly elegant fashion in the synthesis of transition metal-containing phosphazenes. The $(NPCl_2)_x$ compounds date back to 1834 and were first made by Liebig, Wöhler, and Rose [74, 75] by reactions of NH_3 or NH_4Cl with PCl_5, a method which is still used on both laboratory and industrial scales. In the past, researchers have concentrated mainly on substitution reactions of the chlorine atoms bonded to phosphorus. Our approach, however, has been different. We have replaced one, two or three of the phosphorus atoms in the cyclotri- and cyclotetra-phosphazenes by transition metal atoms, as exemplified by products **6-9** [76].

2,4,6-Tris(trifluoromethyl)phenyl derivatives

Another area of research that is being actively investigated by my group concerns the development of highly volatile single-source precursors for the preparation of main group and transition metal based thin films by MOCVD processes. Mention should be made of the successful preparation of CdSe films using 2,4,6-tris(trifluoromethyl)phenyl as

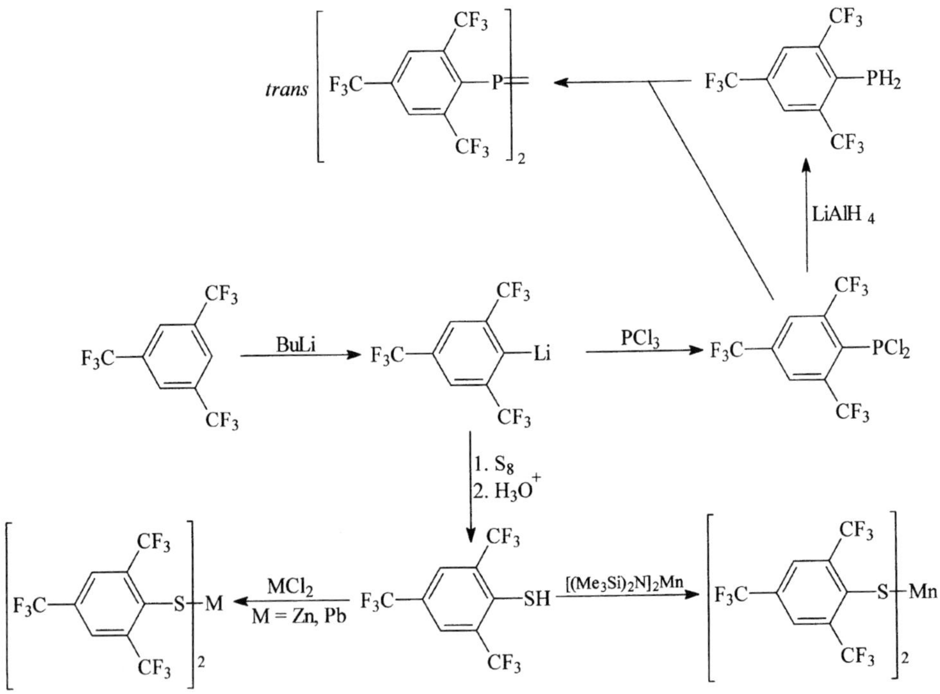

Scheme 24.5. Synthesis of 2,4,6-tris(trifluoromethyl)phenyl derivatives.

a ligand [78]:

$$[(CF_3)_3C_6H_2\text{-Se}]_2Cd \xrightarrow{\text{heat}} CdSe + [(CF_3)_3C_6H_2]_2Se$$

The reaction of $Cd[N(SiMe_3)_2]_2$ with two equivalents of Ar_FSeH [Ar_F = 2,4,6-$(CF_3)_3C_6H_2$] afforded the starting material in good yield [79].

Moreover, the 2,4,6-$(CF_3)_3C_6H_2$ ligand was found capable of stabilizing species otherwise unstable or even unknown under ambient conditions. Obviously, steric crowding, electron-withdrawal, and agostic fluorine interactions stemming from the CF_3 groups enhance the stabilities of the compounds. Some of the 2,4,6-tris(trifluoromethyl)phenyl compounds we have synthesized are shown in Scheme 24.5 [79-85].

Organometallic fluorides of main groups and transition elements

The term organometallic fluoride is used for compounds having both a metal-fluorine and a metal-carbon bond (C-M-F). Compounds of this type have a very polar metal-fluorine bond, resulting often in fluorine bridge formation, and a more covalent rather weak carbon-metal bond that is easily cleaved by oxidative fluorinating agents. This problem has been overcome to a large extent by the introduction of new fluorinating agents.

Trimethyltin fluoride. – In recent years we have introduced Me_3SnF as a fluorinating agent for the preparation of group 4-6 and group 13-15 fluorides. It is easy to make

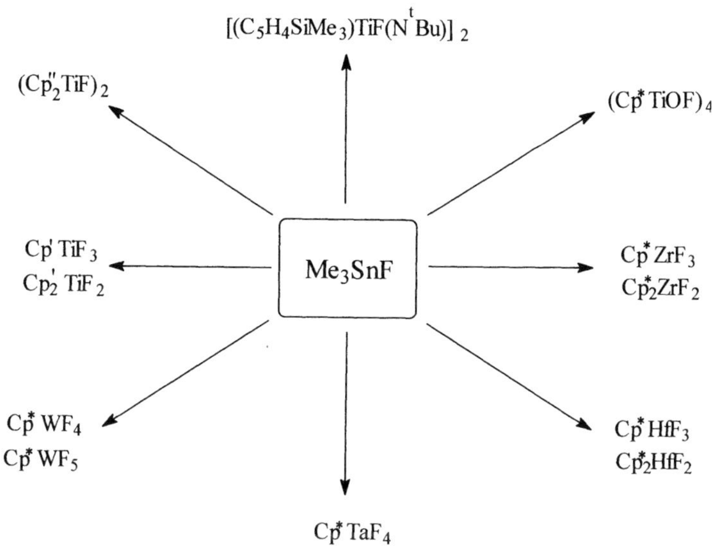

(Cp' = C_5H_4Et; Cp'' = C_5H_4Me; Cp* = C_5Me_5)

Scheme 24.6. Examples of organometallic fluorides prepared by treating the corresponding chlorides with Me_3SnF.

from Me_3SnCl and sodium fluoride, and many starting materials used in the preparation of organometallic fluorides are readily available and conveniently handled as the chloro or bromo derivatives. Consequently, halide metathesis, using Me_3SnF in a stoichiometric amount, has proved to be one of the most reliable routes to the corresponding fluorides. Moreover, the resulting Me_3SnCl can be easily removed from the crude product, in vacuo, and recycled. The insolubility of Me_3SnF in weakly co-ordinating solvents, in combination with the high volatility of Me_3SnCl, makes this an ideal fluorinating system.

Preparation and reactions of group 4 cyclopentadienyl derivatives. – Owing to the higher solubility of the fluorides in organic solvents, substituted cyclopentadienyl compounds of group 4 were preferred substrates. Moreover, these systems give crystalline materials suitable for investigation by X-ray techniques. Scheme 24.6 shows a number of compounds made using Me_3SnF and reveals that mono- and bis-cyclopentadienyl derivatives as well as Ti(III) compounds can be prepared. Mixed fluoro-chloro group 4 compounds can be obtained by treating Cp^*MF_3 ($Cp^* = C_5Me_5$, M = Zr, Hf) with Me_3SiCl, and compounds of composition Cp^*MF_2Cl have been generated in almost quantitative yield. Products obtained by the reduction of Cp^*TiF_3 are shown below (Eqn. 24.1-24.5).

$$6Cp^*TiF_3 + Na/Hg \longrightarrow (Cp^*TiF_2)_6(NaF)_7 \qquad (24.1)$$

$$14Cp^*TiF_3 + Na/Hg \longrightarrow (Cp^*TiF_2)_{12}(TiF_3)_2(NaF)_{18} \qquad (24.2)$$

$$4Cp^*TiF_3 + Mg/Hg \longrightarrow (Cp^*TiF_2)_4(MgF_2)_2 \qquad (24.3)$$

$$6Cp^*TiF_3 + Ca/Hg \longrightarrow (Cp^*TiF_2)_6(CaF_2) \qquad (24.4)$$

$$3Cp_2TiF_2 + Al/Hg \longrightarrow (Cp_2TiF)_3(AlF_3) \qquad (24.5)$$

Reduction of Cp^*TiF_3 and Cp_2TiF_2 using group 1, 2 and 13 metals has led to the structural characterisation of several 'molecular solids' which show a wide range of metal fluoride environments. Furthermore $(Cp_2TiF)_3(AlF_3)$ demonstrates the symbiosis between organometallic systems and Werner co-ordination compounds. The products in equations 24.1-24.5 above are soluble in organic solvents due to the wrapping of the inorganic core by the organic ligands [86].

Of particular interest are the reactions of Cp^*TiF_3 and Cp^*ZrF_3 with $AlMe_3$. Methylation of Cp^*TiF_3 with 2 equivalents of $AlMe_3$ gives Cp^*TiMe_2F at room temperature and the reduced product $Cp^*_2Ti_2(\mu_2\text{-}F)_8Al_4Me_8$ at elevated temperature, with release of methane. The single crystal structure of the latter product shows an octahedral metal core of Ti_2Al_4 with the two Ti atoms occupying *trans* positions [87]. Treatment of Cp^*ZrF_3 with a molar equivalence of $AlMe_3$ results in selective exchange of fluorine for methyl, yielding $[Cp^*Zr(Me)F_3AlMe_2]_2$. However, methane gas evolution is observed when Cp^*ZrF_3 is treated with an excess of $AlMe_3$, yielding a fluorine-free cluster core of composition $Zr_3Al_6C_7$.

An interesting product (**10**) has been isolated from the following reaction:

$$6Cp^*ZrF_3 + 4LiO(2,6\text{-}t Bu_2C_6H_3) \longrightarrow \underset{\mathbf{(10)}}{(Cp^*Zr)_6F_{18}Li_4O_2} + 2(2,6\text{-}t Bu_2C_6H_3)_2O$$

This intercalated lithium oxide is another example of a trapped molecular solid surrounded by soluble Cp^*ZrF_3 molecules.

Aluminium-fluorine compounds

The great differences in melting points and solubilities of $AlCl_3$ (subl. 183 °C; m.p. 193 °C at 1700 mbar) and AlF_3 (subl. 1272 °C; m.p. 1290 °C) indicate the problems associated with the preparation of organoaluminium fluorides.

While monofluorides R_2AlF (R = alkyl, aryl) have been known for many years, compounds of the type $RAlF_2$ have been mentioned only in the patent literature. Recently, however, we were able to prepare $(Me_3Si)_3CAlF_2$ from $(Me_3Si)_3CAlMe_2{\cdot}THF$ and Me_3SnF. The initial product isolated was the solvate $(Me_3Si)_3CAlF_2{\cdot}THF$, which lost THF in vacuo to afford the THF-free aluminium difluoride in the form of its trimer $(Me_3Si)_3CAlF_2$. This contains a puckered six-membered ring which has been characterised by X-ray single crystal structure analysis and contains alternating aluminium and fluorine atoms; the aluminium atoms have distorted tetrahedral geometry with terminal fluorine atoms. The trimer $[(Me_3Si)_3CAlF_2]_3$ has been used to gain access to the novel anions $[(Me_3Si)_3CAlF_3]^-$ and $[\{(Me_3Si)_3CAlF_2\}_2F]^-$.

Of particular interest is the oxidative addition of Ph_2SiF_2 to $(Cp^*Al)_4$ leading to $[(Cp^*AlF)_2SiPh_2]_2$. This product is a rare example of a structurally-characterised compound containing an eight-membered Al_4F_4 ring, and an unprecedented example of one possessing an Al-Si-Al bridge [88].

References

1 O. Glemser, *Angew. Chem.*, *72* (1960) 179.

2 (a) O. Glemser and R. Mews, *Angew. Chem.*, *92* (1980) 904; *Angew. Chem. Int. Ed. Engl.*, *19* (1980) 883; (b) O. Glemser, *Endeavour, 104* (1969) 86.

3 (a) O. Glemser, H. Roesky and K. -H. Hellberg, *Angew. Chem.*, *75* (1963) 346; *Angew. Chem. Int. Ed. Engl.*, *2* (1963) 266; (b) H. Roesky and O. Glemser, *Angew. Chem.*, *75* (1963) 920; *Angew. Chem. Int. Ed. Engl.*, *2* (1963) 626.
4 H. W. Roesky, O. Glemser and K. -H. Hellberg, *Chem. Ber.*, *99* (1966) 459.
5 H. W. Roesky, F. N. Tebbe and E. L. Muetterties, *J. Am. Chem. Soc.*, *89* (1967) 1272; *Inorg. Chem.*, *9* (1970) 831.
6 F. N. Tebbe, H. W. Roesky, W. C. Rode and E. L. Muetterties, *J. Am. Chem Soc.*, *90* (1968) 3578.
7 H. W. Roesky, *Angew. Chem.*, *80* (1968) 844; *Angew. Chem. Int. Ed. Engl.*, *7* (1968) 815.
8 H. W. Roesky, *Chem. Ber.*, *101* (1968) 3679.
9 H. W. Roesky and H. Beyer, *Chem. Ber.*, *102* (1969) 2588.
10 H. W. Roesky and L. F. Grimm, *Chem. Ber.*, *102* (1969) 2319.
11 H. W. Roesky and L. F. Grimm, *Chem. Ber.*, *103* (1970) 1664.
12 H. W. Roesky and L. F. Grimm, *Angew. Chem.*, *82* (1970) 255; *Angew. Chem. Int. Ed. Engl.*, *9* (1970) 244.
13 H. W. Roesky and D. P. Bapp, *Angew. Chem.*, *81* (1969) 705; *Angew. Chem. Int. Ed. Engl.*, *8* (1969) 674.
14 H. W. Roesky and D. P. Bapp, *Angew. Chem.*, *81* (1969) 494; *Angew. Chem. Int. Ed. Engl.*, *8* (1969) 510.
15 H. W. Roesky and D. P. Bapp, *Inorg. Chem.*, *8* (1969) 1733.
16 H. W. Roesky, *Angew. Chem.*, *79* (1967) 724; *Angew. Chem. Int. Ed. Engl.*, *6* (1967) 711.
17 H. W. Roesky, G. Holtschneider and H. Giere, *Z. Naturforsch.*, *25b* (1970) 252.
18 H. W. Roesky and G. Holtschneider, *Z. Anorg. Allg. Chem.*, *378* (1970) 168.
19 H. W. Roesky and H. Wiezer, *Chem. Ber.*, *104* (1971) 2258.
20 H. W. Roesky, M. Diehl and M. Banek, *Chem. Ber.*, *111* (1978) 1503.
21 H. W. Roesky and H. Wiezer, *Chem. Ber.*, *107* (1974) 1153.
22 H. W. Roesky and H. Wiezer, *Chem. Ber.*, *106* (1973) 280.
23 H. W. Roesky and E. Janssen, *Z. Naturforsch.*, *26b* (1971) 679.
24 A. Gieren, B. Dederer, H. W. Roesky and E. Janssen, *Angew. Chem.*, *88* (1976) 853; *Angew. Chem. Int. Ed. Engl.*, *15* (1976) 783.
25 H. W. Roesky and E. Janssen, *Angew. Chem.*, *88* (1976) 24; *Angew. Chem. Int. Ed. Engl.*, *15* (1976) 39
26 H. W. Roesky, G. Holtschneider, H. Wiezer and B. Krebs, *Chem. Ber.*, *109* (1976) 1358.
27 H. W. Roesky, *Angew. Chem.*, *83* (1971) 890; *Angew. Chem. Int. Ed. Engl.*, *10* (1971) 810.
28 H. W. Roesky and M. Aramaki, *Angew. Chem.*, *90* (1978) 127; *Angew. Chem. Int. Ed. Engl.*, *17* (1978) 129.
29 H. W. Roesky, M. Aramaki and L. Schönfelder, *Z. Naturforsch.*, *33b* (1978) 1072.
30 A. Gieren, Ch. Hahn, B. Dederer, H. W. Roesky and N. Amin, *Z. Anorg. Allg. Chem.*, *447* (1978) 179.
31 B. Krebs, M. Hein, M. Diehl and H. W. Roesky, *Angew. Chem.*, *90* (1978) 825; *Angew. Chem. Int. Ed. Engl.*, *17* (1978) 778.
32 H. W. Roesky, M. Diehl, B. Krebs and M. Hein, *Z. Naturforsch.*, *34b* (1979) 814.
33 H. W. Roesky and E. Janssen, *Chem. Ber.*, *108* (1975) 2531.
34 H. W. Roesky and E. Janssen, *Chem. Ztg.*, *98* (1974) 260.
35 I. Rayment, H. M. M. Shearer and H. W. Roesky, *J. Chem. Soc. Dalton Trans.*, (1982) 883.
36 H. W. Roesky, M. Thomas, J. W. Bats and H. Fuess, *J. Chem. Soc. Dalton Trans.*, (1983) 1891.
37 H. W. Roesky, M. Thomas, J. Schimkowiak, M. Schmidt, M. Noltemeyer and G. M. Sheldrick, *J. Chem. Soc. Chem. Commun.*, (1982) 790.
38 H. W. Roesky, M. Thomas, M. Noltemeyer and G. M. Sheldrick, *Angew. Chem.*, *94* (1982) 861; *Angew. Chem. Int. Ed. Engl.*, *21* (1982) 858.
39 H. W. Roesky, M. Thomas, J. W. Bats and H. Fuess, *Inorg. Chem.*, *22* (1983) 2342.
40 H. W. Roesky, M. Thomas, J. Schimkowiak, P. G. Jones, W. Pinkert and G. M. Sheldrick, *J. Chem. Soc. Chem. Commun.*, (1982) 895.
41 H. W. Roesky, M. Thomas, H. G. Schmidt, W. Clegg, M. Noltemeyer and G. M. Sheldrick, *J. Chem. Soc. Dalton Trans.*, (1983) 405.
42 H. W. Roesky, M. Thomas, P. G. Jones, W. Pinkert and G. M. Sheldrick, *J. Chem. Soc. Dalton Trans.*, (1983) 1211.
43 H. W. Roesky, H. Djarrah, M. Thomas, B. Krebs and G. Henkel, *Z. Naturforsch.*, *38b* (1983) 168.
44 H. W. Roesky, H. Hofmann, P. G. Jones, W. Pinkert and G. M. Sheldrick, *J. Chem. Soc. Dalton Trans.*, (1983) 1215.
45 H. Hofmann, P. G. Jones, M. Noltemeyer, E. Peymann, W. Pinkert, H. W. Roesky and G. M. Sheldrick, *J. Organomet. Chem.*, *249* (1983) 97.

46 H. W. Roesky, E. Peymann, J. Schimkowiak, M. Noltemeyer, W. Pinkert and G. M. Sheldrick, *J. Chem. Soc. Chem. Commun.*, (1983) 981.

47 P. G. Jones, T. Gries, H. Grützmacher, H. W. Roesky, J. Schimkowiak and G. M. Sheldrick, *Angew. Chem.*, *96* (1984) 357; *Angew. Chem. Int. Ed. Engl.*, *23* (1984) 376.

48 H. W. Roesky, T. Gries, P. G. Jones, K. -L. Weber and G. M. Sheldrick, *J. Chem. Soc. Dalton Trans.*, (1984) 1781.

49 P. G. Jones, H. W. Roesky, J. Liebermann and G. M. Sheldrick, *Z. Naturforsch.*, *39b* (1984) 1729.

50 H. W. Roesky, H. Hofmann, J. Schimkowiak, P. G. Jones, K. Meyer-Bäse and G. M. Sheldrick, *Angew. Chem.*, *97* (1985) 403; *Angew. Chem. Int. Ed. Engl.*, *24* (1985) 417.

51 P. G. Jones, H. W. Roesky, H. Grützmacher and G. M. Sheldrick, *Z. Naturforsch.*, *40b* (1985) 590.

52 H. W. Roesky, T. Gries, J. Schimkowiak and P. G. Jones, *Angew. Chem.*, *98* (1986) 93; *Angew. Chem. Int. Ed. Engl.*, *25* (1986) 84.

53 (a) H. W. Roesky, J. Schimkowiak, K. Meyer-Bäse and P. G. Jones, *Angew. Chem.*, *98* (1986) 998; *Angew. Chem. Int. Ed. Engl.*, *25* (1986) 1005; (b) P. G. Jones, H. W. Roesky and J. Schimkoviak, *J. Chem. Soc. Chem. Commun.*, (1988) 730.

54 H. W. Roesky, N. K. Homsy, M. Noltemeyer and G. M. Sheldrick, *Angew. Chem.*, *96* (1984) 1002; *Angew. Chem. Int. Ed. Engl.*, *23* (1984) 1000.

55 H. W. Roesky, N. K. Homsy and H. G. Schmidt, *Z. Anorg. Allg. Chem.*, *532* (1986) 131.

56 H. W. Roesky, *J. Fluorine Chem.*, *30* (1985) 123.

57 H. Grützmacher and H. W. Roesky, *Chem. Ber.*, *119* (1986) 2127.

58 H. W. Roesky, V. W. Pogatzki, K. S. Dhathathreyan, A. Thiel, H. G. Schmidt, M. Dyrbusch, M. Noltemeyer and G. M. Sheldrick, *Chem. Ber.*, *119* (1986) 2687.

59 V. W. Pogatzki and H. W. Roesky, *Chem. Ber.*, *119* (1986) 771.

60 N. K. Homsy, H. W. Roesky, M. Noltemeyer and G. M. Sheldrick, *J. Chem. Soc. Dalton Trans.*, (1985) 2205.

61 H. W. Roesky, J. Lucas, K. Keller, K. S. Dhathathreyan, M. Noltemeyer and G. M. Sheldrick, *Chem. Ber.*, *118* (1985) 2659

62 H. W. Roesky, K. S. Dhathathreyan, M. Noltemeyer and G. M. Sheldrick, *Z. Naturforsch.*, *40b* (1985) 240.

63 H. W. Roesky and K. S. Dhathathreyan, *J. Chem. Soc. Chem. Commun.*, (1984) 1053.

64 H. W. Roesky, A. Thiel, M. Noltemeyer and G. M. Sheldrick, *Chem. Ber.*, *118* (1985) 2811.

65 H. W. Roesky and N. Benmohamed, *Rev. Roumaine Chim.*, *31* (1986) 935

66 H. W. Roesky, N. Benmohamed and J. Schimkowiak, *Z. Anorg. Allg. Chem.*, *544* (1987) 209.

67 H. W. Roesky and N. Benmohamed, *Chem.-Zeit.*, *110* (1986) 417.

68 H. W. Roesky and N. Benmohamed, *Z. Anorg. Allg. Chem.*, *545* (1987) 143.

69 H. W. Roesky and A. Thiel, *Chem. Ber.*, *117* (1984) 1980.

70 U. Otten and H. W. Roesky, *Z. Anorg. Allg. Chem.*, *560* (1988) 55.

71 H. W. Roesky and U. Otten, *J. Fluorine Chem.*, *46* (1990) 433.

72 A. May, H. W. Roesky, D. Stalke, F. Pauer and G. M. Sheldrick, *Chem. Ber.*, *123* (1990) 1475.

73 A. May, H. W. Roesky, R. Herbst-Irmer, S. Freitag and G. M. Sheldrick, *Organometallics, 11* (1992) 15.

74 J. Liebig and F. Wöhler, *Ann. Chem.*, *11* (1834) 134.

75 H. Rose, *Ann. Chem.*, *11* (1834) 131.

76 H. W. Roesky, *Synlett,* (1990) 651.

77 H. S. Park, M. Mokhtari and H. W. Roesky, *Chem. Vapor Deposition, 2* (1996) 135.

78 D. Labahn, F. M. Bohnen, R. Herbst-Irmer, E. Pohl, D. Stalke and H. W. Roesky, *Z. Anorg. Allg. Chem.*, *620* (1994) 41.

79 D. Labahn, S. Brooker, G. M. Sheldrick and H. W. Roesky, *Z. Anorg. Allg. Chem.*, *610* (1992) 163.

80 M. Witt and H. W. Roesky, *Progress Inorg. Chem.*, *40* (1992) 353.

81 M. Scholz, H. W. Roesky, D. Stalke, K. Keller and F. T. Edelmann, *J. Organomet. Chem.*, *366* (1989) 73.

82 T. Lübben, H. W. Roesky, H. Gornitzka, A. Steiner and D. Stalke, *Eur. J. Solid State Inorg. Chem.*, *32* (1995) 121.

83 J.-T. Ahlemann, H. W. Roesky, R. Murugavel, E. Parisini, M. Noltemeyer, H.-G. Schmidt, O. Müller, R. Herbst-Irmer, L. N. Markovskii and Y. G. Shermolovich, *Chem. Ber. /Recueil, 130* (1997) 1113.

84 J.-T. Ahlemann, A. Künzel, H. W. Roesky, M. Noltemeyer, L. Markovskii and H. -G. Schmidt, *Inorg. Chem.*, *35* (1996) 6644.

85 M. Scholz, M. Noltemeyer and H. W. Roesky, *Angew. Chem.*, *101* (1989) 1419; *Angew. Chem. Int. Ed. Engl.*, *28* (1989) 1383.

86 E. F. Murphy, R. Murugavel and H. W. Roesky, *Chem. Rev.*, *97* (1997) 3425.
87 P. Yu, P. Müller, M. A. Said, H. W. Roesky, I. Uson, G. Bai and M. Noltemeyer, *Organometallics, 18* (1999) 1669.
88 C. Schnitter, K. Klimek, H. W. Roesky, T. Albers, H.-G. Schmidt, C. Röpken and E. Parisini, *Organometallics, 17* (1998) 2249.

Chapter 25

FLUOROCARBON METAL COMPOUNDS – ROLE MODELS IN ORGANOTRANSITION METAL CHEMISTRY

F. GORDON A. STONE
Department of Chemistry, Baylor University, Waco, Texas 76798-7348, USA

Introduction

Some twenty years have elapsed since studies on metal complexes having fluorinated organic groups as ligands ceased to be a dominant theme for research in my laboratory. I therefore greeted the Editor's invitation to contribute to this book with some trepidation because it is difficult to recall the essence of work done so long ago. However, since I believe there is merit in books which document the history of the sub-areas of our science I am glad to have been given the opportunity to write this chapter.

In 1952 the sandwich structure of the molecule ferrocene $[Fe(\eta^5\text{-}C_5H_5)_2]$ was established [1, 2]. Characterization of this complex initiated a renaissance in organometallic chemistry, followed as it was by the preparation of cyclopentadienyl compounds of most of the transition elements, and by the synthesis of several π-arene complexes of these metals, *e.g.* $[Cr(\eta^6\text{-}C_6H_6)_2]$ [3, 4]. By the early 1950's, however, few metal complexes had been authenticated with alkyl or aryl moieties attached to the metal by conventional two-center two-electron σ bonds [5]. This situation began to change with the discovery of species such as $[TiPh_2(\eta^5\text{-}C_5H_5)_2]$ [6], $[WMe(CO)_3(\eta^5\text{-}C_5H_5)]$ [7, 8] and $[FeMe(CO)_2(\eta^5\text{-}C_5H_5)]$ [8]. A further pivotal result was the synthesis of $[MnMe(CO)_5]$ [9][1] which came after the development at the Ethyl Corporation of a convenient route to its precursor $[Mn_2(CO)_{10}]$ [11]. The successful isolation of compounds like $[FeMe(CO)_2(\eta^5\text{-}C_5H_5)]$ or $[MnMe(CO)_5]$ led to the belief that carbon-metal σ bonds would be stable only if there were also present in the same molecule ligands like CO or $\eta^5\text{-}C_5H_5$ groups possessing both donor and acceptor properties. This idea at the time was seemingly reinforced with the characterization of a plethora of alkyl and aryl platinum complexes of which $[PtI(Me)(PPh_3)_2]$, $[PtPh_2(PEt_3)_2]$, and $[PtI_2(Me)_2(PEt_3)_2]$ are representative [12]. These complexes contained phosphine ligands considered to have both σ-donor and π-acceptor properties. A combination of either carbonyl, cyclopentadienyl, or phosphine ligands with alkyl or aryl groups was considered to favor a synergism between the various bonding modes thereby enhancing stability. These early ideas at the time they were advanced greatly stimulated further research. However, the preparation in much more recent times of isoleptic metal-alkyl and metal-aryl compounds

[1] Methyl(pentacarbonyl)manganese was prepared independently at about the same time by Walter Hieber and his coworkers at the Technischen Hochschule, München. Walter Hieber is generally regarded as the father of metal carbonyl chemistry, and he has given a valuable personal account of his work in ref. [10].

has shown that it is not essential that π bonding groups be present for metal-alkyl or -aryl bonds to exist.[2]

Fluorocarbon metal compounds – early days

It was my good fortune to commence work on organotransition metal chemistry in 1958 when, as I have described elsewhere [15], there was no shortage in this field of interesting problems to address. Moreover, NMR spectroscopy (then at 40 or 60 MHz!) had recently been added to the tools available to gain information on the structure of molecules. The routine use of single-crystal X-ray diffraction for structure determination was to occur some years later. The paucity of compounds of the d-block metals at that time known to have alkyl or aryl groups σ-bonded to the metals was an enigma seeking a solution. Earlier (1948-51) I had carried out my postgraduate research with Harry Emeléus at Cambridge. I believe Norman Greenwood and I were the only persons in his group in this period who were not working with fluorine compounds. I was encouraged to work with diborane as a consequence of contacts Professor Emeléus had with scientists working for the Admiralty. He had been asked to initiate studies with boranes,[3] and I was therefore guided into this area. Nevertheless, I was acutely aware that the main thrust of the group was with fluorine chemistry, one consequence of which had been the discovery of the then very novel mercurial compound $Hg(CF_3)_2$ [16].

In my early years of independent research at Harvard in 1958 I recalled that the properties of $Hg(CF_3)_2$ were very unlike those of $HgMe_2$. Whereas the latter was a highly toxic liquid with the well known ability to transfer Me groups to other metals via reactions with their halides, the former was a white crystalline solid which did not engage in ligand exchange reactions.[4] The property that sets the two molecules apart is the high electronegativity of the CF_3 group, which is comparable with that of a Cl atom. Hence the properties of $Hg(CF_3)_2$ may be qualitatively related to those of $HgCl_2$. I reasoned that a similar situation would exist with the transition elements, with fluoroalkyl-metal derivatives existing with stabilities similar to those of well established metal complex halides. Hence since $[MnCl(CO)_5]$ was stable surely $[Mn(CF_3)(CO)_5]$ would be also. Accordingly we prepared

[2]The much later isolation of molecules such as $[WMe_6]$ or $[Mo_2(CH_2SiMe_3)_6]$ by Wilkinson and his coworkers [13, 14] demonstrated that the facile decomposition of carbon-metal σ-bonds when this occurs is due to kinetic rather than thermodynamic factors. Strategies for the synthesis of 'stable' species must therefore block decomposition pathways.

[3]Towards the end of World War II the American Navy through its Office of Naval Research had sponsored in Herman Schlesinger's group at Chicago extensive studies on metal borohydrides and related species. It was thought that these compounds might be convenient solid sources of hydrogen for use in weather balloons upon hydrolysis. Subsequently both the value of $NaBH_4$ and $LiAlH_4$ as reducing agents in organic chemistry and the possibility of using borane compounds as propellants became apparent. This gave a huge boost to research in boron chemistry in the USA and former Soviet Union which lasted for many years. Scientists at the Admiralty had learned of the work on boron chemistry in the USA and wished to encourage studies in this area in Britain. Unfortunately when I set out to make B_2H_6 neither of the precursors used at that time (BCl_3 and $LiAlH_4$) were commercially available. These had to be made; a character forming experience for a new research student.

[4]Much later perfluoromethyl cadmium compounds were shown to be CF_3 donors and sources of difluorocarbene. For a review see ref. [17].

perfluoroalkylmanganese and -rhenium pentacarbonyls through the intermediacy of their acyl derivatives [18][5]:

$$\mathrm{Na[M(CO)_5] + R_FCOCl \xrightarrow{THF} [M(COR_F)(CO)_5] + NaCl}$$

$$\mathrm{[M(COR_F)(CO)_5] \xrightarrow{heat} [MR_F(CO)_5] + CO}$$

$\mathrm{M = Mn}$ or Re, $\mathrm{R_F = CF_3, C_2F_5}, n\text{-}\mathrm{C_3F_7}$

As we anticipated these species proved to be more robust towards thermal or oxidative decomposition than their alkyl analogs.[6]

Emeléus and Haszeldine had shown that CF_3I would oxidize Hg to CF_3HgI and much of the chemistry they developed reflected the pseudo halide characteristics of perfluoroalkyl groups [21, 22]. This property suggested to me that perfluoroalkyl iodides would similarly oxidize transition metal complexes when the metal was in a low oxidation state, an idea which led to discovery of the important reactions [23-25]:

$$\underset{\mathrm{Fe^0\ (d^8)}}{\mathrm{[Fe(CO)_5]}} + \mathrm{C_3F_7I} \longrightarrow \underset{\mathrm{Fe^{II}\ (d^6)}}{\mathrm{[FeI(C_3F_7)(CO)_4]}} + \mathrm{CO}$$

$$\underset{\mathrm{Co^I\ (d^8)}}{\mathrm{[Co(CO)_2(\eta^5\text{-}C_5H_5)]}} + \mathrm{C_2F_5I} \longrightarrow \underset{\mathrm{Co^{III}\ (d^6)}}{\mathrm{[CoI(C_2F_5)(CO)(\eta^5\text{-}C_5H_5)]}} + \mathrm{CO}$$

These syntheses are now known to be examples of a widespread class of reaction in which the metal's oxidation state and coordination number are both increased as a result of interaction with a substrate molecule. Moreover, the reaction between perfluoroalkyl iodides and iron pentacarbonyl to give $[FeI(R_F)(CO)_4]$ is formally similar to the reaction between the carbonyl complex and I_2 which gives $[FeI_2(CO)_4]$. This similarity in behavior between R_FI and I_2 reflects the pseudo interhalogen like properties of the former [21, 22].

We also observed that $[Fe(CO)_5]$ and $[Co(CO)_2(\eta^5\text{-}C_5H_5)]$ reacted with $CF_2{=}CF_2$ to yield the metallacycles **1** and **2**, respectively [23, 26]. These metallacycles have been described correctly as the first examples of metallacyclopentanes reported in the literature [27]. Metallacycles are now recognized as key intermediates in many syntheses involving transition metals. Similarly, the synthesis of the complexes $[FeI(C_3F_7)(CO)_4]$ and $[CoI(C_2F_5)(CO)(\eta^5\text{-}C_5H_5)]$ demonstrated in these early times the ability of transition met-

[5]Unknown to us, workers at the Ethyl Corporation had earlier reported (First International Conference on Coordination Chemistry held in London in April 1959) the compound $[Mn(CF_3)(CO)_5]$. Abstracts of this meeting were not available to us prior to completion of our study. At this first and relatively informally held ICCC meeting distribution of abstracts of the papers presented was not on the scale of later conferences. In a paper submitted one month later than ref. [18], W. R. McClellan [19] of DuPont also independently reported the synthesis of several cobalt and manganese fluoroalkyl complexes $[CoR_F(CO)_4]$ and $[MnR_F(CO)_5]$. These reports illustrate the increasingly competitive nature of research on organotransition metal compounds which followed the seminal work of Fischer and Wilkinson on the metallocenes.

[6]Following our discoveries, and the subsequent work of others who also made fluoroalkyl metal compounds, the cause of the enhanced stability of these complexes over their hydrocarbon analogs became a matter for controversy revolving around relative bond strengths, back bonding possibilities and electrostatic effects. These arguments have been reviewed by Hughes [20]. It is important to stress that our initial work was intuitive, being based on the known pseudo halogen behavior of CF_3 and other fluoroalkyl groups.

als in low oxidation states to insert into carbon-iodine bonds. Reactions of this type were later widely developed, as in the commercial synthesis of acetic acid from methanol using rhodium or iridium complexes as catalysts, with HI as a co-catalyst.

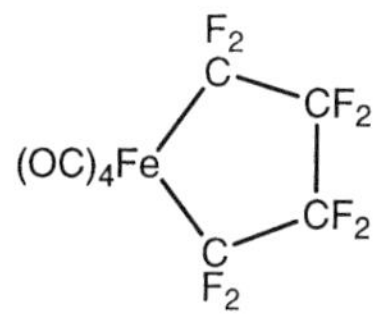

1

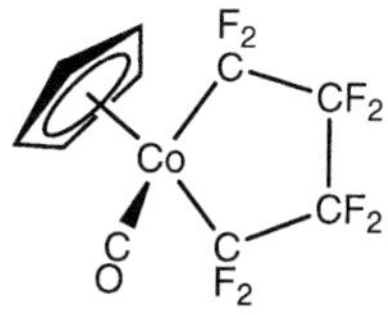

2

Reactions between alkenes, hydrogen and CO to yield aldehydes, with cobalt species as catalysts (hydroformylation), had been known for many years. It had been conjectured that the addition of metal-hydrogen bonds to C=C bonds took place along the reaction pathway. However, well authenticated examples of the addition of metal-hydrogen bonds to C=C or C≡C linkages were not known. The characterization of $[Mn(CF_2CF_2H)(CO)_5]$ and $[Mo(CF_2CF_2H)(CO)_3(\eta^5\text{-}C_5H_5)]$ from reactions between $CF_2{=}CF_2$ and $[MnH(CO)_5]$ and $[MoH(CO)_3(\eta^5\text{-}C_5H_5)]$, respectively [28, 29], as well as the formation of $[Mn\{C(CF_3){=}C(H)CF_3\}(CO)_5]$ from $CF_3C{\equiv}CCF_3$ and $[MnH(CO)_5]$ [29, 30] provided well defined examples of metal-hydride additions to C=C and C≡C bonds.[7] Thus a significant amount of the fluorocarbon-metal chemistry served as a role model for a vast array of hydrocarbon-metal chemistry that later came to light.

The above mentioned similar reactivity patterns shown by $[Fe(CO)_5]$ and $[Co(CO)_2(\eta^5\text{-}C_5H_5)]$ towards certain fluorocarbon substrates led to our drawing attention to other pairs of molecules which behaved similarly in their chemistry [24], *e.g.*, $[Mn_2(CO)_{10}]$ and $[Fe_2(CO)_4(\eta^5\text{-}C_5H_5)_2]$, or $[Co_2(CO)_8]$ and $[Ni_2(CO)_2(\eta^5\text{-}C_5H_5)_2]$. This was an early appreciation of isolobal mapping between molecules, a concept subsequently given a theoretical basis and elegantly developed by Hoffmann [32] to build bridges between inorganic and organic chemistry.

During the period in which the work described in this section was accomplished I trained my coworkers in the use the high vacuum techniques I had learned initially with Harry Emeléus when working with diborane and had developed to a much more sophisticated level when studying with Anton Burg as my postdoctoral mentor. This was important as it allowed us to manipulate compounds in the absence of air and moisture and often to analyze the products we had made without recourse to microanalytical laboratories. A good example of this involved the metallacycle **1**. We established that it contained four carbonyl ligands rather than three by treating the complex with iodine and measuring the CO gas released from weighed samples. In this manner we avoided the error of formulating **1** as a bis(alkene) complex $[Fe(CO)_3(\eta^2\text{-}CF_2{=}CF_2)_2]$.

A further factor in the good progress made in this period was the enthusiasm displayed by the graduate students for research in this new field. Interaction between each and every one was very close to the point where they would join with each other to bring

[7] At about the same time in another demonstration of the addition of metal-hydride to C=C bonds Chatt and Shaw [31] reported a reversible reaction between $[Pt(H)Cl(PEt_3)_2]$ and $CH_2{=}CH_2$ yielding $[PtCl(C_2H_5)(PEt_2)_2]$.

laboratory experiments to a rapid conclusion. An example of this was the synthesis of the important cobaltacycle **2** which was the fruit of work by several collaborators [26].

Pentafluorophenyl metal complexes

Success in preparing perfluoroalkyl metal compounds and the timely coincidence that perfluoroaromatic compounds had become commercially available led naturally to our synthesizing pentafluorophenyl metal species upon my arrival at Queen Mary College in late 1962 [33]. Disruptions associated with moving the research programme were alleviated to a degree by Paul Treichel, one of my Harvard students, who came with me to Queen Mary College as a National Science Foundation postdoctoral fellow. Fortunately his experiences in dealing with the primitive facilities available to us, together with the onset of the last of the London smogs, did not impede his subsequently having a distinguished career as a professor at the University of Wisconsin. Our preparation of molecules like $[Zr(C_6F_5)_2(\eta^5\text{-}C_5H_5)_2]$, $[Re(C_6F_5)(CO)_5]$, and $[Pt(C_6F_5)_2(PEt_3)_2]$ initiated a further sub-area of fluorocarbon metal chemistry,[8] activity in which has continued to this day. In recent years important contributions have been made by Usön and Forniés and their coworkers [36], especially in the area of pentafluorophenyl complexes of palladium and platinum. Novel discoveries from the Zaragoza group include anionic complexes such as $[Pt(C_6F_5)_4]^{2-}$, donor (Pt) acceptor (Ag or Au) molecules like $[PtAg(C_6F_5)_3(SC_4H_8)(PPh_3)]$, and polynuclear metal species such as $[Pt_2Pd_2(C_6F_5)_3(\mu\text{-}PPh_2)_3(CO)(PPh_2C_6F_5)]$ [37].

While studying pentafluorophenyl metal complexes my long time interest in boron compounds, including an earlier preparation of $B(CF{=}CF_2)_3$ [38], led to our synthesizing $B(C_6F_5)_3$and demonstrating the strong Lewis acidity of this compound [39, 40].[9]

After being ignored for some 30 years tris(pentafluorophenyl)boron has re-emerged into the literature as a very important activator component in homogeneous metallocene Ziegler catalyst formation [41-43]. The discovery of $B(C_6F_5)_3$ followed by the very long period before its usefulness became recognized in a different area of chemistry well illustrates the benefit of conducting unfettered research in chemical synthesis.

A new environment

In 1963 I was rescued from the smog and commuting problems associated with living in London by my appointment to a newly created chair of inorganic chemistry at Bristol. Since I had been born in the West Country this was a very welcome move. The studies on fluorocarbon metal compounds were carried over to the new location. Initially I was greatly assisted by Peter Jolly who accompanied me from Queen Mary College. In the syntheses

---

[8] After the article first mentioning C_6F_5 metal derivatives appeared (ref. [33]) it became apparent that others [34, 35] had independently focused on this area, probably as a result of the demonstrated existence of the perfluoroalkyl compounds.

[9] The discovery of $B(C_6F_5)_3$ occurred a few months before I left Queen Mary College. Unfortunately I was unable to take my co-discovers of the compound with me and so left them to bring the study to completion as they have described in ref. [40].

of new metal complexes Peter exploited the susceptibility of highly fluorinated olefinic hydrocarbons to nucleophilic attack by using metal carbonyl anions as nucleophiles [44], *e.g.*

$$[Re(CO)_5]^- + CF_2{=}CF_2 \longrightarrow [Re(CF{=}CF_2)(CO)_5] + F^-$$

$$[Mn(CO)_5]^- + \mathit{cyclo}\text{-}C_4F_6 \longrightarrow [Mn(\overline{C{=}CFCF_2C}F_2)(CO)_5] + F^-$$

This methodology was based on the earlier observations that the anionic complexes $[Mn(CO)_5]^-$ and $[Fe(CO)_2(\eta^5\text{-}C_5H_5)]^-$ react with $CF_2{=}CFCF_2Cl$ to afford $[Mn(CF{=}CFCF_3)(CO)_5]$ [18] and $[Fe(CF{=}CFCF_3)(CO)_2(\eta^5\text{-}C_5H_5)]$ [28], respectively. These reactions involve migration of F^-. That between $CF_2{=}CFCF_2Cl$ and $[Mn(CO)_5]^-$ was independently reported by McClellan [19].

Michael Bruce, a new student who joined my group, showed that $[Na][Re(CO)_5]$ in tetrahydrofuran with C_6F_6 afforded $[Re(C_6F_5)(CO)_5]$ and NaF [45]. My former student Bruce King and one of his coworkers [35] had found that a similar reaction occurred between $[Na][Fe(CO)_2(\eta^5\text{-}C_5H_5)]$ and C_6F_6 to give $[Fe(C_6F_5)(CO)_2(\eta^5\text{-}C_5H_5)]$. In collaboration with Michael Bruce and various students it was subsequently shown that many fluoroaromatic metal complexes could be prepared by reacting carbonyl metal anions with fluoroaromatic compounds [46].

Metallacyclopropanes and metallacycles

My move to Bristol coincided with a substantial expansion of the British University system following adoption by the U.K. government of the recommendations of an independent report (Robbins). It was thus possible to make several new appointments to the inorganic chemistry staff to teach the increasing numbers of students. One of those appointed was Michael Green who came from the University of Manchester Institute of Science and Technology. Michael became actively involved in research with me on the fluorocarbon metal compounds in addition to setting up his own independent research group. He contributed many new ideas, particularly on mechanistic aspects of the chemistry.

In the 1960's tetrafluoroethylene was being used by workers at the Central Research Department of the DuPont Company [47, 48] and by us [49] to probe the ability of d^8 and d^{10} metal complexes to activate small molecules. Recognition of the metallacyclic nature of the product **1** obtained from the reaction between $[Fe(CO)_5]$ and C_2F_4 [23] had led me to appreciate the potential for using the fluoroalkene and other unsaturated fluorocarbons as synthons for preparing new metal complexes. For the early work leading to compounds **1** and **2** we had obtained C_2F_4 by pyrolysis of Teflon powder given to us by DuPont, and had purified the gas by vacuum system techniques. At Bristol we required a steady source of C_2F_4 for our work, and in this regard the ICI laboratory at Runcorn (Cheshire) were very helpful in sending us supplies of the gas at low pressures in balloons shipped by rail in large thick-walled cardboard containers. (It was well known that transporting C_2F_4 under pressure in steel cylinders was hazardous.) These cardboard containers later proved useful for shipping household goods when I moved homes.

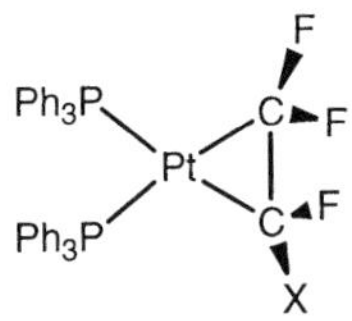

	X
3a	F
3b	Cl
3c	Br
3d	CF_3

At Bristol we focused initially on the reaction of Malatesta's complex [$Pt(PPh_3)_4$] with $CF_2{=}CF_2$ which afforded the very stable compound [$Pt(C_2F_4)(PPh_3)_2$] (**3a**) [49]. Similar products were obtained from $CF_2{=}CFX$ (X = Cl (**3b**), Br (**3c**), or CF_3 (**3d**)). Variable temperature ^{19}F NMR studies conducted by Tony Rest revealed that in solution at ambient temperatures the fluoroalkene ligands in these complexes do not rotate about an axis through the metal atom and midpoint of the C=C bond as do hydrocarbon alkenes in the compounds [$Pt(alkene)(PR_3)_2$]. Moreover, ^{19}F NMR measurements showed that J_{FFgem} values for the CF_2 groups in the fluoroalkene-platinum compounds were similar in value to those for fluorocyclopropanes. These coupling constants were much higher (*ca.* 190 Hz) than expected (*ca.* 60 Hz) for $=CF_2$ groups associated with sp^2 hybridized carbon atoms, a result implying that the complexes are best formulated as metallacyclopropanes with σ bonds between the platinum and the carbon atoms. On the basis of the Dewar-Chatt-Duncanson [50, 51] model for metal-alkene bonding, there is an extreme form of back bonding between the filled d orbitals of platinum and the π^* orbitals of the fluoroalkene ligands in the complexes **3**. This is brought about through the highly electronegative fluorine atoms lowering the energy of the these orbitals. A pivotal study in this area was made by John D. Roberts of the California Institute of Technology in collaboration with workers at DuPont [52]. It was shown by variable temperature NMR measurements that whereas the C_2H_4 ligand in the complex [$Rh(C_2H_4)(C_2F_4)(\eta^5\text{-}C_5H_5)$] underwent rotation about an axis through the rhodium atom and the mid-point of the $H_2C{=}CH_2$ bond, the $F_2C{=}CF_2$ molecule remained rigid behaving in accordance with that expected for a rigid rhodacyclopropane structure $\overline{Rh{-}CF_2{-}CF_2}$. Indeed George Parshall and colleagues at DuPont [47, 48] had earlier prepared the complex [$IrCl(CO)(C_2F_4)(PPh_3)_2$] and had proposed that it was formally an Ir^{III} compound containing a $[C_2F_4]^{2-}$ ligand. Similarly, the complexes **3** are best regarded as Pt^{II} species rather than Pt^0 complexes like [$Pt(C_2H_4)_3$] [15].

As discussed elsewhere [53, 54], the metallacyclopropane nature of molecules like **3b** and **3c** is probably responsible for their facile rearrangement in polar solvents to yield σ-vinylplatinum compounds [$PtX(CF{=}CF_2)(PPh_3)_2$] (X = Cl or Br), a process accelerated by the presence of silver salts. Corresponding reactions with nickel and palladium occur even more readily. Thus treatment of [$Ni(AsMe_2Ph)_4$] with $CF_2{=}CFBr$ in suitable solvents yields [$NiBr(CF{=}CF_2)(AsMe_2Ph)_2$], and [$Pd(CNBu^t)_n$] reacts with $CF_2{=}CFCl$ to give [$PdCl(CF{=}CF_2)(CNBu^t)_2$]; the intermediate fluoroalkene complexes [$M(C_2F_3X)(L)_2$] are not isolated [55, 56]. These reactions provide a good route to perfluorovinyl complexes of Ni, Pd, or Pt.

We were able to widen the scope of our work to probe the reactivity of zerovalent complexes of Ni, Pd and Pt by using $(CF_3)_2C{=}O$ and $(CF_3)_2C{=}NH$ as substrate molecules. In this we were much assisted by the advent of new metal reagents, notably [$Ni(CNBu^t)_4$]

[57], [Ni(cod)$_2$] (cod = cyclo-octa-1,5-diene) [58, 59], and [Pt(cod)$_2$] [60].[10] The synthons [M(cod)$_2$] (M = Ni or Pt) proved especially useful as they could be used to form *in situ* other complexes where the metal is ligated by a variety of phosphines or isocyanide groups. Thus the nucleophilicity of the metal center could be adjusted so as to change its reactivity towards electrophilic fluorocarbon substrate molecules.

Tetrafluoroethylene and the various nickel reagents generally afforded the metallacyclopentanes $[Ni(CF_2CF_2CF_2CF_2)(L_2)]$ (**4**) [62]. Only when [Ni(cdt)] (cdt = *t,t,t*-cyclododeca-1,5,9-triene) was treated with $CF_2{=}CF_2$ was a three-membered ring complex [Ni(C_2F_4)(cdt)] obtained. The latter reacted with PPh_3 to give [Ni(C_2F_4)(PPh_3)$_2$] which then in turn with $CF_2{=}CF_2$ very rapidly yielded the nickelacyclopentane **4a** [63]. This observation implicated metallacyclopropanes as intermediates in the formation of metallacyclopentane ring systems. The resistance of [Ni(C_2F_4)(cdt)] towards further ring expansion with $CF_2{=}CF_2$ is probably due to the cdt ligand blocking access to a coordination site on the metal by another molecule of the fluoroalkene.

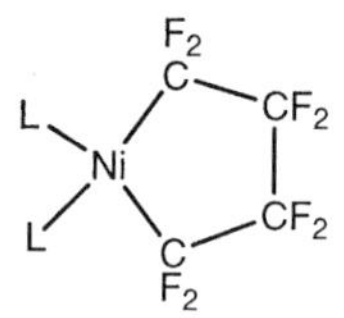

	L
4a	PPh_3
4b	$PMePh_2$
4c	PEt_3
4d	$PBu^n{}_3$
4e	$P(OMe)_3$

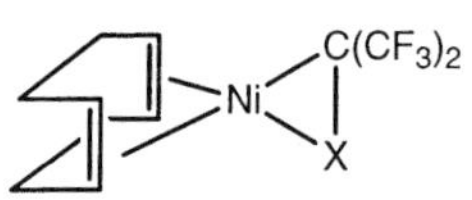

	X
5a	O
5b	NH

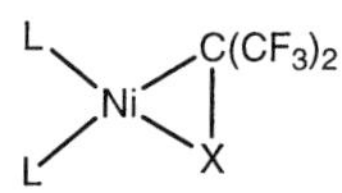

	L	X
5c	PPh_3	O
5d	PPh_3	NH
5e	$PMePh_2$	O
5f	$P(OPh)_3$	O
5g	$CNBu^t$	O
5h	$CNBu^t$	NH
5i	CNPh	O

Further examples of the expansion of three- to five-membered rings came with studies of reactions of Ni^0complexes with the molecules $(CF_3)_2C{=}X$ (X = O or NH) [64-66]. A series of 1:1 adducts **5a-f** were isolated when cyclo-octa-1,5-diene, tertiary phosphines or phosphites were ligating the nickel. However, when [Ni($CNBu^t$)$_4$] was used as the precursor the five-membered ring metallacycles **6** were obtained. These compounds were also obtained from reactions between **5g** or **5h** and $(CF_3)_2C{=}O$ or $(CF_3)_2C{=}NH$, respectively. Compounds **5g** and **5h** required as starting points for these ring expansion reactions were prepared by displacement of cod ligands from **5a** and **5b** with $CNBu^t$. Interestingly ring expansion of **5g** with $(CF_3)_2C{=}NH$ yields exclusively complex **7a**, its isomer **7b** is not formed [66]. Formation of **7a** thus occurs by opening of the carbon-nickel bond in **5g**. Determination of the molecular structure of **7a** by X-ray diffraction by Penfold and Countryman [67] was of pivotal importance. At the time we did not have sufficient capacity at Bristol for X-ray diffraction studies of all the unusual metal complexes being made by the several different groups and so we were helped in this situation by Professor Penfold

[10] We were not the first to prepare [Pt(cod)$_2$] [61]. However, Dr (now Professor) J. L. Spencer in my group was the first to devise a synthesis of the compound that made useful quantities available for its use as a precursor [15].

in Christchurch, New Zealand. In a reaction related to the synthesis of **7a**, the platinum complex **8** reacts with $(CF_3)_2C{=}O$ to give the heterocyclic compound **9**.

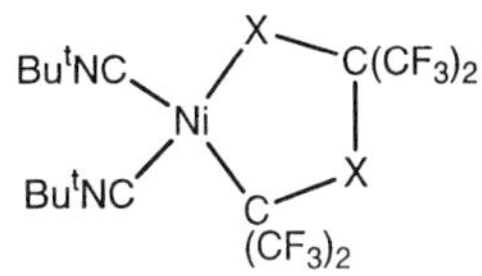

	X
6a	O
6b	NH

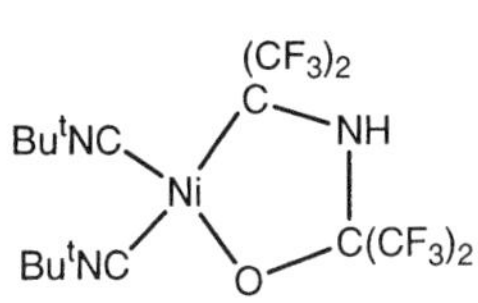

7a

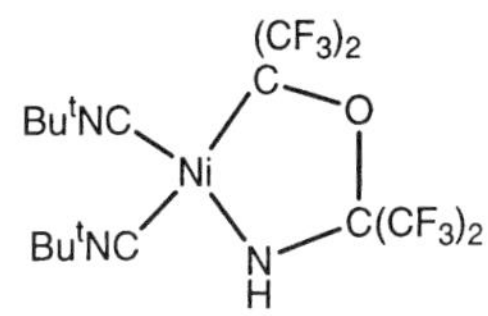

7b

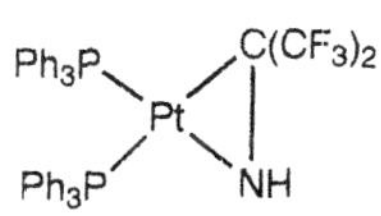

8

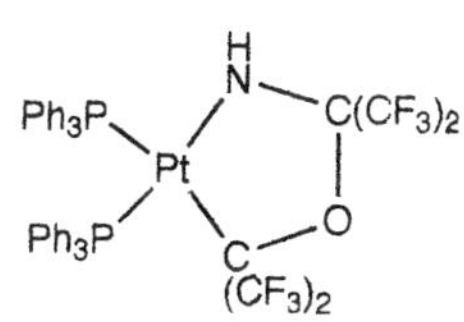

9

It became evident from these and other observations [68] that the formation of three versus five-membered rings is delicately controlled by the metal involved (Ni, Pd or Pt), the relative σ-donor and π-acceptor properties of the ligands on the metals, and the nature of the fluorocarbon substrate molecules. Many features of the chemistry remain unresolved to be solved by others who come later. Thus neither the perfluoropropene-platinum complex **3d** [48] nor more surprisingly the nickel compound $[Ni\{CF_2CF(CF_3)\}(cod)]$ [62] react further with $CF_2{=}CFCF_3$ to afford a five-membered ring heterocycle. Also metallacycles with ring sizes larger than five were not found employing $CF_2{=}CF_2$. Further studies with this fluoroalkene and the M^0 compounds in the presence of cocatalysts are merited to determine if conditions for oligomerization or polymerization can be observed.

Interesting differences between the behavior of analogous Ni^o and Pt^o complexes towards the same fluorocarbon substrates $(CF_3)_2C{=}O$, $CF_2{=}CF_2$, and $CF_2{=}CFCF_3$ became apparent during our work. Thus $[Ni(cod)_2]$ reacts with $(CF_3)_2C{=}O$ to give the three-membered ring complex **5a** as the only product [64]. In contrast, the main product obtained from reactions between $[Pt(cod)_2]$ and $(CF_3)_2C{=}O$ employing a 1:1 ratio of these reagents is the diplatinum complex **10**, with the species **11** and **12** also being formed as minor products [69]. When $(CF_3)_2C{=}O$ is used in excess with $[Pt(cod)_2]$ compounds **11** and **12** are the only products with the former predominating. Compound **13**, the platinum analog of the nickel complex **5a**, was prepared as part of our program by treating $[Pt(C_2H_4)_3]$ with cod followed by addition of $(CF_3)_2C{=}O$.

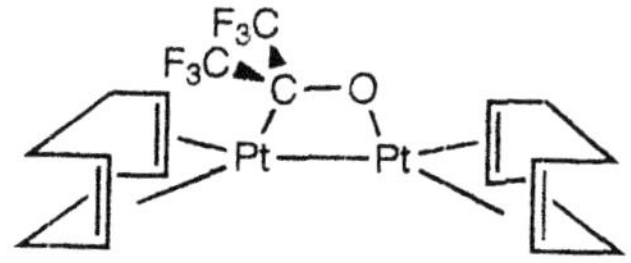

10

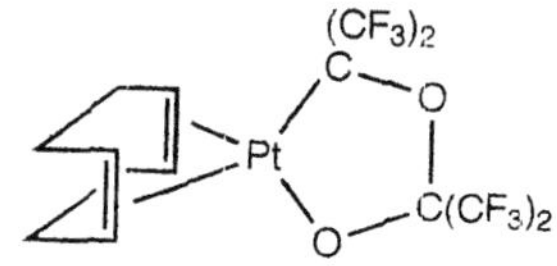

11

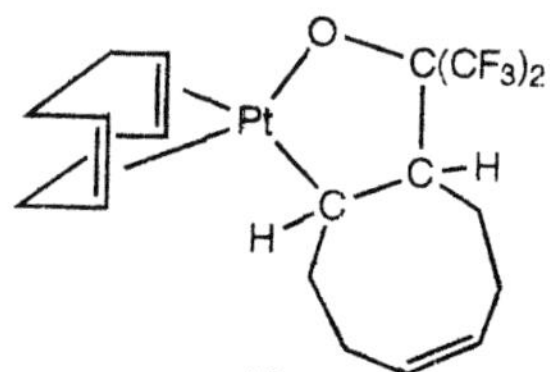

12

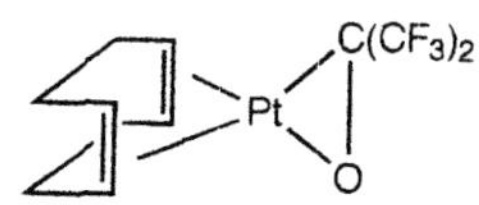

13

Interestingly, whereas Ni^{o} complexes form octafluoronickelacyclopentane structures very readily *e.g.* complexes **4**, the Pt^{o} precursors do not form the corresponding octafluoroplatinacyclopentanes. Thus platinum complexes $[Pt(PR_3)_4]$ or $[Pt(C_2H_4)(PR_3)_2]$ (R = alkyl or aryl) react with $CF_2{=}CF_2$ to afford very stable molecules with platinacyclopropane structures like **3a**. The reagent $[Pt(cod)_2]$ reacts with excess $CF_2{=}CF_2$ to yield the diplatinum complex **14** [70] whereas under the same conditions $[Ni(cod)_2]$ affords a very unstable species possibly $[Ni(C_2F_4)(cod)]$ [62]. If $PMePh_2$ is added to solutions of the latter, before deposition of nickel metal occurs, the nickelacyclopentane complex **4b** is obtained. The cyclo-octa-1,5-diene nickel complex **15** was prepared by an indirect route involving treatment of **4d** with cod in the presence of $ZnBr_2$ as a scavenger for the removal of the PBu^n_3 ligands.

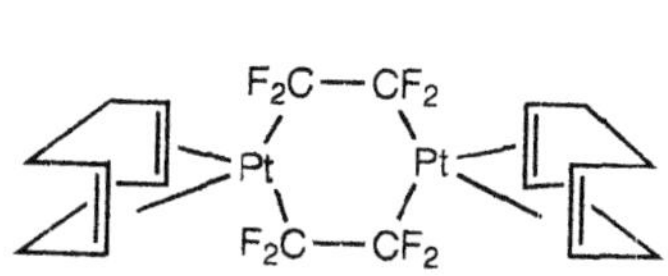

14

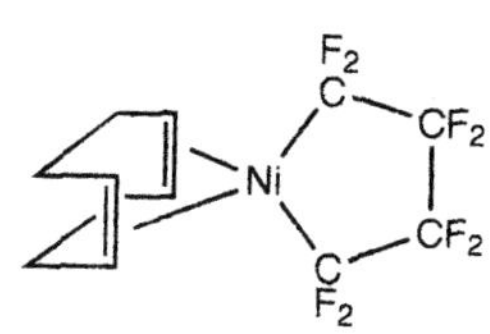

15

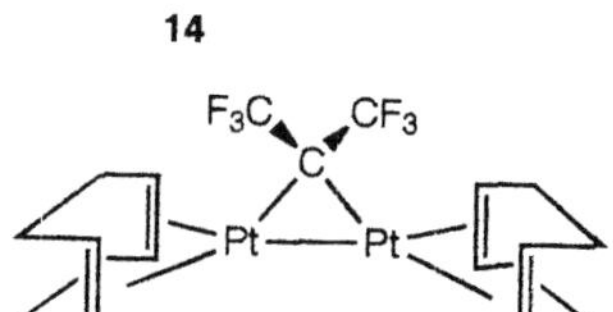

16

Mention was made earlier of the perfluoropropene complex $[Ni\{CF_2CF(CF_3)\}(cod)]$. This labile species forms in the reaction between $[Ni(cod)_2]$ and $CF_2{=}CFCF_3$, and with PPh_3 gives the very stable derivative $[Ni\{CF_2CF(CF_3)\}(PPh_3)_2]$ akin to the platinum complex **3d** [62]. In contrast, $[Pt(cod)_2]$ reacts with $CF_2{=}CFCF_3$ to produce a diplatinum complex **16** [70]. Evidently a fluorine migration reaction occurs in the synthesis of this interesting product, and the intermediacy of a fluoroalkylidene species [Pt{$=C(CF_3)_2$}(cod)], derived from [Pt{$CF_2CF(CF_3)$}(cod)] by F migration, was proposed. Formation of **16** could then result through a combination of the species [Pt{$=C(CF_3)_2$(cod)] and 'Pt(cod)'.

Conclusion

During two decades (1958-1977) the study of fluorocarbon derivatives of the transition elements formed the main focus of my research. However, from 1963 onwards studies in other areas gathered momentum. These were organoruthenium chemistry, following Michael Bruce's discovery of a convenient low pressure synthesis of the carbonyl [$Ru_3(CO)_{12}$] [71], and metallacarboranes [72], through John Spencer's interests in employing Pt^0 complexes for new syntheses of the cage compounds. Work with the fluorocarbon metal compounds seemed less exciting at the time and was therefore terminated. This was especially so following our synthesis of naked platinum complexes where alkenes are the only ligands, *e.g.* [$Pt(C_2H_4)_3$] [15, 73]. The very successful application of isolobal mapping to the synthesis of organometallic compounds [74] diverted us still further from the fluorocarbon studies. Nevertheless, even with the later work the strong bonding properties of $CF_2{=}CF_2$ as a ligand were used to probe the structures and dynamic behavior of some molecules *e.g.* [$Pt(C_2H_4)_2(C_2F_4)$] [75, 76] and [$Pt(C_2H_4)(C_2F_4)\{P(C_6H_{11})_3\}$] [77]. With hindsight I have had some regrets at abandoning studies on fluorocarbon metal compounds as it is easy to envisage much new chemistry that could be developed. Further research in the area would certainly be profitable.

The above work became possible through the aid of many coworkers whom I thank and list approximately in order in which they studied with me, together with their present titles: Professors H. D. Kaesz, R. B. King, Emily Pitcher (Mrs. E. O. Dudek), P. M. Treichel, Drs T. D. Coyle, T. A. Manuel, S. L. Stafford, J. Morris, and Professor P. M. Maitlis of the Harvard group. These were followed in the United Kingdom by Drs. P. W. Jolly, D. T. Rosevear, J. B. Wilford, Professor M. I. Bruce, Drs A. J. Rest, C. S. Cundy, J. Ashley-Smith, Professors J. A. K. Howard, J. L. Spencer, A. J. Mukhedkar, Drs P. K. Maples, Jane Browning (Mrs. D. Berry), H. D. Empsall, A. Greco, Professors A. Laguna and J. Forniés. It is particularly gratifying to me that Antonio Laguna and Juan Forniés in recent years have published so much novel metal complex chemistry involving fluoroaromatic ligands attached to gold and platinum, respectively. The chemistry they have developed could not possibly be envisaged when we began our work in the late 1950's. Finally it should be mentioned that within the confines of this article it was not possible to mention all of our own work with unsaturated fluorocarbons. Many interesting reactions between $CF_3C{\equiv}CCF_3$ and low valent metal complexes were discovered by Professors J. L. Davidson, M. I. Bruce, and Dr B. L. Goodall.

Acknowledgments

I thank the Robert A. Welch Foundation of Houston Texas for supporting my present research and Dr Paul Jelliss for helpful comments.

References

1 G. Wilkinson, M. Rosenblum, M. C. Whiting and R. B. Woodward, *J. Am. Chem. Soc.*, *74* (1952) 2125.
2 E. O. Fischer and W. Pfab, *Z. Naturforsch., B: Anorg. Chem., Org. Chem.*, *7* (1952) 377.
3 E. O. Fischer and W. Hafner, *Z. Naturforsch., B: Anorg. Chem., Org. Chem.*, *10* (1955) 655.
4 H. Zeiss and M. Tsutsui, *J. Am. Chem. Soc.*, *79* (1957) 3062.
5 F. A. Cotton, *Chem. Rev.*, *55* (1955) 551.
6 L. Summers, R. H. Uloth and A. Holmes, *J. Am. Chem. Soc.*, *77* (1955) 3604.
7 E. O. Fischer, *Angew. Chem.*, *67* (1955) 475.
8 G. Wilkinson and T. S. Piper, *J. Inorg. Nucl. Chem.*, *3* (1956) 104.
9 R. D. Closson, J. Kozikowski and T. H. Coffield, *J. Org. Chem.*, *22* (1957) 598.
10 W. Hieber, *Adv. Organomet. Chem.*, *8* (1970) 1.
11 E. O. Brimm, M. A. Lynch and W. J. Sesny, *J. Am. Chem. Soc.*, *76* (1954) 3831.
12 J. Chatt and B. L. Shaw, *J. Chem. Soc.*, (1959) 705, 4020; (1960) 1718.
13 A. J. Shortland and G. Wilkinson, *J. Chem. Soc.*, (1973) 872.
14 F. Huq, W. Mowat, A. Shortland, A. C. Skapski and G. Wilkinson, *J. Chem. Soc., Chem. Commun.*, (1971) 1079.
15 F. G. A. Stone, *Leaving No Stone Unturned – Pathways in Organometallic Chemistry*, American Chemical Society, Washington, DC, 1993.
16 H. J. Emeléus and R. N. Haszeldine, *J. Chem. Soc.*, (1949) 2953.
17 J. A. Morrison, *Adv. Organomet. Chem.*, *35* (1993) 211.
18 H. D. Kaesz, R. B. King and F. G. A. Stone, *Z. Naturforsch., B: Anorg. Chem., Org. Chem.*, *15* (1960) 763.
19 W. R. McClellan, *J. Am. Chem. Soc.*, *83* (1961) 1598.
20 R. P. Hughes, *Adv. Organomet. Chem.*, *31* (1990) 183.
21 H. J. Emeléus, *Angew. Chem. Int. Ed.*, *1* (1962) 129.
22 R. E. Banks and R. N. Haszeldine, *Adv. Inorg. Chem. Radiochem.*, *3* (1961) 337.
23 T. Manuel, S. L. Stafford and F. G. A. Stone, *J. Am. Chem. Soc.*, *83* (1961) 249.
24 R. B. King, P. M. Treichel and F. G. A. Stone, *J. Am. Chem. Soc.*, *83* (1961) 3593.
25 R. B. King, S. L. Stafford, P. M. Treichel and F. G. A. Stone, *J. Am. Chem. Soc.*, *83* (1961) 3604.
26 T. D. Coyle, R. B. King, E. Pitcher, S. L. Stafford, P. M. Treichel and F. G. A. Stone, *J. Inorg. Nucl. Chem.*, *20* (1961) 172.
27 G. Wilke, *Pure Appl. Chem.*, *50* (1978) 677.
28 R. B. King, P. M. Treichel and F. G. A. Stone, *Proc. Chem. Soc.*, (1961) 69.
29 P. M. Treichel, E. Pitcher and F. G. A. Stone, *Inorg. Chem.*, *1* (1962) 511.
30 P. M. Treichel, J. Morris and F. G. A. Stone, *J. Chem. Soc.*, (1963) 720.
31 J. Chatt and B. L. Shaw, *J. Chem. Soc.*, (1962) 5075.
32 R. Hoffmann, *Angew. Chem. Int. Ed.*, *21* (1982) 711 (Nobel Lecture).
33 P. M. Treichel, M. A. Chaudhari and F. G. A. Stone, *J. Organomet. Chem.*, *1* (1963) 98; *2* (1964) 206.
34 M. D. Rausch, *Inorg. Chem.*, *3* (1964) 300.
35 R. B. King and M. B. Bisnette, *J. Organomet. Chem.*, *2* (1964) 38.
36 R. Usön and J. Forniés, *Adv. Organomet. Chem.*, *28* (1988) 219; *Inorg. Chim. Acta*, *198-200* (1992) 165.
37 L. R. Falvello, J. Forniés, C. Fortuño, A. Martín and A. P. Martínez-Sariñana, *Organometallics*, *16* (1997) 5849.
38 S. L. Stafford and F. G. A. Stone, *J. Am. Chem. Soc.*, *82* (1960) 6238.
39 A. G. Massey, A. J. Park and F. G. A. Stone, *Proc. Chem. Soc.*, (1963) 212.
40 A. G. Massey and A. J. Park, *J. Organomet. Chem.*, *2* (1964) 245.
41 S. Döring, G. Erker, R. Fröhlich, O. Meyer and K. Bergander, *Organometallics*, *17* (1998) 2183.
42 X. Yang, C. L. Stern and T. J. Marks, *J. Am. Chem. Soc.*, *116* (1994) 10015.
43 J. Karl, G. Erker and R. Fröhlich, *J. Am. Chem. Soc.*, *119* (1997) 11165.
44 P. W. Jolly and F. G. A. Stone, *J. Chem. Soc., Chem. Commun.*, (1965) 85.
45 P. W. Jolly, M. I. Bruce and F. G. A. Stone, *J. Chem. Soc.*, (1965) 5830.
46 M. I. Bruce and F. G. A. Stone, *Angew. Chem. Int. Ed.*, *7* (1968) 747.
47 R. Cramer and G. W. Parshall, *J. Am. Chem. Soc.*, *87* (1965) 1392.
48 G. W. Parshall and F. N. Jones, *J. Am. Chem. Soc.*, *87* (1965) 5356.

49 M. Green, R. B. L. Osborn, A. J. Rest and F. G. A. Stone, *J. Chem. Soc., Chem. Commun.*, (1966) 502; *J. Chem. Soc. A*, (1968) 2525.
50 M. J. S. Dewar, *Bull. Soc. Chim. Fr.*, *18* (1951) C79; M. J. S. Dewar and G. P. Ford, *J. Am. Chem. Soc.*, *101* (1979) 783.
51 J. Chatt and L. A. Duncanson, *J. Chem. Soc.*, (1953) 2939.
52 R. Cramer, J. B. Kline and J. D. Roberts, *J. Am. Chem. Soc.*, *91* (1969) 2519.
53 A. J. Mukhedkar, M. Green and F. G. A. Stone, *J. Chem. Soc. A*, (1970) 947.
54 P. K. Maples, M. Green and F. G. A. Stone, *J. Chem. Soc., Dalton Trans.*, (1973) 2069.
55 J. Browning, M. Green and F. G. A. Stone, *J. Chem. Soc., Dalton Trans.*, (1971) 453.
56 H. D. Empsall, M. Green, S. K. Shakshooki and F. G. A. Stone, *J. Chem. Soc., Dalton Trans.*, (1971) 3472.
57 S. Otsuka, A. Nakamura and Y. Tatsuno, *J. Am. Chem. Soc.*, *91* (1969) 6994.
58 G. Wilke, *Angew. Chem.*, *72* (1960) 581.
59 B. Bogdanovic, M. Kroner and G. Wilke, *Justus Liebigs Ann. Chem.*, *699* (1966) 1.
60 M. Green, J. A. K. Howard, J. L. Spencer and F. G. A. Stone, *J. Chem. Soc., Chem. Commun.*, (1975) 3; *J. Chem. Soc., Dalton Trans.*, (1977) 271.
61 J. Müller and P. Göser, *Angew. Chem., Int. Ed. Engl.*, *6* (1967) 364.
62 C. S. Cundy, M. Green and F. G. A. Stone, *J. Chem. Soc., Dalton Trans.*, (1970) 1647.
63 A. Greco, M. Green, S. K. Shakshooki and F. G. A. Stone, *J. Chem. Soc., Chem. Commun.*, (1970) 1374.
64 J. Browning, C. S. Cundy, M. Green and F. G. A. Stone, *J. Chem. Soc. A*, (1969) 20.
65 J. Ashley-Smith, M. Green and F. G. A. Stone, *J. Chem. Soc. A*, (1969) 3019.
66 M. Green, S. K. Shakshooki and F. G. A. Stone, *J. Chem. Soc. Dalton Trans.*, (1971) 2828.
67 R. M. Countryman and B. R. Penfold, *Chem. Commun.*, (1971) 1598.
68 F. G. A. Stone, *Pure Appl. Chem.*, *30* (1972) 551.
69 M. Green, J. A. K. Howard, A. Laguna, L. E. Smart and F. G. A. Stone, *J. Chem. Soc., Dalton Trans.*, (1977) 278.
70 M. Green, A. Laguna, J. L. Spencer and F. G. A. Stone, *J. Chem. Soc., Dalton Trans.*, (1977) 1010.
71 M. I. Bruce and F. G. A. Stone, *Angew. Chem., Int. Ed. Engl.*, *7* (1968) 427.
72 F. G. A. Stone, *J. Organomet. Chem.*, *100* (1975) 257.
73 F. G. A. Stone, *Acc. Chem. Res.*, *14* (1981) 318.
74 F. G. A. Stone, *Angew. Chem., Int. Ed. Engl.*, *23* (1984) 89.
75 M. Green, J. A. K. Howard, J. L. Spencer and F. G. A. Stone, *J. Chem. Soc., Chem Commun.*, (1975) 449.
76 J. A. K. Howard, J. L. Spencer and S. A. Mason, *Proc. R. Soc., London A*, *386* (1983) 161.
77 N. C. Harrison, M. Murray, J. L. Spencer and F. G. A. Stone, *J. Chem. Soc., Dalton Trans.*, (1978) 1337.

BIOGRAPHIC NOTE

Gordon Stone

Born in Exeter in 1925 and educated at Exeter School, Gordon Stone entered Christ's College Cambridge as a student in 1945. After obtaining first class honours in Parts 1 and 2 of the Natural Sciences Tripos he carried out graduate work with H. J. Emeléus, receiving his PhD in 1951. He then spent two years as a postdoctoral fellow under the Fulbright programme with Anton Burg at the University of Southern California. He moved to Harvard in 1954 working initially with Gene Rochow but within a few months he was appointed an instructor and subsequently an assistant professor. In 1962 he returned to the United Kingdom to head the inorganic chemistry section at Queen Mary College, University of London. In 1963 he became the first professor of inorganic chemistry at Bristol University remaining there until 1990 when he accepted an invitation to become the Robert A. Welch Foundation Distinguished Professor of Chemistry at Baylor University in Texas where he is able to concentrate on research.

In recognition of his contributions to the advancement of inorganic chemistry, Gordon Stone has received many major awards. He was elected to the Royal Society in 1976, and for services to chemistry in the United Kingdom was appointed CBE in 1990. He is one of very few chemists who have received both the Davy Medal of the Royal Society and the Longstaff Medal of the Royal Society of Chemistry. He has also received the American Chemical Society's award for research in inorganic chemistry, and his RSC awards include the Tilden Lectureship, the Ludwig Mond Medal, and the Frankland Prize Lectureship. He is the author of over 800 primary journal articles, was the founding editor of Advances in Organometallic Chemistry, of which 43 volumes have been published, and co-edited with E. W. Abel and G. Wilkinson the well known series of volumes 'Comprehensive Organometallic Chemistry'. His scientific autobiography 'Leaving No Stone Unturned' was commissioned by the American Chemical Society and published in 1993 as part of a series of scientific autobiographies by eminent chemists. Approximately 60 of the 180 persons who have worked in Gordon's laboratory either as PhD students or postdoctoral assistants now hold permanent academic positions through out the world.

Chapter 26

AROMATIC FLUORINE CHEMISTRY AT SALFORD[1,2]

HANS SUSCHITZKY and BASIL J. WAKEFIELD

Department of Chemistry and Applied Chemistry, University of Salford, Salford M5 4WT, UK

Suschitzky's story: en route to heterocyclic pursuits via organofluorine chemistry

Background

I was born in Linz (Austria) in 1915 but grew up in Vienna, where I also received my education. At secondary school (gymnasium) I was taught very little physics or mathematics and no chemistry at all. However, I acquired an excellent knowledge of Latin and classical Greek – an accomplishment which was hardly a help in the study of the sciences. Yet, surprisingly, this same gymnasium – which has recently celebrated its centenary – can proudly boast to have produced two Nobel Laureates in science, namely Richard Kuhn (Chemistry Prize 1938) and Wolfgang Pauli (Physics Prize 1945). Perhaps the inspiration of Democritus' Atomic Theory was at work here?

Hans Suschitzky

I entered the University of Vienna in 1934 to study in the Faculty of Medicine, and continued my studies until March 1938 when Austria was occupied by Hitler. For racial and political reasons I decided to leave Austria, and after a dramatic escape I managed to reach England shortly before the outbreak of war in Europe. Once there, I was assigned various jobs by the British Government's agricultural authority – 'digging for victory' and other such pursuits. Towards the end of the war, however, I obtained permission to take a job more in line with my qualifications, namely as a chemist in a small pharmaceutical company (Harker, Stagg and Morgan Ltd) and later in the coal tar industry.

Through evening study at the West Ham College of Technology (now the University of East London), I obtained an internal BSc degree in chemistry of the University of London, and in 1949 West Ham College offered me a Lectureship in Chemistry. In 1956 I moved to the then Royal College of Technology, Salford, as a Senior Lecturer, and was

[1] Salford is now part of Greater Manchester, though locals are quick to point out that Salford received its royal charter in 1230, eighty years before Manchester.

[2] For information on fluoro-aliphatic chemistry carried out at Salford by H. Sutcliffe's group, see Chapter 32 (Appendix 32.2).

appointed Professor of Organic Chemistry in the University of Salford in 1967.[3] I retired officially in 1981, according to the rules applying in UK universities, and was granted the status of Professor Emeritus.

Research activities

My first encounter with fluorine chemistry came in the late 1940s when I was working in the coal tar industry. Beckton Gas Works in east London wanted to get rid of twenty tonnes of coal tar residues – which were rich in carbazole – and offered them to my employer (Salamon and Co., Ltd) at an attractive price. The only commercial use for carbazole at that time was as a starting material for fast Hydron Blue dyes,[4] which were very successful at competing with indigo dyes. It was thought that the use of a fluorinated carbazole would improve the fastness of such dyes and thus make them even more competitive. So we prepared a range of fluorinated carbazoles (**1**) and fluorinated tetrahydrocarbazoles (**2**) [1] by means of the famous Balz-Schiemann reaction [2]. This involves a dry and somewhat explosive decomposition of the required diazonium 'borofluorides' (tetrafluoroborates): $ArN_2^+BF_4^- \rightarrow ArF + N_2 + BF_3$ [2].

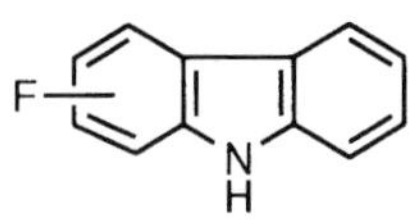

(1)

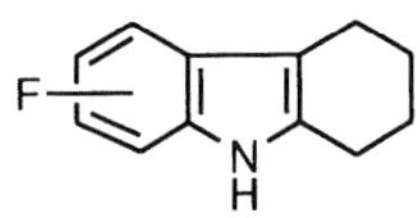

(2)

It was not difficult to convince the chemical engineers at Salamon & Co. by a striking laboratory demonstration that the unruly decomposition of the borofluorides, when carried out on a large scale in the works, would pose a considerable hazard. Thus, so far, fluorinated Hydron Blue dyes have not been produced commercially – possibly because of my concern about the danger. However, the experience of making fluorocarbazoles encouraged me to pursue research on fluorinated heterocyclic compounds. The incentive was twofold: (a) very few fluoroheterocycles had been reported at that time, and (b) I could not resist the temptation to carry out explosive reactions which could be tamed by skilful experimentation. The opportunity to concentrate on this chemistry arose when I moved in 1951 from industry to a teaching post at the West Ham College, and progress accelerated when I took up a Senior Lectureship at the Royal Technical College, Salford, in 1956.[5] Research

[3]The University of Salford has its origins in the Royal Technical Institute, founded in 1896. Chemistry was taught from the outset, and as early as 1911 a research assistant was working on the synthesis of pyrazolidones. In 1921 the Royal Technical Institute became the Royal Technical College, and in 1961 it assumed the rank of a university-level institution as the Royal College of Advanced Technology. It achieved full University status in 1967.

[4]Condensation of carbazole with *p*-nitrosophenol, for instance, leads to Hydron Blue R. Hydron Blue G, a dye of superior fastness, is derived from *N*-ethylcarbazole.

[5]At that time very few if any fluoro-organic compounds were available from suppliers of laboratory chemicals. Realising this, one of my enterprising research students, Philip Koch, sold some of his research F-compounds to L. Light & Co, thus augmenting somewhat his meagre research grant. 'Lights', which was one of the UK's leading laboratory suppliers of the time, eventually merged with Philip Koch's own company to become Koch-Light Laboratories, a firm which for many years offered a whole string of organic F-compounds to researchers.

facilities in Salford at that time left much to be desired, but the resourcefulness and enthusiasm of my research students, many of whom were on release from industry – made up for most of the shortcomings. Somehow they always managed to find sufficient equipment and chemicals – I did not quiz them too much about how they came to be so well supplied! When Salford obtained university status in 1967, the circumstances improved considerably.

Eventually my interest in organic F-compounds was diluted and shifted downwards (in Periodic Table terms) to include polychloro-aromatic and -heteroaromatic compounds [3], which I began to study in collaboration with my departmental colleagues Basil Wakefield and Brian Iddon.[6] This shift to chlorine chemistry was partly due to persuasion from chemical industries, interested in the application of such compounds as pesticides. Also, public interest in these compounds became considerable in 1976 following the Seveso (Italy) disaster which was caused by the accidental escape of the dangerous 2,3,7,8-tetrachlorodibenzo-1,4-dioxin (popularly known as 'dioxin') into the atmosphere.

Since that time my interests have broadened still further and now include heterocyclic compounds in general. Contributions have been made through original researches and by the provision of important review literature, the latter in collaboration with my ex-colleagues Otto Meth-Cohn[7] [4, 5] and Eric Scriven[8] [6]. In 1973 Otto and I arranged a meeting for heterocyclic enthusiasts in Grasmere,[9] along the lines of the Gordon Conferences in the USA. Lectures were given morning and evenings, with afternoons left free for discussion and enjoyment of the countryside (many of the participants took part in quite strenuous hikes up the Lakeland fells). This meeting proved so successful that the Grasmere Heterocyclic Meeting is still running biennially. The lectures are now once again given in the Village Hall, after having been 'demoted' to a hotel location for a few years. It seems that the homely atmosphere of the Village Hall contributes much to the enjoyment of the lectures.

My appointment (dating from 1982) as a Visiting Professor at the University of Heidelberg (Germany) is still in place, so I still attend the Pharmazeutisch-Chemisches Institut there twice a year for short periods, supervising seminars and collaborating in research with Professor Richard Neidlein and his students, as well as acting as an examiner for doctoral students. I feel very privileged to be able to continue to be active still as a heterocyclic chemist.

Details of organofluorine research

My researches in the field of fluorine chemistry at Salford fall into three related areas: the synthesis of novel fluorinated heteroaromatic compounds; modification of the Balz-Schiemann reaction; and the use of fluorine as a chemical label for aromatic compounds.

---

[6]Brian Iddon came to Salford from W. K. R. Musgrave's group at the University of Durham (UK). He managed to combine excellent work in heterocyclic chemistry with political activities in Bolton, and in May 1997 he became a Member of Parliament. He has the unusual distinction of having published papers under the address, 'House of Commons, London SW1A 0AA'.

[7]Now Professor at the University of Sunderland (UK), following posts in South Africa and with Sterling Organics in Newcastle (UK).

[8]Now Research Director of Reilly Industries, Indianapolis (USA).

[9]A village in the English Lake District; home of William Wordsworth and Thomas de Quincey.

Synthesis of fluorinated heteroarenes. – In collaboration with research students at Salford I prepared a range of monofluorinated heteroaromatics and derived compounds [7-11], including fluoroindoles (**3**), fluoroxanthones (**4**), fluoroindazoles (**5**), fluoropyridine-*N*-oxides (**6**), and fluoroquinoline- and fluoroisoquinoline-*N*-oxides (**7**, **8**). Fluorine was invariably introduced, either into the heterocycle or into its precursor, by the Balz-Schiemann reaction.

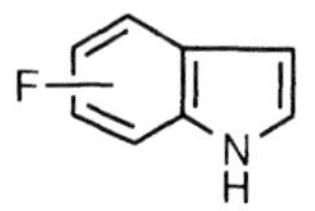

(**3**) 4-F, 5-F, 6-F, 7-F

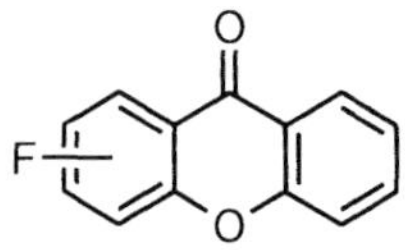

(**4**) 2-F, 3-F, 4-F, 5-F

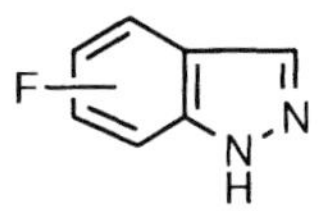

(**5**) 4-F, 5-F, 6-F, 7-F

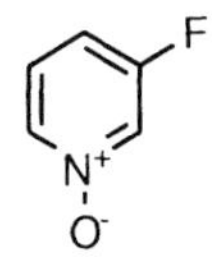

(**6**)

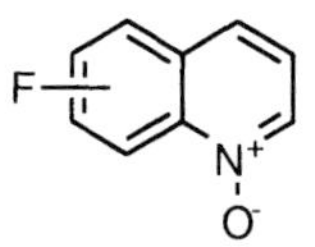

(**7**) 3-F, 5-F, 6-F, 7-F, 8-F

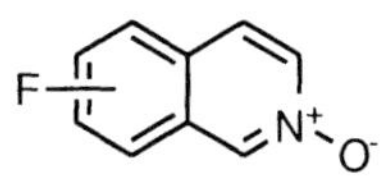

(**8**) 3-F, 4-F, 5-F, 6-F, 7-F, 8-F

Variations of and alternatives to the Balz-Schiemann reaction. – At a chance meeting with Professor Schiemann[10] in 1965 at the international fluorine symposium in Munich he spoke favourably about our work. This encouraged me to continue our work with fluorine-substituted aromatics. In particular I became interested in studying variations of, and alternatives to, the Schiemann method. For instance, we noted that Bergmann's group had described [14] a general method for the preparation of aryl fluorides involving treatment of a diazonium tetrafluoroborate at room temperature with copper powder or cuprous chloride in aqueous or dry acetone; a 52% yield was claimed in the case of *o*-fluoronitrobenzene, which, like other fluoronitrobenzenes, had previously been obtained in only very low yield (10-19%) by a standard Balz-Schiemann procedure. However, we found that Bergmann's procedure always led to deamination when we applied it to the preparation of various fluoronitrobenzenes [15]. Our results confirmed those of Hodgson and Marsden [16], who reported that the diazonium group in stabilised diazonium compounds is replaced by hydrogen on treatment with metallic copper in organic solvents.

In another variation of the Schiemann method, we applied Tedder's procedure [17]. This consists of the direct introduction of the diazonium group into aryl structures. For instance, a tertiary aromatic amine is treated with an aqueous solution of sodium nitrite, ethyl acetate and tetrafluoroboric acid; addition of ether precipitates the diazonium tetrafluoroborate: $R_2NC_6H_5 + 2HNO_2 + HX \rightarrow R_2NC_6H_4N_2^+X^- + 2H_2O + 2[O]$. Again the Schiemann method proved to be preferable because of substantially higher product yields,

[10]Günter Schiemann (1899-1967) was Professor of Technical Chemistry and Director of the Technical University of Hannover (Germany). Together with Günter Balz he pioneered the most important known method for the preparation of aryl fluorides, *i.e.* thermal decomposition of arenediazonium tetrafluoroborates. Their 1927 paper [12] describing the method is one of the most important ever published in the field of organofluorine chemistry. Schiemann himself, with various other co-workers, rapidly extended the scope of the method [13].

although it requires more stages. For example, tetrafluoroborates, 4-$XC_6H_4N_2^+\ BF_4^-$ can be obtained in yields of 16% ($X = Me_2N$), 52% ($X = Et_2N$) and 14% (X = pyrrolidino) using Tedder's method, compared with 52, 72 and 49%, respectively, by the Schiemann procedure ($R_2NC_6H_5 \rightarrow R_2NC_6H_4NO_2 \rightarrow R_2NC_6H_4NH_2 \rightarrow R_2NC_6H_4N_2^+BF_4^- \rightarrow R_2NC_6H_4F$).

As alternatives to the Schiemann procedure we explored the use of arenediazonium hexafluoroantimonates ($ArN_2^+SbF_6^-$) and -arsenates ($ArN_2^+AsF_6^-$) [18]. For instance, fluorobenzene had been obtained in the 1930s by pyrolysis of benzenediazonium hexafluoroantimonate, albeit in low yield [19]. Use of such salts is quite practicable, as they are insoluble and stable to light and air, and their decomposition temperatures are lower than those of the corresponding tetrafluoroborates – which mitigates the severity of their disintegration. Hence in most cases dilution with sand (the most popular moderator in the normal Balz-Schiemann reaction [2]) during their thermal decomposition is not necessary. We also found that electron-withdrawing ring substituents did not lower the yields, as is the case with diazonium tetrafluoroborates. Thus the method proved rewarding for the preparation of the three monofluorobenzoic acids (yields >70%) and *o*-fluoronitrobenzene (40% yield); by contrast, the tetrafluoroborate method provides *o*-, *m*-, *p*-fluorobenzoic acids in yields of only 9.5%, 5% and 40% respectively, while the yield of *o*-fluoronitrobenzene rarely exceeds 10%.

Decomposition of benzenediazonium hexafluoroarsenate did not give fluorobenzene but a deliquescent solid, which, on treatment with water, afforded bis(4-fluorophenyl)arsinic acid (**9**) in 87% yield [18]. This work provided a more convenient route to **9** than the then-current method [20] in view of the high product yield and the ready availability of the starting materials. A rationale for the formation of **9** is outlined in Scheme 26.1. The first step leads to fluorobenzene, as in the Schiemann reaction; however, the arsenic pentafluoride liberated combines with two molecules of fluorobenzene (owing to the para-directing influence of the fluorine towards electrophilic attack) to give the diarylarsenic trifluoride, which on hydrolysis yields the arsinic acid **9**. This scheme is supported by the fact that passing arsenic pentafluoride through boiling fluorobenzene, and then adding water, yields the same bis(fluorophenyl)arsinic acid (**9**).

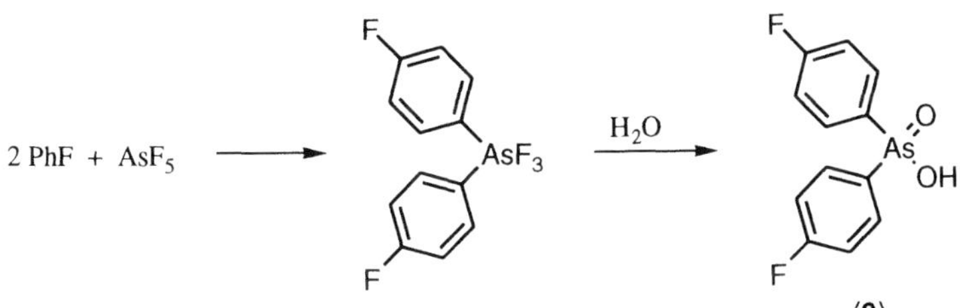

Scheme 26.1.

Aromatic fluorine as a chemical label for investigating reaction mechanisms [21, 22]. – Nucleophilic displacement of 'activated' ring fluorine in aromatic compounds is a well known phenomenon. The order of halogen mobility observed for bimolecular anionic attack activated by electron-withdrawing groups ($-I$, $-M$) runs in the opposite direction, *i.e.* F > Cl > Br > I, to that found in aliphatic S_N2 situations [23]. Foremost in providing

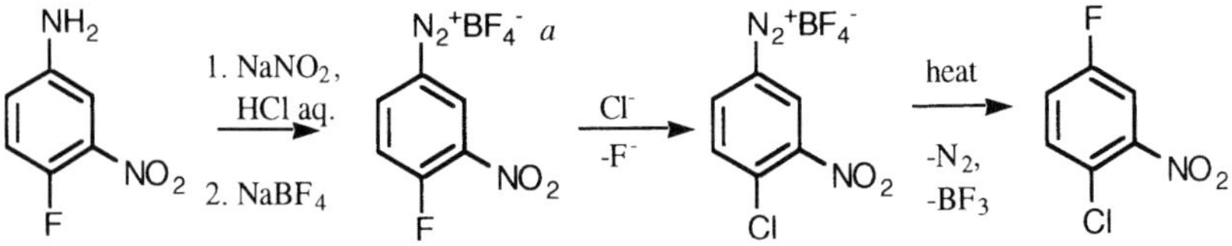

Scheme 26.2. [a] Additional activation of this substrate towards Cl^- attack is provided by the NO_2 group.

such activation in S_NAr reactions is the diazonium ion substituent, $-N_2^+$, which can render aromatic fluorine very labile indeed; thus during a Balz-Schiemann reaction involving diazotisation of a ring-fluorinated arylamine (indeed during any conversion involving an intermediate ring-fluorinated diazonium salt), the aromatic fluorine can (perhaps unexpectedly) be displaced even by a weak nucleophile. This annoying property of a diazonium group has been called the *nuisance effect* by Bunnett and Zahler [24]. The aptness of this description will be appreciated by every chemist who has unintentionally encountered this phenomenon. This nuisance effect is responsible, for instance, for the loss of fluorine from 4-fluoro-3-nitroaniline when concentrated hydrochloric acid is used in the diazotisation stage [25] (Scheme 26.2).

This nuisance effect can sometimes be put to good use when one is studying mechanisms of reactions of aromatic compounds in which the intermediacy of a diazonium group (or other group which can similarly promote nucleofugal behaviour of ring fluorine) needs to be established. In such cases the conversion of aromatic (covalent) fluorine into readily detectable fluoride ion is clearly diagnostic of the mechanism involved.

We have described a number of cases in which aromatic fluorine has been used as a chemical marker [21, 22], studies on the synthesis of 5-fluoroindazole by Jacobsen's method [26] providing a particularly good example. The preparative procedure involved the spontaneous decomposition of the fluoro-*N*-nitrosobenz-*o*-toluidide (**10**) in benzene (see Scheme 26.3). 5-Fluoroindazole (**12**), the expected product, was obtained, but it was accompanied by an equal amount of a fluorine-free compound which we identified as 5-(benzoyloxy)indazole (**14**), and fluoride ion was detected in the reaction mixture.[11] Under the same reaction conditions, fluorine was not replaced in the isomer of **10** with F *meta* to $N(NO)COC_6H_5$. Jacobsen's preparation of indazole (F = H in **12**) involves a rate-determining rearrangement of the corresponding nitroso-compound (F = H in **10**) to the diazoester (F = H in **11**) before ring closure takes place. On this basis, the replacement of the *para*-fluorine in **10** can be explained if the diazoester **11** is in equilibrium with the ion pair **13**: the *para*-fluorine is now liable to replacement by the weakly nucleophilic anion $C_6H_5CO_2^-$ owing to activation stemming from the *para*-diazonium group in **13**. Thus, the aromatic fluorine in **10** behaved as a chemical label, revealing at the time of our experiment the hitherto unsuspected involvement of a diazonium ion pair (*e.g.* **13**) during this type of indazole ring synthesis. Fluorine in this case proved to be specific for providing this information, as the chloro- and bromo-analogues (F = Cl or Br in **10**) gave only the expected halogenoindazoles (F = Cl or Br in **12**).

[11] The presence of F^- is conveniently detected by the zirconium-alizarin spot test [27].

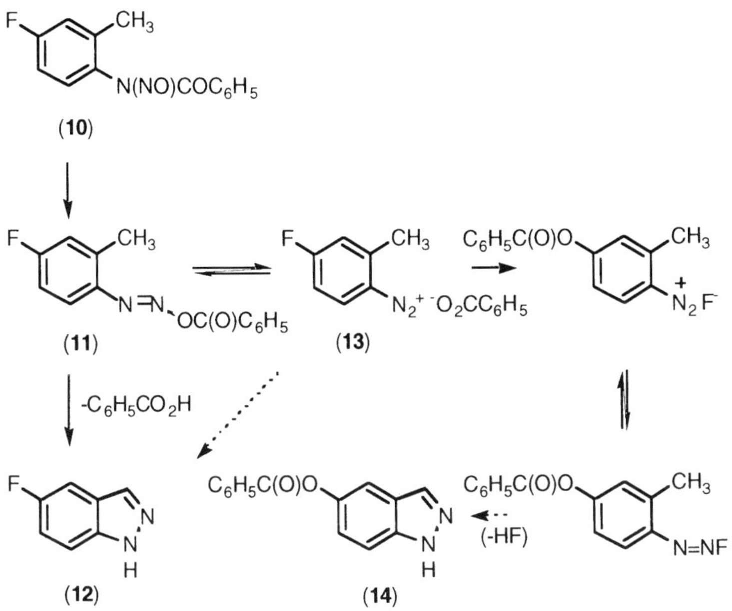

Scheme 26.3.

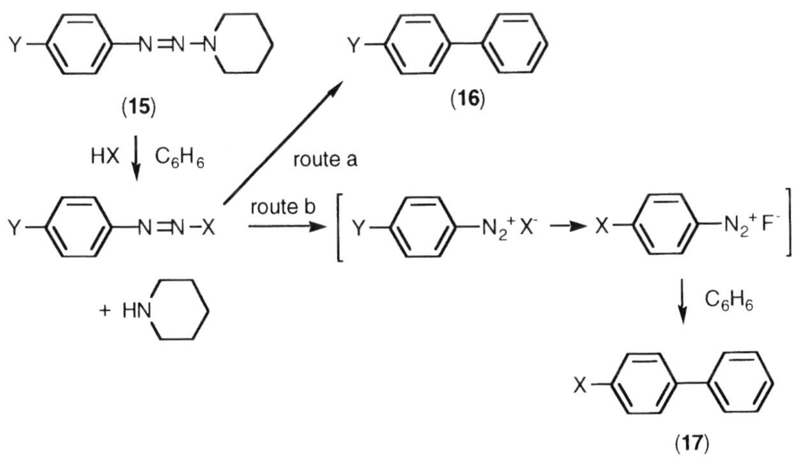

Scheme 26.4.

Details of our work on the use of 'fluorine labelling' in studies on the generation of aryl radicals (*e.g.* see Scheme 26.4) can be found in ref. [21]. An example is our study on the thermolysis of triazenes: when heated in benzene containing an acid, the parent triazene (**15**; Y = H) decomposes homolytically to give nitrogen, biphenyl (**16**; Y = H) and piperidine (Scheme 26.4, route a) [28]; however, when the corresponding fluorotriazene (**15**; Y = F) was decomposed in a mixture of benzene and acetic acid or benzene and hydrochloric acid, the 4-fluorobiphenyl (**16**; Y = F) was accompanied by an additional product (**17**; X = AcO or Cl) and fluoride ion was generated. The inference was that the triazenes could decompose heterolytically (route b) as well as homolytically (route a), as outlined in Scheme 26.4 [21].

Wakefield's story: polyhalogenoaromatic compounds

Following my undergraduate studies and my PhD research at Imperial College, London,[12] then a Postdoctoral Fellowship at Louisville, Kentucky (USA), I worked for three years at Courtaulds Ltd, Coventry (UK). I enjoyed this time in industry, but decided I really wanted to be an academic, and obtained a research post at the University of Reading (UK), where my interest in the chemistry of organolithium and organomagnesium compounds was kindled by Professor Derek Bryce-Smith. In 1964 I was appointed to a lectureship in the Royal College of Advanced Technology, Salford, which was at that time a 'university designate'. The following years were exciting. There was an air of confidence and progress, and the staff already in post were joined by excellent new members, including Brian Iddon, who brought with him his experience of organofluorine chemistry gained in W. K. R. Musgrave's group at Durham. Following the retirement of Professor George Ramage, the Organic Chemistry Section at Salford was under the wing of Hans Suschitzky – sometimes referred to in his absence just as 'HS', but more often as 'Uncle Hans', particularly by his research students.

Basil Wakefield

I was soon introduced to organofluorine chemistry. I was assigned a student, Ted Felstead, for his final year project, and it was customary to invite the industrial supervisors of 'works-based sandwich students'[13] to propose topics for the project. Ted's supervisor at the ICI laboratories situated at The Heath in Runcorn, Cheshire, was the established fluorine chemist Harold Fielding, who said that pentafluorobenzonitrile could well be of interest to his company, so we investigated the reaction of hexafluorobenzene with sodium cyanide in methanol, and isolated a product (**18**) which indicated that pentafluorobenzonitrile had been formed, but was so reactive that it underwent attack by the solvent [29] (Scheme 26.5). Later, I identified a second product – **19**, the result of hexasubstitution even under such mild conditions [30].

Although some work on polyfluoroaromatic compounds continued at Salford for some years (*e.g.* [31-33]), we were aware that this field was becoming crowded; and when large numbers of papers emanating from other UK groups working on fluorinated arenes

[12] During that time, 1952-1958, there were six Nobel Prize winners on the staff of the college, and I felt I was at the centre of the scientific world.

[13] An important course at Salford was the Integrated Course, invariably referred to as the sandwich course, a four-year course which included two six-month periods spent in industry. Most of the students were 'college based', but some already had jobs and were released by their employers: hence 'works based'.

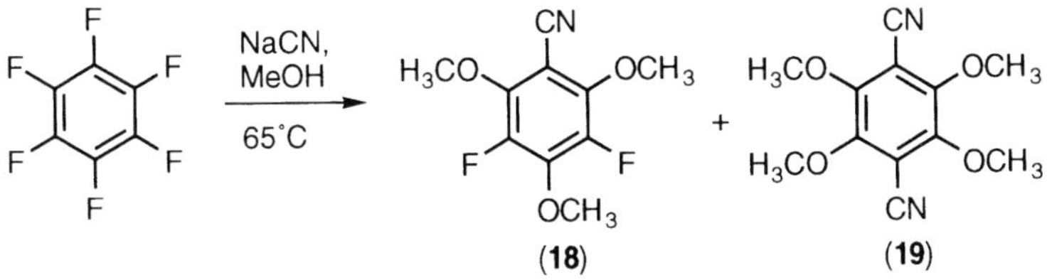

Scheme 26.5.

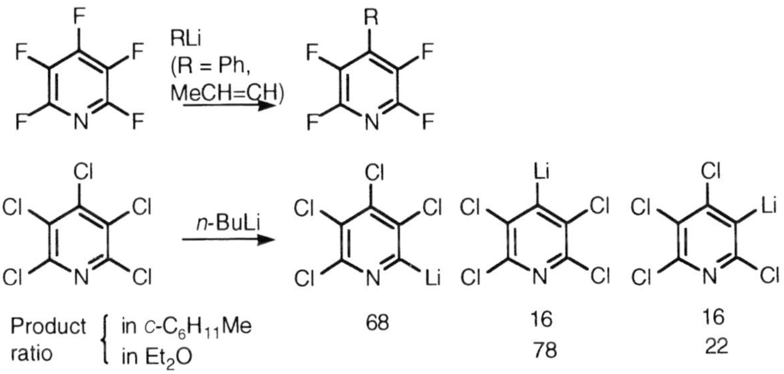

Scheme 26.6.

and hetarenes began to arrive on our desks for refereeing, we decided that it was now *over*crowded.[14] Accordingly, we decided to concentrate on polychloro (and polybromo) aromatic compounds, which were of interest because of their possible biological activity, as well as being of academic interest in their own right. We naturally wished to compare them with their fluorinated analogues. For example, it had been observed that whereas hexafluorobenzene underwent alkylation by organolithium compounds [34], hexachlorobenzene surprisingly undergoes metal-halogen exchange to give pentachlorophenyllithium [35]. Pentafluoropyridine is also alkylated (at the 4-position) [36], and we wondered whether pentachloropyridine, being more susceptible to nucleophilic attack than hexachlorobenzene, would be alkylated, and in which position. In fact, it underwent metal-halogen exchange in all three positions, in proportions depending on the solvent used (Scheme 26.6) [37].

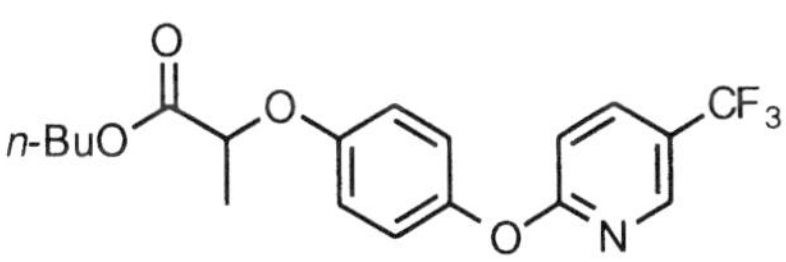

(20)

[14]Rivalry was rumoured to have become so intense that certain groups were unhappy about the prospect of their papers falling prematurely into the hands of rivals via refereeing procedures, and made this plain to editors when submitting publications to journals.

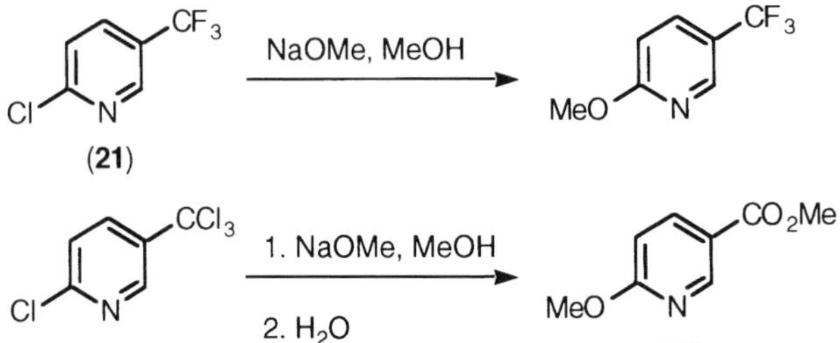

(**21**) (**22**)

Scheme 26.7.

Another remarkable contrast between fluoro-compounds and their chloro-analogues was revealed by work related to the important trifluoromethylpyridyloxy selective herbicides such as fluazifop-butyl [*Fusilade*^{TM} (ICI)] (**20**) [38]. Trifluoromethyl and trichloromethyl substituents in aromatic rings are normally regarded as quite unreactive towards nucleophilic reagents. However, while 2-chloro-5-trifluoromethylpyridine (**21**) was found to undergo replacement of the 2-chlorine by nucleophiles such as methoxide, leaving the trifluoromethyl group unaffected [39], the reaction of 2-chloro-5-trichloromethylpyridine with sodium methoxide in methanol, followed by aqueous work-up, gave the ester **22** [40] (Scheme 26.7).[15] The contrast with the lability of ring fluorine (the 'nuisance effect') compared with ring chlorine is noteworthy.

References

1 F. L. Allen and H. Suschitzky, *J. Chem. Soc.*, (1953) 3854.
2 H. Suschitzky, 'The Balz-Schiemann Reaction', *Adv. Fluorine Chem.*, *4* (1965) 1.
3 H. Suschitzky (ed.), *Polychloroaromatic Compounds,* Plenum, London and New York, 1974.
4 H. Suschitzky and O. Meth-Cohn (eds.), *Specialist Periodical Reports on Aromatic and Heteroaromatic Chemistry*, *7, 8* (1978, 1979).
5 H. Suschitzky and O. Meth-Cohn (eds.), *Specialist Periodical Reports on Heterocyclic Chemistry, 1-5* (1980-1985).
6 H. Suschitzky and E. F. V. Scriven (eds.), *Progress in Heterocyclic Chemistry, 1-7* (1989-1995); H. Suschitzky and G. Gribble (eds.), *Progress in Heterocyclic Chemistry, 8* (1996).
7 F. L. Allen, J. C. Brunton and H. Suschitzky, *J. Chem. Soc.*, (1955) 1283.
8 F. L. Allen, P. Koch and H. Suschitzky, *Tetrahedron Lett.*, (1959) 315.
9 I. K. Barben and H. Suschitzky, *J. Chem. Soc.*, (1960) 672.
10 M. Bellas and H. Suschitzky, *J. Chem. Soc.*, (1963) 4007.
11 M. Bellas and H. Suschitzky, *J. Chem. Soc.*, (1964) 4561.
12 G. Balz and G. Schiemann, *Ber. Dtsch. Chem. Ges.*, *60* (1927) 1186.
13 G. Schiemann, *J. Prakt. Chem.*, *140* (1934) 97.
14 E. D. Bergmann, S. Berkovic and R. T. Kan, *J. Am. Chem. Soc.*, *78* (1956) 6037.
15 I. K. Barben and H. Suschitzky, *Chem. Ind. (London)*, (1957) 1039.
16 H. H. Hodgson and E. D. Marsden, *J. Chem. Soc.*, (1940) 207.
17 J. M. Tedder, *J. Chem. Soc.*, (1957) 4003.
18 C. Sellers and H. Suschitzky, *J. Chem. Soc.*, (1968) 2317.

[15]We also discovered other remarkable reactions of β-trichloromethylpyridines – but that is another story [41].

19 W. Lange and K. Askitopoulos, *Z. Anorg. Chem.*, *223* (1935) 369.

20 O. S. Zeide, S. M. Sherlin and A. B. Bruckner, *Zh. Obshch. Khim.*, *28* (1958) 2404.

21 H. Suschitzky, *Angew. Chem. Int. Edn. Engl.*, *6* (1967) 596.

22 P. Miles and H. Suschitzky, *Tetrahedron, 19* (1963) 385; C. F. Sellers and H. Suschitzky, *Tetrahedron, 25* (1969) 1105.

23 J. Miller, *Aromatic Nucleophilic Substitution*, Elsevier, Amsterdam, 1968.

24 J. F. Bunnett and R. E. Zahler, *Chem. Rev.*, *49* (1951) 273.

25 H. Suschitzky, unpublished results.

26 P. Jacobsen and L. Huber, *Ber. Dtsch. Chem. Ges.*, *41* (1908) 660.

27 F. Feigl, *Spot Tests in Organic Analysis*, Elsevier, New York, 1956.

28 J. Elks and D. H. Hey, *J. Chem. Soc.*, (1934) 441.

29 E. Felstead, H. C. Fielding and B. J. Wakefield, *J. Chem. Soc. (C)*, (1966) 708.

30 B. J. Wakefield, *J. Chem. Soc. (C)*, (1967) 72.

31 M. Bellas, D. Price and H. Suschitzky, *J. Chem. Soc. (C)*, (1967) 1249.

32 D. Price, H. Suschitzky and J. I. Hollies, *J. Chem. Soc. (C)*, (1969) 1967.

33 C. L. Cheong and B. J. Wakefield, *J. Chem. Soc., Perkin Trans. 1*, (1988) 3301.

34 L. Wall, W. J. Pummer, J. E. Fearn and J. M. Antonucci, *J. Res. Nat. Bur. Stand., A, 67* (1963) 481; R. J. Harper, E. J. Soloski and C. Tamborski, *J. Org. Chem.*, *29* (1964) 2385.

35 M. D. Rausch, F. E. Tibbetts and H. B. Gordon, *J. Organometal. Chem.*, *5* (1966) 493.

36 R. D. Chambers, J. Hutchinson and W. K. R. Musgrave, *J. Chem. Soc.*, (1964) 3736; R. E. Banks, J. E. Burgess and R. N. Haszeldine, *J. Chem. Soc.*, (1965) 575.

37 J. D. Cook, B. J. Wakefield and C. J. Clayton, *Chem. Commun.*, (1967) 150; J. D. Cook and B. J. Wakefield, *J. Organometal. Chem.*, *13* (1966) 15.

38 *Weeds Today, 14* (1983) 3.

39 T. Haga, K.-I. Fujikawa, T. Koyonagi, T. Nakajima and K. Hayashi, *Heterocycles, 2* (1984) 117.

40 R. S. Dainter, T. Jackson, A. H. H. Omar, H. Suschitzky, N. Hughes, A. J. Nelson and G. Varvounis, *J. Chem. Soc., Perkin Trans. 1*, (1989) 283.

41 D. Cartwright, J. R. Ferguson, T. Giannopoulos, G. Varvounis and B. J. Wakefield, *Tetrahedron*, *51* (1995), 1279.

Chapter 27

FLUORINE CHEMISTRY AT THE UNIVERSITY OF BIRMINGHAM – A CRADLE OF THE SUBJECT IN THE UK

JOHN COLIN TATLOW
30 Grassmoor Road, Kings Norton, Birmingham B38 8BP, UK

Origins of the Chemistry Department

Chemistry was one of the foundation departments of the Mason Science College in Edmund Street, Birmingham, which opened in 1880. The building and endowments were financed by Sir Josiah Mason (1795-1881), a Birmingham industrialist, whose enterprises included the manufacture of steel pens and processes for silver electroplating. Already a benefactor to the city, having previously established alms-houses and an orphanage, his last great gift was a College for advanced study. The original deed provided for 'instruction without distinction of sex, class or creed,' but excluded 'mere literary pursuits', though it was later widened to permit an arts faculty. Queen's College, an independent teaching foundation established much earlier, became the Medical Faculty in Mason's College in 1892. Full University status was achieved in 1900, after a vigorous campaign led by Joseph Chamberlain, the Birmingham politician, who became Chancellor. Mason had attracted leading academics to be Foundation Professors in his College, and the Chemistry Department established a high reputation from the outset, both in undergraduate teaching and original research. Its early Professors were outstanding chemists, with tenure as follows: Sir William Tilden (1880-1894), P. F. Frankland (1894-1918), and G. T. Morgan (1918-1926) (later Sir Gilbert).

The Birmingham carbohydrate school

In 1926, Walter Norman Haworth was appointed as Professor and Head of Department. Already a leading carbohydrate chemist [1], his work at Birmingham was particularly concerned with ring sizes and the structures of di- and poly-saccharides, and of Vitamin C (ascorbic acid). He and his associates, particularly E. L. (later Sir Edmund) Hirst and S. Peat, made Birmingham an outstanding centre for carbohydrate research. Haworth was the recipient of many honours and awards, including a joint Nobel Prize for Chemistry in 1937 and a knighthood in 1947. An excellent summary of his life and achievements is available [2].

Work during World War II

At the outbreak of war in 1939, Haworth offered the facilities of his department for work of national importance, and some research based in the carbohydrate area and on explosives was undertaken. At that time, the cavity magnetron valve and the early forms of microwave radar were being developed in the Birmingham Physics Department by a team headed by Professor (later Sir Mark) Oliphant. Further, he, Dr (later Professor) P. B. Moon and Haworth were on the British MAUD Committee on nuclear energy, the vital importance of this new departure having been appreciated already. Central to the research on atomic fission was the theoretical work of the Professor of Mathematical Physics, R. (later Sir Rudolph) Peierls. Once it was realized that an atomic bomb based on the fission of uranium-235 was a practical possibility, crash projects were begun on all relevant aspects. The largest of these were in the USA (Manhattan Project) [3], but there was also much effort in the UK (Tube Alloys Project) [4].

The only feasible way to separate the isotopes of uranium on a large scale centred on the diffusion of a gaseous derivative through porous membranes. The only compound with sufficient volatility was uranium hexafluoride and, advantageously, fluorine has only one natural isotope. However, the great reactivity of UF_6 towards organic substances which would otherwise have been used in a gas-diffusion plant as coolants, lubricants, gasketing materials etc. focused attention on perfluorocarbons, examples of which had just been synthesized by J. H. Simons in the US and found to resist attack by the hexafluoride [3]. Clearly, the almost new field of fluorocarbon chemistry had to be developed rapidly.

The Birmingham Chemistry Department was asked to work on the new project, starting with some uranium chemistry. The metal was prepared, and physical properties measured for some of its compounds, such as the hexafluoride. Later, with the recognition that organofluorine compounds had a vital role, this aspect, particularly their preparation, received priority. Techniques entirely new to the department had to be learned and introduced quickly to achieve success.

The names of many of those associated with the earliest work in Birmingham in both areas (uranium and organofluorine) are recorded in the caption to Plate 27.1. Originally in charge of the organic side was F. Smith, an outstanding carbohydrate chemist. When he was seconded to the Manhattan Project in the USA, he was succeeded by M. Stacey, a Meldola Medallist from his work with Haworth on Vitamin C, who later became Head of the Birmingham Chemistry Department. The group's first publications were confidential patents, dated 1944 [5].

The historical development of organofluorine chemistry and its status at that time have been described in detail [6]. In this country, several groups had done some work with mono- or di-fluoroarenes, mostly using normal laboratory facilities, since highly reactive and corrosive fluorinating agents were not needed. However, such reagents, particularly hydrogen fluoride and elemental fluorine, were required to make highly-fluorinated aliphatic compounds. Only a group in the General Chemicals Division of ICI at Runcorn/Widnes was seriously involved in this area, mainly in the preparation of chlorofluoro-methanes and -ethanes (*e.g.* [7]), but cells to generate elemental fluorine had been made and some work on its reactions had begun.

There was always good cooperation between the Runcorn and Birmingham research groups during World War II. Fluorine cells were set up at Birmingham (with assistance from ICI) to generate fluorine, which was used to make perfluorocarbons from hydrocarbons, at first by direct vapour-phase reactions [8], and later via cobaltic fluoride, a superior fluorinating agent [9]. Among the products made were the fluorocarbon derivatives of the homologous series heptane to undecane, of methyl-, dimethyl-, ethyl-, and trimethyl-cyclohexane,

Plate 27.1. The Birmingham Chemistry Research Team, 1941/42 (from 'Mirror to a Mermaid' Birmingham University Centenary Publication, 1975). Left to right, front row: W. J. Hickinbottom (later Professor at Queen Mary College, London, author of a well-known organic textbook), Col. F. W. Pinkard (co-author of Wardlaw and Pinkard, 'Qualitative Analysis'), S. R. Carter (physical chemist), W. N. Haworth, L. L. Bircumshaw (inorganic chemist), M. Stacey, F. Smith; middle row; A. D. Booth (later a computation specialist), G. D. Thomas, W. G. M. Jones, S. P. James (later Reader in Biochemistry), E. Teece, R. Boyle, L. F. Thomas; back row; G. F. Claringbull [later Sir Gordon, Director of the British Museum (Natural History)], H. G. Bray (later Reader in Biochemistry), L. F. Wiggins (later Professor at the University of the West Indies), J. Wilkinson (produced the first 30 lb of metallic uranium; later of ICI), K. F. Chackett, W. K. R. Musgrave: not present was C. B. Amphlett who became a Senior Scientist in the Atomic Energy Authority.

and of decalin. Fluorocarbon oils were made by using elemental fluorine in liquid-phase reactions to di- and tri-merize partially-fluorinated arenes, especially benzotrifluoride, whilst simultaneously fluorinating them further, and then achieving perfluorination using high-valency metal fluorides [10]. Routes to partially-fluorinated precursors needed for these various processes were established [11].

Theses for higher degrees were kept secret at the time. The first PhD degree for work on organofluorides was awarded to W. K. R. Musgrave. I was one of the second wave of research students to earn their PhD's for work in this field, which included R. N. Haszeldine, and the late P. G. Harvey (who became a Director of ICI). The names of the many research workers from the Department, both from these early times and later, who went on to specialize further in some aspect of fluorine chemistry, are listed later in this chapter.

The Birmingham school of fluorine chemistry

Early advances

When hostilities were over, Haworth felt that our hard-won expertise in organofluorine chemistry should not be lost. He retired in 1948 but had established a small team to pursue fundamental research in the field. Smith returned to Birmingham for a short period before emigrating permanently to the USA to pursue carbohydrate chemistry. Before leaving, he wrote a detailed review of the organofluorine field [12]. Stacey then superintended the fluorine team and I became a research fellow and his deputy. Most of the wartime workers had by now moved on and among the new recruits was A. K. Barbour, who began a study of fluorohydrocarbons made using cobalt trifluoride [13]. These hydrogen-containing fluoro-compounds, which had always been difficult to remove from the perfluorocarbons required in the wartime work, seemed well worthy of study in their own right, as was soon confirmed. Despite experimental difficulties, systematic progress was made, justifying a vast team effort, that went on for many years. Some of the compounds made and their typical reactions are recorded in Schemes 27.1 and 27.2.

Also starting research at this time were my wife, C. E. M. Tatlow, and J. M. Tedder who, respectively, initiated work on the use of trifluoroacetic anhydride to make trifluoroacetate esters of carbohydrates [14] and the promotion of acylations and related reactions in general organic chemistry [15]. The trifluoroacetyl group, readily removed from esters and amides, was a useful blocking group; *e.g.* $-CH(OCOCF_3)CH(OCOCF_3)- + MeOH \rightarrow -CH(OH)CH(OCOCF_3)- \rightarrow -CH(OR)CH(OCOCF_3)- + MeOH \rightarrow -CH(OR)CH(OH)-$. The acylation-type reactions were promoted by unsymmetric anhydrides, with reactivities enhanced by the trifluoroacetic acid formed: $RCO_2H + (CF_3CO)_2O \rightarrow RC(O)OCOCF_3 + CF_3CO_2H$. The initial work was followed up by J. Burdon [16]. These developments were the earliest examples of the use of organofluorides in

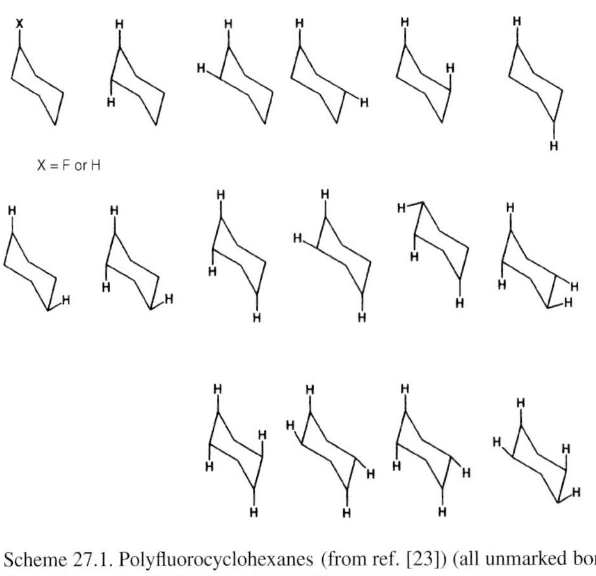

Scheme 27.1. Polyfluorocyclohexanes (from ref. [23]) (all unmarked bonds are to fluorine) produced by fluorination of benzene with CoF_3 at about 150 °C.

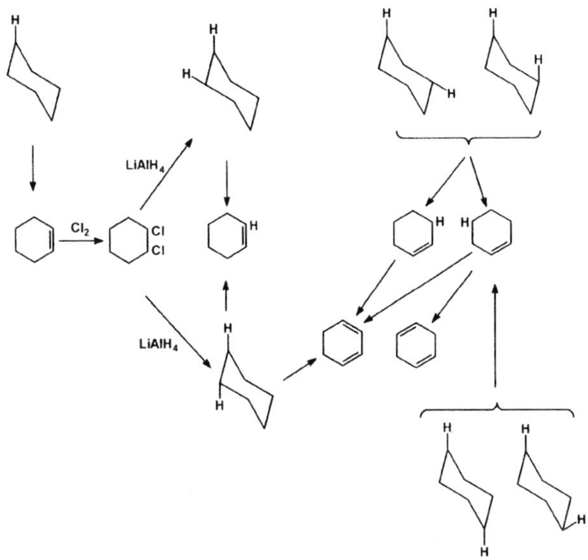

Scheme 27.2. Reactions of decafluorocyclohexanes with aq. KOH (from ref. [23]) (all unmarked bonds are to fluorine).

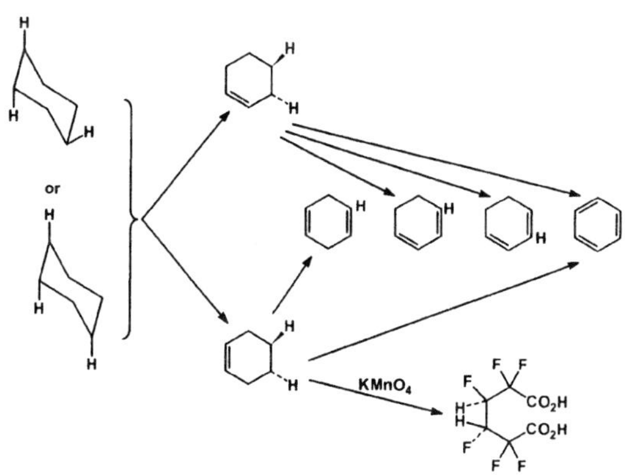

Scheme 27.3. Reactions of nonafluorocyclohexanes with aq. KOH (from ref. [23]) (all unmarked bonds are to fluorine).

general organic chemistry as synthesis aids and both types of application are now of enormous utility. These studies were supervised jointly with my cousin, the late E. J. Bourne, another carbohydrate chemist who later became Professor of Chemistry at the University of London (Royal Holloway College).

Fluorohydrocarbons

These have always been central to much of the work of the Birmingham group. Fluorination of benzene by means of cobalt trifluoride afforded a range of polyfluorocyclo-

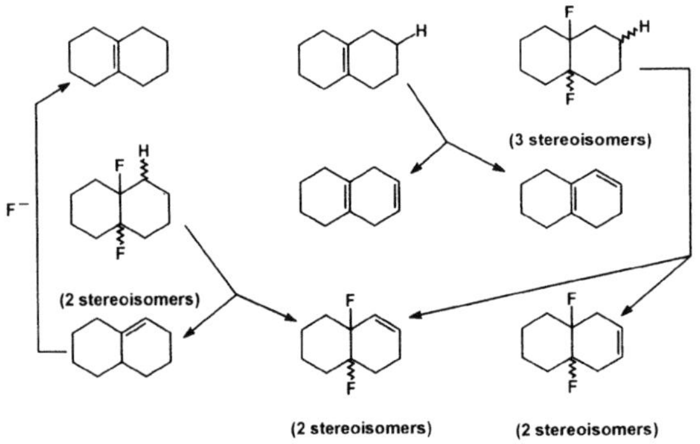

Scheme 27.4. Reactions of heptadecafluorodecalins with aq. KOH (from ref. [23]) (all unmarked bonds are to fluorine).

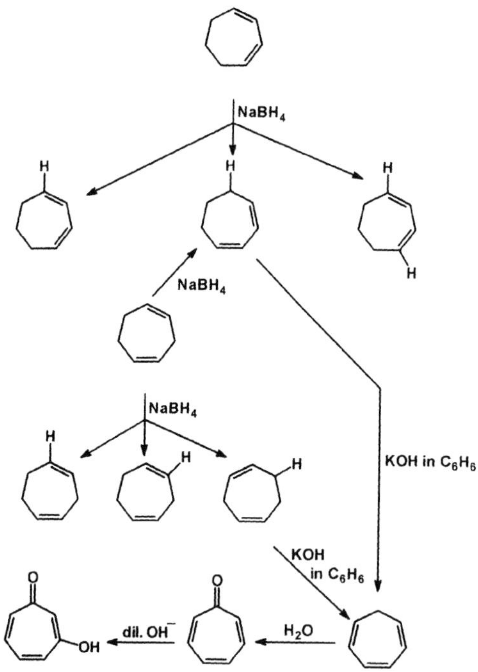

Scheme 27.5. Polyfluorocyclohepta-dienes and -triene (from ref. [23]) (all unmarked bonds are to fluorine).

hexanes, the reactions and stereochemistries of which were fully worked out, following the earliest applications of preparative-scale gas-liquid chromatography; whole new families of polyfluorocyclohex-enes and -dienes were prepared and identified [17] (Schemes 27.2 and 27.3). Later studies gave analogous ranges of derivatives based on the polyfluoromethylcyclohexane, -cyclopentane [18], -decalin (Scheme 27.4), and -cycloheptane systems [19] (Scheme 27.5). Names particularly associated with the work are R. Stephens, D. E. M. Evans, E. Nield and W. J. Feast.

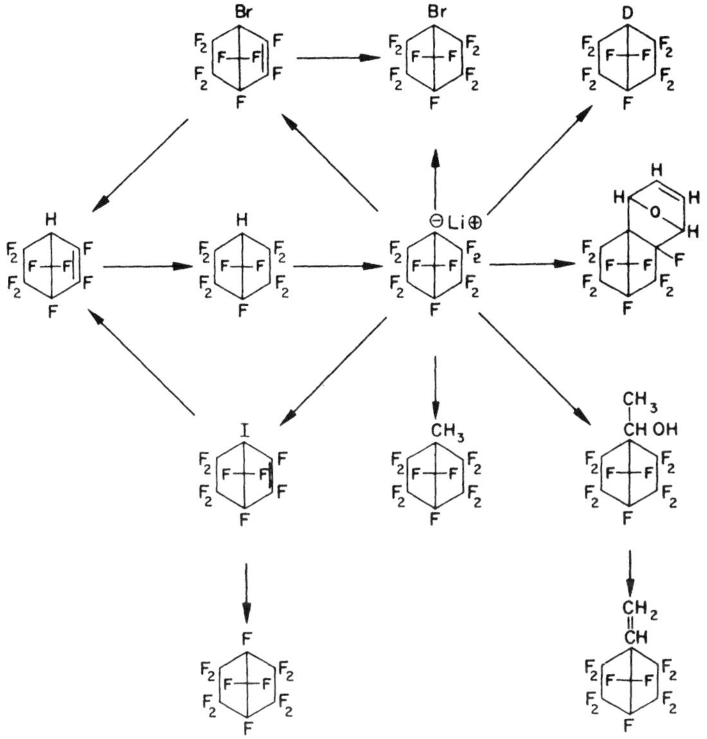

Scheme 27.6. Reactions of polyfluorobicyclo[2.2.1]heptanes [polyfluoronorbornanes] (from ref. [20]).

The studies on polyfluoroalicyclic systems were later extended into a new field, that of highly fluorinated bridgehead compounds (bicyclo-heptanes and -octanes), which showed many novel features, particularly the acidities, and hence reactivities, of C—H bonds at bridgehead positions [20] (Scheme 27.6). Prominently involved with these developments were R. Stephens and S. F. Campbell. The properties of these compounds were instrumental in re-establishing the concept of negative hyperconjugation as a significant feature in organofluorine chemistry [21-23].

Recent review articles describe all our work on fluorohydrocarbons in some detail, from the viewpoint of the polyfluoro-alicyclic systems [23] and of the fluorination processes [24].

Polyfluoro-arenes and -heteroarenes

This branch of the subject was also first entered via reactions of certain polyfluorocyclohexanes. Loss of three pairs of adjacent substituents, either hydrogen and fluorine, or fluorine itself, generated arene rings. Routes were developed to make hexa- and pentafluorobenzene readily available, enabling the vast field of polyfluoroaromatic compounds to be opened up; some of the new compounds synthesized by us are shown in Scheme 27.7. As part of the development of general synthetic methods in the field [25], a necessary and interesting aspect of the early work was establishing, for these compounds, the pattern of

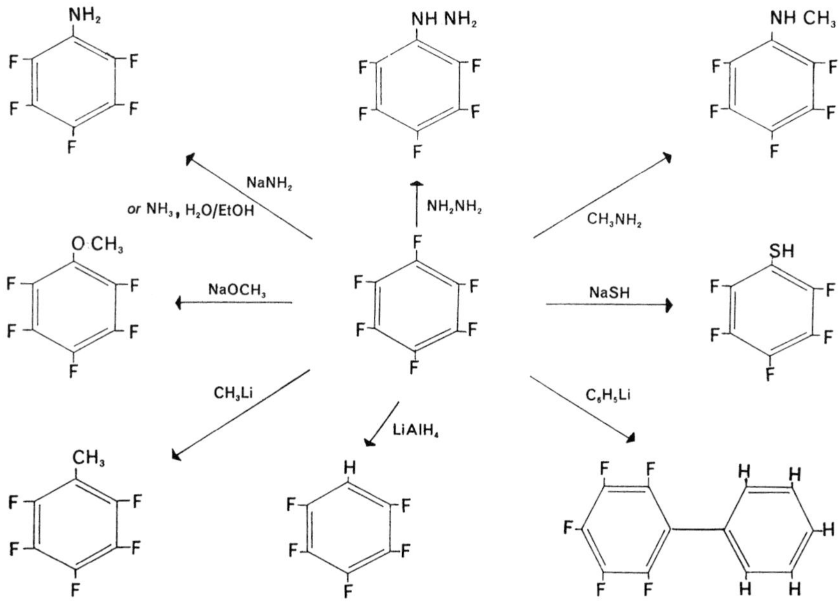

Scheme 27.7. Early reactions of hexafluorobenzene (from ref. [25]).

directional effects that govern attack on per- and poly-fluoroarenes by nucleophilic agents (Scheme 27.8). At that stage, structures of products had to be established by absolute methods. They then became the reference compounds for the newly-developing field of NMR spectroscopy of fluoro-arenes, which led to routine use of this technique for determining the orientation of ring substituents. It emerged that the influence exerted by substituents already present in a polyfluoro-arene ring was by far the major factor determining which fluorine was replaced by an incoming group. This parallels the situation for electrophilic substitution in orthodox aromatic chemistry.

Extensions of these preparative techniques were used later to synthesize highly-fluorinated heterocyclic systems, particularly polyfluoro-furans and -thiophens [23, 26]. Many studies on the effects exerted by fluorine on the reactivities of substituent groups in polyfluoroarenes were undertaken. An interesting transient species generated for the first time was tetrafluorobenzyne, found to react readily with arenes and heteroarenes to give Diels-Alder-type aducts [27].

All the Birmingham staff members whose names are included in the later list, were active in work on various aspects of the polyfluoro-arene field as it was extended. Little known when we began, the area aroused immediate interest [25] and has since grown on a worldwide basis to become one of the most important segments of organo-fluorine chemistry, as testified by a recent authoritative review by a former member of the group, G. M. Brooke [26].

Other work on organofluorides

Investigations into various other areas of specialist importance have been initiated in more recent years, associated particularly with P. L. Coe, who is continuing with the

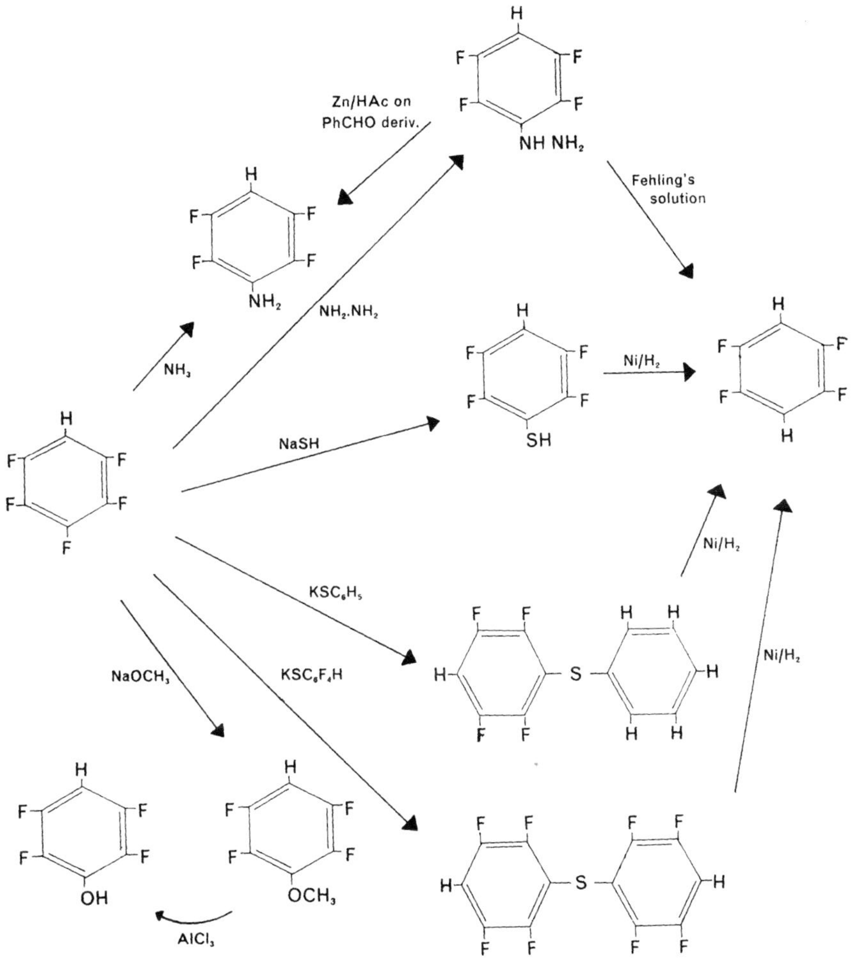

Scheme 27.8. Early reactions of pentafluorobenzene (from ref. [25]).

development of some of them. Selected from among the many topics studied are the following: syntheses of acyclic polyfluoro-ethers [28], *e.g.* $Et_2O \rightarrow (CHF_2CHF)_2O \rightarrow \rightarrow (F_2C{=}CF)_2O$; reactions of oligomers of tetrafluoroethene [29]; epoxidations of the double bonds of acyclic and cyclic polyfluoro-enes, (including oligomers of tetrafluoroethene, and of enes and dienes in the polyfluorocyclo-hexane and -heptane series) and reactions of the oxirane rings generated [30]; and syntheses of fluorinated analogues of anti-cancer agents, including chlorambucil [31] and aminoglutethimide [32].

Analytical, inorganic and physical chemistry involving fluorine compounds

Though our main effort was on synthesis and reactions of organofluorides, important work was done independently in the chemistry department on other branches of fluorine chemistry. For example, considerable effort was devoted by the late Professor R. Belcher and his team to the development of new methods for the analysis of fluorine compounds (*e.g.* [33]). This was not only for the determination of fluorine itself, but also for various other functions in the presence of fluorine, often not easy to measure accurately.

In the field of inorganic fluorine chemistry, R. D. Peacock directed a group, which included J. H. Holloway, that carried out important independent work (*e.g.* [34]). Later, after Peacock's appointment to a chair at the University of Leicester, the work continued under A. J. Edwards (*e.g.* [35]). Their major interests were the development of novel synthetic methods which led to the preparation of new fluorides of transition metals, often in higher valency states, followed by determinations of their structures and their fundamental properties. Complex fluorides involving transition metals were also investigated.

Work on spectroscopic properties of organofluorides was also an important feature, IR, MS and NMR methods being studied, both at a fundamental level and to provide departmental research services. T. A. Hamor's group carried out X-ray crystallographic studies on some of the new compounds synthesized (*e.g.* [36]). Fluorine-containing substrates were also often used in studies of reaction kinetics by colleages in the physical chemistry areas. In short, an integrated approach to extending knowledge in the fluorine chemistry field was adopted whenever possible.

Birmingham-trained fluorine chemists

Highlights of the subsequent careers of some of the members of the Birmingham Research School are indicated below (dates of Birmingham PhD degrees are given in parentheses).

- A. K. Barbour (1951): Research Director of ISC Chemicals – the fluorochemistry subsidiary of the RTZ Group, and later Chief Environmental Scientist for the Group; OBE, 1988.
- G. M. Brooke (1961): Reader in Organic Chemistry, University of Durham.
- J. Burdon (1956): Reader in Organic Chemistry, University of Birmingham.
- M. W. Buxton (1953): Founded a company, Bristol Organics (now part of Sigma Aldrich), for the custom synthesis of organofluorides, especially fluoroaromatics.
- S. F. Campbell (1965): Director of Research, Pfizer (UK); FRS, 1999.
- P. L. Coe (1960): Reader in Organofluorine Chemistry, University of Birmingham.
- A. J. Edwards: Senior Lecturer in Inorganic Chemistry, University of Birmingham.
- W. J. Feast (1963): Professor of Materials Science, University of Durham; FRS, 1998.
- G. Gambaretto (Visiting Research Fellow): Professor, University of Padua, Italy.
- G. B. Hammond (1985): Professor of Chemistry, University of Massachusetts, Dartmouth, USA.
- R. N. Haszeldine (1947); Professor and Head of the Chemistry Department, University of Manchester, Institute of Science and Technology (UMIST), where he founded and directed a leading school of organofluorine chemistry (continued in greatly modified form by Professor R. E. Banks); Meldola Medallist, 1953; Corday-Morgan Medal and Prize, 1960; Tilden Lecturer, 1968; FRS, 1968.

- J. H. Holloway (1963): Professor of Inorganic Chemistry and Head of the Chemistry Department, University of Leicester.
- C.-M. Hu (Visiting Research Fellow): Professor, Institute of Organic Chemistry, Chinese Academy of Sciences, Shanghai, China.
- The late W. G. M. Jones, 1916-1996 (1941, on carbohydrates): ICI Pharmaceuticals Division; *inter alia* worked on fluorinated anaesthetics [37].
- W. K. R. Musgrave (1943): Professor and Head of the Chemistry Department, University of Durham, where he founded and directed a leading school of organofluorine chemistry [38], (continued under Professor R. D. Chambers); later served as Acting Vice-Chancellor.
- I. W. Parsons (1969): Senior Lecturer in Polymer Chemistry, University of Birmingham.
- C. R. Patrick (1954, on kinetics of gas-phase reactions): Reader in Physical Chemistry, University of Birmingham, specializing in physico-chemical aspects of fluorine chemistry (*e.g.* [39]).
- R. D. Peacock: Professor of Inorganic Chemistry and Head of the Chemistry Department, University of Leicester.
- A. E. Pedler: Senior Lecturer in Organic Chemistry, University of Birmingham [40].
- R. G. Plevey (1964): Lecturer in Organic Chemistry, University of Birmingham.
- The late I. N. Rozhkov (Visiting Research Fellow): Professor, Institute of Organoelement Compounds (INEOS), Academy of Sciences, Moscow, Russia.
- P. Sampson (1983): Professor of Chemistry, Kent State University, Ohio, USA.
- S. F. Sellers (1979): Research Director, PCR Inc., USA.
- The late F. Smith, 1911-1965 [41] (1935, on carbohydrates): Oak Ridge National Laboratory, USA; then Professor of Biochemistry, University of Minnesota, USA.
- The late M. Stacey, 1907-1995 [42] (1932, on carbohydrates): Professor of Organic Chemistry (from 1948) and Head of the Chemistry Department (1956-74), University of Birmingham, senior member of the original fluorine research group [43]; FRS, 1950; CBE, 1966.
- R. Stephens (1954 on carbohydrate chemistry): Reader in Organic Chemistry, University of Birmingham.
- J. C. Tatlow (1946): see the Biographic Note below.
- The late J. M. Tedder, 1926-1994 [44] (1951): son of Lord Tedder [Marshall of the Royal Air Force, Deputy Commander of Operation Overlord, (the invasion of Europe, 1944)] and successor to the title (1967); Professor and Head of the Chemistry Department, University of St. Andrews, and leader there of an active group in organofluorine chemistry; FRSE, 1968.
- R. C. Terrell, (Visiting Research Fellow): Senior Scientist, Air Reduction Co., Later Abbot Laboratories Inc., USA.
- L. F. Thomas (1951, after wartime naval service): Senior Lecturer in Physical Chemistry, University of Birmingham, specializing in NMR spectroscopy (*e.g.* [45]).
- E. H. Wiseman (1959): Senior Scientist, Pfizer Inc., USA.

Acknowledgements

My thanks go to Dr P. L. Coe for his recollections and for other generous assistance associated with this article. I am proud to be the co-author of some 340 original papers, patents, reviews etc. arising from work done by the Birmingham fluorine group, and pay tribute to all my colleagues and research students whose names also appear on those publications.

References

1 W. N. Haworth, *The Constitution of Sugars*, Arnold, 1929.
2 E. L. Hirst, 'Obituary Notice for W. N. Haworth (1883-1950)', *J. Chem. Soc.* (1951) 2790.
3 H. Goldwhite, in R. E. Banks, D. W. A. Sharp and J. C. Tatlow (eds.), *Fluorine: The First Hundred Years*, Elsevier Sequoia, Lausanne, 1986, p. 109.
4 *Statements Relating to the Atomic Bomb*, HMSO, London, 1945; R. W. Clark, *The Birth of the Bomb*, Phoenix House, London, 1961.
5 W. N. Haworth, F. Smith and E. V. Appleton, *British Patents 627,685* (1944) [*Chem. Abs.*, *44* (1950) 3524]; *630,606* (1944), [*Chem. Abs.*, *44* (1950) 4029].
6 R. E. Banks and J. C. Tatlow, in R. E. Banks, D. W. A. Sharp and J. C. Tatlow (eds.), *Fluorine: The First Hundred Years*, Elsevier Sequoia, Lausanne, 1986, p. 71.
7 J. P. Baxter, *Brit. Pat. 454,577* (1936) [*Chem. Abs.*, *31* (1937) 1045]; F. D. Leicester, *Brit. Pat. 468,447* (1937) [*Chem. Abs.*, *32* (1938) 587].
8 W. K. R. Musgrave and F. Smith, *J. Chem. Soc.*, (1949) 3021; 3026.
9 R. N. Haszeldine and F. Smith, *J. Chem. Soc.*, (1950) 3617.
10 F. Smith, M. Stacey, J. C. Tatlow, J. K. Dawson and B. R. J. Thomas, *J. Appl. Chem.*, *2* (1952) 97.
11 L. V. Johnson, F. Smith, M. Stacey and J. C. Tatlow, *J. Chem. Soc.*, (1952) 4710.
12 F. Smith, *Ann. Reports Chem. Soc.*, *44* (1947) 86.
13 A. K. Barbour, H. D. Mackenzie, M. Stacey and J. C. Tatlow, *J. Appl. Chem.*, *4* (1954) 341; 347.
14 E. J. Bourne, M. Stacey, Mrs C. E. M. Tatlow and J. C. Tatlow, *J. Chem. Soc.*, (1951) 826.
15 E. J. Bourne, M. Stacey, J. C. Tatlow and J. M. Tedder, *J. Chem. Soc.*, (1949) 2976; J. M. Tedder, *Chem. Rev.*, *55* (1955) 787.
16 E. J. Bourne, J. Burdon and J. C. Tatlow, *J. Chem. Soc.*, (1958) 1274; (1959) 1864.
17 D. E. M. Evans and J. C. Tatlow, *J. Chem. Soc.*, (1955) 1184; D. E. M. Evans, J. A. Godsell, R. Stephens, J. C. Tatlow and E. H. Wiseman, *Tetrahedron*, *2* (1958) 183; E. Nield, R. Stephens and J. C. Tatlow, *J. Chem. Soc.*, (1959) 159; D. E. M. Evans, W. J. Feast, R. Stephens and J. C. Tatlow, *J. Chem. Soc.*, (1963) 4828.
18 J. Burdon, T. M. Hodgins, R. Stephens and J. C. Tatlow, *J. Chem. Soc.*, (1965) 2382.
19 D. J. Dodsworth, C. M. Jenkins, R. Stephens and J. C. Tatlow, *J. Fluorine Chem.*, *24* (1984) 41.
20 S. F. Campbell, R. Stephens and J. C. Tatlow, *Tetrahedron*, *21* (1965) 2997.
21 J. H. Sleigh, R. Stephens and J. C. Tatlow, *J. Fluorine Chem.*, *15* (1980) 411.
22 B. E. Smart, in R. E. Banks, B. E. Smart and J. C. Tatlow (eds.), *Organofluorine Chemistry; Principles and Commercial Applications*, Plenum Press, New York, 1994, p. 77.
23 J. C. Tatlow, *J. Fluorine Chem.*, *75* (1995) 7.
24 J. Burdon, in B. Baasner, H. Hagemann, J. C. Tatlow (eds.), *Organo-Fluorine Compounds*, Vol. E10a, *Methods of Organic Chemistry* (Houben-Weyl), Thieme, Stuttgart, 1999, p. 655.
25 J. C. Tatlow, *Endeavour*, *22* (1963) 89; *New Scientist*, *17* (1963) 236; *Discovery*, (1963) 27.
26 G. M. Brooke, *J. Fluorine Chem.*, *86* (1997) 1.
27 D. D. Callander, P. L. Coe, J. C. Tatlow and A. J. Uff, *Tetrahedron*, *25* (1969) 25.
28 M. Brandwood, P. L. Coe and J. C. Tatlow, *J. Fluorine Chem.*, *6* (1975) 37.
29 P. L. Coe, S. F. Sellers, J. C. Tatlow, H. C. Fielding and G. Whittaker, *J. Chem. Soc., Perkin Trans. I*, (1983) 1957; P. L. Coe, A. Sellars and J. C. Tatlow, *J. Chem. Soc., Perkin Trans. I*, (1985) 2185; P. L. Coe and N. C. Ray, *J. Fluorine Chem.*, *88* (1998) 169.
30 P. L. Coe, A. W. Mott and J. C. Tatlow, *J. Fluorine Chem.*, *49* (1990) 21; P. L. Coe, M. Löhr and C. Rochin, *J. Chem. Soc., Perkin Trans. I*, (1998) 2803.

31 C. W. Buss, P. L. Coe and J. C. Tatlow, *J. Fluorine Chem.*, *34* (1986) 83; P. L. Coe, M. Markou and J. C. Tatlow, *J. Fluorine Chem.*, *89* (1998) 183.
32 G. B. Hammond, R. G. Plevey, P. Sampson and J. C. Tatlow, *J. Fluorine Chem.*, *40* (1988) 81.
33 R. Belcher, E. F. Caldas, S. J. Clark and A. Macdonald, *Mikrochim. Acta*, *3/4* (1953) 283; R. Belcher, M. A. Leonard and T. S. West, *J. Chem. Soc.*, (1959) 3577; A. M. G. Macdonald, in F. A. Smith (ed.), *Handbook of Experimental Pharmacology, New Series, Vol. XX/2*, Springer-Verlag, Berlin, 1970, p. 1.
34 R. D. Peacock, *Adv. Fluorine Chem.*, *4* (1965) 31; A. J. Edwards, R. D. Peacock and R. W. H. Small, *J. Chem. Soc.*, (1962) 4486; J. H. Holloway and R. D. Peacock, *J. Chem. Soc.*, (1963) 527; J. H. Holloway, R. D. Peacock and R. W. H. Small, *J. Chem. Soc.*, (1964) 644.
35 A. J. Edwards, *J. Chem. Soc.*, (1964) 3714; (*A*) (1972) 2325; *Adv. Inorg. Chem. Radiochem.*, *27* (1983) 83; J. C. Tatlow, R. G. Plevey and A. J. Edwards, *British Patent 1,392,571* (1975).
36 M. J. Hamor and T. A. Hamor, *J. Chem. Soc., Perkin Trans II*, (1976) 383; N. Goodhand and T. A. Hamor, *Acta. Cryst.*, *B34* (1978) 513.
37 W. G. M. Jones, in R. E. Banks (ed.), *Preparation, Properties and Industrial Applications of Organofluorine Compounds*, Horwood, Chichester, 1982, p. 157.
38 Various authors, 'Anniversary Issue to Commemorate W. K. R. Musgrave's 70th birthday', *J. Fluorine Chem.*, *40* (1988) 81-434.
39 J. R. Majer and C. R. Patrick, *Trans. Farad. Soc.*, *58* (1962) 17; F. Gozzo and C. R. Patrick, *Tetrahedron*, *22* (1966) 3329.
40 A. M. Doyle, C. R. Patrick and A. E. Pedler, *J. Electroanalytical Chem.*, *33* (1971) 23; G. Gambaretto and A. E. Pedler, *Annali di Chimica*, *64* (1974) 711.
41 R. Montgomery, 'Obituary Notice for F. Smith (1911-1965)', *Adv. Carbohydrate Chem.*, *22* (1967) 1.
42 J. C. Tatlow, 'Obituary Notice for M. Stacey (1907-1994)', *J. Fluorine Chem.*, *73* (1995), vii.
43 Various authors, 'Anniversary Issue to Commemorate M. Stacey's 70th birthday', *J. Fluorine Chem.*, *10* (1977) 437-624.
44 M. Hudlicky, 'Obituary Notice for J. M. Tedder (1926-1994)', *J. Fluorine Chem.*, *90* (1998) 201.
45 J. Battersby, R. Stephens, J. C. Tatlow and L. F. Thomas, *J. Fluorine Chem.*, *15* (1980) 139.

BIOGRAPHIC NOTE

Born in 1923 in Staffordshire, Colin Tatlow went to Rugeley Grammar School and then the University of Birmingham. After getting his PhD degree in 1946 (thesis title: Studies in Fluorination), he married Cherry Joiner, a chemistry student he met whilst supervising undergraduate practical classes. They flew to Canada in a converted Liberator bomber and both worked for a year (1946-47) at the Chalk River Atomic Energy Establishment before sailing home in style, first class, in the newly refitted Queen Mary (with Mae West as a fellow passenger). He took up a Research Fellowship at Birmingham and was deputy leader of the fluorine group. He than progressed through the ranks (Lecturer; Senior Lecturer, Reader) to become Professor of Organic Chemistry (1959) and Head of the Chemistry Department (1974-81); en route he became director of the fluorine research group. In

J. C. Tatlow (late 1970s)

Plate 27.2. Chemical Archway at the wedding of Colin and Cherry Tatlow. Guard of Honour (left to right): ? Smart, B. R. J. Thomas, F. H. Newth, P. G. Harvey, R. N. Haszeldine, P. W. Kent (little was left of a spectacular carpet of vapour from liquid air, poured by Smart, by the time this photo was taken).

1959, the group organized in Birmingham an International Symposium on Fluorine Chemistry, which became the first of the now regular and important familiar series. Since retiring from the University of Birmingham in 1982, Colin has continued to promote fluorine chemistry as a consultant, writer and editor. After being invited to help to establish the Journal of Fluorine Chemistry in 1970, Colin was Joint Editor-in-Chief for more than 25 years (1971-1997) and is now Honorary Editor. He was a joint editor (1960-1973) of the review series *Advances in Fluorine Chemistry* (M. Stacey and R. D. Peacock also served) and of the books *Fluorine: the First Hundred Years* (Elsevier, 1986) and *Organofluorine Chemistry, Principles and Commercial Applications* (Plenum, 1994). In 1997 he was asked to be a co-editor of the new Houben-Weyl issue *Organo-Fluorine Compounds, Vol. E10* in their series, *Methods of Organic Chemistry* a comprehensive work in five volumes, just published. In recognition of the Birmingham group's work, Colin was the SCI Jubilee Lecturer in 1976, a Medallist at the Moissan Centenary Symposium in Paris in 1986, and received the American Chemical Society Award for Creative Work in Fluorine Chemistry in 1990.

On a personal level, Cherry and Colin Tatlow have had great satisfaction as gardeners, particularly from growing most of their own vegetables. Genealogy and family history have also been a major ongoing interest for many years. The earliest Tatlow ancestor for whom they have found records was a farmer in Longford, Derbyshire, and they have a copy of his will, dated 1713. Among the families that married into the Tatlows, one included a bigamist (a direct ancestor) and another a Catholic Saint, Ralph Sherwin, who was executed in 1580, in the reign of Elizabeth I. Cherry and Colin have two daughters, Susan, a lawyer, and Bridget, a medical practitioner, and two grandchildren, Tom and Emma Johnson, and they hope to make a scientist of one of them.

Chapter 28

THE BELATED HEXAFLUOROBENZENE PAPERS OF YVONNE DÉSIRANT

DIRK TAVERNIER

Department of Organic Chemistry, University at Ghent, Krijgslaan 281 (S4), B-9000 Gent, Belgium

Prologue

During the period 1934-36 Miss Yvonne Désirant (born 1899), a co-worker of the famous pioneer of aliphatic fluorine chemistry Frederic Swarts (1866-1940) developed the first synthesis of hexafluorobenzene: pyrolysis ('cracking') of tribromofluoromethane over platinum foil contained in a platinum tube heated to 630-640 °C. Remarkably, this important work was made public as late as 1955, almost twenty years after the experimental work was done. The chronology is as follows. On 15th December 1936, the Belgian Royal Academy (Class of Sciences), accepted the deposition, by Désirant, of a sealed envelope. On 11th June 1955, at the request of Désirant, the sealed envelope was opened by the secretary of the Academy. It contained a brief note of about 200 words, describing the synthesis and some physical and chemical properties of hexafluorobenzene. The note was published in the bulletin of the Academy [1]. Three years later, in 1958, Désirant published the full paper on her hexafluorobenzene work in the *Bulletin of the Belgian Chemical Societies* [2]. Except for a brief reference to Mohler of the National Bureau of Standards, Washington DC, USA, who, in 1954, carried out a mass spectrometric study on a sample of hexafluorobenzene prepared by Désirant, all the information in that paper dates from 1934-36. Neither the 1955 'preliminary communication' nor the 1958 'full paper' state or suggest a reason for the late publication of such important results.

Yvonne Désirant (May 1999)
[Photograph by courtesy of 'F-twee' Publishers, Gent, Belgium.]

Désirant is now a centenarian and lives in Ostend. For the preparation of this chapter, her longtime friend, Mrs Vandepitte (née Hilda Loos) has been to see her in Ostend in order to find out why she delayed publication for such a long time. There was no answer to that question, but Désirant did provide some information concerning hexafluorobenzene-related events dating from the 1930s; unfortunately, however, her memory, otherwise excellent, did not serve to provide information from the 1950s. By writing to and speaking with various people who have been in contact with Désirant, and by consulting the archives of Ghent University, I have pieced together the story related here.

Setting the scene

Some historical background on Belgium and on Ghent University, where Swarts was a professor of chemistry, is necessary. Broadly, Belgium consists of a French-speaking part and a Dutch (or Flemish)-speaking part. At the independence of Belgium in 1830, French became the official language of the new state. The uppermost class in the Dutch-speaking part of Belgium was largely Frenchified, and the other classes acquiesced to the predominance of that language. That acquiescence did not persist. As the nineteenth century progressed, demands grew for the Dutch language to feature in the civil services and in higher education. There were then two state universities in Belgium, one at Liège (French-speaking part of Belgium), and one at Ghent (Dutch-speaking part of Belgium). The conversion of the French-speaking university at Ghent to a Dutch-speaking institution became a key objective of the Flemish movement. This conversion was resisted by the Belgian establishment. It was feared that the formation of a Dutch-speaking educated elite would foster separatism, and the scission of Belgium. Compromise solutions were tried out in the 1920s. However as 1930 – the centennial of Belgian independence – approached, it was feared that demonstrations and perhaps riots about the non-existence of a Dutch-speaking university might spoil the centennial celebrations and tarnish the international image of Belgium. So the establishment gave way, and a law was passed which enabled Ghent University to be converted into a Dutch-speaking institution in 1930. The French-speaking sections were not abolished overnight, enabling students who had enrolled under the French regime to complete their studies in that language, but this was to end with the academic year 1935-36. The new linguistic regime affected the professorial corps of the university. Many knew Dutch, and simply stayed on. Some chose not to teach in Dutch, others were unable to; they were put on the inactive list but kept their titles and salaries. Swarts, a luminary of Ghent University, had fiercely opposed the abolishment of the French linguistic regime. Yet, in a formal sense, his career was not influenced by it because he reached the statutory date of retirement on 30th September 1936, precisely when the French-speaking sections were terminating.

The Désirant story

Yvonne Désirant enrolled at Ghent in 1921 as a chemistry student in the French-speaking sections and became an assistant to Swarts in 1926. With the full support of Swarts, in 1931 she was appointed by the minister of education to a position called in French *chef de travaux* – foreman, if you will. The administrative status of *chef de travaux* apparently has no counterpart in Anglo-Saxon universities. Like a professorship, *chef de travaux* is a tenured position. Unlike a professorship, it is not a 'stand-on-your-own-feet' position. On the contrary, a *chef de travaux* is assigned to a professor, who is his/her superior and who chooses the research activities. The position does not exist anymore. The appointment of Désirant in 1931 was somewhat curious in that she became tenured in the soon (1935-36) to be phased-out French-speaking section of the University. Therefore, prior to her appointment, Désirant was asked by the university administration to state that she could and would, from 1936 on, perform university duties in Dutch. Her affirmative statement was verified by an examination, which took the form of a conversation with the rector of the university, August Vermeylen, a scholar of Dutch literature and a one time novelist.

Upon his retirement in 1936, Swarts was given the full privileges of an important professor emeritus: a private office and a private laboratory space. But he had little or no influence with the men who now held sway in the 'Laboratory of General and Toxicological Chemistry' or in the new (1936) 'Laboratory of Organic Chemistry', namely Professor René Goubau (1886-1976) and Professor Firmin Govaert (1902-1993) respectively. Govaert was appointed to his chair in 1936 – the very year of Swarts' retirement – and this was to have a great influence on Désirant. Neither Goubau nor Govaert were interested in organofluorine chemistry, and with some exaggeration, the Ghent school of fluorine chemistry may be said to have died with the retirement of Swarts. Goubau and Govaert were very different characters. Goubau, the elder man, was not really interested in research. He excelled in commission work, was esteemed for it, and eventually became rector of Ghent University (1939-1944). Govaert, the younger man, was of a different mettle. Ambitious to build up his own research team, he was also eager to introduce himself into the power centre of the university, in which he succeeded soon after his appointment. There he displayed remarkable bureaucratic savvy. He was to establish the 'Laboratory of Organic Chemistry' as one of the university's major research groups.

Désirant, officially severed from Swarts through his retirement, had to be assigned as *chef de travaux* to another professor. She did explore the possibility of joining the research group of Professor Alfred Schoep, a well-known mineralogist, but that plan came to naught. On 14th October 1936 the administrator-general of the university wrote a letter to Désirant, telling her that she had been assigned to Goubau. He sent a copy of that letter to Goubau, and also asked Goubau what research activities he had in mind for Désirant. On 20th November 1936 Goubau replied as follows.

(1) Désirant must supervise the lab classes and the written exercises of my students.
(2) Désirant takes up the apprenticeship of organic microchemical analysis under the guidance of myself *and of colleague Govaert*. (The italics are mine.)
(3) These activities of Désirant will take up all her time.

Which means, I think, that Govaert had seen an opportunity and had grabbed it. Administratively, Désirant was assigned to Goubau, whom she supported in his teaching duties; but her remaining time – available for research – was to profit Govaert, with whom she officially had nothing to do. Govaert had Désirant installed in his lab spaces, and 'the apprenticeship of organic microchemical analysis' meant she did the exacting and time-consuming elemental analysis of the new compounds prepared in Govaert's fledgling research group. He had acquired at no cost the research services of a skilled and meticulous co-worker.

Such is the background of the deposition in the Belgian Academy, by Désirant, of the famous sealed envelope containing the short note describing her synthesis of hexafluorobenzene [1]. Recall the date of deposition: 15th December 1936 (the note itself is dated 14th December). By then, Désirant's fate had been sealed by Goubau and Govaert. There was to be no more fluoro-organic chemistry in her future. In her interview with Hilda Vandepitte, Désirant has been very clear: she deposited the sealed envelope at the request of Swarts himself. She pointed out that Swarts had very high publications standards. Research should only be published when fully rounded off, and in the eyes of Swarts the hexafluorobenzene work was not – according to Désirant, he felt that additional physico-chemical data were required. On the other hand, scientific priority is of paramount importance. The deposition of a sealed envelope in an academy was then an accepted method of establishing priority, and Swarts had used it before. The present-day onlooker remains somewhat puzzled. Who was to round off the hexafluorobenzene work in order to make it – in Swarts' view – publishable? Not Désirant – see Goubau's letter. Did Swarts himself intend to do it? We do not know. Why then not publish this maybe incomplete but certainly innovative and important hexafluorobenzene research and take credit for it? Maybe Swarts had become a prisoner of his reputation as a perfectionist, of his own high publication standards, perhaps also of his declining health, so that he simply could not bear the thought of an incomplete publication coming out of his research group.

We know from Désirant's 1958 publication [2] that Swarts directed the hexafluorobenzene work, and took a great interest in it. Yet the note in the 1936 sealed envelope has Désirant as the sole author [1]. Remarkable, but entirely consistent with Swarts' publication policy. His publications carry his name only, not those of his co-workers, and conversely publications of his co-workers do not carry his name – hardly imaginable today!

Désirant's synthesis of hexafluorobenzene, though unpublished, was certainly not kept secret. It was passed by word of mouth at scientific meetings. Victor Desreux (born 1910), the last PhD student (1935) of Swarts, was provided with samples of the hexafluorobenzene prepared by Désirant, probably for use in Raman spectroscopic studies. In 1938 Desreux gave at Harvard (MA, USA) a speech on fluoro-organic chemistry, wherein he described Désirant's hexafluorobenzene synthesis. In a sense, it even appeared in print, for in 1946 Desreux published [3] an extensive obituary on Swarts wherein Désirant is given full credit for developing a synthesis of hexafluorobenzene by a thermal procedure. Desreux eventually became a professor at Liège University, where he, and others, kept alive the memory of Swarts' discoveries (including the synthesis of hexafluorobenzene), so much so that Desreux came to be known as the man to be contacted by anyone who needed information about Swarts.

Meanwhile, at Ghent, Désirant gradually lost interest in research. With Swarts, she had worked on the frontier of research. On his retirement, she had been relegated to a routine support function in a research group with which she had little contact. Those years cannot have been a happy period of her life. Fortunately, she was given educational responsibilities, mainly as a substitute for Goubau. The students felt a high esteem for her, and some occasionally pay her a visit in Ostend.

Events post 1950

In Belgium

Luc Delbouille (born 1929) a chemist from Liège University, was looking out for a subject for a PhD thesis in late 1954, when the precise molecular structure of polyhalogenobenzenes was still disputed. Professor J. Duchesne, who knew about Désirant's synthesis of hexafluorobenzene, suggested to Delbouille that he should study the Raman and IR spectra of hexafluorobenzene [4]. Desreux agreed to provide Delbouille with the remaining hexafluorobenzene [4] given to him by Désirant before World War II. However the amount available – about 3 ml – was insufficient for Delbouille's purposes, so Desreux suggested to Delbouille that he should contact Désirant. She too had but a very limited supply of hexafluorobenzene left in her collection [4], hence it was viewed necessary to repeat the original synthesis of hexafluorobenzene. Furtunately, the platinum equipment which Désirant had used for the conversion of tribromofluoromethane to hexafluorobenzene still existed. Delbouille then developed a two-pronged attack. First in October 1954 he sent to Dr E. K. Plyler of the US National Bureau of Standards a portion of Desreux's sample of hexafluorobenzene in order to obtain its IR spectrum on a type of spectrometer not then available in Belgium. Second, in early 1955 he went to Ghent, where, with Désirant's help [5], he prepared the required amount of hexafluorobenzene. Delbouille then unequivocally settled its structure: hexafluorobenzene is a planar molecule belonging to point group D_{6h} – just like benzene. He obtained his PhD in 1958, left chemical physics for astrophysics, and eventually became a professor in the subject at Liège University.

There is a story to be told about Swarts' platinum equipment. During World War I, the Germans, starved of raw materials, had emptied Belgium (most of which was in their hands) of nearly all of the metals they could lay their hands on. Even Swarts' platinum research tools had been confiscated, but he had been re-equipped after World War I. After the conquest of Belgium by Germany in 1940, Goubau and Govaert hid Swarts' platinum equipment: they placed it in a zinc container, made by a trusted laboratory technician, and the three of them sank the container in a water well in Govaert's house. After the war, Govaert kept most of this platinum equipment securely locked in his office. Some of it survives to this day, including a platinum tube of 8 mm diameter, possibly the one used by Désirant for the pyrolysis of tribromofluoromethane.

In the US

The action now moved to the United States. The arrival in late 1954 of Delbouille's sample – actually Désirant's sample – of hexafluorobenzene at Plyler's National Bureau of Standards laboratory came to the attention of two more research groups within that institution, the polymer research group led by Dr Leo A.Wall (1918-1972), and the mass spectral group led by F. L. Mohler. The polymer group was interested in perfluoroaromatics as part of an effort to develop heat-resistant polymers for *e.g.* missiles. Hexafluorobenzene interested them more academically as the prototype of the aromatic fluorocarbons, and as a model to assess the chemical reactivity of perfluoro(polyphenylene). No satisfactory synthesis of hexafluorobenzene had yet been reported in the literature, and they were eager to know how that hexafluorobenzene in Plyler's laboratory had been prepared. Max Hellmann from the polymer group did the sleuthing, and contacted Désirant. The latter was glad to be of help and sent a copy of the recipe for the preparation of hexafluorobenzene to Washington [6]. The Americans gallantly insisted that Désirant publish her tribromofluoromethane route to hexafluorobenzene before they themselves inevitably would have to one day. So highly did they think of Désirant's synthetic feat that they suggested *Nature* or *Science* as the appropriate journal. But Désirant simply asked the secretary of the Royal Belgian Academy to open the sealed envelope she had deposited in 1936 and to publish the note contained in it. It must have happened that way. The chronology is right. The formal opening of the sealed envelope was on 11th June 1955, but the archives of the Academy show that the envelope was actually handed over to the secretary of the Academy about four weeks earlier, on 17th May 1955. The letter in which Désirant asked for the opening of the sealed envelope has not been recovered, but must reasonably be assumed to have been written in the early part of May 1955. Delbouille, who was busy in Ghent in early 1955 repeating the synthesis of hexafluorobenzene, does not recall the arrival of Hellmann's letter there, and Désirant probably would have told him. Assuming that Delbouille had resynthesized hexafluorobenzene in February-March 1955, Hellmann's letter or letters must have reached Désirant around April 1955. Leo Wall's group in Washington improved Désirant's method, finding that better yields of hexafluorobenzene can be obtained by pyrolysing tribromofluorobenzene under dinitrogen at 4.5 bar, and also that the outer platinum tubing is not really necessary – graphite tubing is equally satisfactory [6].

Spectroscopists are always interested in novel molecules, and the mass spectral group at the National Bureau of Standards must have been pleased that Désirant's twenty-year-old sample of hexafluorobenzene fell into their laps via Delbouille and Plyler [7]. The mass spectral study of the 'old' hexafluorobenzene sample showed it to be pure, a tribute not only to the stability of hexafluorobenzene, but also to Désirant's experimental skills, and, one guesses, a relief for her, because in the pre-chromatography years 1934-36 doubts about the purity of a new compound must have been a persistent worry.

The 1958 full paper on hexafluorobenzene

It is uncertain what prompted Désirant to publish her full hexafluorobenzene paper in 1958 [2]. In 1957, her professional life was shaken for a second time by a retirement, that of Goubau. Due to the increased number of students in higher education, Goubau's laboratory and teaching load was then rescinded, and Désirant was now assigned to Professor Lucien Massart, a reputed biochemist. This meant moving to a new research group and a new location in another part of the town of Ghent. There, in late 1957 or (more likely) early 1958 an American chemist came to visit her. Nobody remembers his name, or knows where he came from. Professor C. Van Sumeren witnessed their meeting, or at least the initial stages of their meeting. Désirant and the American chemist must have talked about Liège, about polymers, about hexafluorobenzene and about how many research dollars the American chemist would have saved had he known earlier about Désirant's synthesis of hexafluorobenzene. In my view, the American just had to be Leo Wall, but I have not able to confirm that supposition, not even through LeoWall's then co-workers Hellmann and Peters. But consider the dates. Leo Wall visits Désirant in (probably) early 1958, and Désirant submits to the *Bulletin of the Belgian Chemical Societies* her full hexafluorobenzene paper on 6th June, 1958. Maybe, just maybe, it was Leo Wall who caused Désirant to publish that paper, the last from Swarts' research group. Désirant retired on 1st June 1964, aged 65.

References

1 Y. Désirant, *Bull. Classe Sci., Acad. roy. Belg., 41* (1955) 759. For earlier tenuous evidence for the formation of hexafluorobenzene by the pyrolysis of hexafluoroethane, see F. Swarts, *Bull. Soc. Chim. Belg., 42* (1933) 114.

2 Y. Désirant, *Bull. Soc. Chim. Belges, 67* (1958) 676.

3 V. Desreux, *Bull. Soc. Chim. Belg. (No. Spécial), 55* (1948) 21.

4 L. Delbouille, *J. Chem. Phys., 25* (1956) 182.

5 L. Delbouille, *Bull. Classe Sci., Acad. roy. Belg., 44* (1958) 971. See also L. Delbouille, PhD Thesis, 1958, 'Contribution à l'étude de la structure et de la dynamique de la molécule d'hexafluorobenzène', University of Liège, Belgium.

6 M. Hellmann, E. Peters, W. J. Pummer and L. A. Wall, *J. Am. Chem. Soc., 79* (1957) 5654. See also L. A. Wall, J. E. Fearn, W. J. Pummer and R. Lowry, *J. Res. Nat. Bur. Stand., 65A* (1961) 239.

7 V. H. Dibeler, R. M. Reese and F. M. Mohler, *J. Chem. Phys., 26* (1957) 304.

BIOGRAPHIC NOTE

Dirk Tavernier, whose hobby is the history of chemistry, was born in Ghent, Belgium, on the 23rd of August 1941. After studying pharmacy at the University of Ghent, he went on to earn a PhD in Chemistry there in 1973 and is now a Lecturer in Organic Chemistry. His research interests lie in conformational analysis, structure determination by NMR methods, and the synthesis of oligopeptides, and he has served as a Titular Member of the IUPAC Commission on the Nomenclature of Organic Chemistry (1992-95).

Dirk Tavernier

Chapter 29

HIGHLY-TOXIC FLUORINE COMPOUNDS

CHRISTOPHER M. TIMPERLEY

Chemical and Biological Defence Sector, Defence Evaluation and Research Agency, Porton Down, Salisbury, Wiltshire SP4 OJQ, UK

Introduction

In 1915, chemical weapons were used on the World War I battlefield against the Allies. Revenge attacks by the British soon followed, bringing the realization that enthusiasm had to be tempered by scientific study: hence the acquisition in 1916 of Porton Down[1] in Southern England. The emphasis on a retaliatory capability shifted during the inter-war years, with offensive and protective capabilities now required by the British government. Chemical weapons were not used against British forces in World War II, and in the late 1950s all offensive work at Porton was stopped (production capabilities and stockpiles in the UK were destroyed). Since then, the role of Porton Down has been to provide British service personnel with effective protective measures against chemical weapons.

Porton Down was the subject of a special edition of *Chemistry in Britain* in 1988 [1]. For historical coverage, tracing its origins to World War I and describing the changing nature of scientific enquiry from 1916 to 1991, refer to 'Porton Down: 75 Years of Chemical and Biological Research' by Gradon Carter [2]. For an account of chemical defence, and a discussion of the once eminent advisory board connected with the field, read the relevant chapter in 'Cold War, Hot Science: Applied Research in the UK's Defence Research Laboratories, 1945-90' [3]. Both books contain descriptions of the activities (which are little known, often misunderstood and often maligned) of the oldest establishment of its kind in the world.

Background

Research into the synthesis and toxicity of fluorinated molecules in the UK started around 1935 at the Chemical Defence Experimental Station, Porton Down in collaboration with the universities of Cambridge and Oxford, and Imperial College, London (Table 29.1).

[1]Near Salisbury, Wiltshire, on the chalk downs of Salisbury Plain, close to Stonehenge (one of the most important megalithic monuments in Europe). The 7000 acre site of Porton Down provides sanctuary for over 90 species of bird, almost 200 species of spider, and many varieties of fungi, orchids and lichens. It has more species of butterfly than anywhere else in the UK and is one of the top breeding sites for the rare stone curlew. The main reason for the exceptional flora and fauna is that the chalk grassland is untouched by pesticides, fertilisers and the plough; it has accordingly been designated a Site of Special Scientific Interest.

TABLE 29.1
Scientists working for the Ministry of Supply in 1946

Cambridge team [a]		Oxford team	Imperial College team
H. McCombie	B. C. Saunders [c]	A. F. Childs	Prof H. V. A. Briscoe
F. J. Buckle	G. J. Stacey	L. J. Goldsworthy	H. J. Emeléus [f]
N. B. Chapman	F. E. Smith	A. W. Nineham	W. Kocay
H. G. Cook	F. Wild (part-time)	S. G. P. Plant	L. H. Long
R. Heap	I. G. E. Wilding	Prof R. Robinson [e]	A. L. G. Rees
J. D. Ilett	S. J. Woodcock	A. L. Tompsett	F. O. Sporzynski [g]
F. L. M. Pattison [b]	Prof A. R. Todd [d]		

[a] Physiological effects of compounds were examined at the Cambridge Extra Mural Testing Station by Lord E. D. Adrian, K. J. Carpenter, B. A. Kilby and M. Kilby. Sir Rudolph A. Peters, who discovered the mode of action of the fluoroacetates, also worked at Cambridge University. He was chairman of the Chemical Defence Advisory Board from 1952 to 1956.

[b] Pattison joined the Suffield Experimental Station (Defence Research Board of Canada) and later the University of Western Ontario. He is primarily remembered for his work on the fluoroacetates.

[c] Bernard Saunders (1903-1983) wrote with F. G. Mann in the 1930s the book 'Practical Organic Chemistry' which was used for most postwar UK undergraduate courses. His work on phosphorus fluorides has proved of lasting and fundamental importance to organic chemistry, biochemistry and physiology. A biography appears in J. Fluorine Chem., 90 (1998) 193.

[d] Sir Alexander Todd was awarded the Nobel Prize for Chemistry in 1957 for his work on nucleotides and nucleotide co-enzymes. He was chairman of the Chemical Defence Advisory Board from 1949 to 1951.

[e] Sir Robert Robinson was awarded the Nobel Prize for Chemistry in 1947 for investigations on plant products of biological importance, especially the alkaloids. He consulted for the Ministry of Supply.

[f] Harry Emeléus (1903-1993) is remembered for his contributions to physical and inorganic chemistry. At the start of the war, he was involved in the preparation of fluorinated silanes and arsines. The war shaped his work over the following years, culminating in an attachment to the Manhattan Project where his expertise in fluorine chemistry was applied to uranium isotope separation. He was chairman of the Chemical Defence Advisory Board during 1957-1960 and in 1962. An obituary features in J. Fluorine Chem., 67 (1994) vii and 100 (1999) 15.

[g] Sporzynski escaped from Poland to England in 1941 and directed British Intelligence to methyl fluoroacetate (British codename AF-1) which was under development in Poland as a potential chemical warfare agent.

Bernard Saunders directed a team of chemists, biochemists and biologists at Cambridge at the start of World War II. Work focused on the synthesis of toxic species including phosphorus fluorides (nerve agents). Similar work was carried out independently in Germany by Gerhard Schrader. Fortunately, none of the toxic compounds discovered were used in the war. Their first use occurred in 1983 when the nerve agents – tabun and later sarin – were used by Iraq against Iranian troops and Kurdish civilians [4]. In 1995, the Aum Shinrikyo terrorist group released sarin into the Tokyo subway, claiming eleven deaths and five thousand casualties [5]. Such incidences illustrate the extreme dangers posed by some fluorinated substances and reinforce the need to defend against their misuse.

A general survey of the toxic properties of organofluorine compounds by Ulm [6] complements this review, which concentrates solely on highly-toxic fluorinated compounds, many of which have been studied at Porton. The toxic materials discussed include inorganic fluorides, fluorinated arsenic and mercury compounds, fluoroacetates, fluorinated nitrogen mustards, nerve agents, fluorinated carbamates, polyfluorinated alcohols, fluorinated ethers, fluoroalkenes, fluorinated organosulfur compounds, and cage convulsants. The prophylaxis and therapy for poisoning by such compounds is discussed whenever possible (in this article prophylaxis is defined as treatment *before* poisoning and therapy as treatment *after* poisoning). Historical details from the Porton archives are included in the discussions on fluoroacetates and nerve agents. Both topics have been reviewed elsewhere. An account of fluoroacetates appears in a monograph by Pattison entitled 'Toxic Aliphatic Fluorine Compounds' [7]. The wartime work of Saunders was described initially in communications to Porton Down. An overview is given in 'Some Aspects of the Chemistry and Toxic Action of Organic Compounds containing Phosphorus and Fluorine' [8], a book summarising lectures given by Saunders world-wide (including countries then behind the Iron Curtain, such as Czechoslovakia). The chemistry, properties and toxicities of the nerve agents have been reviewed recently by Porton chemists, Robin Black and John Harrison [9]. Only a cursory treatment is possible here.

Note on toxicity measurements

The term LD_{50} is a common measure of acute toxicity. It is the lethal dose that kills 50% of a group of animals and is usually expressed in milligrams of compound per kilogram body weight (mg/kg). When discussing inhalation toxicity, the term LCt_{50} is often used. This is the lethal concentration that kills 50% of a group of animals and is expressed in milligrams of compound per metre cubed multiplied by the exposure time t in minutes (mg min m^{-3}). In cases where a precise figure is not available, an approximate lethal concentration for time t (ALCt) is quoted. This is the concentration of a volatile compound that causes lethal effects. With all toxicity measurements, it is customary to state the animal species and the route of administration. The main routes include intramuscular (im), intraperitoneal (ip), intravenous (iv) and subcutaneous (sc) injection, oral administration, percutaneous application (pc), inhalation, or introduction into the eye.

Inorganic fluorides

Fluorine compounds were reviewed at the start of World War I as possible military gases. At that time, little was known about the physiological effects of fluorinated compounds as few had been prepared. Anhydrous hydrogen fluoride had long been known to be highly corrosive in contact with animal tissue, so it was thought that if it could be introduced into the system, toxic effects might result. Some inorganic fluorides were therefore made by the Imperial College team under Briscoe.

Disulfur decafluoride 'Z' (S_2F_{10}), phosphorus trifluoride (PF_3) and chlorine trifluoride (ClF_3) were studied. The first two were found to possess high toxicity, and the third, in addition to being toxic, had the alarming property as a liquid or concentrated vapour of causing spontaneous inflammation of organic material (*e.g.* asphalt, fabric and hair). From the chemical warfare aspect, however, these fluorides possessed several drawbacks, such as their difficulty and danger of preparation. Considerable attention was given to 'Z' gas (LCt_{50} 1200 mg min m^{-3}) which was several times more toxic than phosgene (LCt_{50} 3200 mg min m^{-3}) to mice exposed for 10 minutes [10]. The main changes in experimental animals were congestion of the lower airways after a delay of 24 hours and a fatal build up of fluid in the lung (pulmonary oedema). In high concentrations, the gas is lethal even after a short period of exposure. The mode of action of disulfur decafluoride (b.p. 29 °C) is

TABLE 29.2
Inhalation toxicities of some inorganic fluorides (1 h exposures) [14]

Common name	Formula	b.p. (°C)	LCt_{50} mice (mg min m^{-3})	LCt_{50} rats (mg min m^{-3})
Oxygen difluoride	OF_2	−145	199	345
Chlorine trifluoride	ClF_3	11	40 157	67 454
Boron trifluoride	BF_3	−100	–	64 552
Boron trichloride	BCl_3	13	–	722 660
Sulfuryl fluoride	SO_2F_2	−50	755 604	933 246
Hydrogen fluoride	HF	19	22 380	47 411
Hydrogen chloride	HCl	−85	98 035	275 558
Hydrogen bromide	HBr	−67	161 660	567 996
Hydrogen cyanide	HCN	26	21 396	32 060
Deuterium fluoride	DF	19	16 699	56 694

uncertain; its toxicity cannot be due to hydrolysis to hydrogen fluoride as it is not attacked by water.

Other Group VI fluorides were examined at Porton but were discounted as chemical warfare agents due to difficulties in synthesis, unfavourable physical properties, or relatively low toxicity. These included the fluorinating agent sulphur tetrafluoride SF_4 (LCt_{50} rat 20 155 mg min m^{-3}, 4 h exposure) [11], and the commercial electrical insulators, sulphur hexafluoride SF_6 (LD_{50} iv rabbit 5790 mg/kg) [12] and selenium hexafluoride SeF_6 (LCt_{50} mice and rats ~14 202 mg min m^{-3}, 3 h exposure) [13].

Chlorine trifluoride (b.p. 11 °C), which reacts violently with water, is used in nuclear reactor fuel processing and as an igniter and propellant for rockets. It was produced on a large scale in Germany as an incendiary agent which at the same time had a toxic effect; tests at Porton revealed the gas to have 1/20th of the toxicity of phosgene [10]. Inhalation of chlorine trifluoride causes distention of the lungs, contraction of the bronchial tubes, corrosion of the upper respiratory tract, and severe bronchitis. High concentrations cause acute coughing and expectoration and lead to serious lung damage which is usually fatal. Skin contact with chlorine trifluoride causes redness, blister formation, abcesses, and necrosis of lower-lying tissues. The eyes are also badly affected.

Other possible military gases were nitrosyl fluoride (NOF), nitryl fluoride (NO_2F) and chromyl fluoride (CrO_2F_2), but these were hydrolysed by water too quickly to be of any use. Others like fluorine nitrate (NO_2OF) were liable to explode spontaneously with great violence. Those of insufficient toxicity included bromine monofluoride (BrF), chlorine monofluoride (ClF), hydrogen fluoride (HF), boron trifluoride (BF_3), sulfuryl fluoride (SO_2F_2), thionyl fluoride (SOF_2) and thiophosphorus fluoride (PSF_3, which inflames in contact with air).

Comparative inhalation data are available for some inorganic fluorides (Table 29.2). Toxicity in rodents decreases in the order: oxygen difluoride $\gg$ hydrogen fluoride $>$ chlorine trifluoride ~boron trifluoride $>$ sulfuryl fluoride. Replacement of fluorine for other halogens tends to lower toxicity; for example, boron trichloride is roughly eleven times less toxic than boron trifluoride, and for hydrogen halides, toxicity follows the order HF $>$ HCl $>$ HBr. Hydrogen fluoride is as toxic as hydrogen cyanide to mice, but less toxic to rats. Deuteration of hydrogen fluoride profoundly alters its biological activity: deuterium fluoride is more toxic to mice than hydrogen fluoride, but less toxic to rats.

Tetrafluorohydrazine (N_2F_4) is also toxic (ALCt rats 51 000 mg min m^{-3}, 4 h exposure) [11]. Signs of poisoning at lethal concentrations include irritation of the eyes and respiratory tract, followed by pulmonary oedema. At lower levels, tetrafluorohydrazine acts on the blood-forming tissues and on the kidney; repeat exposures therefore cause cumulative damage.

Fluorinated arsenic and mercury compounds

Trivalent arsenic compounds are more active than the pentavalent compounds and often have a powerful blister effect if they contain the group $-As(halogen)_2$; activity decreases in the order Cl > Br > I. As little was known about fluorinated arsenic compounds in the 1940s, Long at Imperial College prepared and examined several fluoroarsines. Since they have lower boiling points than the chloro-compounds, it was speculated that they might be more hazardous. However, they were found to be far inferior in this respect.

Arsenic trifluoride (AsF_3), a colourless mobile liquid (b.p. 63 °C) that fumes in moist air, is rapidly absorbed by the skin, causing irritation after several minutes; painful blisters later develop and the absorbed arsenic causes systemic poisoning. The vapour also damages the respiratory tract (LCt_{50} mice 7000 mg min m^{-3}, 10 min exposure) [10]. As the acute toxicity and vesicant activity was about a fifth that of Lewisite (a very potent blister agent) and less than that of arsenic trichloride, the compound was not examined further. Methyl difluoroarsine (**1**), ethyl difluoroarsine (**2**) and (2-chlorovinyl)difluoroarsine (**3**) were obtained by heating the respective chloroarsines to 80 °C with ammonium fluoride; they boiled at 77 and 94 °C and 44 °C at 15 mmHg respectively [10]. Similar chemistry yielded phenyldifluoroarsine (**4**), a solid (m.p. 42 °C) that caused little sensory irritation, but which was toxic by injection (LD_{50} iv rabbit 0.5 mg/kg) [10].

$MeAsF_2$	$EtAsF_2$	$CHCl{=}CHAsF_2$	$PhAsF_2$
(**1**)	(**2**)	(**3**)	(**4**)

Fluoroarsines attack unprotected skin and the lungs. Like chloroarsines, they owe their pharmacological properties to their ability to react with biological thiols, particularly sulfhydryl groups of enzymes. Lipoic acid (**5**), which belongs to the pyruvic acid oxidase system, reacts with difluoroarsines to form the cyclic adduct (**6**), which impairs tissue respiration (Scheme 29.1). The antidote 2,3-dimercaptopropanol, $HSCH_2CH(SH)CH_2OH$ (British Anti-Lewisite, BAL),[2] chelates more effectively with the arsenic atom, giving the cyclic adduct (**7**) and freeing the lipoic acid. Thus, the damage done to the lipoate receptor can be reversed.

[2]The toxic blister agent Lewisite ($CHCl{=}CHAsCl_2$) had been developed in America towards the end of World War I, and its reappearance in the 1940s was greatly feared. The antidote BAL was discovered by Robert Thompson while he was a biochemistry demonstrator at Oxford University (UK) during World War II. It was one of the first drugs to be developed not by chance, but from a logically pursued programme to fit chemical properties to a particular task. Professor R. H. S. Thompson CBE, FRS (1912-1998) served on several advisory commitees at the UK Ministry of Defence, and was chairman of the UK Chemical Defence Advisory Board from 1968 to 1975.

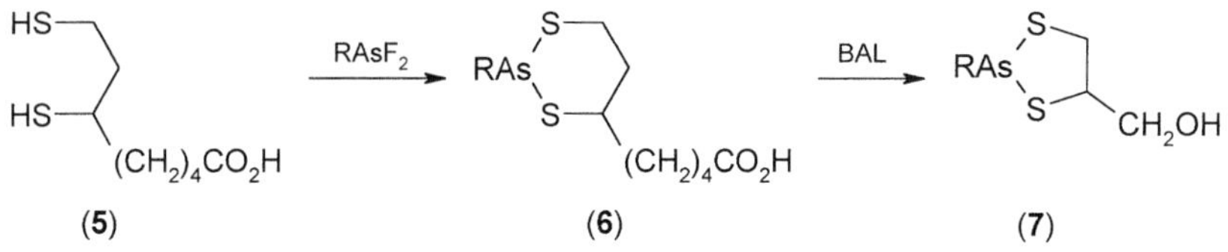

Scheme 29.1.

Some fluorinated arsenic and mercury compounds cause temporary incapacitation when inhaled. Dimethyl(trifluoromethylthio)arsine, Me_2AsSCF_3, irritates the eyes, nose and throat at a concentrations as low as 0.05 ppm [15]; rapid recovery of test animals occurs a few minutes after exposure. The related compound, bis(trifluoromethylthio)mercury, $Hg(SCF_3)_2$, a solid with a sweet odour (m.p. 38 °C), damages the skin and the gastrointestinal tract [16]. Bis(trifluoromethyl)mercury, $(CF_3)_2Hg$, a solid with a pungent odour (m.p. 165 °C), renders skin temporarily insensitive and causes a strong headache when its vapour is inhaled. Bis(perfluoro-*t*-butyl)mercury, $[(CF_3)_3C]_2Hg$, a volatile compound, has similar toxic properties [16].

Fluoroacetates

The development of toxic fluorocarbon compounds starts with fluoroacetic acid (FCH_2CO_2H) and derivatives, known collectively as fluoroacetates [17]. Fluoroacetic acid was first prepared by Swarts in Belgium in 1896 [18] but no mention was made of its toxicity. Many years later, Gryszkiewicz-Trochimowski and colleagues [19, 20] working at the Warsaw Polytechnic during the period 1935-1939, carried out investigations into the preparation and properties of fluoroacetates but, due to the war, publication of the work was delayed until 1947. Their discovery of the toxicity of fluoroacetates stemmed from work on iodoacetates and their lachrymatory effects. When a bomb filled with methyl iodoacetate exploded, a purple cloud resulted due to iodine vapour, warning field troops. While examining related compounds which would not have this undesirable property, methyl fluoroacetate was prepared and assessed for lachrymatory action. The substance was applied to the eye of a rabbit. No lachrymation was observed but the rabbit died, prompting a systematic study of fluoroacetates. During pharmacological investigations, it became apparent that fluoroacetates were highly toxic to a range of animals by all routes of administration [20]. The cause of death after a delayed action was cardiac arrest, respiratory failure and exhaustion from convulsions. From consideration of many compounds, the Polish workers concluded that fluoroacetic acid was responsible for toxicity. Compounds which could produce it by hydrolysis or oxidation were toxic whilst those that could not produce it were non-toxic.

Independent work on organic fluorine compounds was carried out in 1934 by Schrader in Germany (Fig. 29.1). Little of this work featured in the literature at the time, and came to light only after the war through interviews with Schrader by British Intelligence [21]. The purpose of the work was to develop new pesticides. Acid fluorides, RCOF and RSO_2F, were the first compounds examined, of which methanesulfonyl fluoride was an

Fig. 29.1. Dr Gerhard Schrader pioneered the chemistry of fluoroacetates and organophosphorus compounds which led to the discovery of many new pesticides. In August 1945, an investigating team that included Porton scientists S. A. Mumford and E. A. Perren interviewed him in Germany. The team requested an account of his research, including preparative details, his ideas on the relationship between chemical constitution and toxicity towards insects and mammals, and his suggestions for future work. In response, several voluminous reports were received. Schrader was handicapped in writing these by the destruction of many of his records by IG Farben, and there were certain gaps, especially in the insecticidal data and preparative methods. However, it was still possible after translation to piece together a coherent account of his work [21]. There is little doubt that Gerhard Schrader visited Porton Down in 1946 or 1947. (Photograph reproduced by courtesy of DERA.)

outstanding fumigant.[3] Other sulfonyl fluorides and related substances were then studied. In 1935, Schrader prepared 2-fluoroethanol, FCH_2CH_2OH and 2-fluoro-2′-hydroxydiethyl ether, $FCH_2CH_2OCH_2CH_2OH$, both of which were toxic to mammals; 2-fluoroethanol was patented as a rodenticide [21].

A great achievement of Schrader was the discovery of the first systemic insecticides. These are substances which are absorbed by a plant and are subsequently distributed

[3]Methanesulfonyl fluoride, CH_3SO_2F (b.p. 124 °C), a volatile liquid formerly used as an insecticide, is toxic to mammals (LD_{50} sc mice 3.5, iv rabbit 0.3-1.0, and ip rat 3 mg/kg) [12, 22]. Accidental inhalation by man resulted in severe damage to the upper respiratory tract and death from spasm and pulmonary oedema [12]. The compound is also an inhibitor of cholinesterase [23, 24] and disrupts the central nervous system in a manner similar to organophosphorus nerve agents.

throughout its system. The absorbed agents or metabolic products kill sucking or chewing insects without harming the plant. The most active of the new compounds were acetals (**8**) and (**9**), made from paraformaldehyde, the fluoroalcohol and a trace of sulfuric acid [21].

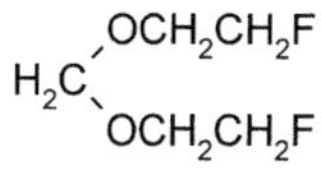

(**8**)

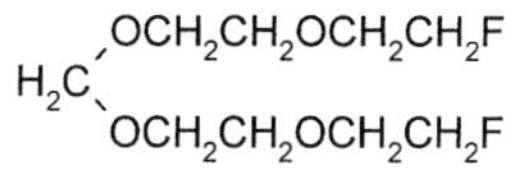

(**9**)

In summary, the major advances in fluoroacetate chemistry up to 1939 had been made in Belgium, Poland and Germany. A new development occurred in 1941 when Sporzynski [20] defected to England (see Table 29.1) and directed the attention of British Intelligence to methyl fluoroacetate, $FCH_2CO_2CH_3$, a potentially new military agent. The Cambridge group under Saunders was requested to study the fluoroacetates in secret. Methyl fluoroacetate was the first compound to be investigated in detail and extensive work was carried out to select the best method of preparation. By heating methyl chloroacetate and potassium fluoride together for four hours at 220 °C, a 60% yield of methyl fluoroacetate was obtained [8]. This method formed the basis of its production and that of related substances on a large scale.

Methyl fluoroacetate, a mobile liquid (b.p. 104 °C), with an extremely faint odour, is toxic by injection (LD_{50} iv rabbits 0.25 mg/kg) and by inhalation (LCt_{50} rabbits 1000 mg min m^{-3}, 10 min exposure) [8]. Even at high doses, however, animals did not show any symptoms while being exposed to lethal concentrations of the vapour, and no obvious signs were evident until some 30 to 60 minutes after exposure; violent convulsions then set in and death followed within several hours. The toxicities of many derivatives of fluoroacetic acid such as the ethyl, propyl and isopropyl esters were similar to that of methyl fluoroacetate [25-29]. Compounds missing the fluoroacetate group, for example CF_2HCO_2H, CF_3CO_2H and CR_2FCO_2R lacked toxicity [30].[4]

2-Fluoroethanol, a stable colourless liquid (b.p. 101 °C) that is completely miscible with water and practically odourless, is a convulsant poison like methyl fluoroacetate, and is equally toxic. As it produced a toxic effect comparable to that of fluoroacetic acid, the Cambridge team decided to synthesise a compound in which the 'active' groups of these molecules were combined with the hope of obtaining a compound of increased potency. 2-Fluoroethyl fluoroacetate, $FCH_2CO_2CH_2CH_2F$, was therefore prepared in 1943 by treatment of fluoroacetyl chloride with 2-fluoroethanol. In accordance with expectation, the compound possessed enhanced toxicity (LCt_{50} rabbit 500 mg min m^{-3}, 10 min exposure), being twice as toxic to rabbits by inhalation than methyl fluoroacetate. The conclusion that only compounds that can form fluoroacetic acid by hydrolysis and/or oxidation are toxic was confirmed. While research in England proceeded, Kharasch and his colleagues in Chicago worked on related problems, and reports were exchanged between the two groups.

---

[4]Acute toxicity by intravenous injection to mice for the three fluoroacetic acids decreases with increasing fluorine substitution; *i.e.* FCH_2CO_2H (LD_{50} 6.6 mg/kg) [31] > F_2CHCO_2H (LD_{50} 180 mg/kg) [7] > CF_3CO_2H (LD_{50} 1200 mg/kg) [32].

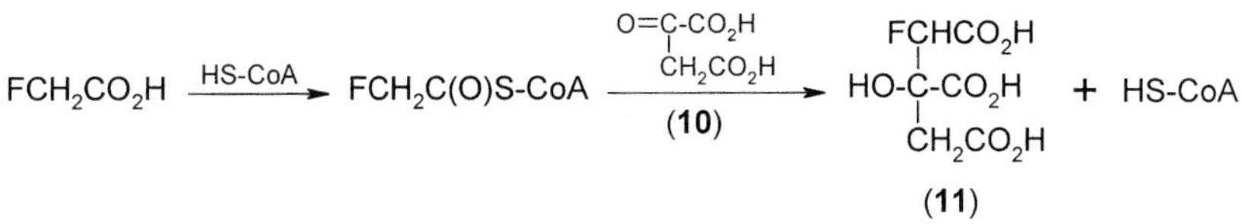

Scheme 29.2.

Towards the end of the war, several developments occurred. In 1944, Marais announced that fluoroacetic acid was the toxic principle of 'gifblaar' (*Dichapetalum cymosum*) [33], a poisonous plant from South Africa. Later fluoroacetic acid was found in another toxic plant, 'gidyea' (*Acacia georginae*) from Australia [34], and 18-fluoro-oleic acid, *cis*-$F(CH_2)_8CH{=}CH(CH_2)_7CO_2H$ was found in the shrub 'ratsbane' (*Dichapetalum toxicarium*) from Sierra Leone [36].[5]

The discovery of fluoroacetic acid in nature, and the use of sodium fluoroacetate (compound 1080) and fluoroacetamide (compound 1081) as rodenticides,[6] inspired scientists to investigate its biochemistry. The mode of action was deduced by Sir Rudolph Peters at Cambridge who showed that it could enter the Krebs tricarboxylic acid cycle, where activation with acetyl coenzyme A and condensation with oxaloacetic acid (**10**) gave rise to fluorocitric acid (**11**) [37-39] (Scheme 29.2). Toxicity was ascribed to inhibition of the enzyme aconitase by fluorocitrate. Subsequent studies have shown, however, that a mitochondrial enzyme responsible for formation of a citrate-glutathione ester, is inhibited irreversibly by fluorocitrate [40, 41]. The toxic effects of fluorocitrate are therefore related to inhibition of mitochondrial citrate transport; note that only one diastereoisomer is toxic, namely 2*R*,3*R*-2-fluorocitrate [42].

Examination of ω-fluorocarboxylic acids, $F(CH_2)_nCO_2H$ [43], led to the discovery of alternation in toxicity while ascending the homologous series: if the total number of carbon atoms in the acid moiety was even, the compound was toxic and produced symptoms similar to those produced by fluoroacetic acid, but if the number was odd, the compound was non-toxic (Table 29.3).

The alternation in toxicity correlated with the β-oxidation theory of fatty acid metabolism [7]. Toxic acids can be degraded to fluoroacetic acid, whereas non-toxic acids can be oxidised only to non-toxic 3-fluoropropanoic acid or its metabolites. The increase in toxicity of the higher members may be due to their greater lipid solubility. Because of the interesting toxicology of the ω-fluorocarboxylic acids, a variety of related compounds were prepared and examined by Pattison and workers [7], who attempted to correlate the toxicity pattern with metabolic detoxification processes. Since the biological oxidation of alcohols to acids was well known, the homologous series of ω-fluoroalcohols, $F(CH_2)_nOH$,

[5]The fluoro-oleic acid-containing extract of *Dichapetalum toxicarium* exerts its toxic effects by severely reducing cardiac function, which can lead to death. The lethal dose that killed 100% of a group of rats injected intraperitoneally with the extract was 10 mg/kg [35].

[6]Sodium fluoroacetate (m.p. 200 °C) and fluoroacetamide (m.p. 108 °C) are white, odourless, tasteless solids whose high toxicity has restricted their use to prepared baits. Both agents are well absorbed from the gastrointestinal tract. Acute oral toxicity of fluoroacetate in the rat is about 0.1 mg/kg, whereas that of fluoroacetamide is 4-15 mg/kg [12]. These chemicals are uniquely effective in mice and rats because of the high metabolic rate in tissues that are susceptible to inhibition.

TABLE 29.3
Toxicities of ω-fluorocarboxylic acids and ω-fluoroalcohols [7]

Acid	LD_{50} ip mice (mg/kg)	Alcohol	LD_{50} ip mice (mg/kg)
FCH_2CO_2H	6.6	FCH_2CH_2OH	10
$F(CH_2)_2CO_2H$	60	$F(CH_2)_3OH$	47
$F(CH_2)_3CO_2H$	0.7[a]	$F(CH_2)_4OH$	0.9
$F(CH_2)_4CO_2H$	>100	$F(CH_2)_5OH$	>100
$F(CH_2)_5CO_2H$	1.4	$F(CH_2)_6OH$	1.2
$F(CH_2)_6CO_2H$	40	$F(CH_2)_7OH$	80
$F(CH_2)_7CO_2H$	0.6	$F(CH_2)_8OH$	0.6
$F(CH_2)_8CO_2H$	>100	$F(CH_2)_9OH$	32
$F(CH_2)_9CO_2H$	1.5[a]	$F(CH_2)_{10}OH$	1.0
$F(CH_2)_{10}CO_2H$	58	$F(CH_2)_{11}OH$	>100
$F(CH_2)_{11}CO_2H$	1.3	$F(CH_2)_{12}OH$	1.5
$F(CH_2)_{17}CO_2H$	5.7	$F(CH_2)_{18}OH$	4.0

[a] Sodium salt used.

TABLE 29.4
Toxicities of ω-fluoro compounds, $F(CH_2)_nX$ (LD_{50} ip mice, mg/kg) [7]

X =	H	F	Cl	Br	I	CN	NO_2	NH_2	OR[a]	SCN	CHO
$n = 2$			>100	>100	28	10			15-70	15	
$n = 3$				>100		16	92	46	>100	18	2
$n = 4$		3	1	8	5	1	11		0.8-6	3	81
$n = 5$		18	32	11	9	50	90	50	90	30	0.6
$n = 6$	2		6	13	5	3	13	0.9	4	5	>100
$n = 7$	35	21	>100	>100		>100		50			2
$n = 8$	3	2	2	20				0.8			53

[a] R = alkyl or aryl.

was investigated. An alternation in toxicity comparable to that for the ω-fluorocarboxylic acids was found, with the same generalisations regarding the odd and even members, confirming the conversion of alcohols to acids in mammals. Similar alternation in toxicities has been noted for other monofluorinated compounds (Table 29.4). Even-number carbon compounds are generally more toxic than their odd-carbon homologues, but exceptions do exist, *e.g.* in the amino acid series $F(CH_2)_nCH(NH_2)CO_2H$ [44].

Because of the resistance of the carbon-fluorine bond to hydrolysis, it was anticipated that the ω,ω'-difluoroalkanes, $F(CH_2)_nF$, would be excreted unchanged. The members listed in Table 29.4 were submitted to routine testing and were found to be surprisingly toxic [45]; even-carbon members of this series caused citric acid accumulation [46]. 1,2-Difluoroethane, FCH_2CH_2F (b.p. 26 °C), is toxic by inhalation (LCt_{50} for mice 42 120 and for rats 48 600 mg min m^{-3}, 4 h exposure) [47], and clinical and spectroscopic evidence suggests that it is metabolised to fluorocitrate; some of its analogues are also very toxic by inhalation, *e.g.* the 1-halo-2-fluoroethanes FCH_2CH_2X, X = Cl or Br, and 1-chloro-1,2-difluoroethane $FClCHCH_2F$ [47].

The ω-fluoroalkylamines $F(CH_2)_nNH_2$ are perhaps the most hazardous compounds metabolisable to fluoroacetates. They are nearly as toxic through the skin as when injected. Approximate percutaneous toxicities of 6-fluorohexylamine (b.p. 54 °C/13 mmHg) are: rabbit 0.25, rat 1.5, guinea pig 1.4 mg/kg [7, 48]. Fluoroamines, like unsubstituted alkylamines, cause a characteristic red patch on the skin.

Among the compounds prepared at Cambridge, several are particularly interesting. Sesquifluoro-H (**12**) was non-toxic and did not produce fluoroacetate symptoms, suggesting metabolic stability [28, 30, 49]. It is the fluorine analogue of sesqui-H (**13**), a potent member of the mustard group of blister agents; the lack of activity shows that the vesicant action of the mustards depends on reactive halogens. Also, triethyllead fluoroacetate (**14**) (LD_{50} ip mice 15 mg/kg) combined the sternutatory properties (irritation of nose, throat and chest) of organolead salts with the convulsant properties of the fluoroacetates [50, 51].

$CH_2SCH_2CH_2F$ | $CH_2SCH_2CH_2F$ (**12**)

$CH_2SCH_2CH_2Cl$ | $CH_2SCH_2CH_2Cl$ (**13**)

$FCH_2CO_2Pb(C_2H_5)_3$ (**14**)

No reliable antidote to fluoroacetate poisoning has yet been discovered. Sodium acetate [52] and ethanol [53] have been found to be effective in certain animals. In one case of fluoroacetate poisoning in man [7], an hourly drink of 100 proof whiskey mixed with sugar and water apparently provided limited relief! Entry of fluoroacetate into the citric acid cycle is minimised by co-administration of ethanol and sodium acetate [54], consistent with the reduced toxicity of the mixture [55]. Fluoroacetamide poisoning is partially alleviated by one-hour pretreatment with phenyl saligenin cyclic phosphonate or glutathione, apparently by slowing the rate of fluoroacetate liberation [55]. No antidotes have been found that will remove or inactivate the fluorocitrate once it has been formed. Barbiturates may be used to control convulsions, but such treatment may accelerate death in some animals by increasing the respiratory depression [56]. Of the antidotes examined, only glycerol monoacetate [57] and acetamide [58] are outstanding, but by no means universally effective. Their protective action is thought to arise from liberation of acetyl coenzyme A. It is unsurprising therefore that the most effective antagonist of 4-fluorobutyric acid is glycerol monobutyrate [59, 60], presumably for the same reason. Despite the difficulty of developing an effective prophylaxis, a recent study showed that sheep innoculated with genetically-modified ruminal bacteria (*Butyrivibrio fibrisolvens*) were protected from the toxic effects of fluoroacetate [61].

Fluorinated nitrogen mustards

Pharmacological research into halogenated tertiary amines, carried out after World War II, was aimed at the use of these substances as anti-cancer drugs. However, during the war, some of the so-called nitrogen mustards were considered for use as chemical warfare agents. A necessity for the skin damaging effect of this class is the presence of reactive 2-haloalkyl groups in the molecule. The most dangerous compounds are those of general formula, $RN(CH_2CH_2Hal)_2$ where R = CH_3, CH_3CH_2 or CH_2CH_2Hal. Lengthening the haloalkyl groups, or putting the halogens in a place other than the 2-position, leads to a

TABLE 29.5
Effect of fluorination on the toxicology of nitrogen mustards [10, 62]

Compound	Vesicant activity	LD_{50} sc mice (mg/kg)	LD_{50} iv rabbit (mg/kg)
$HN(CH_2CH_2F)_2$	none	17	0.2
$CH_3N(CH_2CH_2Cl)_2$	potent	2.6	–
$CH_3N(CH_2CH_2F)_2$	none	27	0.1
$CH_3CH_2N(CH_2CH_2Cl)_2$	potent	1.2	–
$ClCH_2CH_2N(CH_2CH_2Cl)_2$	potent	2	–
$FCH_2CH_2N(CH_2CH_2Cl)_2$	moderate	–	–
$ClCH_2CH_2N(CH_2CH_2F)_2$	none	–	–

reduction in toxicity. As nothing was known about the effects of fluorination, the Imperial College team prepared several fluorine analogues. These were much less toxic than the chlorine compounds and did not cause skin blisters (Table 29.5). An exception was $FCH_2CH_2N(CH_2CH_2Cl)_2$ which retained some vesicant activity due to the presence of two reactive chlorine atoms; the difluoride, $ClCH_2CH_2N(CH_2CH_2F)_2$, with a sole chlorine atom, had no effect on the skin.

Organophosphorus nerve agents

The biological activity of organophosphorus fluorides[7] was discovered in 1932 by German chemists, Lange and von Krüger [64]. They observed that dimethyl and diethyl phosphorofluoridates, $(MeO)_2P(O)F$ and $(EtO)_2P(O)F$, caused breathlessness and affected vision when inhaled in tiny amounts. This observation led to the synthesis and examination of a series of dialkyl phosphorofluoridates in the UK at Cambridge during World War II [8, 65]. In general these compounds are colourless, stable and faintly scented liquids. The eye-effect was confirmed by Saunders and his colleagues who voluntarily entered gas chambers containing low concentrations of phosphorofluoridates! Exposure resulted in persistent miosis, *i.e.* contraction of the pupil of the eye, causing blurred vision (Fig. 29.2).

Tests on animals by Lord Adrian and his team [66, 67] showed the phosphorofluoridates to be highly toxic by inhalation, having a lethal action at higher concentrations. Low-level exposure is characterised by miosis. Exposure to higher levels causes breathing difficulties, cramp, vomiting, and involuntary defecation and urination. Muscular symptoms become more pronounced with higher doses, leading to convulsions and unconsciousness. Death may occur within several minutes from paralysis of the respiratory muscles.

[7] Organophosphorus nomenclature is full of inconsistencies and it is uncommon for chemists to agree on the same name for a particular compound! Fortunately, as the types of compound covered by this review are limited, only a few rules will suffice. Important P^{III} precursors are trialkyl phosphites $(RO)_3P$ and dialkyl phosphites $(RO)_2POH$ [usually depicted as $(RO)_2P(O)H$]. The P^{V} nerve agents are named after the acids from which they are derived: $(RO)_2P(O)F$ are dialkyl phosphorofluoridates [they are diesters of phosphorofluoridic acid, $(HO)_2P(O)F$] and ROP(O)MeF are *O*-alkyl methylphosphonofluoridates [they are esters of methylphosphonofluoridic acid HOP(O)MeF]. Tabun derivatives, $ROP(O)(NMe_2)CN$, are *O*-alkyl *N*,*N*-dimethylphosphoramidocyanidates [they are esters of dimethylphosphoramidocyanidic acid $ROP(O)(NMe_2)CN$]. The term phosphylated, as suggested by Hudson and Keay [63], is used to denote any group covalently bound through phosphorus.

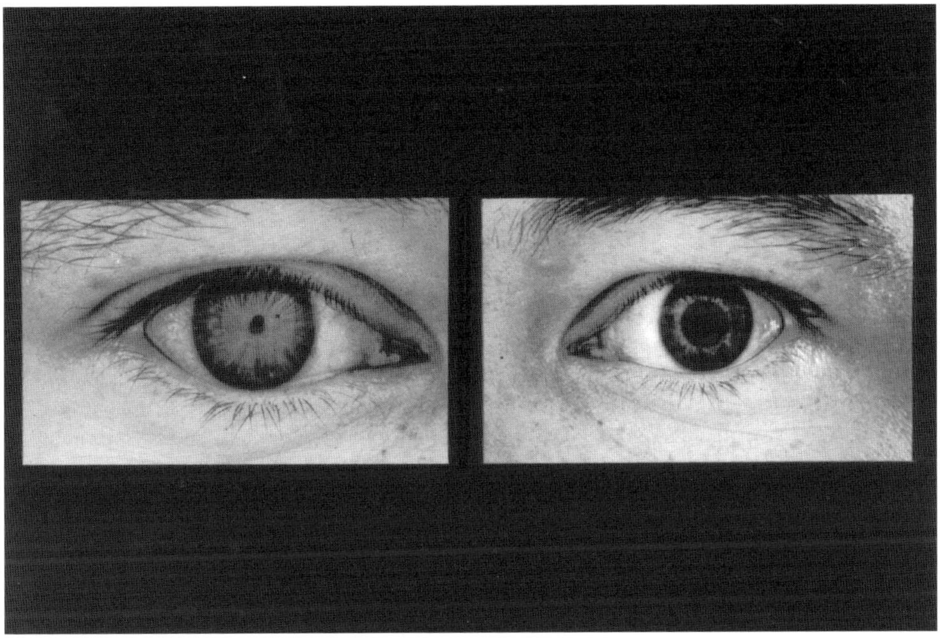

Fig. 29.2. The eye on the left has been exposed to a nerve agent and has developed miosis, in contrast to the eye on the right which is normal. Miosis is a constriction of the pupil to a pin-point size. The amount of light entering the eye and the powers of accommodation are greatly reduced. The condition results in photophobia, headaches, and a pain experienced in changing from bright to dull light. (Photograph reproduced by courtesy of DERA.)

Tests on enzymes in 1942 by Dixon showed that toxicity was due to inhibition of acetylcholinesterase (AChE) [68], an enzyme that controls nerve impulse transmission by hydrolysing acetylcholine to acetic acid and choline. Nerve agents react with cholinesterase by a bimolecular displacement reaction, in which a serine hydroxyl group in the active site makes a nucleophilic attack on the electrophilic phosphorus atom, displacing fluoride and phosphylating the enzyme [8, 9] (Scheme 29.3). The reactivity of the nerve agent, and hence its toxicity, is determined by the electrophilic character of the phosphorus atom and the affinity for the active site as determined by substituents R^1 and R^2.

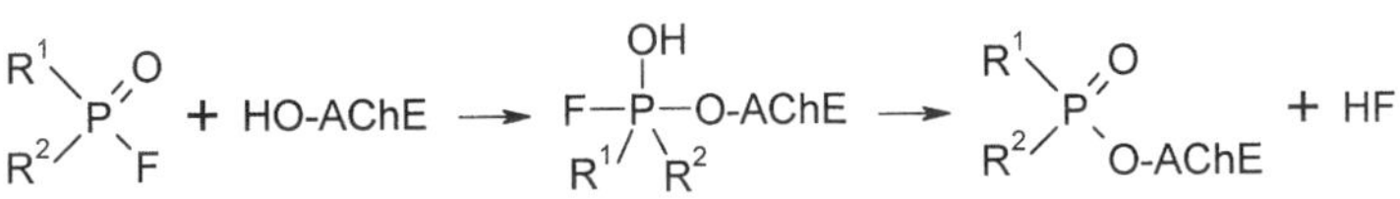

Scheme 29.3.

One of the most potent compounds made at Cambridge in 1941 was diisopropyl phosphorofluoridate (DFP), a volatile liquid with a fruity odour, which was found to be highly toxic by injection (LD_{50} iv rabbit 0.5 mg/kg) and by inhalation (LCt_{50} mice 4400 mg min m^{-3}, 10 min exposure) [69-71]. DFP containing radioactive phosphorus was also prepared [72] and shown to combine with cholinesterase [73].

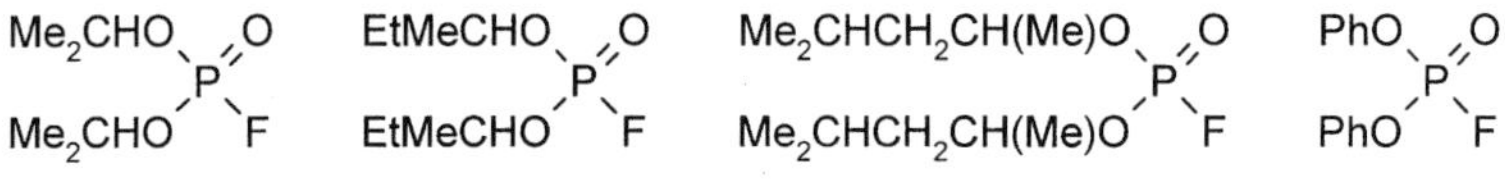

DFP (15) (16) (17)

Investigations on the relationship between structure and biological activity showed that the most toxic phosphorofluoridates are derived from secondary alcohols [74]; *e.g.* the isopropyl and isobutyl esters were more potent than the respective *n*-propyl and *n*-butyl esters. The diisobutyl ester (**15**) was as toxic as DFP. The dicyclohexyl ester was very toxic by injection (LD_{50} iv rabbit 0.1 mg/kg) and by inhalation (LCt_{50} mice 1400 mg min m^{-3}, 10 min exposure) [8]. Branching of the carbon chain next to oxygen gave compounds of higher potency than those with terminal branching; *e.g.* diisoamyl phosphorofluoridate (**16**) caused miosis, yet di-*n*-amyl phosphorofluoridate did not. Also, toxicity was low for the aromatic series; *e.g.* diphenyl phosphorofluoridate (**17**), had no miotic action and was relatively non-toxic. Ethyl phosphorodifluoridate, $EtOP(O)F_2$, similarly had neither miotic nor toxic action [75].

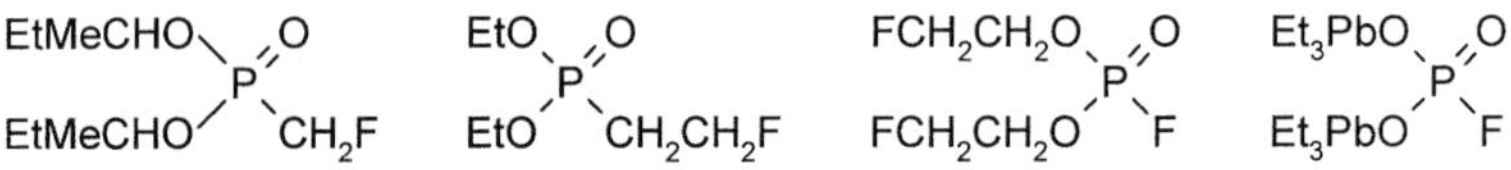

(18) (19) (20) (21)

To determine whether fluorine must be bonded to phosphorus for high toxicity, diisobutyl fluoromethylphosphonate (**18**) was prepared and found to be only slightly toxic. Compound (**19**), the first recorded example of a compound with a 2-fluoroethyl group attached to phosphorus, showed neither the toxic symptoms of the fluoroacetates or the phosphorofluoridates; the animal body was evidently unable to break the P-C link [70]. Bis(2-fluoroethyl) phosphorofluoridate (**20**) was prepared with the idea of combining the toxic actions of the fluoroacetates and the phosphorofluoridates. The compound caused miosis but the toxicity was lower than anticipated. At a concentration of 0.5 mg/l (10 min exposure) it produced in rats a state of hyperactivity followed by unusual convulsions leading to coma and death [76]. Efforts to combine the powerful anticholinesterase action of the nerve agents with the irritant effect of the lead trialkyls led to the synthesis of bis(triethyllead) phosphorofluoridate (**21**), a crystalline substance (m.p. >260 °C). At concentrations of only $1{:}10^8$ it exerted an irritant effect but did not cause miosis; concentrations of $1{:}10^6$ were intolerable [77-79]. In substances of structure $(RO)_2P(O)X$, toxicity was high when X is fluorine, but low when X was other functional groups (*e.g.* H, alkyl or aryl, OH, OR, CN, SCN) [8]. Substitution of fluorine by chlorine or bromine results in more reactive molecules that hydrolyse more easily and are unable to penetrate the nervous system intact [80].

Two routes (Scheme 29.4) to dialkyl phosphorofluoridates were developed at Cambridge: (A) conversion of phosphorus trichloride to a dialkyl phosphorochloridate, followed by halogen exchange using sodium fluoride; and (B) treatment of phosphorus oxychloride with antimony fluoride to give the oxydichlorofluoride, followed by alcoholysis. The first

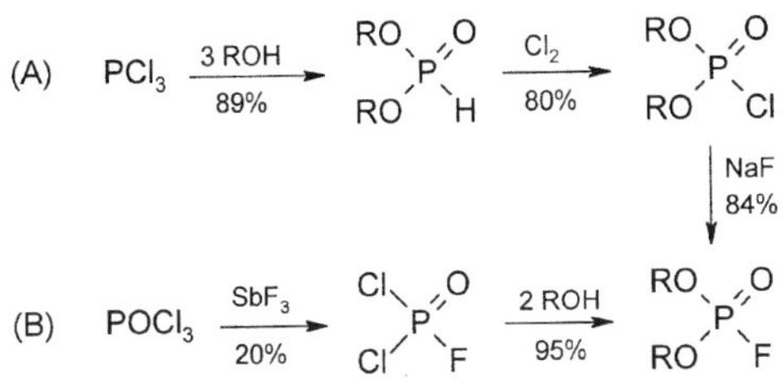

Scheme 29.4.

(A) was later improved by the Porton chemist, Arthur Ford-Moore [81]. The second (B) is only useful for small-scale work.

Arthur Ford-Moore (1896-1958)

Arthur Ford-Moore

Arthur Ford-Moore was a brilliant experimentalist, whose working life was spent at the bench or at the laboratory table writing up the results of his own work. His output was prodigious and there is no doubt that, were it not for the secrecy surrounding his work, he would have been recognised as one of the leading researchers of his day. His outstanding ability for research was recognised in 1952 by special promotion to the UK government grade of Senior Principal Scientific Officer.

Ford-Moore was born at Ealing, Middlesex in England, on 11 October 1896, and died suddenly at his home in Salisbury on 29 May 1958. After service as a Captain with the London Regiment in Palestine from 1915 to 1919, he graduated in 1921 with a BSc degree from Trinity College, Cambridge (UK), and that year joined the Chemical Defence Experimental Station, Porton Down. He was employed there until his death, except between 1942-1944 when he collaborated with H. R. Ing at the Dyson-Perrins Laboratory, Oxford. Although security considerations restricted publication of much of his work, Ford-Moore was internationally accepted as an authority on chemical defence problems, and in particular on the chemistry of organic compounds containing phosphorus and sulphur. At the time of his death he was preparing a series of papers describing his latest results.

In his early researches, Ford-Moore gained recognition for his work on the chemistry of mustard gas, $S(CH_2CH_2Cl)_2$, and related compounds. He devoted his later years to the chemistry of organophosphorus compounds, and it is for work in this field, where his superb laboratory technique found ample scope, that he will be primarily remembered. He carried out a comprehensive examination of the chemistry of insecticidal and toxic organophosphorus compounds. This included the elucidation of chemical reactions of importance in the preparation of these compounds, the establishment of synthetic routes for producing them in high purity, and the effect of variations in molecular structure on activity.

Arthur Ford-Moore also made lesser but valuable contributions to other branches of synthetic chemistry. Particularly noteworthy is the result of his collaboration with

H. R. Ing. Together they prepared some forty alkamine esters, mostly of benzylic acids. Their testing as mydriatics (drugs that cause dilation of the pupil of the eye) was carried out in 1947. An outcome of this work was the manufacture and use in medicine of the mydriatic drug lachesine, $Ph_2C(OH)CO_2CH_2CH_2N^+Me_2Et\ Br^-$. (Photograph reproduced by courtesy of DERA.)

In 1942, the Cambridge group prepared a new type of fluorine compound via the action of amines on phosphorus oxydichlorofluoride [82]. The P—F bond remained intact and the chlorine atoms were replaced (Scheme 29.5). The reaction was found to be general and was extended to a series of phosphorodiamidofluoridates which were isolated as liquid or crystalline solids. Many of these compounds were toxic by injection or inhalation; for example, one of the most poisonous was N,N,N',N'-tetramethyl phosphorodiamidofluoridate, later marketed as the systemic insecticide dimefox (LD_{50} sc mice 1 mg/kg). Unlike the phosphorofluoridates, however, they did not induce miosis. The phosphorodiamidofluoridates are very stable and are not easily hydrolysed by water. They are extremely neurotoxic and produce irreversible nerve damage in doses below their LD_{50} values [83].

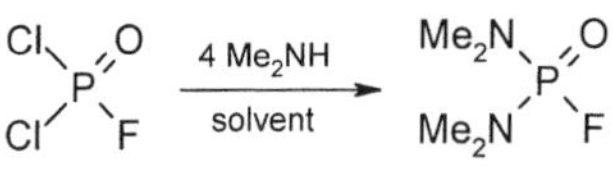

Scheme 29.5.

The Cambridge team decided to 'combine' the toxicities of a phosphorofluoridate with that of a phosphorodiamidofluoridate in a hybrid molecule. The resulting O-alkyl N,N-alkylphosphoroamidofluoridates[8] were more toxic than the diamido compounds and, in contrast to the latter, induced miosis. Special interest was taken in fluoride (**22**), the properties of which were investigated both in England and in Germany. This water-soluble and volatile liquid of high toxicity (LD_{50} sc mice 2.5 mg/kg) is slightly more poisonous than DFP. Monoalkylamino compounds were also studied. The methylamino analogue (**23**) had been prepared by Schrader in Germany in 1939 and was found to have high toxicity (LD_{50} sc mice 0.4 mg/kg) [21], but was of no military value due to its low volatility. The phenylamino analogue (**24**) had a much lower toxicity (LD_{50} sc mice 10 mg/kg) [8].

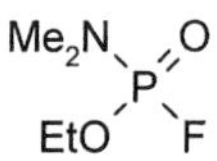

(22)

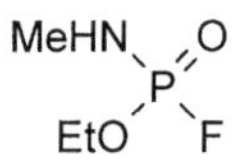

(23)

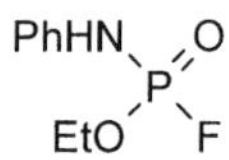

(24)

In 1945, munition dumps in western Germany were captured which contained high explosive chemical shells that appeared to be filled with a new substance. Samples sent to Porton not only confirmed this supposition, but also indicated that the substance had a toxicity and rapidity of action much greater than anything previously examined. Not only was the

[8] The precursor N,N-dimethylphosphoramidic difluoride, $Me_2NP(O)F_2$, is unusally toxic for a phosphorus difluoride (LD_{50} iv mice 1-2.5 mg/kg) [10]. Its reactivity towards various nucleophiles was investigated at Porton [84].

substance toxic by the intravenous route, it was also highly effective on skin or eye contact, or when inhaled as a vapour. At very low concentrations, it possessed miotic properties to a greater degree than the dialkyl phosphorofluoridates. The identity of the substance was established to be that of *O*-ethyl *N,N*-dimethylphosphoramidocyanidate, known as tabun. Pure tabun is a colourless liquid with a fruit-like odour. It is very soluble in water and is soluble in most organic solvents except hydrocarbons [9].

Tabun was discovered accidentally by Schrader while he was working on pesticides in the Leverkusen laboratories of the IG Farben company. On 23 December 1936, he treated dichloride (**25**) with sodium cyanide in ethanol in an attempt to make the corresponding dicyanide. Instead, he isolated tabun (Scheme 29.6) and fell prey to its toxic effects: 'I made the observation that the new substance, besides its contact action, exercised an extremely unpleasant effect on man. The first symptom noticed was an inexplicable action causing the power of sight to be much weakened in artificial light. In the darkness of early January, it was hardly possible to read by electric light, or after working hours to reach my home by car' [21].

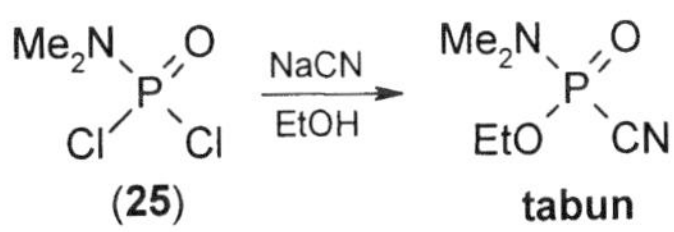

Scheme 29.6.

Schrader had miosis and realized this was a rather persistent effect: 'By stopping working on the substance for many days, the symptoms vanished, proving that the physiological action was due to the new cyanide' [10]. Other symptoms soon became apparent: 'By further, and now very carefully carried out, invesigations the observation was made that the smallest quantity of the substance dropped by inattention on the bench caused strong irritation of the cornea, and a very strong feeling of oppression in the chest. All these symptoms disappeared in 1 to 3 days in the fresh air, with the exception of the narrowing of the pupil, which causes sight trouble, which is unpleasantly noticible, and which only slowly improves' [21].

In 1939, Schrader and his collaborator, Dr Kükenthal, applied for a patent for tabun and related pesticides. The patent was classified Top Secret and Schrader was instructed to demonstrate the preparation of tabun in the Army Anti-gas Laboratories, Berlin, where the military significance of the new substance was quickly recognised. From then on, Schrader had little association with the development of tabun as a chemical warfare agent. He transferred from IG Leverkusen to IG Elberfeld in 1937 to study the preparation and properties of related substances, which led to the discovery of more potent nerve agents.

After World War II, interrogation of German scientists revealed that they knew of another phosphorus compound having greater toxicity than tabun but, unlike tabun, its preparation had advanced only as far as the pilot plant stage. This substance contained fluorine instead of nitrogen and was first synthesised by Schrader in 1938. It was isopropyl methylphosphonofluoridate, commonly called sarin – a name derived from those of the scientists involved in its discovery: *S*chrader, *A*mbros, *R*üdiger and van der L*in*de. Pure sarin is a colourless, odourless liquid. It is hygroscopic and miscible with water in all proportions, very soluble in organic solvents, and is the most volatile of the commonly encountered

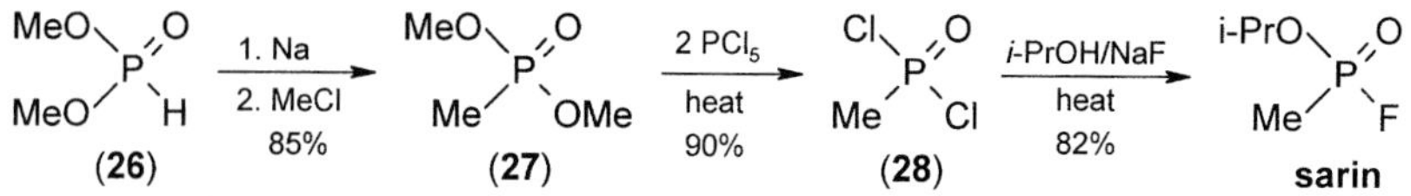

Scheme 29.7.

nerve agents [9]. The German workers developed a high-yield route to sarin starting from dimethyl phosphite (**26**), which first was methylated to yield dimethyl methylphosphonate (**27**); chlorination of this ester with phosphorus pentachloride, followed by treatment of the resulting methylphosphonic dichloride (**28**) with sodium fluoride and isopropanol gave sarin (Scheme 29.7). The Germans found that the action of sarin as a toxic war substance was 'in comparison with hitherto-known substances, astonishingly high'; consequently, the reaction details were forwarded to Berlin [21].

The work of Schrader was also passed to Richard Kuhn, head of the Kaiser Wilhelm Institute for Medical Research, Berlin, and the 1938 recipient of the Nobel Prize for Chemistry (awarded in 1939) for work on carotenoids and vitamins. His colleague Henkel synthesised about ten compounds by esterification of methylphosphonic difluoride, $MeP(O)F_2$. Reaction with 3,3-dimethylbutanol yielded a cholinesterase inhibitor (**29**) that was more toxic than tabun, but as the alcohol was too difficult to produce, pinacolyl alcohol was used as a replacement. This gave a more toxic compound in 1944 called soman, from the Greek verb 'to sleep' or the Latin verb 'to bludgeon'. Pure soman is a colourless liquid, supposedly with a pineapple-like smell, and of intermediate volatility. It has a low solubility in water and a high solubility in organic solvents [9].

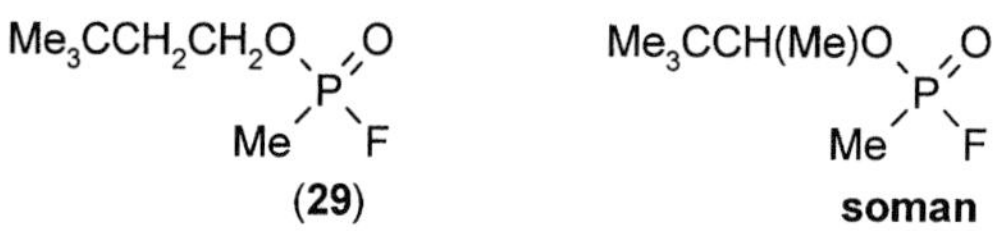

The new phosphorus compounds, referred to as the German agents (G agents), possessed the properties required for an 'assault gas' which was in general demand during World War II. The properties required for such a substance were that it should be rapidly and intensely incapacitating (or lethal), and contaminate large tracts of land only for short periods. Sarin, by virtue of its volatility and high toxicity, was particularly valuable in this respect. It was therefore clear that all major effort in chemical warfare research would have to be directed towards a study of the new compounds. Comparatively little was known about the organic chemistry of phosphorus and it was obvious that a study would have to be undertaken not only of sarin and analogues, but also of phosphorus chemistry in general. From the chemical warfare standpoint, a survey of sarin and analogues had to predominate. For several reasons, it was necessary for this to be carried out at Porton Down (and at other UK sites that have since been decommisioned – Sutton Oak in Lancashire and Nancekuke in Cornwall). Firstly, the dangerous nature of the substances made it undesirable, if not impossible, to carry out the work extramurally. Secondly, in an attempt to correlate chemical structure with toxicity, it was essential that the compounds should be in a high state of purity at the time of assessment. Carrying out the work in the Chemistry Division at Porton Down ensured that the toxicity was determined on-site in the Medical Division with little delay and under identical conditions, thereby eliminating errors that may have arisen owing to decomposition on storage or other factors. The work at Porton on sarin and tabun-type compounds was carried out between 1945 and 1952 by a team of chemists led by Ford-Moore.

Of the methods available for the synthesis of sarin, some were tedious or gave impure products. Two improved processes were therefore developed by Ford-Moore. In the first, treatment of diisopropyl methylphosphonate (**30**) [85] with phosgene yielded isopropyl methylphosphonochloridate (**31**), which was converted into sarin by sodium fluoride in hot dichloromethane [86] (Scheme 29.8). In the second, an equimolar mixture of methylphosphonic dichloride (**28**) and difluoride (**32**) was treated with isopropanol in a warm inert solvent (Scheme 29.9; the difluoride was easily prepared by treating the dichloride with hydrogen fluoride) [86].

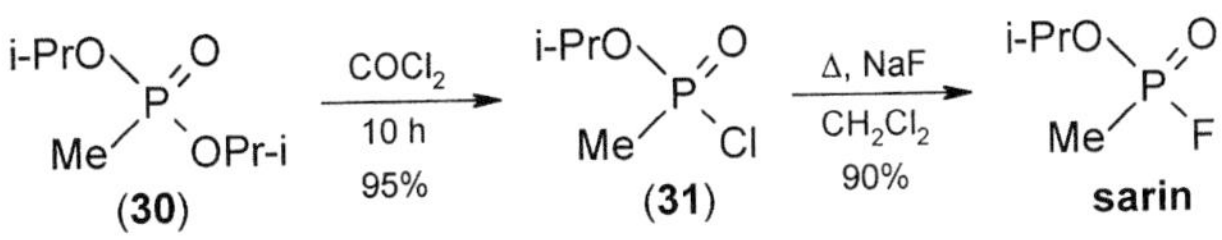

Scheme 29.8.

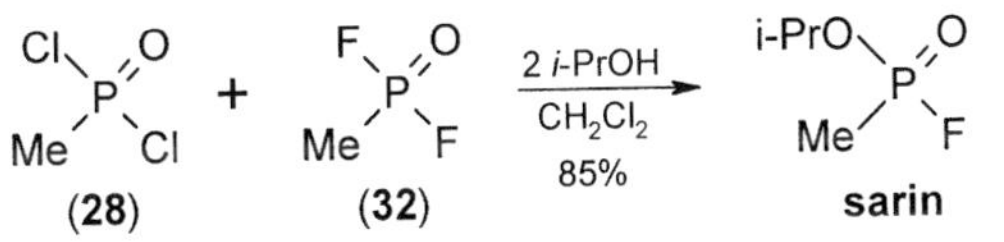

Scheme 29.9.

Maximum toxicity in the phosphonofluoridate series is associated with compounds having both a *P*-methyl substituent and a highly-branched C_4-C_6 alkoxy substituent [87] (Table 29.6). Inhibition rates of acetylcholinesterase follow the same trend [88]. In the G agent series, tabun is abbreviated to GA, sarin to GB, and soman to GD. Another highly toxic relative is *O*-cyclohexyl methylphosphonofluoridate, or cyclosarin, which is abbrievated to GF.

Shortly after research into the synthesis of sarin-type compounds at Porton had been wound down, another potent anticholinesterase was discovered by Ghosh at ICI laboratories in Manchester, UK [89]. It contained sulphur and nitrogen but no fluorine, and was unusually toxic by percutaneous administration. The substance, under the ICI tradename Amiton® [90, 91], was brought to the attention of Porton scientists. In structure, it closely resembled the natural substrate acetylcholine.

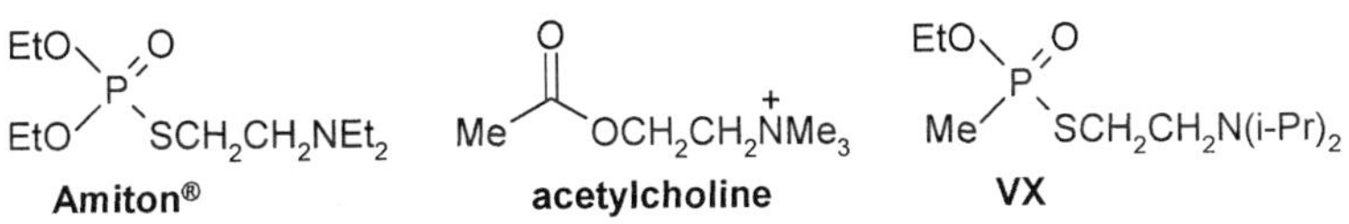

The histories of Amiton® and tabun are similar in that both were discovered in a search for insecticides, both were found to be too toxic to be used safely for this purpose, and both were adopted at one time by the services of the two countries of origin as potential war gases. At Porton, modification of Amiton® by Ford-Moore and his workers led to the

TABLE 29.6
Toxicities of *O*-alkyl methylphosphonofluoridates, $CH_3P(O)(F)OR$ [9]

C_x [a]	R group	Name	LD_{50} iv rabbit (mg/kg)
1	CH_3-	methyl	0.04
2	CH_3CH_2-	ethyl	0.05
3	$CH_3(CH_2)_2$-	*n*-propyl	0.03
3	$(CH_3)_2CH$-	*i*-propyl	0.02 [b]
4	$CH_3(CH_2)_3$-	*n*-butyl	0.05
4	$C_2H_5CH(CH_3)$-	*s*-butyl	0.01
4	$(CH_3)_2CHCH_2$-	*i*-butyl	0.19
5	$C_3H_7CH(CH_3)$-	2-pentyl	0.02
5	$(CH_3)_2CHCH(CH_3)$-	3-methyl-2-butyl	0.01
5	$(CH_3)_3CCH_2$-	neopentyl	0.01
6	$CH_3(CH_2)_5$-	*n*-hexyl	0.15
6	$(CH_3)_2CHCH_2CH(CH_3)$-	4-methyl-2-pentyl	0.02
6	$(CH_3)_3CCH(CH_3)$-	3,3-dimethyl-2-butyl	0.01 [c]
6	C_6H_{11}-	cyclohexyl	0.02 [d]

[a] No. of carbon atoms in R group.
[b] Sarin (GB).
[c] Soman (GD).
[d] Cyclosarin (GF).

TABLE 29.7
Amiton® and fluorinated analogues, $(EtO)_2P(X)SCH_2CH_2R$ [92]

R	pK_a water	pI_{50} [a] red cell	LD_{50} ip mice (mg/kg)
Oxono species (X = O)			
$N(C_2H_5)_2$ [Amiton®]	8.5	8.1	0.3
$N(C_2H_5)(CH_2CH_2F)$	6.5	7.0	0.7
$N(CH_2CH_2F)_2$	4.7	5.2	1.5
Thiono species (X = S)			
$N(C_2H_5)_2$	8.2	–	2.5
$N(C_2H_5)(CH_2CH_2F)$	6.2	<3	2.5
$N(CH_2CH_2F)_2$	4.2	–	3.0

[a] Negative logarithms of concentration inhibiting red cell cholinesterase by 50%.

discovery of the so-called venomous agents (V agents). Similar work in the United States led to the synthesis of VX, which was described by Ford-Moore as having 'quite fantastic toxicity'. Pure VX is a colourless, involatile liquid that has a negligible vapour pressure. It is hygroscopic and moderately soluble in water [9].

The toxicity of Amiton® and related oxono compounds (P=O) is partly due to the basicity of the amino group. The pK_a of the nitrogen atom in Amiton® is 8.5, consequently it is 97% protonated at physiological pH. Introduction of fluorine atoms into the amino group lowers basicity, reducing anticholinesterase activity and toxicity (Table 29.7). The

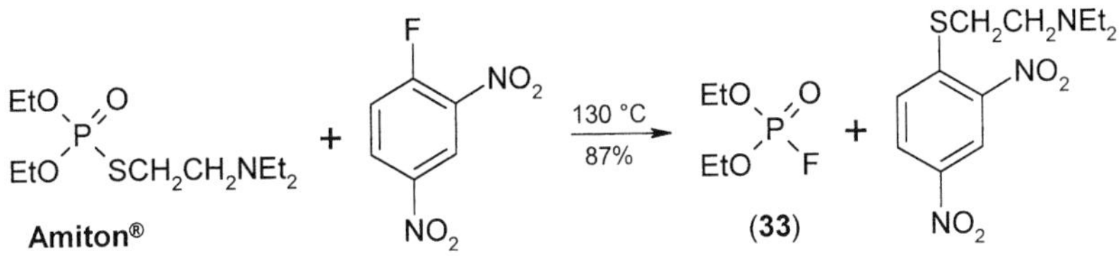

Scheme 29.10.

thiono compounds (P=S) are without anticholinesterase activity, as is almost invariably the case with thiono pesticides. They are toxic because of metabolic oxidation to the oxono form, but the relationship of pK_a to LD_{50} is more complex.

An unusual route to dialkyl fluorophosphates was discovered at Porton by Bebbington and Ley in the mid-1960s [93]. Heating Amiton® with 1-fluoro-2,4-dinitrobenzene converted it into diethyl fluorophosphate (**33**) in high yield (Scheme 29.10).

Two series of phosphorus compounds which combine structural features of G and V agents were researched but not developed as chemical weapons. The first series, studied in Sweden in the 1950s [94, 95], are known as Tammelin esters (after Lars-Erik Tammelin who discovered them) and possess a side-chain with a quaternary nitrogen atom, as in acetylcholine. Some of these compounds are as toxic as V agents but are solids of low stability. The second series, often called GV agents, were studied in former Czechoslovakia [96]. They are liquids that have the volatility of G agents and a percutaneous toxicity approaching that of V agents. They contain a tertiary nitrogen, as in the V agents, and the P-NMe_2 and P-F groups, as in tabun and sarin, respectively.

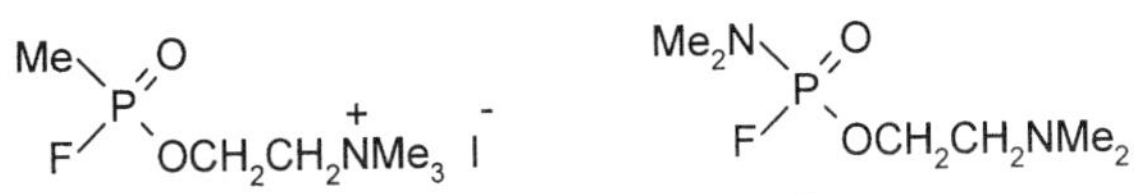

Structure-activity data for organophosphorus nerve agents have been reported [9]. They are much more toxic than sodium fluoroacetate. For the fluorine-containing nerve agents, toxicity increases in the order DFP $<$ sarin $<$ soman $<$ Tammelin ester $\sim$ GV (Table 29.8). With the exception of Tammelin esters, which are solids, DFP and G agents pose an inhalation hazard due to their volatility. Soman is the only G agent that possesses high percutaneous toxicity. Amiton and VX have low volatility and present mainly a percutaneous hazard.

Two rational lines of treatment for nerve agent poisoning are available. The first, essentially prophylactic, is to find a compound that can be phosphylated as rapidly as acetylcholinesterase and which, if introduced into the body, might protect the enzyme from inhibition by competing with it for the nerve agent. The second is to find a compound which might restore the activity of the inhibited enzyme by dephosphylating it. Compounds with either or both these properties are *o*-dihydroxybenzene derivatives, metal complexes, hydroxylamine, hydroxamic acids and oximes [97].

Antidotes for nerve agent poisoning

The search for antidotes to nerve agent poisoning was pioneered by a host of Porton scientists that included Askew, Berry, Brown, Davies, Green, Holmes, Ladell and Rutland. An early study in 1953 examined the effect of sixteen amino acids on the inhibition

TABLE 29.8
Physical properties and toxicities of some organophosphorus nerve agents [31]

Compound	Estimated b.p.[b] (°C)	Vapour pressure ($\times 10^{-3}$ mmHg)	LD_{50} rat sc (mg/kg)	LD_{50} mice sc (mg/kg)	LD_{50} mice ip (mg/kg)	LCt_{50} mice inh (mg min m^{-3})	LD_{50} rabbit sc (mg/kg)	LD_{50} rabbit iv (mg/kg)	LD_{50} rabbit pc (mg/kg)
FCH_2CO_2Na	solid	low	5		7			0.5	
DFP	183	579	1.44	3.00	2.45	4400 [c]	1.00	0.30	
Tabun (GA)	248	57	0.16	0.25	0.60	450 [d]	0.38	0.06	63
Sarin (GB)	158	2900	0.07	0.04	0.39	30 [d]	0.02	0.02	38
Soman (GD)	198	400	0.10	0.06	0.22	150 [d]	0.03	0.01	1.5
Amiton	320		0.15	0.19	0.30		0.13	0.08	0.2
VX	298	0.6	0.01	0.02	0.05		0.01	0.01	0.03
Tammelin ester [a]	solid	low	0.15		0.15			0.01	low
GV	225	49	0.02	0.03				0.02	0.2

[a] $CH_3P(O)(F)OCH_2CH_2\overset{+}{N}Me_3\ Br^-$.
[b] Estimated boiling point at 760 mmHg.
[c] Exposure time of 10 minutes.
[d] Exposure time of 30 minutes.

of horse-serum cholinesterase by sarin [98]. Only dopa (3,4-dihydroxyphenylalanine) was able, in high concentration, to abolish inhibition *in vitro* [99]. Indeed, catechol derivatives protected cholinesterases *in vitro* from inhibition by DFP, sarin and tabun; the basis of the protective mechanism was attributed to reaction between catechol and the nerve agent.

In the 1950s it was found that sarin was decomposed fairly slowly by hydroxylamine [97]. In an attempt to improve the reactivity, many substituted hydroxylamines were examined at Porton, which eventually led to the discovery that oximes have a high affinity for nerve agents [100]. Several hydroxamic acids were tested as antidotes to poisoning by sarin [101], but they were only effective in doses too high to be of practical significance. Studies of the effect of structure on their reactivity towards nerve agents showed that the deciding factor was the ionisation constant and that the reactivity could be predicted reasonably accurately from the ionisation constant alone [102]. It was also possible to show that for reaction at physiological pH (7.4), there was an optimum ionisation constant equivalent to a pK_a of about 8. To test this principle, many hydroxylamine derivatives were synthesised with pK_as ranging from 5 to 12 [103]. Most of them did reactivate *in vitro* and the best compounds at pH 7.4 had a pK_a of around 8. With some exceptions, oximes with this pK_a tended to be more active than analogous hydroxamic acids. In particular, pyridine-2-aldoxime methiodide (PAM) and the methanesulphonate salt (P2S) were more active than other compounds tested (Table 29.9) [104-110]. Their effectiveness was almost certainly due to a nucleophilic oxime group and a quarternised nitrogen atom that could interact strongly with the active site of the inhibited enzyme. In conjunction with the anticholinergic drug atropine, PAM or P2S were effective against many of the more common organophosphorus pesticides, although when given alone their activity was generally low. Oximes alleviate nerve agent poisoning by attacking the phosphorus atom of the inhibited enzyme, thus releasing the bound inhibitor and restoring enzyme activity (Scheme 29.11).

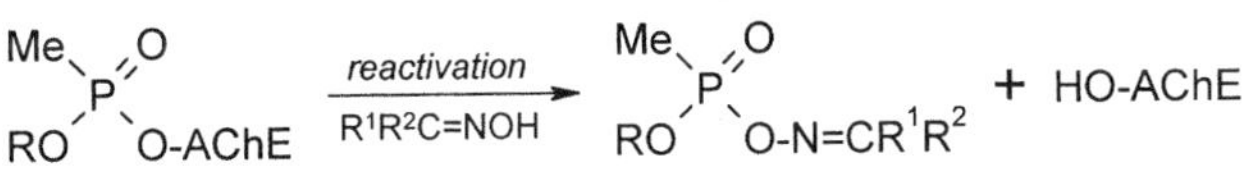

Scheme 29.11.

Both PAM and P2S were relatively harmless in therapeutically active doses; P2S administered to humans produced no ill-effects on oral doses of over 5 g, or as a 15% aqueous solution, in intramuscular doses of over 1 g. Other oximes were explored in detail, but their early promise was not maintained. The rates of reaction of some fluorinated oximes (**34**)-(**37**) made by Haszeldine[9] *et al.* (under contract to Porton) with sarin were measured (Table 29.9). The tetrafluoro-oxime (**36**) had a K_{ox}- approaching that of the good reactivator P2S, but suffered from the disadvantage that even four fluorine atoms had reduced the pK_a to only 9.1. Other studies of the reactivities of oximes towards nerve agents and inhibited cholinesterase have been published by Porton scientists [112-120].

When inactivated enzyme is kept at room temperature for a time, the proportion of reactivatable enzyme becomes progressively less [121, 122]. This transformation of inhibited

---

[9] Professor Robert N. Haszeldine, FRS, was chairman of the UK Defence Advisory Board from 1981 to 1983. The chemistry committee was founded in 1946 and disbanded in 1994. Its members provided scientific advice and did much to foster the research programmes in chemical defence and civil research in the UK.

TABLE 29.9
Second-order rate constants for reaction of oxime anions with sarin [111]

Oxime		pK_a	K_{ox^-} (l mol^{-1} min^{-1})
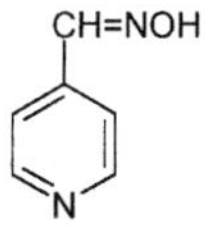	**(34)**	10.1	–
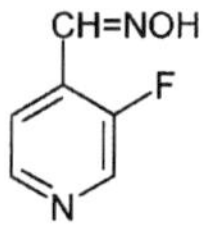	**(35)**	9.8	1800
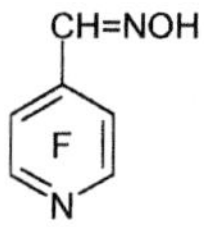	**(36)**	9.1	310
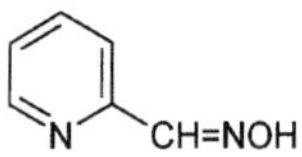	**P2A**	10.1	1690
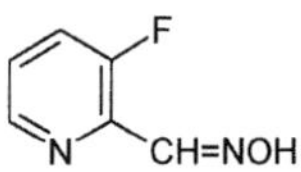	**(37)**	10.1	1300
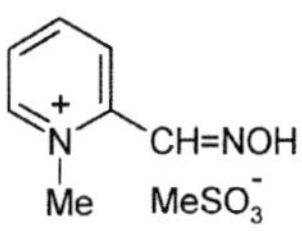	**P2S**	7.8	120

but reactivatable enzyme to a form that is resistant to the action of nucleophilic reagents was termed 'aging' [123]. Aging describes cleavage of the PO–C bond of the inhibited enzyme with loss of a carbenium ion (Scheme 29.12). The negative charge on the bound inhibitor renders the phosphorus atom resistant to attack by nucleophiles such as oximes.

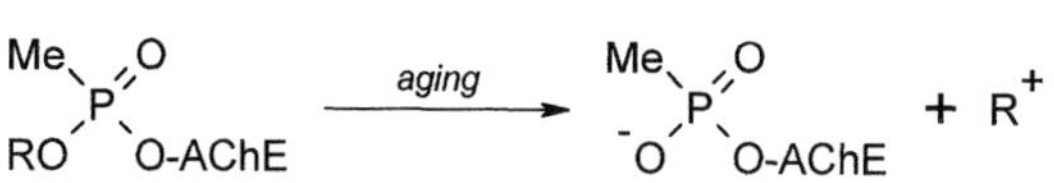

Scheme 29.12.

Aging in enzyme inhibited by G agents is slow when the ester group is derived from a primary alcohol, but is much more rapid if the alkyl group is a secondary or cyclic alcohol as the latter produce relatively stable carbenium ions; aging is also accelerated by increase of temperature or a decrease of pH [124]. Aging proceeds at a finite rate which is sometimes very rapid; *e.g.* the half-life of soman-inhibited enzyme at 35 °C is probably less than 2 min [124]. Clearly the effect of treatment with oximes and atropine in poisoning depends on the nerve agent. Enzyme inhibited by DFP, although initially reactivatable, is difficult to reactivate after a time [121, 122]. Pretreatment with PAM and atropine raised the LD_{50} of DFP 16 times, but of sarin only 2 times [125]. Enzyme inhibited by soman was resistant to treatment within minutes of poisoning, the effect of oxime plus atropine being no better than would be expected from atropine alone. The therapy of GV intoxication is also very difficult [126]. Diazepam or avizophone are used as anticonvulsant drugs for the treatment of convulsions arising from nerve agent poisoning [9]. When the threat of exposure to nerve agents is anticipated, high-affinity carbamate cholinesterase inhibitors, such as pyridostigmine or physostigmine, provide effective pretreatment for ensuring survival [9].

Fluorinated carbamates

Fluorinated carbamates are anticholinesterases like the nerve agents. Several have been assessed as pesticides but none have found practical use. Dimethylcarbamoyl fluoride, $(CH_3)_2NCOF$, was prepared by Schrader in 1947 by the action of antimony trifluoride on the corresponding chloride, and found to be a good insecticide [21]. It is however highly toxic to mammals (LD_{50} ip mice 0.2 mg/kg) [127], irreversibly inhibiting cholinesterase, the fluorine atom departing and the dimethylcarbamoyl group 'carbamoylating' the active site of the enzyme [128]. The chloro analogue, $(CH_3)_2NCOCl$, is acutely non-toxic (LD_{50} ip mice 350 mg/kg) [12], but is a potent carcinogen [129].

Polyfluorinated alcohols

The toxicity of monofluorinated alcohols has already been discussed. Few fluorinated alcohols containing more than one fluorine atom have been investigated and these are of moderate toxicity [130]. However, the highly acidic (pK_a 5.95) perfluoropinacol, $(CF_3)_2C(OH)C(OH)(CF_3)_2$ (b.p. 129 °C), is acutely toxic by skin contact (one drop on the skin has been found sufficient to kill a guinea pig [131]). It is also toxic by inhalation (ALCt in rats: 164 400 mg min m^{-3}, 4 h exposure) [131], although its mode of action is unknown. Related compounds are of lower toxicity, *e.g.* hexafluoroacetone, $(CF_3)_2C{=}O$ (LCt_{50} in rats: 488 880 mg min m^{-3}, 4 h exposure), is three times less toxic than perfluoropinacol, and hexafluoroisopropanol, $(CF_3)_2CHOH$ (ALCt in rats: 5 276 160 mg min m^{-3}, 4 h exposure), is also much less toxic [132]. Both hexafluoroacetone sesquihydrate and hexafluoroisopropanol cause permanent eye damage in rabbits [132].

Another toxic polyfluorinated alcohol is 2,2-difluoro-2-nitroethanol, $CF_2(NO_2)CH_2OH$, a colourless liquid (b.p. 64 °C/15 mmHg) which causes temporary insensitiveness of the skin [16]. Difluoronitroacetic acid, $CF_2(NO_2)CO_2H$, a colourless solid (m.p. 37 °C), produces similar symptoms on skin contact [16].

Fluorinated ethers

The biological activities of fluoroethers are unpredictable [133]. At present, there is generally no physical or chemical property that will allow biological activity to be predicted. Toxicity results for monofluorinated ethers can be explained by simple rupture of the ether link, $F(CH_2)_nOR$ to $F(CH_2)_nOH$. This biochemical cleavage is strikingly confirmed

TABLE 29.10
Toxicities of monofluoro and difluoro dibutyl ethers [7]

Compound		b.p. (°C/mmHg)	LD_{50} ip mice (mg/kg)
$Cl(CH_2)_4O(CH_2)_4Cl$	(**38**)	129/13	>100
$F(CH_2)_4O(CH_2)_4CN$	(**39**)	135/10	1.5
$F(CH_2)_4O(CH_2)_4Cl$	(**40**)	101/10	1.3
$F(CH_2)_4O(CH_2)_4F$	(**41**)	74/10	0.8

TABLE 29.11
Toxicities of some polyfluorinated ethers

Compound	Bp (°C)	Biological activity	Ref.
$FCH_2OCHFCl$	54	Toxic[a]	134
$CH_3OCF{=}CCl_2$	102	Toxic[b]	133
$CH_3OCH_2CF_2CF_2CF_3$	71	Convulsions, delayed death[c]	135
CH_3OCF_2CHFCl	64	Convulsions, delayed death[d]	136
$CH_2{=}CHOCH_2CF_2CF_3$	58	Pulmonary oedemagen[c]	135
$CH_2ClOCF_2CHF_2$	77	Delayed death[a]	133
$CF_2ClOCF_2CFCl_2$	–	Convulsions, delayed death[b]	136
$CH_3CH_2OCF_2CHFCl$	88	Anaesthesia, delayed death[d]	137
$(CH_3)_2CHOCF_2CHFCl$	100	Anaesthesia, delayed death[d]	137
$CF_3CH_2OCFClCF_3$	–	Convulsions, delayed death[b]	136
$CF_3CH(OCF_3)CF_2OCF_3$	42	Highly toxic (see text)[c]	138
$CF_3OCH_2CF_2CF_2H$	46	Highly toxic (see text)[d]	139
$CF_3OCF{=}CF_2$	–	Extremely toxic (see text)	6

[a] Dog.
[b] Animal species not stated – presumably rat or mouse.
[c] Rat.
[d] Mouse.

by the results obtained with four dibutyl ethers (Table 29.10). The dichloro ether **38**, containing no fluorine, is non-toxic. The toxicities of the fluorocyano and fluorochloro ethers, **39** and **40** respectively, are almost identical, and very similar to that of 4-fluorobutanol on a molar basis. The difluoro ether **41** can theoretically give rise to twice the amount of 4-fluorobutanol on hydrolytic fission, and, according to expectation, is roughly twice as toxic as the two mono-fluoro ethers.

Toxicity results for polyfluorinated ethers are difficult to explain at present. While methoxyflurane ($CH_3OCF_2CHCl_2$), fluroxene ($CF_3CH_2OCH{=}CH_2$), and the more recent agents enflurane (CHF_2OCF_2CHFCl), isoflurane ($CHF_2OCHClCF_3$), sevoflurane [$CH_2FOCH(CF_3)_2$], and desflurane ($CHF_2OCHFCF_3$), are well-known inhalation anaesthetics, other polyfluorinated ethers can cause convulsions and/or delayed death after inhalation (Table 29.11). Convulsive seizures usually start in less than 1 min after the inhalation of 4-6 breaths. Flurothyl ($CF_3CH_2OCH_2CF_3$) produces seizures in rats at concentrations as low as 30 ppm and has been used in the treatment of mentally ill patients [140, 141]; tests at Porton confirmed its low inhalation toxicity (LCt_{50} mice >50 000 mg min m^{-3}, 10 min exposure) [10].

Recent studies have shown that diethers $CF_3CH(OCH_2F)CF_2OR$ (R = CH_3, CH_2F) sedate rats when administered intravenously [138]. However, the related diether $CF_3CH(OCF_3)CF_2OCF_3$ proved to be highly toxic: the lethal dose for rats by iv injection was 15-25 μl, resulting in rapid cessation of respiration and death within 2 min [138]; and inhalation of 0.5% v/v of vapour for 30 minutes proved fatal to rats during the exposure or shortly after its cessation [138]. Other ethers containing the trifluoromethoxy group are toxic; for example, $CF_3OCH_2CF_2CF_2H$ at a vapour concentration of 0.5% caused violent convulsions in mice and death within 30-120 seconds [139]. Perfluoro(2-methoxy)ethene $CF_3OCF{=}CF_2$ is extremely toxic, having an LCt_{50} to mice of 5.5 mg min m^{-3} for a 2 h exposure [8].

Fluoroalkenes

The toxicities of fluoroalkenes, which generally are greater than those of related fluoroalkanes [142], have attracted attention because of the importance of members of this fluoro-organic class as commercial monomers [143] and synthons [144]. Most information concerns fluoroalkenes of moderate inhalation toxicity that react with lung thiols, such as glutathione [145-151] and cysteine [152-157], to give saturated or unsaturated products which are degraded enzymatically to toxic species that can cause liver or kidney damage [158].

$(CF_3)_2C{=}CF_2$

PFIB

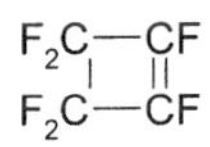

HFCB

Not all fluoroalkenes behave this way and those of high inhalation toxicity specifically target the lung. Best known of these is perfluoroisobutene (PFIB, b.p. 6 °C)[10] which is unusually toxic [160, 161]. It is formed as a by-product during the manufacture of tetrafluoroethylene [162], during pyrolysis of some perfluorinated monomers and polymers (notably PTFE) [163], and possibly during the extinguishment of fires by some Halon replacements [164]. Industrial accidents involving human exposure to PFIB have been reported [143, 165].[11] Related fluoroalkenes are toxic upon inhalation [166, 167], but less so than PFIB; however, hexafluorocyclobutene (HFCB, b.p. 5 °C) is reported to possess a similar order of toxicity [168].

During the mid-1980s, it was recognised that fluoroalkenes might pose a risk not only to workers in the chemical industry but also to military personnel involved in fire-fighting operations. Porton scientists studied the synthesis and toxic properties of PFIB and related alkenes with the aim of elucidating their mechanism of action. It was found that PFIB caused pulmonary oedema (similar to that associated with phosgene) after a latent period of 6-15 hours, depending on the inhaled dose [169]. It was clear that the

[10] A fascination with this electrophile led me to a review of the chemistry of PFIB by the distinguished Russian fluorine chemist Ivan Ludvigovich Knunyants and his co-workers [159]. Reading this inspired me to investigate the reactions of PFIB with thiols, an area largely neglected by other researchers.

[11] Five workers accidentally exposed to a gas containing 2% PFIB sufferred irritation of the respiratory tract (cough, shortness of breath and wheezing) within 24 h of exposure; two of the workers died at 11 and 13 days after exposure, and pathlogical examination revealed congestion of the lungs consistent with oedema [143].

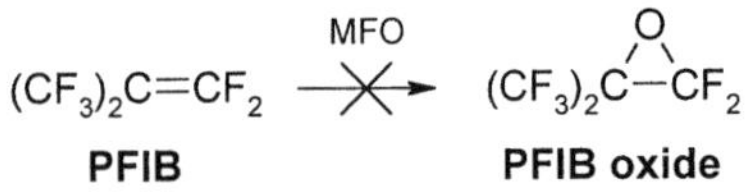

Scheme 29.13.

alkene had a direct action on the lung since damage was confined to this organ [170] and metabolic activators/inhibitors did not modify toxicity: the LCt_{50} in rats was not affected by pretreatment with mixed function oxidase (MFO) inducing or inhibiting drugs, indicating that PFIB oxide was not formed (Scheme 29.13).[12]

Hydrolysis was not responsible for toxicity as reaction of PFIB with water gave less toxic products; studies at Porton showed that PFIB hydrolysed to hexafluoroisobutyric acid, $(CF_3)_2CHCO_2H$, which decarboxylated to give a mixture of the hexafluoropropane $CF_3CH_2CF_3$ and the pentafluoropropene $CF_3CH{=}CF_2$ [175]. There were no signs of poisoning or deaths in mice exposed to 20 000 mg min m^{-3} of the fluoropropane or 50 000 mg min m^{-3} of the fluoropropene (5 min exposure, 7 day observation). Another hydrolysis product, the nonafluoropropane, $(CF_3)_3CH$, formed by addition of liberated HF to perfluoroisobutene, was considered unlikely to possess appreciable toxicity.

The breakthough in understanding the mechanism of action of PFIB was the finding that it lowered lung thiol levels in rats [169]. Conversely, cysteine esters administered as pretreatments increased thiol levels in the rat lung and protected against lethal doses of PFIB [176, 177]. Therefore it appeared that reactions with biological thiols were responsible for toxicity. It was later shown that in aqueous neutral buffer, in the presence of other nucleophilic centres, thiols reacted with PFIB to give ketene dithioacetals, $(CF_3)_2C{=}C(SR)_2$, while cysteine isopropyl ester reacted to give thiazole **42** (Scheme 29.14) [178]. Phosgene combines similarly with cysteine *in vivo* to yield thiazolidinone **43** [179]. The hexafluoroisobutenylidene group of perfluoroisobutene, $(CF_3)_2C{=}C$, therefore acts as an analogue of the carbonyl group, O=C, of phosgene. Studies at Porton showed that PFIB was three to four times more toxic than phosgene to mice (10 min exposures);[13] this difference might, in part, arise from the higher reactivity of the former towards nucleophiles.

The inhalation toxicity of PFIB is probably enhanced by its low solubility in water, which allows it to penetrate further into the lungs than more soluble gases such as hydrogen fluoride.[14] The upper respiratory tract is protected against fluoroalkenes by a thick

[12] The claim by Russian workers that 'the toxicity of perfluoroisobutene oxide is higher than that of perfluoroisobutene' and 'the quantity of oxides and peroxides that are formed are in fact the products that produce damage to lung tissue' [171] was refuted at Porton: PFIB oxide, made by treatment of perfluoroisobutene with hydrogen peroxide [172], had an LCt_{50} to mice of 17 500 mg min m^{-3} for a 10 min exposure (the comparable LCt_{50} for perfluoroisobutene is 880 mg min m^{-3}) [10]. By analogy to reactions with alcohols and amines [173, 174], perfluoroisobutene oxide will probably combine with thiols to give disubstituted products, $RSC(CF_3)_2C(O)SR$.

[13] In the 1970s, the Porton scientists Nash and Pattle [180] demonstrated that the high toxicity of phosgene was due to acylation of biological nucleophiles, rather than to hydrolytic production of hydrochloric acid in the lung. The reactivity of nucleophiles towards phosgene was found to decrease in the order amines > alcohols > thiols. Consequently, it was soon discovered that animals pretreated with amines could be protected from the lethal effects of phosgene.

[14] The high water solubility of HF causes irritation of all tissues with which it comes into contact. The larynx and throat are attacked and the lungs become oedematous.

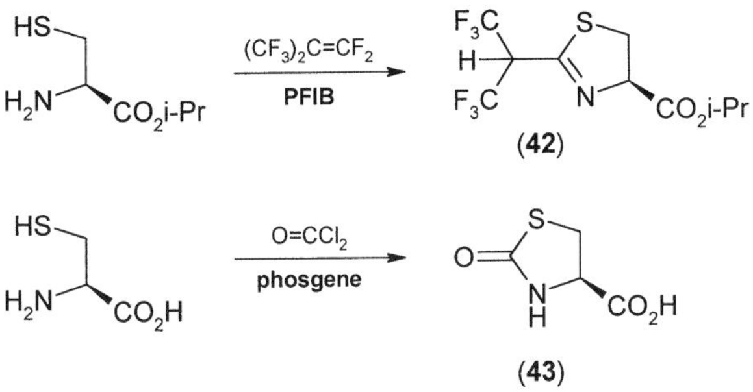

(42)

(43)

Scheme 29.14.

mucus layer which prevents hydrolysis; however, due to a high affinity for water, hydrogen fluoride can penetrate this layer and attack tissues. In the deep lung, the damage caused by a fluoroalkene is greatest as this region is overlayed only by a very thin layer of surfactant through which the gas can penetrate. Once absorbed, the fluoroalkene is able to attack important biological nucleophiles much faster than it can attack the surrounding water. Fluoroalkenes generally react with thiols in preference to other nucleophilic centres. For example, the biological thiols cysteine and glutathione react with tetrafluoroethylene, $CF_2{=}CF_2$, and hexafluoropropene, $CF_3CF{=}CF_2$, to give addition products CF_2HCF_2SR [145] and CF_3CFHCF_2SR [147], respectively. These perfluoroalkenes are much less reactive than perfluoroisobutene, and they combine with only one molar equivalence of a thiol, hence they are much less toxic [166, 168]. The toxicity value for a given fluoroalkene may vary not only for a different species, but also for different strains and even for different laboratories. Clearly, there was a need for strictly comparable values to be determined under as nearly similar conditions as possible. Inhalation toxicities were therefore measured at Porton for PFIB and a number of its structural analogues in an attempt to discern structure-activity relationships (Table 29.12).

The LCt_{50} values obtained correlate with the electrophilicities of the fluoroalkenes and their reactivities towards model biological thiols, RSH, *e.g.* *n*-propanethiol. Thus, fluoroalkenes **44** and **45**, the least toxic alkenes of the group, are of low reactivity compared to PFIB, and give only mono-adducts of structure $(CF_3)_2CHCHXSR$ ($X = Br$ or F) plus the related olefin $(CF_3)_2C{=}CHSR$. Alkene **46** is at least as reactive as fluoroalkenes **44** and **45**, but, having no vinylic halogen substituents, can only react with a molar equivalence of a thiol to give a mono-adduct, $(CF_3)_2CHC(CF_3)_2SR$. Alkene **47** also combines with a molar equivalence of thiol to give the product of vinylic fluoride loss, $(CF_3)_2C{=}C(SR)(OMe)$.[15] Dichlorohexafluorobutene **48**, a minor by-product in the manufacture of the important inhalation anaesthetic Fluothane® (ICI) which has to be carefully removed [181], can react with thiols to give a disubstituted product, $(CF_3)(RS)C{=}C(SR)(CF_3)$. The toxicities of fluoroalkenes that can react with thiols to form

[15]The substitution of one of PFIB's vinylic fluorines with a methoxy group converts a bis-alkylating agent into the less toxic mono-alkylating agent $(CF_3)_2C{=}CF(OMe)$. Similiarly, the substitution of one chlorine atom in phosgene with a methoxy group, as in chlorocarbonic acid methyl ester O=CCl(OMe), produces an eye irritant which has only a slight lung-damaging effect. This is a further example of the analogy between the hexafluoroisobutylidene group of perfluoroisobutene and the carbonyl group of phosgene.

TABLE 29.12
Inhalation toxicities of perfluoroisobutene (PFIB) and its analogues [10]

Fluoroalkene		LCt_{50} mice[a] (mg min m^{-3})
$(CF_3)_2C{=}CHBr$	**(44)**	>10 000
$(CF_3)_2C{=}CHF$	**(45)**	>50 000
$(CF_3)_2C{=}C(CF_3)_2$	**(46)**	>50 000
$(CF_3)_2C{=}CF(OMe)$	**(47)**	>15 000
$CF_3CCl{=}CClCF_3$	**(48)**	30 000
$(CF_3)_2C{=}CFCl$	**(49)**	28 000
$(CF_3)_2C{=}CCl_2$	**(50)**	>25 000
$(CF_3)_2C(CF_3)C{=}CF_2$	**(51)**	<10 000
$(ClCF_2)CF_3C{=}CFCl$	**(52)**	2400
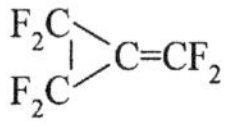	**(53)**	1750
$(CF_3)_2C{=}CF_2$	**PFIB**	880
$(ClCF_2)CF_3C{=}CF_2$	**(54)**	460
$(ICF_2)CF_3C{=}CF_2$	**(55)**	460

[a] Based on deaths occurring up to 14 days following exposures of 10 min duration.

disubstituted products increase in the order **(48)** ~ **(49)** ~ **(50)** < **(51)** < **(52)** < **(53)** < PFIB < **(54)** ~ **(55)**, and correlate with increasing alkene electrophilicity. The greater toxicity of dichloride **52** over monochloride **49** can be explained by the possibility of the former giving rise to trisubstituted products, $(RSCF_2)CF_3C{=}C(SR)_2$, via substitution of allylic chlorine. Similarly, fluoroalkenes **54** and **55** are almost twice as toxic as PFIB, their impressive toxicity presumably stemming from high electrophilicity coupled with ease of conversion to trisubstituted products, $(RSCF_2)CF_3C{=}C(SR)_2$. The observation by Knunyants *et al.* [182] that alkenes **49**, **50** and **52** are less active than PFIB in reactions with nucleophiles provides further evidence that the toxicities of fluoroalkenes are related to their alkylating abilities.

The high electrophilicity and toxicity of PFIB is due partly to the strongly electron-withdrawing geminal trifluoromethyl groups. Substitution of one CF_3 group for a less electronegative atom or group results in molecules of much lower electrophilicity and toxicity; *e.g.* the LCt_{50} values in mice exposed to $CF_3CH{=}CF_2$ and $CF_3CI{=}CF_2$ were 20 000 and >40 000 mg min m^{-3} respectively (10 min exposures) [10]. Removal of the geminal difluoromethylene group of PFIB also results in loss of toxicity; *e.g.* bis(trifluoromethyl)ketene, $(CF_3)_2C{=}C{=}O$, and tetrakis(trifluoromethyl)allene, $(CF_3)_2C{=}C{=}C(CF_3)_2$, had LCt_{50} values to mice of 30 000 and >50 000 mg min m^{-3} respectively (10 min exposures) [10]. These unsaturated systems are mono-alkylating agents of lower reactivity than PFIB which combine with one equivalence of a thiol, giving $(CF_3)_2CHC(O)SR$ or $(CF_3)_2CHC(SR){=}C(CF_3)_2$, respectively. In addition to toxicity studies, the ability of carbon filters to remove PFIB [175] and $(ClCF_2)CF{=}CF_2$ [183] was studied at Porton by Lawston *et al.*; Watts *et al.* showed that ion mobility spectroscopy has potential for detecting and monitoring PFIB [184].

Tests on candidate anaesthetics in the 1960s had revealed that certain fluorinated cycloalkenes were toxic to mice [185, 186]. Comparative toxicity data were however unavail-

TABLE 29.13
Inhalation toxicities of hexafluorocyclobutene (HFCB) and its analogues [10, 190]

Fluoroalkene		Bp (°C)	LCt_{50} mice[a] (mg min m^{-3})
	(**56**)	27	>100 000
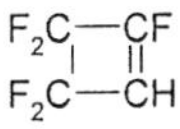	(**57**)	25	10 000-20 000
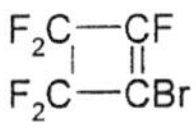	(**58**)	54	>6750
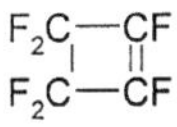	**HFCB**	5	6000
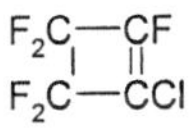	(**59**)	36	6000
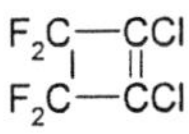	(**60**)	67	>5000
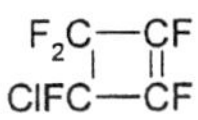	(**61**)	32	1930
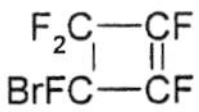	(**62**)	47	1890
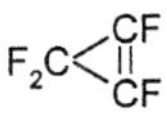	(**63**)	−13	1000

[a] Based on deaths occurring up to 14 days following exposures of 10 min duration.

able, therefore inhalation toxicities were measured at Porton for hexafluorocyclobutene (HFCB) and some of its analogues (Table 29.13); the retention of inhaled HFCB in the rat was also studied [187].

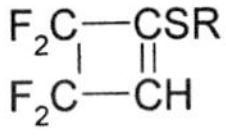

(A)

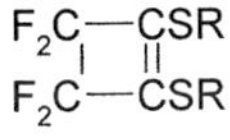

(B)

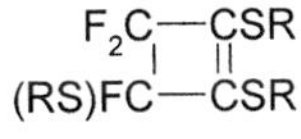

(C)

Again, toxicities correlate with substrate electrophilicities and their reactivities towards biological thiols. The reactivities of perfluorocycloalkenes increase with decreasing ring size [188, 189], therefore toxicity increases in the order: perfluorocyclo-pentene (**56**) < -butene (HFCB) < -propene (**63**). The first two cycloalkenes can react with thiols to give disubstituted products [190] in a manner analogous to $CF_3CCl{=}CClCF_3$. HFCB's 1-hydro analogue (**57**), which reacts with thiols to give only mono-substituted adducts (**A**)

[190], is the least toxic cyclobutene. The similar toxicities of HFCB and butenes **58-60** can be rationalised on the basis of their similar electrophilicities and the fact that they are all bis-alkylating agents, giving disubstituted products (**B**) with thiols [190]. The greater toxicities of the 3-chloro- and 3-bromopentafluorocyclobutenes (**61** and **62**) are ascribable to their ability to act as tris-alkylating agents in a manner analogous to $(ClCF_2)CF_3C{=}CF_2$ (**54**) and $(ICF_2)CF_3C{=}CF_2$ (**55**), giving trisubstituted products (**C**) [190]. Perfluorocyclopropene (**63**) is the most toxic cycloalkene we have examined to date, its toxicity approaching that of PFIB. It is highly electrophilic and reacts violently with nucleophiles, often with ring opening, to give polysubstituted adducts [191, 192]. The conclusion that toxicity results from addition-elimination cascades involving cellular thiols should aid assessment of the inhalation hazards posed by known or novel fluoroalkenes and hopefully reduce the need for animal experimentation.

Low concentrations of PFIB are tolerated by rats and humans when inhaled for several minutes and may cause short-lived lung irritation [143], but high concentrations, even for short periods, may cause a potentially fatal oedema. There is no current antidote for fluoroalkene poisoning and because the alkylation reactions are likely to be largely irreversible, it is unlikely that an effective therapy will be found. However it is possible to design prophylaxes that are effective in animals [193, 194]. Studies at Porton by Lailey *et al.* showed that PFIB depleted lung thiol levels in the rat, and that pretreatment with a supplementary thiol reduced its toxicity, and in some cases provided complete protection [176]. Further studies showed that intraperitoneal or oral administration of *N*-acetylcysteine, cysteine, or some cysteine esters, protected rats against PFIB [195]. Animals pretreated with *N*-acetylcysteine by the oral route were protected for up to 4 hours and to a lesser extent for up to 8 hours [195], and those pretreated by the intratracheal route were protected for up to 3 hours [196]. The thiols apparently exert a protective effect by reacting with the fluoroalkene and preventing it from alkylating essential cellular components.

Fluorinated organosulfur compounds

Bistrifluoromethyl disulfide, CF_3SSCF_3 (b.p. 34 °C), is an industrial fumigant [197] that like the trisulfide, CF_3SSSCF_3, is a pulmonary oedemagen [198, 199]. The disulfide (LCt_{50} rats 1900 mg/kg) is about half as toxic as PFIB (LCt_{50} rats 1000 mg/kg) for a 10 minute exposure [200, 201]. Replacement of the trifluoromethyl groups on sulfur gives compounds of reduced toxicity; *e.g.* bisperfluorocyclohexyl disulfide, $C_6F_{11}SSC_6F_{11}$, which is relatively non-toxic [6]. Other fluorinated organosulfur compounds are toxic by inhalation. Tetrafluoro-1,3-dithietane (**64**) produces irritation to the upper and lower respiratory tract of rats and also nervous system effects, such as incoordination and convulsions, at lethal doses [11].

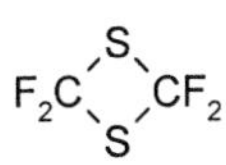

(64)

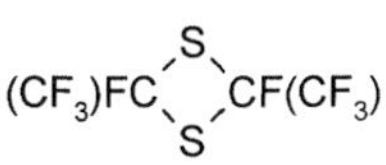

(65)

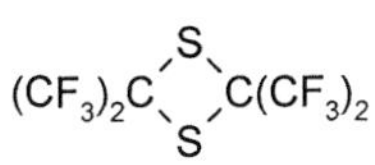

(66)

Trifluorothioacetyl fluoride dimer (**65**) is a lung irritant of moderate toxicity (LCt_{50} rats $>5\,000\,000$ mg min m^{-3}, 3 h exposure) [11]. Hexafluorothioacetone dimer (**66**) is however much more toxic (LCt_{50} rats 14 000 mg min m^{-3}, 4 h exposure) [11]. Comparative tests at Porton on mice exposed for 10 minutes showed that dimer (**64**) (LCt_{50} 30 000 mg min m^{-3}) was about three times less toxic than dimer (**66**) (LCt_{50} $<$10 000 mg min m^{-3}). The mode of action of the fluorinated organosulfur compounds, like that of disulfur decafluoride (S_2F_{10}), is poorly understood.

Fluorinated cage convulsants

While investigating the biological properties of the alkene $(F_3C)_2C{=}C(CN)_2$, DuPont scientists discovered in 1982 that various cyclopentadiene adducts (Scheme 29.15) were toxic by oral administration to rodents [202]. Norbornene **67** (LD_{50} ip for mice, 1 mg/kg) and norbornane **68** (LD_{50} ip for mice, 0.1 mg/kg) are also highly toxic by injection; death occurs rapidly from convulsive seizures [203].

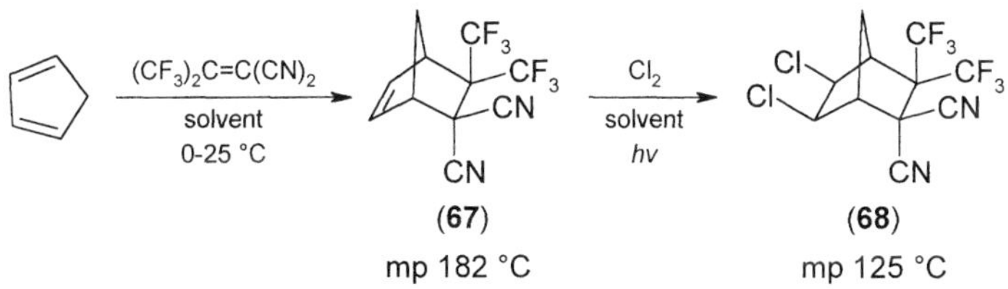

Scheme 29.15.

The mode of action of these 'cage convulsants' remained unknown for ten years until workers at the Institute of Physiologically Active Substances in Moscow found that they blocked gamma-aminobutyric acid (GABA) induced chloride ion currents in the central nervous system [203, 204]. The CF_2 groups mimic the lactone C=O groups of the naturally-occurring convulsant picrotoxin, which acts on the same gamma-aminobutyric acid receptor, *i.e.* subtype A ($GABA_A$) [205]. Until structure-activity relationships are available for the fluorinated norbornanes and norbornenes, great care should be taken in any experimental work in this area.

Other potent antagonists of the $GABA_A$ receptor are 1-(fluorinated phenyl)-4-substituted bicyclo-orthocarboxylates (Fig. 29.3) [206]. Their biological activity depends on the location and number of fluorine atoms on the phenyl ring, with high toxicity being obtained by *ortho* or *para* fluorination (as in **69** and **70** respectively); the corresponding hydro or chloro compounds are of reduced toxicity. Perfluorination results in a slight fall in potency (see **71**), while a *para* trifluoromethyl group causes it to be lost completely (see **72**). The nature of the 4-bridgehead group is also important, with toxicity generally decreasing in the order *t*-butyl > phenyl > cyclohexyl [206].

Conclusion

Several classes of fluorinated compounds rank among the most toxic chemicals known. In these materials fluorine acts in one or more of three ways: (1) as an atom which mimics hydrogen but does not participate in a vital enzymic reaction, *e.g.* as with fluoroacetates; (2) as an activated leaving group which either permits compounds to react with a

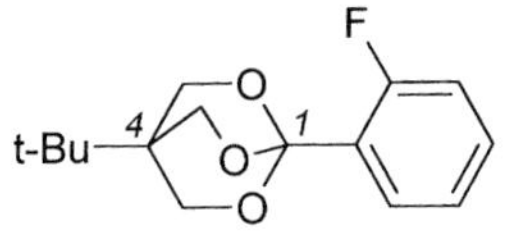

(69)
0.7 mg/kg

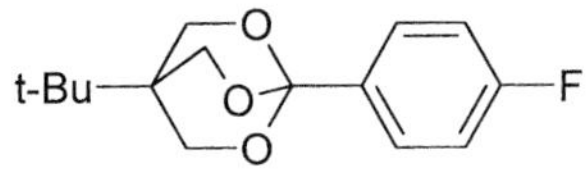

(70)
0.8 mg/kg

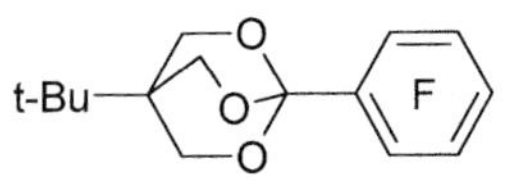

(71)
1.1 mg/kg

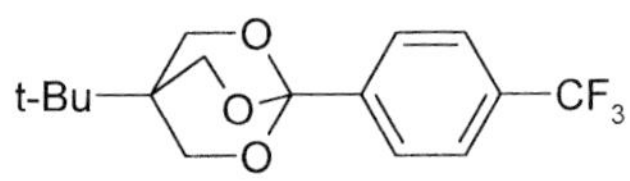

(72)
38 mg/kg

Fig. 29.3. Toxic fluorinated bicyclo-orthodicarboxylates and their LD_{50} values (ip mice).

vital enzyme, *e.g.* as with nerve agents, or facilitates single or multiple attack by biological thiols, *e.g.* as with fluoroalkenes; (3) as a typical halogen providing a stable lipophilic group attached to an aliphatic or aromatic nucleus, *e.g.* as with cage convulsants. The greatest revelation is the profound changes in biological properties caused by the introduction of atoms of fluorine into a hydrocarbon skeleton.

During World Wars I and II, intensive research on toxic chemicals was carried out in many of the belligerent countries. Much of this research had, and still has, a rather restricted interest, but the studies on fluoro-organic compounds, initiated in the UK by scientists at Porton Down have had broad consequences. Not only have these fluorine compounds influenced biochemistry, toxicology and even clinical medicine (in the case of prophylaxis and therapy), but they have achieved industrial importance in pest control. The development of systemic insecticides, arising from independent wartime work by Saunders in England and Schrader in Germany, is an exciting story that is not as well known as it deserves to be. The secrecy of some of the early work at Porton, in collaboration with the institutions listed in Table 29.1, has previously prevented its disclosure; the present review serves to remedy this.

Recent work at Porton has assessed the hazards to UK troops and civilian personnel from other toxic fluorine compounds (*e.g.* fluoroalkenes). The research carried out on perfluoroisobutene by the author and his many colleagues led to the discovery of its mode of action and that of related fluoroalkenes. The visualisation that toxicity was related to their alkylating abilities, which explained results that could not be previously rationalised, was an exhilarating moment for the author. However, there is still a need for information on metabolism and other biological mechanisms of many of the materials covered in this review. Whilst the modes of action of some compounds are well understood (*e.g.* fluoroacetates, nerve agents and fluorinated carbamates), those of the fluorinated ethers, fluorinated organosulfur compounds, and the cage convulsants await clarification during the next millennium. Scientists at Porton Down will continue to study highly-toxic fluorinated compounds with a view to developing defensive measures against them. It should, however, be pointed out that an antidote for scientists fascinated by fluorine has yet to be discovered!

Acknowledgements

I am very grateful to my wife, Helen, for her constant encouragement and patience. I also wish to thank Porton colleagues, Robin Black, Gradon Carter, Rick Hall, Ian Lawston and David Upshall for useful discussions, and Eric Banks for his invitation to write this article.

References

1 *Chemistry in Britain, 24* (1988) 657-688.
2 G. B. Carter, *Porton Down: 75 Years of Chemical and Biological Research*, HMSO, London, 1992.
3 G. B. Carter and B. Balmer, 'Chemical and Biological Warfare and Defence', in *Cold War, Hot Science: Applied Research in the UK's Defence Research Laboratories, 1945-90*, R. Bud and P. Gummett (eds.), Harwood Academic Publishers, 1999, in press.
4 R. M. Black and G. S. Pearson, *Chemistry in Britain, 29* (1993) 584.
5 D. Van Biema, *Time Magazine*, 3 April (1995) 21.
6 K. Ulm, 'Toxicity', in Houben-Weyl (B. Baasner, H. Hagemann and J. C. Tatlow (eds.), *Organofluorine Compounds, Vol. E10*), Georg Thieme Verlag, Stuttgart, 1999, Ch. 5, pp. 33-58.
7 F. L. M. Pattison, *Toxic Aliphatic Fluorine Compounds*, Elsevier, London, 1959.
8 B. C. Saunders, *Some Aspects of the Chemistry and Toxic Action of Organic Compounds containing Phosphorus and Fluorine*, Cambridge University Press, 1957.
9 R. M. Black, J. M. Harrison, 'The Chemistry of Organophosphorus Chemical Warfare Agents', in F. R. Hartley (ed.), *The Chemistry of Organophosphorus Compounds*, Vol. 4, John Wiley & Sons, Chichester, 1996, Ch. 10, pp. 781-840.
10 Unpublished data from Porton archives.
11 J. W. Clayton, *J. Occup. Med., 4* (1962) 262.
12 M. L. Richardson (ed.), *The Dictionary of Substances and their Effects*, Royal Society of Chemistry, Clays Ltd., St. Ives Plc., 1993 and references therein.
13 G. Kimmerle, *Arch. Toxicol., 18* (1960) 140.
14 E. H. Vernot, J. D. MacEwen, C. C. Haun and E. R. Kinkead, *Toxicol. Appl. Pharmacol., 42* (1977) 417.
15 S. Munavalli, D. I. Rossman, D. K. Rohrbaugh, C. P. Ferguson and L. Buettner, *J. Fluorine Chem., 65* (1993) 15.
16 I. L. Knunyants and G. G. Yacobson (eds.), *Syntheses of Fluoroorganic Compounds*, Springer-Verlag, New York, 1985.
17 M. B. Chenoweth, *Pharmacol. Rev., 1* (1949) 383.
18 F. Swarts, *Bull. Acad. Roy. Belg., 31* (1896) 675.
19 E. Gryszkiewicz-Trochimowski, *Rec. Trav. Chim., 66* (1947) 427, 430.
20 E. Gryszkiewicz-Trochimowski, A. Sporzynski and J. Wnuk, *Rec. Trav. Chim., 66* (1947) 413, 419.
21 G. Schrader, 'The Development of New Insecticides and Chemical Warfare Agents', in British Intelligence Objectives Sub-Committee (BIOS) Final Report 714 (1945), Item No. 8 (reported by S. A. Mumford and E. A. Perren of the Chemical Defence Experimental Station, Porton Down).
22 *The Pesticide Manual*, British Crop Protection Council, Worcestershire, *1* (1968) 287.
23 D. E. Moss, H. Kobayashi, G. Pacheco, R. Palacios and R. G. Perez, *Current Res. Alzheimer Ther. Cholinesterase Inhib.*, Taylor and Francis, New York, 1988, p. 305.
24 D. E. Fahrney and A. M. Gold, *J. Am. Chem. Soc., 85* (1963) 997.
25 B. C. Saunders and G. J. Stacey, *J. Chem. Soc.*, (1948) 1773.
26 B. C. Saunders, G. J. Stacey and I. G. E. Wilding, *J. Chem. Soc.*, (1949) 773.
27 F. J. Buckle, R. Heap and B. C. Saunders, *J. Chem. Soc.*, (1949) 912.
28 F. J. Buckle, R. Heap and B. C. Saunders, *J. Chem. Soc.*, (1949) 1471.
29 H. McCombie and B. C. Saunders, *Nature, 158* (1946) 382.
30 B. C. Saunders, *Endeavour, 19* (1960) 36.
31 Registry of toxic effects of chemical substances (RTECS).

32 M. M. Airaksinen and T. Taministo, *Ann. Med. Exp. Biol. Fenn., 46* (1968) 242; *Chem. Abs., 69* (1968) 65967.

33 J. S. C. Marais, *J. Vet. Sci. Animal Ind., 20* (1944) 67.

34 P. B. Oelrichs and T. McEwan, *Nature, 190* (1961) 808.

35 A. Tosaki and D. J. Hearse, *Basic Res. Cardiol., 83* (1988) 158.

36 R. A. Peters, R. J. Hall, P. F. V. Ward and N. Sheppard, *Biochem. J., 77* (1960) 17, 22.

37 P. Buffa, R. A. Peters and R. W. Wakelin, *Biochem. J., 48* (1951) 467.

38 R. A. Peters, R. W. Wakelin, P. Buffa and L. C. Thomas, *Proc. Roy. Soc. B, 140* (1953) 497.

39 F. L. M. Pattison and R. A. Peters, in F. A. Smith (ed.), *Handbook of Experimental Pharmacology*, Vol. XX/1, *Pharmacology of Fluorides*, Part 1, Springer-Verlag, New York, 1966, Ch. 8, pp. 387-458.

40 E. Kun, E. Kirsten and M. L. Sharma, *Proc. Natl. Acad. Sci. USA, 74(11)* (1977) 4942.

41 E. Kirsten, M. L. Sharma and E. Kun, *Mol. Pharmacol., 14* (1977) 172.

42 D. D. Clarke, *Neurochem. Res., 16* (1991) 1055.

43 F. L. M. Pattison, S. B. D. Hunt and J. B. Stothers, *J. Org. Chem., 21* (1956) 883.

44 M. S. Raasch, *J. Org. Chem., 23* (1958) 1567.

45 F. L. M. Pattison and J. J. Norman, *J. Am. Chem. Soc., 79* (1957) 2311.

46 I. G. Walker and J. M. Parker, *Can. J. Biochem. Physiol., 36* (1958) 339.

47 P. H. Lieder and D. A. Keller, *Chem. & Eng. News*, 11 May (1992) 2.

48 J. M. Parker and I. G. Walker, *Can. J. Biochem. Physiol., 35* (1957) 407.

49 B. C. Saunders and G. J. Stacey, *J. Chem. Soc.*, (1949) 916.

50 B. C. Saunders, *J. Chem. Soc.*, (1949) 1279.

51 B. C. Saunders, in M. Stacey, J. C. Tatlow, A. G. Sharpe (eds.), *Adv. in Fluorine Chem.*, Vol. 2, Butterworths, London, 1961, p. 183.

52 W. W. Tourtellotte and J. M. Coon, *J. Pharmacol. Expt., 101* (1951) 82.

53 J. O. Hutchens, H. Wagner, B. Podolsky and T. M. McMahon, *J. Pharmacol. Expt. Ther., 95* (1949) 62.

54 M. B. Chenoweth, *J. Pharmacol. Exp. Ther. II, 97* (1949) 383.

55 B. Tecle and J. E. Casida, *Chem. Res. Toxicol., 2* (1989) 429.

56 R. A. Peters, *Adv. in Enzymol. Relat. Subjects Biochem.*, Vol. XVIII, Interscience Publishers Inc., New York, 1957, p. 113.

57 M. B. Chenoweth, A. Kandel, L. B. Johnson and D. R. Bennett, *J. Pharm. Expt. Ther., 102* (1951) 31.

58 S. Gitter, *Biochem. J., 63* (1956) 182.

59 L. C. Hendershot and M. B. Chenoweth, *J. Pharm. Expt. Ther., 113* (1955) 160.

60 A. Kandel and M. B. Chenoweth, *J. Pharm. Expt. Ther., 104* (1952) 234.

61 K. Gregg, B. Hamdorf, K. Henderson, J. Kopecny and C. Wong, *Appl. Environ. Microbiol., 64* (1998) 3496.

62 S. Franke, *Textbook of Military Chemistry*, 2nd Edition, Vol. 1, Military Publishing House, Berlin, 1977.

63 R. F. Hudson and L. Keay, *J. Chem. Soc.*, (1960) 1859.

64 W. Lange and G. von Krüger, *Chem. Ber., 65* (1932) 1598.

65 B. C. Saunders, 'The Chemistry and Toxicology of Organic Fluorine and Phosphorus Compounds', The Royal Institute of Chemistry, Lectures, Monographs and Reports, No. 1, 1953.

66 E. D. Adrian, W. Feldberg and B. A. Kilby, *Brit. J. Pharmacol., 2* (1947) 56.

67 J. F. Mackworth and E. C. Webb, *Biochem. J., 42* (1948) 91.

68 H. McCombie and B. C. Saunders, *Nature, 157* (1946) 776.

69 H. McCombie and B. C. Saunders, *Nature, 157* (1946) 287.

70 B. C. Saunders and G. J. Stacey, *J. Chem. Soc.*, (1948) 695.

71 H. McCombie, B. C. Saunders and C. L. Wheeler, *British Patent 601,210* (1948).

72 B. C. Saunders and T. S. Worthy, *J. Chem. Soc.*, (1950) 1320.

73 J. C. Boursnell and E. C. Webb, *Nature, 164* (1949) 875.

74 H. G. Cook, B. C. Saunders and F. E. Smith, *J. Chem. Soc.*, (1949) 635.

75 B. C. Saunders, G. J. Stacey, F. Wild and I. G. E. Wilding, *J. Chem. Soc.*, (1948) 699.

76 N. B. Chapman and B. C. Saunders, *J. Chem. Soc.*, (1948) 1010.

77 B. C. Saunders and G. J. Stacey, *J. Chem. Soc.*, (1949) 919.

78 R. Heap and B. C. Saunders, *J. Chem. Soc.*, (1949) 2983.

79 B. C. Saunders, *J. Chem. Soc.*, (1950) 684.

80 D. R. Davies, P. Holland and M. J. Rumens, *Brit. J. Pharmacol., 15* (1960) 271.

81 A. H. Ford-Moore, L. J. Lermit and C. Stratford, *J. Chem. Soc.*, (1953) 1776.

82 R. Heap and B. C. Saunders, *J. Chem. Soc.*, (1948) 1313.
83 D. R. Davies, P. Holland and M. Rumens, *Biochem. Pharmacol., 15* (1966) 1783.
84 J. Amos, R. A. Chittenden, G. H. Cooper and R. L. Rickard, *Phosphorus, 6* (1975) 35.
85 A. H. Ford-Moore and B. J. Perry, *Org. Synth., 31* (1951) 33.
86 P. J. R. Bryant, A. H. Ford-Moore, B. J. Perry, A. W. H. Wardrop and T. F. Watkins, *J. Chem. Soc.*, (1960) 1553.
87 R. H. Rohrbaugh, P. C. Jurs, W. P. Ashman, E. G. Davis and J. H. Lewis, *Chem. Res. Toxicol., 1* (1988) 123.
88 M. Maksimovic and M. Stevanovic, *Acta Pharm. Jugosl., 25* (1975) 19.
89 R. Ghosh, *British Patent 738,839* (1955).
90 J. I. G. Cadogan and L. C. Thomas, *J. Sci. Instrum., 37* (1960) 60.
91 T. R. Fukuto and E. M. Stafford, *J. Am. Chem. Soc.*, (1957) 6083.
92 R. D. O'Brian and B. D. Hilton, *J. Agr. Food Chem., 12(1)* (1964) 53.
93 A. Bebbington and R. V. Ley, *J. Chem. Soc.*, (1966) 1410.
94 L. E. Tammelin, *Acta Chem. Scand., 11* (1957) 859.
95 L. E. Tammelin, *Ark. Kemi., 12* (1958) 287.
96 J. Bajgar, J. Fusek, V. Hrdina, J. Patocka and J. Vachek, *Physiol. Res., 41* (1992) 399.
97 D. R. Davies and A. L. Green, *Brit. J. Industr. Med., 16* (1959) 128.
98 W. K. Berry, *Proc. Roy. Soc. Med., 46* (1953) 801.
99 W. K. Berry, K. P. Fellowes, P. J. Fraser, J. P. Rutland and A. Todrick, *Biochem. J., 59* (1955) 1.
100 A. L. Green and B. Saville, *J. Chem. Soc.*, (1956) 3887.
101 B. M. Askew, *Brit. J. Pharmacol., 11* (1956) 417.
102 A. L. Green, G. L. Sainsbury, B. Saville and M. Stansfield, *J. Chem. Soc.*, (1958) 1583.
103 A. F. Childs, D. R. Davies, A. L. Green and J. P. Rutland, *Brit. J. Pharmacol., 10* (1955) 462.
104 D. R. Davies and A. L. Green, *Disc. Faraday. Soc., 20* (1955) 269.
105 R. Holmes and E. L. Robins, *J. Brit. Pharmacol., 10* (1955) 490.
106 D. R. Davies, A. L. Green and G. L. Willey, *Brit. J. Pharmacol., 14* (1959) 5.
107 W. S. S. Ladell, *Brit. Med. J., 2* (1958) 141.
108 R. V. Brown, *J. Physiol., 149* (1959) 49-50P.
109 R. V. Brown, *Brit. J. Pharmacol., 15* (1960) 170.
110 N. H. Creasey and A. L. Green, *J. Pharm. Pharmacol., 11* (1959) 485.
111 R. E. Banks, R. N. Haszeldine and I. M. Young, unpublished results (1966).
112 B. M. Askew, *Brit. J. Pharmacol., 12* (1957) 340.
113 A. L. Green and H. J. Smith, *Biochem. J., 68* (1958) 28, 32.
114 J. P. Rutland, *Brit. J. Pharmacol., 13* (1958) 399.
115 B. M. Askew, D. R. Davies, A. L. Green and R. Holmes, *Brit. J. Pharmacol., 11* (1956) 424.
116 B. M. Askew, D. R. Davies and A. L. Green, *Biochem. J., 66* (1957) 43P.
117 D. R. Davies and G. L. Willey, *Brit. J. Pharmacol., 13* (1958) 202.
118 B. W. Ford and P. Watts, *J. Chem. Soc., Perkin Trans. II*, (1974) 1009.
119 B. Harvey, D. J. Sellers and P. Watts, *Biochem. Pharmacol., 33* (1984) 3499.
120 B. Harvey, R. P. Scott, D. J. Sellers and P. Watts, *Biochem. Pharmacol., 35* (1986) 737, 745.
121 D. R. Davies and A. L. Green, *Biochem. J., 63* (1956) 529.
122 F. Hobbiger, *Brit. J. Pharmacol., 11* (1956) 295.
123 B. J. Jandorf, H. O. Michel, N. K. Schaffer, R. Egan and W. H. Sommerson, *Disc. Faraday Soc., 20* (1955) 134.
124 W. K. Berry and D. R. Davies, *Biochem. J., 100* (1966) 572.
125 W. K. Berry, *Biochem. Pharm., 20* (1971) 1333.
126 J. Bajgar, J. Fusek and V. Vachek, *ASA Newsletter*, 94-4, 10.
127 J. E. Casida, K. B. Augustinsson and G. Jonsson, *J. Econ. Entomol., 53* (1960) 205.
128 I. B. Wilson, M. A. Harrison and S. Ginsburg, *J. Biol. Chem., 236* (1961) 1498.
129 A. R. Sellakumar, S. Laskin, M. Kuschner, G. Rusch, G. V. Katz, C. A. Snyder and R. E. Albert, *J. Environ. Path. Toxicol., 4* (1980) 107.
130 J. W. Clayton, in F. A. Smith (ed.), *Handbook of Experimental Pharmacology*, Vol. XX/1, *Pharmacology of Fluorides, Part 1*, Springer-Verlag, New York, 1966, Ch. 9, pp. 459-500.
131 W. J. Middleton and R. V. Lindsey, *J. Am. Chem. Soc., 86* (1964) 4948.
132 C. G. Krespan and W. J. Middleton, *Fluorine Chem. Rev., 1(1)* (1967) 145.

133 E. R. Larsen, *Fluorine Chem. Rev., 3* (1969) 1.
134 A. Van Posnak and J. F. Artusio, *Toxicol. Appl. Pharmacol., 2* (1960) 374.
135 G. Lu, S. L. Johnson, M. S. Ling and J. C. Krantz, *Anaesthesiology, 14* (1953) 466.
136 J. C. Krantz, F. G. Rudo, in F. A. Smith (ed.), *Handbook of Experimental Pharmacology*, Vol. XX/1, *Pharmacology of Fluorides, Part 1*, Springer-Verlag, New York, 1966, Ch. 10, pp. 501-564.
137 T. H. S. Burns, J. M. Hall, A. Bracken, G. Gouldstone and D. S. Newland, *Anaesthesia, 16* (1961) 3.
138 J. A. Ruzicka, S. D. Pedersen and M. T. Baker, *J. Fluorine Chem., 76* (1996) 213.
139 P. E. Aldrich and W. A. Sheppard, *J. Org. Chem., 29* (1964) 11.
140 J. C. Krantz, E. B. Truitt, A. S. C. Ling and A. A. Kurland, *Toxicol. Appl. Pharmacol., 1* (1959) 362.
141 J. C. Krantz, A. Esquibel, E. B. Truitt and A. S. C. Ling, *JAMA, 166* (1958) 1555.
142 J. W. Clayton, *Fluorine Chem. Rev., 1* (1967) 197.
143 G. L. Kennedy, *Crit. Rev. Toxicol., 21* (1990) 149.
144 R. E. Banks, B. E. Smart and J. C. Tatlow (eds.), *Organofluorine Chemistry: Principles and Commercial Applications*, Plenum Press, London, 1994.
145 J. Odum and T. Green, *Toxicol. Appl. Pharmacol., 76* (1984) 306.
146 S. Vamvakas, E. Kremling and W. Dekant, *Biochem. Pharm., 38* (1989) 2297.
147 M. Koob and W. Dekant, *Drug Metab. Dispos., 18* (1990) 911.
148 M. B. Finkelstein, R. B. Baggs and M. W. Anders, *J. Pharm. Exp. Therap., 261* (1992) 1248.
149 S. J. Hargus, M. E. Fitzsimmons, Y. Aniya and M. W. Anders, *Biochem., 30* (1991) 717.
150 D. R. Dohn, A. J. Quebbemann, R. F. Borch and M. W. Anders, *Biochem., 24* (1985) 5137.
151 L. Jin, M. R. Davis, E. D. Kharasch, G. A. Doss and T. A. Baillie, *Chem. Res. Toxicol., 9* (1996) 555.
152 H. Wark, J. Earl, D. D. Chau, J. Overton and H. T. A. Cheung, *Brit. J. Anaesth., 64* (1990) 469.
153 T. Green and J. Odum, *Chem.-Biol. Interactions, 54* (1985) 15.
154 M. B. Finkelstein, W. Dekant, A. S. Kende and M. W. Anders, *J. Am. Chem. Soc., 117* (1995) 9590.
155 E. A. Lock and J. Ishmael, *Arch. Toxicol., 72* (1998) 347.
156 J. N. M. Commandeur, R. A. J. Oostendorp, P. R. Schoofs, B. Xu and N. P. E. Vermeulen, *Biochem. Pharm., 36* (1987) 4229.
157 J. W. Harris, W. Dekant and M. W. Anders, *Chem. Res. Toxicol., 5* (1992) 34.
158 M. W. Anders and W. Dekant, *Annu. Rev. Pharmacol. Toxicol., 38* (1998) 501 and references therein.
159 I. L. Knunyants, Yu. V. Zeifman, E. G. Ter-Gambrielyan and N. P. Gambaryan, *Usp. Khim., 53* (1984) 431 [see *Russian Chem. Rev., 53* (1984) 431].
160 L. W. Smith, R. J. Gardner and G. L. Kennedy, *Drug and Chem. Toxicol., 5* (1982) 295.
161 B. D. Karpov, *Tr. Leningrad San. Gig. Med. Inst., 111* (1975) 30.
162 G. Paulet and G. P. Bernard, *Biol. Med. (Paris), 57* (1968) 247.
163 H. Arito and R. Soda, *Ann. Occup. Hyg., 20* (1977) 247.
164 R. E. Banks, *J. Fluorine Chem., 67* (1994) 193.
165 I. D. Makulova, *Gig. Trud. Prof. Zabol., 9* (1965) 20.
166 E. W. Cook and J. S. Pierce, *Nature*, 30 March (1973) 337.
167 K. Ulm, *J. Fluorine Chem., 35* (1987) 255.
168 H. Reichelt, *Z. Ges. Hyg., 30* (1984) 204.
169 M. P. Maidment and D. G. Upshall, *J. Appl. Toxicol., 12(6)* (1992) 393.
170 R. F. R. Brown and P. Rice, *Int. J. Exp. Path., 72* (1991) 437.
171 A. I. Korbakova, I. D. Makulova, Y. N. Marchenko and T. K. Mikitenko, in *Toxicology of Organofluorine Compounds and Industrial Hygiene in their Production*, Meditsina, Moscow, 1975.
172 F. J. Pavlik, *US Patent 3,385,904* (1968).
173 I. L. Knunyants, V. V. Shokina, V. V. Tyuleneva, Yu. A. Cheburkov and Yu. E. Aronov, *Izv. Akad. Nauk. SSSR, Ser. Khim.*, (1966) 1831.
174 I. L. Knunyants, V. V. Shokina and V. V. Tyuleneva, *Izv. Akad. Nauk. SSSR, Ser. Khim.*, (1968) 415.
175 C. R. Hall, I. W. Lawston and A. M. Tinsley, *Chem. and Ind. (London)*, (1989) 145.
176 A. F. Lailey, L. Hill, I. W. Lawston, D. Stanton and D. G. Upshall, *Biochem. Pharmacol., 42* (1991) S47.
177 M. J. Hobbs, M. Butterworth, G. M. Cohen and D. G. Upshall, *Biochem. Pharmacol., 45* (1993) 1605.
178 C. M. Timperley, *J. Fluorine Chem., 94* (1999) 37.
179 D. Mansuy, P. Beaune, T. Cresteil, M. Lange and J. P. Leroux, *Biochem. Biophys. Res. Commun., 79* (1977) 513.
180 T. Nash and R. E. Pattle, *Ann. Occup. Hyg., 14* (1971) 227.

181 E. N. Cohen, H. W. Brewer, J. W. Bellville and R. Sher, *Anaethesiology, 26* (1965) 140.
182 V. V. Tyuleneva, L. A. Rozov, Y. V. Zeifman and I. L. Knunyants, *Izv. An. SSSR, Ser. Khim., 5* (1975) 1042 (English transl.).
183 I. W. Lawston, M. W. Smith, N. Stansfield and D. Stanton, *J. Fluorine Chem., 54* (1992) 170.
184 A. J. Bell, C. J. Hayhurst, C. A. Mayhew and P. Watts, *Int. J. Mass. Spec. Ion Processes., 140* (1994) 133.
185 T. H. S. Burns, J. M. Hall, A. Bracken and G. Gouldstone, *Anaesthesia, 19(2)* (1964) 167.
186 T. H. S. Burns, J. M. Hall, A. Bracken, G. Gouldstone and D. S. Newland, *Anaesthesia, 16* (1961) 3.
187 M. P. Maidment, P. Rice and D. G. Upshall, *J. Appl. Toxicol., 14* (1994) 395.
188 M. R. Bryce, R. D. Chambers and G. Taylor, *J. Chem. Soc. Perkin I*, (1984) 509.
189 J. D. Park, R. J. McMurtry and J. H. Adams, *Fluorine Chem. Rev., 2* (1968) 55.
190 M. P. Maidment, S. K. Patel, D. G. Upshall and C. M. Timperley, *J. Appl. Toxicol., 19* (1999) 113.
191 P. B. Sargeant and C. G. Krespan, *J. Am. Chem. Soc., 91* (1969) 415.
192 R. L. Soulen and D. W. Paul, *J. Fluorine. Chem., 10* (1977) 261.
193 M. J. Hobbs, N. E. Williams, S. K. Patel and D. G. Upshall, *Biochem. Pharmacol., 55* (1998) 1573.
194 A. F. Lailey and D. G. Upshall, *Human & Exp. Toxicol., 13* (1994) 776.
195 A. F. Lailey, *Human & Exp. Toxicol., 16* (1997) 212.
196 A. F. Lailey, *Human & Exp. Toxicol.*, (1999) in press.
197 R. E. A. Dear and E. E. Gilbert, *US Patent 3505459* (1968).
198 S. Munavalli, A. J. Muller, D. I. Rossman, D. K. Rohrbaugh and C. P. Ferguson, *J. Fluorine Chem., 67* (1994) 37.
199 S. Munavalli, D. I. Rossman, D. K. Rohrbaugh and C. P. Ferguson, *J. Fluorine Chem., 61* (1993) 147.
200 J. B. Nold, J. P. Petrali, H. G. Wall and D. H. Moore, *Inhal. Toxicol., 3* (1991) 123.
201 D. M. Stavert, D. Archuleta and B. E. Lehnert, *The Toxicologist, 11* (1991) 104.
202 W. J. Middleton and E. M. Bingham, *J. Fluorine Chem., 20* (1982) 397.
203 V. I. Fetisov, A. A. Maslov, V. A. Panarin and N. V. Trefilov, *J. Fluorine Chem., 58* (1992) 366.
204 V. I. Fetisov, A. A. Maslov and N. V. Trefilov, *J. Fluorine Chem., 58* (1992) 195.
205 V. I. Fetisov, *J. Fluorine Chem., 58* (1992) 199.
206 J. E. Casida, C. J. Palmer and L. M. Cole, *Mol. Pharmacol., 28* (1985) 246.

BIOGRAPHIC NOTE

Chris Timperley was born in Cheshire, England in 1970. He received a BSc degree from the University of Sheffield in 1991 and a PhD degree from the University of Newcastle upon Tyne in 1995. In 1996, he joined the Chemical and Biological Defence Establishment at Porton Down, where, as a Scientific Leader, he manages a small team of synthetic organic chemists. His research interests include all aspects of phosphorus chemistry, the reactions of fluoroalkenes with nucleophiles, and the toxicology of fluorinated molecules.

Chris Timperley

Chapter 30

FROM COMPLEX FLUORIDES TO CFC ALTERNATIVES – AN ACCOUNT OF FLUORINE CHEMISTRY AT GLASGOW

JOHN M. WINFIELD

Department of Chemistry, University of Glasgow, Glasgow G12 8QQ, UK

Introduction

A. G. Sharpe

The author's first significant encounter with an inorganic fluoride occurred in 1958, not in a laboratory or lecture theatre, but during an inorganic chemistry tutorial given to four or five young first-year undergraduate students by David Sharp, who at that time was a junior member of staff at Imperial College, London. The chemical in question was BrF_3, home-made and contained in a mild steel pot. David Sharp occupied a small lab-office (the organisation of which left something to be desired!) in the Chemistry Department at I.C., and the BrF_3 container was inadvertently disturbed by one of us during a tutorial; luckily it did not fall over, otherwise this account would probably not have been written! Shortly afterwards, in 1961, I was one of several postgraduate students who moved with David from London to Glasgow, to what was then the Royal College of Science and Technology – a well-respected institution which became Strathclyde University[1] in 1964. The following account describes some of the fluorine chemistry carried out in the City of Glasgow that, at least in my opinion, can reasonably be described as 'landmarks'. Where appropriate, I have placed this work in context by reference to earlier studies carried out either at Cambridge University, where David Sharp was a student, or at Imperial College.

All UK fluorine chemists of the author's vintage have been influenced profoundly by the famous fluorine school founded by H.J. Emeléus[2] at Cambridge University in the

[1] Strathclyde was one of the ancient Scottish Kingdoms and was used to name the region of local government covering the Western part of Scotland.

[2] Harry Julius Emeléus, CBE, FRS (1903-1993).

immediate post-World War II period.[3,4] In particular the work of A. G. (Alan) Sharpe, David Sharp's PhD supervisor at Cambridge, exemplifies the importance of synthesis coupled with physico-chemical studies. This has been a recurring principle in much of the Glasgow work, which has regularly involved 'physical' and 'organic' chemistry as well as 'inorganic'. A thematic approach has been adopted in this account, and while the work is described in an approximate chronological order, the overlaps, both in time and location, are considerable.

Complex fluorides

The first general synthetic route to ternary fluoride complexes was developed by the Cambridge School and reported by Emeléus, Sharpe and Woolf in a series of papers published in the late '40s and early '50s [1-7]. Solvolysis of halides or oxides in liquid BrF_3 was rationalised [2] on the basis of the postulated self-ionization $2BrF_3 \rightleftarrows [BrF_2]^+ + [BrF_4]^-$ with acid-base reactions, for example, between $A[BrF_4]$, A = K or Ag, and $[BrF_2][SbF_6]$ or $[BrF_2]_2[MF_6]$, M = Sn [2] or Ti [4], leading to complex fluoride or oxo-fluoride products. A large number of ternary fluorides, particularly those containing d-block metals have been prepared in this way. In some cases, decomposition of $[BrF_2]_2[MF_6]$ adducts yields the parent tetrafluoride, *e.g.* PtF_4 [4], but preparation of binary fluorides by this route may result in impure products that contain bromine. The oxidizing power (sometimes uncontrollable) of liquid BrF_3 is illustrated by its reaction with a silver-gold metallic mixture to give $Ag[AuF_4]$ [3], an example that the author has sometimes found useful as an 'unseen' problem question in undergraduate examinations.

Despite its disadvantages, the BrF_3 route to ternary fluorides, A_2BF_6 or ABF_6, remained popular for a considerable period. Reagents that were used as later alternatives include SF_4 [8], particularly useful for mixtures of Group 1 fluorides with metal oxides, sulfides or carbonyls. A third general route involves the reduction of d-block hexafluorides with iodide anion in liquid SO_2, a route exploited by R. D. Peacock and co-workers which has been widely applied.

Although a brief collaboration between A. G. Sharpe and R. S. (Ron) Nyholm resulted in an early magnetochemical study of binary and ternary d-block fluorides [9], the main physical investigation of these compounds involved the determination of lattice parameters in the series A_2BF_6 and ABF_6 by X-ray powder diffraction. Early mapping of the A_2BF_6, A = Group 1 cation, B^{IV} = Si, Ge, Ti, Mn, Ni, Pd and Pt, [10, 11] by Cox and Sharpe was followed by an investigation of ABF_6 structural types [12] and culminated in an exhaustive study of $A^IB^VF_6$ compounds, work carried out by R. D. W. Kemmitt, D. R. Russell and D. W. A. Sharp at Imperial College and Strathclyde [13]. Further XRD

[3] Note that Emeléus' influence spread globally. The tradition of enhancing the activities of the inorganic school through overseas research fellows and other visiting researchers that he established in the very early days at Cambridge, continued for many years. Those who contributed to fluorine chemistry included Ang How Ghee (Singapore), R. G. Cavell (Canada), H. C. Clark (Canada), A. F. Clifford (USA), W. I. Cullen (New Zealand), J. Grobe (Germany), A. Haas (Germany), A. G. MacDiarmid (USA), T. A. O'Donnell (Australia), R. C. Paul (India), J. M. Shreeve (USA), L. E. Smythe (Australia), E. G. Wazaschewski (Poland) and S. M. Williamson (USA).

[4] See Appendix 30.1 for A. A. (Alf) Woolf's memoir on those early days.

TABLE 30.1
Structures adopted by $A^{I}B^{V}F_6$ complex fluorides,[a,b] reproduced with permission from ref. [13]

B^{V}	A^{I}						
	Li	Na	Ag	K	Tl	Rb	Cs
P	R_1	C_1	C_2	C_2, R_2	C_2	C_2	C_2
As	R_1	R_1	C_2	R_2	R_2	R_2	R_2
V	R_1	R_1	T	R_2	R_2	R_2	R_2
Ru	R_1	R_1	T	R_2	R_2	R_2	R_2
Ir	R_1	R_1	T	R_2		R_2	R_2
Os	R_1	R_1	T	R_2		R_2	R_2
Re	R_1	C_1		T		R_2	R_2
Mo	R_1	C_1		T	R_2	R_2	R_2
W	R_1	C_1		T		R_2	R_2
Sb	R_1	C_1	T	T	R_2	R_2	R_2
Nb	R_1	C_1	T	T	R_2	R_2	R_2
Ta	R_1	C_1	T	T	R_2	R_2	R_2

[a] C_1 = cubic $NaSbF_6$ structure; C_2 = cubic $CsPF_6$ structure; R_1 = rhombohedral $LiSbF_6$ structure; R_2 = rhombohedral $KOsF_6$ structure; T = tetragonal $KNbF_6$ structure.
[b] Ordering of A^{I} (l → r) and B^{V} (t → b) according to increasing size.

powder work [14] on some $Na_2B^{IV}F_6$ compounds, extending the earlier study [11], followed shortly afterwards. As a result of these papers, appearing over the period 1953-65, the crystal chemistry of ternary fluorides, particularly those containing d-block elements, was placed on a secure basis. The work, most of which is still regarded as correct, has informed subsequent studies of the spectra of these compounds and a selection of the results for $A^{I}B^{V}F_6$ complexes is presented in Table 30.1. It is ironic that this type of very detailed physico-chemical examination which had no obvious immediate 'use' would probably be very difficult to carry out in 1998 because of present UK funding policy.

Reliable recording CW IR spectrometers became available commercially during the mid-1950s and this spectroscopic method attracted David Sharp's attention. He applied the technique to various weakly-basic, fluorine-containing anions such as $[SO_3F]^-$, $[BF_4]^-$ and hexafluorometallates(V), an interesting example being $[Ph_3C]^+$ salts of these anions in which a planar, propeller-like D_3 configuration was established for the cation [15]. A paper co-authored by R. D. (Ray) Peacock and Sharp from Imperial College proved to be a classic, judged by the number of citations it received subsequently, and was one of the early attempts to apply simple symmetry considerations to the IR spectra of a variety of O_h and T_d fluorometallate anions in the solid state [16]. Subsequent work, at Strathclyde, treated IR spectra (now routinely available down to 250 cm^{-1}) in terms of site symmetry [17] and factor group [18] analyses and culminated in a detailed spectral examination by Andrew Lane of cubic and orthorhombic perovskite fluorides, ABF_3, including a study of the low temperature cubic → orthorhombic (or tetragonal) phase change in $KMnF_3$ [19].

This section would be incomplete without an acknowledgement of the contribution made by A. G. Sharpe to the development of a thermodynamic basis to inorganic fluorine chemistry, being applied not only to complex fluorides [20] but also to the energetics of halogen-exchange reactions involving ionic binary fluorides [21] and to the solvation of the F^- anion [22].

Substituted derivatives of high oxidation state fluorides

The move from England to Scotland in 1961 coincided with an expansion of Sharp's research interests, to include molecular chemistry of non-metal fluorides, and the period 1961-79 was characterized by a steady research output, initially from the Royal College (Strathclyde) and later from both of the chemistry departments at Glasgow and Strathclyde. However the first person in Scotland to carry out research in this area was probably D. S. Payne at Glasgow University. Douglas, a graduate of Imperial College and Cambridge where he was a student of Emeléus, was primarily a phosphorus chemist, but his interests included mixed halide derivatives [23], for example the preparation of salts such as $[PCl_4][PCl_5F]$ [24].

To some extent this new research was prompted by the increasing availability of fluorides such as SF_4 and PF_5, the former available as gifts from ICI or DuPont, the latter generated from *Phosflorogen A* (Ozark Mahoning Co.), an aryldiazonium $[PF_6]^-$ salt. A 60-Amp fluorine cell loaned by ICI was in regular use and the DuPont synthesis of SF_4, from $SCl_2 + NaF$ in acetonitrile, often employed. The advent of ^{19}F NMR spectroscopy was crucial to this work, the department taking delivery of its first spectrometer, 40 MHz with a permanent magnet, in *ca* 1963. I spent many happy (!) hours, usually in the evenings when instrument performance was more reliable, recording ^{19}F spectra of both perfluorocarbon derivatives and transition metal fluorides. My other abiding memory of the early sixties is being asked to construct a vacuum line suitable for the manipulation of volatile fluorides without the services of a professional glass- blower. As with many self-taught glass-blowers, the results were not always pretty but I was very lucky that David Russell (a 2nd year PhD student) was accomplished and he often helped me out. Our studies of substituted derivatives ranged over both p- and d-block fluorides and were carried out during the period 1961-79. A selection of the molecules prepared is contained in Fig. 30.1.

Phosphorus-fluorine chemistry received a significant boost with the arrival of George Fraser as a postgraduate student. He argued that there were analogies to be drawn between boron halides and PF_5, and proceeded to demonstrate this by synthesising secondary amine-PF_5 adducts which could be decomposed thermally to give dialkylaminofluorophosphoranes such as $PF_4(NMe_2)$ and $PF_3(NMe_2)_2$ [25]. Complex formation between PF_5 and various Lewis bases proved to be a very productive area for study; several P$-$P bonded adducts were characterized, $P(NMe_2)_3$ being an alternative reagent for the introduction of the $-NMe_2$ ligand to P^V [26]. Aminolysis of PF_5 or PF_3 using primary amines was studied subsequently with similar results [27]. A highlight from this period is work on the $(MeO)_3P \cdot PF_5$ adduct, which proved too unstable thermally to be characterized, but whose decomposition led to a rich chemistry, the main elements of which can be rationalised by the equations: $PF_5 + P(OMe)_3 \rightarrow PF_4(OMe) + PF(OMe)_2$; $2PF_4(OMe) \rightarrow (MeO)P(O)F_2 + \{Me^+PF_6^-\}$; $P(OMe)_3 + \{Me^+PF_6^-\} \rightarrow [(MeO)_3PMe][PF_6]$; $[(MeO)_3PMe]^+ \rightarrow MeP(O)(OMe)_2 + \{Me^+\}$; ref. [28]. The last reaction proceeds by a second-order autocatalytic process and corresponds to the decay of the intermediate of the Michaelis-Arbuzov rearrangement of $P(OMe)_3$. An additional product, $(MeO)(Me)P(O)F$, is believed to be formed by a similar rearrangement of $PF(OMe)_2$. An analogous investigation based on $P(OEt)_3$ was terminated abruptly after George Fraser and a fellow student, Donald Barclay, were forced to pay an

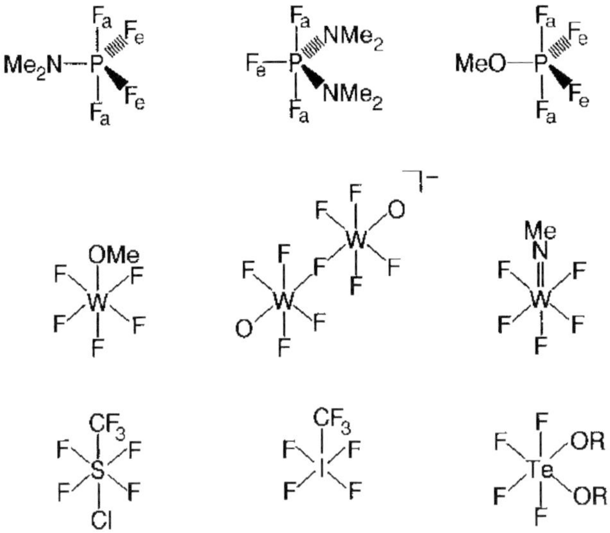

Fig. 30.1. Some of the substituted derivatives prepared at Glasgow in the '60s and '70s (identified by NMR spectroscopy).

unexpected visit to Glasgow's Royal Infirmary with symptoms attributable to nerve gas exposure.

Other P-F substituted derivatives prepared included $PF_4(SMe)$, using Me_3SiSMe [29], $F_2P(S)OP(S)F_2$ [30], $PF_4N(Me)PF_2$ [31] and $F_2(O)PN(R)P(O)F_2$ [32]. Sharp's group made a substantial contribution therefore to the explosion in phosphorus-fluorine chemistry that occurred in the '60s and early '70s, and numerous links with other fluorine groups world-wide were established.

In 1965, when I was appointed as a junior lecturer at the University of Glasgow, molecular chemistry involving d-block high-oxidation-state fluorides was relatively undeveloped, and the possibilities in this area of chemistry, using aprotic organic solvents as reaction media were immediately explored. A reaction carried out between dimethyl sulfite and WF_6, designed to prepare the unknown $(MeO)_2SF_2$ by utilising the capacity of WF_6 to form WOF_4, led instead to the isolation of WF_5OMe and very soon afterwards to most of the members of the series of methoxo-tungsten(VI) fluorides, $WF_{6-n}(OMe)_n$ [33], using Me_3Si-reagents, a route that was then in vogue for the preparation of substituted derivatives of many binary fluorides. ^{19}F NMR spectroscopy, including ^{183}W measurements made by using double resonance techniques developed by W. McFarlane and his coworkers, were crucial in making unambiguous structural assignments [34]. The other notable study in this area involved the tungsten(VI) chlorofluorides, $WF_{6-n}Cl_n$, work that was initiated by Mary Mercer[5] and Ray Peacock at Birmingham University and carried out by George Fraser and his wife Joy (née Gibbs) after Peacock had been appointed to the Chair of Inorganic Chemistry at Leicester. The two series of compounds make an interesting contrast, the chlorofluorides readily undergoing redistribution reactions in solution

[5] See Appendix 30.2 for a memoir by Mary Mercer concerning events in Birmingham.

below ambient temperature, while the methoxofluorides, which do not show this propensity, exhibit marked preferences for the *cis* configuration in $WF_4(OMe)_2$ and $WF_2(OMe)_4$. The differences were rationalised on the basis of the good π-donor properties of the MeO-ligand towards W^{VI}, an important factor also for the effect of the oxo-ligand in the NMR spectra of $[WOF_5]^-$ and the F-bridged dimer $[W_2O_2F_9]^-$ [34].

David Sharp moved from Strathclyde to Glasgow University in 1968 to occupy the newly-established Ramsay Chair and we began to collaborate in exploring further the chemistry of d- block fluoride substituted derivatives, notably diethylamido derivatives of WF_6 [35, 36] NbF_5, TaF_5 [37] and VOF_3 [38], pentafluorophenoxo derivatives of WF_6 [36] and alkylamidotungsten(VI) fluorides with their associated anions. This last group of compounds yielded superb ^{19}F NMR spectra, since the effects of ^{14}N coupling in the linear Me$-$N$=$W moiety were readily observed [39].

This activity in the d block did not mean that p-block fluorides were neglected. Although an early attempt to prepare methoxosulfur(IV) fluorides using $(MeO)_2SO$ was unsuccessful due to their thermal instability, this was not a problem for phenoxo-analogues, hence the series $SF_{4-n}(OPh)_n$ [40], $(ArO)_nS(O)F_{4-n}$, $n = 1$-3, Ar = aryl [41] and analogous compounds based on CF_3SF_3 [42] were all characterized. Exploitation of the oxidative chlorofluorination of S^{IV} by CsF/Cl_2 led to the isolation of *trans*-CF_3SF_4Cl, free-radical addition of which to several alkenes and alkynes was demonstrated [43]. The heavier p-block elements, iodine and tellurium respectively, were investigated at Glasgow (by me) and at Strathclyde (by Fraser, who was a lecturer there for a short time after Sharp's departure).

The work on iodine(V) derivatives was initiated by an attempt to extend the scope of a reaction reported by DuPont chemists who had demonstrated that n-C_4F_9I could be oxidized to n-$C_4F_9IF_4$ using ClF_3 below room temperature. Oxidation of CF_3I to CF_3IF_4 proved more problematic and was punctuated by numerous small explosions before we found the correct conditions [44]. Luckily Gerry Oates, the research student involved, was a very phlegmatic character! Decomposition of CF_3IF_4 appears to be catalysed by traces of I_2 and IF_5, although the compound is more stable in this respect than the I^{III} analogue, CF_3IF_2. However substitution of F in CF_3IF_4 by OMe can be achieved, and the series $CF_3IF_{4-n}(OMe)_n$ was identified by ^{19}F NMR spectroscopy [44]. Other compounds of the R_FIF_4 type are more stable thermally, particularly where $R_F = C_6F_5$ [45], and evidence was obtained also for all members of the series $IF_{5-n}(OMe)_n$ except where $n = 2$ [46]. A characteristic of all the methoxo-iodine(V) fluorides studied is the lability of the fluorine, which is in marked contrast to the W^{VI} analogues.

George Fraser's interest in TeF_6 originated from his period with Ray Peacock at Leicester University, during which time they published several papers dealing with chloro-, dialkylamido- and methoxo-tellurium(VI) fluorides, and he continued work on the last class of compound during his second period at Strathclyde. The general synthetic route to $TeF_{6-n}(OR)_n$, R = various alkyl groups, involves reactions between TeF_6 and the parent alcohol, ROH [47-49], although the extent to which substitution of F occurs depends on the electronic properties of the R group [48]. *Cis* configurations for $TeF_4(OR)_2$ are preferred over *trans,* though π-donor effects are less likely to be important here than with W^{VI}, and TeF_5OR compounds are powerful alkylating agents, better than $(MeO)_2SO_2$ [50], a property shared by WF_5OMe [33]. Hydrolysis of TeF_6, if carefully controlled, leads to

$TeF_{6-n}(OH)_n$, $n = 1$-4 [51]; Si—O rather than O—H bond cleavage is observed also in reactions with Me_3SiOH or Ph_3SiOH.

The work during this very productive period was enriched enormously by discussions with visitors from other fluorine labs throughout the world. We enjoyed the friendly rivalry with Ray Peacock and his group at Leicester and the frequent visits from Reinhard Schmutzler (DuPont, thence Loughborough and on to Braunschweig), who not only gave us the experience of his 'rapid-fire' lecture delivery but also shared his NMR results.

Perfluorocarbon ligand complexes of d-block metals

Organometallic chemistry carried out during the 1970s by David Sharp and his students, notably Jack Davidson (presently a member of the Chemistry Department at Heriot-Watt University, Edinburgh), was centered on the behaviour of CF_3SSCF_3 and $CF_3C{\equiv}CCF_3$ towards a range of metal carbonyl and organometallic compounds. Photolytic generation of the $CF_3S\cdot$ radical was exploited as a general route to metal-SCF_3 derivatives, for example $CpMo(CO)_3(SCF_3)$, $Cp = \eta^5$-C_5H_5, $[M(CO)_4(SCF_3)]_2$, M = Mo, W, and various complexes originating from $CpCo(CO)_2$. Analogous reactions were carried out also with $C_6F_5SSC_6F_5$, yielding products such as $[CpCo(SC_6F_5)]_2$ and $CpCo(CO)(SC_6F_5)_2$ [52]. In parallel with this work, the photochemical addition of CF_3SSCF_3 to a range of olefins and fluoro-olefins was explored, the products ranging from 1:1 adducts to medium-range telomers [53].

Sharp's interest in $CF_3C{\equiv}CCF_3$ as a potential ligand in organometallic chemistry originated at Imperial College [54]. Reactions between $CF_3C{\equiv}CCF_3$ or related alkynes and the thiolate complexes mentioned above, lead to a variety of products, which include metallothiacyclobutenes and metallated vinyls; the $16e^-$ complexes, $CpMo(R_FCCR_F)_2X$, X = Cl, Br, I, result from reactions of $CpMo(CO)_3X$ [55]. The compounds $[CpM(E)(RCCR)(SR_F)]$, M = Mo, W; E = O or CO; R = CF_3, Me, Ph; $R_F = CF_3$, C_6F_5, which result from reactions of $CpM(CO)_3(SR_F)$ with alkynes exist in two different geometries (Fig. 30.2) which was demonstrated using low temperature NMR spectroscopy

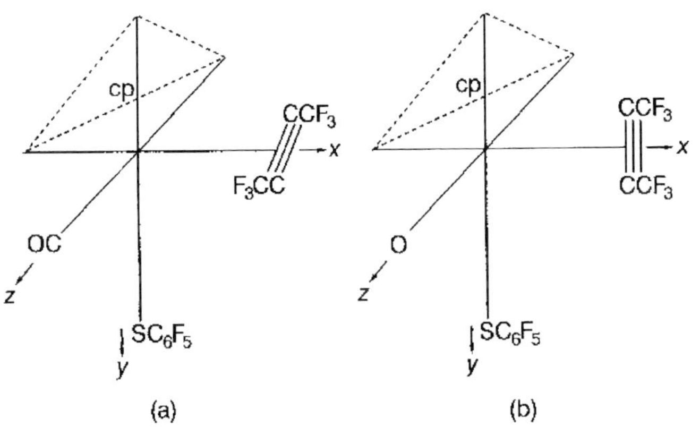

Fig. 30.2. Structures of (a) $[CpMo(CO)(CF_3C_2CF_3)(SC_6F_5)]$ and (b) $[CpMo(O)(CF_3C_2CF_3)(SC_6F_5)]$ ($Cp = \eta^5$-C_5H_5). (Reproduced with permission, from ref. [56].)

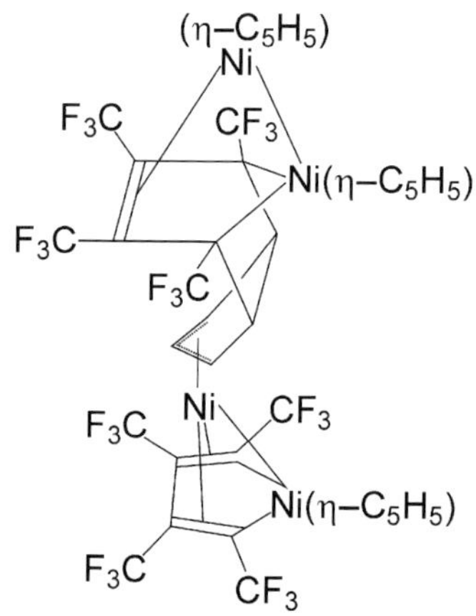

Fig. 30.3. Structure of the black crystalline tetramer $[Ni(Cp)CF_3C_2CF_3)]_4$ ($Cp = \eta^5$-C_5H_5). (Reproduced, with permission, from ref. [57].)

[56] and confirmed subsequently by X-ray crystallography. The versatility of $CF_3C{\equiv}CCF_3$ and its analogues is evident from these examples, and progress would not have been possible without structural determinations by X-ray crystallography for definitive identifications. A tetrameric Ni complex, $[NiCp(CF_3C_2CF_3)]_4$ [57] that was characterized in this way is shown in Fig. 30.3.

Binary fluorides in organic media – complexes, redox reactions and solvated cations

A distinctive feature of fluorine chemistry at Glasgow has been the use of dipolar aprotic organic solvents for solution studies on binary fluorides rather than the more obvious solvents anhydrous hydrogen fluoride, halogen fluorides or sulfur dioxide. The importance of cation solvation in determining the solubility of fluoroacid salts in organic solvents was recognised very early by Sharp and Sharpe, solvation of Ag^+ and Cu^+ by aromatic hydrocarbons being pertinent examples [58]. The approach was extended later to prepare solutions of Ag^I or Cu^{II} $[BF_4]^-$ or $[CF_3CO_2]^-$ salts in Et_2O or $MeNO_2$ [59]. A very early report (in 1948) of a red solution formed by WF_6 in C_6H_6 attracted the author's attention during his PhD work (I found out much later that Mary Mercer had been similarly intrigued during her doctoral work at Birmingham!) and this led to a systematic study of interactions between high-oxidation-state binary fluorides and π- or n-donor organic molecules that was carried out over several years. A range of interactions is possible, from very weak, of the contact charge-transfer type, which is observed for WF_6 and MoF_6 with both π- and σ-donors [60, 61], to the conventional case of isolable Lewis acid-base complex formation between the pentafluorides of Mo [62, 63] Nb, Ta [63, 64] or U [65] and σ-donors such as pyridine (py), Me_2O or MeCN. The behaviour of IF_5 in MeCN is intermediate between these two extremes, since although an adduct is not isolated, there is Raman spectroscopic evidence for the perturbation of the IF_5 liquid structure by MeCN [66].

Our original objective in examining redox behaviour of high-oxidation fluorides such as WF_6 and MoF_6 in organic solvents was to develop the solution chemistry of fluorometallate anions without recourse to the specialized equipment that would be required for spectroscopic studies in anhydrous HF. Preliminary experiments established that both nitromethane and acetonitrile were suitable solvents to replicate the I^- anion reduction of WF_6 or MoF_6, reactions which were known to occur in liquid SO_2. Using R_4N^+ counter-cations produced very soluble salts, but side-reactions involving attack of the fluoride on the alkyl groups were a limitation. This led us to contemplate metals as reducing agents with the possibility that solvation of the metal cation formed by the solvent would provide an additional driving force. Acetonitrile proved to be an almost ideal solvent for the purpose. It is relatively easily purified, has a convenient liquid range and a relatively high dielectric constant. Anions, even F^- [22], are poorly solvated in this solvent, but a range of d-block and post-d cations, particularly those with a d^{10} electronic configuration, are solvated moderately well.

Our initial success owed much to Ann Prescott, who came to Glasgow in 1971 (from Leeds University) as a research fellow with a background in non-aqueous solution chemistry. Her initial survey focussed on the oxidation of 3d and post-d metals by MoF_6 or WF_6 in MeCN. Both hexafluorides are $1e^-$ oxidants in MeCN, but the behaviour of WF_6 is complicated by its ability to react with $[WF_6]^-$ generating the $[WF_7]^-$ anion [67].

$$MF_6 + \text{metal} \rightarrow [MF_6]^- + \text{solvated metal cation } (M = \text{Mo, W})$$

$$WF_6 + [WF_6]^- \rightleftarrows [WF_7]^- + \text{`}WF_5\text{'}$$

Although salts of $[WF_7]^-$ were isolated with Tl^I or Cu^{II} counter-cations, as part of this study the complex $WF_5(NCMe)$ was not satisfactorily identified until much later [68]. The actinide hexafluoride UF_6 in MeCN has similar redox and F^- ion acceptor properties, although it is slowly reduced by MeCN to give U^V [69, 70].

These observations were placed on a firm thermodynamic basis by Graham Heath (University of Stirling then Edinburgh) and David Sharp, who undertook a comprehensive examination of redox behaviour of 4d and 5d Group 1 hexafluorometallates(V) by cyclic voltammetry. Successive E^0 values for the couples $[MF_6]^{z/z-1}$, $z = 0, -1, -2$, when M = Ta, W, Re, Os and Ir or Nb, Mo and Ru, show a remarkable linear progression. Oxidizing power increases with increasing d^n configuration, the 4d member being significantly more oxidizing than its 5d counterpart. The linear relationships appear to be the result of central-metal core charge and configuration and deviations (Fig. 30.4) are observed at d^3/d^4 couples due to spin-pairing effects [71]. The pleasingly-simple pattern is replicated for analogous chloro-metallates in CH_2Cl_2 [72]. This work, together with cyclic voltammetry measurements of hexafluorometallates(V) having redox active cations [73], led to the order of oxidizing ability in MeCN, $UF_6 > MoF_6 > [NO]^+ >$ solvated $Cu^{2+} \geqslant WF_6$ and to the identification of redox and F^- ion transfer equilibria in the system Cu metal/solvated Cu^{n+} ($n = 1$ or 2) /WF_6/$[WF_6]^-$. The quantitative difference (*ca.* 1.1 V) between the thermodynamic oxidizing abilities of MoF_6 and WF_6 in MeCN is virtually identical with these fluorides' properties in anhydrous HF, work undertaken by Tom O'Donnell's group in Melbourne,[6] and is consistent with solvation effects being relatively unimportant in these redox couples.

[6] Tom O'Donnell and his group kept the fluorine flag flying in Australia (University of Melbourne) throughout the '60s, '70s and '80s. His physical-inorganic studies of solute behaviour in anhydrous HF are regarded

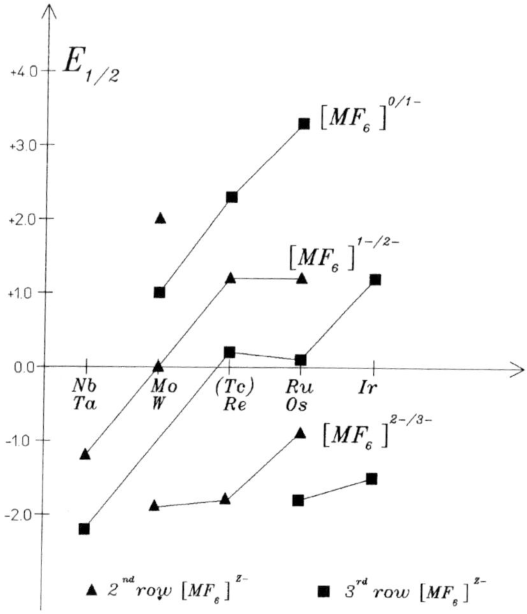

Fig. 30.4. $E_{1/2}$/V vs. S.C.E. data for second and third row d block $[MF_6]^{z/z-1}$ couples in MeCN. (Redrawn, with permission, from data in ref. [71].)

There were two immediate consequences of the work described above: firstly studies to determine whether UF_6, MoF_6 and WF_6 could be used to oxidize non-metallic elements in MeCN, and secondly, explorations of the coordination chemistry of selected metals using the solvated- metal-cation fluoroanion salts as starting materials. Elemental iodine is oxidized by UF_6 or MoF_6, but not by WF_6, in MeCN below room temperature to give the $[I(NCMe)_2]^+$ cation [74], shown to have a linear N−I−N skeleton by I K-edge EXAFS [75]. This redox behaviour is in contrast to the situation in IF_5, where I_2 is oxidized by UF_6 or ReF_6, but not MoF_6, to give the I_2^+ cation [65]. Dibromine reacts with UF_6 but not with MoF_6, in MeCN under similar conditions but evidently Br^+ is too electrophilic to exist as a simple cation solvated by MeCN and oligomerization of MeCN occurs to give a N−Br heterocyclic cation [76]. Similarly, attempts to prepare the solvated Te^{IV} cation, by oxidation of Te by UF_6 or MoF_6, resulted in $[TeF_3(NCMe)_2]^+$ the formation of which is rationalized by F^- transfer from $[MF_6]^-$, M = Mo, U to Te^{4+} [77]. Similar behaviour is observed for other heavy p-block elements; for example the solvated Bi^{III} cation appears to have marginal kinetic stability, but fluoride ion transfer can be prevented by coordination of the macrocyclic N_4 ligand, 1,4,8,11-tetraazacyclotetradecane to Bi^{III} [78].

Our interests in coordination chemistry in MeCN derived from solvated metal cations have centred on thallium [73, 79] and copper [73, 80], for which two oxidation states are

highly. Irene Irvine, a Glasgow graduate, undertook some of the early work on WF_6 and MoF_6 in HF as part of her PhD studies with Tom. He and his redoubtable wife, Pat are inveterate world travellers, so we still have frequent contacts.

accessible and where oxidation reactions by Cu^{II} or Tl^{III} can be studied [81, 82] and on the stepwise ligand replacement reactions of coordinated MeCN, NH_3 or py by $P(OMe)_3$ at the Fe^{II} centre [83-85]. Solvated cation/fluoroanion salts were prepared either by direct oxidation of the metal using MoF_6 and WF_6, or, for Cu^{II} and Fe^{II}, by acid-base reactions between the anhydrous metal fluorides and PF_5. More recently, replacement of weakly-coordinated MeCN by macrocycles at Cu^{II}, Fe^{II}, Fe^{III}, I^{I} and Bi^{III} centres has been studied [78, 86]. As expected, replacement of the weak MeCN ligand by a macrocycle has profound effects. Stabilization of Bi^{III} by this means was referred to above [78]; a second example is that the Fe^{III} oxidation state, which is not accessible from $[Fe(NCMe)_6]^{2+}$ in MeCN by hexafluoride oxidation [83], is formed by oxidation of $[FeL(NCMe)_2]^{2+}$, L = [12]aneN_4 or [14]aneN_4, using the weak oxidant PF_5 [86]. Similarly, oxidation of $[Agpy_4]^+$ or $[Agpy_2]^+$ cations by MoF_6 or UF_6 respectively, leads to Ag^{III} species, whereas Ag^{I} solvated by MeCN is not oxidized [87].

Fluorine-18 radiotracer studies: catalysis by fluoride ion and the lability of covalently-bound fluoride ligands

Glasgow is associated with the pioneering work by Soddy and Cranston during the period 1904-12 which led to the first isolation of the element protactinium; the term 'isotope' was coined by Soddy at a dinner party held in a house on University Avenue. Since then, the use of isotopes for the investigation of various chemical phenomena has been a Glasgow tradition. The only feasible fluorine isotope for use as a tracer is fluorine-18, a β^+ emitter whose $t_{1/2}$ is 110 min. Our use of the isotope dates from the early 1970s following conversations between the author and Geoff Webb, a Glasgow colleague whose research interests included the application of [^{14}C] labelling for the study of heterogeneous catalysis by oxide-supported transition metals. Enthused by G. H. Cady's work on heterogeneous catalytic fluorination of CO and F_2CO, we chose to apply the [^{18}F] tracer method to heterogeneous catalysis by F^- anion.

In the initial phase of the work it was established that the extent of the interaction between solid Group 1 fluorides and [^{18}F]-labelled SF_4, F_2CO or $CF_3C(O)F$ followed the order Cs > Rb > K > Na, Li and, in the case of F_2CO, was enhanced by the presence of acetonitrile or diglyme [88]. These observations are in harmony with the behaviour of Group 1 fluorides as catalytic sources of F^- ion, however extension of the work to other ionic fluorides [89] indicated that simple correlations, for example with cation size, were not appropriate. Progress resulted by examining the effect of Group I fluoride activation on their B.E.T. areas, particularly that of CsF [90, 91], and by studying Lewis base-acid interactions between activated CsF and SF_4, F_2CO and CO_2 using both [^{18}F] and [^{35}S] or [^{14}C] (both β^- emitters) as appropriate [91]. Pretreatment of CsF by thermal decomposition of its 1:1 adducts with $(CF_3)_2CO$ or F_2CO leads to increased B.E.T. areas, from 0.31-0.19 to 3.01-2.09 $m^2\ g^{-1}$, possibly due to the development of some degree of meso- or macro-porous structure. Reactions between activated CsF and SF_4 or F_2CO involve both weakly adsorbed and strongly bound species, and can be described as follows [91]:

$$CsF(s) + SF_4(g) \rightarrow CsF \cdot SF_4(ad) \rightarrow Cs^+[SF_5]^-(s)$$

$$CsF(s) + F_2CO(g) \rightarrow CsF \cdot F_2CO(ad) \rightleftarrows Cs^+[OCF_3]^-(s)$$

Although adsorption of $^{14}CO_2$ on activated CsF was detected, no complex was isolated. Other solid F^- anion sources that were characterised in this way are CsF or KF supported on calcined γ-alumina [92] or on fluorinated γ-alumina [93], the latter being useful when a non-hydroxylic material of relatively high B.E.T. area is required.

Using a combination of [^{36}Cl] and [^{35}S] radiotracers enabled the progress of the room- temperature chlorofluorination of SF_4 by ClF to be followed and the mechanism of the catalysis involving CsF [94] or Cs^+ or K^+ fluorides supported on fluorinated γ-alumina [95] to be determined. Caesium fluoride supported on fluorinated γ-alumina, loading 5.5 mmol g^{-1}, has catalytic activity which is comparable with that of unsupported activated CsF; supported KF although less active, has acceptible activity. The mechanism is identical in all cases [94, 95] and can be described as follows:

$$SF_4(g) + MF(s) \rightarrow MF \cdot SF_4(ad)$$
$$ClF(g) + MF(s) \rightarrow MF \cdot ClF(ad)$$
$$MF \cdot ClF(ad) + MF \cdot SF_4(ad) \rightarrow SF_5Cl(g) + 2MF(s)$$
$$MF \cdot SF_4(ad) \rightarrow M^+[SF_5]^-(s)$$
$$MF \cdot ClF(ad) \rightarrow Cs^+[ClF_2]^-(s)$$

in which formation of $[SF_5]^-$ and, particularly, $[ClF_2]^-$ result in catalyst poisoning.

The radioisotope [^{18}F] has proved to be a useful complementary technique to ^{19}F NMR spectroscopy to study fluorine exchange, being particularly suited to situations in solution where exchange may be slow on the NMR time scale and to the study of weak interactions occurring under heterogeneous conditions. The extent of [^{18}F] exchange in solution between $Me_3Si^{18}F$ and various substituted derivatives of high-oxidation-state fluorides depends both on the identity of the central atom and on the number and nature of the substituents; for example, with W^{VI} and the MeO-substituent, exchange rates vary in the order $WF_6 < WF_5(OMe) \geqslant$ *cis*-$WF_4(OMe)_2 > WF_3(OMe)_3 >$ *cis*-$WF_2(OMe)_4$; associative mechanisms have been formulated [96]. Exchange was also observed with MoF_6, IF_5, $IF_4(OMe)$ [96] UF_6 and $UF_5(OMe)$ [97], although not with TeF_6 or $TeF_5(OMe)$ [96]. [^{18}F] Exchange reactions between the hexafluorides of Mo, W and U and various fluoroanions have established that fluorine in $[SbF_6]^-$ and to some extent in $[AsF_6]^-$ is less labile than fluorine in $[BF_4]^-$ or in other hexafluorometallates(V) [98]. The isotope [^{18}F] has been used also to demonstrate that $[MoF_7]^-$ and $[WF_7]^-$ anions co-exist with their hexafluorides in MeCN at room temperature [99]. Surface complexation of SF_4, labelled with [^{18}F] or [^{35}S], with solid AlF_3, CrF_3 and MF_5, M = Nb, Ta or U is rationalised on the basis of $[SF_3]^+$ formation [100], although isolable complexes were not formed under the conditions used (*cf.* its behaviour towards CsF described above [91]).

Chlorofluorocarbons and their alternatives

Our interest in heterogeneous catalytic fluorination of a C—Cl bond to a C—F bond dates from a request from ICI in the early 1980s to determine whether using [^{18}F] labelling would throw any light on the processes occurring during the HF pretreatment of chromia catalysts prior to their use for the large scale production of chlorofluoroethanes, $C_2Cl_{6-n}F_n$, particularly where $n = 3$, 4, or 5. We found that labelling experiments using

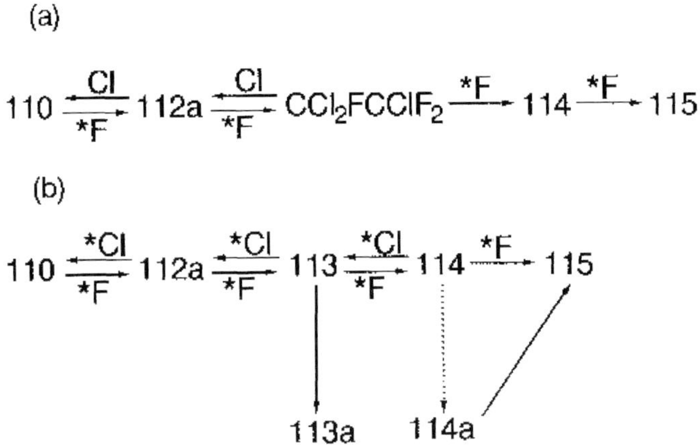

Fig. 30.5. Halogen exchange and isomerization reaction scheme for chlorofluoroethane behaviour on fluorinated chromia at 700 K. (a) Partial model deduced using CCl_2FCClF_2 as feedstock and catalyst labelled with [^{18}F]. (b) Model deduced using all experimental data. F and Cl are catalytically active surface halogen species, * = radioactive. Key to compounds: 110 = C_2Cl_6, 112a = CCl_3CClF_2,113 = CCl_2FCClF_2, 113a = CCl_3CF_3, 114 = $CClF_2CClF_2$, 114a = CCl_2FCF_3, 115 = $CClF_2CF_3$. (Reproduced, with permission, from ref. [104].)

$H^{18}F$ were ideal for the purpose and by [^{18}F] exchange between HF vapour and [^{18}F]-labelled fluorinated chromia catalysts, proposed the existence of three types of surface fluorine-containing species: weakly adsorbed, catalytically active, and inert. The differing behaviour observed over a catalyst's lifetime was rationalized on the basis of a slow replacement of $Cr^{III}-O$ by $Cr^{III}-F$ bonds [101]. Catalytic fluorination of $C_2Cl_4 + Cl_2$ ($\equiv C_2Cl_6$) by HF over fluorinated chromia at $\geqslant$ 623 K characteristically produces mixtures of $C_2Cl_{6-n}F_n$ (n = 1-5) even when an excess of HF is present. [^{18}F] Labelling experiments established that the radioisotope was incorporated into the products from the fluorinated catalyst [102], and [^{36}Cl]-labelling showed that a parallel phenomenon occurred, even though the chlorine-containing species was present on the catalyst to a very small extent [103]. This indicates that the reaction should be viewed as two distinct processes, fluorination of C−Cl to C−F and chlorination of C−F to C−Cl, a suggestion first made by Professor J. C. (Colin) Tatlow in 1974, rather than the more conventional dismutation processes. Further work on this system in which the [^{18}F] or [^{36}Cl] activities in individual products were determined by radio-G.C., resulted in a refinement of the halogen-exchange proposal and a model for the dynamic behaviour of chlorofluoroethanes on fluorinated chromia [104] (Fig. 30.5). In this model, catalytic fluorination and catalytic chlorination are both viewed as intermolecular processes in which the catalytically-active Cl originates from C−Cl as a result of the fluorination step. There is no requirement however for chlorination and fluorination to be concerted, as would be the case for a dismutation. In contrast, isomerizations of CCl_2FCClF_2 to CCl_3CF_3 and $CClF_2CClF_2$ to CCl_2FCF_3 are intramolecular, and exchange with surface halogen-containing species is not involved.

The advent of the Montreal and subsequent protocols has resulted in increased interest in heterogeneous catalytic fluorination for the synthesis of hydrochlorofluorocarbons (HCFCs) and hydrofluorocarbons (HFCs), notably CF_3CH_2F, as replacements for CFC-based refrigerants. Rather unexpectedly, we discovered a room temperature heterogeneous catalytic halogen- exchange process whereby hydrochloroethanes, for example CH_3CCl_3, are converted to HCFCs [105]. The catalysts involved are based on γ-alumina, fluorinated with SF_4 to promote surface activity and to provide a 'pool' of labile fluorine [106] then treated with CH_3CCl_3 (which undergoes dehydrochlorination with subsequent oligomerization of the $CH_2{=}CCl_2$ formed) to give an organic-layer supported on the fluorinated oxide surface [105]. Our model for the behaviour of these oxide-supported organic layer catalysts in reactions such as $CH_3CCl_3 + HF \rightarrow CH_3CCl_2F + HCl$ is that the organic-layer provides a quasi-liquid medium to trap the reactant, CH_3CCl_3; the latter undergoes dehydrochlorination at strongly Lewis-acidic sites on the fluorinated oxide surface, and $CH_2{=}CCl_2$ so formed is hydrofluorinated to give CH_3CCl_2F. Labile fluoride on the catalyst is replenished by HF [107]. Catalytic fluorination at room temperature using fluorinated Fe_3O_4 and Co_3O_4 occurs in a similar manner, but where the organic substrate does not undergo dehydrochlorination readily, radical intermediates are believed to be involved [108].

Conclusion – current research activities

Our interest in CFC alternatives continues and is increasingly focussed on catalysis involving CF_3CH_2F, both its formation via the heterogeneous catalytic fluorination of CF_3CH_2Cl, and the ways in which it might be activated catalytically to generate new chemistry. In this context, calculations to determine its Brønsted acidity have been carried out [109].

Chemical-mechanical polishing, the process whereby surface roughness of electronic, optical or electro-optic materials is removed prior to the use of these materials in device manufacture, is usually regarded as the province of engineers; however, chemists do have a role to play in characterization of the surface etching processes that are involved. Halogen-containing compounds are important etchants and much of our work has been concerned with the role of the $[HF_2]^-$ anion, using $[^{18}F]$ to probe the surface events. Polishing of the ferroelectric $LiNbO_3$ is enhanced, both in terms of time and surface quality, by incorporation of $[HF_2]^-$ in the conventional silica sol polishing reagent. The reason is that $LiNbOF_4$ is formed as an intermediate and is readily hydrolysed under the aqueous medium polishing conditions used to give sparingly-soluble Nb_2O_5 as one product [110]. The $[HF_2]^-$ anion is also implicated in a process for the rapid polishing of silica using an $[HF_2]^-$ ceria mixture at low pH. In this process the intermediate is a K_2SiF_6 particle which is coated with a thin silica layer [111]. Silica wafers polished in this way are excellent substrates for the deposition of thin films of polycyclic hydrocarbons, which in turn may be modified chemically by adsorption of MoF_6, WF_6 or AsF_5 [112].

Theoretical studies of small, highly endothermic, non-metal species, including fluorides, became a new Glasgow interest with the arrival of T. M. (Thomas) Klapoetke from Berlin in 1995 to occupy the Ramsay Professorship of Chemistry. Since azides are his major interest, we became quite used to the occurrence of small (controlled!) explosions from his laboratory [113]; his work includes the study of cations such as $[ICNI]^+$, as its $[AsF_6]^-$ salt [114] and intermediates such as $FXe(N_3)$ [115]. More conventional chemistry includes a modified laboratory synthesis of AuF_3 [116], a 'Cambridge' compound prepared using BrF_3 [3]. Collaborative work with Tom Klapoetke continues since his return to Germany in 1997 to become Head of the Institute for Inorganic Chemistry at the University of Munich.

Fluorine research at Glasgow today, although very different in character than that of the '60s, continues to survive and thrive in the uncertain climate of funding that is prevelant in UK universities. The work that has been reviewed here would not have been possible without the enormous contributions made by talented and enthusiastic students and research assistants. Many have gone on to pursue academic careers, either continuing with fluorine chemistry or changing to other fields. Others have pursued radically different careers; David Sharp's former students, for example, include two ordained members of the Churches of England and Scotland. Sadly, the careers of four former colleagues, John Fuggle, Gerry Oates, Matt Baillie and Andrew Lane, were cut short by their untimely deaths. There is no telling where a 'fascination with fluorine' will lead.

Acknowledgements

I am very grateful to those who have helped in compiling this account of Fluorine Chemistry at Glasgow, particularly Drs M. Mercer and A. A. Woolf who provided personal memoirs and Prof. D. W. A. Sharp, with whom I have reminisced much about the 'early days' and who cast a critical eye over the manuscript.

References

1 A. G. Sharpe and H. J. Emeléus, *J. Chem. Soc.*, (1948) 2135.
2 A. A. Woolf and H. J. Emeléus, *J. Chem. Soc.*, (1949) 2865.
3 A. G. Sharpe, *J. Chem. Soc.*, (1949) 2901.
4 A. G. Sharpe, *J. Chem. Soc.*, (1950) 2907, 3444.
5 A. G. Sharpe and A. A. Woolf, *J. Chem. Soc.*, (1951) 798.
6 A. G. Sharpe, *J. Chem. Soc.*, (1952) 4538.
7 A. G. Sharpe, *J. Chem. Soc.*, (1953) 197.
8 R. D. W. Kemmitt and D. W. A. Sharp, *J. Chem. Soc.*, (1961) 2496.
9 R. S. Nyholm and A. G. Sharpe, *J. Chem. Soc.*, (1952) 3579.
10 B. Cox and A. G. Sharpe, *J. Chem. Soc.*, (1953), 1783.
11 B. Cox, *J. Chem. Soc.*, (1954) 3251.
12 B. Cox, *J. Chem. Soc.*, (1956) 876.
13 R. D. W. Kemmitt, D. R. Russell and D. W. A. Sharp, *J. Chem. Soc.*, (1963) 4408.
14 D. H. Brown, K. R. Dixon, R. D. W. Kemmitt and D. W. A. Sharp, *J. Chem. Soc.*, (1965) 1559.
15 D. W. A. Sharp and N. Sheppard, *J. Chem. Soc.*, (1957) 674.
16 R. D. Peacock and D. W. A. Sharp, *J. Chem. Soc.*, (1959) 2762.
17 D. H. Brown, K. R. Dixon, C. M. Livingston, R. H. Nuttall and D. W. A. Sharp, *J. Chem. Soc. (A)*, (1967) 100.
18 A. P. Lane and D. W. A. Sharp, *J. Chem. Soc. (A)*, (1969) 2942.
19 A. P. Lane, D. W. A. Sharp, J. M. Barraclough, D. H. Brown and D. A. Patterson, *J. Chem. Soc. (A)*, (1971) 94.
20 A. G. Sharpe, *Adv. Fluorine Chem., 1* (1960) 29.
21 A. G. Sharpe, *Inorganic Chemistry*, 3rd edn, Longman, 1992, pp. 231 and 409.
22 A. G. Sharpe, *J. Chem. Educ., 67* (1990) 309.
23 D. S. Payne, *Quart. Rev., 15* (1961) 173.
24 T. Kennedy and D. S. Payne, *J. Chem. Soc.*, (1960) 4126.
25 D. H. Brown, G. W. Fraser and D. W. A. Sharp, *J. Chem. Soc. (A)*, (1966) 171.
26 D. H. Brown, K. D. Crosbie, G. W. Fraser and D. W. A. Sharp, *J. Chem. Soc. (A)*, (1969) 551.
27 J. S. Harman and D. W. A. Sharp, *J. Chem. Soc. (A)*, (1970) 1138.
28 D. H. Brown, K. D. Crosbie, G. W. Fraser and D. W. A. Sharp, *J. Chem. Soc. (A)*, (1969) 872.

29 D. H. Brown, K. D. Crosbie, J. I. Darragh, D. S. Ross and D. W. A. Sharp, *J. Chem. Soc. (A)*, (1970) 914.
30 C. B. Colburn, W. E. Hill and D. W. A. Sharp, *J. Chem. Soc. (A)*, (1970) 2221.
31 J. S. Harman, M. E. McCartney and D. W. A. Sharp, *J. Chem. Soc. (A)*, (1971) 1547.
32 M. E. Harman, R. Keat and D. W. A. Sharp, *J. Inorg. Nucl. Chem., Supplement* (1976) 49.
33 A. M. Noble and J. M. Winfield, *J. Chem. Soc. (A)*, (1970) 501; 2574.
34 W. McFarlane, A. M. Noble and J. M. Winfield, *J. Chem. Soc. (A)*, (1971) 948.
35 A. Majid, R. R. McLean, D. W. A. Sharp and J. M. Winfield, *Z. Anorg. Allg. Chem., 385* (1971) 85.
36 A. Majid, D. W. A. Sharp, J. M. Winfield and I. Hanley, *J. Chem. Soc., Dalton Trans.*, (1973) 1876.
37 J. C. Fuggle, D. W. A. Sharp and J. M. Winfield, *J. Chem. Soc., Dalton Trans.*, (1972) 1766.
38 G. A. Kolta, D. W. A. Sharp and J. M. Winfield, *J. Fluorine Chem., 14* (1979) 153.
39 O. R. Chambers, M. E. Harman, D. S. Rycroft, D. W. A. Sharp and J. M. Winfield, *J. Chem. Res. (S)*, (1977) 150; *J. Chem. Res. (M)*, (1977) 1849.
40 J. I. Darragh and D. W. A. Sharp, *Angew. Chem., 82* (1970) 45.
41 D. S. Ross and D. W. A. Sharp, *J. Chem. Soc., Dalton Trans.*, (1972) 34.
42 J. I. Darragh, S. F. Hossain and D. W. A. Sharp, *J. Chem. Soc., Dalton Trans.*, (1975) 218.
43 J. I. Darragh, G. Haran and D. W. A. Sharp, *J. Chem. Soc., Dalton Trans.*, (1973) 2289.
44 G. Oates and J. M. Winfield, *J. Chem. Soc., Dalton Trans.*, (1974) 119.
45 J. A. Berry, G. Oates and J. M. Winfield, *J. Chem. Soc., Dalton Trans.*, (1974) 509.
46 G. Oates, J. M. Winfield and O. R. Chambers, *J. Chem. Soc., Dalton Trans.*, (1974) 1380.
47 G. W. Fraser and J. B. Millar, *J. Chem. Soc., Dalton Trans.*, (1974) 2029.
48 G. W. Fraser and G. D. Meikle, *J. Chem. Soc., Perkin Trans. 2*, (1975) 312.
49 G. W. Fraser and G. D. Meikle, *J. Chem. Soc., Dalton Trans.*, (1977) 1985.
50 G. W. Fraser and G. D. Meikle, *J. Chem. Soc., Perkin Trans. 1*, (1979) 2005.
51 G. W. Fraser and G. D. Meikle, *J. Chem. Soc., Chem. Commun.*, (1974) 624; see also G. W. Fraser and J. B. Millar, *ibid.*, (1972) 1113.
52 J. L. Davidson and D. W. A. Sharp, *J. Chem. Soc., Dalton Trans.*, (1972) 107; (1973) 1957; (1975) 813.
53 G. Haran and D. W. A. Sharp, *J. Chem. Soc., Perkin Trans. 1*, (1972) 34.
54 J. L. Boston, D. W. A. Sharp and G. Wilkinson, *J. Chem. Soc.*, (1962) 3488.
55 J. L. Davidson and D. W. A. Sharp, *J. Chem. Soc., Dalton Trans.*, (1975) 2283, 2531.
56 P. S. Braterman, J. L. Davidson and D. W. A. Sharp, *J. Chem. Soc., Dalton Trans.*, (1976) 241.
57 J. L. Davidson and D. W. A. Sharp, *J. Chem. Soc., Dalton Trans.*, (1976) 1123.
58 D. W. A. Sharp and A. G. Sharpe, *J. Chem. Soc.*, (1956) 1855, 1858.
59 M. J. Baillie, D. H. Brown, K. C. Moss and D. W. A. Sharp, *J. Chem. Soc. (A)*, (1968) 104.
60 H. J. Clase, A. M. Noble and J. M. Winfield, *Spectrochim Acta, Part A, 25* (1969) 293.
61 R. R. McLean, D. W. A. Sharp and J. M. Winfield, *J. Chem. Soc., Dalton Trans.*, (1972) 676.
62 M. Mercer, T. J. Ouellette, C. T. Ratcliffe and D. W. A. Sharp, *J. Chem. Soc. (A)*, (1969) 2532.
63 J. C. Fuggle, D. W. A. Sharp and J. M. Winfield, *J. Fluorine Chem., 1* (1971/72) 427.
64 J. C. Fuggle, D. A. Tong, D. W. A. Sharp, J. M. Winfield and J. H. Holloway, *J. Chem. Soc., Dalton Trans.*, (1974) 205.
65 J. A. Berry, A. Prescott, D. W. A. Sharp and J. M. Winfield, *J. Fluorine Chem., 10* (1977) 247.
66 J. A. Berry, D. W. A. Sharp and J. M. Winfield, *J. Chem. Res. (S)*, (1978) 377; *J. Chem. Res. (M)*, (1978) 4664.
67 A. Prescott, D. W. A. Sharp and J. M. Winfield, *J. Chem. Soc., Dalton Trans.*, (1975) 934, 936.
68 N. Bao and J. M. Winfield, *J. Fluorine Chem., 50* (1990) 339.
69 J. A. Berry, R. T. Poole, A. Prescott, D. W. A. Sharp and J. M. Winfield, *J. Chem. Soc., Dalton Trans.*, (1976) 272.
70 D. K. Sanyal, D. W. A. Sharp and J. M. Winfield, *J. Fluorine Chem., 19* (1981/82) 55.
71 S. Brownstein, G. A. Heath, A. Sengupta and D. W. A. Sharp, *J Chem. Soc., Chem. Commun.*, (1983) 669.
72 G. A. Heath, K. A. Moock, D. W. A. Sharp and L. J. Yellowlees, *J. Chem. Soc., Chem. Commun.*, (1985) 1503.
73 G. M. Anderson, J. Iqbal, D. W. A. Sharp, J. M. Winfield, J. H. Cameron and A. G. McLeod, *J. Fluorine Chem., 24* (1984) 303.
74 G. M. Anderson and J. M. Winfield, *J. Chem. Soc., Dalton Trans.*, (1986) 337.
75 C. D. Garner, L. McGhee, A. Steel and J. M. Winfield, *J. Fluorine Chem., 69* (1994) 73.
76 L. McGhee, D. S. Rycroft and J. M. Winfield, *J. Fluorine Chem., 36* (1987) 351.

77 L. McGhee and J. M. Winfield, *J. Fluorine Chem., 57* (1992) 147.
78 S. I. Ajiboye, J. Iqbal, R. D. Peacock, N. Prouff, G. J. Taylor, J. M. Winfield and X. Liu, *J. Fluorine Chem., 91* (1998) 213.
79 R. M. Siddique and J. M. Winfield, *J. Fluorine Chem., 40* (1988) 71.
80 A. C. Baxter, J. H. Cameron, A. McAuley, F. M. McLaren and J. M. Winfield, *J. Fluorine Chem., 10* (1977) 289.
81 G. M. Anderson, J. H. Cameron, A. G. Lappin, J. M. Winfield and A. McAuley, *Polyhedron, 1* (1982) 467.
82 R. M. Siddique and J. M. Winfield, *Can. J. Chem., 67* (1989) 1780.
83 C. J. Barbour, J. H. Cameron and J. M. Winfield, *J. Chem. Soc., Dalton Trans.*, (1980) 2001.
84 J. H. Cameron, A. G. Lappin, J. M. Winfield and A. McAuley, *J. Chem. Soc., Dalton Trans.*, (1981) 2172.
85 L. McGhee, R. M. Siddique and J. M. Winfield, *J. Chem. Soc., Dalton Trans.*, (1988) 1309.
86 S. H. Wang, S. I. Ajiboye, G. Haining, L. McGhee, R. D. Peacock, G. Peattie, R. M. Siddique and J. M. Winfield, *J. Chem. Soc., Dalton Trans.*, (1995) 3837.
87 J. Iqbal, D. W. A. Sharp and J. M. Winfield, *J. Chem. Soc., Dalton Trans.*, (1989) 461.
88 C. J. W. Fraser, D. W. A. Sharp, G. Webb and J. M. Winfield, *J. Chem. Soc., Dalton Trans.*, (1972) 2226; (1974) 112.
89 C. J. W. Fraser, D. W. A. Sharp, R. A. Sule, G. Webb and J. M. Winfield, *J. Chem. Res. (S)*, (1978) 2; *J. Chem. Res. (M)*, (1978) 311.
90 G. A. Kolta, G. Webb and J. M. Winfield, *J. Fluorine Chem., 14* (1979) 331.
91 K. W. Dixon and J. M. Winfield, *J. Chem. Soc., Dalton Trans.*, (1989) 937.
92 T. Baird, A. Bendada, G. Webb and J. M. Winfield, *J. Mater Chem., 1* (1991) 1071.
93 T. Baird, A. Bendada, G. Webb and J. M. Winfield, *J. Fluorine Chem., 66* (1994) 117.
94 G. A. Kolta, G. Webb and J. M. Winfield, *Appl. Catal., 2* (1982) 257.
95 T. Baird, A. Bendada, M. Selougha, G. Webb and J. M. Winfield, *J. Fluorine Chem., 69* (1994) 109.
96 R. T. Poole and J. M. Winfield, *J. Chem. Soc., Dalton Trans.*, (1976) 1557.
97 D. K. Sanyal and J. M. Winfield, *J. Fluorine Chem., 24* (1984) 75.
98 M. F. Ghorab and J. M. Winfield, *J. Fluorine Chem., 49* (1990) 367.
99 M. F. Ghorab and J. M. Winfield, *J. Fluorine Chem., 62* (1993) 101.
100 K. W. Dixon, M. F. Ghorab and J. M. Winfield, *J. Fluorine Chem., 37* (1987) 357.
101 J. Kijowski, G. Webb and J. M. Winfield, *Appl. Catal., 27* (1986) 181.
102 J. Kijowski, G. Webb and J. M. Winfield, *J. Fluorine Chem., 27* (1985) 213.
103 L. Rowley, G. Webb, J. M. Winfield and A. McCulloch, *Appl. Catal., 52* (1989) 69.
104 L. Rowley, J. Thomson, G. Webb, J. M. Winfield and A. McCulloch, *Appl. Catal. A, 79* (1991) 89.
105 J. Thomson, G. Webb, J. M. Winfield, D. Bonniface, C. Shortman and N. Winterton, *Appl. Catal. A, 97* (1993) 67.
106 A. Bendada, G. Webb and J. M. Winfield, *Eur. J. Solid State Inorg. Chem., 33* (1996) 907.
107 A. Bendada, D. W. Bonniface, F. McMonagle, R. Marshall, C. Shortman, R. R. Spence, J. Thomson, G. Webb, J. M. Winfield and N. Winterton, *Chem. Commun.*, (1996) 1947.
108 J. Thomson, *J. Chem. Soc., Faraday Trans., 90* (1994) 3585.
109 T. M. Klapoetke and J. M. Winfield, *J. Fluorine Chem., 88* (1998) 19.
110 M. Beveridge, L. McGhee, S. G. McMeekin, M. I. Robertson, A. Ross and J. M. Winfield, *J. Mater Chem., 4* (1994) 119.
111 D. S. Boyle and J. M. Winfield, *J. Mater. Chem., 6* (1996) 227.
112 D. S. Boyle and J. M. Winfield, *J. Mater. Chem., 7* (1997) 2039.
113 H. Holfter, T. M. Klapoetke and A. Schulz, *Propellants, Explosives, Pyrotechnics, 22* (1997) 51.
114 T. M. Klapoetke, *J. Chem. Soc., Dalton Trans.*, (1997) 553.
115 A. Schulz and T. M. Klapoetke, *Inorg. Chem., 36* (1997) 1929.
116 I. C. Tornieporth-Oetting and T. M. Klapoetke, *Chem. Ber., 128* (1995) 957.

BIOGRAPHIC NOTES

David Sharp has recently retired from directing the Office of International Programmes at the University of Glasgow, a post which he has held for the past 22 years. Prior to that he was the first occupant of the Ramsay Chair of Chemistry, and at various periods was Head of Department and Dean of Science. He maintains his interest in fluorine chemistry as one of the Editors of *J. Fluorine Chem.* and is a regular attendee at fluorine chemistry symposia.

David Sharp

John Winfield holds a Personal Professorship in Inorganic Chemistry at the University of Glasgow and is currently Head of Department. His BSc and PhD degrees were awarded from the Universities of London (Imperial College) and Glasgow (Royal College of Science and Technology) respectively in 1961 and 1964. He was appointed a lecturer at the University of Glasgow in 1965 following postdoctoral work with R. J. (Ron) Gillespie, McMaster University, Hamilton, Ontario, and has progressed through the ranks as Senior Lecturer (1974), Reader (1985) and Professor (1995).

John Winfield

Appendix 30.1

Recollections of early days in the Cambridge inorganic lab

A. A. WOOLF

Faculty of Applied Science, University of the West of England, Bristol, BS16 1QY, UK

Emeléus was appointed Professor of Inorganic Chemistry in Cambridge in 1945 and commenced work there during a period of considerable austerity, albeit one of great expectations (1945-7). The student body at that time was reinforced by a more mature and experienced set of wartime returnees who had left after the Part I Tripos (2nd year examinations) and were keen to make up for lost time and complete their courses. Inorganic research was a very junior partner to Todd's organic activities and was allocated only half a teaching laboratory and a few off-rooms. Emeléus started up with three students from Imperial College (A. A. Banks, A. G. Maddock and D. S. Payne), two ICI employees from Runcorn (V. Kerrigan and Davies), an experienced fluorine chemist from Birmingham University (R. N. Haszeldine), and the first Cambridge graduate (A. G. Sharpe).

Present day essentials such as IR, MS, NMR, GC, HPLC etc. were absent, weighing was by swings rather than push-button, and the most advanced instrument was a new Beckmann UV-VIS spectrometer. As far as fluorine chemistry was concerned, a high-temperature fluorine cell which occasionally lifted off its base as H_2 and F_2 mixed explosively was in use until ICI donated a tamer low-temperature cell. A few quartz vessels served for reactions since fabricated fluoroplastic tubes and valves were unavailable. (Actually the use of a quartz flask was advantageous, enabling the key compound CF_3I to be made from CI_4 and IF_5; previous use of copper vessels caused Ullmann coupling to C_2F_6.) The main innovation introduced by Emeléus was the use of vacuum systems for most of the work. The ones made for Si and B hydride chemistry employed Stock mercury float valves, and Emeléus relaxed in the summer by cutting down systems, cleaning and rejoining them all in the difficult-to-work soft glass. Other students set up their own Pyrex systems with conventional taps. Volatile compounds were purified by trap-to-trap vacuum distillation using a series of slush (cooling) baths and identified by vapour pressure measurements and molecular weight determination using gas density bulbs (Regnault's method). Fluorine analyses were by long-winded CaF_2 gravimetry, or the less precise $Th(NO_3)_4$ titrimetry after Willard-Winter distillation, since the fluoride electrode had yet to be invented. Most of the fluorochemicals were not obtainable commercially and had to be home-made. No doubt the laboratory would have failed today's stringent safety regulations: liquid oxygen was used as a coolant and this caused a few bangs when one was making up slush baths with organic solvents; the extensive use of mercury in manometers and valves probably resulted in an appreciable level in the lab atmosphere; and waste fluorine was not trapped too efficiently, causing a penetrating odour reminiscent of hypochlorous acid or ozone in the cell room, especially when the wind was in the wrong direction.

The build up of inorganic research student numbers continued with Cambridge graduates (me and J. Banus in 1947, followed by F. W. Bennett and F. G. A. Stone in 1948) and an overseas contingent [R. L. Martin, N. N. Greenwood (Australia), V. Gutmann (Austria), Gambroya (Spain), and G. Brandt (Finland)] of whom six were involved in fluorine chemistry. Supervision was more of a term-by-term rather than a daily event, since Emeléus would suggest a starting project and leave the student to get on with it. Only if results were not forthcoming would he suggest an alternative or persuade another student to transfer a working project (this was quite feasible in the co-operative and relaxed atmosphere which prevailed in the lab). Researchers were also allowed to develop their own projects, within available resources. Looking back it is remarkable that so much could be achieved without the sophisticated equipment and facilities available today; and of course it was much more affordable then.

Appendix 30.2

Chemistry at Birmingham in the 1960s

MARY MERCER

Advice on how to select a research supervisor came from a very young Professor in Ottawa: "You start at the top and work your way downwards" he stated, handing me a list of British inorganic chemists in what he claimed was their order of merit. Twelve applications and eight offers later, more advice was required: "You have to choose between being happy and going to Oxford," was his second effort, followed by a thumb-nail sketch – unprintable – of what each of the potential candidates for my services would be like to work for. It was on this promising note that in 1961 I found myself in the laboratories of a certain Dr. R. D. Peacock at Birmingham University, with the proximity of the Welsh mountains the main week-end attraction.

Ray Peacock had gone into fluorine chemistry in its early days, when passing fluorine over a new element was all that was needed to get a PhD: anything that came out the other end was a new compound. By the time I arrived on the scene it was getting slightly harder. George Hargreaves was before my time, but copies of his thesis dealing mainly with complex fluorides – a truly massive tome – remained behind in every lab, supposedly to act as a guide to the next generation on the quantity of work we, too, were expected to produce. Each new student was given a compound, with instructions to determine its chemical properties. John Holloway had RuF_5; Dave Hugill had TcF_6, and was locked up in a little room alone with it, with instructions not to let it out; John Holmes was working on tin compounds; Vijayan Nayar had $TiCl_4$ to start with; and I was presented with WF_6, a substance chiefly remarkable for not being as reactive as MoF_6. Tony Edwards, then a research fellow from Imperial College, spent his days painstakingly growing single crystals by subliming volatile solids such as MoF_5, VF_5, NbF_5, TaF_5, WOF_4, $MoOF_4$ and $ReOF_4$. These were coaxed up very fragile glass tubes in the faint hope that one of them would be a perfect crystal and land obligingly at least a millimetre away from anything else.

In the spring of 1964, with my thesis soon due in, my grant was running out and my landlady was pressing for the rent (Bernie Cohen, an American research fellow, lent me £50, which kept me going for three months), and I needed to do something novel with WF_6. One day as I sat cross-legged on the bench watching my neighbour Vijayan Nayar doing chemistry Indian style with $TiCl_4$, I wondered aloud what would happen if our two compounds were mixed together, "Absolutely nothing", opined by labmates unanimously, "They're both d^0 configurations so they can't possibly react." My despised reaction went with violent enthusiasm, and Bernie came on the scene to help design the apparatus for measuring the physical properties of the product WF_5Cl, the first chlorofluoride of tungsten.

With Ray's departure for Leicester the whole group broke up. Tony stayed at Birmingham. John followed Ray to Leicester, after some years in Aberdeen. I ended up in Glasgow, working for David Sharp after satisfying my yearning to see the world via Australia. Chemistry moved into the age of the computer, and safety regulations stifled the enterprise of the more exuberant fluorine chemists.

Chapter 31

DEVELOPMENT OF INORGANIC FLUORINE CHEMISTRY IN SLOVENIA

BORIS ŽEMVA

Jožef Stefan Institute, 1000 Ljubljana, Slovenia

Beginnings of fluorine chemistry in Slovenia

Work on fluorine chemistry in Slovenia dates back to the year 1953 when the Laboratory for Fuel and Uranium Hexafluoride was established at the Jožef Stefan Physical Institute[1] as part of planned research and development in the field of nuclear technology in the then Yugoslavia. The Director of the institute, Professor Anton Peterlin, appointed Professor Branko Brčić (who at that time also chaired the Department of Inorganic Chemistry at the University of Ljubljana) to head up this new laboratory, the main purpose of which was to master the production of elemental fluorine and the synthesis of uranium hexafluoride. Since fluorine was classified as a strategic material on account of its military uses, it was impossible at the time for us to buy fluorine or the special equipment required for its generation and use.

Boris Žemva

In the beginning, therefore, the activities of Brčić's group centred on the development and construction of an electrolytic cell for the production of elemental fluorine – work which necessitated mastering the demanding experimental techniques well known to be associated with this halogen. Important among Brčić's first coworkers were Jože Slivnik, who was studying chemistry at the University of Ljubljana, and the technician Anton Zemljič.

Development of electrolytic cells for production of elemental fluorine

The first electrolytic cell which was constructed was a copy of von Wartenberg's cylindrical cell dating from the year 1930,[2] which allowed only a relatively low electric

[1] Jožef Stefan (1835-1893) was one of the most distinguished physicists of the nineteenth century. Born to Slovenian parents in Sveti Peter near Celovec, he graduated in Mathematics and Physics from the University of Vienna. He originated together with Ludwig Boltzmann the law that the total radiation from a black body is proportional to the 4th power of its absolute temperature. The Institute of Physics, a predecessor of the Jožef Stefan Institute, was established within the Slovenian Academy of Sciences and Arts in 1949.

[2] See Chapter 24 for information on Hans von Wartenberg's career.

current to be used – up to six amperes. On the basis of that experience, a new 20 A cell was constructed with an anode made of nickel and an electrolyte of composition approximately KF·2HF [1]. Operated at 393 K, to keep the electrolyte molten, this cell provided the group with its first usable quantities of elemental fluorine and enabled the first few grams of uranium hexafluoride to be prepared in 1958 via the well-established route $UO_2 \rightarrow$ (with HF) $UF_4 \rightarrow$ (with F_2) UF_6. Work with this F_2 generator was not simple: its cover, carrying the bell which separated the anodic space from the cathodic one, was not fastened to the top of the cell but was only laid upon it; therefore, since the cell was not hermetically sealed, the opportunity existed for moist air to enter it and cause explosions during the production of fluorine [2]. One particular explosion caused the massive cover weighing several kilograms to be blown to the top of the fume hood together with the HF traps and all other apparatus connected to the fluorine outlet tube.

Since working with bigger cells is less difficult than working with smaller ones, in the late fifties a 250 A electrolytic cell was designed and constructed. This worked better than the 20 A cell, although some of the old problems remained and some new ones were encountered. This cell was also not completely airtight, so the electrolyte absorbed some water during long periods of non-operation; additionally, the amount of fluorine produced was too great to be used immediately. The solution to the latter problem was seen as storage of the fluorine produced in cylinders under pressure, but at that time this did not prove easy to achieve. The fluorine (b.p. 85 K) had to be condensed in metal traps cooled in liquid nitrogen (77 K) and then expanded under pressure into a steel cylinder. The main problem with this procedure was how to regulate the flow of fluorine during its condensation. When the condensation of fluorine was too quick, molten electrolyte was drawn into the cell exit line, where it cooled, solidified and caused a blockage; when the rate of condensation was too slow, a pressure of fluorine built up. In both cases, fluorine flowed under the bell into the cathodic space where an explosive reaction with hydrogen took place. Normally helium is used in the liquefaction of fluorine to maintain the appropriate pressure, but unfortunately in those days it was not possible for us to buy helium and therefore it was necessary to find another solution.

In 1960 Andrej Šmalc joined the group and constructed a new fluorine generator – an 80 A electrolytic cell which had a tightly-sealed lid, hence the electrolyte stayed dry even after a long period of non-operation [3]. Together with Anton Zemljič he tackled the problem of storing fluorine in cylinders under pressure. In order to maintain atmospheric pressure in the anode compartment, they used a specially-designed control gauge connected to a magnetic valve to regulate the flow of fluorine into the liquid nitrogen trap. Fluorine thus condensed was transferred under a pressure of 30 bar into special cylinders which were made from oxygen cylinders by replacing the usual valves with valves suitable for use with fluorine (constructed in our departmental workshop) and then passivating the inner surface of the cylinders with fluorine. On the basis of this work, a pilot plant was built for the production, purification and filling of cylinders with elemental fluorine from a 120 A electrolytic cell [4] (Fig. 31.1). The system employed for the purification and loading of fluorine into 10-litre cylinders at a pressure of 30 bar is shown in Fig. 31.2; we used this equipment for about 25 years, until it finally became possible to buy what we needed. We are ready at any time to start up our plant again if the situation changes. For example, after the Slovenian War of Independence in 1991 there was an embargo on the importation of fluorine into Slovenia.

Fig. 31.1. 120 Ampere electrolytic cell for the production of elemental fluorine (built in 1966).

This pioneering period was distinguished by extraordinary devotion to research by each member of our group, and by a unique collective eagerness to overcome difficulties. Jože Slivnik was not just the Head of the group but also the centre of inspiration. In spite of the general lack of apparatus and spare parts, there was a strong belief that everything could be made in 'house'. Possessed of unshakeable confidence in their own professional expertise, group members adopted the attitude that no task would prove too difficult and that all obstacles would be surmounted. The know-how obtained during this period proved to be very advantageous later.

Synthesis of xenon(VI) fluoride and other xenon chemistry

Although the Jožef Stefan Nuclear Institute was under the authority of the Federal Commission for Nuclear Energy at the beginning of the 1960s, the then Department of Fluorine Chemistry at the institute was also doing fundamental research in the general field of fluorine chemistry, in addition to research on nuclear materials. The first of these

Fig. 31.2. System for the purification of elemental fluorine and its transfer into cylinders (built in 1966).

endeavours concerned complexes of metal fluorides with hydrazine and hydrazinium fluorides. This research area was initiated when the possibility of reducing uranium(VI) to uranium(IV) with hydrazine was tested [5]. A special impetus to fundamental research in the field of fluorine chemistry was provided by the discovery of the first noble gas compound by Neil Bartlett in 1962 [6][3], and the subsequent synthesis of the first binary compound of xenon, XeF_4, in the Argonne National Laboratory, USA [7]. The Department of Fluorine Chemistry here, headed by Jože Slivnik, was immediately engaged in this newly created field of research – the chemistry of noble gases. Since Slivnik's group had the experience and the necessary equipment for work with elemental fluorine at high pressures (up to 200 bar) and temperatures (up to 1000 K), they were immediately able to repeat the synthesis of XeF_4 [8], samples of which were used in NMR studies by physicists at the Jožef Stefan Institute; using higher pressures and a higher molar ratio of xenon to fluorine (1:20), they went on to become the first in the world to succeed in preparing XeF_6, xenon(VI) fluoride, the synthesis of which was published in *Croat. Chem. Acta* in December 1962 [9].

At the beginning of 1963, three independent publications appeared on the synthesis of xenon(VI) fluoride from American laboratories [10-12]; this shows that during the pioneering period of noble gas chemistry time was a most important factor. Senior staff here still remember the night when an entire experiment went wrong. Everybody was tired and wanted to go home except Slivnik, who asked the others to help him set up a new experiment before they left. Of course, once the new experiment was under way, nobody

[3]See Chapter 3 for details of Neil Bartlett's contributions to fluorine chemistry.

wanted to miss the outcome. This was in fact successful, and the next day the very first batch of xenon(VI) fluoride became available. When they were convinced that they had synthesized a new binary fluoride of xenon, Slivnik and his team naturally wanted to publish this achievement as soon as possible. At that time all publications had to be approved by the Director of the institute before they could be sent to journals, and when Slivnik was on his way to the Director with the XeF_6 paper, he met a physicist who explained to him that theoreticians have already calculated that the synthesis of XeF_6 should not be possible. Fortunately, the Director believed the experimental results and the paper was immediately sent to the journal *Croatica Chemica Acta* [9].

At that time I was a student in the 4th year of chemistry studies at the University of Ljubljana, supported by a grant from the textile industry. However, I became fascinated by fluorine and its compounds at a lecture given by Slivnik, who explained how XeF_6 had been synthesized by scientists in the Ljubljana group and provided some details about the experiment. With Slivnik's help, I started working at the Jožef Stefan Institute in the autumn of 1965, and ever since then the chemistry of noble gases has been one of my major interests.

The synthesis of XeF_6 put us on the world map of fluorine chemistry groups. As a result of the work with noble gases, a new technique for synthesis under high pressure was developed and new pressure vessels were constructed, better valves for work with fluorine were acquired, transmitters for the measurement of high pressures of fluorine were developed, etc. (This equipment, in practically unchanged form, is still in use here.) Coworkers of the group also developed high-vacuum and dry-box techniques for work with reactive and volatile fluorides, and also complete methodology for the characterisation of binary and complex fluorides, ranging from measurements of magnetic susceptibility to X-ray diffraction analyses. Note that during these formative years of fluorine chemistry in Slovenia, Boris Frlec, Andrej Šmalc and Bogdan Volavšek made invaluable contributions to the development of equipment and experimental techniques for work with elemental fluorine and fluorides. For the last three and a half decades, Borka Sedej was responsible for the indispensable support provided by the analytical laboratory to all fundamental and applied research performed in the Department.

Following the synthesis of XeF_6, a study was made of the kinetics of the reaction between xenon and fluorine under conditions suitable for the synthesis of xenon fluorides in larger quantities; also the catalytic effect of the first-row transition metal fluorides on the rate of reaction between xenon and fluorine was investigated [13]. During this research, a series of new xenon(II) and xenon(VI) fluorometalates was isolated [14], *e.g.* $n XeF_2 \cdot TiF_4$ ($n = 1.5; 1; 0.5$), $XeF_2 \cdot VF_5$, $n XeF_2 \cdot CrF_4$ ($n = 1; 0.5$), $n XeF_2 \cdot MnF_4$ ($n = 1; 0.5$), $n XeF_6 \cdot TiF_4$ ($n = 4; 1; 0.5$), $n XeF_6 \cdot VF_5$ ($n = 1; 0.5$), $XeF_6 \cdot CrF_4$, $n XeF_6 \cdot MnF_4$ ($n = 4; 2; 1; 0.5$), $XeF_6 \cdot FeF_3$ and $XeF_6 \cdot CoF_3$. Owing to our expertise in the large-scale synthesis of xenon fluorides, the International Atomic Energy Agency funded a project dealing with the separation of waste radioactive krypton and xenon from nuclear power plants and their storage in solid form.

Photochemical syntheses

In the seventies we turned our attention to the field of photochemical reactions involving elemental fluorine. Initially the influence of the reaction conditions on rates for some

Fig. 31.3. Photochemical synthesis of KrF_2 in liquefied fluorine at 77 K.

gaseous systems (*e.g.* the AsF_5-O_2-F_2 system) [15] was studied. Similar studies, extending also to the influence of catalysts, were also performed in the case of the synthesis of XeF_2 and, later, of other xenon fluorides [16, 17] and chlorine(V) fluoride [18]. For the benefit of organic chemists, a new method was developed for the preparation of larger quantities of XeF_2 (up to 500 g in one batch). In 1972 Jože Slivnik invited Alfred Pollak, an Assistant

Professor at the University of Ljubljana, to study the fluorination of organic compounds with XeF_2. During December of that year Alfred Pollak and Marko Zupan, a junior research assistant at the University, performed their first successful fluorination with XeF_2. This was the beginning of organic fluorine chemistry in Slovenia. By mid-1998, nearly 2 kg of XeF_2 had been prepared for use by organic chemists.

Our investigations of photochemical reactions were later also extended to systems with liquefied fluorine, and a special apparatus was developed for fluorination at 77 K which was used successfully for the first time to synthesise KrF_2 (Fig. 31.3). This method is still one of the most efficient methods for producing KrF_2, yielding up to 1 g of material per hour [19] (this is about ten times more than the previous most efficient method – electrical discharge). Thus the chance arose to do some krypton chemistry, and, later on, also to specialize in the high-oxidation-state chemistry of transition elements.

Besides this, a photochemical method for the production of O_2F_2 by irradiation of a liquefied mixture of oxygen and fluorine was developed [20]. This method is also used for the preparation of large quantities of spectroscopically-pure fluorine [20] (up to a hundred litres at STP of pure fluorine per batch). The main impurities in fluorine are usually hydrogen fluoride, carbon fluorides and oxygen. The last of these is the most difficult to remove. With this photochemical method, oxygen was converted to O_2F_2 which is a solid at 100 K so fluorine could be distilled over. By repeating this procedure four times, very pure fluorine is obtained.

Binary and complex fluorides with transition metals in high oxidation states

In the 1980s we studied the preparation of XeF_5^+ salts with metals in higher oxidation states using KrF_2 as an oxidant, XeF_6 as a moderately strong base, and anhydrous hydrogen fluoride as solvent. Using this synthetic route the following XeF_5^+ salts were isolated: XeF_5AgF_4 [21, 22], $(Xe_2F_{11})_2NiF_6$ [23] and $(XeF_5)_2NiF_6$ [24]. These salts are excellent starting materials for further research in the field of thermodynamically unstable binary fluorides.

This research in the field of binary fluorides in higher oxidation states was a joint project with Neil Bartlett of the University of California at Berkeley, USA, supported by the US National Science Foundation. During this collaboration a general method for the isolation of thermodynamically unstable binary fluorides was developed [25]. The main idea for the preparation of binary fluorides is fluoride ion capture from their anionic relatives in anhydrous hydrogen fluoride solution by strong fluoride ion acceptors (*e.g.* AsF_5). The method is especially advantageous in the syntheses of highest-oxidation-state transition-metal polymeric fluorides insoluble in anhydrous hydrogen fluoride. Using this method, a whole series of new thermodynamically unstable binary fluorides was isolated *e.g.* AgF_3 [26], Ag_3F_8 [26], Ag_2F_5 [26], CuF_3 and NiF_4 [27], and different crystallographic forms of NiF_3 [27]. All these compounds are high-energy oxidants able to oxidize xenon to xenon fluorides at room temperature, and some of them even at lower temperatures (*e.g.* 200 K). During the study of the oxidizing capability of higher silver fluorides it was found that even cationic Ag(II) is able to oxidize xenon to XeF_2 [28]. It was further shown that cationic $Ag(III)_{solv}$ and cationic $Ni(IV)_{solv}$ are the strongest known oxidants and surpass even KrF^+ salts, up to now the strongest known oxidants [29].

Present activities of the Department of Inorganic Chemistry and Technology

In 1992, the Department of Fluorine Chemistry was renamed the Department of Inorganic Chemistry and Technology. The new name more clearly expresses today's activities, which are no longer centred solely on fluorine chemistry and are much broader in scope. In recent years the Department has been active mainly in the fields of inorganic fluorine chemistry, chemical technology and the environment, and chemical education. The inorganic research mainly concerns high-energy oxidizers, chemistry in superacids, lanthanoid metals, inorganic materials with interesting electrical, optical and magnetic properties, binary fluorides as supercritical fluids, and technologically important halogenated compounds. In the chemical technology/environmental area, many projects are devoted to the optimization of technological processes in Slovenian industry and to ecological problems, especially to investigations related to the environmental impact of fossil fuel combustion and the treatment of gaseous and liquid industrial wastes. In this context, part of our activity is directed towards solving the problems of volatile waste fluorides from different industries. Recently, pollution by volatile fluorides has been detected by biomonitoring. Finally, the Department's role in the education of undergraduate and postgraduate students should not be overlooked: 101 BSc theses, 29 MSc theses and 21 PhD theses have been completed as a result of research studies since 1955. Besides this, the education of secondary school students in our School of Experimental Chemistry is taking place, and each year around 30 groups are involved in one-week courses.

We have been particularly fortunate in obtaining long-term support from the Ministry of Science and Technology of the Republic of Slovenia; however, the level of support has been too small to cover all activities and a considerable proportion of the funds (30-50%) has been obtained from Slovenian and foreign industries.

References

1 J. Slivnik, A. Zemljič, *Reports J. Stefan Institute, 5* (1958) 49.
2 G. H. Cady, in J. H. Simons (ed.), *Fluorine Chemistry*, Vol. 1, Academic Press, New York, 1950, p. 304.
3 J. Slivnik, A. Šmalc and A. Zemljič, *Vestn. Slov. Kem. Drus., 9* (1962) 61.
4 J. Slivnik, A. Šmalc and A. Zemljič, *Vestn. Slov. Kem. Drus., 12* (1965) 17.
5 B. Frlec, B. S. Brčić and J. Slivnik, *Croat. Chem. Acta, 36* (1964) 173.
6 N. Bartlett, *Proc. Chem. Soc.*, (1962) 218.
7 H. H. Claassen, H. Selig and J. G. Malm, *J. Am. Chem. Soc., 84* (1962) 3593.
8 J. Slivnik, B. S. Brčić, B. Volavšek, A. Šmalc, B. Frlec, A. Zemljič and A. Anžur, *Croat. Chem. Acta, 34* (1962) 187.
9 J. Slivnik, B. S. Brčić, B. Volavšek, J. Marsel, V. Vrščaj, A. Šmalc, B. Frlec and A. Zemljič, *Croat. Chem. Acta, 34* (1962) 253.
10 J. G. Malm, I. Sheft and C. L. Chernick, *J. Am. Chem. Soc., 85* (1963) 110.
11 F. B. Dudley, G. Gard and G. H. Cady, *Inorg. Chem., 2* (1963) 228.
12 E. E. Weaver, B. Weinstock and C. P. Knop, *J. Am. Chem. Soc., 85* (1963) 111.
13 B. Žemva and J. Slivnik, *J. Inorg. Nucl. Chem., H.H. Hyman Memorial Issue*, (1976) 173.
14 B. Žemva, *Croat. Chem. Acta, 61* (1988) 163.
15 A. Šmalc and K. Lutar, *J. Fluorine Chem., 9* (1977) 399.
16 A. Šmalc, K. Lutar and J. Slivnik, *J. Fluorine Chem., 8* (1976), 95.
17 K. Lutar, A. Šmalc and J. Slivnik, *Vestn. Slov. Kem. Drus., 26* (1979) 435.
18 A. Šmalc, B. Žemva, J. Slivnik and K. Lutar, *J. Fluorine Chem., 17* (1981) 381.
19 J. Slivnik, A. Šmalc, K. Lutar, B. Žemva and B. Frlec, *J. Fluorine Chem., 5* (1975) 273.

20 A. Šmalc, K. Lutar and J. Slivnik, *J. Fluorine Chem.*, *6* (1975) 287.
21 K. Lutar, A. Jesih and B. Žemva, *Rev. Chim. Miner.*, *23* (1986) 565.
22 K. Lutar, A. Jesih, I. Leban, B. Žemva and N. Bartlett, *Inorg. Chem.*, *28* (1989) 3467.
23 A. Jesih, K. Lutar, I. Leban and B. Žemva, *Inorg. Chem.*, *28* (1989) 2911.
24 A. Jesih, K. Lutar, I. Leban and B. Žemva, *Eur. J. Solid State Inorg. Chem.*, *28* (1991) 829.
25 B. Žemva, K. Lutar, A. Jesih, W. J. Casteel, Jr. and N. Bartlett, *J. Chem. Soc., Chem. Commun.*, (1989) 346.
26 B. Žemva, K. Lutar, A. Jesih, W. J. Casteel, Jr., A. P. Wilkinson, D. E. Cox, R. B. Von Dreele, H. Borrmann and N. Bartlett, *J. Am. Chem. Soc.*, *113* (1991) 4192.
27 B. Žemva, K. Lutar, L. Chacon, M. Fele-Beuermann, J. Allman, C. Shen and N. Bartlett, *J. Am. Chem. Soc.*, *117* (1995) 10025.
28 B. Žemva, R. Hagiwara, W. J. Casteel, Jr., K. Lutar, A. Jesih and N. Bartlett, *J. Am. Chem. Soc.*, *112* (1990) 4846.
29 G. Lucier, C. Shen, W. J. Casteel, Jr., L. Chacon and N. Bartlett, *J. Fluorine Chem.*, *72* (1995) 157.

Chapter 32

GOING WITH THE FLUO

RONALD ERIC BANKS

Chemistry Department, UMIST, Manchester M60 1QD, UK

Prologue

My own enduring fascination with fluorine stems from finding out about *perfluorocarbon chemistry* when I was a novice in W. K. R. Musgrave's research group in the 1950s (it was termed *fluorocarbon* chemistry then, of course). That event and just how I came to be classed as an *organofluorine chemist*, plus information concerning some of my activities in the fluorine field over the years are documented here. I've sectioned my story to cover six contiguous eras: 1932-1953 (birth to graduation); 1953-56 (serving my apprenticeship in fluorine chemistry); 1956-58 (a spell in the fluorochemicals industry); 1958-74 (becoming steeped in fluorine chemistry); 1974-94 (flying solo); 1994-2000 (tapering off). And I've mentioned as many of the people I've 'interacted' with while pursuing fluorine chemistry as possible. Mostly I regret not having been able to record the names and achievements of every research student it has been my privilege to work with – those many foot soldiers of fluorine chemistry who enabled me to advance so far. It has also been my good fortune to receive much help from skilful technical staff in UMIST's Chemistry Department; I'm truly grateful to all those involved, and particularly to Dr Roger Perry and his assistants in the Analytical Laboratory for elemental analyses (with F values a priority, of course) and to our HF Technician, Mr E. ('Ted') A. Laing, who made sure that our F_2 generators never failed us and that our Simons ECF cells always functioned properly.

School and undergraduate years

Auspiciously perhaps, I was born on Gunpowder Plot Day[1] (5 November) 1932 in the Potteries (Stoke-on-Trent of Wedgwood fame). In common with Moissan [1], my father was a railwayman and my mother a seamstress; talented, God-fearing, working-class people, they strove hard to ensure that both my elder sister (Margaret, who became an accomplished pianist and introduced me to orchestral concerts and Gilbert and Sullivan comic operettas) and I were given every chance to enjoy good educations within the British

[1]Guy Fawkes Day – the anniversary of the discovery of the Gunpowder Plot, an unsuccessful conspiracy to blow up King James I and Parliament in Westminster, London, in 1605. This event is still widely celebrated in Britain every 5th of November with fireworks and bonfires.

state system. This enabled me to enter Bede College[2] in the Durham Colleges Division of the University of Durham in 1950, armed with my food ration book and a Staffordshire County Scholarship, well qualified to commence studies in the university's honours school of chemistry, and – as the crow flies – about 120 miles north-east of my home in central England.

I showed a marked aptitude for science at grammar school and chose to major in chemistry rather than physics at university because just being in a chemistry lab gave me a buzz. In fact, if it had been allowed, I would have spent the major part of every school day (except for sports days) working at the bench when I reached the sixth form (pre-university level at Leek High School[3]). As it was, we did a substantial amount of practical chemistry overall, and I became very proficient at qualitative and quantitative inorganic analysis and also gained experience in both analytical (Lassaigne's test – 'sodium fusion') and preparative (making chloroform, aniline etc.) organic chemistry. Physical chemistry didn't feature much, although we did a fair amount of practical physics. As for mathematics... well, I got by okay. My maths master coached field hockey, and I'm eternally grateful to him for introducing me to that sport, which I played at top club level (including an unbroken 6-year spell in the Durham Colleges 1st XI – a record) for many years until wear-and-tear on my right knee and left hip called a halt.[4]

Well before my last year at school, I'd decided to train to become a grammar school chemistry teacher. On paper, Bede College[2] seemed to be an ideal institution to attend – especially in view of my religiousness; and that breathtaking view of the cathedral and castle one gets as northbound trains pull into Durham railway station clinched it for me. The icing

[2]Bede College (founded 1839) was a Diocesan Training College of the Church of England for men teachers, and also a recognised College of the University of Durham. Bede (?673-735 AD) was the greatest Anglo-Saxon scholar of his generation – a saintly monk and priest at the monastery in Jarrow (Northumbria) who devoted his life to study, teaching and writing; his prodigious literary output included the Latin *Ecclesiastical History of the English People* (publ. 731), which apparently became the international bestseller of the early Middle Ages [2]. In my day, Durham's collegiate system extended 14 miles northwards to Newcastle upon Tyne to include King's College, where Neil Bartlett started on his life's work in fluorine chemistry. Although our research years overlapped (1953-56 vs. 1954-57), I don't recall meeting Neil before 1962 (at the 2nd ISFC in Estes Park, Colorado); that 14 miles was quite a barrier.

[3]I entered the grammar school system at Longton High School in Stoke-on-Trent and transferred to Leek High School when my father became stationmaster at Rudyard Lake, a North Staffordshire beauty spot near the mill-and-market town of Leek. My first encounter with HF occurred at Leek: I was helping the chemistry master to reorganize his stores and came across a gutta-percha bottle containing about 200 ml of hydrofluoric acid. 'Steer clear of that', he ordered, 'the contents cause terrible burns'.

[4]In 1989, while anaesthetized with isoflurane ($CF_3CHClOCHF_2$), I was fitted with a 'plastic' hip by Kevin Hardinge, MCh Orth, FRCS, a consultant orthopaedic surgeon from the internationally famous Centre for Hip Surgery at Wrightington Hospital near Wigan (Lancashire), established by the late Sir John Charnley (1911-1982). Charnley – who trained for surgery in the Medical School of the Victoria University of Manchester – provided me with the photograph of a polytetrafluoroethylene-based arthroplasty which can be seen in the 1st edition (1964) of my monograph *Fluorocarbons and their Derivatives*. By the time I composed the 2nd edition, that photograph had to be removed because Charnley had abandoned PTFE 'sockets' (owing to adverse tissue reactions caused by wear debris) in favour of cups made from high-density polyethylene (similar to the one I carry now). Little did I think when Haydn Sutcliffe and I visited John Charnley at Wigan in the early '60s that in later life I would join the ranks of the many thousands of 'hip-problems' sufferers who owe so much to his dedication and skill. According to Kevin Hardinge's latest paper [D. H. Sochart and K. Hardinge, *J. Bone and Joint Surg.*, 80-B (1998) 577], my hip implant should be good for at least another five years.

on the cake came when I arrived at Bede for an interview in early March 1950 and discovered that not only did the college overlook the university sports fields, but also the Principal (the Reverend Canon G. E. Brigstocke) was President of the college hockey club. A few days later, a letter arrived from Canon Brigstocke offering me 'a place in October, 1950, as a candidate for a Degree followed by a year's course for the Diploma in Education'.

Bedemen breakfasted early (7.30 am) on weekdays so that they could attend morning prayers in chapel before lectures commenced. This proved crucial in determining whether I took the honours course in chemistry or the ordinary (or even the honours course in physics), for when I arrived at the University Science Laboratories on registration day 1950 I was in time to get the only unassigned place in the 1st year honours chemistry course. Dr G. H. Christie, the senior organic lecturer who administered the chemistry department because Professor F. A. Paneth spent a lot of time elsewhere,[5] told me that Canon Brigstocke had failed to consult him about me following my college interview seven months earlier!

Otherwise, everything went fairly smoothly during my undergraduate days, probably because I became a workaholic, although I certainly played a lot of hockey (sometimes 3 matches per week in the Michaelmas and Lent Terms), which meant that I had to forgo cricket in the Summer Terms in order to catch up on my studies.[6] Final examinations came and went, leaving me holding a first-class honours degree (BSc). This success enabled me to make my apologies to the university's Institute of Education and register instead as a DSIR-maintained (*D*epartment of *S*cientific and *I*ndustrial *R*esearch) PhD student in Dr W. K. R. Musgrave's laboratory, with my military service deferred. A downside was that I had to vacate my room in Bede and find lodgings. Gone forever were those cosy Christmas and Easter breaks – catching up on lecture notes and enjoying home comforts – and those frustratingly long summer vacations, reading around my chosen subject and discovering the real world by (of economic necessity!) taking on 'vac jobs'. Gone also was my plan to become a schoolteacher: I now set my sights on working towards a post in the higher education sector where I could specialize in organic chemistry – an amazingly information-rich field I'd been drawn towards through some excellent lectures by Dr Christie and his younger colleague, Dr W. K. R. Musgrave.

[5]George Hallatt Christie (*d* 1965), well known for his contributions to the discovery of 'biphenyl isomerism' (atropisomerism) in the 1920s, played a major role in the administration of the chemistry department, often standing in for its head, the late Professor Friedrich Adolf Paneth (1887-1958). Paneth, a well known radiochemist, meteorite expert, atmospheric scientist, alkyl radical pioneer (via the Paneth-Hofeditz 'lead mirror test' for methyl radicals, 1929), and member of the joint British-Canadian atomic energy team in Montreal during WWII [3], was a friend of Harry Emeléus (note the source of information quoted in ref. [3]).

[6]This also eliminated a serious potential hazard: perhaps having to face the bowling of 'Typhoon' Tyson (Hatfield College; 1950-55). Frank Tyson became a national cricketing hero in 1954-55 for his part in winning the Ashes test series on Australian soil, and was acclaimed as the fastest bowler in the history of cricket after he had taken 7 wickets for 27 runs in Australia's second innings against England in Melbourne.

Aggressive fluorine chemistry with Musgrave in Durham: getting to know the ropes (1953-56)

Discovering organofluorine chemistry

Although Musgrave pioneered in organofluorine chemistry at Birmingham University[7] during WWII and had been extending his work in that area since he arrived in Durham in 1945, I detected no sign of his research interests in his lectures; indeed, if it had not been for conversations I'd had with his research students (like me later, they supplemented meagre maintenance grants by teaching – 'demonstrating' – in undergraduate labs), I would have classed him as a natural products chemist (his lectures on carbohydrates[8] and steroids were impressive). Overall, though, I did pick up information on fluorine chemistry from other parts of the degree course, and particularly so in the final year (1952-53) by reading (heavy scanning would be a more appropriate description) the second edition (1952) of *Modern Aspects of Inorganic Chemistry* by H. J. Emeléus and J. S. Anderson. The short section entitled *Fluorides of Carbon* caught my eye, so I became aware of the existence of a large group of compounds 'known collectively as fluorocarbons' (C_nF_{2n+2}) 'which parallel the hydrocarbons in structure and, to a certain extent, in physical properties'. Three now-classical methods for achieving exhaustive fluorination of hydrocarbons were defined: liquid-phase and vapour-phase ('catalytic') direct fluorination (with F_2-N_2 blends) and the cobalt fluoride process. Soon afterwards, I met Professor Emeléus in person for the first of many times: he was the External Examiner in Chemistry at Durham when I sat my finals.

During my compulsory oral exam (a one-to-one affair), Emeléus indicated that I was heading for a 'first' and then asked about my future. After commenting favourably on Musgrave's fluorine research, he remarked casually that one of his young colleagues at Cambridge, a Dr Haszeldine, had recently become interested in perfluorinated amines. After writing out the structure of one of them (perfluoro-*N*-fluoropiperidine), he turned up the heat by asking me to deduce something about its reactivity. Afterwards, he inquired whether I would consider joining Haszeldine's group at Cambridge. Emeléus also tried to poach the other student in my year who got a 'first', but the Durham staff objected to losing PhD students in this way, and the matter was dropped. Perhaps Emeléus was psychic, because I did catch up with Haszeldine later and since then have carried out a lot of research involving perfluoro- *N*-fluoropiperidine (**1**; see Scheme 32.1).

---

[7]Now an Emeritus Professor of Organic Chemistry, William Kenneth Rodgerson Musgrave (*b.* 1918) received his degrees (BSc, PhD, DSc) from the University of Birmingham, UK. Following his PhD work, he spent a year (1944-45) at the nuclear facility at Chalk River, Canada where the first reactor outside the US provided the experimental tool around which Britain's postwar nuclear work was to grow. After meeting Paneth in Canada, he joined the teaching staff at Durham and rose through the ranks to become the university's Acting Vice-Chancellor (1979) [4].

[8]No doubt Musgrave was good on sugar chemistry because during his student days Professor Sir Norman Haworth strode the corridors of the Chemistry department in Birmingham [Haworth shared the 1937 Nobel Prize for chemistry (with Paul Karrer, Zurich) for his work on carbohydrates and Vitamin C]. Haworth's team included the young Maurice Stacey (1907-1994) – of whom more later. It was through Haworth's volition that Musgrave (1941) conducted some of the first research on perfluorocarbons in the UK.

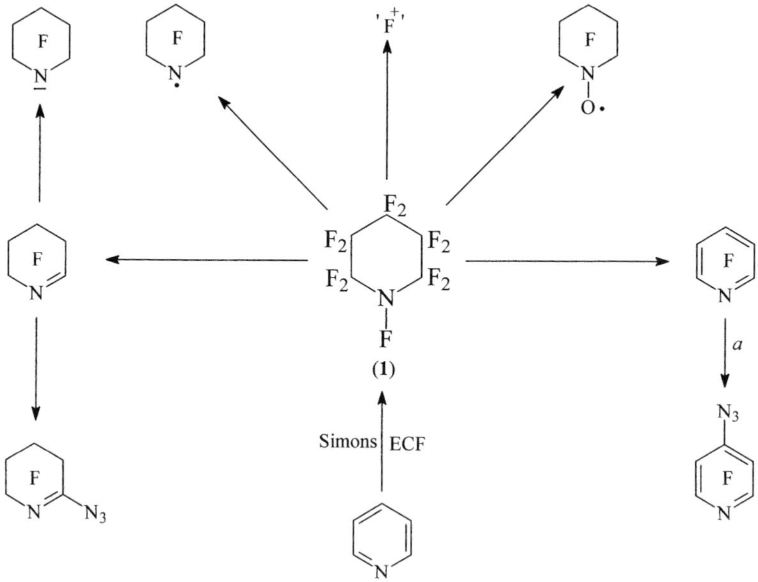

Scheme 32.1. An indication of the author's activities based on perfluoro-*N*-fluoropiperidine (**1**). [a] Actually, sodium azide was not among the reagents we used initially to investigate nucleophilic substitution in pentafluoropyridine [5]: by the time 4-azidotetrafluoropyridine was needed for studies on perfluoroarylnitrenes, production of pentfluoropyridine from (**1**) had long been abandoned in favour of halex perfluorination of pentachloropyridine.

Chlorofluorinating benzene

Musgrave did not define my research topic until I arrived back in Durham at the beginning of the September following my graduation (BSc) in June 1953. During the interim he had sent me a copy of Weissberger's *Technique of Organic Chemistry, Vol IV, Distillation* to peruse, but nothing on fluorine chemistry. I found out why when he told me that our ultimate objective was to synthesize hexafluorobenzene via a route involving the vapour-phase chlorofluorination of benzene with chlorine trifluoride, a notoriously reactive incendiary agent which he had acquired in quantity (a 10 lb cylinder) from ICI via Jack Rudge. By using concentric-tube distillation columns he hoped that I would be able to isolate potential hexafluorobenzene precursors (Scheme 32.2) from the complex ill-defined material he knew[9] could be obtained from benzene and ClF_3-N_2 blends in an all-metal apparatus of the type he had used at Birmingham to study the direct fluorination of hydrocarbons [6]. That hope was not fulfilled: after using a more advanced, packed (Cu clippings) reactor incorporating a Bigelow 'cool flame' burner [7] to produce nearly 5 kg of chlorofluorocyclohexane material at 220-260 °C [$ClF_3 + C_6H_6$ (both diluted with N_2) $\rightarrow C_6H_xF_yCl_z$ ($x = 0$-6; y

[9] I inherited my topic from Mr R. Sowler, who abandoned the work prematurely and had left by the time I arrived on the scene, leaving Musgrave's lab unoccupied. Sowler is mentioned in Musgrave's review 'The Halogen Fluorides – Their Preparation and Uses in Organic Chemistry' [*Adv. Fluorine Chem.*, *1* (1960) 1]. Note that Sowler had no knowledge of gas-liquid chromatography (GLC), the first account of which was published by James and Martin towards the end of 1952.

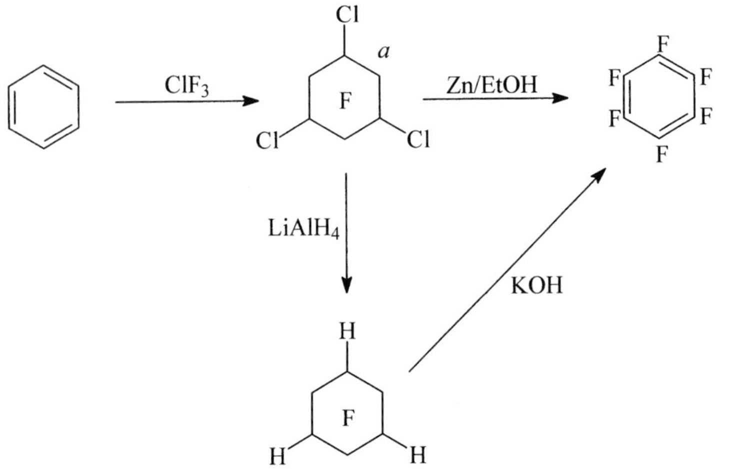

Scheme 32.2. An illustration of W. K. R. Musgrave's strategy. [a] In fact, the complex product of thermal chlorofluorination of benzene was shown by the author to contain trichlorononafluorocyclohexane material: C_6H_6 + ClF_3 → $C_6F_{10}Cl_2$, $C_6HF_9Cl_2$, $C_6H_2F_8Cl_2$, $C_6H_3F_7Cl_2$, **$C_6F_9Cl_3$**, $C_6H_6F_5Cl$, C_6H_5Cl, C_6H_4FCl + unidentified products (R. E. Banks, PhD Thesis, University of Durham, 1956).

and $z = 0\text{-}12$; $x + y + z = 12$)], then employing tedious precise distillation methodology followed by large-scale gas-liquid chromatography, I had to admit defeat and write my PhD thesis (submitted in June 1956) without having prepared even a sniff of hexafluorobenzene. Also, I hadn't addressed my secondary objective, mentioned in passing by Musgrave at the very outset, namely to develop a chlorofluorination route to pentafluoropyridine – an unknown compound then, unlike hexafluorobenzene. 'Once you've done that' (made C_6F_6), he'd called after me, as I was leaving his office, 'try applying the method to pyridine'. Will PhD supervisors ever change?

Details of my chlorofluorination work were not published until loose ends had been tied up by Peter Johncock and Dick Mobbs [8], research students contemporary with Dick Chambers, who took the photograph of Musgrave, Forster Cuthbertson and me reproduced here as Fig. 32.1. Dick was working on a final-year undergraduate research project in our lab at the time, and one afternoon managed to blow up one of my semi-micro concentric-tube distillation columns, thereby suffering some nasty cuts. The incident certainly didn't put him off fluorine chemistry. Amazingly, my full-size concentric-tube columns survived (relatively unscathed) an explosion which occurred between midnight and 6 am during 1954. At that juncture I was developing a new test mixture with which to determine the efficiencies (plate equivalencies) of my concentric-tube stills under reduced pressure; this entailed determining vapour-liquid equilibrium data for the methyl benzoate-methyl caprylate system at 50 mmHg, six hours being allowed for my modified Williams still to come to equilibrium before samples of distillate and distilland were analysed by refractometry. Thus, I was on a 12 noon → 6 pm → 12 midnight → 6 am schedule for a week, hence my certainty regarding the period during which the explosion happened. It took place in a fume cupboard containing a fair-sized piece of glass apparatus M. C. (Charles) Tanner was using to study the action of ClF_3 on CCl_3CO_2H [9]. When I arrived at the lab at about

Fig. 32.1. Left to right: W. K. R. Musgrave, R. E. Banks and F. Cuthbertson enjoying a tea break in 1956.

5.45 am to perform analyses, only the metal frame on which Charlie's apparatus had been clamped remained in the fume cupboard, and the blast screen he'd left in front of a Pyrex flask containing a reaction mixture had been blown across the room. One of the lab's lights was hanging down – suspended from only one of its two chains – and there was powdered glass and glass 'shot' everywhere. The bulbs on my vacuum system had been perforated in several places, some projectiles obviously having gone in at one side and out the other. I telephoned Musgrave, who told me to start cleaning up immediately so that there was no chance of the office cleaners or anyone else seeing what had happened ('otherwise our work might be closed down'); about 20 minutes later he appeared with a coat over his pyjamas to check out the situation. You will deduce how lucky I was not to have been in the lab when the detonation occurred.

The paper containing details of my vapour-liquid equilibrium study and the subsequent testing of my concentric-tube columns appeared in 1956 [10]. The distillation curve it contains for the portion of my complex chlorofluorination product boiling between room temperature and 160 °C reveals a lot about my problems. With hindsight, I should have pyrolyzed each fraction over hot iron in my quest for fluoroaromatics before bringing my home-made preparative GLC apparatus into play.

Further distillation studies connected with my work resulted in a paper dealing with vapour-liquid equilibrium data for the C_6H_6-C_6H_5F system [11]. Again, that was not my first publication, because Musgrave viewed it necessary to develop an improved method of elemental analysis for fluoro-organic materials before I could start tackling the separation and identification of products from the ClF_3-C_6H_6 reaction. Done in collaboration with Forster Cuthbertson, who was known affectionately as 'Flash' by everyone, that piece of research (the semi-micro determination of fluorine, chlorine and nitrogen in organic compounds [12]) is something I'm still very proud of.

Tanner (son of ICI's Dr C. C. Tanner), Forster Cuthbertson and I (all from the class of '53) surely became one of the most skilful groups of young technicians ever seen in an academic fluoro-organic laboratory. Musgrave's lab was noticeably underfunded, so we had virtually to make everything on the apparatus front.[10] Charlie Tanner even fabricated vacuum stopcocks and ground-glass joints; I learned the basic principles of glass-blowing from him, which set me up to become admirably skilled in that art. In fact, my arduous stint in Musgrave's lab set me up for life as an *organofluorine chemist* – respectful but not afraid of HF or ClF_3 (or related 'aggressives'), fully conversant with metal equipment and electrical accessories, a distillation 'king', an expert (for those early days) in gas-liquid chromatography (we designed and built both analytical and preparative machines, including katharometers) and skilled in F analysis. If only Musgrave had possessed an infrared spectrophotometer my repertoire would have been complete for those days; however, I was not totally ignorant of IR analysis since I spent a few days utilizing Dr D. H. Whiffen's home-made grating instrument at Birmingham University while looking for evidence of material bearing C−H and C=C bonds in my chlorofluorination products. That was done by courtesy of Professor Maurice Stacey, who later examined my PhD thesis, and with practical help from Dr (then Mr) James ('Jim') Burdon, whose career in organofluorine chemistry has mirrored mine to some extent.

On the theoretical side of organofluorine chemistry, I didn't learn a lot from my own research but did pick up a considerable amount through reading appropriate literature;[11] this also greatly expanded my general knowledge of the subject and revealed precisely where perfluorocarbon chemistry – a completely man-made subject straddling organic and inorganic chemistry – fitted into the scheme of things. This led me to cherish the hope that I would find an opportunity to specialize further in this sub-field of organic chemistry. I was very lucky where this ambition was concerned: Musgrave had connections with the Imperial Smelting Corporation (ISC) in Avonmouth (Bristol), and arranged for me to visit the company for an interview with Dr A. K. Barbour. He needed an organofluorine chemist to work on a government-sponsored research project and liked what he saw, so with a lovely wife[12] on my arm, deferment of military service assured yet again, and very little money in my bank account, I set off at the end of July 1956 to honeymoon in a West Country seaside resort before starting work in Avonmouth with a salary of £650 per annum.

---

[10]Even electrical heaters for flasks and a great number of other things, including a sizeable metal tube which contained our novel preparative GLC column. This tube was made from a length of cast iron pipe I found some workman digging up outside the Science Laboratories one morning. Fortunately, the superintendent of the mechanical workshop was away on holiday, so I was able to persuade his young apprentice to help me cut the piping to length then clean it up, using a lathe.

[11]Including Musgrave's excellent 'Chem Soc' review *The Reactions of Organic Fluorine Compounds* [13] [published in the year (1954) I was elected to the Associateship of the Royal Institute of Chemistry (ARIC) and also became a Fellow of The Chemical Society (London)], that great little (17 × 11 × 1 cm) 1951 Methuen monograph *Fluorine and its compounds* written by the Cambridge duo R. N. Haszeldine and A. G. Sharpe (from Harry Emeléus' group), and, not least, the £5 book prize *Fluorine Chemistry (Volume I)* (ed. J. H. Simons, Acad. Press, 1950) I picked up just after joining Musgrave's group for having been awarded a first-class degree.

[12]Linda (née Raine) was a local girl who had been educated at Durham Girls' Grammar School and then trained to be a teacher at St Hild's College (the women's counterpart of Bede College; the two colleges merged in 1975 – too late for me!).

An industrial interlude (1956-58): hexafluorobenzene by courtesy of Mistress Fluorine[13]

Chemistry is a noble science and becomes useful in many sorts of business as well as a lasting source of amusement.

LUKE or ROBERT HOWARD [15]

Prelude

Three primary factors are responsible for the super-substituent status accorded to fluorine in the field of organic chemistry [16]: the great strength of the C—F bond (*fluorine forms the strongest single bond to carbon encountered in organic chemistry*); the small size of bound fluorine (*fluorine is the smallest substituent after hydrogen*); and fluorine is extremely electronegative (*fluorine is the most electronegative of all the combinative chemical elements*). The first two '*F-factors*' made it possible more than fifty years ago to contemplate *total* replacement of C—H bonds by C—F in all hydrocarbons and their functionalized derivatives – something not possible in the case of substitution of hydrogen by chlorine. Furthermore, it was realized that when *stepwise* conversion of hydrocarbons or their derivatives to perfluorocarbon analogues was taken into account (*e.g.* $CH_3CO_2H \rightarrow CH_2FCO_2H \rightarrow CHF_2CO_2H \rightarrow CF_3CO_2H$), vast numbers of new organic compounds were there for the making, among which one was certain to find commercially-useful materials. For example, in the famous March 1947 issue of *Industrial and Engineering Chemistry* dealing with the prodigious development of fluorine chemistry during World War II, Grosse and Cady (War Research Laboratories, Columbia University, New York) pointed out that about a trillion organofluorine compounds could be patterned structurally on the million or so organic compounds then known [17]; and in the same journal, McBee (Purdue University and Research Foundation, Indiana) pronounced: 'Peacetime products containing fluorine are expected to include new and useful dyes, plastics, pharmaceuticals, lubricants, tanning agents, metal fluxes, fumigants, insecticides, fungicides, fire extinguishers, solvents, fireproofing compounds, heat transfer media, and other products of benefit to Society' [18]. All this has come to pass, of course: fluorine chemistry nowadays touches all our lives in more ways than even many chemists are aware of – and all thanks to the impressive commercial development of the subject during my lifetime [19-22].

My own awareness of the immediate postwar burgeoning of the fluorochemicals industry came through a two-year period of National Service spent working on a Ministry

[13]This is the name the New York poet Vernon Newton used when giving me permission to reproduce his clever quatrain about element 9 ('*Fervid Fluorine, though just Nine, ...*) in my Moissan centennial article about the isolation of fluorine [14]. I had discovered the quatrain (belonging to Vernon's clever collection of verses *Adam's Atoms: making light of the Elements*) in Truman Schwartz's superb book *Chemistry: Imagination and Implication* (Academic Press, New York, 1973, p. 100), where it is accompanied by the remark 'So seductive is fluorine that even the stern and stalwart stuff of the rare gases finally succumbed to her [fluorine's] electric charms'. I tried unsuccessfully in 1985 to persuade The Royal Society of Chemistry to publish a second edition of Vernon's book (by then out of print). It was rejected on economic grounds. Vernon had intended to change '*Fervid Fluorine*' to '*Perfervid Fluorine*' in a second edition. I've lost touch with him now (my letter sent in September 1998 to his old address in New York came back labelled 'return to sender; no forward order on file').

of Supply contract in ISC's research department, where I got to grips with fluoropolymers – and never let go. That interlude, during which the zest of applied chemistry contributed much to the feast of fluorine chemistry I enjoyed, led me to consider somewhat seriously whether I should abandon my ambition to become an academic.

ISC at Avonmouth, Bristol

In stark contrast to the University science labs in Durham, ISC's research department in Avonmouth was located in a docks/factory area some 200 metres from the main entrance to the National Smelting Company (NSC) – one of ISC's operating companies, where, in May 1948, the UK's first large-scale continuous-production HF plant came on stream [23, 24]. The considerable compensations were that admirable facilities existed for pursuing all manner of fluorine chemistry, technical help was on hand from skilful Assistants who were studying chemistry part-time at Bristol Technical College, and within reasonable bounds, I was allowed to manage my own team's consumables/apparatus budget.

A less attractive feature was that I had to spend a fair amount of time on NSC territory running a relatively small stand-alone brick-metal-and-concrete multi-kilo lab (the 'Monomer Laboratory'); this was sited next to the fluorine plant and piped-up to it so that we could refuel our large cobalt fluoride reactors (Fig. 32.2). Nearby was kept a large supply of complimentary cool fresh milk which workers with HF or F_2 were advised to consume at the rate of two pints per day to 'settle the fluoride'; we certainly did our best to overcomply, despite the cost of instant coffee.

Two commercial (Pennsalt) 2000-amp medium-temperature ($KF{\cdot}2HF$ electrolyte) F_2 generators were available for use. Fitters from the HF plant serviced these when necessary, but otherwise we switched them on ourselves. Waste fluorine was burned in coal gas and the effluent scrubbed with aqueous alkali before discharge to atmosphere. Lighting the burner was a risky business, particularly when my senior assistant Mr A. ('Tony') E. Tipping used a blazing rolled-up sheet of newspaper as the source of ignition. Tony had a neat way, though, of testing the quality of fluorine entering the pipework in the monomer lab: determine how long it took to ignite a cigarette. Tony was an expert when it came to running the CoF_3 reactors since he had worked with my predecessor Dr R. ('Bob') Stephens on the Ministry Contract (which had commenced in 1954; see Appendix 32.1). Even Tony wasn't prepared, however, when the external pipework on the scrubbing system's circulator fractured while we were on night-shift; we thought at first that it was raining heavily!

We used the cobalt fluoride reactors to fluorinate benzene ($C_6H_6 \rightarrow C_6H_xF_y$; $x + y = 12$, $x = 0, 1, 2, 3 \ldots$), the 'hot spot' which travelled down each reactor as fluorination proceeded being controlled by a set of water sprays fixed above the horizontal copper body (when I arrived, Tony was using a hosepipe for this purpose!). I also greatly improved the benzene feed and insisted on having a proper ventilation system installed (see Fig. 32.2).

Fluorinated monomers and polymers

My immediate boss at ISC was ex-Birmingham fluorine chemist Dr A. K. Barbour (see Appendix 32.1), the Section Leader in organic chemistry. Known to us all as Joe, he ran a happy and productive group at ISC (Fig. 32.3) and allowed me much freedom of action within the remit of my tasks. These centred on providing fluoromonomers for use by Joe's polymerization expert, Mr W. ('Bill') Hopkin, who was responsible for the synthesis

Fig. 32.2. Paired (for semi-continuous operation), horizontal, stirred-bed, gas-heated, water-cooled, cobalt fluoride reactors (*ca.* 7 kg CoF_3 in each) used by the author at the NSC Co., Ltd., Avonmouth.

and testing of high-performance fluorinated fluids, lubricants and elastomers of potential use in high-speed aircraft and rockets.

By far my most important and demanding task was to provide Bill Hopkin with substantial amounts of perfluorocyclohexa-1,3-diene (**2**), made via the 'Birmingham route'

Fig. 32.3. The 'Ministry' component of Joe Barbour's research group at Avonmouth (1957): (left to right) W. Hopkin, W. I. Bevan, A. K. Barbour, A. E. Tipping, B. D. Joyner and R. E. Banks.

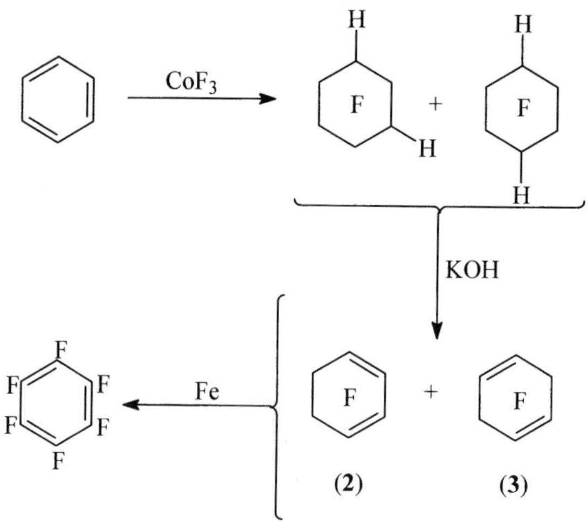

Scheme 32.3. PhD goal achieved: a workable fluorination-defluorination route from benzene to hexafluorobenzene [28, 29].

[25][14] outlined in Scheme 32.3. Also shown there is a pivotal conversion which Tony Tipping and I stumbled on while attempting to boost our output of 1,3-C_6F_8 (**2**) via thermal isomerization of its 1,4-isomer (**3**), namely defluorination of these dienes to hexafluo-

[14]Ref. [25] is a recent review by Colin Tatlow of his group's massive contributions to the chemistry of per- and poly-fluorinated cycloaliphatic compounds.

robenzene. This discovery opened the way for the production of hexafluorobenzene and its derivatives and congeners on a commercial scale by ISC (and subsequently ISC Chemicals) [26, 27], thereby enabling great progress to be made in polyfluoroaromatic chemistry. The rapture I experienced through realizing my PhD objective in this manner was modified later when Joe Barbour informed me that the credit for the discovery of this route to C_6F_6 must be shared with the Birmingham group, as reflected in the publications subsequently issued [28, 29]. Related work from that era on pyrolytic [30, 31] and base-induced [32, 33] dehydrofluorination of polyfluorocyclohexanes was also published. Work on the latter method stemmed from Musgrave's suggestion when I was working on chlorofluorocyclohexanes in Durham that related hydrofluorocyclohexanes (Scheme 32.2) might be easier to dehydrofluorinate cleanly using strongly basic ion-exchange resins rather than hot concentrated KOH aq; this proved to be so, undecafluorocyclohexane, for example, undergoing smooth dehydrofluorination when shaken with Amberlite IR-400(OH) in tetrachloroethylene at 25 °C for 1.5 h [*cyclo*-C_6HF_{11} → *cyclo*-C_6F_{10} (96%)] [32]. Similar dehydrofluorination of the trihydro-compound 1*H*,2*H*,4*H*-$C_6H_3F_9$ at 50 °C (no solvent; 4 h) gave a complex mixture of mono-enes ($C_6H_2F_8$), dienes (C_6HF_7) and some hexafluorobenzene [33]. Previously, detailed work at Birmingham on the KOH-dehydrofluorination of this trihydro-compound had led to the isolation of hexafluorobenzene, as reported in *Nature* in July 1956 [34]; this important note contained the first information on nucleophilic displacement (S_NAr) of fluorine from hexafluorobenzene [C_6F_6 + MeONa → C_6F_5OMe + $C_6F_4(OMe)_2$ (orientation not determined)] and thereby heralded much activity (synthetic and mechanistic) worldwide in the field of polyfluoroaromatic chemistry.

NSC's 1967 *Catalogue of Highly Fluorinated Aromatic and Alicyclic Compounds* listed more than 60 polyfluoroaromatic compounds available in development quantities. Details of the development and potential applications of fluoroaromatic chemistry up to that time can be found in the informative review [26] by Joe Barbour (by then Chemical Research Manager at ISC; see Appendix 32.1) and one of his marketing colleagues. Published in January 1966, this review ('Highly Fluorinated Aromatic and Alicyclic Compounds') deals in critical fashion with the two major routes to perfluoroaromatics which had been established: fluorination/defluorination of arenes [*e.g.* $C_6H_5CH_3$ → (with CoF_3) $C_6F_{11}CF_3$ → (with Fe or Fe_3O_4) $C_6F_5CF_3$], and halex (halogen exchange) fluorination of perchloroarenes [*e.g.* C_6Cl_6 → (with KF) C_6F_6 + C_6F_5Cl + 1,3-$C_6F_4Cl_2$ + 1,3,5-$C_6F_3Cl_3$]. Also mentioned is C_6F_6 formation via pyrolysis of some simple fluoroaliphatic compounds, the most famous example of which is the 'lost reaction of Yvonne Désirant' ($6CFBr_3$ → C_6F_6 + $9Br_2$) – discovered in the mid-1930s but not revealed to the world until 1955, as explained by Dirk Tavernier in Chapter 28 here. Halex fluorination of *lightly* chlorinated aromatics carrying groups which encourage attack of F^- is of vital importance nowadays to fluorochemical businesses associated with the provision of fluoroaromatic intermediates for use in the manufacture of agrochemicals, pharmaceuticals, advanced polymers, dyestuffs and liquid crystals [22].

Interviewed by stealth

Early in 1958 the most charismatic of ISC's consultants, Professor R. N. Haszeldine, made an unscheduled visit to the office I shared with Bill Hopkin, annoyingly interrupting my work on a report for the Ministry of Supply. Bill slipped out of the room, and Haszeldine, somehow casually yet insistently, quizzed me about various matters, including what

I'd learned at ISC, what I really knew about gas-liquid chromatography, and what I had in mind to do if the Ministry Contract was terminated later in the year. He then revealed that he had interviewed me for a junior teaching post (Assistant Lecturer) at the Manchester College of Science and Technology (known locally there as 'Tech'), adding that if I was interested he'd need to receive a two-page (foolscap size) hand-written statement giving my reasons. Apparently, he was looking for someone to help him assemble the biggest fluorine research group in the world! And he hinted that if I lived up to my reputation, he would recommend me for promotion to a full Lectureship after one year in post. In view of his reputation as an organic/organoelemental fluorine chemist gained through his work in Emeléus' labs at Cambridge, I was quite flattered; and he seemed to bear me no grudge for showing earlier (at ISC) that all was not well with his personal work [35] on the addition reaction between ICI and $CF_2{=}CFCl$.[15]

What a stroke of luck, I thought. Here was a chance not only to achieve my ambition to teach at university level, but also to couple that with both *instant* and *sizeable* involvement with research in organofluorine chemistry. The research opportunity really excited me because clearly it would allow me to indulge my fast-growing fascination with the seemingly unparalleled synthetic and mechanistic challenges offered by *per*fluorocarbon chemistry in a manner which well suited my temperament. Note that I was not an absolute novice where lecturing was concerned because I'd deputized for a late-arriving organic lecturer at Durham in 1956 and also taught evening classes (physical chemistry and inorganic analysis) at Bristol Technical College in order to supplement my ISC salary.

Having checked first with Linda that she would not object strongly to a move to 'murky' (as it was then) Manchester, I posted a job application to Haszeldine. An official notification that I had been appointed Assistant Lecturer (salary £700 per annum) in 'Tech', the Faculty of Technology of the University of Manchester, arrived at our flat in Bristol at the end of March 1958, by which time the Ministry of Labour & National Service had given me permission ('in view of the very great importance of university teaching') to leave ISC a few weeks before my 26th birthday ('call up' cut-off day!) in order to join Haszeldine on October 1st. Linda would have liked me to return to Durham, but was not impressed that Musgrave appeared able to offer me only the possibility of a postdoctoral fellowship.

Linda went ahead to Manchester to occupy a house we'd managed to get a mortgage for and to take up a new teaching post. I moved from our flat in Bristol to a homely guesthouse, where, for almost two months, I spent many 'leisure' hours reading reprints of Haszeldine's numerous papers (*his* parcel was totally different from the one Musgrave had sent me in the summer of '53!) and trying to get to grips (at Haszeldine's request) with the new-to-me NMR method of analysis through literature obtained via ISC's library (the lady librarians played hockey, so I was a privileged borrower; NSC had a good men's hockey team, too, happily for me).

Amazingly, three of the ISC chemists who, though in celebrative condition, made sure that I caught the 'midnight-to-Manchester' train from Bristol at the end of September

[15] We used the Haszeldine/Henne-Postelnek route [35, 36], *i.e.* $CF_2{=}CFCl + ICl \rightarrow CF_2ClCFClI \rightarrow (CF_2ClCFCl)_2 \rightarrow (CF_2{=}CF)_2$, to procure perfluorobutadiene for Bill Hopkin. With the aid of GLC, which Haszeldine did not have access to, we found that under his reaction conditions [35] (as far as we could judge) the first stage is not regioselective but gives an approximately 1:1 mixture of $CF_2ClCFClI$ and CF_2ICFCl_2.

1958 subsequently followed me to Haszeldine's department as PhD students: Haydn Sutcliffe, Tony Tipping and Bill Bevan. There was more to it than that, though, because Haydn and I were old schoolmates! (see Appendix 32.2).

Megafluor in Manchester (1958-74)

Background

It seems a pity that when Bill Bryson stopped over in Manchester recently during his valedictory tour of Britain (see his 1996 bestseller *Notes from a Small Island*) he did not resolve his uncertainty about 'where to go next' as he stood on the edge of the city centre's Piccadilly gardens by taking the short walk from that location to UMIST's campus. Once there, he could have marvelled at the splendid developments on the 28-acre site where the Russian revolution began,[16] noted for future use the fact that a viaduct carrying a main railway line cuts through the campus en route to Piccadilly Station, and seen Professor R. N. Haszeldine's contribution to the Manchester skyline: the impressive new-in-1966 chemistry building.

UMIST's origins date back to 1824, when a group of businessmen and industrialists founded the Manchester Mechanics' Institution. Since then several name changes have occurred – each marking a step in the evolution of this educational establishment and culminating, in 1966, with a title so long (The University of Manchester Institute of Science and Technology) that it was soon abandoned in favour of the now widely understood acronym UMIST. This title finally convinced all and sundry that the Institute was part of the Victoria University of Manchester, as indeed it had been since the 1905 Concordat established its position as the university's Faculty of Technology. UMIST became an independent university in 1994, but retained its name to indicate the special links that still connect it with the University of Manchester, under whose academic aegis it enjoyed and attained many historic successes.

Gary S. Messinger, Bill Bryson's fellow American, mentions only the Manchester Mechanics' Institution in his book *Manchester in the Victorian Age* (Manchester University Press, 1985), and avoids the important events leading up to its staged transformation to the Manchester Municipal School of Technology, opened in 1902; but then, even John Dalton (1766-1844), a Vice-President (1839-1841) of the Mechanics' Institution and lecturer there, receives scant attention. However, Messinger's book does provide an informative account of the origins of the University of Manchester, which began life as Owens College[17] in 1851 and benefited greatly from the services of Henry Enfield Roscoe who, aged 24, succeeded Edward Frankland as professor of chemistry in 1857. To come to the point, fluorine chemistry might well have prospered in the University of Manchester long before

[16]Writing about UMIST in 1974, the then Principal (the late Lord Bowden of Chesterfield) referred to the fact that the Institute's post-50s expansion programme replaced slums by 'splendid buildings', then continued [37]: 'It was only after we had abolished all these slums that we discovered that the area had at one time been immortalised by Engels in his account of the condition of the working classes that inspired so much of the *Communist manifesto*. Few people realise that the Russian revolution began on what is now the campus of UMIST in a bend of the Medlock which disappeared when we built our culvert and hid the river for ever'.

[17]John Owens (1790-1845) was a textiles/furs magnate who bequeathed £100K to establish a college in Manchester. Students and faculty still tend to refer to the University of Manchester as 'Owens' to distinguish it from UMIST ('Tech').

Haszeldine arrived if research in Roscoe's department by Sydney Young into the synthesis of alkyl fluorides [38], or by Bohuslav Brauner on the generation of fluorine [39, 40], had come to fruition.

Boom time

UMIST was known as the Manchester College of Science and Technology when Haszeldine's embryonic Manchester Fluorine Team[18] arrived from Cambridge in time for the start of the university's academic year at the beginning of October 1957. I arrived a year later, by which time Haszeldine and his band of pioneers were well established on E and F floors of the Main Building on Sackville Street. Aged 32 at the time and already established internationally as a perfluorocarbon chemist, 'Bob' (Robert Neville) Haszeldine was back on home territory (he was educated at Stockport Grammar School, not 10 miles from UMIST) and in charge of a university chemistry department that needed to lie closer to industry than most conventional departments without sacrificing academic standards. I felt privileged to have been given the chance to participate in his highly ambitious plans regarding the revitalization of the chemistry department he had 'inherited' from Professor H. N. Rydon (1952-57).

Through Haszeldine's amazing drive, professionalism and fund-raising know-how, not only had the chemistry department become the largest in the Faculty by academic year 1964-65 but also one of the biggest in the UK in terms of numbers. And the new chemistry building then rising at the edge of the campus furthest from the railway viaduct was of a size to match. I moved into the research tower of this £3 $\times$ 10^6 building ('Fort Haszeldine') in August/September 1966; HRH the Prince Philip, Duke of Edinburgh, KG, performed the official opening ceremony on 2 May 1968.

The Manchester Evening News covered the opening ceremony in style, reproducing inter alia a photograph bearing the caption 'the Prince "met" this life-size figure wearing safety equipment as protection against acid' (the mannequin was dressed after the style of our HF technician), and one showing Professor Haszeldine presenting the Prince with a commemorative Blue John chalice. The background to the beautiful portrait of Haszeldine (Fig. 32.4) presented to the chemistry department to mark the realization of his 1957 vision of a new chemistry building was inspired by the unmistakable colours and patterns associated with Blue John, a unique variety of fluorspar found at Castleton (Derbyshire), 29 miles from Manchester in the direction of Sheffield [14]. When Prince Philip attends the opening ceremony for the UK's Millennium Dome (Greenwich, London) with its impressively large PTFE-coated glass fibre roof, he may perhaps recall a few things I told him about PTFE (polytetrafluoroethylene) as he paused by our display of fluorochemicals during his UMIST tour in 1968. Bryan Willoughby, one of my research students who helped with that exhibition on our opening day, has recently acquired an interesting view of PTFE in art form (see Appendix 32.3).

---

[18] One of its members (Colin Young, studying polyfluoroalkyl derivatives of silicon) was responsible for my return to playing 1st XI university hockey (for 'Owens'[17]), but I moved on to a large local club (Bramhall, Cheshire) after one season because being absent from the labs on Wednesday afternoons became a problem; and in any case, I was not a student. Colin and I played in the University of Manchester team which represented the UK universities at the 1959 international hockey festival in Groningen (Netherlands). Although billed as 'Manchester United' towards the end, we didn't make the semi-finals; perhaps if I'd not missed that tackle on the Dutch centre-forward...

Fig. 32.4. Portrait composed by Fred Deane from photographs of Professor R. N. Haszeldine demonstrating a classical Moissan discovery that iron burns like a firework in a stream of fluorine. (Photo kindly provided by Professor R. N. Haszeldine FRS; the portrait was presented originally by Mr. G. S. J. White, Chairman of the Chemistry Building Planning Committee and of the Construction Committee, and now hangs in the chemistry department's library at UMIST.)

Watershed: dissolution of 'the old firm'

Once the dust had settled after the opening of our new building, I put the finishing touches to the second (enlarged) edition of my 1964 monograph on perfluorocarbon chemistry, *Fluorocarbons and their Derivatives* [41] and braced myself to start work on the first of a related three-volume series of biennial specialist periodical reports – *Fluorocarbon and Related Chemistry* [42]. Through the help and support of only my good friend and ex-PhD student Dr M. ('Mike') G. Barlow, volume 1, covering the years 1969 and 1970, appeared on time; realizing that the subject was about to boom, we coped for the next four years (volumes 2 and 3) by enlisting as reporters 'fellow officers' in Professor Haszeldine's F team (Drs J. M. Birchall, R. Fields, W. T. Flowers, D. R. Taylor and A. E. Tipping). On economic grounds, The Chemical Society discontinued the publication of the series after volume 3. Not long afterwards, I took on the editorship of two books concerned with industrial applications of organofluorine compounds [20, 21], the first of which proved particularly popular. By that time (1978) I was ploughing my own fluorocarbon furrow at UMIST, having resigned my commission in Haszeldine's officer corps in the early '70s.

Detailed discussion of the dissolution of 'the old firm', as Haszeldine used to refer to our professional relationship, has little place in this *Festschrift*, a decision which I trust will not disappoint too many readers. The straw which finally broke the camel's back concerned his ongoing failure to deal efficiently with research papers I'd written carrying his name as the senior author. Other 'lieutenants' had long suffered the same irritation and continued to do so, but none eventually took the sort of action which led to my watershed: I submitted a 'joint' paper directly to the editor of *J. Chem. Soc.*

Details of my joint research with Haszeldine featured prominently in the DSc thesis I submitted to the Victoria University of Manchester in 1970; with that degree under my belt, I was promoted in 1973 from Senior Lecturer to Reader in Chemistry in the University ('post to be held at UMIST'). Understandably, I continued to write papers based on research associated with 'the old firm' of Banks & Haszeldine (although to the chagrin, I'm sure, of some of the PhD students involved, I didn't complete that task), hence some outsiders still viewed me as the crown prince when Haszeldine succeeded Lord Bowden as Principal of UMIST in 1976; however, the 'natural products' expert R. ('Bob') Ramage (University of Liverpool) was appointed to the Chair of organic chemistry, in part to help promote biosciences at UMIST. As an 'unnatural products' chemist myself,[19] I was well aware – as was Haszeldine, especially on the basis of his own experience – that I would have to very lucky to acquire a chair of organic chemistry in a 'conventional' establishment of UMIST's rank. After two abortive attempts to appear on 'shortlists', I settled for what I already had (excellent fluorination, workshop and high-pressure facilities) and might well struggle to achieve elsewhere on cost and/or safety grounds: the opportunity to pursue fluorine chemistry in almost any direction that outside funding required. This decision eventually served me well.

[19]Naturally-occurring organofluorine compounds are rare and none are isolated for utilization [16]; and of the twelve I'm aware of, only CF_4 [16] and CF_3CO_2H (see later) are perfluorocarbon in nature. Only the lightly-fluorinated varieties (monofluorides) were known when Bob Ramage arrived at UMIST.

Pushing back the frontiers of fluorocarbon chemistry

During the 15 years I spent as a senior collaborator in Professor Haszeldine's very large F team, I tackled fluorocarbon chemistry (including organoelemental and polymer chemistry) on a broad front, with emphasis on potential industrial applications. It was the busiest period of my life where day-to-day supervision of research students is concerned, and resulted in several commendable achievements, including: development of the first routes to perfluoropyridine [43], perfluorocyclopentadiene [44], perfluoronorbornadiene [45], and perfluoropropyne [46]; discovery of a method for cross-linking the original *nitroso rubber* (a 1:1 alternating semi-inorganic copolymer of $CF_2{=}CF_2$ and the famous blue gas $CF_3N{=}O$ [47-49]); introduction of the prototypical electrophilic fluorinating agent of the N$-$F class [50]; and unearthing the first tangible evidence that thermal ring-expansion of arylnitrenes gives azacycloheptatrienylidenes [51, 52]. Schemes 32.4-32.8 provide basic information on these, except for the somewhat complicated cross-linking method. Scheme 32.9 shows a good example taken from our Diels-Alder studies [53] of the 'think negative' approach all devotees of perfluorocarbon and related chemistry must quickly learn to adopt.

For the record, the majority of my research papers from the Banks-Haszeldine period were published under the following series titles: (1) Heterocyclic Polyfluoro-compounds; (2) Studies in Azide Chemistry; (3) *N*-Fluoro-compounds; (4) Perfluoroalkyl Derivatives of Nitrogen; (5) Nitroxide Chemistry; (6) Polyhalogeno-allenes; (7) Polyfluorocyclopentadienes; (8) Fluoro-olefins; (9) Perfluoroalkyl Derivatives of Sulphur; (10) Perfluorinated Carboxylic Acids and Their Derivatives. Quite a few patents were filed, including two arising from work by Peter N. Preston associated with the treatment of coal with HF [54, 55]. Reviews I co-authored during that period covered fluoropolymers [56, 57], polyfluoroalkyl derivatives of metalloids and nonmetals [58, 59], and an overview of fluorine chemistry for Frank Smith's *Pharmacology of Fluorides* [60].

Series 4, 8 and 9 in the list above had been started by Haszeldine at Cambridge; the azide series (2) was solely mine, and series 1 and 3 stemmed from our independent interests in the chemistry of the N$-$F compound **1** (Scheme 32.4) Emeléus quizzed me about during

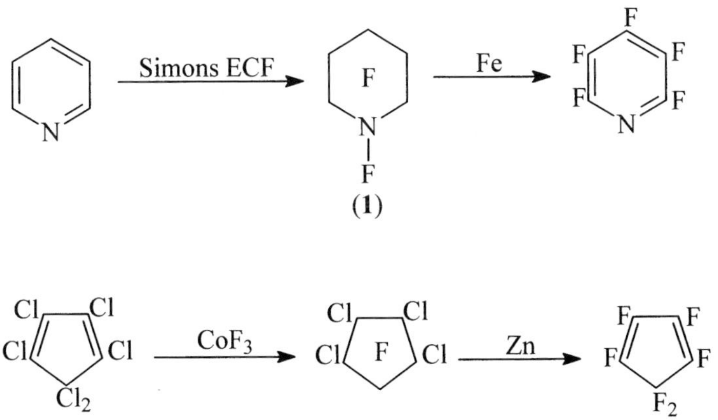

Scheme 32.4. The first routes to pentafluoropyridine [43] and hexafluorocyclopentadiene [44].

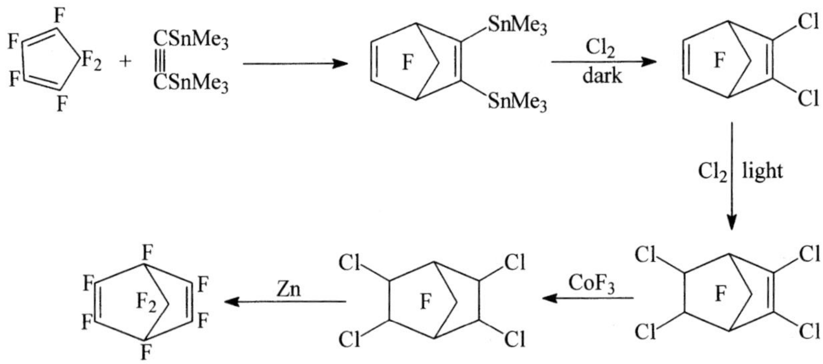

Scheme 32.5. Diels-Alder route to perfluoronorbornadiene [45].

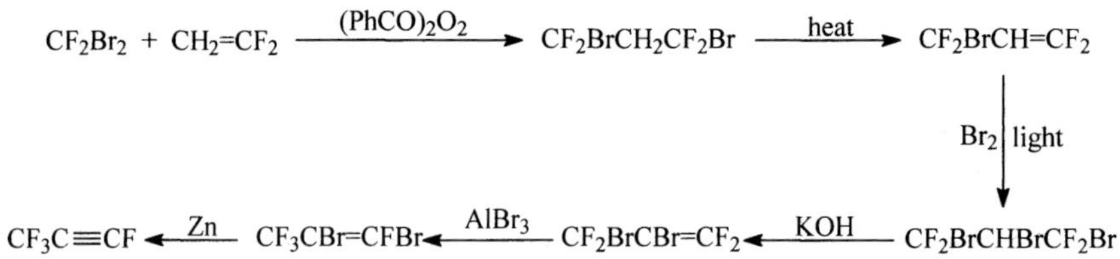

Scheme 32.6. Synthesis of the first perfluoro(monoalkylacetylene), tetrafluoropropyne [46].

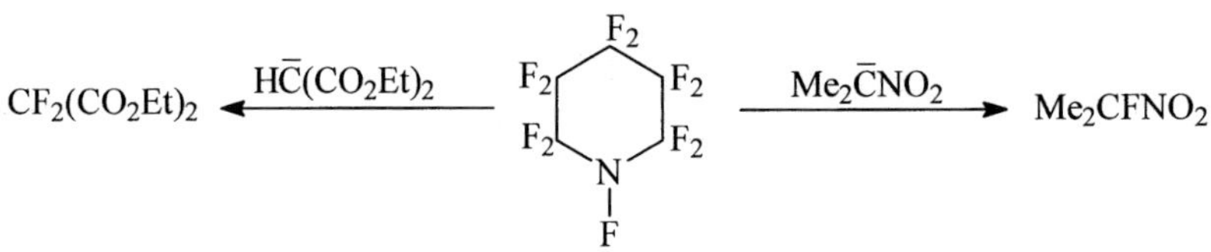

(1)

Scheme 32.7. The first examples of electrophilic fluorination involving an N—F reagent [50].

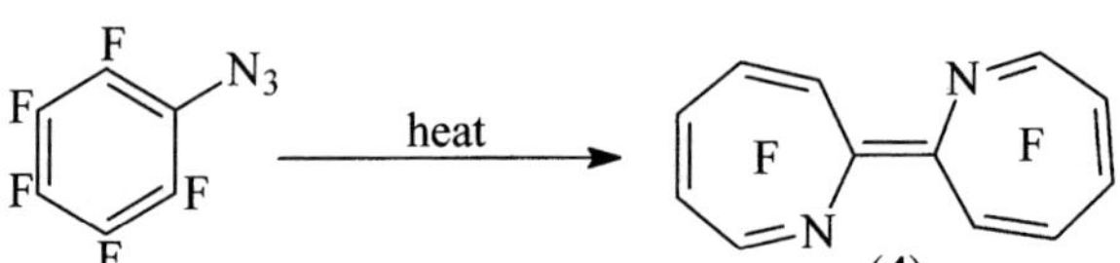

(4)

Scheme 32.8. Synthesis of perfluoro-2,7′-diazaheptafulvalene (**4**; 'pentafluorophenylnitrene dimer'): the first concrete evidence for ring-expansion of an arylnitrene to an azacycloheptatrienylidene [51, 52].

my BSc viva in 1953. Not unexpectedly, competition in the pentafluoropyridine area came from both Tatlow's group and Musgrave's, almost immediately and short-lived in the former case [61] and somewhat later, but in extenso, where the Durham F team was concerned [62]. As I mentioned earlier, achieving a synthesis of pentafluoropyridine was already on Musgrave's 'wish list' when he briefed me regarding my PhD topic in September 1953;

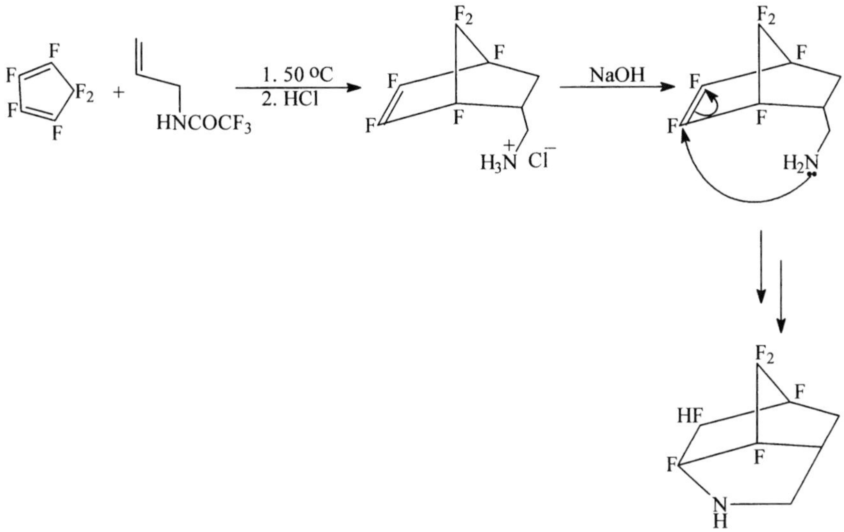

Scheme 32.9. A fluorocarbon analogue of the classical bromo/iodo-lactonization method used to establish cyclopentadiene's compliance with Alder's '*endo*-rule' [53].

perhaps he might have beaten us to this area if he had had Simons ECF equipment (as did Haszeldine and Tatlow), and hence access to perfluoro-*N*-fluoropiperidine (**1**). As it was, the halex route to pentafluoropyridine and its synthetically-useful chlorofluoro analogues (C_5Cl_5N + xs. KF $\rightarrow$ C_5F_5N + 3-ClC_5F_4N + 3,5-$Cl_2C_5F_3N$) was reported independently from Durham and UMIST in 1964 [63, 64]; this development enabled both groups to progress in the area, and also led to commercialization of these compounds.

Haszeldine was well into his blue period on arrival in Manchester. I refer, of course, to his work on trifluoronitrosomethane, a deep blue, toxic, monomeric gas (b.p. −84.6 °C) which condenses to a deep blue, unassociated liquid. The failure of this nitroso-compound to isomerize ($CF_3N{=}O \not\rightarrow CF_2{=}NOF$) or form a colourless dimer of the di-*N*-oxide type is a feature common to perfluoronitrosoalkanes and sets them apart from their hydrocarbon analogues, nicely exemplifying the fascination of fluorine. My involvement with CF_3NO stemmed from Haszedline's vigorous extension in Manchester of the Barr-Haszeldine discovery in Cambridge that tetrafluoroethylene combines smoothly with this nitroso-compound to yield a 1:1 cycloadduct, perfluoro(2-methyl-1,2-oxazetidine), and the novel elastomer $[N(CF_3)OCF_2CF_2]_n$. This alternating copolymer, a unique example then of polymerization via addition across the N=O bond, was found to be insoluble in common solvents, unaffected by hot concentrated acids and aqueous alkalis, and blessed with good low-temperature properties that placed it in the 'Arctic rubber' class. This triggered military and aerospace interest in fluorinated elastomers of the nitroso class, so the field boomed [65]. Research at UMIST, financed by the UK's Ministry of Supply (I was working for the Government again!), culminated in the development of a cure-site approach to the vulcanization of Haszeldine's prototypical nitroso rubber based on our work on perfluorinated nitroxides [47-49].

Regarding perfluorinated nitroxides, my extensive work with Haszeldine on the 'magic radical' $(CF)_2NO\cdot$ (bistrifluoromethyl nitroxide)[20] – a *purple* gas made from a *blue* gas [$2CF_3NO + NH_3 \rightarrow (CF_3)_2NOH{\cdot}H_2O \rightarrow$ (with $KMnO_4$) $(CF_3)_2NO\cdot$] commenced after its isolation had been reported independently in the mid 1960s by American and Russian investigators.[21] We concentrated on the reactions of this persistent radical with unsaturated fluorinated compounds and with hydrocarbon derivatives containing a range of common functional groups. Emeléus and his associates researched inorganic aspects, developing, for example, the mercurial $[(CF_3)_2NO]_2Hg$ as a bis(trifluoromethyl)nitroxylating agent. We extended that approach and also developed *N,N*-bistrifluoromethylhydroxylamine and its sodium salt $(CF_3)_2NO^-Na^+$ as reagents.

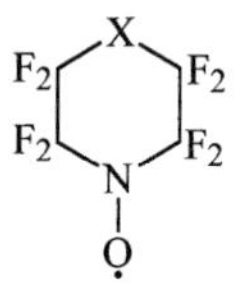

(5) X = CF_2
(6) X = O

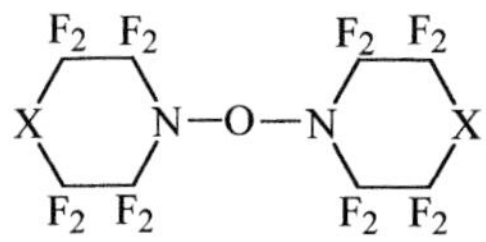

(8) X = CF_2
(9) X = O

The first perfluorinated nitroxide I encountered was perfluoropiperidine *N*-oxyl (**5**), a deep purplish-blue liquid (b.p. 60-62 °C), discovered accidentally during work by Keith Mullen and Eddie Williamson [66] on photochemical reactions (see Scheme 32.10) of the corresponding N—F compound (**1**) (we're back to that conversation with Emeléus again!). This gave me my first taste of esr spectroscopy, which was extended through work by Gordon Smith on the morpholino analogue (**6**) of **5** [67] and by Mahmood Nickkho-Amiry on its 2,6-bis(trifluoromethyl) derivative [68]. Like $(CF_3)_2NO\cdot$ [$2(CF_3)_2NO\cdot \not\rightarrow (CF_3)_2NOF + CF_3N^+(O^-)=CF_2$], none of these nitroxides suffer the characteristic disproportionation of some of their hydrocarbon cousins. Again the great strength of the C—F bond, so useful in the study of radical reactions, comes into play. The N—F bond is a different proposition, of course, in terms not only of homolytic (*e.g.* Scheme 32.10) but also heterolytic fission, and I shall return later to the Banks-Williamson (Scheme 32.7) [50] discovery that perfluoro-*N*-fluoropiperidine (**1**) will deliver fluorine in positive mode to overt and covert carbanionic sites.

On a lighter note, I dubbed the N—O—N compounds (**8** and **9**) we encountered during the UV photolysis of our cyclic N—F compounds **1** and **7** in the presence of oxygen (Scheme 32.10, [66, 67]) and via UV irradiation of the 'magic radical', *i.e.* $2(CF_3)_2NO\cdot \rightarrow (CF_3)_2NON(CF_3)_2$ [69], *de Gaulle compounds*, the etymology of which lies in the structural connection with President de Gaulle's rejection in November 1967 of Britain's application to join the EEC. The term didn't catch on though, and the nickname 'magic

[20] The origin of the identification 'magic radical' is the exclamation 'that's magic!' uttered by Phillip Carson during a conversation I was having with him and Barry Justin in 1966 about choice of PhD topics. Phillip was reacting to my assertion that one couldn't fail to get results on a topic involving $(CF_3)_2NO\cdot$ because it even attacks filter paper, neckties, hair and suede jackets.

[21] For detailed information about $(CF_3)_2NO\cdot$ and its analogues, see my reviews in ref. [42] (all three volumes).

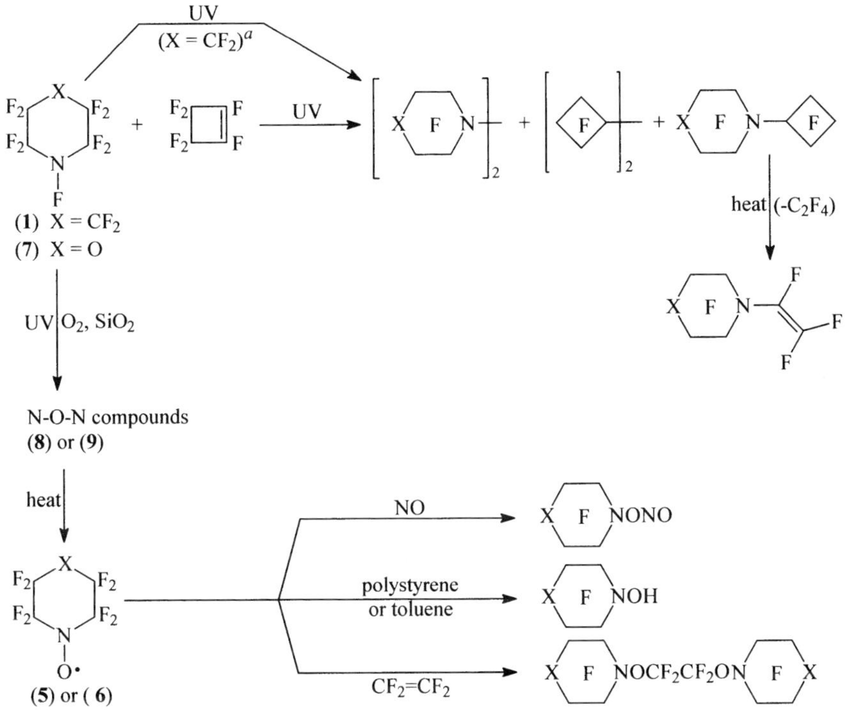

Scheme 32.10. Photochemical reactions of perfluoro-*N*-fluoropiperidine (**1**) and its morpholino analogue (**7**), and some conversions involving the derived nitroxides (*N*-oxyls) **5** and **6**. [a]Perfluoro-*N*-fluoropiperidine (**1**) is colourless, but it was noticed that unchanged material recovered by distillation of products from UV experiments was very pale blue; this colour desappeared rapidly when the material was shaken with toluene (due to benzylic H˙ abstraction). It was this observation which prompted us to carry out preparative experiments leading to the isolation of nitroxides **5** [66] and **6** [67].

radical' has not been used much outside UMIST, so neither are included in that delightful 1987 book *Organic Chemistry: The Name Game* [70].

Further information on research activities in Haszeldine's laboratories can be found here in Appendices 32.2-32.4, and in refs [41] and [42]; Appendix 32.4 reveals the excitement engendered in undergraduates by our activities in fluorine chemistry. Even though Haszeldine's ability to lay his hands on research funding was phenomenal, not every aspiring PhD student could be accommodated; also, the research activities of other members of staff needed to be considered. Even so, Haszeldine (who was in charge of postgraduate admissions) took the lion's share of successful applicants, directing them to one or other of his 'lieutenants' for supervision. I benefited enormously from this initially, even being allocated five research students immediately upon arrival in October 1958 to team up with J. M. ('Mike') Birchall and Dr Harold Goldwhite. Harold had earned his PhD degree under the direction of B. C. Saunders at Cambridge University, then crossed the Atlantic to pursue postdoctoral work on reaction mechanisms involving fluoro-olefins with W. ('Bill') T. Miller at Cornell University, so he was ideally equipped to collaborate with Haszeldine on fluorocarbon derivatives of phosphorus. Mike, who had taken part in the exodus of

Haszeldine's group from Cambridge, was already established as a research supervisor even though he had not yet completed his own doctoral work (on hexafluorobenzene chemistry); clearly he was indispensable to Haszeldine, and remained so for many years. Harold, with his American wife, Marie, returned to the States in 1962 to join the faculty of California State University, Los Angeles. Subsequently Dr Michael Green came and went (to Bristol, to join F. G. A. Stone), as did Dr A. B. P. Lever (to Canada).

Home-grown lieutenants soon started to emerge from the graduate school, and one of those, Dr D. R. Taylor, took over from me as supremo of the departmental GC and distillation services. David and I produced a 3-part videotaped course of lecture-demonstrations on distillation for undergraduates that is still in use today! We also collaborated with Versal Glass of Manchester to produce a commercial range of semi-micro one-piece distillation units and various interchangeable fractional distillation columns (plus accessories) of the type research students were finding invaluable. The pièce de résistance, though, was a semi-automatic low-temperature unit built and fine-tuned by Haydn Sutcliffe [71]; it was based on fully-automatic apparatus Haydn and I had seen used in the CFC lab at ISC Avonmouth, and when integrated with a large 'Cambridge type' vacuum system provided a powerful tool for work with gaseous fluorocarbon compounds.

The lighter side

On the social front, Professor Haszeldine was great company: a lover of fine food and wine, he seemed to know all the best restaurants and also entertained superbly at his home, with the help of his remarkable late wife, Peggy. And when he partied, it was in style, as Wojciech Dmowski clearly remembers (see Chapter 12). He was admirably considerate towards disadvantaged students, faculty and technicians, especially when health problems were involved; and I still meet ex-PhD students from the early days who remark on how much they owe him for giving them the opportunity to better themselves at UMIST when many other universities looked askance at their degree-equivalent professional qualifications earned part-time while working in industry or their modest university degrees backed up by several years experience of industrial R & D. Tony Tipping, Haydn Sutcliffe (see Appendix 32.2) and Bill Bevan who all followed me to UMIST from ISC are perfect examples of how well this policy worked out in terms of mutual benefit. Tony, of course, stayed on at UMIST, becoming a Reader in Organic Chemistry and producing some 200 papers concerning fluorine chemistry. He retired in 1998.

Given space, I could regale the reader with amusing stories about my 'social' and 'chemical' interactions with Professor Haszeldine, both within UMIST and elsewhere: in the early days there was never a dull moment. A highlight of the early years was my first trip to America in 1962, specifically to attend the 2nd ISFC in Estes Park, Colorado. To save money, Mike Birchall and I flew out overnight from a military air base in Suffolk, by courtesy of the US Airforce and through our research connections with the UK's Ministry of Aviation. Haszeldine was travelling in style, of course, using commercial flights to reach our rendezvous in New York. Our noisy turboprop transport plane (I've suffered from tinnitus ever since) was packed to the gunnels with US military personnel and their families, and on landing it was found that some of the children had developed a red rash. We were quarantined and had visions of not making the rest of the tour. However, the malady turned out to be nothing more serious than prickly heat, so eventually we set off, tired and hungry, for the main gate of the air base where we had landed. The armed guard at the barrier glanced at my papers and let me through without demur; but as I wandered off, zombie-like, in the early morning light I heard a shout from Mike. Clearly he needed my help: the guard had demanded the name of someone in the US who could vouch for him and Mike had named me! The guard accepted my statement and we headed smartly for the shuttle (bus) to New York before he changed his mind. One would have thought that a 20-hour journey from Manchester of the type we'd endured would dull the mind, not sharpen it! Perhaps that doesn't apply to Cambridge graduates.

On the next leg of the journey to Denver, the plane lost hydraulic fluid and Haszeldine had an interesting altercation with a stewardess concerning surrender of his bulky briefcase for storage at the rear before we adopted crash-landing postures. In the event, the plane landed smoothly, and I would have been one of the first to know otherwise since the flight engineer had lifted an inspection hatch by my feet, so I could see the undercarriage assembly he had been hitting previously with a large wrench as we flew around 'losing' fuel. If things had gone very badly the UMIST F team would have been shortlived indeed! In fact, I've not experienced a smoother landing since, nor encountered such jubilation from fellow passengers: the cheering was deafening, and as we left the plane under the gaze of fire, ambulance and mass media crews, several passengers kissed the tarmac.

Eventually going with the flow (1974-1993)

My ebullient nature and addiction to fluorine chemistry, coupled with my enjoyment of teaching organic chemistry (particularly applied and polymer aspects), saw me through some trying events during the first ten years of my flying-solo period. Departmentally, fluorine chemistry took a nosedive when Professor Haszeldine was elevated in 1976 to the top executive post (Principal) in UMIST: he modified his fluorine team, replacing it with a much smaller group (known officially as the Principal's Research Group, but as his private army by the likes of me), and appointed Bob Ramage Professor of Organic Chemistry, leaving the inorganic chair vacant.[22] Not without good reason, Bob Ramage set about normalizing the organic section of the chemistry department, and establishing his own excellent research group. The upshot was that by the time (1985) he had been replaced by Professor R. J. Stoodley from the University of Newcastle upon Tyne, a specialist in the same area of organic chemistry but noticeably different otherwise, only Tony Tipping and I were viewed as dedicated organofluorine chemists.

On track again

Knowing the ropes well, from 1976 I was gradually able to recover my laboratories and establish a research group I could be proud of. This was the life I'd always wanted, and publications started to flow at a rate which indicated that I had picked up speed again. The announcement of my death in *Chemistry in Britain* in 1980 was greatly exaggerated [72], but I really appreciated the letters of condolence received from home and abroad.[23]

[22] Professor Haszeldine's official status had been Head of Department (HOD) and Professor of Chemistry (Organic and Inorganic). He was succeeded as HOD by the Professor of Physical Chemistry, P. G. ('Sandy') Ashmore.

[23] The Editor of *Chemistry in Britain* explained that 'About once a year either the computer or the file search goes wrong, and, if not caught in time, results in a wrong announcement'. The Dr Banks who had actually died was *Robert* Eric Banks (University Chemical Laboratory, Cambridge), sadly only 27 years old.

Fig. 32.5. The author – still smiling several days after promotion!

Owing to unfortunate circumstances, Professor Haszeldine left UMIST in 1982 and his fluorine group was disbanded. In mid-1984, the new Principal of UMIST asserted that I was to be considered for promotion to a personal chair of fluorine chemistry. Following a series of events which beggar belief, I was formally appraised on 16 November 1989. Having passed muster, I emerged from UMIST's Council Chamber smiling (Fig. 32.5).

Under the conditions of my promotion, I could not ensure that fluorine chemistry would survive at UMIST – even in minifluor form – once I'd reached normal retirement age (65), so when I felt the need to take early retirement at the end of 1993, and knowing that I would be allowed to run down my research activities gradually, I concluded a deal with UMIST's authorities regarding enhancement of my university pension to the 40 years mark. Part of that deal, which enabled Dr Alan K. Brisdon from the Leicester fluorine team to be appointed to a lectureship in inorganic chemistry at UMIST, involved the release of £19 000 from my personal departmental accounts. An appropriate sum well spent, I judged.

Research highlights

Papers carrying Haszeldine's name gradually became less prominent from 1974 onwards in my list of publications and had disappeared altogether by 1983, by which time my

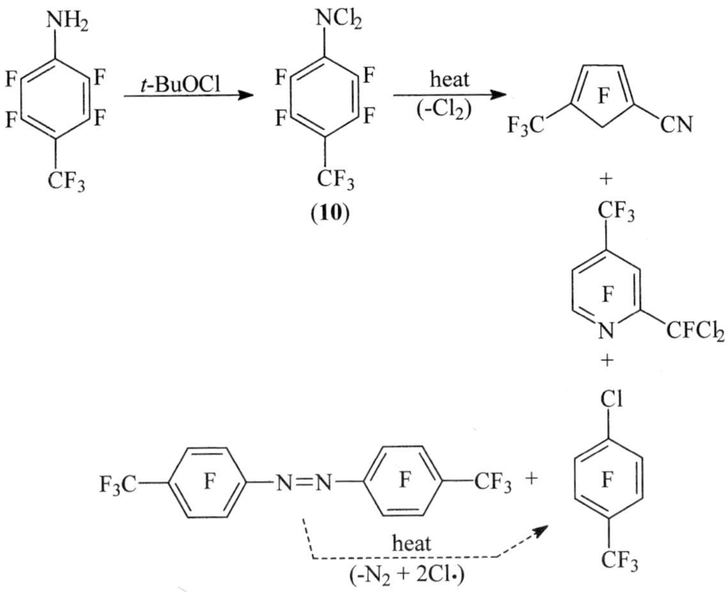

Scheme 32.11. First evidence for nitrene formation via pyrolysis of an *N,N*-dihalogenoaniline (**10**) [76].

own F team was nicely established. Also, I was enjoying collaboration at times with Tony Tipping and often with Mike Barlow. Mike ('Mr Megahertz'), both a fluorine chemist and an NMR expert – just like Roy Fields, with whom I've also collaborated to my advantage – participated in my work on novel fluoroaromatic *N,N*-dichloroamines. This stemmed from my interest in $C_6F_5NF_2$ as both an electrophilic fluorinating agent and a source of pentafluorophenylnitrene, previously accessed via the corresponding azide, $C_6F_5N_3$. Tim Noakes initiated my work in this area through his studies on $C_6F_5NCl_2$ [73], and it was carried forward by John Hornby [74], Manouchehr Mamaghani [74], Mohamad Saleh [75], Balkis Al-Saleh [76], Dev Venayak [77], and Sharique Zuberi (see Appendix 32.5). Our work on the pyrolysis of *N,N*-dichloroperfluoro-*p*-toluidine (**10**) was a highlight, since it established at last that nitrenes are implicated in the thermal decomposition of *N,N*-dichloroanilines (Scheme 32.11) [76]. We used the toluidine derivative **10** owing to its reluctance to rearrange 'spontaneously' to the corresponding *N*,4-dichlorocyclohexa-2,5-dienylideneamine, as found with $C_6F_5NCl_2$ [73]. The only *N,N*-dichloroarylamine capable of manipulation in comfort until compounds of the fluorocarbon class became available [73] was the perchloro compound $C_6Cl_5NCl_2$.

My long-standing involvement with fluorinated azides as nitrene precursors continued. Following studies in the early 1960s on C_2-C_4 acyclic compounds by Geoff Moore, Mike McGlinchey and David Berry [78], I switched my attention to fluoroaryl azides when we moved to the new chemistry building because Mike was experiencing too many explosions. (Absorbing flying glass fragments is one thing, but taking aboard HF and toxic fluoro-organics simultaneously is quite another.) George Sparkes [79], followed by Ajai Prakash [80], provided the springboard for work in the fluoroaromatic area by Dev Ve-

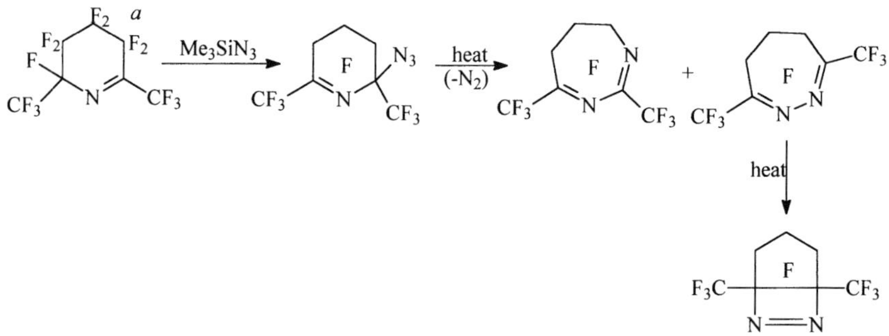

Scheme 32.12. First example of ring expansion involving a cyclic perfluorinated azide in which the N_3 moiety is attached to sp^3-hybridized carbon [84].
[a]Prepared by the route:

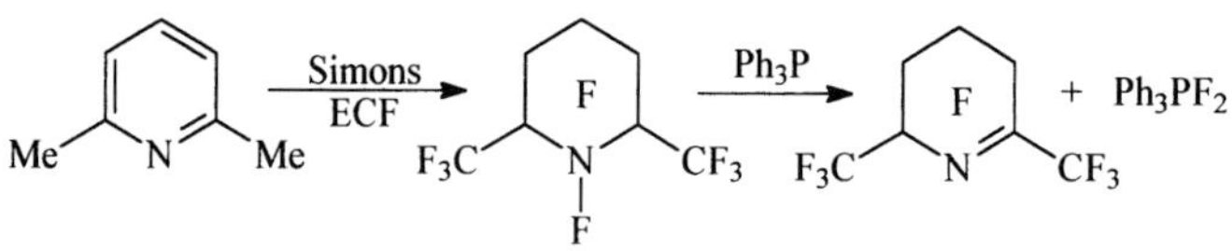

nayak [52, 81] and Ismail Madany [82]. Allan Bailey then took us into the perfluoro-heteroalicyclic azide field by mastering the nucleophilic mono-azidation of perfluoro-1-azacyclohexane (Scheme 32.12) without blowing the apparatus to smithereens, as happened when McGlinchey tried to accomplish the same conversion years previously [83]; this paved the way for Mohammad Abed-Rabboh to demonstrate an azidopiperideine → dihydrodiazepine ring expansion of the perfluorocarbon class [84] (Scheme 32.12).

Stephen Hitchen, who initiated my researches (see Scheme 32.13) on building-block approaches to partially-fluorinated *N*-heterocyclic compounds via pyridinium-*N*-imines and related methylides [85],[24] Neil Dickinson, who [88], like Adrian Richards [89], kept me in touch with fluorocarbon nitroso-compounds and nitroxides, and Adam Alty, who discovered [90] by chance a novel versatile synthesis of diareno-1,2-diazepines exemplified in Scheme 32.14, had a tremendous influence on life in my laboratory over several years, and I can't thank them enough for that. Adam went on to a postdoc position in Bill Dolbier's group, then returned to my lab for a spell to study surface fluorination of polyolefins with 'magic radical'; missing the Florida lifestyle, he flew west again to take up employment with PCR, joining Rick Du Boisson there. Rick, who had been in Haszeldine's private army at one time, earned his PhD through work on fluorinated nitrones with Tony Tipping and myself.

My addiction to synthetic polymer chemistry was satisfied by work on polymer-supported fluorodehydroxylating agents [92], *e.g.* **14** (with Abdul-Karim Barrage, Ezzatol-

[24]Continued by Julie Thomson [86], who until recently worked at F2 Chemicals Ltd. (UK) following a spell at Durham with Dick Chambers, and Suad Mohialdin [87], who later did postdoctoral work on the development of electrophilic fluorinating agents (see later).

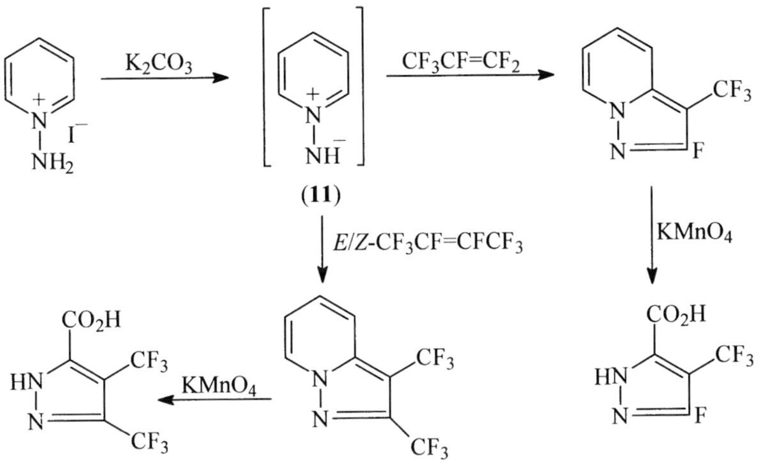

Scheme 32.13. Examples of the use of perfluoro-olefin building blocks for the synthesis of fluorinated pyrazole-carboxylic acids (via 3-azaindolizines derived from *N*-iminopyridinium ylide **11** generated in situ) [85].

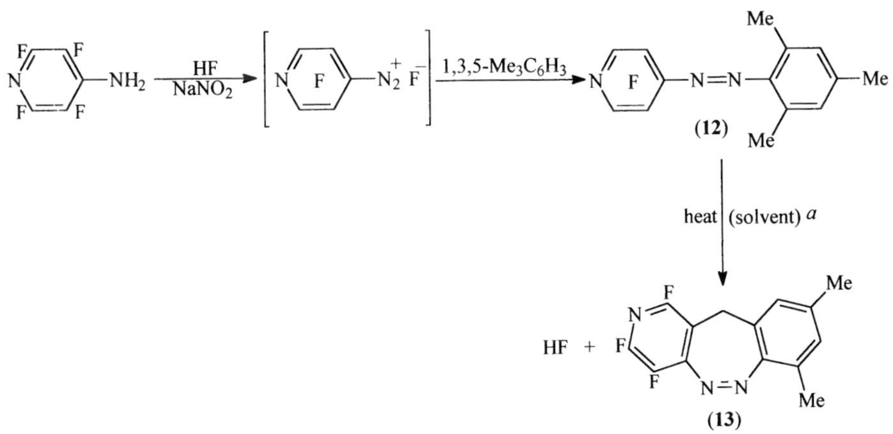

Scheme 32.14. The first example of a novel synthesis of diareno-1,2-diazepines via intramolecular dehydrofluorination of 2,4,6-trimethylphenylazo-derivatives of fluoroaromatic compounds [90, 91]. Adam Alty, who wanted to research dyestuffs chemistry for his PhD, was studing the diazotization and azo-coupling of weakly-basic aromatic amines in anhydrous hydrogen fluoride (work sponsored by ICI) when he happened to recrystallize azo-compound **12** from acetic acid and discovered that partial conversion of **12** to the pyridobenzo-1,2-diazepine **13**, a bright orange solid, had occurred. Clean efficient conversion of **12** to **13** occurs in the boiling non-protic 'inert' solvents identified below.
[a]Boiling 1,3,5-trimethylbenzene (b.p. 165 °C) or *o*-dichlorobenzene (b.p. 180.5 °C).

lah Khoshdel and Jian-An Jiang), on polymeric electrophilic fluorinating agents of the N−F class, *e.g.* **15** (with Efthimios Tsiliopoulos [93]), on piezoelectric materials derived from fluoroethylenes (with Frank Moss [94]), and on a fluorine-free topic concerning polymer-anchored amines useful for curing epoxy resins (with John Otaigbe [95]). My good friend

Dr Sydney Smith, who was Haszeldine's polymer expert and had collaborated with me earlier on the vulcanization of nitroso rubbers, played a major rôle in the last two ventures.

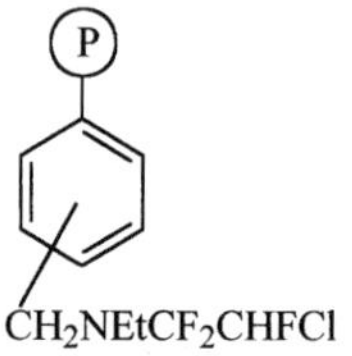

(14)

(P) = polystyrene type 'support'

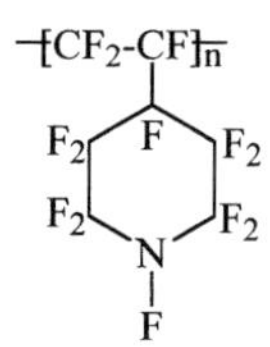

(15)
polymeric analogue of **1**

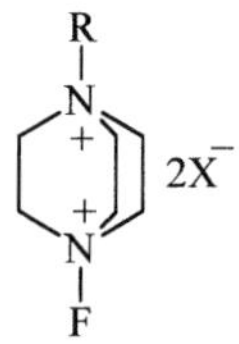

(16) R = Me, CH_2Cl, Et, CF_3CH_2, C_8H_{17};

$X^- = BF_4^-$, PF_6^-, TfO^- ($CF_3SO_3^-$)

Tsiliopoulos' research on polymeric analogues (**15**) of perfluoro-*N*-fluoropiperidine (**1**), which was matched by an extension (with some input by Vincent Murtagh [96]) of Williamson's original early 1960s discovery of the 'F^+' nature of (**1**) itself [50], was part of the drive internationally during the 1980s to address the shortcomings of electrophilic fluorinating agents available for site-selective fluorination (including ^{18}F placement) of bioactive organic molecules. The renaissance of activity in the N−F area, triggered by reports in 1983-1984 on the synthesis and 'F^+' delivery capabilities of *N*-fluoropyridin-2(1*H*)-one (Purrington's reagent) and *N*-fluoro-*N*-alkylsulfonamides (Barnette reagents), led to today's healthy list of reagents [97, 98] designed to be more generally acceptable (less aggressive, nonexplosive, less toxic, relatively inexpensive) than perchloryl fluoride, O-F compounds like trifluoromethyl hypofluorite and caesium fluoroxysulfate, xenon difluoride or fluorine itself. Our most notable contribution to that list is a range of easily-handled 1-alkyl-4-fluoro-1,4-diazoniabicyclo[2.2.2]octane salts (**16**) [99 - 101], particularly the chloromethylated tetrafluoroborate ($R = CH_2Cl$, $X^- = BF_4^-$) known as F-TEDA-BF_4 (TEDA = triethylenediamine), which is manufactured nowadays on a multi-tonne scale by Air Products (USA), primarily for use in the pharmaceutical industry.

The background to, mechanism of 'F^+' transfer from, and uses in synthesis of 1-chloromethyl-4-fluoro-1,4-diazoniabicyclo[2.2.2]octane bis(tetrafluoroborate) are detailed in my recent review 'Selectfluor™ reagent F-TEDA-BF_4 in action: tamed fluorine at your service' [16]. Details of the process employed by Air Products to manufacture Selectfluor™ reagent F-TEDA-BF_4 at its Hometown facility in Pennsylvania (US) are proprietary. Routes developed at UMIST are indicated in Scheme 32.15; Scheme 32.16 shows a few of the selective fluorinations of 'model' electron-rich substrates we have demonstrated, some in collaboration with Dr Nick Lawrence [102 - 104], whose labs are just down the corridor from my office. Everything came together so nicely: Air Products was a major producer of F_2, and TEDA – a relatively common amine used in the polyurethane foam industry – also happened to be made and manufactured by the same company.

Air Products funded much of my group's work on electrophilic fluorinating agents of the N−F class during the past decade. My gratitude for that support, and for the friendship and collaboration I've enjoyed with Dr Guido P. Pez of Air Products & Chemicals, Inc., in Allentown (PA) cannot be overstated. Guido and his colleagues Drs Robert G. Syvret and G. Sankar Lal played vitally important rôles in bringing F-TEDA-BF_4 to the marketplace [16] and hence great joy to me.

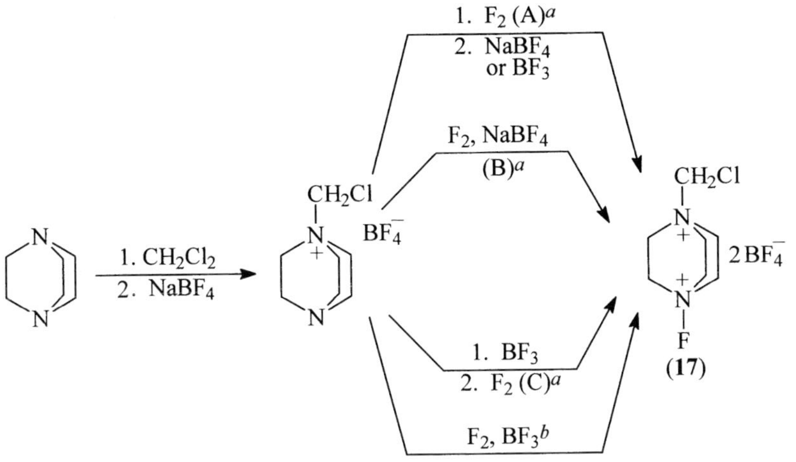

Scheme 32.15. UMIST routes to F-TEDA-BF_4 (**17**) [100]. [a] 'Flow fluorination' reactions A, B and C were carried out with F_2-N_2 blends (1:10 by vol.) at *ca.* −35 °C in acetonitrile, i.e. under electrophilic fluorination conditions. [b] In dry CH_3CN at −35°C with a 1:1 molar mixture of F_2 and BF_3 in a closed reactor at 10-20 mmHg.

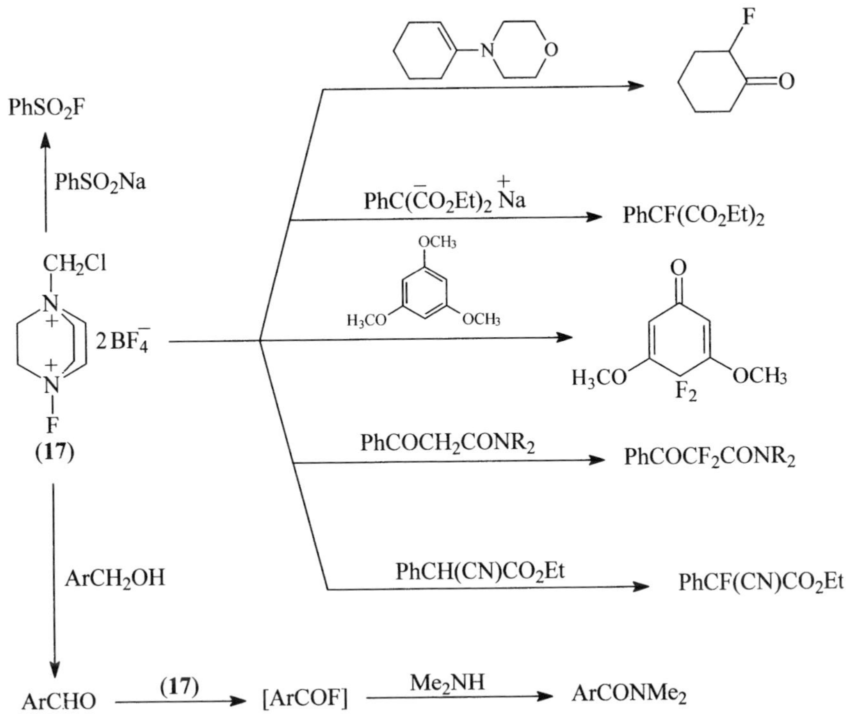

Scheme 32.16. Some of the conversions involving F-TEDA-BF_4 (**17**) demonstrated at UMIST.

CFC alternatives and all that

In 1987, the year in which the Montreal Protocol on Substances that Deplete the Ozone Layer was promulgated, I participated in the discussions of an international committee of experts from industry and academia assembled by Professor Richard ('Dick') J. Lagow (University of Texas at Austin) at the request of the US Environmental Protection Agency (EPA). The committee's mandate was to identify the most promising 'in-kind' substitutes for CFCs 11 ($CFCl_3$), 12 (CF_2Cl_2) and 113 ($CF_2ClCFCl_2$), and to establish and consider all factors which would influence commercialization of those substitutes. Two 3-day workshops were held in the US in the early part of 1987 (Washington DC in February; Yountville CA in April) but the findings were not published until April 1988 [105]. I was on board to bring my expertise in organofluorine synthesis to bear on matters under discussion, but it turned out that my knowledge of reaction mechanisms, hence ability to proffer the opinion that release of substitutes CF_3CH_2F (HFC-134a) and CF_3CHCl_2(HCFC-123) to the troposphere might cause contamination of the biosphere by trifluoroacetic acid (TFA) [106], made a greater impact. Professor F. S. (Sherry) Rowland attended the Yountville meeting and seemed happy with my reaction schemes.

Those readers who have followed the abundant literature stemming from extensive industrial and governmental programmes dealing with tropospheric fates (via oxidation) of 'alternative fluorocarbons' will know well that my concern (shared by many others), especially in the case of HFC-134a, which nowadays is a hugely important refrigerant, has been alleviated [107, 108]. TFA is indeed formed via tropospheric breakdown of HFC-134a, HCFC-123 and HCFC-124 (CF_3CHFCl) [107]; surprisingly, though, environmental measurements in many diverse global locations have revealed that current levels of TFA cannot be explained by known industrial sources and 'imply a long-term, possibly pre-industrial source' [108].

My involvement as a voyeur in environmental aspects (including global warming) of industrial organofluorine compounds extended also to 'in-kind' replacements for Halon fire extinguishants [109]. On the practical side, in mid-1993 I initiated work at UMIST with the considerable help of Dr Ziad El-Koussa on the destruction of 'environmentally-offensive waste halocarbons' with molten sodium. This project, which lasted until the end of 1995 (hence into my semi-retirement period) was funded by a substantial European Community grant, and involved a consortium led by EA Technology (Capenhurst, UK). Basically, Ziad's courageous work (which harked back to the days when I analysed my chlorofluorination products in Musgrave's lab) enabled us to scale up the old Lassaigne test (*ca.* 0.05 g of Na) to the point where mineralization of organic entities like CF_3Br and $CFCl_3$ could be achieved safely and with excellent efficiency (>99% utilization of sodium) at the 200 g Na level. The method works well with a range of perfluorinated compounds (*e.g.* CF_3SF_5, C_5F_{12}, and – of course! – perfluoro-*N*-fluoropiperidine), and we are still actively pursuing commercial exploitation of our technology.

Tapering off (1994 onwards)

Collecting the 1993 ACS Award for Creative Work in Fluorine Chemistry (an honour I'll always treasure) near the beginning of the year which saw an end to my having to juggle research, teaching and administrative duties at UMIST was a fantastic bonus. However, though able to concentrate from 1 January 1994 on running down my research group in a structured fashion, I was not completely free from deadlines and duties. Importantly, I was committed to directing UMIST's contribution to the EA Technology-led project concerning the destruction of halocarbons with sodium (the Final Report was dated January 1996) and to at least seeing out the current UMIST/Air Products research contract. Also,

Fig. 32.6. Now that the time has come for the author to consider taking six weeks vacation... (Cartoon and caption reproduced by permission of Grace Johns; the copyright is held by *Cartoons by JOHNS*.)

the UMIST Chemserve conference *Fluorine in Medicine* (April 1994 [110]) was looming large on the horizon, and plans were already well advanced for its fluorine-in-the-service-of-man sister symposium *Fluorine in Agriculture* (January 1995 [111]). The latter meeting was a joint Chemserve-Fluorine Technology Ltd. (FTL) affair, FTL being the small company established in 1988 by F. R. (Roger) Benn (then Director of UMIST's Chemserve Unit within the Chemistry department), M. ('Mike') J. Stevenson (then Director of Fluorochem Ltd., Derbyshire) and myself. My contribution to *Organofluorine Chemistry: Principles and Commercial Applications* [22] was more or less finished, and I thought the same applied to a piece on nomenclature I'd penned for the 'forthcoming' mammoth Houben-Weyl

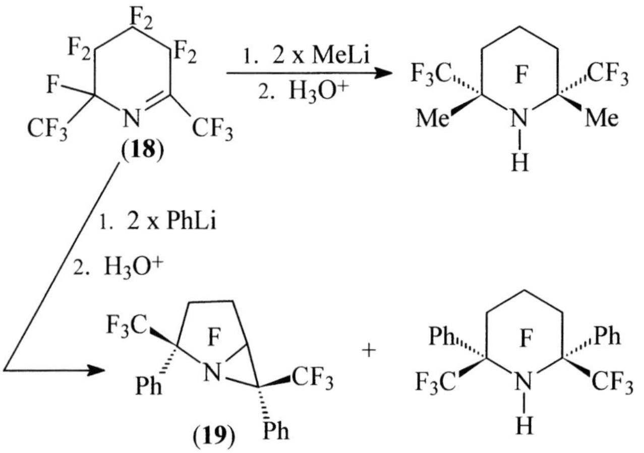

Scheme 32.17. Novel conversion of perfluoro(2,6-dimethyl-l-azacyclohexene) **18** to a l-azabicyclo[3.1.0]hexane (**19**) [112].

volumes on organofluorine compounds (see ref. [98]). Ongoing also, I sincerely hoped, would be my contribution to FTL's Bureau – a unique Manchester-based 'club' for anyone with interests centred on fluorine.

Thing have worked out quite well, so by the time *Fascinated by Fluorine* appears in print, I shall have left Alan Brisdon, Nick Lawrence and Dick Powell to push back the frontiers of fluorine chemistry at UMIST. I intend to continue chipping away at my publications backlog and helping Roger Benn and Vic Garner to run Fluorine Technology Bureau; should I lose access to the non-fictional literature (See Fig. 32.6 – one of my favourite cartoons[25]), maybe I'll start work on that novel!

Epilogue

Fairly recently, my last PhD student (David Tovell) chanced on the ring-contraction shown in Scheme 32.17 [112], reminding me of the enduring truth of Harry Emeléus' assertion in the Foreword to *Fluorine: The First Hundred Years (1886-1986)*, that 'Fortunately, too, for those who have yet to enter this fascinating field, another of his [Moissan's] sayings still holds true: "... L'étude des composés fluorés resérve encore bien des surprises" '. It's no wonder I've enjoyed every moment of being fascinated by fluorine.

References

1 J. Flahaut and C. Viel, in R. E. Banks, D. W. A. Sharp and J. C. Tatlow (eds.), *Fluorine: The First Hundred Years (1886-1986)*, Elsevier Sequoia, Lausanne and New York, 1986, p. 27.

[25] I came across this JOHNS cartoon nearly twenty years ago in the August 1980 issue of *CHEMTECH* (p. 481) and have kept it in mind ever since. With the help of Arleen Courtney and Marcia Dresner of the ACS Publications Division, I was able to contact Grace Johns in Pebble Beach, California.

2 M. Wood, *In Search of the Dark Ages*, British Broadcasting Corp., 1981, p. 8.
3 E. T. Williams and H. M. Palmer (eds.), *Dictionary of National Biography: 1951-1960*, Oxford Univ. Press, 1971, p. 788.
4 See *Who's Who 1986*, A & C Black Ltd., London.
5 R. E. Banks, J. E. Burgess, W. M. Cheng and R. N. Haszeldine, *J. Chem. Soc.*, (1965) 575.
6 W. K. R. Musgrave and F. Smith, *J. Chem. Soc.*, (1949) 3021 and 3026.
7 E. A. Tyczkowski and L. A. Bigelow, *J. Am. Chem. Soc.*, *75* (1953) 3523.
8 R. E. Banks, P. Johncock, R. H. Mobbs and W. K. R. Musgrave, *Ind. Eng. Chem., Process Design and Development*, *1* (1962) 262.
9 F. Cuthbertson, T. F. Holmes, W. K. R. Musgrave and M. C. Tanner, *J. Appl. Chem.*, *8* (1958) 390.
10 R. E. Banks and W. K. R. Musgrave, *J. Appl. Chem.*, *6* (1956) 214.
11 R. E. Banks and W. K. R. Musgrave, *J. Chem. Soc.*, (1956) 4682.
12 R. E. Banks, F. Cuthbertson and W. K. R. Musgrave, *Anal. Chim. Acta*, *13* (1955) 442.
13 W. K. R. Musgrave, *Quart. Rev.*, *8* (1954) 331.
14 R. E. Banks, 'Isolation of fluorine by Moissan: setting the scene', in R. E. Banks, D. W. A. Sharp and J. C. Tatlow (eds.), *Fluorine: The First Hundred Years (1886-1986)*, Elsevier Sequoia, Laussane and New York, 1986, p. 3.
15 Quoted in the editorial article 'Quinine trade begun', *Chem. and Ind. (London)*, (1971) 1467.
16 R. E. Banks, *J. Fluorine Chem.*, *87* (1998) 1 and references cited there.
17 A. V. Grosse and G. H. Cady, *Ind. Eng. Chem.*, *39* (1947) 367.
18 E. T. McBee, *Ind. Eng. Chem.*, *39* (1947) 236.
19 M. Howe-Grant (ed.), *Fluorine Chemistry: A Comprehensive Treatment*, John Wiley, New York, 1995 (reprinted from the Kirk-Othmer Encyclopedia of Chemical Technology).
20 R. E. Banks (ed.), *Organofluorine Chemcals and Their Industrial Applications*, Ellis Horwood Publishers, Chichester, 1979.
21 R. E. Banks (ed.), *Preparation, Properties, and Industrial Applications of Organofluorine Compounds*, Ellis Horwood Publishers/John Wiley, 1982.
22 R. E. Banks, B. E. Smart, and J. C. Tatlow (eds.), *Organofluorine Chemistry: Principles and Commercial Applications*, Plenum Press, New York, 1994.
23 W. G. Hiscock, in *Acid Handling: The transport and handling of sulphuric and hydrofluoric acids*, (2nd edition), Imperial Smelting Corporation (Research Department Publication), London, 1958.
24 'The production of Anhydrous Hydrofluoric Acid', *The Industrial Chemist*, December 1948, 801.
25 J. C. Tatlow, *J. Fluorine Chem.*, *75* (1995) 7.
26 A. K. Barbour and P. Thomas, *Ind. Eng. Chem.*, *58* (1966) 48.
27 W. Prescott, *Chem. and Ind. (London)*, (1978) 56.
28 R. E. Banks, A. K. Barbour, B. Gething, C. R. Patrick, J. C. Tatlow and A. E. Tipping, *Nature*, *183* (1959) 586.
29 R. E. Banks, A. K. Barbour, C. R. Patrick and J. C. Tatlow, *US Patent 3,004,077* (1961).
30 R. E. Banks and A. E. Tipping, *Chem. and Ind. (London)*, (1959) 1491.
31 R. E. Banks, *British Patent 920,796* (1963).
32 R. E. Banks, W. I. Bevan and W. K. R. Musgrave, *Chem. and Ind. (London)*, (1959) 296.
33 R. E. Banks, *British Patent 922,610* (1963).
34 J. A. Godsell, M. Stacey and J.C. Tatlow, *Nature*, 178 (1956) 199.
35 R. N. Haszeldine, *J. Chem. Soc.*, (1952) 4423.
36 A. L. Henne and W. Postelnek, *J. Am. Chem. Soc.*, *77* (1955) 2334.
37 B. V. Bowden, in D. S. L. Cardwell (ed.), *Artisan to Graduate*, Manchester University Press, 1974, pp. 248-257.
38 S. Young, *J. Chem. Soc.*, *39* (1881) 489.
39 B. Brauner, *J. Chem. Soc.*, *41* (1882) 68.
40 B. Brauner, *J. Chem. Soc.*, *65* (1894) 393.
41 R. E. Banks, *Fluorocarbons and their Derivatives*, Oldbourne Press, London, 1964 (1st edn.); Macdonald & Co., London, 1970 (2nd edn.).
42 R. E. Banks and M. G. Barlow (senior reporters), *Fluorocarbon and Related Chemistry*, The Chemical Society, London, Vol. 1 (1971); Vol. 2 (1974); Vol. 3 (1976).

43 R. E. Banks, A E. Ginsberg and R. N. Haszeldine, *Proc. Chem. Soc. (London)*, (1960) 211 [full paper: *J. Chem. Soc.*, (1961) 1740].
44 R. E. Banks, R. N. Haszeldine and J. B. Walton, *J. Chem. Soc.*, (1963) 5581.
45 R. E. Banks, R. N. Haszeldine and A. Prodgers, *J. Chem. Soc., Perkin Trans. 1*, (1973) 596.
46 R. E. Banks, M. G. Barlow, W. D. Davies, R. N. Haszeldine and D. R. Taylor, *J. Chem. Soc. (C)*, (1969) 1104.
47 R. E. Banks, P. A. Carson and R. N. Haszeldine, *J. Chem. Soc.*, (1973) 1111.
48 R. E. Banks, K. C. Eapen, R. N. Haszeldine, P. Mitra, T. Myerscough and S. Smith, *J.C.S. Chem. Comm.*, (1972) 833.
49 R. E. Banks, R. N. Haszeldine, P. Mitra, T. Myerscough and S. Smith, *J. Macromol. Sci.-Chem., (A8)* (1974) 1325.
50 R. E. Banks and G. E. Williamson, *Chem. and Ind. (London)*, (1964) 1864.
51 R. E. Banks and A. Prakash, *Tetrahedron Lett.*, (1973) 99 (see also ref. [80]).
52 R. E. Banks, N. D. Venayak and T. A. Hamor, *J.C.S. Chem. Comm.*, (1980) 900.
53 R. E. Banks, L. E. Birks and R. N. Haszeldine, *J. Chem. Soc. (C)*, (1970) 201.
54 R. N. Haszeldine and R. E. Banks, *British Patent 1,118,629* (1968) [to Coal Industry (Patents) Ltd.].
55 R. N. Haszeldine and R. E. Banks, *British Patent 1,120,228* (1968) [to Coal Industry (Patents) Ltd.].
56 R. E. Banks and R. N. Haszeldine, *J. Oil Colour Chem. Assoc.*, *42* (1959) 591.
57 R. E. Banks, J. M. Birchall and R. N. Haszeldine, *Soc. Chem. Ind.*, Monograph No. 13 (1961) 270.
58 R. E. Banks and R. N. Haszeldine, *Adv. Inorg. Chem. Radiochem.*, *3* (1961) 337.
59 R. E. Banks and R. N. Haszeldine, in N. Kharasch and C. Y. Myers (eds.), *The Chemistry of Sulfur Compounds*, Vol. 2, Pergamon Press, Oxford, 1966, p. 137.
60 R. E. Banks and H. Goldwhite, in O. Eichler, A. Farah, H. Herken and A. D. Welch (eds.), *Handbook of Experimental Pharmacology: Vol. XX/1, Pharmacology of Fluorides* (Part 1; sub-ed. F. A. Smith), Springer-Verlag, New York, 1966, p. 1.
61 J. Burdon, D. J. Gilman, C. R. Patrick, M. Stacey and J. C. Tatlow, *Nature*, *186* (1960) 231.
62 R. D. Chambers and C. R. Sargent, *Adv. Heterocyclic Chem.*, *28* (1981) 1.
63 R. D. Chambers, J. Hutchinson and W. K. R. Musgrave, *Proc. Chem. Soc. (London)*, (1964) 83 [full paper: *J. Chem. Soc.*, (1964) 3573].
64 R. E. Banks, R. N. Haszeldine, J. V. Latham and I. M. Young, *Chem. Ind. (London)*, (1964) 835 [full paper: *J. Chem. Soc.*, (1965) 594].
65 M. C. Henry, C. G. Griffis and E. C. Stump, *Fluorine Chem. Rev.*, *1* (1967) 1.
66 R. E. Banks, K. Mullen and G. E. Williamson, *J. Chem. Soc. (C)*, (1968) 2608.
67 R. E. Banks, A. J. Parker, M. J. Sharp and G. F. Smith, *J. Chem. Soc., Perkin Trans. 1*, (1973) 5.
68 M. Nickkho-Amiry, R. E. Banks, A. J. Parker and M. J. Parrott, *J. Fluorine Chem.*, *75* (1995) 205.
69 R. E. Banks, R. N. Haszeldine and T. Myerscough, *J. Chem. Soc., Perkin Trans. 1*, (1972) 1449.
70 A. Nickon and E. F. Silversmith, *Organic Chemistry: The Name Game*, Pergamon Press, New York, 1987.
71 R. E. Banks and H. Sutcliffe, *Chem. and Ind. (London)*, (1962) 979.
72 *Chem. Brit.*, *16* (1980) 296 (corrigendum: p. 349).
73 R. E. Banks and T. J. Noakes, *J. Chem. Soc., Perkin Trans. 1*, (1976) 143.
74 R. E. Banks, M. G. Barlow, J. C. Hornby and M. Mamaghani, *J. Chem. Soc., Perkin Trans. 1*, (1980) 817.
75 R. E. Banks, M. G. Barlow, T. J. Noakes and M. M. Saleh, *J. Chem. Soc., Perkin Trans. 1*, (1977) 1746.
76 B. A. Al-Saleh, R. E. Banks and M. G. Barlow, *J.C.S. Chem. Comm.*, (1980) 997.
77 R. E. Banks, M. G. Barlow and N. D. Venayak, *J.C.S. Chem. Comm.*, (1980) 151.
78 R. E. Banks and M. J. McGlinchey, *J. Chem. Soc. (C)*, (1971) 3971 and the previous 3 papers in this series locatable from there.
79 R. E. Banks and G. R. Sparkes, *J. Chem. Soc., Perkin Trans. 1*, (1972) 2964.
80 R. E. Banks and A. Prakash, *J. Chem. Soc., Perkin Trans. 1*, (1974) 1365.
81 R. E. Banks, A. Prakash and N. D. Venayak, *J. Fluorine Chem.*, *16* (1980) 325.
82 R. E. Banks and I. M. Madany, *J. Fluorine Chem.*, *30* (1985) 211 and 413.
83 A. R. Bailey and R. E. Banks, *J. Fluorine Chem.*, *24* (1984) 117.
84 M. Abed-Rabboh, R. E. Banks and B. Beagley, *J. Chem. Soc., Chem. Commun.*, (1983) 1117.
85 R. E. Banks and S. M. Hitchen, *J. Chem. Soc., Perkin Trans. 1*, (1982) 1593.
86 R. E. Banks, R. G. Pritchard and J. Thomson, *J. Chem. Soc., Perkin Trans. 1*, (1986) 1769 and previous papers listed there.

87 R. E. Banks and S. N. Mohialdin, *J. Fluorine Chem.*, *51* (1991) 407.
88 R. E. Banks and N. Dickinson, *J. Chem. Soc., Perkin Trans. 1*, (1982) 685.
89 R. E. Banks and A. Richards, *J. Chem. Soc., Chem Commun.*, (1985) 205.
90 A. C. Alty, R. E. Banks, B. R. Fishwick, R. G. Pritchard and A. R. Thompson, *J. Chem. Soc., Chem. Commun.*, (1984) 832.
91 R. E. Banks, Y. Djebli, R. Fields, N. O. Olawore, R. G. Pritchard, E. Tsiliopoulos and J. Mason, *J. Chem. Soc., Perkin Trans. 1*, (1989) 1117.
92 R. E. Banks, A.-K. Barrage and E. Khoshdel, *J. Fluorine Chem.*, *17* (1981) 93.
93 R. E. Banks and E. Tsiliopoulos, *J. Fluorine Chem.*, *34* (1986) 281.
94 P. A. Payne, Q. X. Chen, R. E. Banks, S. Smith, F. Moss and M. Nicholls, in C. Brook and P. D. Hanstead (eds.), *Impact of Non-Destructive Testing (NDT-89)*, Pergamon, Oxford, 1989, p. 93.
95 J. O. E. Otaigbe, R. E. Banks and S. Smith, *Brit. Polymer J.*, *20* (1988) 53.
96 R. E. Banks, V. Murtagh and E. Tsiliopoulos, *J. Fluorine Chem.*, *52* (1991) 389.
97 G. S. Lal, G. P. Pez and R. G. Syvret, *Chem. Rev.*, *96* (1996) 1737.
98 G. G. Furin, in B. Baasner, H. Hagemann and J. C. Tatlow (eds.), *Methods of Organic Chemistry (Houben-Weyl): Vol. E10a, Organo-Fluorine Compounds*, Georg Thieme Verlag, Stuttgart, 1999, p. 432.
99 R. E. Banks, *US Patent 5,086,178* (1992) (to Air Products & Chemicals, Inc.).
100 R. E. Banks, I. Sharif and R. G. Pritchard, *Acta Cryst.*, *C49* (1993) 492.
101 R. E. Banks, M. K. Besheesh and S. N. Mohialdin-Khaffaf, *J. Chem. Soc.*, *Perkin Trans. 1*, (1996) 2069.
102 R. E. Banks, N. J. Lawrence and A. L. Popplewell, *J. Chem. Soc.*, *Chem. Commun.*, (1994) 343.
103 R. E. Banks, N. J. Lawrence and A. Popplewell, *Synlett.*, (1994) 831.
104 R. E. Banks, M. K. Besheesh, R. W. Görski, N. J. Lawrence and A. J. Taylor, *J. Fluorine Chem.*, *96* (1999) 129.
105 '*Findings of the Chlorofluorocarbon Chemical Substitutes International Committee*', Publ. No. EPA-600/9-88-009, NTI Service, Springfield, VA 22161, USA.
106 See *Chem. Brit.*, *26* (1990) 217 and O. Tickell, *New Scientist*, *128* No. 1739 (1990) 41.
107 A. McCulloch, *J. Fluorine Chem.*, *100* (1999) 163.
108 J. C. Boutennet *et al.*, *Human and Ecological Risk Assessment*, *5* (1999) 59.
109 R. E. Banks, *J. Fluorine Chem.*, *67* (1994) 193.
110 See R. E. Banks and K. C. Lowe (eds.), *Fluorine in Medicine in the 21st Century*, RAPRA Technology, Shawbury (UK), 1994.
111 See R. E. Banks (ed.), *Fluorine in Agriculture*, RAPRA Technology, Shawbury (UK), 1995.
112 R. E. Banks, M. K. Besheesh, N. J. Lawrence, R. G. Pritchard and D. J. Tovell, *J.C.S. Chem. Commun.*, (1999) 47.

Appendix 32.1

Recollections of fluorochemical research at Avonmouth[1]

ANTHONY K. BARBOUR
7 Pitch & Pay Park, Sneyd Park, Bristol BS9 1NJ, UK

My initiation (1947) into organic fluorine chemistry at the University of Birmingham came initially through the late Fred Smith and then as Colin Tatlow's first research student (along with Cherry Tatlow) under the general guidance of Maurice Stacey. Those were the days of medium-temperature, open, nickel-anode fluorine cells, utilized – when they could be persuaded to work! – for liquid-phase fluorination and for vapour-phase fluorination using the cobaltic fluoride and metallic 'catalysts' (Bigelow) procedures. I utilized the CoF_3 method, which had been worked on extensively at Birmingham by Bob Haszeldine, but benefited from the use of a rotary reactor developed mainly by our workshop superintendent Bill Massingham. The finale of the programme on completely fluorinated compounds at Birmingham was the use of this method to make the bicyclohexyl series of perfluorocarbons. These 'dumbbell' compounds did not possess the improved viscosity-temperature characteristics which some had predicted; in fact, several were crystalline solids at room temperature. My own work was therefore directed away from complete fluorination (associated with products of high chemical stability) to partial fluorination, hence providing access to chemically reactive species. With benzene as the substrate, partial fluorination with CoF_3 was developed to provide the hydrofluorocyclohexane series of compounds which Colin Tatlow, Bob Stephens and many of their research students at Birmingham used in numerous stereochemical and mechanistic studies.

Those early days were notable for the absence of significant spectroscopic or chromatographic techniques, though the latter problem was partly compensated for by the massive (to us) distillation columns largely handmade by Stan Jacobs, who seemed to prefer this activity to bench chemistry! Others around at that time in the fluorine group (*i.e.* those who worked with fluorine in the famous 'large-scale' laboratory) were Ralph Worthington and Brian Barlow. Contemporaneously, equally significant work on reactions promoted by trifluoroacetic acid and its anhydride was proceeding at the hands of Cherry Tatlow, Bob Cartwright, John Tedder and H. D. MacKenzie under the joint supervision of Colin Tatlow and Ted Bourne.

In November 1950 I transferred to Imperial Smelting Corporation at Avonmouth, where a Pennsalt/Ozark-Mahoning absorption plant for anhydrous HF production existed despite the fact that essentially no market existed for the product. After unsuccessful attempts to licence the DuPont liquid-phase elevated-pressure process to manufacture CF_2Cl_2 (aerosols had not arrived then in the UK, so there was no market for $CFCl_3$), research bosses Ken Morgan and Arnold Edwards decided to develop a vapour-phase process

[1] Adapted, with permission, from *Fluorine Technology Bulletin*, No. 21 (1996). Joe still regularly attends meetings of Fluorine Technology Bureau, the Manchester-based 'club' for fluorine chemists and technologists.

in the laboratory (I worked on the fluorination of CCl_4, while Stamford Green and Leon Belf concentrated on methane-based processes). Subsequently, successful pilot-planting was directed by Dick Kingdom, and so ISC's ISCEON® range came into being. All of this work benefited greatly from Stamford Green's skills in first designing a Podbielniak low-temperature still to analyse the products, and then leading the development of home-made gas-liquid chromatography units which expedited progress enormously. Commercialization of ISCEON® started in 1957-8, and the process employed was highly successful through many capacity extensions until the Montreal Protocol essentially halted production in 1994-5. Unfortunately, the process could not be used satisfactorily to manufacture either the hydrochlorofluorocarbon or the two-carbon CFC series, so a liquid-phase route had to be worked out for products such as CHF_2Cl (HCFC-22) and $CF_2ClCFCl_2$ (CFC-113).

In 1954, research started on thermally-stable copolymers based on isomeric cyclo-C_6F_8 monomers; this effort was funded by the then Ministry of Supply, and co-ordinated by Jim Fear of RAE Farnborough. These perfluorinated cyclohexadienes were synthesised *via* dehydrofluorination of $C_6H_2F_{10}$ isomers produced by partial CoF_3-fluorination of benzene, which connected the work to my previous researches at Birmingham. The Avonmouth team was reinforced by Bob Stephens (a carbohydrate PhD from Birmingham) and Bill Hopkin (an organic chemist from Glasgow with strong polymer interests). Bill, who retired a few years ago, subsequently became influential in RTZ Services through his work on the technico-economic analysis of metallurgical processes and the safe disposal of arsenic. Bob became greatly interested in the environmental toxicology of lead after he returned to Birmingham (in 1956) to join Colin Tatlow's fluorine group; he was replaced at Avonmouth by Eric Banks, fresh from his PhD researches on chlorine trifluoride under Ken Musgrave at Durham University. The Ministry of Supply team was then supported by an active group of embryonic high-fliers, including Tony Tipping, Bill Bevan, Brian Joyner, Malcolm Sainsbury and David Brown. What turned out to be a milestone of this period was the observation by Eric Banks of a 'ghost' GLC peak in material he had obtained while attempting to effect thermal isomerization of 1,4-C_6F_8 to its conjugated 1,3-isomer in an iron tube. This was shown to arise from the defluorination product hexafluorobenzene and, quite quickly, this important breakthrough provided the starting point for company-funded programmes to develop fluoroaromatic chemistry commercially at Avonmouth. Eric Banks left to join Bob Haszeldine's team in September 1958; Tony Tipping followed shortly afterwards.

In the early sixties, organic fluorine research at Avonmouth followed three main lines: (i) fluoro-aromatics; (ii) further development of routes to fluorohalogenoalkanes, including inhalation anaesthetics; and (iii) development of a new family of novel thermally-stable copolymers, carried out jointly with Monsanto and funded by the USAAF from Wright Field, Dayton, Ohio, under the technical supervision of Christ Tamborski. Mike Buxton, David Wotton, George Fuller and Dick Mobbs were additional members of the Avonmouth research group during this time. Throughout the whole of this period (starting in 1950) – and subsequently – the work was guided and supported by wise and incisive advice from consultants Bob Haszeldine, Colin Tatlow and Ken Musgrave, with less frequent interventions from Geoff Coates and Maurice Stacey, Maurice having provided the initial driving force for the entire concept.

The early 1960s saw the curtailment of my fluorochemical activities because promotions meant that I had to take more and more interest in inorganic chemicals and the metallurgy of zinc, lead and cadmium; also, both Departmental and Company administrative matters took up much time. Virtually complete formal severance from responsibility for research management then came through the general upsurge of interest in environmental and pollution matters in the early 1970s. I was transferred to the Group Company RTZ Services as Group Environmental Scientist, responsible for advising member companies worldwide on problems arising from mining, minerals extraction, smelting, chemical production and light engineering. In this role I pioneered the now widely-accepted technique of environmental auditing, did much trade association work, and travelled widely to assess both current and potential RTZ sites globally. My connection with fluorochemicals was restricted to visiting manufacturing plants, occasional refereeing for the *Journal of Fluorine Chemistry* and discussions with many long-standing friends – even the CFC/stratospheric ozone issue was excluded, being handled most ably by Brian Joyner.

Leadership of the Avonmouth research group passed, as part of RTZ Chemicals, to David Robertson and Nigel Simpson, who were ably supported by Stam Green, David Slinn, David Wotton, David Cozens and, for a few years, George Fuller, who conducted much valuable work on the production of fluoroaromatics by halogen exchange. George eventually joined Borax Consolidated (a major member of the RTZ Group); previously, Mike Buxton had left to start his own successful enterprise, Bristol Organics Ltd (later sold to Aldrich) which specialized in producing fluoroaromatics. During this period, the emphasis at Avonmouth returned to the 'FLUTEC' series of completely fluorinated alkanes and cycloalkanes produced by an advanced type of cobaltic fluoride process (initiated by Dick Kingdom), halex processes for fluoroaromatics production, and the commericalization of fluorine-containing intermediates for use in the pharmaceutical industry.

As a relatively early worker in organofluorine chemistry, it is great to note that many others have been fascinated by its potential for the unusual. Significantly for me too, fluorine-containing entities are now part of the everyday armoury of synthetic organic chemists generally – and also of some physical chemists. Let us hope that innovative chemistry and scientific regard for environmental management will continue to provide satisfactory and enjoyable appointments for chemists in both a prosperous chemical industry and a fully appreciated and funded academic world.

BIOGRAPHIC NOTE

A. K. Barbour

A. K. Barbour, 'Joe' to the legion of friends he has acquired worldwide during his long career in the chemical industry, celebrated his 73rd birthday on 27 June, 1999 (one day earlier would have meant that he was born *exactly* 40 years from the date on which Moissan isolated fluorine). His hands-on contribution to fluorine chemistry ceased in the late 1960s but the group he established at ISC continued to flourish throughout the seventies as part of RTZ Chemicals, which was sold to Rhône-Poulenc (now Rhodia) when the parent company 'returned to basics' to concentrate on mining and minerals extraction in 1989 (the year in which Joe joined the noble company of RTZ pensioners). Simultaneously with his work post-1970 as Group Environmental Scientist to the RTZ Services Group, Joe was a non-executive board member of Wessex Water (both Authority and plc) from 1980 to 1994, doing much environmental auditing for them during that period. His work on the diagnosis and amelioration of environmental and pollution issues, a field he has remained active in since retirement from RTZ, has brought him important honours: OBE (Order of the British Empire), 1988; RSC (The Royal Society of Chemistry) Distinguished Environmental Guest Lectureship, 1994; SCI (Society of Chemical Industry, London) Environmental Medal, 1995; RSC Award in Environmental Chemistry, 1996; Honorary DSc, University of the West of England, 1998.

Joe still lives in Bristol, where his wife, Audrey, is still very active as a magistrate. Their daughter Elizabeth operates the administrative side of her husband's electrical contracting business in Aldershot; their son Neil is a partner in a legal practice in Bath and specializes in environmental and land issues.

Appendix 32.2

Fluorine chemistry: keeping my hand in

HAYDN SUTCLIFFE

School of Environment and Life Sciences,[1] *University of Salford, Salford M6 6PU, UK*

H. Sutcliffe

My first contact with fluorine chemistry was in Dr A. K. ('Joe') Barbour's team in the Research Department of the then National Smelting Company (NSC) in Avonmouth. There I worked with Dr Mike Buxton synthesizing a stock of methylene fluoride ready for a study of its bromination. Under Mike's watchful eye, I learnt about gas handling, storage of liquefied gases, and the use of autoclaves in synthesis. Halogen exchange was widely used at NSC in those days and I soon learnt about the various catalysts that one could use, in both gas and liquid phase. Finally, I was 'let loose' on my own to examine the gas-phase bromination of methylene fluoride. This proved more difficult than was at first thought, especially where obtaining a consistent reactant ratio was concerned.

It was about this time that I re-established contact with Eric Banks, who, unknown to me, was working on the fluorine cells in a remote part of the plant. Eric and I had previously been at Leek High School together during the '40s, then we lost touch. [I graduated (BSc, External London) at University College, Hull, then did my army service (1954-56) before finding employment]. I also met Bob Haszeldine at ISC/NSC, since he visited the group as a consultant from time to time. Then things began to happen rapidly. Eric, who had always wanted an academic life, left Avonmouth to help Bob build up a new fluorine team in Manchester. I soon followed and began work as a research student at Manchester with Bob and Eric in November 1958. This started a 'haemorrhage' from Avonmouth to Manchester (or should I say transfusion?), since we were followed by Tony Tipping and Bill Bevan, who also joined the UMIST fluorine group as research students. I recall a wry comment from Joe one day: 'It looks as if industry is training people for academia instead of vice versa.'

At Manchester, I worked on the reaction of CF_3NO with a variety of partially-fluorinated olefins and one or two commercial hydrocarbon monomers [1]. This was a particularly enjoyable part of my life and, apart from one incident, the work proceeded reasonably smoothly. The incident in question was an explosive reaction between CF_3NO and C_2H_4. A few grams of these two gases, in a small autoclave, detonated at 40 atm while

[1] Currently, Dr Sutcliffe holds a part-time appointment in this Department, having retired from the Department of Chemistry and Applied Chemistry at Salford University in 1992.

I was holding the autoclave in a sink full of water, checking for leaks. The net result was a straight Bourdon tube, a broken sink and the end of one of my fingers hanging off. However, I recovered and graduated with a PhD in December 1961. I stayed on in the fluorine group at Manchester for a further year as a postdoctoral fellow before moving to the then Royal College of Advanced Technology, Salford, as a Lecturer in Inorganic Chemistry.

At Salford I initiated some research on the effect of gamma radiation on the simple fluorinated molecules CF_3I and CF_3Br in both the gas and liquid phase. In this work, I was aided by Ian McAlpine, my first postgraduate student. Ian was a meticulous experimentalist and we published five papers [2-6] on our studies, plus a review of this field [7]. During the course of our work we achieved the first synthesis of CF_2I_2 and studied its spectroscopic (IR) properties [8].

I then turned my attention to the chemistry of CCl_3NO, thinking that, as with CF_3NO, it might be possible to synthesize some interesting nitroso rubbers, then go on to use the CCl_3 group for crosslinking purposes. In the event, CCl_3NO, a deep blue liquid, proved to be less stable than CF_3NO owing to the lability of the C—Cl and C—N bonds. However, we did discover the presence of two novel free radicals, namely the nitroxide $(CCl_3)_2NO\cdot$ [9] and the perchloronitrone-related species $CCl_2{=}N^+(O^-)CCl_2\cdot$ [10] in liquid CCl_3NO using esr techniques. The discovery of the nitroxide was serendipity personified. I was working on the synthesis of CCl_3NO in a laboratory next door to that of my colleague, Harry Wardale, an electron spin resonance expert, and I asked him to 'have a look' at a sample of the nitroso compound in his spectrometer. The result was a very pretty 57 line spectrum which matched perfectly the structure of the chlorinated nitroxide [9]. Appropriately, we were using blue ink at the time!! The second radical was detected later in CCl_3NO solution [10].

At this juncture, I returned to fluorine chemistry and began to build a continuous-flow Simons-type ECF cell; in this I was helped by another research student, David Lines. The major problem with such a cell is ensuring that there is a uniform flow of the AHF-based electrolyte. We had a 'mock-up' built from Perspex^{TM} and studied a variety of different inlet and outlet geometries, using water and a blue dye to follow flow patterns. I believed this work to be an important contribution to the design of 'flow cells' [11], a claim soon confirmed when the company we were associated with began to run a flow cell without considering such design features and encountered the formation of localised, heavy deposits of carbonaceous material in the electrode pack: more haste, less progress! Unfortunately, towards the end of our work, the company abandoned cell development, so we had to discontinue our study owing to lack of funding.

The final phase of my fluorine research involved the synthesis of perfluoroalkanecarboxylates of zirconium. We hoped that the affinity of zirconium for cellulosic materials would enable us to stick perfluoroalkyl groups onto materials via an intermediate zirconium atom, and did indeed manage to render filter paper oil- and water-repellent; however, the tendency for the compounds to hydrolyse slowly in air was worrisome. The work was continued on a purely academic basis with the help of another student, Robert Holmes, and a number of structurally very interesting zirconium perfluoroalkanecarboxylates were synthesized, such as $Zr_6O_4(OH)_4(C_3F_7CO_2)_{12}$, which was prepared by the partial hydrolysis of $ZrO(C_3F_7CO_2)_2$. The structures of such complexes (determined by X-ray methods) represent a big step forward in understanding the structural chemistry of zirconium carboxylates, and I really ought to find time to publish our results.

References

1 R. E. Banks, R. N. Haszeldine, H. Sutcliffe and C. J. Willis, 'The reaction of trifluoronitrosomethane with trifluoroethylene, vinylidene fluoride, vinyl fluoride, and ethylene', *J. Chem. Soc.*, (1965) 2506.

2 I. McAlpine and H. Sutcliffe, 'The radiolysis of trifluoroiodomethane in the gas phase', *J. Phys. Chem.*, *73* (1969) 3215.

3 I. McAlpine and H. Sutcliffe, 'The radiolysis of gaseous trifluoroiodomethane in the presence of nitric oxide', *J. Phys. Chem.*, *74* (1970) 848.

4 I. McAlpine and H. Sutcliffe, 'The radiolysis of liquid trifluoroiodomethane', *J. Phys. Chem.*, *74* (1970) 1422.

5 I. McAlpine and H. Sutcliffe, 'A comparison of the radiolysis of liquid bromotrifluoromethane with the radiolysis of liquid trifluoroiodomethane', *J. Phys. Chem.*, *76* (1972) 2070.

6 H. Sutcliffe, 'The reactivity and electronic structure of excited trifluoromethyl radicals formed during radiolysis', *Int. J. Radiat. Phys. Chem.*, *4* (1972) 499.

7 H. Sutcliffe and I. McAlpine, 'The radiation chemistry of polyfluorinated organic compounds', *Fluorine Chemistry Reviews*, *6* (1973) 1.

8 I. McAlpine and H. Sutcliffe, 'The infrared absorption spectra and vibrational assignment of difluorodiiodomethane', *Spectrochimica Acta*, *25A* (1969) 1723.

9 H. Sutcliffe and H. W. Wardale, 'Bis(trichloromethyl) nitroxide. A novel electron spin resonance spectrum', *J. Am. Chem. Soc.*, *89* (1967) 5487.

10 V. Astley and H. Sutcliffe, '$\alpha\alpha$-Dichloro-*N*-(dichloromethyl)nitrone radical, a novel free radical', *J.C.S. Chem. Comm.*, (1971) 1303.

11 D. Lines and H. Sutcliffe, 'The electrochemical fluorination of octanoyl fluoride with electrolyte circulation', *J. Fluorine. Chem.*, *17* (1981) 423.

Appendix 32.3

Chemistry – but not in black and white

BRYAN G. WILLOUGHBY

Rapra Technology Limited, Shawbury, Shrewsbury, Shropshire SY4 4NR, UK

I joined the fluorine group at UMIST almost immediately after first graduating (BSc) there in the summer of 1966. To become part of such an active group was a major inducement, as was the opportunity to study derivatives of an element so extreme in its behaviour as fluorine. What has never failed to amaze me since that time is that fluorine in its compounds can display such contrasting character – some materials showing exceptional stability while others are so reactive that even containment is made difficult. Always there are surprises in store with such 'all-or-nothing' chemistry.

One such surprise lay in the uncharacteristic (for the class) infrared spectrum of trifluoromethyl isocyanide (CF_3NC). I was the first person in the research group to isolate trifluoromethyl isocyanide [1], a feat not helped by its 'spontaneous' polymerization. Obtaining the infrared spectrum of the gaseous monomer at low pressure presented no problem, but great difficulty was experienced in convincing others at UMIST that this was indeed the spectrum of an isocyanide. The unusual hot bands [2] met a cool reception from many, so I was left to confirm the identity of my product via something more believable. There was quite a celebration therefore when the diagnostic ^{19}F NMR spectrum was seen – ever so briefly – when a solution of CF_3NC in trichlorofluoromethane was examined as it warmed up from -80 °C to room temperature [2].

Cryogenics and vacuum techniques figured prominently in the work of the UMIST laboratories at that time since they provided the means by which volatile product mixtures could be fractionated. How well I remember a colleague proudly displaying a trap-to-trap separation of bistrifluoromethyl nitroxide (purple gas, yellow solid) from trifluoronitrosomethane (blue gas, deep blue liquid) and its photo-dimer (orange-red gas, reddish-brown liquid) and exclaiming 'none of your black-and-white chemistry for me!'. Shortly afterwards I was able to add to the decoration of the laboratory through the accidental synthesis of a telomer (thought to be a tetramer) of pentafluorophenyl isocyanide (C_6F_5NC) which was a brilliant cherry red [3]. The tetramer of phenyl isocyanide (C_6H_5NC) is indigodianil, which is blue and a member of the class of indigo dyes. Given that a related member of this class is woad, the warpaint of the ancient Britons, we felt we had compelling evidence to suggest that the perfluoro-analogue of woad would be red and not blue! Other colours, all red to a degree, were seen in the 'spontaneous' polymerization of pentafluorophenyl isocyanide. Although visually attractive, they were scientifically unwelcome when the task before me was to stabilize the monomer.

My seemingly irrepressible skill to turn anything into a polymer was viewed by Professor R. N. Haszeldine as a natural calling for work on polymers. Hence a research fellowship spent studying fluoro-olefin insertion into C—H bonds of commodity polymers followed my PhD research. Expectations that new low-friction graft polymers might result were dispelled when I was able to place a fistful of poly(ethylene-*graft*-hexafluoropropene) on Professor Haszeldine's office wall, where it remained for some time, defying gravity by virtue of its own tack. Polarity is indeed one area where fluoropolymers can show remarkable extremes of behaviour. Thus, while PTFE has the lowest dielectric constant of any known polymer except for the new amorphous fluoropolymers (such as Teflon® AF), and polyethylene is not far behind, PVDF ($[CH_2CF_2]_n$ which represents the 'half-way house' between the two) has a dielectric constant about four times that of PTFE – possibly the highest of any uncompounded polymer.

Bryan G. Willoughby

Polymer chemistry has figured prominently in my career since leaving academia, and I have been a Principal Consultant at the (UK) Shropshire-based rubber and plastics consultancy Rapra Technology Ltd since 1976. In this role, I expect to handle the majority of queries put to Rapra on fluoropolymers, both plastic and rubber. This has brought me into contact with a diversity of issues, from production problems to consumer

Plate 3.1. 'Polytetrafluoroethylene' by Timothy Willoughby (acrylic; 1996). (Reproduced by courtesy of John N. MacSween.)

concerns. Perhaps the most unusual of the questions I have received was a request to show what it would be like to be inside a fluoropolymer molecule! Always ready to tackle the unusual, I accepted this invitation (from the journal *Plastics in Engineering*) and sought inspiration from a classical ball-and-stick model of polytetrafluoroethylene. Given the brief, I chose to look into the section rather than view it sideways on. What I saw surprised me: the spiral of fluorines in PTFE, which is commonly perceived as a smooth enveloping sheath, takes on quite a different character when viewed in section. Rather than closing in around the carbons, the fluorines are seen as thrusting out to meet the viewer. In a quite different, and certainly unusual, context those contrasting characteristics of fluorine were again emerging.

Of course, finding examples of such character from the real world of fluoropolymers is not at all difficult: there are fluoropolymer molecules which are perfectly safe when left intact, but which show a vicious streak when they eventually degrade [4]. But what a scientist may think is only one thing – the perception of an artist is something different. Does structural chemistry therefore offer any possibilities as an art form?

This question was addressed by another Willoughby, namely my younger son Timothy, who (when I wrote this article) was studying Fine Art at Reading University, despite having a pre-university qualification (UK A-level) in Chemistry. Having looked at my efforts and built a ball-and-stick model of PTFE for himself, he produced two pictures – without a green ball to be seen anywhere! In fact his avoidance of the conventional colouring was quite deliberate: an attempt to let the viewer see a geometric arrangement without preconceptions as to what that arrangement represents. Both paintings are in acrylic, and one was displayed at the Newtown (Powys, Wales) Summer Exhibition of 1996. This painting (shown here) is now owned by John N. MacSween of W. L. Gore & Associates (UK) Ltd, and the other, which decorates the cover of this book, is owned by Rapra in Shawbury. The pictures have proved thought-provoking, especially amongst those artists who have queried the extent to which any art should be dictated by the representational forms of science. On this basis we may not see many more pictures of their ilk – perhaps 'polyfluorism' is not to be the next art movement!

References

1 R. E. Banks, R. N. Haszeldine, M. J. Stevenson and B. G. Willoughby, *J. Chem. Soc. (C)*, (1969) 2119.

2 J. Lee and B. G. Willoughby, *Spectrochim. Acta*, *33A* (1977) 395.

3 R. E. Banks, R. N. Haszeldine and B. G. Willoughby, *J. Chem. Soc., Perkin Trans. 1*, (1975) 2451.

4 B. G. Willoughby, 'Fluoropolymer Emissions and the Environment', *Fluoropolymers '92*, Conference paper 10 (publ. Rapra Technology Ltd., 1991). Fluoropolymers '92 was a joint UMIST Chemserve-RSC/SCI Macro Group UK Conference held in Manchester, 6-8 January 1992.

Appendix 32.4

Fluorine: fascination, frustration, and fulfilment

RUSSELL P. HUGHES

Chemistry Department, Dartmouth College, 6128 Burke Laboratory, Hanover, NH 03755-3564, USA

Russell P. Hughes

As a sixteen-year-old schoolboy I knew I was going to be a research chemist, although I wasn't quite sure what that really meant. I was doing my Advanced level examinations at Ecclesfield Grammar School, near Sheffield (UK), and my interest in chemistry was due in large part to a very enthusiastic teacher, Mr Stanley Spencer. I particularly enjoyed experimental aspects of the subject. Qualitative analysis of metal salts proved fascinating, using flame tests and abundant hydrogen sulphide generated in a Kipp's apparatus; and the sulphonation and nitration of benzene were accomplished with great vigour and enjoyment using fuming acids. Moreover, each afternoon in the lab culminated with a rigorous cleansing of young hands by washing up with benzene! (Fluorine was nowhere to be seen, of course.)

I came to understand that in order to become a research chemist, and perhaps actually get paid for doing exciting lab work, one had to go to university and get a Bachelor's degree, and maybe even go on to do postgraduate work. My applications for entrance were made to various universities discreetly distanced from Sheffield itself – far enough to be beyond immediate parental scrutiny, but close enough to accommodate necessary 'laundry runs' when things got critical in that department. Manchester was ideally located, and a lasting impression was generated when I visited the University's Faculty of Technology (soon to become UMIST) for an interview. The whole Chemistry Department there seemed to vibrate with excitement about research, particularly the Haszeldine-led effort in organofluorine chemistry. I was fortunate to be taken on a tour of some research labs in the old building where I encountered a graduate student who flourished before my eyes a tube filled with CF_3NO! I was fascinated by its beautiful blue colour, and also by the dark mutterings of the student about its rumoured toxicity! Immediately I wanted to be part of this great endeavour. Happily, my application was approved and I entered UMIST in the autumn of 1964. During my undergraduate days there I especially enjoyed lectures in organic chemistry given by Haszeldine, Banks, Barlow and Tipping, and liberally sprinkled with examples of organofluorine compounds. I could hardly wait to get involved in research in this area. Then came final exams.

I got an upper-second class honours degree, good enough to go on for the PhD; but when I was summoned to Haszeldine's office to discuss research options, it became frustratingly clear that the project on fluorinated heterocycles for which I yearned had been assigned to another student. If I were to stay at UMIST to do research, it would have to be in a different field, one in which I was relatively ignorant. So I was given into the care of a new research supervisor, John Powell, a freshly-minted organometallic PhD from Leeds University, and when the Department moved to its new labs further down Sackville Street I quickly became interested in his area of chemistry. However, in lab F108, where my bench was located, everyone else worked on fluorine chemistry. My fascination and frustration increased as I observed Dave Woodward and Ron Hubbard carrying out occasionally nerve-racking vacuum transfers of perfluoro 'Dewar benzene', and the great excitement all round when the first examples of perfluoroalkylated valence isomers of benzene were observed. After a year in Manchester, Powell moved to the University of Toronto in Canada, and I accompanied him. My entire PhD project on reactions of allylic complexes of palladium contained only a smattering of fluorine in the form of ancillary hexafluoroacetylacetonato ligands. By that time I had decided that the academic life was definitely for me, so doing a 'postdoc' was going to be necessary in order to get a faculty job. Perhaps, I mused, I could get a shot at actually doing fluorine chemistry in this capacity!

Research in organometallic chemistry using fluorinated ligands was dominated at that time by the groups of Gordon Stone and Michael Green at the University of Bristol, and in 1973 I was fortunate enough to get a fellowship to work in Mike Green's group. On my arrival I was assigned to tie up a couple of loose ends involving some trifluoroacetonitrile and hexafluoro-2-butyne chemistry, but clearly the heyday of fluorine chemistry at Bristol was past. I had missed the boat again! However, Mike Green quickly engaged my interest in the organometallic chemistry of small ring compounds, and I had a most productive and enjoyable stay at Bristol. But when the time came to seek employment, jobs were very scarce in the UK, so I decided to return to Canada as a postdoc with John Harrod at McGill University in Montréal. As a result I was ideally placed geographically to interview for an academic job at Dartmouth College in the USA, where a particularly attractive feature was the presence on the faculty of Dave Lemal, actively engaged in organofluorine chemistry. Coincidentally, Dave had been in the process of co-discovering the perfluoroalkylated valence isomers of benzene at Dartmouth at the very same time that I was watching the same compounds being made in Manchester!

I got the job, but since a young faculty member must make his own mark and not risk the criticism of unseemly overcollaboration with his colleagues, my initial research thrusts quite deliberately avoided fluorine altogether. Nevertheless, one day in 1978 Dave Lemal showed me a beautifully crystalline sample of octafluorocyclo-octatetraene (OFCOT) and the NMR spectrum of its iron tricarbonyl derivative; it was clearly quite different from that of the well-known hydrocarbon ligand complex. Ray Davis at the University of Texas solved the structure and my enthusiasm for organofluorine compounds was reawakened. Exploration of the organometallic chemistry of OFCOT led to a large number of new kinds of organometallic complexes. The presence of fluorine, rather than hydrogen, on the carbon skeletons of the ligands imbued the resultant complexes with significantly higher thermal and air stability, and often led to significantly different structural and bonding patterns. Furthermore, the chemistry of OFCOT appeared to be dominated by its tendency to rearrange or to react to give a ligand with as many sp^3 carbon-to-fluorine bonds as possible. (Contributions to this area were reviewed in 1990 [1]). The air stability of these compounds led us subsequently to explore synthetic routes to a variety of fluorinated analogues of hydrocarbon ligands, including tetrafluoroethylene [2], hexafluorobutadiene [3], and the long-sought-after pentafluorocyclopentadienyl ligand [4]. Recent gas-phase ionization and electronic structural studies of the latter complexes and their partially-fluorinated analogues indicate that fluorine is indeed a net σ-electron withdrawing substituent, but that this effect is strongly attenuated by π-donation [5], thereby providing no surprises at all for experienced fluorine chemists!

More recently we have turned our attention to organometallic compounds containing perfluoroalkylated ligands as part of two specific projects: to make transition metal compounds soluble in saturated perfluorocarbon and supercritical CO_2 solvents [6], thereby allowing catalytic and stoichiometric chemistry to be performed in these media; and to investigate methodology for increasing the chemical vulnerability of the normally inert carbon-fluorine bond [7] in order to develop new ways of functionalizing perfluorocarbons.

Finally, after an initial period of frustrated flirtation, the carbon-fluorine bond has proven to be a fascinating companion and a challenging adversary. It has been well worth the wait!

References

1 R. P. Hughes, *Adv. Organomet. Chem.*, *31* (1990) 183.

2 O. J. Curnow, R. P. Hughes and A. L. Rheingold, *J. Am. Chem. Soc.*, *114* (1992) 3153; O. J. Curnow, R. P. Hughes, E. N. Mairs and A. L. Rheingold, *Organometallics*, *12* (1993) 3102.

3 R. P. Hughes, P. R. Rose and A. L. Rheingold, *Organometallics*, *12* (1993) 3109; R. P. Hughes, P. R. Rose, X. Zheng and A. L. Rheingold, *Organometallics*, *14* (1995) 2407.

4 O. J. Curnow and R. P. Hughes, *J. Am. Chem. Soc.*, *114* (1992) 5895; R. P. Hughes, P. R. Rose, X. Zheng and A. L. Rheingold, *Organometallics*, *14* (1995) 2407.

5 D. E. Richardson, M. F. Ryan, W. E. Geiger, T. T. Chin, R. P. Hughes and O. J. Curnow, *Organometallics*, *12* (1993) 613; D. E. Richardson, L. Lang, J. R. Eyler, S. R. Kircus, X. Zheng, C. A. Morse and R. P. Hughes, *Organometallics*, *16* (1997) 149; D. L. Lichtenberger, Y. Elkadi, N. E. Gruhn, R. P. Hughes, O. J. Curnow and X. Zheng, *Organometallics*, *16* (1997) 5209; R. P. Hughes, X. Zheng, C. A. Morse, O. J. Curnow, J. R. Lomprey, A. L. Rheingold and G. P. A. Yap, *Organometallics*, *17* (1998) 457.

6 R. P. Hughes and H. A. Trujillo, *Organometallics*, *15* (1996) 286; R. P. Hughes, T. L. Husebo, A. L. Rheingold, L. M. Liable-Sands and G. P. A. Yap, *Organometallics*, *16* (1997) 5; R. P. Hughes, S. M. Maddock, A. L. Rheingold and L. M. Liable-Sands, *J. Am. Chem. Soc.*, *119* (1997) 5988; R. P. Hughes, T. L. Husebo, B. J. Holliday, A. L. Rheingold and L. M. Liable-Sands, *J. Organomet. Chem.*, *548* (1997) 109.

7 R. P. Hughes, D. C. Lindner, A. L. Rheingold and L. M. Liable-Sands, *J. Am. Chem. Soc.*, *119* (1997) 11544; R. P. Hughes, T. L. Husebo, S. M. Maddock, A. L. Rheingold and I. A. Guzei, *J. Am. Chem. Soc.*, *119* (1997) 10231.

Appendix 32.5

India to the USA via UMIST: a fluorine chemistry trail

SHARIQUE S. ZUBERI

Aerojet Fine Chemicals, Building 05025/Dept. 9651, P.O. Box 1718, Rancho Cordova, CA 95741, USA

I set off on my journey in fluorine chemistry in the spring of 1980 at UMIST (Manchester, UK) under the direction of Professor R. E. Banks, who assigned me a short research project on perhalogenated aromatic azo-compounds as part of a Masters programme. The objective was to probe the possibility of using azoarenes like decafluoroazobenzene and octafluoro-4,4′-azopyridine (**1**) as thermal sources of radicals for use in synthesis: $Ar_FN{=}NAr_F$ (heat) $\rightarrow 2Ar_F^{\bullet} + N_2$. Pyrolysis of the azohetarene (**1**), prepared via electrophilic chlorination of 4-aminotetrafluoropyridine with *t*-BuOCl, provided a new route to octafluoro-4,4′-bipyridyl (**2**): 4-$(H_2N)C_5F_4N \rightarrow$ 4-$(Cl_2N)C_5F_4N \rightarrow$ (at 160 °C/1 atm) $2Cl_2$ + 4-$(C_5F_4N)N{=}N(C_5F_4N)$-4′ (**1**) $\rightarrow$ (at 550 °C/0.2 mmHg) 4,4′-$(C_5F_4N)_2$ (**2**) + numerous byproducts.

By the time I graduated (October 1980) [1], I had become highly enthused by this 'new area' of chemistry despite hearing stories about the dangers associated with hydrogen fluoride and elemental fluorine! And soon I found myself not only having to work with both of these 'nasties' but also with azides – a real baptism of fire, which I came through unscathed and which set me up for life as a practicising chemist. I refer to my adventures as a PhD student in Professor Banks' group during the period 1980-1983 [2]. Anhydrous hydrogen fluoride was used to extract kerogen from source rock samples, and elemental fluorine to fluorinate kerogen and coal in order to obtain structural information. Strange as it might seem, the expertise gained was employed to probe the structure of a novel fluoropolymer, $(C_5F_8N_2)_x$, obtained by one of my senior fellow chemists at UMIST, Allan Bailey, via pyrolysis of perfluoro-2-azido-1-azacyclohexene (**3**) [3].

Problems with a P-GC-MS instrument, coupled with the oft-found 'time-did-not-allow' situation, precluded a positive result; however, the novel piperidine **4**, distinguished by containing both an N−F bond and an azide function, was synthesized via lethargic fluorination (treatment with F_2 in a stainless steel reactor for 48 days at room temperature) of perfluoro-(6-azido-2,6-dimethyl-1-azacyclohexene) (**5**). This experiment was prompted by Schack's preparation of CF_3NF_2 from CF_3N_3 and F_2 at 70 °C [4]; the duration of my experiment was not intended: I was called home to India just after I'd loaded the s.s. reaction vessel, and I left for the airport without turning the reactor's heater on.

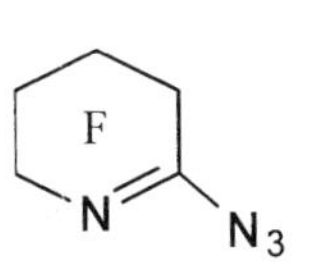

(**3**)

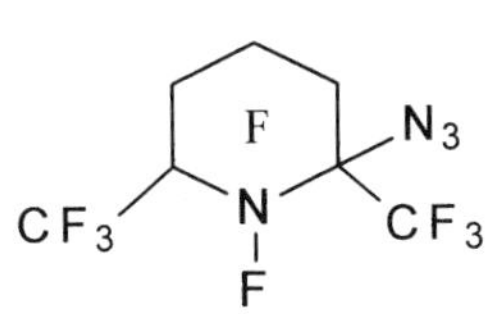

(**4**)

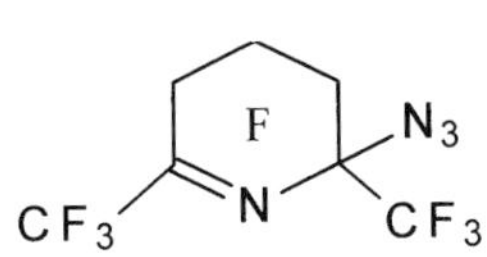

(**5**)

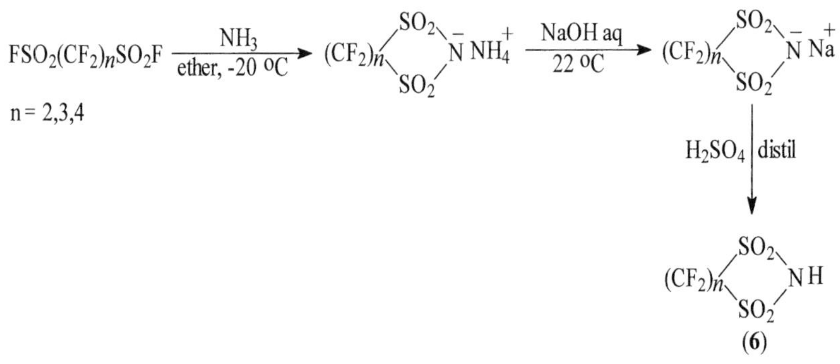

Scheme 5.1.

After completing my doctorate at UMIST [2], I spent several years back home in India working for the Oil & Natural Gas Commission as a Scientific Pool officer. The opportunity to work in the fluorine field came again in 1985 in the form of a postdoctoral position in Professor Darryl D. DesMarteau's group at Clemson University, South Carolina, USA. My work there centred on the synthesis and chemistry of fluorinated sulfonimides, $(R_FSO_2)_2NH$, which are extremely acidic and have proved to be excellent candidates for electrolytes in fuel cells. This research brought me in contact with N—Cl and N—F compounds, and involved using elemental fluorine, hence it linked nicely with my studies at UMIST.

The strategy developed for the preparation of cyclic perfluoroalkanesulfonimides (**6**) is shown in Scheme 5.1 [5]. *N*-Halogeno derivatives of the sulfonimides were obtained as shown in Scheme 5.2 [6]; like $(CF_3SO_2)_2NF$ (the prototypical DesMarteau reagent), the cyclic N—F compounds were powerful electrophilic

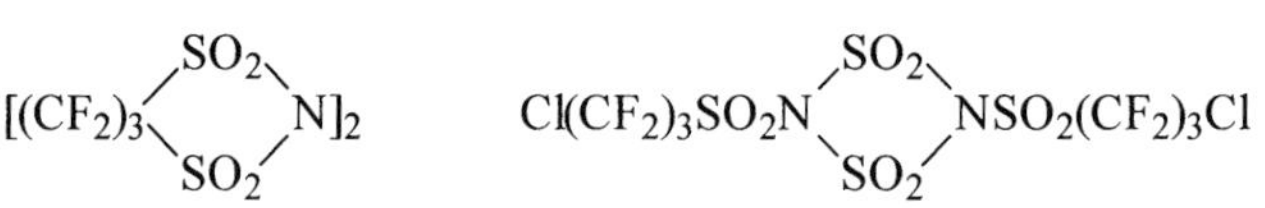

(**8**) (**9**)

fluorinating agents [7]. Interestingly, UV photolysis of the N—Cl derivative (**7**) ($n = 3$) did not yield the expected hydrazine (**8**) but the dimer **9**.

In 1987, I moved from Clemson to Professor James L. Adcock's group at Knoxville, where, as a postdoctoral research associate, I worked on the aerosol direct fluorination of organic molecules [8-10]. This work was very exciting since molecules were recovered having undergone perfluorination with little or no skeletal fragmentation. The most noteworthy advance involved the synthesis of 1- and 2-substituted hydryl-, methyl-, and (difluoromethyl)-*F*-adamantanes (*e.g.* Scheme 5.3 [11]).

By 1990 I had received permanent US resident status and was hired by Great Lakes Chemical Corporation in the February of that year to work in the Fluorine Chemicals Division. Great Lakes Chemical Corporation is a specialty chemicals company, headquartered in Indiana, USA. There are more than 7000 employees worldwide involved in research, sales, service, production and distribution activities. The corporation is organized into different business groups, which are driven by the needs of its customers. Each business group

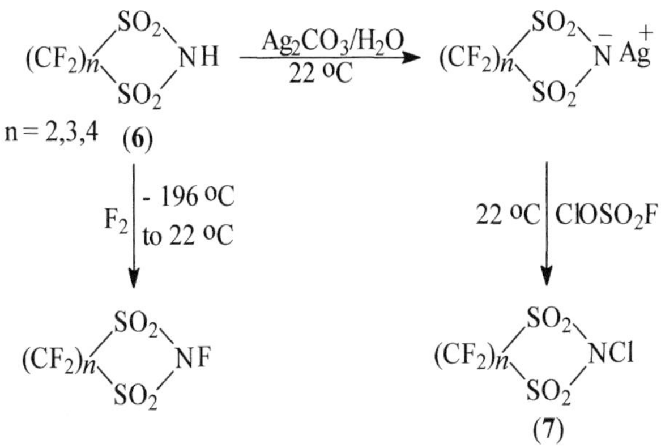

Scheme 5.2.

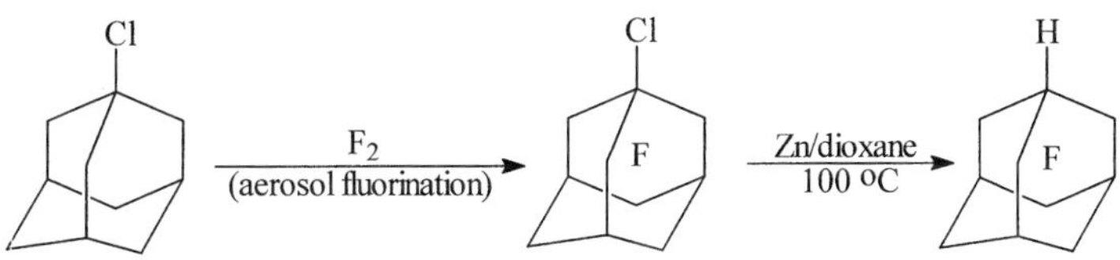

Scheme 5.3.

has resources to conduct research and development, manufacturing, marketing and technical service. Great Lakes stock is traded on the New York Stock Exchange under the symbol GLK.

The Fluorine Chemicals Division is part of the Performance Chemicals Business group (1997 sales topped $58m) and produces a wide array of compounds, *e.g.*, 2*H*-heptafluoropropane (FM-200^{TM}), 3,3,3-trifluoropropene, 1-bromo-2-fluoroethane, trifluoromethane (R-23), and difluoromethane (R-32). In the past, Great Lakes has also produced the brominated fire-extinguishants Halons 1301 (trifluoromethyl bromide), 1202 (dibromodifluoromethane), 2402 (1,2-dibromo-tetrafluoroethane), and 1211 (bromochlorodifluoromethane), which, being partly responsible for stratospheric ozone depletion, fell into disfavour under the 1987 Montreal Protocol. I gained tremendous experience in research and development, trouble shooting, and, to some extent, scale-up while working on most of these products, and this will stand me in good stead as I move on now to GenCorp Aerojet (CA).

References

1 S. S. Zuberi, MSc Dissertation, Victoria University of Manchester, UK, 1980.
2 S. S. Zuberi, PhD Thesis, Victoria University of Manchester, UK, 1983.
3 A. R. Bailey, PhD Thesis, Victoria University of Manchester, UK, 1981.
4 C. J. Schack, *J. Fluorine Chem.*, *18* (1981) 583.
5 S. S. Zuberi and D. D. DesMarteau, to be published.
6 D. D. DesMarteau, S. S. Zuberi, W. T. Pennington and B. B. Randolph, *Eur. J. Solid. State Inorg. Chem.*, *29* (1992) 777.

7 S. Singh, D. D. DesMarteau, S. S. Zuberi, M. Witz and H.-N. Huang, *J. Amer. Chem. Soc.*, *109* (1987) 7194.
8 J. L. Adcock, in *Chemistry of Organic Fluorine Compounds II*, ACS Monograph, *187* (1995) 97.
9 J. L. Adcock, in G. A. Olah, R. D. Chambers and G. K. S. Prakash (eds.), *Synthetic Fluorine Chemistry*, John Wiley, New York, 1992, pp. 127-141.
10 J. L. Adcock, in R. E. Banks, D. W. A. Sharp and J. C. Tatlow (eds.), *Fluorine: The First Hundred Years (1886-1986)*, Elsevier Sequoia, Lausanne and New York, 1986, pp. 327-330.
11 J. L. Adcock, H. Luo and S. S. Zuberi, *J. Org. Chem.*, *57* (1992) 4749.

BIOGRAPHIC NOTE

Sharique Zuberi graduated [BSc (1974), MSc (1976)] from Aligarh University, India before entering the Chemistry Department at UMIST (Manchester, UK) where he earned an MSc degree (1980) before taking a doctorate (PhD, 1983) in R. E. Banks' fluorine team. After doing postdoctoral work in the US with D. D. DesMarteau (Clemson University, SC; 1985-87) and then J. L. Adcock (University of Tennessee, Knoxville, TN; 1987-90), he worked in the Fluorine Chemicals Division of Great Lakes Chemical Corporation, leaving in mid-1998 to become a Senior Chemist at GenCorp Aerojet (CA).

Sharique S. Zuberi

NAME INDEX[1]

[1] Only names appearing in the main text are listed here; (b) = biographic note; (p) = photograph.

SUBJECT INDEX

ESTABLISHMENT INDEX[1]

[1]UK Universities appear under Universities (British); others appear in the form, for example, Baylor University or under the heading Universities (other).